The Electrical Engineering Handbook
Third Edition

Circuits, Signals, and Speech and Image Processing

The Electrical Engineering Handbook Series

Series Editor

Richard C. Dorf

University of California, Davis

Titles Included in the Series

The Electrical Engineering Handbook
Third Edition

Edited by
Richard C. Dorf

Circuits, Signals, and Speech and Image Processing

Electronics, Power Electronics, Optoelectronics, Microwaves, Electromagnetics, and Radar

Sensors, Nanoscience, Biomedical Engineering, and Instruments

Broadcasting and Optical Communication Technology

Computers, Software Engineering, and Digital Devices

Systems, Controls, Embedded Systems, Energy, and Machines

The Electrical Engineering Handbook
Third Edition

Circuits, Signals, and Speech and Image Processing

Edited by

Richard C. Dorf
University of California
Davis, California, U.S.A.

Taylor & Francis
Taylor & Francis Group
Boca Raton   London   New York

A CRC title, part of the Taylor & Francis imprint, a member of the
Taylor & Francis Group, the academic division of T&F Informa plc.

Published in 2006 by
CRC Press
Taylor & Francis Group
6000 Broken Sound Parkway NW, Suite 300
Boca Raton, FL 33487-2742

International Standard Book Number-10: 0-8493-7337-9 (Hardcover)
International Standard Book Number-13: 978-0-8493-7337-4 (Hardcover)
Library of Congress Card Number 2005054346

Library of Congress Cataloging-in-Publication Data

Circuits, signals, and speech and image processing / edited by Richard C. Dorf.
    p. cm.
  Includes bibliographical references and index.
  ISBN 0-8493-7337-9 (alk. paper)
  1. Signal processing. 2. Electric circuits. I. Dorf, Richard C. II. Title.

TK5102.9.C495 2005
621.3--dc22   2005054346

Taylor & Francis Group
is the Academic Division of Informa plc.

Preface

Purpose

The purpose of *The Electrical Engineering Handbook, 3rd Edition* is to provide a ready reference for the practicing engineer in industry, government, and academia, as well as aid students of engineering. The third edition has a new look and comprises six volumes including:

Circuits, Signals, and Speech and Image Processing
Electronics, Power Electronics, Optoelectronics, Microwaves, Electromagnetics, and Radar
Sensors, Nanoscience, Biomedical Engineering, and Instruments
Broadcasting and Optical Communication Technology
Computers, Software Engineering, and Digital Devices
Systems, Controls, Embedded Systems, Energy, and Machines

Each volume is edited by Richard C. Dorf, and is a comprehensive format that encompasses the many aspects of electrical engineering with articles from internationally recognized contributors. The goal is to provide the most up-to-date information in the classical fields of circuits, signal processing, electronics, electromagnetic fields, energy devices, systems, and electrical effects and devices, while covering the emerging fields of communications, nanotechnology, biometrics, digital devices, computer engineering, systems, and biomedical engineering. In addition, a complete compendium of information regarding physical, chemical, and materials data, as well as widely inclusive information on mathematics is included in each volume. Many articles from this volume and the other five volumes have been completely revised or updated to fit the needs of today and many new chapters have been added.

The purpose of this volume (*Circuits, Signals, and Speech and Image Processing*) is to provide a ready reference to subjects in the fields of electric circuits and components, analysis of circuits, and the use of the Laplace transform. We also discuss the processing of signals, speech, and images using filters and algorithms. Here we provide the basic information for understanding these fields. We also provide information about the emerging fields of text-to-speech synthesis, real-time processing, embedded signal processing, and biometrics.

Organization

The information is organized into three sections. The first two sections encompass 27 chapters and the last section summarizes the applicable mathematics, symbols, and physical constants.

Most articles include three important and useful categories: defining terms, references, and further information. *Defining terms* are key definitions and the first occurrence of each term defined is indicated in boldface in the text. The definitions of these terms are summarized as a list at the end of each chapter or article. The *references* provide a list of useful books and articles for follow-up reading. Finally, *further information* provides some general and useful sources of additional information on the topic.

Locating Your Topic

Numerous avenues of access to information are provided. A complete table of contents is presented at the front of the book. In addition, an individual table of contents precedes each section. Finally, each chapter begins with its own table of contents. The reader should look over these tables of contents to become familiar with the structure, organization, and content of the book. For example, see Section II: Signal Processing, then Chapter 18: Multidimensional Signal Processing, and then Chapter 18.2: Video Signal Processing. This tree-and-branch table of contents enables the reader to move up the tree to locate information on the topic of interest.

Two indexes have been compiled to provide multiple means of accessing information: subject index and index of contributing authors. The subject index can also be used to locate key definitions. The page on which the definition appears for each key (defining) term is clearly identified in the subject index.

The Electrical Engineering Handbook, 3rd Edition is designed to provide answers to most inquiries and direct the inquirer to further sources and references. We hope that this handbook will be referred to often and that informational requirements will be satisfied effectively.

Acknowledgments

This handbook is testimony to the dedication of the Board of Advisors, the publishers, and my editorial associates. I particularly wish to acknowledge at Taylor & Francis Nora Konopka, Publisher; Helena Redshaw, Editorial Project Development Manager; and Susan Fox, Project Editor. Finally, I am indebted to the support of Elizabeth Spangenberger, Editorial Assistant.

Richard C. Dorf
Editor-in-Chief

Editor-in-Chief

Richard C. Dorf, Professor of Electrical and Computer Engineering at the University of California, Davis, teaches graduate and undergraduate courses in electrical engineering in the fields of circuits and control systems. He earned a Ph.D. in electrical engineering from the U.S. Naval Postgraduate School, an M.S. from the University of Colorado, and a B.S. from Clarkson University. Highly concerned with the discipline of electrical engineering and its wide value to social and economic needs, he has written and lectured internationally on the contributions and advances in electrical engineering.

Professor Dorf has extensive experience with education and industry and is professionally active in the fields of robotics, automation, electric circuits, and communications. He has served as a visiting professor at the University of Edinburgh, Scotland; the Massachusetts Institute of Technology; Stanford University; and the University of California, Berkeley.

Professor Dorf is a Fellow of The Institute of Electrical and Electronics Engineers and a Fellow of the American Society for Engineering Education. Dr. Dorf is widely known to the profession for his *Modern Control Systems, 10th Edition* (Addison-Wesley, 2004) and *The International Encyclopedia of Robotics* (Wiley, 1988). Dr. Dorf is also the co-author of *Circuits, Devices and Systems* (with Ralph Smith), *5th Edition* (Wiley, 1992), and *Electric Circuits, 7th Edition* (Wiley, 2006). He is also the author of *Technology Ventures* (McGraw-Hill, 2005) and *The Engineering Handbook, 2nd Edition* (CRC Press, 2005).

Advisory Board

Contributors

Taan El Ali
Benedict College
Columbia, South Carolina

Nick Angelopoulos
Hoffman Engineering
Scarborough, Canada

A. Terry Bahill
University of Arizona and
    BAE Systems
Tucson, Arizona

Norman Balabanian
University of Florida
Gainesville, Florida

Sina Balkir
University of Nebraska
Lincoln, Nebraska

Glen Ballou
Ballou Associates
Guilford, Connecticut

Mahamudunnabi Basunia
University of Dayton
Dayton, Ohio

Stella N. Batalama
State University of New York
Buffalo, New York

Peter Bendix
LSI Logic Corp.
Milpitas, California

Theodore A. Bickart
Michigan State University
East Lansing, Michigan

Bill Bitler
InfiMed
Liverpool, New York

Theodore F. Bogart, Jr.
University of Southern Mississippi
Hattiesburg, Mississippi

Bruce W. Bomar
University of Tennessee Space Institute
Tullahoma, Tennessee

N.K. Bose
Pennsylvania State University
University Park, Pennsylvania

John E. Boyd
Cubic Defense Systems
San Diego, California

Marcia A. Bush
Xerox Palo Alto Research Center
Palo Alto, California

James A. Cadzow
Vanderbilt University
Nashville, Tennessee

Yu Cao
Queen's University
Kingston, Ontario, Canada

Shu-Park Chan
International Technological University
Santa Clara, California

Rulph Chassaing
Roger Williams University
Bristol, Rhode Island

Rama Chellappa
University of Maryland
College Park, Maryland

Chih-Ming Chen
National Taiwan University
    of Science and Technology
Taipei, Taiwan

Wai-Kai Chen
University of Illinois
Chicago, Illinois

Michael D. Ciletti
University of Colorado
Colorado Springs, Colorado

J.R. Cogdell
University of Texas
Austin, Texas

Israel Cohen
Israel Institute of Technology
Haifa, Israel

Reza Derakhshani
University of Missouri
Kansas City, Missouri

Hui Dong
University of California
Santa Barbara, California

Richard C. Dorf
University of California
Davis, California

Yingzi Du
Indiana University/Purdue University
Indianapolis, Indiana

Gary W. Elko
Agere Systems
Murray Hill, New Jersey

Yariv Ephraim
George Mason University
Fairfax, Virginia

Delores M. Etter
United States Naval Academy
Annapolis, Maryland

Jesse W. Fussell
U.S. Department of Defense
Fort Meade, Maryland

Jerry D. Gibson
University of California
Santa Barbara, California

Lawrence Hornak
West Virginia University
Morgantown, West Virginia

Jerry L. Hudgins
University of Nebraska
Lincoln, Nebraska

Mohamed Ibnkahla
Queen's University
Kingston, Ontario, Canada

J. David Irwin
Auburn University
Auburn, Alabama

Robert W. Ives
United States Naval Academy
Annapolis, Maryland

W. Kenneth Jenkins
The Pennsylvania State University
University Park, Pennsylvania

David E. Johnson
Birmingham-Southern College
Birmingham, Alabama

Dimitri Kazakos
University of Idaho
Moscow, Idaho

William J. Kerwin
The University of Arizona
Tucson, Arizona

Yong Deak Kim
Ajou University
Suwon, South Korea

Allan D. Kraus
Naval Postgraduate School
Pacific Grove, California

Dean J. Krusienski
Wadsworth Center
Albany, New York

Pradeep Lall
Auburn University
Auburn, Alabama

Kartikeya Mayaram
Oregon State University
Corvallis, Oregon

Michael G. Morrow
University of Wisconsin-Madison
Madison, Wisconsin

Paul Neudorfer
Seattle University
Seattle, Washington

Norman S. Nise
California State Polytechnic University
Orange, California

Keshab K. Parhi
University of Minnesota
Minneapolis, Minnesota

Sujan T.V. Parthasaradhi
Bioscrypt, Inc.
Markham, Ontario, Canada

Clayton R. Paul
Mercer University
Macon, Georgia

Michael Pecht
University of Maryland
College Park, Maryland

S. Unnikrishna Pillai
Polytechnic University
Brooklyn, New York

Alexander D. Poularikas
University of Alabama
Huntsville, Alabama

Jose C. Principe
University of Florida
Gainesville, Florida

Sarah A. Rajala
North Carolina State University
Raleigh, North Carolina

J. Gregory Rollins
Technology Modeling Associates Inc.
Sunnyvale, California

Amit K. Roy-Chowdhury
University of California
Riverside, California

C. Sankaran
Electro-Test
Seattle, Washington

Juergen Schroeter
AT&T Labs
Florham Park, New Jersey

Stephanie A.C. Schuckers
Clarkson University
Potsdam, New York

Yun Q. Shi
New Jersey Institute of Technology
Newark, New Jersey

Theodore I. Shim
Polytechnic University
Brooklyn, New York

L.H. Sibul
Pennsylvania State University
University Park, Pennsylvania

L. Montgomery Smith
University of Tennessee Space Institute
Tullahoma, Tennessee

M. Mohan Sondhi
AT&T Labs
Florham Park, New Jersey

Wei Su
U.S. Army RDECOM CERDEC
Fort Monmouth, New Jersey

Ahmad Iyanda Sulyman
Queen's University
Kingston, Ontario, Canada

David D. Sworder
University of California
San Diego, California

Ferenc Szidarovszky
University of Arizona
Tucson, Arizona

Ronald J. Tallarida
Temple University
Philadelphia, Pennsylvania

Charles W. Therrien
Naval Postgraduate School
Monterey, California

Vyacheslav Tuzlukov
Ajou University
Suwon, South Korea

Bo Wei
Southern Methodist University
Dallas, Texas

Thad B. Welch
United States Naval Academy
Annapolis, Maryland

Lynn D. Wilcox
Rice University
Houston, Texas

Cameron H.G. Wright
University of Wyoming
Laramie, Wyoming

Ping Xiong
State University of New York
Buffalo, New York

Won-Sik Yoon
Ajou University
Suwon, South Korea

Shaohua Kevin Zhou
University of Maryland
College Park, Maryland

Contents

SECTION II   Signal Processing

SECTION III Mathematics, Symbols, and Physical Constants

Indexes

I

Circuits

Passive Components

Michael Pecht
University of Maryland

Pradeep Lall
Auburn University

Glen Ballou
Ballou Associates

C. Sankaran
Electro-Test

Nick Angelopoulos
Hoffman Engineering

1.1 Resistors

Michael Pecht and Pradeep Lall

The resistor is an electrical device whose primary function is to introduce resistance to the flow of electric current. The magnitude of opposition to the flow of current is called the resistance of the resistor. A larger resistance value indicates a greater opposition to current flow.

The resistance is measured in ohms. An ohm is the resistance that arises when a current of one ampere is passed through a resistor subjected to one volt across its terminals.

The various uses of resistors include setting biases, controlling gain, fixing time constants, matching and loading circuits, voltage division, and heat generation. The following sections discuss resistor characteristics and various resistor types.

Resistor Characteristics

Voltage and Current Characteristics of Resistors

The resistance of a resistor is directly proportional to the **resistivity** of the material and the length of the resistor and inversely proportional to the cross-sectional area perpendicular to the direction of current flow. The resistance R of a resistor is given by

$$R = \frac{\rho l}{A} \tag{1.1}$$

where ρ is the resistivity of the resistor material ($\Omega \cdot$cm), l is the length of the resistor along direction of current flow (cm), and A is the cross-sectional area perpendicular to current flow (cm^2) (Figure 1.1). Resistivity is an inherent property of materials. Good resistor materials typically have resistivities between 2×10^{-6} and 200×10^{-6} $\Omega \cdot$cm.

The resistance can also be defined in terms of sheet resistivity. If the sheet resistivity is used, a standard sheet thickness is assumed and factored into resistivity. Typically, resistors are rectangular in shape; therefore the length *l* divided by the width *w* gives the number of squares within the resistor (Figure 1.2). The number of squares multiplied by the resistivity is the resistance.

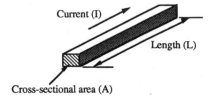

FIGURE 1.1 Resistance of a rectangular cross-section resistor with cross-sectional area *A* and length *L*.

$$R_{sheet} = \rho_{sheet} \frac{l}{W} \qquad (1.2)$$

where ρ_{sheet} is the sheet resistivity (Ω/square), *l* is the length of resistor (cm), *w* is the width of the resistor (cm), and R_{sheet} is the sheet resistance (Ω).

The resistance of a resistor can be defined in terms of the **voltage drop** across the resistor and current through the resistor related by Ohm's law:

$$R = \frac{V}{I} \qquad (1.3)$$

where *R* is the resistance (Ω), *V* is the voltage across the resistor (V), and *I* is the current through the resistor (A). Whenever a current is passed through a resistor, a voltage is dropped across the ends of the resistor. Figure 1.3 depicts the symbol of the resistor with the Ohm's law relation.

All resistors dissipate power when a voltage is applied. The power dissipated by the resistor is represented by

$$P = \frac{V^2}{R} \qquad (1.4)$$

where *P* is the power dissipated (W), *V* is the voltage across the resistor (V), and *R* is the resistance (Ω). An ideal resistor dissipates electric energy without storing electric or magnetic energy.

Resistor Networks

Resistors may be joined to form networks. If resistors are joined in series, the effective resistance (R_T) is the sum of the individual resistances (Figure 1.4).

$$R_T = \sum_{i=1}^{n} R_i \qquad (1.5)$$

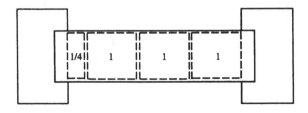

THE ABOVE RESISTOR IS 3.25 SQUARES
IF $\rho = 100\ \Omega/\square$, THEN R = $3.25\square \times 100\ \Omega/\square = 325\ \Omega$

FIGURE 1.2 Number of squares in rectangular resistor.

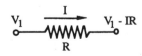

FIGURE 1.3 A resistor with resistance *R* having a current *I* flowing through it will have a voltage drop of *IR* across it.

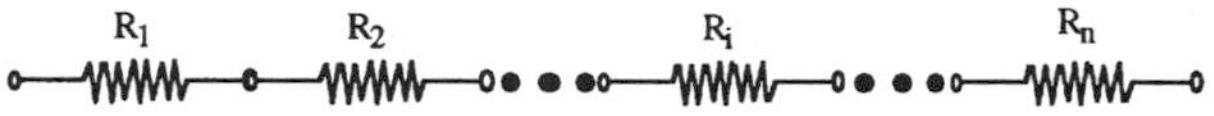

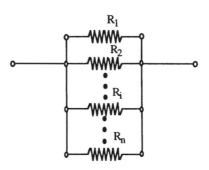

FIGURE 1.4   Resistors connected in series.

If resistors are joined in parallel, the effective resistance (R_T) is the reciprocal of the sum of the reciprocals of individual resistances (Figure 1.5).

$$\frac{1}{R_T} = \sum_{i=1}^{n} \frac{1}{R_i} \tag{1.6}$$

Temperature Coefficient of Electrical Resistance

The resistance for most resistors changes with temperature. The temperature coefficient of electrical resistance is the change in electrical resistance of a resistor per unit change in temperature. The **temperature coefficient of resistance** is measured in $\Omega/°C$. The temperature coefficient of resistors may be either positive or negative. A positive temperature coefficient denotes a rise in resistance with a rise in temperature; a negative temperature coefficient of resistance denotes a decrease in resistance with a rise in temperature. Pure metals typically have a positive temperature coefficient of resistance, while some metal alloys such as constantin and manganin have a zero temperature coefficient of resistance. Carbon and graphite mixed with binders usually exhibit negative temperature coefficients, although certain choices of binders and process variations may yield positive temperature coefficients. The temperature coefficient of resistance is given by

FIGURE 1.5   Resistors connected in parallel.

$$R(T_2) = R(T_1)[1 + \alpha_{T1}(T_2 - T_1)] \tag{1.7}$$

where α_{T1} is the temperature coefficient of electrical resistance at reference temperature T_1, $R(T_2)$ is the resistance at temperature T_2 (Ω), and $R(T_1)$ is the resistance at temperature T_1 (Ω). The reference temperature is usually taken to be 20°C. Because the variation in resistance between any two temperatures is usually not linear as predicted by Equation (1.7), common practice is to apply the equation between temperature increments and then to plot the resistance change versus temperature for a number of incremental temperatures.

High-Frequency Effects

Resistors show a change in their resistance value when subjected to ac voltages. The change in resistance with voltage frequency is known as the *Boella effect*. The effect occurs because all resistors have some inductance and capacitance along with the resistive component and thus can be approximated by an equivalent circuit shown in Figure 1.6. Even though the definition of useful frequency range is application dependent, typically, the useful range of the resistor is the highest frequency at which the impedance differs from the resistance by more than the tolerance of the resistor.

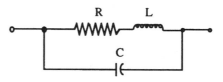

FIGURE 1.6   Equilavent circuit for a resistor.

The frequency effect on resistance varies with the resistor construction. Wire-wound resistors typically exhibit an increase in their impedance with frequency. In composition resistors the capacitances are formed by the many conducting particles which are held in contact by a dielectric binder. The ac impedance for film resistors remains constant until 100 MHz (1 MHz $= 10^6$ Hz) and then decreases at higher frequencies (Figure 1.7).

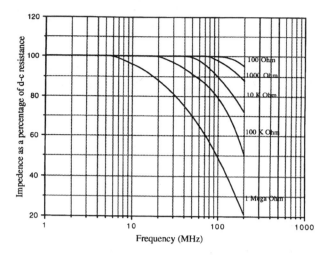

FIGURE 1.7   Typical graph of impedance as a percentage of dc resistance versus frequency for film resistors.

For film resistors, the decrease in dc resistance at higher frequencies decreases with increase in resistance. Film resistors have the most stable high-frequency performance.

The smaller the diameter of the resistor the better is its frequency response. Most high-frequency resistors have a length to diameter ratio between 4:1 and 10:1. Dielectric losses are kept to a minimum by proper choice of base material.

Voltage Coefficient of Resistance

Resistance is not always independent of the applied voltage. The **voltage coefficient of resistance** is the change in resistance per unit change in voltage, expressed as a percentage of the resistance at 10% of rated voltage. The voltage coefficient is given by the relationship

$$\text{voltage coefficient} = \frac{100(R_1 - R_2)}{R_2(V_1 - V_2)} \tag{1.8}$$

where R_1 is the resistance at the rated voltage V_1, and R_2 is the resistance at 10% of rated voltage V_2.

Noise

Resistors exhibit electrical noise in the form of small ac voltage fluctuations when dc voltage is applied. Noise in a resistor is a function of the applied voltage, physical dimensions, and materials. The total noise is a sum of Johnson noise, current flow noise, noise due to cracked bodies, and loose end caps and leads. For variable resistors the noise can also be caused by the jumping of a moving contact over turns and by an imperfect electrical path between the contact and resistance element.

The Johnson noise is temperature-dependent thermal noise (Figure 1.8). Thermal noise is also called "white noise" because the noise level is the same at all frequencies. The magnitude of thermal noise, E_{RMS} (V), is dependent on the resistance value and the temperature of the resistance due to thermal agitation:

$$E_{\text{RMS}} = \sqrt{4kRT\Delta f} \tag{1.9}$$

where E_{RMS} is the root-mean-square value of the noise voltage (V), R is the resistance (Ω), K is the Boltzmann constant (1.38×10^{-23} J/K), T is the temperature (K), and Δf is the bandwidth (Hz) over which the noise energy is measured.

Figure 1.8 shows the variation in current noise versus voltage frequency. Current noise varies inversely with frequency and is a function of the current flowing through the resistor and the value of the resistor.

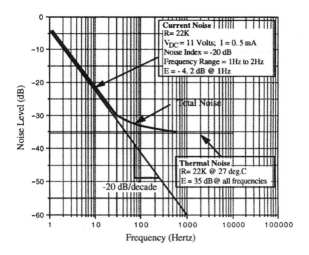

FIGURE 1.8 The total resistor noise is the sum of current noise and thermal noise. The current noise approaches the thermal noise at higher frequencies. (*Source:* Phillips Components, Discrete Products Division, *1990–91 Resistor/Capacitor Data Book,* 1991. With permission.)

The magnitude of current noise is directly proportional to the square root of current. The current noise magnitude is usually expressed by a noise index given as the ratio of the root-mean-square current noise voltage (E_{RMS}) over one decade bandwidth to the average voltage caused by a specified constant current passed through the resistor at a specified hot-spot temperature (Phillips, 1991).

$$\text{N.I.} = 20 \log_{10}\left(\frac{\text{noise voltage}}{\text{dc voltage}}\right) \tag{1.10}$$

$$E_{RMS} = V_{dc} \times 10^{\text{N.I.}/20} \sqrt{\log\left(\frac{f_2}{f_1}\right)} \tag{1.11}$$

where N.I. is the noise index, V_{dc} is the dc voltage drop across the resistor, and f_1 and f_2 represent the frequency range over which the noise is being computed. Units of noise index are μV/V. At higher frequencies, the current noise becomes less dominant compared to Johnson noise.

Precision film resistors have extremely low noise. Composition resistors show some degree of noise due to internal electrical contacts between the conducting particles held together with the binder. Wire-wound resistors are essentially free of electrical noise unless resistor terminations are faulty.

Power Rating and Derating Curves

Resistors must be operated within specified temperature limits to avoid permanent damage to the materials. The temperature limit is defined in terms of the maximum power, called the *power rating*, and the derating curve. The power rating of a resistor is the maximum power in watts which the resistor can dissipate. The maximum power rating is a function of resistor material, maximum voltage rating, resistor dimensions, and maximum allowable hot-spot temperature. The maximum hot-spot temperature is the temperature of the hottest part on the resistor when dissipating full-rated power at rated ambient temperature.

The maximum allowable power rating as a function of the ambient temperature is given by the derating curve. Figure 1.9 shows a typical power rating curve for a resistor. The derating curve is usually linearly drawn from the full-rated load temperature to the maximum allowable no-load temperature. A resistor may be operated at ambient temperatures above the maximum full-load ambient temperature if operating at lower than full-rated power capacity. The maximum allowable no-load temperature is also the maximum storage temperature for the resistor.

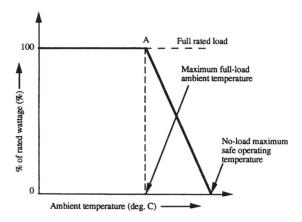

FIGURE 1.9   Typical derating curve for resistors.

Voltage Rating of Resistors

The maximum voltage that may be applied to the resistor is called the **voltage rating** and is related to the power rating by

$$V = \sqrt{PR} \qquad (1.12)$$

where V is the voltage rating (V), P is the power rating (W), and R is the resistance (Ω). For a given value of voltage and power rating, a critical value of resistance can be calculated. For values of resistance below the critical value, the maximum voltage is never reached; for values of resistance above the critical value, the power dissipated is lower than the rated power (Figure 1.10).

Color Coding of Resistors

Resistors are generally identified by color coding or direct digital marking. The color code, given in Table 1.1, is commonly used in composition resistors and film resistors, and essentially consists of four bands of different colors. The first band is the most significant figure, the second band is the second significant figure, the third band is the multiplier or the number of zeros that have to be added after the first two significant figures, and the fourth band is the tolerance on the resistance value. If the fourth band is not present, the resistor tolerance is the standard 20% above and below the rated value. When the color code is used on fixed wire-wound resistors, the first band is applied in double width.

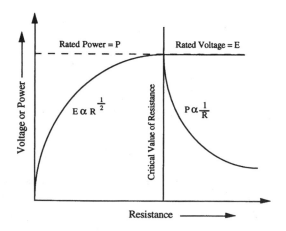

FIGURE 1.10   Relationship of applied voltage and power above and below the critical value of resistance.

TABLE 1.1   Color Code Table for Resistors

Color	First Band	Second Band	Third Band	Fourth Band Tolerance, %
Black	0	0	1	
Brown	1	1	10	
Red	2	2	100	
Orange	3	3	1,000	
Yellow	4	4	10,000	
Green	5	5	100,000	
Blue	6	6	1,000,000	
Violet	7	7	10,000,000	
Gray	8	8	100,000,000	
White	9	9	1,000,000,000	
Gold			0.1	5%
Silver			0.01	10%
No band				20%

Blanks in the table represent situations which do not exist in the color code.

Resistor Types

Resistors can be broadly categorized as fixed, variable, and special-purpose. Each of these resistor types is discussed in detail with typical ranges of their characteristics.

Fixed Resistors

The fixed resistors are those whose value cannot be varied after manufacture. Fixed resistors are classified into composition resistors, wire-wound resistors, and metal-film resistors. Table 1.2 outlines the characteristics of some typical fixed resistors.

Wire-Wound Resistors. Wire-wound resistors are made by winding wire of nickel–chromium alloy on a ceramic tube covering with a vitreous coating. The spiral winding has inductive and capacitive characteristics that make it unsuitable for operation above 50 kHz. The frequency limit can be raised by noninductive winding so that the magnetic fields produced by the two parts of the winding cancel.

Composition Resistors. Composition resistors are composed of carbon particles mixed with a binder. This mixture is molded into a cylindrical shape and hardened by baking. Leads are attached axially to each end, and the assembly is encapsulated in a protective encapsulation coating. Color bands on the outer surface indicate the resistance value and tolerance. Composition resistors are economical and exhibit low noise levels for resistances above 1 MΩ. Composition resistors are usually rated for temperatures in the neighborhood of 70°C for power ranging from 1/8 to 2 W. Composition resistors have end-to-end shunted capacitance that may be noticed at frequencies in the neighborhood of 100 kHz, especially for resistance values above 0.3 MΩ.

Metal-Film Resistors. Metal-film resistors are commonly made of nichrome, tin-oxide, or tantalum nitride, either hermetically sealed or using molded-phenolic cases. Metal-film resistors are not as stable as the

TABLE 1.2   Characteristics of Typical Fixed Resistors

Resistor Types	Resistance Range	Watt Range	Operating Temperature Range	α, ppm/°C
Wire-wound resistor				
Precision	0.1 to 1.2 MΩ	1/8 to 1/4	−55 to 145	10
Power	0.1 to 180 kΩ	1 to 210	−55 to 275	260
Metal-film resistor				
Precision	1 to 250 MΩ	1/20 to 1	−55 to 125	50–100
Power	5 to 100 kΩ	1 to 5	−55 to 155	20–100
Composition resistor				
General purpose	2.7 to 100 MΩ	1/8 to 2	−55 to 130	1500

wirewound resistors. Depending on the application, fixed resistors are manufactured as precision resistors, semiprecision resistors, standard general-purpose resistors, or power resistors. Precision resistors have low voltage and power coefficients, excellent temperature and **time stabilities**, low noise, and very low reactance. These resistors are available in metal-film or wire constructions and are typically designed for circuits having very close resistance tolerances on values. Semiprecision resistors are smaller than precision resistors and are primarily used for current-limiting or voltage-dropping functions in circuit applications. Semiprecision resistors have long-term temperature stability. General-purpose resistors are used in circuits that do not require tight resistance tolerances or long-term stability. For general-purpose resistors, initial resistance variation may be in the neighborhood of 5% and the variation in resistance under full-rated power may approach 20%. Typically, general-purpose resistors have a high coefficient of resistance and high noise levels. Power resistors are used for power supplies, control circuits, and voltage dividers where operational stability of 5% is acceptable. Power resistors are available in wire-wound and film constructions. Film-type power resistors have the advantage of stability at high frequencies and have higher resistance values than wire-wound resistors for a given size.

Variable Resistors

Potentiometers. The potentiometer is a special form of variable resistor with three terminals. Two terminals are connected to the opposite sides of the resistive element, and the third connects to a sliding contact that can be adjusted as a voltage divider.

Potentiometers are usually circular in form with the movable contact attached to a shaft that rotates. Potentiometers are manufactured as carbon composition, metallic film, and wire-wound resistors available in single-turn or multiturn units. The movable contact does not go all the way toward the end of the resistive element, and a small resistance called the *hop-off* resistance is present to prevent accidental burning of the resistive element.

Rheostat. The rheostat is a current-setting device in which one terminal is connected to the resistive element and the second terminal is connected to a movable contact to place a selected section of the resistive element into the circuit. Typically, rheostats are wire-wound resistors used as speed controls for motors, ovens, and heater controls and in applications where adjustments on the voltage and current levels are required, such as voltage dividers and bleeder circuits.

Special-Purpose Resistors

Integrated Circuit Resistors. Integrated circuit resistors are classified into two general categories: semiconductor resistors and deposited film resistors. Semiconductor resistors use the bulk resistivity of **doped** semiconductor regions to obtain the desired resistance value. Deposited film resistors are formed by depositing resistance films on an insulating substrate which are etched and patterned to form the desired resistive network. Depending on the thickness and dimensions of the deposited films, the resistors are classified into thick-film and thin-film resistors.

Semiconductor resistors can be divided into four types: diffused, bulk, pinched, and ion-implanted. Table 1.3 shows some typical resistor properties for semiconductor resistors. Diffused semiconductor resistors use resistivity of the diffused region in the semiconductor substrate to introduce a resistance in the circuit. Both n-type and p-type diffusions are used to form the diffused resistor.

A bulk resistor uses the bulk resistivity of the semiconductor to introduce a resistance into the circuit. Mathematically the sheet resistance of a bulk resistor is given by

$$R_{\text{sheet}} = \frac{\rho_e}{d} \tag{1.13}$$

where R_{sheet} is the sheet resistance in (Ω/square), ρ_e is the sheet resistivity (Ω/square), and d is the depth of the n-type **epitaxial layer.**

Pinched resistors are formed by reducing the effective cross-sectional area of diffused resistors. The reduced cross section of the diffused length results in extremely high sheet resistivities from ordinary diffused resistors.

TABLE 1.3   Typical Characteristics of Integrated Circuit Resistors

Resistor Type	Sheet Resistivity (per square)	Temperature Coefficient (ppm/°C)
Semiconductor		
Diffused	0.8 to 260 Ω	1100 to 2000
Bulk	0.003 to 10 kΩ	2900 to 5000
Pinched	0.001 to 10 kΩ	3000 to 6000
Ion-implanted	0.5 to 20 kΩ	100 to 1300
Deposited resistors		
Thin-film		
Tantalum	0.01 to 1 kΩ	∓100
SnO$_2$	0.08 to 4 kΩ	−1500 to 0
Ni–Cr	40 to 450 Ω	∓100
Cermet (Cr–SiO)	0.03 to 2.5 kΩ	∓150
Thick-film		
Ruthenium–silver	10 Ω to 10 MΩ	∓200
Palladium–silver	0.01 to 100 kΩ	−500 to 150

Ion-implanted resistors are formed by implanting ions on the semiconductor surface by bombarding the silicon lattice with high-energy ions. The implanted ions lie in a very shallow layer along the surface (0.1 to 0.8 μm). For similar thicknesses ion-implanted resistors yield sheet resistivities 20 times greater than diffused resistors. Table 1.3 shows typical properties of diffused, bulk, pinched, and ion-implanted resistors. Typical sheet resistance values range from 80 to 250 Ω/square.

Varistors. Varistors are voltage-dependent resistors that show a high degree of nonlinearity between their resistance value and applied voltage. They are composed of a nonhomogeneous material that provides a rectifying action. Varistors are used for protection of electronic circuits, semiconductor components, collectors of motors, and relay contacts against overvoltage.

The relationship between the voltage and current of a varistor is given by

$$V = kI^{\beta} \tag{1.14}$$

where V is the voltage (V), I is the current (A), and k and β are constants that depend on the materials and manufacturing process. The electrical characteristics of a varistor are specified by its β and k values.

Varistors in Series. The resultant k value of n varistors connected in series is nk. This can be derived by considering n varistors connected in series and a voltage nV applied across the ends. The current through each varistor remains the same as for V volts over one varistor. Mathematically, the voltage and current are expressed as

$$nV = k_1 I^{\beta} \tag{1.15}$$

Equating the expressions (1.14) and (1.15), the equivalent constant k_1 for the series combination of varistors is given as

$$k_1 = nk \tag{1.16}$$

Varistors in Parallel. The equivalent k value for a parallel combination of varistors can be obtained by connecting n varistors in parallel and applying a voltage V across the terminals. The current through the varistors will still be n times the current through a single varistor with a voltage V across it. Mathematically the current and voltage are related as

$$V = k_2 (nI)^{\beta} \tag{1.17}$$

From Equation (1.14) and Equation (1.17) the equivalent constant k_2 for the series combination of varistors is given as

$$k_2 = \frac{k}{n^\beta} \tag{1.18}$$

Thermistors. Thermistors are resistors that change their resistance exponentially with changes in temperature. If the resistance decreases with increase in temperature, the resistor is called a negative temperature coefficient (NTC) resistor. If the resistance increases with temperature, the resistor is called a positive temperature coefficient (PTC) resistor.

NTC thermistors are ceramic semiconductors made by sintering mixtures of heavy metal oxides such as manganese, nickel, cobalt, copper, and iron. The resistance temperature relationship for NTC thermistors is

$$R_T = Ae^{B/T} \tag{1.19}$$

where T is temperature (K), R_T is the resistance (Ω), and A, B are constants whose values are determined by conducting experiments at two temperatures and solving the equations simultaneously.

PTC thermistors are prepared from $BaTiO_3$ or solid solutions of $PbTiO_3$ or $SrTiO_3$. The resistance temperature relationship for PTC thermistors is

$$R_T = A + Ce^{BT} \tag{1.20}$$

where T is temperature (K), R_T is the resistance (Ω), and A, B are constants determined by conducting experiments at two temperatures and solving the equations simultaneously. Positive thermistors have a PTC only between certain temperature ranges. Outside this range the temperature is either zero or negative. Typically, the absolute value of the temperature coefficient of resistance for PTC resistors is much higher than for NTC resistors.

Defining Terms

Doping: The intrinsic carrier concentration of semiconductors (e.g., Si) is too low to allow controlled charge transport. For this reason some impurities called dopants are purposely added to the semiconductor. The process of adding dopants is called doping. Dopants may belong to group IIIA (e.g., boron) or group VA (e.g., phosphorus) in the periodic table. If the elements belong to the group IIIA, the resulting semiconductor is called a p-type semiconductor. On the other hand, if the elements belong to the group VA, the resulting semiconductor is called an n-type semiconductor.

Epitaxial layer: Epitaxy refers to processes used to grow a thin crystalline layer on a crystalline substrate. In the epitaxial process the wafer acts as a seed crystal. The layer grown by this process is called an epitaxial layer.

Resistivity: The resistance of a conductor with unit length and unit cross-sectional area.

Temperature coefficient of resistance: The change in electrical resistance of a resistor per unit change in temperature.

Time stability: The degree to which the initial value of resistance is maintained to a stated degree of certainty under stated conditions of use over a stated period of time. Time stability is usually expressed as a percent or parts per million change in resistance per 1000 h of continuous use.

Voltage coefficient of resistance: The change in resistance per unit change in voltage, expressed as a percentage of the resistance at 10% of rated voltage.

Voltage drop: The difference in potential between the two ends of the resistor measured in the direction of flow of current. The voltage drop is $V = IR$, where V is the voltage across the resistor, I is the current through the resistor, and R is the resistance.

Voltage rating: The maximum voltage that may be applied to the resistor.

References

Phillips Components, Discrete Products Division, *1990–91 Resistor/Capacitor Data Book,* 1991.
C.C. Wellard, *Resistance and Resistors,* New York: McGraw-Hill, 1960.

Further Information

IEEE Transactions on Electron Devices and *IEEE Electron Device Letters:* Published monthly by the Institute of Electrical and Electronics Engineers.
IEEE Components, Hybrids and Manufacturing Technology: Published quarterly by the Institute of Electrical and Electronics Engineers.
G.W.A. Dummer, *Materials for Conductive and Resistive Functions,* New York: Hayden Book Co., 1970.
H.F. Littlejohn and C.E. Burckel, *Handbook of Power Resistors,* Mount Vernon, NY: Ward Leonard Electric Company, 1951.
I.R. Sinclair, *Passive Components: A User's Guide,* Oxford: Heinemann Newnes, 1990.

1.2   Capacitors and Inductors

Glen Ballou

Capacitors

If a potential difference is found between two points, an electric **field** exists that is the result of the separation of unlike charges. The strength of the field will depend on the amount the charges have been separated.

Capacitance is the concept of energy storage in an electric field and is restricted to the area, shape, and spacing of the **capacitor** plates and the property of the material separating them.

When electrical current flows into a capacitor, a force is established between two parallel plates separated by a **dielectric**. This energy is stored and remains even after the input is removed. By connecting a **conductor** (a resistor, hard wire, or even air) across the capacitor, the charged capacitor can regain electron balance, that is, discharge its stored energy.

The value of a parallel-plate capacitor can be found with the equation

$$C = \frac{x\epsilon[(N-1)A]}{d} \times 10^{-13} \tag{1.21}$$

where C = capacitance, F; ϵ = dielectric constant of insulation; d = spacing between plates; N = number of plates; A = area of plates; and x = 0.0885 when A and d are in centimeters, and x = 0.225 when A and d are in inches.

The work necessary to transport a unit charge from one plate to the other is

$$e = kg \tag{1.22}$$

where e = volts expressing energy per unit charge, g = coulombs of charge already transported, and k = proportionality factor between work necessary to carry a unit charge between the two plates and charge already transported. It is equal to $1/C$, where C is the capacitance, F.

The value of a capacitor can now be calculated from the equation

$$C = \frac{q}{e} \tag{1.23}$$

where q = charge (C) and e is found with Equation (1.22).

The energy stored in a capacitor is

$$W = \frac{CV^2}{2} \tag{1.24}$$

where W = energy, J; C = capacitance, F; and V = applied voltage, V.

The **dielectric constant** of a material determines the electrostatic energy which may be stored in that material per unit volume for a given voltage. The value of the dielectric constant expresses the ratio of a capacitor in a vacuum to one using a given dielectric. The dielectric of air is 1, the reference unit employed for expressing the dielectric constant. As the dielectric constant is increased or decreased, the capacitance will increase or decrease, respectively. Table 1.4 lists the dielectric constants of various materials.

The dielectric constant of most materials is affected by both temperature and frequency, except for quartz, Styrofoam, and Teflon, whose dielectric constants remain essentially constant.

The equation for calculating the *force of attraction* between two plates is

$$F = \frac{AV^2}{k(1504S)^2} \tag{1.25}$$

where F = attraction force, dyn; A = area of one plate, cm^2; V = potential energy difference, V; k = dielectric coefficient; and S = separation between plates, cm.

The Q for a capacitor when the resistance and capacitance is in series is

$$Q = \frac{1}{2\pi fRC} \tag{1.26}$$

where Q = ratio expressing the factor of merit; f = frequency, Hz; R = resistance, Ω; and C = capacitance, F.

When capacitors are connected in *series*, the total capacitance is

$$C_{\mathrm{T}} = \frac{1}{1/C_1 + 1/C_2 + \cdots + 1/C_n} \tag{1.27}$$

and is always less than the value of the smallest capacitor.

When capacitors are connected in *parallel*, the total capacitance is

$$C_{\mathrm{T}} = C_1 + C_2 + \cdots + C_n \tag{1.28}$$

and is always larger than the largest capacitor.

When a voltage is applied across a group of capacitors connected in series, the voltage drop across the combination is equal to the applied voltage. The drop across each individual capacitor is inversely proportional to its capacitance:

$$V_{\mathrm{C}} = \frac{V_A C_X}{C_{\mathrm{T}}} \tag{1.29}$$

TABLE 1.4 Comparison of Capacitor Dielectric Constants

Dielectric	K (Dielectric Constant)
Air or vacuum	1.0
Paper	2.0–6.0
Plastic	2.1–6.0
Mineral oil	2.2–2.3
Silicone oil	2.7–2.8
Quartz	3.8–4.4
Glass	4.8–8.0
Porcelain	5.1–5.9
Mica	5.4–8.7
Aluminum oxide	8.4
Tantalum pentoxide	26
Ceramic	12–400,000

Source: G. Ballou, *Handbook for Sound Engineers, The New Audio Cyclopedia,* Carmel, Ind.: Macmillan Computer Publishing Company, 1991. With permission.

where V_C = voltage across the individual capacitor in the series (C_1, C_2,...,C_n), V; V_A = applied voltage, V; C_T = total capacitance of the series combination, F; and C_X = capacitance of individual capacitor under consideration, F.

In an ac circuit, the **capacitive reactance**, or the **impedance**, of the capacitor is

$$X_C = \frac{1}{2\pi f C} \tag{1.30}$$

where X_C = capacitive reactance, Ω; f = frequency, Hz; and C = capacitance, F. The current will lead the voltage by 90° in a circuit with a pure capacitor.

When a dc voltage is connected across a capacitor, a time t is required to charge the capacitor to the applied voltage. This is called a **time constant** and is calculated with the equation

$$t = RC \tag{1.31}$$

where t = time, sec; R = resistance, Ω; and C = capacitance, F.

In a circuit consisting of pure resistance and capacitance, the *time constant t* is defined as the time required to charge the capacitor to 63.2% of the applied voltage.

During the next time constant, the capacitor charges to 63.2% of the remaining difference of full value, or to 86.5% of the full value. The charge on a capacitor can never actually reach 100% but is considered to be 100% after five time constants. When the voltage is removed, the capacitor discharges to 63.2% of the full value.

Capacitance is expressed in microfarads (μF, or 10^{-6} F) or picofarads (pF, or 10^{-12} F) with a stated accuracy or tolerance. Tolerance may also be stated as GMV (guaranteed minimum value), sometimes referred to as MRV (minimum rated value).

All capacitors have a *maximum working voltage* that must not be exceeded and is a combination of the dc value plus the peak ac value which may be applied during operation.

Quality Factor (Q)

Quality factor is the ratio of the capacitor's **reactance** to its resistance at a specified frequency and is found by the equation

$$Q = \frac{1}{2\pi f C R} = \frac{1}{PF} \tag{1.32}$$

where Q = quality factor; f = frequency, Hz; C = value of capacitance, F; R = internal resistance, Ω; and PF = power factor.

Power Factor (PF)

Power factor is the preferred measurement in describing capacitive losses in ac circuits. It is the fraction of input volt-amperes (or power) dissipated in the capacitor dielectric and is virtually independent of the capacitance, applied voltage, and frequency.

Equivalent Series Resistance (ESR)

Equivalent series resistance is expressed in ohms or milliohms (Ω, mΩ) and is derived from lead resistance, termination losses, and dissipation in the dielectric material.

Equivalent Series Inductance (ESL)

The *equivalent series inductance* can be useful or detrimental. It reduces high-frequency performance; however, it can be used in conjunction with the internal capacitance to form a resonant circuit.

Dissipation Factor (DF)

The **dissipation factor** in percentage is the ratio of the effective series resistance of a capacitor to its reactance at a specified frequency. It is the reciprocal of *quality factor* (*Q*) and an indication of power loss within the capacitor. It should be as low as possible.

Insulation Resistance

Insulation resistance is the resistance of the dielectric material and determines the time a capacitor, once charged, will hold its charge. A discharged capacitor has a low insulation resistance; however once charged to its rated value, it increases to megohms. The leakage in electrolytic capacitors should not exceed

$$I_{\mathrm{L}} = 0.04C + 0.30 \tag{1.33}$$

where $I_{\mathrm{L}} =$ leakage current, μA, and $C =$ capacitance, μF.

Dielectric Absorption (DA)

The *dielectric absorption* is a reluctance of the dielectric to give up stored electrons when the capacitor is discharged. This is often called "memory" because if a capacitor is discharged through a resistance and the resistance is removed, the electrons that remained in the dielectric will reconvene on the electrode, causing a voltage to appear across the capacitor. DA is tested by charging the capacitor for 5 min, discharging it for 5 sec, then having an open circuit for 1 min after which the recovery voltage is read. The percentage of DA is defined as the ratio of recovery to charging voltage $\times$ 100.

Types of Capacitors

Capacitors are used to filter, couple, tune, block dc, pass ac, bypass, shift phase, compensate, feed through, isolate, store energy, suppress noise, and start motors. They must also be small, lightweight, reliable, and withstand adverse conditions.

Capacitors are grouped according to their dielectric material and mechanical configuration.

Ceramic Capacitors

Ceramic capacitors are used most often for bypass and coupling applications (Figure 1.11). Ceramic capacitors can be produced with a variety of *K* values (dielectric constant). A high *K* value translates to small size and less stability. High-*K* capacitors with a dielectric constant >3000 are physically small and have values between 0.001 to several microfarads.

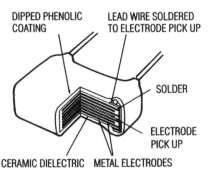

Voltage Ratings: 50 and 100 WVDC
Capacitance Range: 1.0 pF to 4.7 µF
Size Range: 0.150″ x 0.150″ 0.100″ to 0.500″ x 0.500″ x 0.125″
Primary Applications: Used where capacitors with EIA
    Characteristics Z5U, X7R, and COG must be selected
    to meet specific requirements.

FIGURE 1.11   Monolythic® multilayer ceramic capacitors. (Courtesy of Sprague Electric Company.)

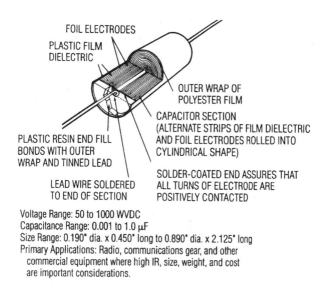

FOIL ELECTRODES
PLASTIC FILM
DIELECTRIC
OUTER WRAP OF
POLYESTER FILM
CAPACITOR SECTION
(ALTERNATE STRIPS OF FILM DIELECTRIC
AND FOIL ELECTRODES ROLLED INTO
CYLINDRICAL SHAPE)
PLASTIC RESIN END FILL
BONDS WITH OUTER
WRAP AND TINNED LEAD
SOLDER-COATED END ASSURES THAT
ALL TURNS OF ELECTRODE ARE
POSITIVELY CONTACTED
LEAD WIRE SOLDERED
TO END OF SECTION

Voltage Range: 50 to 1000 WVDC
Capacitance Range: 0.001 to 1.0 μF
Size Range: 0.190" dia. x 0.450" long to 0.890" dia. x 2.125" long
Primary Applications: Radio, communications gear, and other
   commercial equipment where high IR, size, weight, and cost
   are important considerations.

FIGURE 1.12   Film-wrapped film capacitors. (Courtesy of Sprague Electric Company.)

Good temperature stability requires capacitors to have a K value between 10 and 200. If high Q is also required, the capacitor will be physically larger. Ceramic capacitors with a zero temperature change are called **negative-positive-zero (NPO)** and come in a capacitance range of 1.0 pF to 0.033 μF.

An N750 temperature-compensated capacitor is used when accurate capacitance is required over a large temperature range. The 750 indicates a 750-ppm decrease in capacitance with a 1°C increase in temperature (750 ppm/°C). This equates to a 1.5% decrease in capacitance for a 20°C temperature increase. N750 capacitors come in values between 4.0 and 680 pF.

Film Capacitors

Film capacitors consist of alternate layers of metal foil and one or more layers of a flexible plastic insulating material (dielectric) in ribbon form rolled and encapsulated (see Figure 1.12).

Mica Capacitors

Mica capacitors have small capacitance values and are usually used in high-frequency circuits. They are constructed as alternate layers of metal foil and mica insulation, which are stacked and encapsulated, or are silvered mica, where a silver electrode is screened on the mica insulators.

Paper-Foil-Filled Capacitors

Paper-foil-filled capacitors are often used as motor capacitors and are rated at 60 Hz. They are made of alternate layers of aluminum and paper saturated with oil that are rolled together. The assembly is mounted in an oil-filled, hermetically sealed metal case.

Electrolytic Capacitors

Electrolytic capacitors provide high capacitance in a tolerable size; however, they do have drawbacks. Low temperatures reduce performance, while high temperatures dry them out. The **electrolytes** themselves can leak and corrode the equipment. Repeated surges above the rated working voltage, excessive ripple currents, and high operating temperature reduce performance and shorten capacitor life.

Electrolytic capacitors are manufactured by an electrochemical formation of an oxide film on a metal surface. The metal on which the oxide film is formed serves as the **anode** or positive terminal of the capacitor; the oxide film is the dielectric, and the **cathode** or negative terminal is either a conducting liquid or a gel.

The equivalent circuit of an electrolytic capacitor is shown in Figure 1.13, where A and B are the capacitor terminals, C is the effective capacitance, and L is the self-inductance of the capacitor caused by terminals, electrodes, and geometry.

The shunt resistance (insulation resistance) R_s accounts for the dc leakage current. Heat is generated in the ESR from ripple current and in the shunt resistance by voltage. The ESR is due to the spacer–electrolyte–oxide system and varies only slightly except at low temperature, where it increases greatly.

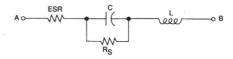

FIGURE 1.13   Simplified equilavent circuit of an electrolytic capacitor.

The *impedance* of a capacitor (Figure 1.14) is frequency-dependent. The initial downward slope is caused by the capacitive reactance X_C. The trough (lowest impedance) is almost totally resistive, and the upward slope is due to the capacitor's self-inductance X_L. An ESR plot would show an ESR decrease to about 5–10 kHz, remaining relatively constant thereafter.

Leakage current is the direct current that passes through a capacitor when a correctly polarized dc voltage is applied to its terminals. It is proportional to temperature, becoming increasingly important at elevated **ambient temperatures**. Leakage current decreases slowly after voltage is applied, reaching steady-state conditions in about 10 min.

If a capacitor is connected with reverse polarity, the oxide film is forward-biased, offering very little resistance to current flow. This causes overheating and self-destruction of the capacitor.

The total heat generated within a capacitor is the sum of the heat created by the $I_{leakage} \times V_{applied}$ and the I^2R losses in the ESR.

The ac **ripple current** rating is very important in filter applications because excessive current produces temperature rise, shortening capacitor life. The maximum permissible rms ripple current is limited by the internal temperature and the rate of heat dissipation from the capacitor. Lower ESR and longer enclosures increase the ripple current rating.

Capacitor life expectancy is doubled for each 10°C decrease in operating temperature, so a capacitor operating at room temperature will have a life expectancy 64 times that of the same capacitor operating at 85°C (185°F).

The *surge voltage* specification of a capacitor determines its ability to withstand high transient voltages that generally occur during the starting up period of equipment. Standard tests generally specify a short on and long off period for an interval of 24 h or more, and the allowable surge voltage levels are generally 10% above the rated voltage of the capacitor.

Figure 1.15 shows how temperature, frequency, time, and applied voltage affect electrolytic capacitors.

Aluminum Electrolytic Capacitors. *Aluminum electrolytic capacitors* use aluminum as the base material (Figure 1.16). The surface is often etched to increase the surface area as much as 100 times that of unetched foil, resulting in higher capacitance in the same volume.

Aluminum electrolytic capacitors can withstand up to 1.5 V of reverse voltage without detriment. Higher reverse voltages, when applied over extended periods, lead to loss of capacitance. Excess reverse voltages applied for short periods cause some change in capacitance but not to capacitor failure.

Large-value capacitors are often used to filter dc power supplies. After a capacitor is charged, the rectifier stops conducting and the capacitor discharges into the load, as shown in Figure 1.17, until the next cycle. Then the capacitor recharges again to the peak voltage. The Δe is equal to the total peak-to-peak ripple voltage and

is a complex wave containing many harmonics of the fundamental ripple frequency, causing the noticeable heating of the capacitor.

Tantalum Capacitors. *Tantalum electrolytics* are the preferred type where high reliability and long service life are paramount considerations.

Tantalum capacitors have as much as three times better capacitance per volume efficiency than aluminum electrolytic capacitors, because tantalum pentoxide has a dielectric constant three times greater than that of aluminum oxide (see Table 1.4).

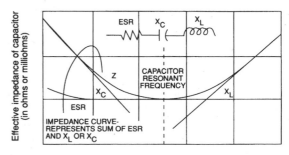

FIGURE 1.14   Impedance characteristics of a capacitor.

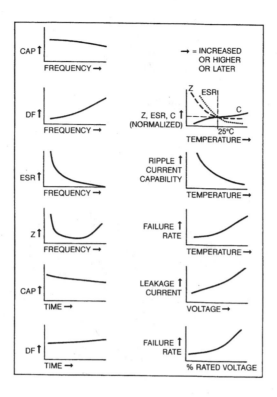

FIGURE 1.15 Variations in aluminum electrolytic characteristics caused by temperature, frequency, time, and applied voltage. (Courtesy of Sprague Electric Company.)

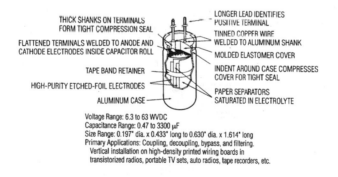

FIGURE 1.16 Verti-lytic® miniature single-ended aluminum electrolytic capacitor. (Courtesy of Sprague Electric Company.)

The capacitance of any capacitor is determined by the surface area of the two conducting plates, the distance between the plates, and the dielectric constant of the insulating material between the plates (see Equation (1.21)).

In tantalum electrolytics, the distance between the plates is the thickness of the tantalum pentoxide film, and since the dielectric constant of the tantalum pentoxide is high, the capacitance of a tantalum capacitor is high.

Tantalum capacitors contain either liquid or solid electrolytes. The liquid electrolyte in wet-slug and foil capacitors, generally sulfuric acid, forms the cathode (negative) plate. In solid-electrolyte capacitors, a dry material, manganese dioxide, forms the cathode plate.

Foil Tantalum Capacitors. *Foil tantalum capacitors* can be designed to voltage values up to 300 V dc. Of the three types of tantalum electrolytic capacitors, the foil design has the lowest capacitance per unit volume and is best suited for the higher voltages primarily found in older designs of equipment. It is expensive and used only where neither a solid-electrolyte (Figure 1.18) nor a wet-slug (Figure 1.19) tantalum capacitor can be employed.

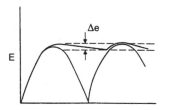

FIGURE 1.17   Full wave capacitor charge and discharge.

Foil tantalum capacitors are generally designed for operation over the temperature range of −55 to +125°C (−67 to +257°F) and are found primarily in industrial and military electronics equipment.

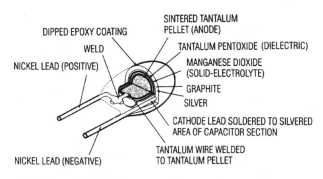

Voltage Range: 3 to 50 WVDC
Capacitance Range: 0.10 to 680 μF
Size Range: 0.175" dia. x 0.280" high to 0.400" dia. x 0.750" high
Primary Applications: For printed wiring boards applications where
   low cost, small size, high stability, low d-c leakage , and low
   dissipation factor are important.

FIGURE 1.18   Tantalex® solid electrolyte tantalum capacitor. (Courtesy of Sprague Electric Company.)

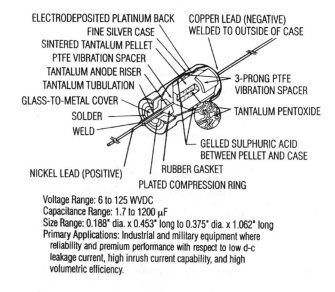

Voltage Range: 6 to 125 WVDC
Capacitance Range: 1.7 to 1200 μF
Size Range: 0.188" dia. x 0.453" long to 0.375" dia. x 1.062" long
Primary Applications: Industrial and military equipment where
   reliability and premium performance with respect to low d-c
   leakage current, high inrush current capability, and high
   volumetric efficiency.

FIGURE 1.19   Hermetically sealed sintered-anode tantalum capacitor. (Courtesy of Sprague Electric Company.)

Solid-electrolyte sintered-anode tantalum capacitors differ from the wet versions in their electrolyte, which is manganese dioxide.

Another variation of the solid-electrolyte tantalum capacitor encases the element in plastic resins, such as epoxy materials offering excellent reliability and high stability for consumer and commercial electronics with the added feature of low cost.

Still other designs of "solid tantalum" capacitors use plastic film or sleeving as the encasing material, and others use metal shells that are backfilled with an epoxy resin. Finally, there are small tubular and rectangular molded plastic encasements.

Wet-electrolyte sintered-anode tantalum capacitors, often called "wet-slug" tantalum capacitors, use a pellet of sintered tantalum powder to which a lead has been attached, as shown in Figure 1.19. This anode has an enormous surface area for its size.

Wet-slug tantalum capacitors are manufactured in a voltage range to 125 V dc.

Use Considerations. Foil tantalum capacitors are used only where high-voltage constructions are required or where there is substantial reverse voltage applied to a capacitor during circuit operation.

Wet sintered-anode capacitors, or "wet-slug" tantalum capacitors, are used where low dc leakage is required. The conventional "silver can" design will not tolerate reverse voltage. In military or aerospace applications where utmost reliability is desired, tantalum cases are used instead of silver cases. The tantalum-cased wet-slug units withstand up to 3 V reverse voltage and operate under higher ripple currents and at temperatures up to 200°C (392°F).

Solid-electrolyte designs are the least expensive for a given rating and are used where their very small size is important. They will typically withstand a reverse voltage up to 15% of the rated dc working voltage. They also have good low-temperature performance characteristics and freedom from corrosive electrolytes.

Inductors

Inductance is used for the storage of magnetic energy. Magnetic energy is stored as long as current keeps flowing through the inductor. In a perfect **inductor**, the current of a sine wave lags the voltage by 90°.

Impedance

Inductive reactance X_L, the impedance of an inductor to an ac signal, is found by the equation

$$X_L = 2\pi f L \tag{1.34}$$

where X_L = inductive reactance, Ω; f = frequency, Hz; and L = inductance, H.

The type of wire used for its construction does not affect the inductance of a **coil**. Q of the coil will be governed by the resistance of the wire. Therefore coils wound with silver or gold wire have the highest Q for a given design.

To increase inductance, inductors are connected in series. The total inductance will always be greater than the largest inductor:

$$L_T = L_1 + L_2 + \cdots + L_n \tag{1.35}$$

To reduce inductance, inductors are connected in parallel:

$$L_T = \frac{1}{1/L_1 + 1/L_2 + \cdots + 1/L_n} \tag{1.36}$$

The total inductance will always be less than the value of the lowest inductor.

Mutual Inductance

Mutual inductance is the property that exists between two conductors carrying current when their magnetic lines of force link together.

The mutual inductance of two coils with fields interacting can be determined by the equation

$$M = \frac{L_A - L_B}{4} \tag{1.37}$$

where M = mutual inductance of L_A and L_B, H; L_A = total inductance, H, of coils L_1 and L_2 with fields aiding; and L_B = total inductance, H, of coils L_1 and L_2 with fields opposing.

The *coupled inductance* can be determined by the following equations. In parallel with fields aiding:

$$L_T = \frac{1}{\dfrac{1}{L_1 + M} + \dfrac{1}{L_2 + M}} \tag{1.38}$$

In parallel with fields opposing:

$$L_T = \frac{1}{\dfrac{1}{L_1 - M} + \dfrac{1}{L_2 - M}} \tag{1.39}$$

In series with fields aiding:

$$L_T = L_1 + L_2 + 2M \tag{1.40}$$

In series with fields opposing:

$$L_T = L_1 + L_2 - 2M \tag{1.41}$$

where L_T = total inductance, H; L_1 and L_2 = inductances of the individual coils, H; and M = mutual inductance, H.

When two coils are inductively coupled to give transformer action, the coupling coefficient is determined by

$$K = \frac{M}{\sqrt{L_1 L_2}} \tag{1.42}$$

where K = coupling coefficient; M = mutual inductance, H; and L_1 and L_2 = inductances of the two coils, H.

An inductor in a circuit has a reactance equal to $j2\pi fL$ Ω. Mutual inductance in a circuit has a reactance equal to $j2\pi fL$ Ω. The operator j denotes that the reactance dissipates no energy; however, it does oppose current flow.

The energy stored in an inductor can be determined by the equation

$$W = \frac{LI^2}{2} \tag{1.43}$$

where W = energy, J (W·s); L = inductance, H; and I = current, A.

Coil Inductance

Inductance is related to the turns in a coil as follows:

1. The inductance is proportional to the square of the turns.
2. The inductance increases as the length of the **winding** is increased.
3. A shorted turn decreases the inductance, affects the frequency response, and increases the insertion loss.
4. The inductance increases as the permeability of the core material increases.
5. The inductance increases with an increase in the cross-sectional area of the core material.
6. Inductance is increased by inserting an iron core into the coil.
7. Introducing an air gap into a choke reduces the inductance.

A conductor moving at any angle to the lines of force cuts a number of lines of force proportional to the sine of the angles. Thus

$$V = \beta L v \sin\theta \times 10^{-8} \tag{1.44}$$

where β = flux density; L = length of the conductor, cm; and v = velocity, cm/sec, of conductor moving at an angle θ.

The maximum voltage induced in a conductor moving in a magnetic field is proportional to the number of magnetic lines of force cut by that conductor. When a conductor moves parallel to the lines of force, it cuts no lines of force; therefore, no current is generated in the conductor. A conductor that moves at right angles to the lines of force cuts the maximum number of lines per inch per second, therefore creating a maximum voltage. The right-hand rule determines direction of the induced electromotive force (emf). The emf is in the direction in which the axis of a right-hand screw, when turned with the velocity vector, moves through the smallest angle toward the flux density vector.

The **magnetomotive force** (mmf) in **ampere-turns** produced by a coil is found by multiplying the number of turns of wire in the coil by the current flowing through it:

$$\text{ampere-turns} = T\left(\frac{V}{R}\right) = TI \tag{1.45}$$

where T = number of turns; V = voltage, V; and R = resistance, Ω.

The inductance of a single layer, a spiral, and multilayer coils can be calculated by using either Wheeler's or Nagaoka's equations. The accuracy of the calculation will vary between 1 and 5%. The inductance of a single-layer coil can be calculated using Wheeler's equation:

$$L = \frac{B^2 N^2}{9B + 10A} \mu H \tag{1.46}$$

For the multilayer coil:

$$L = \frac{0.8B^2 N^2}{6B + 9A + 10C} \mu H \tag{1.47}$$

For the spiral coil:

$$L = \frac{B^2 N^2}{8B + 11C} \mu H \tag{1.48}$$

where B = radius of the winding, N = number of turns in the coil, A = length of the winding, and C = thickness of the winding.

Q

Q is the ratio of the inductive reactance to the internal resistance of the coil and is affected by frequency, inductance, dc resistance, inductive reactance, the type of winding, the core losses, the distributed capacity, and the permeability of the core material.

The Q for a coil where R and L are in series is

$$Q = \frac{2\pi fL}{R} \tag{1.49}$$

where f = frequency, Hz; L = inductance, H; and R = resistance, Ω.

The Q of the coil can be measured using the circuit of Figure 1.20 for frequencies up to 1 MHz. The voltage across the inductance (L) at resonance equals $Q(V)$ (where V is the voltage developed by the oscillator); therefore, it is only necessary to measure the output voltage from the oscillator and the voltage across the inductance.

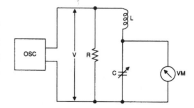

FIGURE 1.20 Circuit for measuring the Q of a coil.

The oscillator voltage is driven across a low value of resistance, R, about 1/100 of the anticipated rf resistance of the LC combination, to assure that the measurement will not be in error by more than 1%. For most measurements, R will be about 0.10 Ω and should have a voltage of 0.1 V. Most oscillators cannot be operated into this low impedance, so a step-down matching transformer must be employed. Make C as large as convenient to minimize the ratio of the impedance looking from the voltmeter to the impedance of the test circuit. The LC circuit is then tuned to resonate and the resultant voltage measured. The value of Q may then be equated:

$$Q = \frac{\text{resonant voltage across } C}{\text{voltage across } R} \tag{1.50}$$

The Q of any coil may be approximated by the equation

$$Q = \frac{2\pi fL}{R} = \frac{X_L}{R} \tag{1.51}$$

where f = the frequency, Hz; L = the inductance, H; R = the dc resistance, Ω (as measured by an ohmmeter); and X_L = the inductive reactance of the coil.

Time Constant

When a dc voltage is applied to an RL circuit, a certain amount of time is required to change the circuit [see text with Equation (1.31)]. The time constant can be determined with the equation

$$T = \frac{L}{R} \tag{1.52}$$

where R = resistance, Ω; L = inductance, H; and T = time, sec.

The *right-hand rule* is used to determine the direction of a magnetic field around a conductor carrying a direct current. Grasp the conductor in the right hand with the thumb extending along the conductor pointing in the direction of the current. With the fingers partly closed, the finger tips will point in the direction of the magnetic field.

Maxwell's rule states, "If the direction of travel of a right-handed corkscrew represents the direction of the current in a straight conductor, the direction of rotation of the corkscrew will represent the direction of the magnetic lines of force."

Impedance

The total impedance created by resistors, capacitors, and inductors in circuits can be determined with the following equations.

For resistance and capacitance in series:

$$Z = \sqrt{R^2 + X_C^2} \tag{1.53}$$

$$\theta = \arctan\frac{X_C}{R} \tag{1.54}$$

For resistance and inductance in series:

$$Z = \sqrt{R^2 + X_L^2} \tag{1.55}$$

$$\theta = \arctan\frac{X_L}{R} \tag{1.56}$$

For inductance and capacitance in series:

$$Z = \begin{cases} X_L - X_C & \text{when } X_L > X_C \\ X_C - X_L & \text{whcn } X_C > X_L \end{cases} \tag{1.57} \tag{1.58}$$

For resistance, inductance, and capacitance in series:

$$Z = \sqrt{R^2 + (X_L - X_C)^2} \tag{1.59}$$

$$\theta = \arctan\frac{X_L - X_C}{R} \tag{1.60}$$

For capacitance and resistance in parallel:

$$Z = \frac{RX_C}{\sqrt{R^2 + X_C^2}} \tag{1.61}$$

For resistance and inductance in parallel:

$$Z = \frac{RX_L}{\sqrt{R^2 + X_L^2}} \tag{1.62}$$

For capacitance and inductance in parallel:

$$Z = \begin{cases} \dfrac{X_L X_C}{X_L - X_C} & \text{when } X_L > X_C \tag{1.63} \\[4mm] \dfrac{X_C X_L}{X_C - X_L} & \text{when } X_C > X_L \tag{1.64} \end{cases}$$

For inductance, capacitance, and resistance in parallel:

$$Z = \frac{R X_L X_C}{\sqrt{X_L^2 X_C^2 + R^2 (X_L - X_C)^2}} \tag{1.65}$$

$$\theta = \arctan \frac{R(X_L - X_C)}{X_L X_C} \tag{1.66}$$

For inductance and series resistance in parallel with resistance:

$$Z = R_2 \sqrt{\frac{R_1^2 + X_L^2}{(R_1 + R_2)^2 + X_L^2}} \tag{1.67}$$

$$\theta = \arctan \frac{X_L R_2}{R_1^2 + X_L^2 + R_1 R_2} \tag{1.68}$$

For inductance and series resistance in parallel with capacitance:

$$Z = X_C \sqrt{\frac{R^2 + X_L^2}{R^2 + (X_L - X_C)^2}} \tag{1.69}$$

$$\theta = \arctan \frac{X_L(X_C - X_L) - R^2}{R X_C} \tag{1.70}$$

For capacitance and series resistance in parallel with inductance and series resistance:

$$Z = \sqrt{\frac{(R_1^2 + X_L^2)(R_2^2 + X_C^2)}{(R_1 + R_2)^2 + (X_L - X_C)^2}} \tag{1.71}$$

$$\theta = \arctan \frac{X_L(R_2^2 + X_C^2) - X_C(R_1^2 + X_L^2)}{R_1(R_2^2 + X_C^2) - R_2(R_1^2 + X_L^2)} \tag{1.72}$$

where $Z =$ impedance, Ω; $R =$ resistance, Ω; $L =$ inductance, H; $X_L =$ inductive reactance, Ω; $X_C =$ capacitive reactance, Ω; and $\theta =$ **phase** angle, degrees, by which current leads voltage in a capacitive circuit or lags voltage in an inductive circuit ($0°$ indicates an in-phase condition).

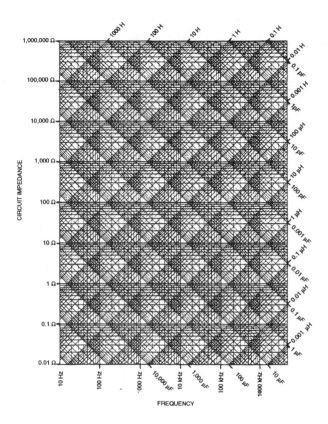

FIGURE 1.21   Reactance chart. (Courtesy AT&T Bell Laboratories.)

Resonant Frequency

When an inductor and capacitor are connected in series or parallel, they form a resonant circuit. The **resonant frequency** can be determined from the equation

$$f = \frac{1}{2\pi\sqrt{LC}} = \frac{1}{2\pi C X_C} = \frac{X_L}{2\pi L} \tag{1.73}$$

where f = frequency, Hz; L = inductance, H; C = capacitance, F; and X_L, X_C = impedance, Ω.

The resonant frequency can also be determined through the use of a reactance chart developed by the Bell Telephone Laboratories (Figure 1.21). This chart can be used for solving problems of inductance, capacitance, frequency, and impedance. If two of the values are known, the third and fourth values may be found with its use.

Defining Terms

Air capacitor: A fixed or variable capacitor in which air is the dielectric material between the capacitor's plates.

Ambient temperature: The temperature of the air or liquid surrounding any electrical part or device. Usually refers to the effect of such temperature in aiding or retarding removal of heat by radiation and convection from the part or device in question.

Ampere-turns: The magnetomotive force produced by a coil, derived by multiplying the number of turns of wire in a coil by the current (A) flowing through it.

Anode: The positive electrode of a capacitor.

Capacitive reactance: The opposition offered to the flow of an alternating or pulsating current by capacitance measured in ohms.

Capacitor: An electrical device capable of storing electrical energy and releasing it at some predetermined rate at some predetermined time. It consists essentially of two conducting surfaces (electrodes) separated by an insulating material or dielectric. A capacitor stores electrical energy, blocks the flow of direct current, and permits the flow of alternating current to a degree dependent essentially upon capacitance and frequency. The amount of energy stored, $E = 0.5\ CV^2$.

Cathode: The capacitor's negative electrode.

Coil: A number of turns of wire in the form of a spiral. The spiral may be wrapped around an iron core or an insulating form, or it may be self-supporting. A coil offers considerable opposition to ac current but very little to dc current.

Conductor: A bare or insulated wire or combination of wires not insulated from one another, suitable for carrying an electric current.

Dielectric: The insulating (nonconducting) medium between the two electrodes (plates) of a capacitor.

Dielectric constant: The ratio of the capacitance of a capacitor with a given dielectric to that of the same capacitor having a vacuum dielectric.

Disk capacitor: A small single-layer ceramic capacitor with a dielectric insulator consisting of conductively silvered opposing surfaces.

Dissipation factor (DF): The ratio of the effective series resistance of a capacitor to its reactance at a specified frequency measured in percent.

Electrolyte: Current-conducting solution between two electrodes or plates of a capacitor, at least one of which is covered by a dielectric.

Electrolytic capacitor: A capacitor solution between two electrodes or plates of a capacitor, at least one of which is covered by a dielectric.

Equivalent series resistance (ESR): All internal series resistance of a capacitor concentrated or "lumped" at one point and treated as one resistance of a capacitor regardless of source, i.e., lead resistance, termination losses, or dissipation in the dielectric material.

Farad: The basic unit of measure in capacitors. Acapacitor charged to 1 volt with a charge of 1 coulomb (1 ampere flowing for 1 sec) has a capacitance of 1 farad.

Field: A general term referring to the region under the influence of a physical agency such as electricity, magnetism, or a combination produced by an electrical charged object.

Impedance (Z): Total opposition offered to the flow of an alternating or pulsating current measured in ohms. (Impedance is the vector sum of the resistance and the capacitive and inductive reactance, i.e., the ratio of voltage to current.)

Inductance: The property which opposes any change in the existing current. Inductance is present only when the current is changing.

Inductive reactance (X_L): The opposition to the flow of alternating or pulsating current by the inductance of a circuit.

Inductor: A conductor used to introduce inductance into a circuit.

Leakage current: Stray direct current of relatively small value which flows through a capacitor when voltage is impressed across it.

Magnetomotive force: The force by which the magnetic field is produced, either by a current flowing through a coil of wire or by the proximity of a magnetized body. The amount of magnetism produced in the first method is proportional to the current through the coil and the number of turns in it.

Mutual inductance: The property that exists between two current-carrying conductors when the magnetic lines of force from one link with those from another.

Negative-positive-zero (NPO): An ultrastable temperature coefficient (± 30 ppm/°C from -55 to 125°C) temperature-compensating capacitor.

Phase: The angular relationship between current and voltage in an ac circuit. The fraction of the period which has elapsed in a periodic function or wave measured from some fixed origin. If the time for one period is represented as 360° along a time axis, the phase position is called phase angle.

Polarized capacitor: An electrolytic capacitor in which the dielectric film is formed on only one metal electrode. The impedance to the flow of current is then greater in one direction than in the other. Reversed polarity can damage the part if excessive current flow occurs.

Power factor (PF):   The ratio of effective series resistance to impedance of a capacitor, expressed as a percentage.

Quality factor (Q):   The ratio of the reactance to its equivalent series resistance.

Reactance (X):   Opposition to the flow of alternating current. Capacitive reactance (X_C) **is the opposition offered by capacitors at a specified frequency and is measured in ohms.**

Resonant frequency: The frequency at which a given system or object will respond with maximum amplitude when driven by an external sinusoidal force of constant amplitude.

Reverse leakage current: A nondestructive current flowing through a capacitor subjected to a voltage of polarity opposite to that normally specified.

Ripple current: The total amount of alternating and direct current that may be applied to an electrolytic capacitor under stated conditions.

Temperature coefficient (TC):   A capacitor's change in capacitance per degree change in temperature. May be positive, negative, or zero and is usually expressed in parts per million per degree Celsius (ppm/°C) if the characteristics are linear. For nonlinear types, TC is expressed as a percentage of room temperature (25°C) capacitance.

Time constant: In a capacitor-resistor circuit, the number of seconds required for the capacitor to reach 63.2% of its full charge after a voltage is applied. The time constant of a capacitor with a capacitance (C) in farads in series with a resistance (R) in ohms is equal to $R \times C$ seconds.

Winding: A conductive path, usually wire, inductively coupled to a magnetic core or cell.

References

Exploring the capacitor, *Hewlett-Packard Bench Briefs*, September/October 1979. Sections reprinted with permission from *Bench Briefs*, a Hewlett-Packard service publication.

Capacitors, *1979 Electronic Buyer's Handbook,* vol. 1, November 1978. Copyright 1978 by CMP Publications, Inc. Reprinted with permission.

W.G. Jung and R. March, "Picking capacitors," *Audio,* March 1980.

"Electrolytic capacitors: Past, present and future," and "What is an electrolytic capacitor," *Electron. Des.,* May 28, 1981.

R.F. Graf, "Introduction To Aluminum Capacitors," Sprague Electric Company. Parts reprinted with permission. "Introduction To Aluminum Capacitors," Sprague Electric Company. Parts reprinted with permission. *Handbook of Electronics Tables and Formulas,* 6th ed., Indianapolis: Sams, 1986.

1.3   Transformers

C. Sankaran

The electrical transformer was invented by an American electrical engineer, William Stanley, in 1885 and was used in the first ac lighting installation at Great Barrington, Massachusetts. The first transformer was used to step up the power from 500 to 3000 V and transmitted for a distance of 1219 m (4000 ft). At the receiving end the voltage was stepped down to 500 V to power street and office lighting. By comparison, present transformers are designed to transmit hundreds of megawatts of power at voltages of 700 kV and beyond for distances of several hundred miles.

Transformation of power from one voltage level to another is a vital operation in any transmission, distribution, and utilization network. Normally, power is generated at a voltage that takes into consideration

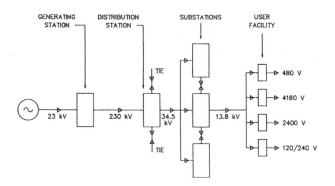

FIGURE 1.22   Power flow line diagram.

the cost of generators in relation to their operating voltage. Generated power is transmitted by overhead lines many miles and undergoes several voltage transformations before it is made available to the actual user. Figure 1.22 shows a typical power flow line diagram.

Types of Transformers

Transformers are broadly grouped into two main categories: dry-type and liquid-filled transformers. Dry-type transformers are cooled by natural or forced circulation of air or inert gas through or around the transformer enclosure. Dry-type transformers are further subdivided into ventilated, sealed, or encapsulated types depending upon the construction of the transformer. Dry transformers are extensively used in industrial power distribution for rating up to 5000 kVA and 34.5 kV.

Liquid-filled transformers are cooled by natural or forced circulation of a liquid coolant through the windings of the transformer. This liquid also serves as a **dielectric** to provide superior voltage-withstand characteristics. The most commonly used liquid in a transformer is a mineral oil known as transformer oil that has a continuous operating temperature rating of 105°C, a flash point of 150°C, and a fire point of 180°C. A good grade transformer oil has a **breakdown strength** of 86.6 kV/cm (220 kV/in.) that is far higher than the breakdown strength of air, which is 9.84 kV/cm (25 kV/in.) at atmospheric pressure.

Silicone fluid is used as an alternative to mineral oil. The breakdown strength of silicone liquid is over 118 kV/cm (300 kV/in.) and it has a flash point of 300°C and a fire point of 360°C. Silicone-fluid-filled transformers are classified as less flammable. The high dielectric strengths and superior thermal conductivities of liquid coolants make them ideally suited for large high-voltage power transformers that are used in modern power generation and distribution.

Principle of Transformation

The actual process of transfer of electrical power from a voltage of V_1 to a voltage of V_2 is explained with the aid of the simplified transformer representation shown in Figure 1.23. Application of voltage across

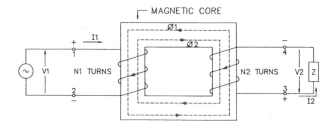

FIGURE 1.23   Electrical power transfer.

the primary winding of the transformer results in a **magnetic field** of ϕ_1 Wb in the magnetic core, which in turn induces a voltage of V_2 at the secondary terminals. V_1 and V_2 are related by the expression $V_1/V_2 = N_1/N_2$, where N_1 and N_2 are the number of turns in the primary and secondary windings, respectively. If a load current of I_2 A is drawn from the secondary terminals, the load current establishes a magnetic field of ϕ_2 Wb in the core and in the direction shown. Since the effect of load current is to reduce the amount of primary magnetic field, the reduction in ϕ_1 results in an increase in the primary current I_1 so that the net magnetic field is almost restored to the initial value and the slight reduction in the field is due to leakage **magnetic flux**. The currents in the two windings are related by the expression $I_1/I_2 = N_2/N_1$. Since $V_1/V_2 = N_1/N_2 = I_2/I_1$, we have the expression $V_1 \cdot I_1 = V_2 \cdot I_2$. Therefore, the voltamperes in the two windings are equal in theory. In reality, there is a slight loss of power during transformation that is due to the energy necessary to set up the magnetic field and to overcome the losses in the transformer core and windings. Transformers are static power conversion devices and are therefore highly efficient. Transformer efficiencies are about 95% for small units (15 kVA and less), and the efficiency can be higher than 99% for units rated above 5 MVA.

Electromagnetic Equation

Figure 1.24 shows a magnetic core with the area of cross-section $A = W \cdot D$ m^2. The transformer primary winding that consists of N turns is excited by a sinusoidal voltage $v = V\sin(\omega t)$, where ω is the angular frequency given by the expression $\omega = 2\pi f$ and f is the frequency of the applied voltage waveform. ϕ is magnetic field in the core due to the excitation current i:

$$\phi = \Phi\sin\left(\omega t - \frac{\pi}{2}\right) = -\Phi\cos(\omega t)$$

Induced voltage in the winding:

$$e = -N\frac{d\phi}{dt} = N\frac{d[\Phi\cos(\omega t)]}{dt} = -N\omega\Phi\sin(\omega t)$$

Maximum value of the induced voltage:

$$E = N\omega\Phi$$

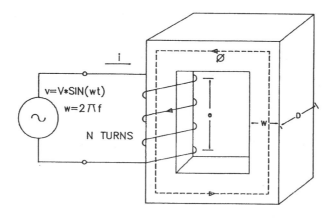

FIGURE 1.24 Electromagnetic relation.

The root-mean-square value:

$$E_{rms} = \frac{E}{\sqrt{2}} = \frac{2\pi f N \Phi}{\sqrt{2}} = 4.44f\,NBA$$

where flux Φ (webers) is replaced by the product of the flux density B (teslas) and the area of cross-section of the core.

This fundamental design equation determines the size of the transformer for any given voltage and frequency. Power transformers are normally operated at flux density levels of 1.5 T.

Transformer Core

The transformer core is the medium that enables the transfer of power from the primary to the secondary to occur in a transformer. In order that the transformation of power may occur with the least amount of loss, the magnetic core is made up of laminations which have the highest permeability, permeability being a measure of the ease with which the magnetic field is set up in the core.

The magnetic field reverses direction every one half cycle of the applied voltage and energy is expended in the core to accomplish the cyclic reversals of the field. This loss component is known as the hysteresis loss P_h:

$$P_h = 150.7 V_e f B^{1.6}\,W$$

where V_e is the volume of the core in cubic meters, f is the frequency, and B is the maximum flux density in teslas.

As the magnetic field reverses direction and cuts across the core structure, it induces a voltage in the laminations known as eddy voltages. This phenomenon causes eddy currents to circulate in the laminations. The loss due to eddy currents is called the eddy current loss P_e:

$$P_e = 1.65 V_e B^2 f^2 t^2 / r$$

where V_e is the volume of the core in cubic meters, f is the frequency, B is the maximum flux density in teslas, t is thickness of the laminations in meters, and r is the resistivity of the core material in ohm-meters.

Hysteresis losses are reduced by operating the core at low flux densities and using core material of high permeability. Eddy current losses are minimized by low flux levels, reduction in thickness of the laminations, and high resistivity core material.

Cold-rolled, grain-oriented silicon steel laminations are exclusively used in large power transformers to reduce core losses. A typical silicon steel used in transformers contains 95% iron, 3% silicon, 1% manganese, 0.2% phosphor, 0.06% carbon, 0.025% sulphur, and traces of other impurities.

Transformer Losses

The heat developed in a transformer is a function of the losses that occur during transformation. Therefore, the transformer losses must be minimized and the heat due to the losses must be efficiently conducted away from the core, the windings, and the cooling medium. The losses in a transformer are grouped into two categories: (1) no-load losses and (2) load losses. The no-load losses are the losses in the core due to excitation and are mostly composed of hysteresis and eddy current losses. The load losses are grouped into three categories: (1) winding I^2R losses, (2) winding eddy current losses, and (3) other stray losses. The winding I^2R losses are the result of the flow of load current through the resistance of the primary and secondary windings. The winding eddy current losses are caused by the magnetic field set up by the winding current, due to formation of eddy voltages in the conductors. The winding eddy losses are proportional to the square of the rms value of the current and to the square of the frequency of the current. When transformers are required to supply loads that are rich in **harmonic frequency** components, the eddy loss factor must be given extra consideration. The other stray loss component is the result of induced currents in the buswork, core clamps, and tank walls by the magnetic field set up by the load current.

Transformer Connections

A single-phase transformer has one input (primary) winding and one output (secondary) winding. A conventional three-phase transformer has three input and three output windings. The three windings can be connected in one of several different configurations to obtain three-phase connections that are distinct. Each form of connection has its own merits and demerits.

Y Connection (Figure 1.25)

In the Y connection, one end of each of the three windings is connected together to form a Y, or a neutral point. This point is normally grounded, which limits the maximum potential to ground in the transformer to the line to neutral voltage of the power system. The grounded neutral also limits transient overvoltages in the transformer when subjected to lightning or switching surges. Availability of the neutral point allows the transformer to supply line to neutral single-phase loads in addition to normal three-phase loads. Each phase of the Y-connected winding must be designed to carry the full line current, whereas the phase voltages are only 57.7% of the line voltages.

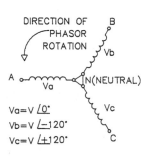

FIGURE 1.25 Y connection.

Delta Connection (Figure 1.26)

In the delta connection, the finish point of each winding is connected to the start point of the adjacent winding to form a closed triangle, or delta. A delta winding in the transformer tends to balance out unbalanced loads that are present on the system. Each phase of the delta winding only carries 57.7% of the line current, whereas the phase voltages are equal to the line voltages.

Large power transformers are designed so that the high-voltage side is connected in Y and the low-voltage side is connected in delta. Distribution transformers that are required to supply single-phase loads are designed in the opposite configuration so that the neutral point is available at the low-voltage end.

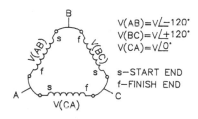

FIGURE 1.26 Delta connection.

Open-Delta Connection (Figure 1.27)

An open-delta connection is used to deliver three-phase power if one phase of a three-phase bank of transformers fails in service. When the failed unit is removed from service, the remaining units can still supply three-phase power but at a reduced rating. An open-delta connection is also used as an economical means to deliver three-phase power using only two single-phase transformers. If P is the total three-phase kVA, then each transformer of the open-delta bank must have a rating of $P/\sqrt{3}$ kVA. The disadvantage of the open-delta connection is the unequal **regulation** of the three phases of the transformer.

FIGURE 1.27 Open-delta connection.

T Connection (Figure 1.28)

The T connection is used for three-phase power transformation when two separate single-phase transformers with special configurations are available. If a voltage transformation from V_1 to V_2 volts is required, one of the units (main transformer) must have a voltage ratio of V_1/V_2 with the midpoint of each winding brought out. The other unit must have a ratio of $0.866V_1/0.866V_2$ with the neutral point brought out, if needed.

The Scott connection is a special type of T connection used to transform three-phase power to two-phase power for operation of electric furnaces and two-phase motors. It is shown in Figure 1.29.

Zigzag Connection (Figure 1.30)

This connection is also called the interconnected star connection where the winding of each phase is divided into two halves and interconnected to form a zigzag configuration. The zigzag connection is mostly used to derive a neutral point for grounding purposes in three-phase, three-wire systems. The neutral point can be used to (1) supply single-phase loads, (2) provide a safety ground, and (3) sense and limit ground fault currents.

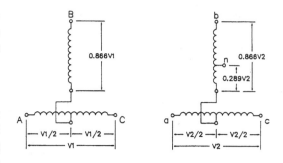

FIGURE 1.28   T connection.

Transformer Impedance

Impedance is an inherent property in a transformer that results in a voltage drop as power is transferred from the primary to the secondary side of the power system. The impedance of a transformer consists of two parts: resistance (R) and reactance (X). The resistance component is due to the resistance of the material of the winding and the percentage value of the voltage drop due to resistance becomes less as the rating of the transformer increases. The reactive component, which is also known as leakage reactance, is the result of incomplete linkage of the magnetic field set up by the secondary winding with the turns of the primary winding, and vice versa. The net impedance of the transformer is given by $Z = \sqrt{R^2 + X^2}$. The impedance value marked on the transformer is the percentage voltage drop due to this impedance under full-load operating conditions:

$$\% \text{ impedance } z = IZ\left(\frac{100}{V}\right)$$

where I is the full-load current of the transformer, Z is the impedance in ohms of the transformer, and V is the voltage rating of the transformer winding. It should be noted that the values of I and Z must be referred to the same side of the transformer as the voltage V.

Transformers are also major contributors of impedance to limit the fault currents in electrical power systems.

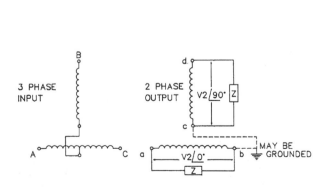

FIGURE 1.29   Three-phase–two-phase transformation.

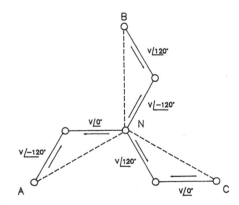

FIGURE 1.30   Zigzag connection.

Defining Terms

Breakdown strength: Voltage gradient at which the molecules of medium break down to allow passage of damaging levels of electric current.

Dielectric: Solid, liquid, or gaseous substance that acts as an insulation to the flow of electric current.

Harmonic frequency: Integral multiples of fundamental frequency. For example, for a 60-Hz supply the harmonic frequencies are 120, 180, 240, 300, . . .

Magnetic field: Magnetic force field where lines of magnetism exist.

Magnetic flux: Term for lines of magnetism.

Regulation: The change in voltage from no-load to full-load expressed as a percentage of full-load voltage.

References and Further Information

Bean, Chackan, Moore, and Wentz, *Transformers for the Electric Power Industry*, New York: McGraw-Hill, 1966.

General Electric, *Transformer Connections*, 1960.

A. Gray, *Electrical Machine Design*, New York: McGraw-Hill.

IEEE, *C57 Standards on Transformers*, New York: IEEE Press, 1992.

IEEE Transactions on Industry Applications.

R.R. Lawrence, *Principles of Alternating Current Machinery*, New York: McGraw-Hill, 1920.

Power Engineering Review.

C. Sankaran, *Introduction to Transformers*, New York: IEEE Press, 1992.

S.A. Stigant, and A.C. Franklin, *The J & P Transformer Book*, London: Newnes-Butterworths, 1973.

1.4   Electrical Fuses

Nick Angelopoulos

The fuse is a simple and reliable safety device. It is second to none in its ease of application and its ability to protect people and equipment.

The fuse is a current-sensitive device. It has a conductor with a reduced cross-section (element) normally surrounded by an arc-quenching and heat-conducting material (filler). The entire unit is enclosed in a body fitted with end contacts. A basic fuse element design is illustrated in Figure 1.31.

Ratings

Most fuses have three electrical ratings: ampere rating, voltage rating, and **interrupting rating**. The ampere rating indicates the current the fuse can carry without melting or exceeding specific temperature rise limits. The voltage rating, ac or dc, usually indicates the maximum system voltage that can be applied to the fuse. The interrupting rating (I.R.) defines the maximum short-circuit current that a fuse can safely interrupt. If a fault current higher than the interrupting rating causes the fuse to operate, the high internal pressure may cause the fuse to rupture. It is imperative, therefore, to install a fuse, or any other type of protective device, that has an interrupting rating not less than the available short-circuit current. A violent explosion may occur if the interrupting rating of any protective device is inadequate.

A fuse must perform two functions. The first, the "passive" function, is one that tends to be taken for granted. In fact, if the fuse performs the passive function well, we tend to forget that the fuse exists at all. The passive function simply entails that the fuse can carry up to its normal load current without aging

FIGURE 1.31   Basic fuse element.

or overheating. Once the current level exceeds predetermined limits, the "active" function comes into play and the fuse operates. It is when the fuse is performing its active function that we become aware of its existence.

In most cases, the fuse will perform its active function in response to two types of circuit conditions. The first is an overload condition, for instance, when a hair dryer, teakettle, toaster, and radio are plugged into the same circuit. This overload condition will eventually cause the element to melt. The second condition is the overcurrent condition, commonly called the short circuit or the fault condition. This can produce a drastic, almost instantaneous, rise in current, causing the element to melt usually in less than a quarter of a cycle. Factors that can lead to a fault condition include rodents in the electrical system, loose connections, dirt and moisture, breakdown of insulation, foreign contaminants, and personal mistakes. Preventive maintenance and care can reduce these causes. Unfortunately, none of us is perfect and faults can occur in virtually every electrical system—we must protect against them.

Fuse Performance

Fuse performance characteristics under overload conditions are published in the form of *average melting time–current characteristic curves,* or simply *time–current curves.* Fuses are tested with a variety of currents, and the melting times are recorded. The result is a graph of time versus current coordinates that are plotted on log-log scale, as illustrated in Figure 1.32.

Under short-circuit conditions the fuse operates and fully opens the circuit in less than 0.01 sec. At 50 or 60 Hz, this represents operation within the first half cycle. The current waveform let-through by the fuse is the shaded, almost triangular, portion shown in Figure 1.33(a). This depicts a fraction of the current that would have been let through into the circuit had a fuse not been installed.

Fuse short-circuit performance characteristics are published in the form of peak let-through (I_p) graphs and I^2t graphs. I_p (peak current) is simply the peak of the shaded triangular waveform, which increases as the fault current increases, as shown in Figure 1.33(b). The electromagnetic forces, which can cause mechanical damage to equipment, are proportional to I_p^2.

I^2t represents heat energy measured in units of A^2 sec (ampere squared seconds) and is documented on I^2t graphs. These I^2t graphs, as illustrated in Figure 1.33(c), provide three values of I^2t: minimum melting I^2t, arcing I^2t, and total clearing I^2t. I^2t and I_p short-circuit performance characteristics can be used to coordinate fuses and other equipment. In particular, I^2t values are often used to selectively coordinate fuses in a distribution system.

Selective Coordination

In any power distribution system, selective coordination exists when the fuse immediately upstream from a fault operates, leaving all other fuses further upstream unaffected. This increases system reliability by isolating the faulted branch while maintaining power to all other branches. Selective coordination is easily assessed by comparing the I^2t characteristics for feeder and branch circuit fuses. The branch fuse should have a total clearing I^2t value that is less than the melting I^2t value of the feeder or upstream fuse. This ensures that the branch fuse will melt, arc, and clear the fault before the feeder fuse begins to melt.

Standards

Overload and short-circuit characteristics are well documented by fuse manufacturers. These characteristics are standardized by product standards written in most cases by safety organizations such as CSA (Canadian Standards Association) and UL (Underwriters Laboratories). CSA standards and UL specify product designations, dimensions, performance characteristics, and temperature rise limits. These standards are used in

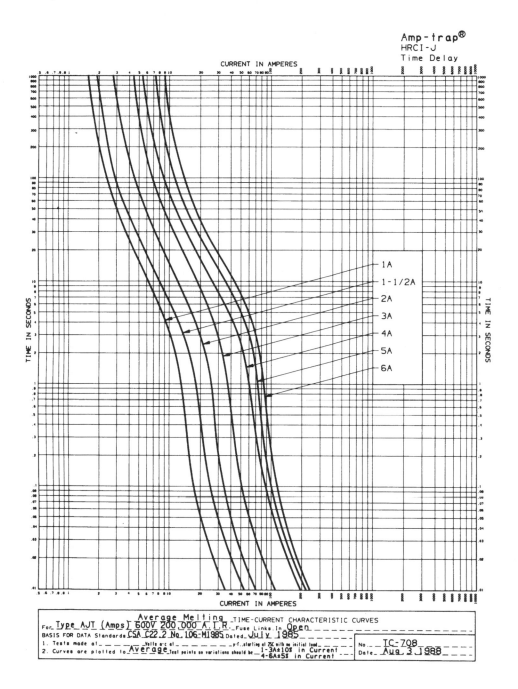

FIGURE 1.32 Time–current characteristic curves.

conjunction with national code regulations such as CEC (Canadian Electrical Code) and NEC (National Electrical Code) that specify how the product is applied.

IEC (International Electrotechnical Commission, Geneva, Switzerland) was founded to harmonize electrical standards to increase international trade in electrical products. Any country can become a member and participate in the standards-writing activities of IEC. Unlike CSA and UL, IEC is not a certifying body that certifies or approves products. IEC publishes consensus standards for national standards authorities such as CSA (Canada), UL (USA), BSI (UK), and DIN (Germany) to adopt as their own national standards.

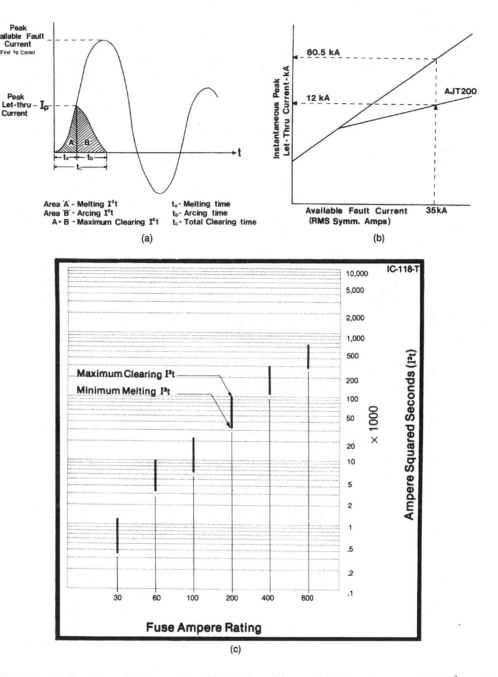

FIGURE 1.33   (a) Fuse short-circuit operation. (b) Variation of fuse peak let-through current I_p. (c) I^2t graph.

Products

North American low-voltage distribution fuses can be classified under two types: Standard or Class H, as referred to in the United States, and **HRC (high rupturing capacity)** or current-limiting fuses, as referred to in Canada. It is the interrupting rating that essentially differentiates one type from the other.

Most Standard or Class H fuses have an interrupting rating of 10,000 A. They are not classified as HRC or current-limiting fuses, which usually have an interrupting rating of 200,000 A. Selection is often based on the calculated available short-circuit current.

In general, short-circuit currents in excess of 10,000 A do not exist in residential applications. In commercial and industrial installations, short-circuit currents in excess of 10,000 A are very common. Use of HRC fuses usually means that a fault current assessment is not required.

Standard—Class H

In North America, Standard or Class H fuses are available in 250- and 600-V ratings with ampere ratings up to 600 A. There are primarily three types: one-time, time-delay, and renewable. Rating for rating, they are all constructed to the same dimensions and are physically interchangeable in standard-type fusible switches and fuse blocks.

One-time fuses are not reusable once blown. They are used for general-purpose resistive loads such as lighting, feeders, and cables.

Time-delay fuses have a specified delay in their overload characteristics and are designed for motor circuits. When started, motors typically draw six times their full load current for approximately 3 to 4 sec. This surge then decreases to a level within the motor full-load current rating. Time-delay fuse overload characteristics are designed to allow for motor starting conditions.

Renewable fuses are constructed with replaceable links or elements. This feature minimizes the cost of replacing fuses. However, the concept of replacing fuse elements in the field is not acceptable to most users today because of the potential risk of improper replacement.

HRC

HRC or current-limiting fuses have an interrupting rating of 200 kA and are recognized by a letter designation system common to North American fuses. In the United States they are known as Class J, Class L, Class R, etc., and in Canada they are known as HRCI-J, HRC-L, HRCI-R, and so forth. HRC fuses are available in ratings up to 600 V and 6000 A. The main differences among the various types are their dimensions and their short-circuit performance (I_p and I^2t) characteristics.

One type of HRC fuse found in Canada, but not in the United States, is the HRCII-C or Class C fuse. This fuse was developed originally in England and is constructed with bolt-on-type blade contacts. It is available in a voltage rating of 600 V with ampere ratings from 2 to 600 A. Some higher ampere ratings are also available but are not as common. HRCII-C fuses are primarily regarded as providing short-circuit protection only. Therefore, they should be used in conjunction with an overload device.

HRCI-R or Class R fuses were developed in the United States. Originally constructed to Standard or Class H fuse dimensions, they were classified as Class K and are available in the United States with two levels of short-circuit performance characteristics: Class K1 and Class K5. However, they are not recognized in Canadian Standards. Under fault conditions, Class K1 fuses limit the I_p and I^2t to lower levels than do Class K5 fuses. Since both Class K1 and K5 are constructed to Standard or Class H fuse dimensions, problems with interchangeability occur. As a result, a second generation of these K fuses was therefore introduced with a rejection feature incorporated in the end caps and blade contacts. This rejection feature, when used in conjunction with rejection-style fuse clips, prevents replacement of these fuses with Standard or Class H 10-kA I.R. fuses. These rejection style fuses are known as Class RK1 and Class RK5. They are available with time-delay or nontime-delay characteristics and with voltage ratings of 250 or 600 V and ampere ratings up to 600 A. In Canada, CSA has only one classification for these fuses, HRCI-R, which have the same maximum I_p and I^2t current-limiting levels as specified by UL for Class RK5 fuses.

HRCI-J or Class J fuses are a more recent development. In Canada, they have become the most popular HRC fuse specified for new installations. Both time-delay and nontime-delay characteristics are available in ratings of 600 V with ampere ratings up to 600 A. They are constructed with dimensions much smaller than HRCI-R or Class R fuses and have end caps or blade contacts which fit into 600-V Standard or Class H-type fuse clips.

However, the fuse clips must be mounted closer together to accommodate the shorter fuse length. Its shorter length, therefore, becomes an inherent rejection feature that does not allow insertion of Standard or HRCI-R fuses. The blade contacts are also drilled to allow bolt-on mounting if required. CSA and UL specify these fuses to have maximum short-circuit current-limiting I_p and I^2t limits lower than those specified for HRCI-R and HRCII-C fuses. HRCI-J fuses may be used for a wide variety of applications. The time-delay type is commonly used in motor circuits sized at approximately 125 to 150% of motor full-load current.

HRC-L or Class L fuses are unique in dimension but may be considered as an extension of the HRCI-J fuses for ampere ratings above 600 A. They are rated at 600 V with ampere ratings from 601 to 6000 A. They are physically larger and are constructed with bolt-on-type blade contacts. These fuses are generally used in low-voltage distribution systems where supply transformers are capable of delivering more than 600 A.

In addition to Standard and HRC fuses, there are many other types designed for specific applications. For example, there are medium- or high-voltage fuses to protect power distribution transformers and medium-voltage motors. There are fuses used to protect sensitive semiconductor devices such as diodes, SCRs, and triacs. These fuses are designed to be extremely fast under short-circuit conditions. There is also a wide variety of dedicated fuses designed for protection of specific equipment requirements such as electric welders, capacitors, and circuit breakers, to name a few.

Trends

Ultimately, it is the electrical equipment being protected that dictates the type of fuse needed for proper protection. This equipment is forever changing and tends to get smaller as new technology becomes available. Present trends indicate that fuses also must become smaller and faster under fault conditions, particularly as available short-circuit fault currents are tending to increase.

With free trade and the globalization of industry, a greater need for harmonizing product standards exists. The North American fuse industry is taking big steps toward harmonizing CSA and UL fuse standards, and at the same time is participating in the IEC standards process. Standardization will help the electrical industry to identify and select the best fuse for the job—anywhere in the world.

Defining Terms

HRC (high rupturing capacity):   A term used to denote fuses having a high interrupting rating. Most low-voltage HRC-type fuses have an interrupting rating of 200 kA rms symmetrical.

I^2t (ampere squared seconds):   A convenient way of indicating the heating effect or thermal energy which is produced during a fault condition before the circuit protective device has opened the circuit. As a protective device, the HRC or current-limiting fuse lets through far less damaging I^2t than other protective devices.

Interrupting rating (I.R.):   The maximum value of short-circuit current that a fuse can safely interrupt.

References

R.K. Clidero and K.H. Sharpe, *Application of Electrical Construction*, Ontario, Canada: General Publishing Co. Ltd., 1982.
Gould Inc., *Shawmut Advisor*, Newburyport, MA: Circuit Protection Division.
C.A. Gross, *Power Systems Analysis*, 2nd ed., New York: Wiley, 1986.
E. Jacks, *High Rupturing Capacity Fuses*, New York: Wiley, 1975.
A. Wright and P.G. Newbery, *Electric Fuses*, London: Peter Peregrinus Ltd., 1984.

Further Information

For greater detail the "Shawmut Advisor" (Gould, Inc., 374 Merrimac Street, Newburyport, MA 01950) or the "Fuse Technology Course Notes" (Gould Shawmut Company, 88 Horner Avenue, Toronto, Canada M8Z 5Y3) may be referred to for fuse performance and application.

2
Voltage and Current Sources

Richard C. Dorf
University of California

Clayton R. Paul
Mercer University

J.R. Cogdell
University of Texas at Austin

2.1 Step, Impulse, Ramp, Sinusoidal, Exponential, and DC Signals

Richard C. Dorf

The important signals for circuits include the step, impulse, ramp, sinusoid, and dc signals. These signals are widely used and are described here in the time domain. All of these signals have a Laplace transform.

Step Function

The **unit-step** function $u(t)$ is defined mathematically by

$$u(t) = \begin{cases} 1, & t \geq 0 \\ 0, & t < 0 \end{cases}$$

Here *unit step* means that the amplitude of $u(t)$ is equal to 1 for $t \geq 0$. Note that we are following the convention that $u(0) = 1$. From a strict mathematical standpoint, $u(t)$ is not defined at $t = 0$. Nevertheless, we usually take $u(0) = 1$. If A is an arbitrary nonzero number, $Au(t)$ is the step function with amplitude A for $t \geq 0$. The unit step function is plotted in Figure 2.1.

The Impulse

The **unit impulse** $\delta(t)$, also called the *delta function* or the *Dirac distribution*, is defined by

$$\delta(t) = 0, \quad t \neq 0$$
$$\int_{-\varepsilon}^{\varepsilon} \delta(\lambda)\mathrm{d}\lambda = 1, \quad \text{for any real number } \varepsilon > 0$$

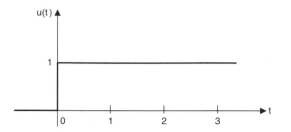

FIGURE 2.1   Unit-step function.   **FIGURE 2.2**   Graphical representation of the impulse $K\delta(t)$.

The first condition states that $\delta(t)$ is zero for all nonzero values of t, while the second condition states that the area under the impulse is 1, so $\delta(t)$ has unit area. It is important to point out that the value $\delta(0)$ of $\delta(t)$ at $t = 0$ is not defined; in particular, $\delta(0)$ is not equal to infinity. For any real number K, $K\delta(t)$ is the impulse with area K. It is defined by

$$K\delta(t) = 0, \quad t \neq 0$$

$$\int_{-\varepsilon}^{\varepsilon} K\delta(\lambda)\,d\lambda = K, \quad \text{for any real number } \varepsilon > 0$$

The graphical representation of $K\delta(t)$ is shown in Figure 2.2. The notation K in the figure refers to the area of the impulse $K\delta(t)$.

The unit-step function $u(t)$ is equal to the integral of the unit impulse $\delta(t)$; more precisely, we have

$$u(t) = \int_{-\infty}^{t} \delta(\lambda)\,d\lambda, \quad \text{all } t \text{ except } t = 0$$

Conversely, the first derivative of $u(t)$ with respect to t is equal to $\delta(t)$ except at $t = 0$, where the derivative of $u(t)$ is not defined.

Ramp Function

The *unit-ramp function* $r(t)$ is defined mathematically by

$$r(t) = \begin{cases} t, & t \geq 0 \\ 0, & t < 0 \end{cases}$$

Note that for $t \geq 0$, the slope of $r(t)$ is 1. Thus, $r(t)$ has *unit slope*, which is the reason $r(t)$ is called the unit-ramp function. If K is an arbitrary nonzero scalar (real number), the ramp function $Kr(t)$ has slope K for $t \geq 0$. The unit-ramp function is plotted in Figure 2.3.

The unit-ramp function $r(t)$ is equal to the integral of the unit-step function $u(t)$; that is:

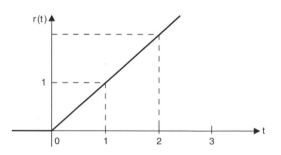

FIGURE 2.3   Unit-ramp function.

$$r(t) = \int_{-\infty}^{t} u(\lambda)\,d\lambda$$

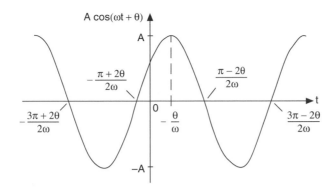

FIGURE 2.4 The sinusoid $A \cos(\omega t + \theta)$ with $-\pi/2 < \theta < 0$.

Conversely, the first derivative of $r(t)$ with respect to t is equal to $u(t)$ except at $t = 0$, where the derivative of $r(t)$ is not defined.

Sinusoidal Function

The sinusoid is a continuous-time signal: $A \cos(\omega t + \theta)$.

Here A is the amplitude, ω is the frequency in radians per second (rad/sec), and θ is the phase in radians. The frequency f in cycles per second, or hertz (Hz), is $f = \omega/2\pi$. The sinusoid is a periodic signal with period $2\pi/\omega$. The sinusoid is plotted in Figure 2.4.

Decaying Exponential

In general, an exponentially decaying quantity (Figure 2.5) can be expressed as

$$a = Ae^{-t/\tau}$$

where a = instantaneous value

A = amplitude or maximum value

e = base of natural logarithms = 2.718...

τ = time constant in seconds

t = time in seconds

The current of a discharging capacitor can be approximated by a decaying exponential function of time.

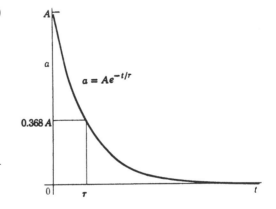

FIGURE 2.5 The decaying exponential.

Time Constant

Since the exponential factor only *approaches* zero as t increases without limit, such functions theoretically last forever. In the same sense, all radioactive disintegrations last forever. In the case of an exponentially decaying current, it is convenient to use the value of time that makes the exponent −1. When $t = \tau =$ the *time constant,* the value of the exponential factor is

$$e^{-t/\tau} = e^{-1} = \frac{1}{e} = \frac{1}{2.718} = 0.368$$

In other words, after a time equal to the time constant, the exponential factor is reduced to approximately 37% of its initial value.

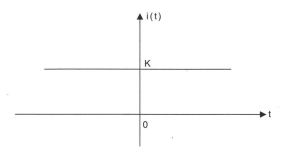

FIGURE 2.6   The dc signal with amplitude *K*.

DC Signal

The direct current signal (dc signal) can be defined mathematically by

$$i(t) = K \qquad -\infty < t < +\infty$$

Here, *K* is any nonzero number. The dc signal remains a constant value of *K* for any $-\infty < t < \infty$. The dc signal is plotted in Figure 2.6.

Defining Terms

Ramp: A continually growing signal such that its value is zero for $t \leq 0$ and proportional to time *t* for $t > 0$.
Sinusoid: A periodic signal $x(t) = A \cos(\omega t + \theta)$ where $\omega = 2\pi f$ with frequency in hertz.
Unit impulse: A very short pulse such that its value is zero for $t \mid 0z$ and the integral of the pulse is 1.
Unit step: Function of time that is zero for $t < t_0$ and unity for $t > t_0$. At $t = t_0$ the magnitude changes from zero to one. The unit step is dimensionless.

References

R.C. Dorf, *Introduction to Electric Circuits,* 6th ed., New York: Wiley, 2004.
R.C. Dorf, *The Engineering Handbook*, 2nd ed., Boca Raton, FL: CRC Press, 2004.

Further Information

IEEE Transactions on Circuits and Systems
IEEE Transactions on Education

2.2   Ideal and Practical Sources

Clayton R. Paul

A *mathematical model* of an electric circuit contains *ideal models* of physical circuit elements. Some of these ideal circuit elements (e.g., the resistor, capacitor, inductor, and transformer) were discussed previously. Here we will define and examine both *ideal* and *practical voltage and current sources*. The terminal characteristics of these models will be compared to those of actual sources.

Ideal Sources

The *ideal independent voltage source* shown in Figure 2.7 constrains the terminal voltage across the element to a prescribed function of time, $v_S(t)$, as $v(t) = v_S(t)$. The polarity of the source is denoted by $\pm$ signs within the circle which denotes this as an ideal *independent* source. Controlled or *dependent* ideal voltage sources will be discussed in Section "Controlled Sources." The current through the element will be determined by the circuit that is attached to the terminals of this source.

The *ideal independent current source* in Figure 2.8 constrains the terminal current through the element to a prescribed function of time, $i_S(t)$, as $i(t) = i_S(t)$. The polarity of the source is denoted by an arrow within the circle which also denotes this as an ideal *independent* source. The voltage across the element will be determined by the circuit that is attached to the terminals of this source.

Numerous functional forms are useful in describing the source variation with time. These were discussed in Section "The Step, Impulse, Ramp, Sinusoidal, and dc Signals." For example, an ideal independent dc voltage source is described by $v_S(t) = V_S$, where V_S is a constant. An ideal independent sinusoidal current source is described by $i_S(t) = I_S \sin(\omega t + \phi)$ or $i_S(t) = I_S \cos(\omega t + \phi)$, where I_S is a constant, $\omega = 2\pi f$ with f the *frequency* in hertz, and ϕ is a phase angle. Ideal sources may be used to model actual sources such as temperature transducers, phonograph cartridges, and electric power generators. Thus usually the time form of the output cannot generally be described with a simple, basic function such as dc, sinusoidal, ramp, step, or impulse waveforms. We often, however, represent the more complicated waveforms as a linear combination of more basic functions.

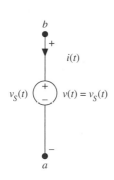
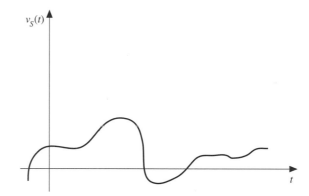

FIGURE 2.7   Ideal independent voltage source.

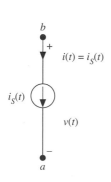
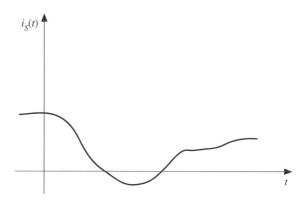

FIGURE 2.8   Ideal independent current source.

Practical Sources

The preceding ideal independent sources constrain the terminal voltage or current to a *known* function of time *independent of the circuit that may be placed across its terminals.* Practical sources, such as batteries, have their terminal voltage (current) dependent upon the terminal current (voltage) caused by the circuit attached to the source terminals. A simple example of this is an automobile storage battery. The battery's terminal voltage is approximately 12 V when no load is connected across its terminals. When the battery is applied across the terminals of the starter by activating the ignition switch, a large current is drawn from its terminals. During starting, its terminal voltage drops as illustrated in Figure 2.9(a). How shall we construct a *circuit model* using the ideal elements discussed thus far to model this nonideal behavior? A model is shown in Figure 2.9(b) and consists of the *series* connection of an ideal resistor, R_S, and an ideal independent voltage source, $V_S = 12$ V.

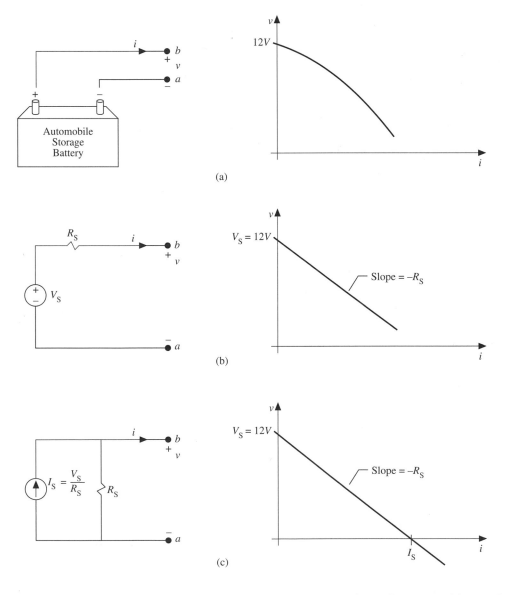

FIGURE 2.9 Practical sources. (a) Terminal v–i characteristic; (b) approximation by a voltage source; (c) approximation by a current source.

To determine the terminal voltage–current relation, we sum Kirchhoff's voltage law around the loop to give

$$v = V_S - R_S i \tag{2.1}$$

This equation is plotted in Figure 2.9(b) and approximates that of the actual battery. The equation gives a straight line with slope $-R_S$ that intersects the v axis ($i = 0$) at $v = V_S$. The resistance R_S is said to be the *internal resistance* of this nonideal source model. It is a fictitious resistance but the model nevertheless gives an equivalent *terminal behavior*.

Although we have derived an approximate model of an actual source, another equivalent form may be obtained. This alternative form is shown in Figure 2.9(c) and consists of the *parallel* combination of an ideal independent current source, $I_S = V_S/R_S$, and the same resistance, R_S, used in the previous model. Although it may seem strange to model an automobile battery using a current source, the model is completely equivalent to the series voltage source–resistor model of Figure 2.9(b) *at the output terminals a–b*. This is shown by writing Kirchhoff's current law at the upper node to give

$$i = I_S - \frac{1}{R_S} v \tag{2.2}$$

Rewriting this equation gives

$$v = R_S I_S - R_S i \tag{2.3}$$

Comparing Equation (2.3) to Equation (2.1) shows that

$$V_S = R_S I_S \tag{2.4}$$

Therefore, we can convert from one form (voltage source in series with a resistor) to another form (current source in parallel with a resistor) very simply.

An ideal voltage source is represented by the model of Figure 2.9(b) with $R_S = 0$. An actual battery therefore provides a close approximation of an ideal voltage source since the source resistance R_S is usually quite small. An ideal current source is represented by the model of Figure 2.9(c) with $R_S = \infty$. This is very closely represented by the bipolar junction transistor (BJT).

Defining Term

Ideal source: An ideal model of an actual source that assumes that the parameters of the source, such as its magnitude, are independent of other circuit variables.

Reference

C.R. Paul, *Analysis of Linear Circuits,* New York: McGraw-Hill, 1989.

2.3   Controlled Sources

J.R. Cogdell

When the analysis of electronic (nonreciprocal) circuits became important in circuit theory, controlled sources were added to the family of circuit elements. Table 2.1 shows the four types of controlled sources. In this section, we will address the questions: What are controlled sources? Why are controlled sources important? How do controlled sources affect methods of circuit analysis?

TABLE 2.1   Names, Circuit Symbols, and Definitions for the Four Possible Types of Controlled Sources

Name	Circuit Symbol	Definition and Units
Current-controlled voltage source (CCVS)	i_1 $r_m\,i_1$ $+$ v_2 $-$	$v_2 = r_m i_1$ $r_m =$ transresistance units, ohms
Current-controlled current source (CCCS)	i_1 i_2 βi_1	$i_2 = \beta i_1$ β, current gain, dimensionless
Voltage-controlled voltage source (VCVS)	$+$ $+$ v_1 μv_1 v_2 $-$ $-$	$v_2 = \mu v_1$ μ, voltage gain, dimensionless
Voltage-controlled current source (VCCS)	$+$ i_2 v_1 $g_m v_1$ $-$	$i_2 = g_m v_1$ g_m, transconductance units, Siemans (mhos)

What Are Controlled Sources?

By *source* we mean a voltage or current source in the usual sense. By *controlled* we mean that the strength of such a source is controlled by some circuit variable(s) elsewhere in the circuit. Figure 2.10 illustrates a simple circuit containing an (independent) current source, i_s, two resistors, and a controlled voltage source, whose magnitude is controlled by the current i_1. Thus, i_1 determines two voltages in the circuit, the voltage across R_1 via Ohm's law and the controlled voltage source via some unspecified effect.

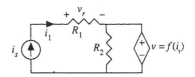

FIGURE 2.10 A simple circuit containing a controlled source.

A controlled source may be controlled by more than one circuit variable, but we will discuss those having a single controlling variable since multiple controlling variables require no new ideas. Similarly, we will deal only with resistive elements, since inductors and capacitors introduce no new concepts. The controlled voltage or current source may depend on the controlling variable in a linear or nonlinear manner. When the relationship is nonlinear, however, the equations are frequently linearized to examine the effects of small variations about some dc values. When we linearize, we will use the customary notation of small letters to represent general and time-variable voltages and currents and large letters to represent constants such as the dc value or the peak value of a sinusoid. On subscripts, large letters represent the total voltage or current and small letters represent the **small-signal** component. Thus, the equation $i_B = I_B + I_b \cos \omega t$ means that the total base current is the sum of a constant and a small-signal component, which is sinusoidal with an amplitude of I_b.

To introduce the context and use of controlled sources we will consider a circuit model for the bipolar junction transistor (BJT). In Figure 2.11 we show the standard symbol for an *npn* BJT with base (B), emitter (E), and collector (C) identified, and voltage and current variables defined. We have shown the common emitter configuration, with the emitter terminal shared to make input and output terminals. The base current, i_B, ideally depends upon the base-emitter voltage, v_{BE}, by the relationship

$$i_B = I_0 \left\{ \exp\left[\frac{v_{BE}}{V_T}\right] - 1 \right\} \qquad (2.5)$$

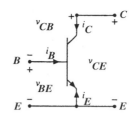

FIGURE 2.11 An *npn* BJT in the common emitter configuration.

where I_0 and V_T are constants. We note that the base current depends on the base-emitter voltage only, but in a nonlinear manner. We can represent this current by a voltage-controlled current source, but the more common representation would be that of a nonlinear conductance, $G_{BE}(v_{BE})$, where

$$G_{BE}(v_{BE}) = \frac{i_B}{v_{BE}}$$

Let us model the effects of small changes in the base current. If the changes are small, the nonlinear nature of the conductance can be ignored and the circuit model becomes a linear conductance (or resistor). Mathematically this conductance arises from a first-order expansion of the nonlinear function. Thus, if $v_{BE} = V_{BE} + v_{be}$, where v_{BE} is the total base-emitter voltage, V_{BE} is a (large) constant voltage and v_{be} is a (small) variation in the base-emitter voltage, then the first two terms in a Taylor series expansion are

$$i_B = I_0 \left\{ \exp\left[\frac{V_{BE} + v_{be}}{V_T}\right] - 1 \right\} \cong I_0 \left\{ \exp\left[\frac{V_{BE}}{V_T}\right] - 1 \right\} + \frac{I_0}{V_T} \exp\left[\frac{V_{BE}}{V_T}\right] v_{be} \qquad (2.6)$$

We note that the base current is approximated by the sum of a constant term and a term that is first order in the small variation in base-emitter voltage, v_{be}. The multiplier of this small voltage is the linearized conductance, g_{be}. If we were interested only in small changes in currents and voltages, only this conductance would be required in the model. Thus, the input (base-emitter) circuit can be represented for the small-signal base variables, i_b and v_{be}, by either equivalent circuit in Figure 2.12.

The voltage-controlled current source, $g_{be}v_{be}$, can be replaced by a simple resistor because the small-signal voltage and current associate with the same branch. The process of **linearization** is important to the modeling of the collector-emitter characteristic, to which we now turn.

The collector current, i_C, can be represented by one of the Eber and Moll equations:

$$i_C = \beta I_0 \left\{ \exp\left[\frac{v_{BE}}{V_T}\right] - 1 \right\} - I_0' \left\{ \exp\left[\frac{v_{BC}}{V_T}\right] - 1 \right\} \qquad (2.7)$$

where β and I_0' are constants. If we restrict our model to the amplifying region of the transistor, the second term is negligible and we may express the collector current as

$$i_C = \beta I_0 \left\{ \exp\left[\frac{v_{BE}}{V_T}\right] - 1 \right\} = \beta i_B \qquad (2.8)$$

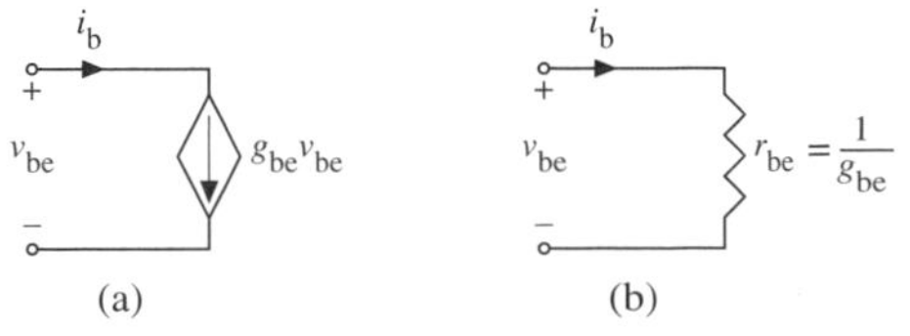

FIGURE 2.12 Equivalent circuits for the base circuit: (a) uses a controlled source and (b) uses a resistor.

Thus, for the ideal transistor, the collector-emitter circuit may be modeled by a current-controlled current source, which may be combined with the results expressed in Equation (2.5) to give the model shown in Figure 2.13.

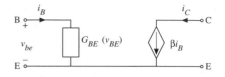

Using the technique of small-signal analysis, we may derive either of the small-signal equivalent circuits shown in Figure 2.14.

The small-signal characteristics of the *npn* transistor in its amplifying region is better represented by the equivalent circuit shown in Figure 2.15. Note we have introduced a voltage-

FIGURE 2.13   Equivalent circuit for BJT.

controlled voltage source to model the influence of the (output) collector-emitter voltage on the (input) base-emitter voltage, and we have placed a resistor, r_{ce}, in parallel with the collector current source to model the influence of the collector-emitter voltage on the collector current.

The four parameters in Figure 2.15 (r_{be}, h_{re}, β, and r_{ce}) are the hybrid parameters describing the transistor properties, although our notation differs from that commonly used. The parameters in the small-signal equivalent circuit depend on the operating point of the device, which is set by the time-average voltages and currents (V_{BE}, I_C, etc.) applied to the device. All of the parameters are readily measured for a given transistor and operating point, and manufacturers commonly specify ranges for the various parameters for a type of transistor.

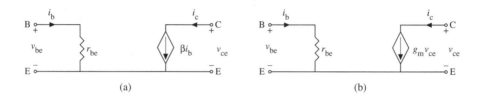

FIGURE 2.14   Two BJT small-signal equivalent circuits ($g_m = \beta/r_{be}$): (a) uses a CCCS and (b) uses a VCCS.

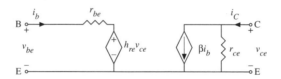

FIGURE 2.15   Full hybrid parameter model for small-signal BJT.

What Is the Significance of Controlled Sources?

Commonplace wisdom in engineering education and practice is that information and techniques that are presented visually are more useful than abstract mathematical forms. Equivalent circuits are universally used in describing electrical engineering systems and devices because circuits portray interactions in a universal, pictorial language. This is true generally, and it is doubly necessary when circuit variables interact through the mysterious coupling modeled by controlled sources. This is the primary significance of controlled sources: that they represent unusual couplings of circuit variables in the universal visual language of circuits.

A second significance is illustrated by our equivalent circuit of the *npn* bipolar transistor, namely, the characterization of a class of similar devices. For example, the parameter β in Equation (2.8) gives important information about a single transistor, and similarly for the range of β for a type of transistor. In this connection, controlled sources lead to a vocabulary for discussing some property of a class of systems or devices, in this case the current gain of an *npn* BJT.

How Does the Presence of Controlled Sources Affect Circuit Analysis?

The presence of nonreciprocal elements, which are modeled by controlled sources, affects the analysis of the circuit. Simple circuits may be analyzed through the direct application of Kirchhoff's laws to branch circuit variables. Controlled sources enter this process similar to the constitutive relations defining R, L, and C, i.e., in defining relationships between branch circuit variables. Thus, controlled sources add no complexity to this basic technique.

The presence of controlled sources negates the advantages of the method that uses series and parallel combinations of resistors for voltage and current dividers. The problem is that the couplings between circuit variables that are expressed by controlled sources make all the familiar formulas unreliable.

When superposition is used, the controlled sources are left on in all cases as independent sources are turned on and off, thus reflecting the kinship of controlled sources to the circuit elements. In principle, little complexity is added; in practice, the repeated solutions required by superposition entail much additional work when controlled sources are involved.

The classical methods of nodal and loop (mesh) analysis incorporate controlled sources without great difficulty. For purposes of determining the number of independent variables required, that is, in establishing the topology of the circuit, the controlled sources are treated as ordinary voltage or current sources. The equations are then written according to the usual procedures. Before the equations are solved, however, the controlling variables must be expressed in terms of the unknowns of the problem. For example, let us say we are performing a nodal analysis on a circuit containing a current-controlled current source. For purposes of counting independent nodes, the controlled current source is treated as an open circuit. After equations are written for the unknown node voltages, the current source will introduce into at least one equation its controlling current, which is not one of the nodal variables. The additional step required by the controlled source is that of expressing the controlling current in terms of the nodal variables.

The parameters introduced into the circuit equations by the controlled sources end up on the left side of the equations with the resistors rather than on the right side with the independent sources. Furthermore, the symmetries that normally exist among the coefficients are disturbed by the presence of controlled sources.

The methods of Thévenin and Norton equivalent circuits continue to be very powerful with controlled sources in the circuits, but some complications arise. The controlled sources must be left on for calculation of the Thévenin (open-circuit) voltage or Norton (short-circuit) current and also for the calculation of the output impedance of the circuit. This usually eliminates the method of combining elements in series or parallel to determine the output impedance of the circuit, and one must either determine the output impedance from the ratio of the Thévenin voltage to the Norton current or else excite the circuit with an external source and calculate the response.

Defining Terms

Controlled source (dependent source): A voltage or current source whose intensity is controlled by a circuit voltage or current elsewhere in the circuit.

Linearization: Approximating nonlinear relationships by linear relationships derived from the first-order terms in a power series expansion of the nonlinear relationships. Normally the linearized equations are useful for a limited range of the voltage and current variables.

Small-signal: Small-signal variables are those first-order variables used in a linearized circuit. A small-signal equivalent circuit is a linearized circuit picturing the relationships between the small-signal voltages and currents in a linearized circuit.

References

E. J. Angelo, Jr., *Electronic Circuits*, 2nd ed., New York: McGraw-Hill, 1964.

N. Balabanian and T. Bickart, *Linear Network Theory*, Chesterland, OH: Matrix Publishers, 1981.

L. O. Chua, *Introduction to Nonlinear Network Theory*, New York: McGraw-Hill, 1969.

B. Friedland, O. Wing, and R. Ash, *Principles of Linear Networks*, New York: McGraw-Hill, 1961.

L. P. Huelsman, *Basic Circuit Theory*, 3rd ed., Englewood Cliffs, NJ: Prentice-Hall, 1981.

3
Linear Circuit Analysis

Michael D. Ciletti
University of Colorado

J. David Irwin
Auburn University

Allan D. Kraus
Naval Postgraduate School

Norman Balabanian
University of Florida

Theodore A. Bickart
Michigan State University

Shu-Park Chan
International Technological University

Norman S. Nise
California State Polytechnic University

3.1 Voltage and Current Laws

Michael D. Ciletti

Analysis of linear circuits rests on two fundamental physical laws that describe how the voltages and currents in a circuit must behave. This behavior results from whatever voltage sources, current sources, and energy storage elements are connected to the circuit. A voltage source imposes a constraint on the evolution of the voltage between a pair of nodes; a current source imposes a constraint on the evolution of the current in a branch of the circuit. The energy storage elements (capacitors and inductors) impose initial conditions on currents and voltages in the circuit; they also establish a dynamic relationship between the voltage and the current at their terminals.

Regardless of how a linear circuit is stimulated, every node voltage and every branch current, at every instant in time, must be consistent with Kirchhoff's voltage and current laws. These two laws govern even the

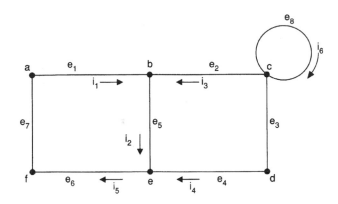

FIGURE 3.1 Graph representation of a linear circuit.

most complex linear circuits. (They also apply to a broad category of nonlinear circuits that are modeled by point models of voltage and current.)

A circuit can be considered to have a topological (or graph) view, consisting of a labeled set of nodes and a labeled set of edges. Each edge is associated with a pair of nodes. A node is drawn as a *dot* and represents a connection between two or more physical components; an edge is drawn as a *line* and represents a path, or branch, for current flow through a component (see Figure 3.1).

The edges, or branches, of the graph are assigned current labels, i_1, i_2,..., i_m. Each current has a designated direction, usually denoted by an *arrow* symbol. If the arrow is drawn toward a node, the associated current is said to be *entering* the node; if the arrow is drawn away from the node, the current is said to be *leaving* the node. The current i_1 is entering node b in Figure 3.1; the current i_5 is leaving node e.

Given a branch, the pair of nodes to which the branch is attached defines the convention for measuring voltages in the circuit. Given the ordered pair of nodes (a, b), a voltage measurement is formed as follows:

$$v_{ab} = v_a - v_b$$

where v_a and v_b are the absolute electrical potentials (voltages) at the respective nodes, taken relative to some reference node. Typically, one node of the circuit is labeled as *ground,* or reference node; the remaining nodes are assigned voltage labels. The measured quantity, v_{ab}, is called the *voltage drop* from node a to node b. We note that

$$v_{ab} = -v_{ba}$$

and that

$$v_{ba} = v_b - v_a$$

is called the *voltage rise* from a to b. Each node voltage implicitly defines the voltage drop between the respective node and the ground node.

The pair of nodes to which an edge is attached may be written as (a, b) or (b, a). Given an ordered pair of nodes (a, b), a *path from a to b* is a directed sequence of edges in which the first edge in the sequence contains node label a, the last edge in the sequence contains node label b, and the node indices of any two adjacent members of the sequence have at least one node label in common. In Figure 3.1, the edge sequence $\{e_1, e_2, e_4\}$ is not a path, because e_2 and e_4 do not share a common node label. The sequence $\{e_1, e_2\}$ is a path from node a to node c.

A path is said to be *closed* if the first node index of its first edge is identical to the second node index of its last edge. The following edge sequence forms a closed path in the graph given in Figure 3.1: $\{e_1, e_2, e_3, e_4, e_6, e_7\}$. Note that the edge sequences $\{e_8\}$ and $\{e_1, e_1\}$ are closed paths.

Kirchhoff's Current Law

Kirchhoff's current law (KCL) imposes constraints on the currents in the branches that are attached to each node of a circuit. In simplest terms, KCL states that the sum of the currents that are entering a given node must equal the sum of the currents that are leaving the node. Thus, the set of currents in branches attached to a given node can be partitioned into two groups whose orientation is away from (into) the node. The two groups must contain the same net current. Applying KCL at node b in Figure 3.1 gives

$$i_1(t) + i_3(t) = i_2(t)$$

A connection of water pipes that has no leaks is a physical analogy of this situation. The net rate at which water is flowing into a joint of two or more pipes must equal the net rate at which water is flowing away from the joint. The joint itself has the property that it only connects the pipes and thereby imposes a structure on the flow of water, but it cannot store water. This is true regardless of when the flow is measured. Likewise, the nodes of a circuit are modeled as though they cannot store charge. (Physical circuits are sometimes modeled for the purpose of simulation as though they store charge, but these nodes implicitly have a capacitor that provides the physical mechanism for *storing* the charge. Thus, KCL is ultimately satisfied.)

KCL can be stated alternatively as: "the algebraic sum of the branch currents entering (or leaving) any node of a circuit at any instant of time must be zero." In this form, the label of any current whose orientation is away from the node is preceded by a minus sign. The currents *entering* node b in Figure 3.1 must satisfy

$$i_1(t) - i_2(t) + i_3(t) = 0$$

In general, the currents entering or leaving each node *m* of a circuit must satisfy

$$\sum i_{km}(t) = 0$$

where $i_{km}(t)$ is understood to be the current in branch k attached to node m. The currents used in this expression are understood to be the currents that would be measured in the branches attached to the node, and their values include a magnitude and an algebraic sign. If the measurement convention is oriented for the case where currents are entering the node, then the actual current in a branch has a positive or negative sign depending on whether the current is truly flowing toward the node in question.

Once KCL has been written for the nodes of a circuit, the equations can be rewritten by substituting into the equations the voltage–current relationships of the individual components. If a circuit is resistive, the resulting equations will be algebraic. If capacitors or inductors are included in the circuit, the substitution will produce a differential equation. For example, writing KCL at the node for v_3 in Figure 3.2 produces

$$i_2 + i_1 - i_3 = 0$$

and

$$C_1 \frac{dv_1}{dt} + \frac{v_4 - v_3}{R_2} - C_2 \frac{dv_2}{dt} = 0$$

KCL for the node between C_2 and R_1 can be written to eliminate variables and lead to a solution describing the capacitor voltages. The capacitor voltages, together with the applied voltage source, determine the remaining voltages and currents in the circuit. Nodal analysis (see Section "Node and Mesh Analysis") treats the systematic modeling and analysis of a circuit under the influence of its sources and energy storage elements.

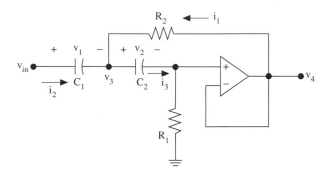

FIGURE 3.2 Example of a circuit containing energy storage elements.

Kirchhoff's Current Law in the Complex Domain

Kirchhoff's current law is ordinarily stated in terms of the real (time-domain) currents flowing in a circuit, because it actually describes physical quantities, at least in a macroscopic, statistical sense. It also applied, however, to a variety of purely mathematical models that are commonly used to analyze circuits in the so-called complex domain.

For example, if a linear circuit is in the sinusoidal steady state, all of the currents and voltages in the circuit are sinusoidal. Thus, each voltage has the form

$$v(t) = A\sin(\omega t + \phi)$$

and each current has the form

$$i(t) = B\sin(\omega t + \theta)$$

where the positive coefficients A and B are called the magnitudes of the signals, and ϕ and θ are the phase angles of the signals. These mathematical models describe the physical behavior of electrical quantities, and instrumentation, such as an oscilloscope, can display the actual waveforms represented by the mathematical model. Although methods exist for manipulating the models of circuits to obtain the magnitude and phase coefficients that uniquely determine the waveform of each voltage and current, the manipulations are cumbersome and not easily extended to address other issues in circuit analysis.

Steinmetz (Smith and Dorf, 1992) found a way to exploit complex algebra to create an elegant framework for representing signals and analyzing circuits when they are in the steady state. In this approach, a model is developed in which each physical sign is replaced by a "complex" mathematical signal. This complex signal in polar, or exponential, form is represented as

$$v_c(t) = Ae^{(j\omega t + \phi)}$$

The algebra of complex exponential signals allows us to write this as

$$v_c(t) = Ae^{j\phi}e^{jwt}$$

and Euler's identity gives the equivalent rectangular form:

$$v_c(t) = A\big[\cos(\omega t + \phi) + j\sin(\omega t + \phi)\big]$$

So we see that a physical signal is either the real (cosine) or the imaginary (sine) component of an abstract, complex mathematical signal. The additional mathematics required for treatment of complex numbers allows us to associate a phasor, or complex amplitude, with a sinusoidal signal. The time-invariant phasor associated with $v(t)$ is the quantity

$$\mathbf{V}_c = Ae^{j\phi}$$

Notice that the phasor $\mathbf{v}_c$ is an algebraic constant and that it incorporates the parameters A and ϕ of the corresponding time-domain sinusoidal signal.

Phasors can be thought of as being vectors in a two-dimensional plane. If the vector is allowed to rotate about the origin in the counterclockwise direction with frequency ω, the projection of its tip onto the horizontal (real) axis defines the time-domain signal corresponding to the real part of $v_c(t)$, i.e., $A\cos[\omega t + \phi]$, and its projection onto the vertical (imaginary) axis defines the time-domain signal corresponding to the imaginary part of $v_c(t)$, i.e., $A\sin[\omega t + \phi]$.

The composite signal $v_c(t)$ is a mathematical entity; it cannot be seen with an oscilloscope. Its value lies in the fact that when a circuit is in the steady state, its voltages and currents are uniquely determined by their corresponding phasors, and these in turn satisfy Kirchhoff's voltage and current laws! Thus, we are able to write

$$\sum I_{km} = 0$$

where I_{km} is the phasor of $i_{km}(t)$, the sinusoidal current in branch k attached to node m. An equation of this form can be written at each node of the circuit. For example, at node b in Figure 3.1 KCL would have the form

$$I_1 - I_2 + I_3 = 0$$

Consequently, a set of linear, algebraic equations describes the phasors of the currents and voltages in a circuit in the sinusoidal steady state, i.e., the notion of time is suppressed (see Section "Node and Mesh Analysis"). The solution of the set of equations yields the phasor of each voltage and current in the circuit from which the actual time-domain expressions can be extracted.

It can also be shown that KCL can be extended to apply to the Fourier transforms and the Laplace transforms of the currents in a circuit. Thus, a single relationship between the currents at the nodes of a circuit applies to all of the known mathematical representations of the currents (Ciletti, 1988).

Kirchhoff's Voltage Law

Kirchhoff's voltage law (KVL) describes a relationship among the voltages measured across the branches in any closed, connected path in a circuit. Each branch in a circuit is connected to two nodes. For the purpose of applying KVL, a path has an orientation in the sense that in "walking" along the path one would enter one of the nodes and exit the other. This establishes a direction for determining the voltage across a branch in the path: the voltage is the difference between the potential of the node entered and the potential of the node at which the path exits. Alternatively, the voltage drop along a branch is the difference of the node voltage at the entered node and the node voltage at the exit node. For example, if a path includes a branch between node a and node b, the voltage drop measured along the path in the direction from node a to node b is denoted by v_{ab} and is given by $v_{ab} = v_a - v_b$. Given v_{ab}, branch voltage along the path in the direction from node b to node a is $v_{ba} = v_b - v_a = -v_{ab}$.

Kirchhoff's voltage law, like Kirchhoff's current law, is true at any time. KVL can also be stated in terms of voltage rises instead of voltage drops.

KVL can be expressed mathematically as "the algebraic sum of the voltages drops around any closed path of a circuit at any instant of time is zero." This statement can also be cast as an equation:

$$\sum v_{km}(t) = 0$$

where $v_{km}(t)$ is the instantaneous voltage drop measured across branch k of path m. By convention, the voltage drop is taken in the direction of the edge sequence that forms the path.

The edge sequence $\{e_1, e_2, e_3, e_4, e_6, e_7\}$ forms a closed path in Figure 3.1. The sum of the voltage drops taken around the path must satisfy KVL:

$$v_{ab}(t) + v_{bc}(t) + v_{cd}(t) + v_{de}(t) + v_{ef}(t) + v_{fa}(t) = 0$$

Since $v_{af}(t) = -v_{fa}(t)$, we can also write

$$v_{af}(t) = v_{ab}(t) + v_{bc}(t) + v_{cd}(t) + v_{de}(t) + v_{ef}(t)$$

Had we chosen the path corresponding to the edge sequence $\{e_1, e_5, e_6, e_7\}$ for the path, we would have obtained

$$v_{af}(t) = v_{ab}(t) + v_{be}(t) + v_{ef}(t)$$

This demonstrates how KCL can be used to determine the voltage between a pair of nodes. It also reveals the fact that the voltage between a pair of nodes is independent of the path between the nodes on which the voltages are measured.

KVL in the Complex Domain

KVL also applies to the phasors of the voltages in a circuit in steady state and to the Fourier transforms and Laplace transforms of the voltages in a circuit.

Importance of KVL and KCL

KCL is used extensively in nodal analysis because it is amenable to computer-based implementation and supports a systematic approach to circuit analysis. Nodal analysis leads to a set of algebraic equations in which the variables are the voltages at the nodes of the circuit. This formulation is popular in CAD programs because the variables correspond directly to physical quantities that can be measured easily.

KVL can be used to completely analyze a circuit, but it is seldom used in large-scale circuit simulation programs. The basic reason is that the currents that correspond to a loop of a circuit do not necessarily correspond to the currents in the individual branches of the circuit. Nonetheless, KVL is frequently used to troubleshoot a circuit by measuring voltage drops across selected components.

Defining Terms

Branch: A symbol representing a path for current through a component in an electrical circuit.
Branch current: The current in a branch of a circuit.
Branch voltage: The voltage across a branch of a circuit.
Independent source: A voltage (current) source whose voltage (current) does not depend on any other voltage or current in the circuit.
Node: A symbol representing a physical connection between two electrical components in a circuit.
Node voltage: The voltage between a node and a reference node (usually ground).

References

M.D. Ciletti, *Introduction to Circuit Analysis and Design*, New York: Holt, Rinehart and Winston, 1988.
R.H. Smith and R.C. Dorf, *Circuits, Devices and Systems*, New York: Wiley, 1992.

Further Information

Kirchhoff's laws form the foundation of modern computer software for analyzing electrical circuits. The interested reader might consider the use of determining the minimum number of algebraic equations that fully characterizes the circuit. It is determined by KCL, KVL, or some mixture of the two.

3.2 Node and Mesh Analysis

J. David Irwin

In this section, Kirchhoff's current law (KCL) and Kirchhoff's voltage law (KVL) will be used to determine currents and voltages throughout a network. For simplicity, we will first illustrate the basic principles of both node analysis and mesh analysis using only dc circuits. Once the fundamental concepts have been explained and demonstrated, we will employ them to determine the currents and voltages in an ac circuit. Since the application of these techniques to large circuits requires the use of computer-aided analysis tools, the use of MATLAB as a solution method will be demonstrated, together with PSPICE, for purposes of comparison.

Node Analysis

In a node analysis, the node voltages are the variables in a circuit, and KCL is the vehicle used to determine them. One node in the network is selected as a reference node, and then all other node voltages are defined with respect to that particular node. This reference node is typically referred to as ground using the symbol ($\perp$), indicating that it is at ground-zero potential.

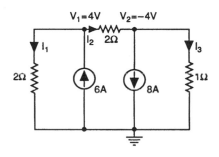

FIGURE 3.3   A three-node network.

Consider the network shown in Figure 3.3. The network has three nodes, and the node at the bottom of the circuit has been selected as the reference node. Therefore the two remaining nodes, labeled V_1 and V_2, are measured with respect to this reference node.

Suppose that the node voltages V_1 and V_2 have somehow been determined, i.e., $V_1 = 4$ V and $V_2 = -4$ V. Once these node voltages are known, Ohm's law can be used to find all branch currents. For example:

$$I_1 = \frac{V_1 - 0}{2} = 2A$$

$$I_2 = \frac{V_1 - V_2}{2} = \frac{4 - (-4)}{2} = 4A$$

$$I_3 = \frac{V_2 - 0}{1} = \frac{-4}{1} = -4A$$

Note that KCL is satisfied at every node, i.e.:

$$I_1 - 6 + I_2 = 0$$

$$-I_2 + 8 + I_3 = 0$$

$$-I_1 + 6 - 8 - I_3 = 0$$

Therefore, as a general rule, if the node voltages are known, all branch currents in the network can be immediately determined.

In order to determine the node voltages in a network, we apply KCL to every node in the network except the reference node. Therefore, given an N-node circuit, we employ $N-1$ linearly independent simultaneous equations to determine the $N-1$ unknown node voltages. Graph theory, which is covered in Section "Graph Theory", can be used to prove that exactly $N-1$ linearly independent KCL equations are required to find the $N-1$ unknown node voltages in a network.

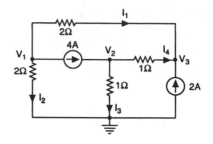

FIGURE 3.4   A four-node network.

Let us now demonstrate the use of KCL in determining the node voltages in a network. For the network shown in Figure 3.4, the bottom node is selected as the reference and the three remaining nodes, labeled V_1, V_2, and V_3, are measured with respect to that node. All unknown branch currents are also labeled. The KCL equations for the three nonreference nodes are

$$I_1 + 4 + I_2 = 0$$

$$-4 + I_3 + I_4 = 0$$

$$-I_1 - I_4 - 2 = 0$$

Using Ohm's law, these equations can be expressed as

$$\frac{V_1 - V_3}{2} + 4 + \frac{V_1}{2} = 0$$

$$-4 + \frac{V_2}{1} + \frac{V_2 - V_3}{1} = 0$$

$$-\frac{(V_1 - V_3)}{2} - \frac{V_2 - V_3}{1} - 2 = 0$$

Solving these equations, using any convenient method, yields $V_1 = -8/3$ V, $V_2 = 10/3$ V, and $V_3 = 8/3$ V. Applying Ohm's law we find that the branch currents are $I_1 = -16/6$ A, $I_2 = -8/6$ A, $I_3 = 20/6$ A, and $I_4 = 4/6$ A. A quick check indicates that KCL is satisfied at every node.

The circuits examined thus far have contained only current sources and resistors. In order to expand our capabilities, we next examine a circuit containing voltage sources. The circuit shown in Figure 3.5 has three non-reference nodes labeled V_1, V_2, and V_3. However, we do not have three unknown node voltages. Since known voltage sources exist between the reference node and nodes V_1 and V_3, these two-node voltages are known, i.e., $V_1 = 12$ V and $V_3 = -4$ V. Therefore, we have only one unknown node voltage, V_2. The equations for this network are then

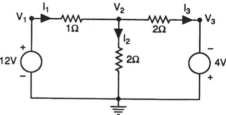

FIGURE 3.5   A four-node network containing voltage sources.

$$V_1 = 12$$

$$V_3 = -4$$

and

$$-I_1 + I_2 + I_3 = 0$$

The KCL equation for node V_2 written using Ohm's law is

$$-\frac{(12 - V_2)}{1} + \frac{V_2}{2} + \frac{V_2 - (-4)}{2} = 0$$

Solving this equation yields $V_2 = 5$ V, $I_1 = 7$ A, $I_2 = 5/2$ A, and $I_3 = 9/2$ A. Therefore, KCL is satisfied at every node.

Thus, the presence of a voltage source in the network actually simplifies a node analysis. In an attempt to generalize this idea, consider the network in Figure 3.6. Note that in this case $V_1 = 12$ V and the difference between node voltages V_3 and V_2 is constrained to be 6 V. Hence, two of the three equations needed to solve for the node voltages in the network are

$$V_1 = 12$$

$$V_3 - V_2 = 6$$

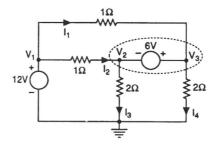

FIGURE 3.6　A four-node network used to illustrate a supernode.

To obtain the third required equation, we form what is called a supernode, indicated by the dotted enclosure in the network. Just as KCL must be satisfied at any node in the network, it must be satisfied at the supernode as well. Therefore, summing all the currents leaving the supernode yields the equation

$$\frac{(V_2 - V_1)}{1} + \frac{V_2}{2} + \frac{V_3 - V_1}{1} + \frac{V_3}{2} = 0$$

The three equations yield the node voltages $V_1 = 12$ V, $V_2 = 5$ V, and $V_3 = 11$ V, and therefore $I_1 = 1$ A, $I_2 = 7$ A, $I_3 = 5/2$ A, and $I_4 = 11/2$ A.

Mesh Analysis

In a mesh analysis, the mesh currents in the network are the variables and KVL is the mechanism used to determine them. Once all the mesh currents have been determined, Ohm's law will yield the voltages anywhere in a circuit. If the network contains N independent meshes, then graph theory can be used to prove that N independent linear simultaneous equations will be required to determine the N mesh currents.

The network shown in Figure 3.7 has two independent meshes. They are labeled I_1 and I_2, as shown. If the mesh currents are known to be $I_1 = 7$ A and $I_2 = 5/2$ A, then all voltages in the network can be calculated. For example, the voltage V_1, i.e., the voltage across the

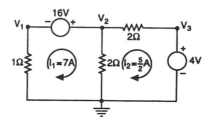

FIGURE 3.7　A network containing two independent meshes.

1-Ω resistor, is $V_1 = -I_1 R = -(7)(1) = -7$ V. Likewise, $V_2 = (I_1 - I_2)R = (7 - 5/2)(2) = 9$ V. Furthermore, we can check our analysis by showing that KVL is satisfied around every mesh. Starting at the lower left-hand corner and applying KVL to the left-hand mesh, we obtain

$$-(7)(1) + 16 - (7 - 5/2)(2) = 0$$

where we have assumed that increases in energy level are positive and decreases in energy level are negative.

Consider now the network in Figure 3.8. Once again, if we assume that an increase in energy level is positive and a decrease in energy level is negative, the three KVL equations for the three meshes defined are

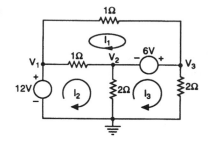

FIGURE 3.8   A three-mesh network.

$$-I_1(1) - 6 - (I_1 - I_2)(1) = 0$$

$$+12 - (I_2 - I_1)(1) - (I_2 - I_3)(2) = 0$$

$$-(I_3 - I_2)(2) + 6 - I_3(2) = 0$$

These equations can be written as

$$2I_1 - I_2 = -6$$

$$-I_{12} + 3I_2 - 2I_3 = 12$$

$$-2I_2 + 4I_3 = 6$$

Solving these equations using any convenient method yields $I_1 = 1$ A, $I_2 = 8$ A, and $I_3 = 5.5$ A. Any voltage in the network can now be easily calculated, e.g., $V_2 = (I_2 - I_3)(2) = 5$ V and $V_3 = I_3(2) = 11$ V.

Just as in the node analysis discussion, we now expand our capabilities by considering circuits that contain current sources. In this case, we will show that for mesh analysis, the presence of current sources makes the solution easier.

The network in Figure 3.9 has four meshes which are labeled I_1, I_2, I_3, and I_4. However, since two of these currents, i.e., I_3 and I_4, pass directly through a current source, two of the four linearly independent equations required to solve the network are

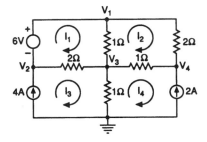

$$I_3 = 4$$
$$I_4 = -2$$

The two remaining KVL equations for the meshes defined by I_1 and I_2 are

FIGURE 3.9   A four-mesh network containing current sources.

$$+6 - (I_1 - I_2)(1) - (I_1 - I_3)(2) = 0$$

$$-(I_2 - I_1)(1) - I_2(2) - (I_2 - I_4)(1) = 0$$

Solving these equations for I_1 and I_2 yields $I_1 = 54/11$ A and $I_2 = 8/11$ A. A quick check will show that KCL is satisfied at every node. Furthermore, we can calculate any node voltage in the network. For example, $V_3 = (I_3 - I_4)(1) = 6$ V and $V_1 = V_3 + (I_1 - I_2)(1) = 112/11$ V.

An AC Analysis Example

Both node analysis and mesh analysis have been presented and discussed. Although the methods have been presented within the framework of dc circuits with only independent sources, the techniques are applicable to ac analysis and circuits containing dependent sources.

To illustrate the applicability of the two techniques to ac circuit analysis, consider the network in Figure 3.10. All voltages and currents are phasors and the impedance of each passive element is known.

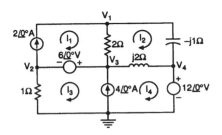

In the node analysis case, the voltage V_4 is known and the voltage between V_2 and V_3 is constrained. Therefore, two of the four required equations are

$$V_4 = 12\underline{/0^\circ}$$

$$V_2 + 6\underline{/0^\circ} = V_3$$

FIGURE 3.10 A network containing five nodes and four meshes.

KCL for the node labeled V_1 and the supernode containing the nodes labeled V_2 and V_3 is

$$\frac{V_1 - V_3}{2} + \frac{V_1 - V_4}{-j1} = 2\underline{/0^\circ}$$

$$\frac{V_2}{1} + 2\underline{/0^\circ} + \frac{V_3 - V_1}{2} + \frac{V_3 - V_4}{-j2} = 2\underline{/0^\circ}$$

Solving these equations yields the remaining unknown node voltages:

$$V_1 = 11.9 - j0.88 = 11.93\underline{/-4.22^\circ}\,V$$

$$V_2 = 3.66 - j1.07 = 3.91\underline{/-16.34^\circ}\,V$$

$$V_3 = 9.66 - j1.07 = 9.72\underline{/-6.34^\circ}\,V$$

In the mesh analysis case, the currents I_1 and I_3 are constrained to be

$$I_1 = 2\underline{/0^\circ}$$

$$I_4 - I_3 = -4\underline{/0^\circ}$$

The two remaining KVL equations are obtained from the mesh defined by mesh current I_2 and the loop that encompasses the meshes defined by mesh currents I_3 and I_4:

$$-2(I_2 - I_1) - (-j1)I_2 - j2(I_2 - I_4) = 0$$

$$-I_3 + 6\underline{/0^\circ} - j2(I_4 - I_2) - 12\underline{/0^\circ} = 0$$

Solving these equations yields the remaining unknown mesh currents

$$I_2 = 0.88\underline{/-6.34^\circ}\,A$$

$$I_3 = 3.91\underline{/163.66^\circ}\,A$$

$$I_4 = 1.13\underline{/72.35^\circ}\,A$$

As a quick check we can use these currents to compute the node voltages. For example, if we calculate

$$V_2 = -1(I_3)$$

and

$$V_1 = -j1(I_2) + 12\underline{/0°})$$

we obtain the answers computed earlier.

Since both node and mesh analyses will yield all currents and voltages in a network, which technique should be used? The answer to this question depends upon the network to be analyzed. If the network contains more voltage sources than current sources, node analysis might be the easier technique. If, however, the network contains more current sources than voltage sources, mesh analysis may be the easiest approach.

Computer Simulation of Networks

While any network can be analyzed using mesh or nodal techniques, the required calculations are cumbersome for more than three loops or nodes. In these cases, computer simulation is an attractive alternative. As an example, we will solve for the voltage V_0 in the circuit in Figure 3.11 using first MATLAB and then PSPICE.

MATLAB requires a matrix representation of the network. Using mesh analysis yields the equations

$$I_1 = 2I_X$$

$$-I_1 + (2 - j1)I_2 + (j1)I_4 = 0$$

$$(j1)I_2 + (j1)I_3 + (2 - j1)I_4 - (j1)I_5 - 2I_6 = 6\underline{/30°}$$

$$I_3 - I_4 = 3\underline{/-60°}$$

$$-(j1)I_3 + (4 + j1)I_6 - 2I_6 = 0$$

$$-2I_4 - 2I_5 + 4I_6 + 4V_X = 0$$

$$I_4 - I_6 = I_X$$

$$(I_4 - I_1)(1) = V_X$$

Note that a "super-mesh" path around the I_3 and I_4 meshes has been used to avoid the 3-A current source. This is analogous to the supernode in Figure 3.6. Eliminating I_X and V_X, the equations are put into matrix format:

$$
\begin{bmatrix}
1 & 0 & 0 & -2 & 0 & 2 \\
-1 & (2-j1) & 0 & j1 & 0 & 0 \\
0 & j1 & j1 & 2-j1 & -j1 & -2 \\
0 & 0 & 1 & -1 & 0 & 0 \\
0 & 0 & -j1 & 0 & 4+j1 & -2 \\
-4 & 4 & 0 & -2 & -2 & 4
\end{bmatrix}
\begin{bmatrix}
I_1 \\
I_2 \\
I_3 \\
I_4 \\
I_5 \\
I_6
\end{bmatrix}
=
\begin{bmatrix}
0 \\
0 \\
5.196 + j3 \\
1.5 - j2.598 \\
0 \\
0
\end{bmatrix}
$$

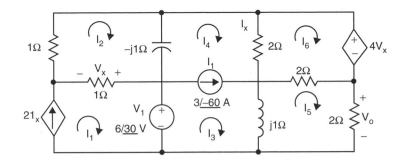

FIGURE 3.11 A six mesh–five non-reference node network.

Solving for the currents:

$$
\begin{bmatrix} \mathbf{I}_1 \\ \mathbf{I}_2 \\ \mathbf{I}_3 \\ \mathbf{I}_4 \\ \mathbf{I}_5 \\ \mathbf{I}_6 \end{bmatrix} =
\begin{bmatrix}
1 & 0 & 0 & -2 & 0 & 2 \\
-1 & (2-j1) & 0 & j1 & 0 & 0 \\
0 & j1 & j1 & 2-j1 & -j1 & -2 \\
0 & 0 & 1 & -1 & 0 & 0 \\
0 & 0 & -j1 & 0 & 4+j1 & -2 \\
-4 & 4 & 0 & -2 & -2 & 4
\end{bmatrix}^{-1}
\begin{bmatrix} 0 \\ 0 \\ 5.196+j3 \\ 1.5-j2.598 \\ 0 \\ 0 \end{bmatrix}
$$

In MATLAB, we first specify the sources, $\mathbf{V}_1$ and $\mathbf{I}_1$:

$$>>\text{V1} = 5.196 + 3\text{i}$$

$$>>\text{I1} = 1.5 - 2.598\text{i}$$

(In MATLAB, "i" is the imaginary operator $\sqrt{-1}$.) Next, we enter the impedance matrix and voltage vector:

```
>> Z = [1 0 0 −2 0 2; −1 2 −1i 0 1i 0 0; 0 1i 1i 2 −1i −1i −2; 0 0 1
        −1 0 0; 0 0 −1i 0 4+1i −2; −4 4 0 −2 −2 4]

>>V = [0; 0; V1; I1; 0; 0]
```

To solve for the current vector, we enter

$$>>\text{I} = \text{inv}(\text{Z}) * \text{V}$$

The results are

```
I = 1.3713 + 0.5786i
    1.7369 + 0.4299i
    2.9556 − 0.0655i
    1.4556 + 2.5325i
    0.8155 + 1.6566i
    0.7700 + 2.2432i
```

Finally, $\mathbf{V}_0 = 2\mathbf{I}_5$. Since $\mathbf{I}_5 = 0.8155 + j1.6566$ A, or $1.8464\underline{/63.79°}$ in polar form, we find $\mathbf{V}_0 = 3.6928\underline{/63.79°}$ V.

In PSPICE, we draw the circuit using one of the accompanying schematic entry tools, either *Schematics* or *Capture*. The resulting *Schematics* file is shown in Figure 3.12. Note that capacitors and inductors must be specified in Farads and Henries, respectively. Therefore, any excitation frequency can be chosen with the L and C values calculated from the known impedances. The most convenient frequency is $\omega = 1$ rad./sec or 0.1591 Hz. Figure 3.13 shows the AC Sweep settings to produce a single frequency analysis at 0.1591 Hz. Also, the VPRINT1 part is required to load the simulation results into the OUTPUT file.

Finally, to preserve the clarity of the schematic, the controlling connections on both dependent sources are reversed with respect to Figure 3.12. To accommodate, we set the gains of the sources at -4 and -2. From the

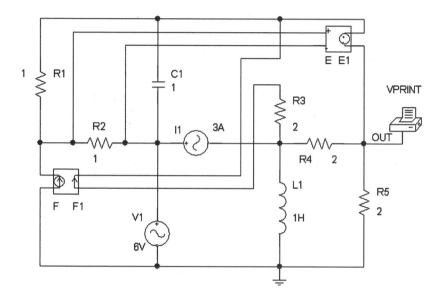

FIGURE 3.12   The *Schematics* file for the network in Figure 3.11 ready for simulation.

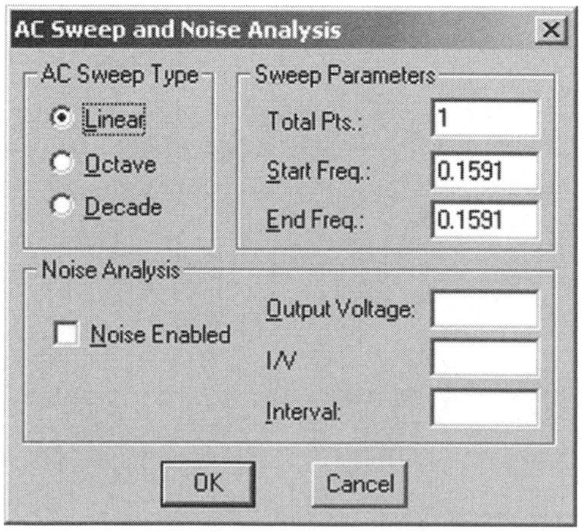

FIGURE 3.13   The AC Sweep attribute window in PSPICE.

OUTPUT file, the simulation results are

$$\text{FREQ} \quad \text{VM(OUT)} \quad \text{VP(OUT)}$$

$$0.1591 \quad 3.692 \quad 63.79$$

or $\mathbf{V}_0 = 3.692 \underline{/63.79°}$ V. The PSPICE and MATLAB results match to four significant digits.

Defining Terms

ac: An abbreviation for alternating current.

dc: An abbreviation for direct current.

Kirchhoff's current law (KCL): This law states that the algebraic sum of the currents either entering or leaving a node must be zero. Alternatively, the law states that the sum of the currents entering a node must be equal to the sum of the currents leaving that node.

Kirchhoff's voltage law (KVL): This law states that the algebraic sum of the voltages around any loop is zero. A loop is any closed path through the circuit in which no node is encountered more than once.

MATLAB and PSPICE: Computer-aided analysis techniques.

Mesh analysis: A circuit analysis technique in which KVL is used to determine the mesh currents in a network. A mesh is a loop that does not contain any loops within it.

Node analysis: A circuit analysis technique in which KCL is used to determine the node voltages in a network.

Ohm's law: A fundamental law which states that the voltage across a resistance is directly proportional to the current flowing through it.

Reference node: One node in a network that is selected to be a common point, and all other node voltages are measured with respect to that point.

Supernode: A cluster of nodes, interconnected with voltage sources, such that the voltage between any two nodes in the group is known.

Reference

J.D. Irwin, *Basic Engineering Circuit Analysis*, 7th ed., New York: John Wiley & Sons, 2002.

3.3   Network Theorems

Allan D. Kraus

Linearity and Superposition

Linearity

Consider a system (which may consist of a single network element) represented by a block, as shown in Figure 3.14, and observe that the system has an input designated by *e* (for excitation) and an output designated by *r* (for response). The system is considered to be *linear* if it satisfies the *homogeneity* and *superposition* conditions.

The *homogeneity condition*: If an arbitrary input to the system, *e*, causes a response, *r*, then if *ce* is the input, the output is *cr* where *c* is some arbitrary constant.

The *superposition condition*: If the input to the system, e_1, causes a response, r_1, and if an input to the system, e_2, causes a response, r_2, then a response, $r_1 + r_2$, will occur when the input is $e_1 + e_2$.

If neither the homogeneity condition nor the superposition condition is satisfied, the system is said to be *nonlinear*.

FIGURE 3.14 A simple system.

The Superposition Theorem

While both the homogeneity and superposition conditions are necessary for linearity, the superposition condition, in itself, provides the basis for the superposition theorem:

If cause and effect are linearly related, the total effect due to several causes acting simultaneously is equal to the sum of the individual effects due to each of the causes acting one at a time.

Example 3.1

Consider the network driven by a current source at the left and a voltage source at the top, as shown in Figure 3.15(a). The current phasor indicated by $\hat{I}$ is to be determined. According to the superposition theorem, the current $\hat{I}$ will be the sum of the two current components $\hat{I}_V$ due to the voltage source acting alone as shown in Figure 3.15(b) and $\hat{I}_C$ due to the current source acting alone shown in Figure 3.15(c):

$$\hat{I} = \hat{I}_V + \hat{I}_C$$

Figure 3.15(b) and (c) follow from the methods of removing the effects of independent voltage and current sources. Voltage sources are nulled in a network by replacing them with short circuits and current sources are nulled in a network by replacing them with open circuits.

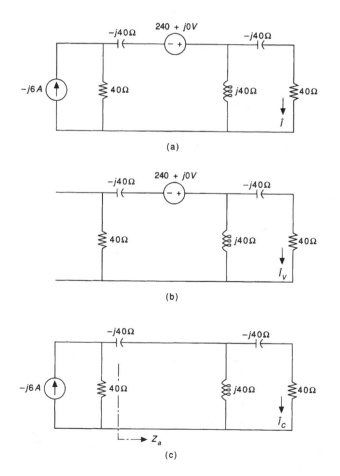

FIGURE 3.15   (a) A network to be solved by using superposition; (b) the network with the current source nulled; and (c) the network with the voltage source nulled.

The networks displayed in Figure 3.15(b) and (c) are simple ladder networks in the phasor domain, and the strategy is to first determine the equivalent impedances presented to the voltage and current sources. In Figure 3.15(b), the group of three impedances to the right of the voltage source are in series-parallel and possess an impedance of

$$Z_P = \frac{(40 - j40)(j40)}{40 + j40 - j40} = 40 + j40 \, \Omega$$

and the total impedance presented to the voltage source is

$$Z = Z_P + 40 - j40 = 40 + j40 + 40 - j40 = 80 \, \Omega$$

Then $\hat{I}_1$, the current leaving the voltage source, is

$$\hat{I}_1 = \frac{240 + j0}{80} = 3 + j0 \, \text{A}$$

and by a current division

$$\hat{I}_V = \left[\frac{j40}{40 - j40 + j40}\right](3 + j0) = j(3 + j0) = 0 + j3 \, A$$

In Figure 3.15(b), the current source delivers current to the 40Ω resistor and to an impedance consisting of the capacitor and Z_p. Call this impedance Z_a so that

$$Z_a = -j40 + Z_P = -j40 + 40 + j40 = 40 \, \Omega$$

Then, two current divisions give $\hat{I}_C$:

$$\hat{I}_C = \left[\frac{40}{40 + 40}\right]\left[\frac{j40}{40 - j40 + j40}\right](0 - j6) = \frac{j}{2}(0 - j6) = 3 + j0 \, \text{A}$$

The current $\hat{I}_C$ in the circuit of Figure 3.15(a) is

$$\hat{I} = \hat{I}_V + \hat{I}_C = 0 + j_3 + (3 + j0) = 3 + j3 \, \text{A}$$

The Network Theorems of Thévenin and Norton

If interest is to be focused on the voltages and across the currents through a small portion of a network such as network B in Figure 3.16(a), it is convenient to replace network A, which is complicated and of little interest, by a simple equivalent. The simple equivalent may contain a single, equivalent, voltage source in series with an equivalent impedance in series as displayed in Figure 3.16(b). In this case, the equivalent is called a *Thévenin equivalent*. Alternatively, the simple equivalent may consist of an equivalent current source in parallel with an equivalent impedance. This equivalent, shown in Figure 3.16(c), is called a *Norton equivalent*. Observe that as long as Z_T (subscript T for Thévenin) is equal to Z_N (subscript N for Norton), the two equivalents may be obtained from one another by a simple source transformation.

Conditions of Application

The Thévenin and Norton network equivalents are only valid at the terminals of network A in Figure 3.16(a) and they do not extend to its interior. In addition, there are certain restrictions on networks A and B. Network A may contain only linear elements but may contain both independent and dependent sources. Network B, on

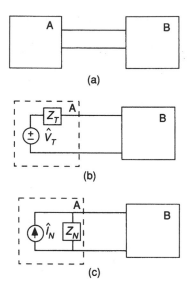

FIGURE 3.16   (a) Two one-port networks; (b) the Thévenin equivalent for network A; and (c) the Norton equivalent for network A.

the other hand, is not restricted to linear elements; it may contain nonlinear or time-varying elements and may also contain both independent and dependent sources. Together, there can be no controlled source coupling or magnetic coupling between networks A and B.

The Thévenin Theorem

The statement of the Thévenin theorem is based on Figure 3.16(b):

> Insofar as a load which has no magnetic or controlled source coupling to a one-port is concerned, a network containing linear elements and both independent and controlled sources may be replaced by an ideal voltage source of strength, $\hat{V}_T$, and an equivalent impedance, Z_T, in series with the source. The value of $\hat{V}_T$ is the open-circuit voltage, $\hat{V}_{OC}$, appearing across the terminals of the network and Z_T is the driving point impedance at the terminals of the network, obtained with all independent sources set equal to zero.

The Norton Theorem

The Norton theorem involves a current source equivalent. The statement of the Norton theorem is based on Figure 3.16(c):

> Insofar as a load which has no magnetic or controlled source coupling to a one-port is concerned, the network containing linear elements and both independent and controlled sources may be replaced by an ideal current source of strength, $\hat{I}_N$, and an equivalent impedance, Z_N, in parallel with the source. The value of $\hat{I}_N$ is the short-circuit current, $\hat{I}_{SC}$, which results when the terminals of the network are shorted and Z_N is the driving point impedance at the terminals when all independent sources are set equal to zero.

The Equivalent Impedance, $Z_T = Z_N$

Three methods are available for the determination of Z_T. All of them are applicable at the analyst's discretion. When controlled sources are present, however, the first method cannot be used.

The first method involves the direct calculation of $Z_{eq} = Z_T = Z_N$ by looking into the terminals of the network after all independent sources have been nulled. Independent sources are nulled in a network by replacing all independent voltage sources with a short circuit and all independent current sources with an open circuit.

The second method, which may be used when controlled sources are present in the network, requires the computation of both the Thévenin equivalent voltage (the open-circuit voltage at the terminals of the network) and the Norton equivalent current (the current through the short-circuited terminals of the network). The equivalent impedance is the ratio of these two quantities:

$$Z_T = Z_N = Z_{eq} = \frac{\hat{V}_T}{\hat{I}_N} = \frac{\hat{V}_{OC}}{\hat{I}_{SC}}$$

The third method may also be used when controlled sources are present within the network. A test voltage may be placed across the terminals with a resulting current calculated or measured. Alternatively, a test current may be injected into the terminals with a resulting voltage determined. In either case, the equivalent resistance can be obtained from the value of the ratio of the test voltage $\hat{V}_0$ to the resulting current $\hat{I}_0$:

$$Z_T = \frac{\hat{V}_0}{\hat{I}_0}$$

Example 3.2

The current through the capacitor with impedance $-j35\ \Omega$ in Figure 3.17(a) may be found using Thévenin's theorem. The first step is to remove the $-j35\ \Omega$ capacitor and consider it as the load. When this is done, the network in Figure 3.17(b) results.

The Thévenin equivalent voltage is the voltage across the 40-Ω resistor. The current through the 40-Ω resistor was found in Example 3.1 to be $I = 3 + j3\ \Omega$. Thus

$$\hat{V}_T = 40(3 + j3) = 120 + j120\ \text{V}$$

The Thévenin equivalent impedance may be found by looking into the terminals of the network in Figure 3.17(c). Observe that both sources in Figure 3.17(a) have been nulled and that, for ease of computation, impedances Z_a and Z_b have been placed on Figure 3.17(c). Here

$$Z_a = \frac{(40 - j40)(j40)}{40 + j40 - j40} = 40 + j40\ \Omega$$

$$Z_b = \frac{(40)(40)}{40 + 40} = 20\ \Omega$$

and

$$Z_T = Z_b + j15 = 20 + j15\ \Omega$$

Both the Thévenin equivalent voltage and impedance are shown in Figure 3.17(d); and when the load is attached, as in Figure 3.17(d), the current can be computed as

$$\hat{I} = \frac{\hat{V}_T}{20 + j15 - j35} = \frac{120 + j120}{20 - j20} = 0 + j6\ \text{A}$$

The Norton equivalent circuit is obtained via a simple voltage-to-current source transformation and is shown in Figure 3.18. Here it is observed that a single current division gives

$$\hat{I} = \left[\frac{20 + j15}{20 + j15 - j35} \right](6.72 + j0.96) = 0 + j6\ \text{A}$$

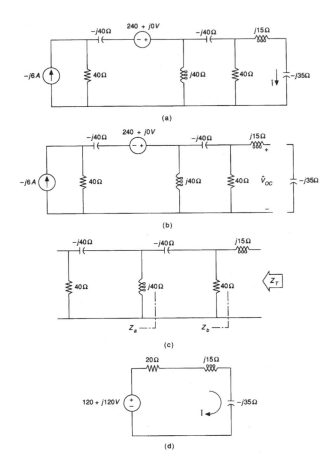

FIGURE 3.17 (a) A network in the phasor domain; (b) the network with the load removed; (c) the network for the computation of the Thévenin equivalent impedance; and (d) the Thévenin equivalent.

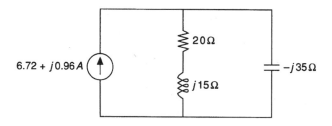

FIGURE 3.18 The Norton equivalent of Figure 3.17(d).

Tellegen's Theorem

Tellegen's theorem states:

> In an arbitrarily lumped network subject to KVL and KCL constraints, with reference directions of the branch currents and branch voltages associated with the KVL and KCL constraints, the product of all branch currents and branch voltages must equal zero.

Tellegen's theorem may be summarized by the equation

$$\sum_{k=1}^{b} v_k j_k = 0$$

where the lower case letters v and j represent instantaneous values of the branch voltages and branch currents, respectively, and where b is the total number of branches. A matrix representation employing the branch current and branch voltage vectors also exists. Because V and J are column vectors

$$\mathbf{V} \cdot \mathbf{J} = \mathbf{V}^T \mathbf{J} = \mathbf{J}^T \mathbf{V}$$

The prerequisite concerning the KVL and KCL constraints in the statement of Tellegen's theorem is of crucial importance.

Example 3.3

Figure 3.19 displays an oriented graph of a particular network in which there are six branches labeled with numbers within parentheses and four nodes labeled by numbers within circles. Several known branch currents and branch voltages are indicated. Because the type of elements or their values is not germane to the construction of the graph, the other branch currents and branch voltages may be evaluated from repeated applications of KCL and KVL. KCL may be used first at the various nodes:

$$\text{node 3:} \quad j_2 = j_6 - j_4 = 4 - 2 = 2\,\text{A}$$
$$\text{node 1:} \quad j_3 = -j_1 - j_2 = -8 - 2 = -10\,\text{A}$$
$$\text{node 2:} \quad j_5 = j_3 - j_4 = -10 - 2 = -12\,\text{A}$$

Then KVL gives

$$v_3 = v_2 - v_4 = 8 - 6 = 2\,\text{V}$$
$$v_6 = v_5 - v_4 = -10 - 6 = -16\,\text{V}$$
$$v_1 = v_2 + v_6 = 8 - 16 = -8\,\text{V}$$

The transpose of the branch voltage and current vectors are

$$\mathbf{V}_T = \begin{bmatrix} -8 & 8 & 2 & 6 & -10 & -16 \end{bmatrix}\,\text{V}$$

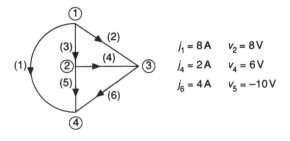

$$j_1 = 8\,\text{A} \qquad v_2 = 8\,\text{V}$$
$$j_4 = 2\,\text{A} \qquad v_4 = 6\,\text{V}$$
$$j_6 = 4\,\text{A} \qquad v_5 = -10\,\text{V}$$

FIGURE 3.19   An oriented graph of a particular network with some known branch currents and branch voltages.

and

$$\mathbf{J}_T = [8 \quad 2 \quad -10 \quad 2 \quad -12 \quad 4]\ V$$

The scalar product of $\mathbf{V}$ and $\mathbf{J}$ gives

$$-8(8) + 8(2) + 2(-10) + 6(2) + (-10)(-12) + (-16)(4) = -148 + 148 = 0$$

and Tellegen's theorem is confirmed.

Maximum Power Transfer Theorem

The maximum power transfer theorem pertains to the connections of a load to the Thévenin equivalent of a source network in such a manner as to transfer maximum power to the load. For a given network operating at a prescribed voltage with a Thévenin equivalent impedance

$$Z_T = |Z_T|\underline{/\theta_T}$$

the real power drawn by any load of impedance

$$Z_0 = |Z_0|\underline{/\theta_0}$$

is a function of just two variables, $|Z_0|$ and θ_0. If the power is to be a maximum, there are three alternatives to the selection of $|Z_0|$ and θ_0:

(1) Both $|Z_0|$ and θ_0 are at the designer's discretion and both are allowed to vary in any manner in order to achieve the desired result. In this case, the load should be selected to be the complex conjugate of the Thévenin equivalent impedance

$$Z_0 = Z_T^*$$

(2) The angle θ_0 is fixed but the magnitude $|Z_0|$ is allowed to vary. For example, the analyst may select and fix $\theta_0 = 0°$. This requires that the load be resistive (Z is entirely real). In this case, the value of the load resistance should be selected to be equal to the magnitude of the Thévenin equivalent impedance

$$R_0 = |Z_T|$$

(3) The magnitude of the load impedance $|Z_0|$ can be fixed, but the impedance angle θ_0 is allowed to vary. In this case, the value of the load impedance angle should be

$$\theta_0 = \arcsin\left[-\frac{2|Z_0||Z_T|\sin\theta_T}{|Z_0|^2 + |Z_T|^2}\right]$$

Example 3.4

Figure 3.20(a) is identical to Figure 3.17(b) with the exception of a load, Z_0, substituted for the capacitive load. The Thévenin equivalent is shown in Figure 3.20(b). The value of Z_0 to transfer maximum power is to be found if its elements are unrestricted, if it is to be a single resistor, or if the magnitude of Z_0 must be 20 Ω but its angle is adjustable.

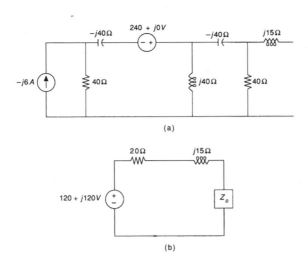

FIGURE 3.20 (a) A network for which the load, Z_0, is to be selected for maximum power transfer, and (b) the Thévenin equivalent of the network.

For maximum power transfer to Z_0 when the elements of Z_0 are completely at the discretion of the network designer, Z_0 must be the complex conjugate of Z_T:

$$Z_0 = Z_T^* = 20 - j15\,\Omega$$

If Z_0 is to be a single resistor, R_0, then the magnitude of $Z_0 = R_0$ must be equal to the magnitude of Z_T. Here

$$Z_T = 20 + j15 = 25\,\underline{/36.87^\circ}$$

so that

$$R_0 = |Z_0| = 25\,\Omega$$

If the magnitude of Z_0 must be 20 Ω but the angle is adjustable, the required angle is calculated from

$$\theta_0 = \arcsin\left[-\frac{2|Z_0||Z_T|}{|Z_0|^2 + |Z_T|^2}\sin\theta_T\right]$$

$$= \arcsin\left[-\frac{2(20)(25)}{(20)^2 + (25)^2}\sin\underline{/36.87^\circ}\right]$$

$$= \arcsin(-0.585) = -35.83^\circ$$

This makes Z_0:

$$Z_0 = 20\,\underline{/-35.83^\circ} = 16.22 - j11.71\,\Omega$$

The Reciprocity Theorem

The reciprocity theorem is a useful general theorem that applies to all linear, passive, and bilateral networks. However, it applies only to cases where current and voltage are involved:

> The ratio of a single excitation applied at one point to an observed response at another is invariant with respect to an interchange of the points of excitation and observation.

The reciprocity principle also applies if the excitation is a current and the observed response is a voltage. It will not apply, in general, for voltage–voltage and current–current situations, and, of course, it is not applicable to network models of nonlinear devices.

Example 3.5

It is easily shown that the positions of v_s and i in Figure 3.21(a) may be interchanged as in Figure 3.21(b) without changing the value of the current i.

In Figure 3.21(a), the resistance presented to the voltage source is

$$R = 4 + \frac{3(6)}{3+6} = 4 + 2 = 6 \; \Omega$$

Then

$$i_a = \frac{v_s}{R} = \frac{36}{6} = 6 \, \text{A}$$

and by current division

$$i_a = \frac{6}{6+3} i_a = \left(\frac{2}{3}\right) 6 = 4 \, \text{A}$$

In Figure 3.21(b), the resistance presented to the voltage source is

$$R = 3 + \frac{6(4)}{6+4} = 3 + \frac{12}{5} = \frac{27}{5} \; \Omega$$

Then

$$i_b = \frac{v_s}{R} = \frac{36}{27/5} = \frac{180}{27} = \frac{20}{3} \, \text{A}$$

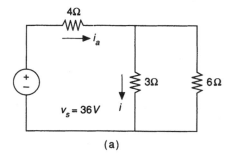

(a)

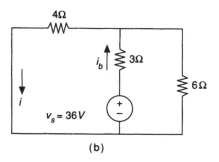

(b)

FIGURE 3.21   Two networks that can be used to illustrate the reciprocity principle.

and again, by current division

$$i = \frac{6}{4+6} i_{\mathrm{b}} = \left(\frac{3}{5}\right)\frac{20}{3} = 4\,\text{A}$$

The network is reciprocal.

The Substitution and Compensation Theorems

The Substitution Theorem

Any branch in a network with branch voltage, v_k, and branch current, i_k, can be replaced by another branch provided it also has branch voltage, v_k, and branch current, i_k.

The Compensation Theorem

In a linear network, if the impedance of a branch carrying a current $\hat{I}$ is changed from Z to $Z + \Delta Z$, then the corresponding change of any voltage or current elsewhere in the network will be due to a compensating voltage source, $\Delta Z\hat{I}$, placed in series with $Z + \Delta Z$ with polarity such that the source, $\Delta Z\hat{I}$, is opposing the current $\hat{I}$.

Defining Terms

Linear network: A network in which the parameters of resistance, inductance, and capacitance are constant with respect to voltage or current or the rate of change of voltage or current and in which the voltage or current of sources is either independent of or proportional to other voltages or currents or their derivatives.

Maximum power transfer theorem: In any electrical network that carries direct or alternating current, the maximum possible power transferred from one section to another occurs when the impedance of the section acting as the load is the complex conjugate of the impedance of the section that acts as the source. Here, both impedances are measured across the pair of terminals in which the power is transferred with the other part of the network disconnected.

Norton theorem: The voltage across an element that is connected to two terminals of a linear, bilateral network is equal to the short-circuit current between these terminals in the absence of the element divided by the admittance of the network looking back from the terminals into the network with all generators replaced by their internal admittances.

Principle of superposition: In a linear electrical network, the voltage or current in any element resulting from several sources acting together is the sum of the voltages or currents from each source acting alone.

Reciprocity theorem: In a network consisting of linear, passive impedances, the ratio of the voltage introduced into any branch to the current in any other branch is equal in magnitude and phase to the ratio that results if the positions of the voltage and current are interchanged.

Thévenin theorem: The current flowing in any impedance connected to two terminals of a linear, bilateral network containing generators is equal to the current flowing in the same impedance when it is connected to a voltage generator whose voltage is the voltage at the open-circuited terminals in question and whose series impedance is the impedance of the network looking back from the terminals into the network, with all generators replaced by their internal impedances.

References

J.D. Irwin, *Basic Engineering Circuit Analysis*, 7th ed., New York: Wiley, 2003.
A.D. Kraus, *Circuit Analysis*, St. Paul: West Publishing, 1991.
R.C. Dorf, *Introduction to Electric Circuits*, New York: Wiley, 2004.

3.4 Power and Energy

Norman Balabanian and Theodore A. Bickart

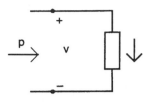

FIGURE 3.22 Power delivered to a circuit.

The concept of the voltage, v, between two points was introduced in Section "Voltage and Current Laws" as the energy, w, expended per unit charge in moving the charge between the two points. Coupled with the definition of current, i, as the time rate of charge motion and that of power, p, as the time rate of change of energy, this leads to the following fundamental relationship between the power delivered to a two-terminal electrical component and the voltage and current of that component, with standard references (meaning that the voltage reference plus is at the tail of the current reference arrow) as shown in Figure 3.22:

$$p = vi \tag{3.1}$$

Assuming that the voltage and current are in volts and amperes, respectively, the power is in *watts*. This relationship applies to any two-terminal component or network, whether linear or nonlinear.

The power delivered to the basic linear resistive, inductive, and capacitive elements is obtained by inserting the v–i relationships into this expression. Then, using the relationship between power and energy (power as the time derivative of energy, and energy, therefore, as the integral of power), the energy stored in the capacitor and inductor is also obtained:

$$p_R = v_R i_R = Ri^2$$

$$p_C = v_C i_C = Cv_C \frac{dv_C}{dt} \qquad w_C(t) = \int_0^t Cv_C \frac{dv_C}{dt} dt = \frac{1}{2} Cv_C^2(t)$$

$$p_L = v_L i_L = Li_L \frac{di_L}{dt} \qquad w_L(t) = \int_0^t Li_L \frac{di_L}{dt} dt = \frac{1}{2} Li_L^2(t) \tag{3.2}$$

where the origin of time ($t=0$) is chosen as the time when the capacitor voltage (respectively, the inductor current) is zero.

Tellegen's Theorem

A result that has far-reaching consequences in electrical engineering is Tellegen's theorem. It will be stated in terms of the networks shown in Figure 3.23. These two are said to be topologically equivalent; that is, they are represented by the same graph but the components that constitute the branches of the graph are not necessarily the same in the two networks. They can even be nonlinear, as illustrated by the diode in one of the networks. Assuming all branches have standard references, including the source branches, Tellegen's theorem states that

$$\sum_{\text{all } j} v_{bj} i_{aj} = 0$$

$$v'_b i_a = 0 \tag{3.3}$$

In the second line, the variables are vectors and the prime stands for the transpose. The a and b subscripts refer to the two networks.

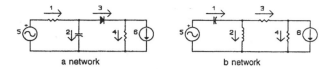

FIGURE 3.23 Topologically equivalent networks.

This is an amazing result. It can be easily proved with the use of Kirchhoff's two laws. The products of *v* and *i* are reminiscent of power as in Equation (3.1). However, the product of the voltage of a branch in one network and the current of its topologically corresponding branch (which may not even be the same type of component) in another network does not constitute power in either branch. Furthermore, the variables in one network might be functions of time, while those of the other network might be steady-state phasors or Laplace transforms.

Nevertheless, some conclusions about power can be derived from Tellegen's theorem. Since a network is topologically equivalent to itself, the b network can be the same as the a network. In that case each *vi* product in Equation (3.3) represents the power delivered to the corresponding branch, including the sources. The equation then says that if we add the power delivered to all the branches of a network, the result will be zero.

This result can be recast if the sources are separated from the other branches and one of the references of each source (current reference for each *v*-source and voltage reference for each *i*-source) is reversed. Then the *vi* product for each source, with new references, will enter Equation (3.3) with a negative sign and will represent the power supplied by this source. When these terms are transposed to the right side of the equation, their signs are changed. The new equation will state in mathematical form that:

> In any electrical network, the sum of the power supplied by the sources is equal to the sum of the power delivered to all the nonsource branches.

This is not very surprising since it is equivalent to the law of conservation of energy, a fundamental principle of science.

AC Steady-State Power

Let us now consider the ac steady-state case, where all voltages and currents are sinusoidal. Thus, in the two-terminal circuit of Figure 3.22:

$$v(t) = \sqrt{2}|V|\cos(\omega t + \alpha) \leftrightarrow V = |V|e^{j\alpha}$$

$$i(t) = \sqrt{2}|I|\cos(\omega t + \beta) \leftrightarrow I = |I|e^{j\beta}$$

(3.4)

The capital *V* and *I* are phasors representing the voltage and current, and their magnitudes are the corresponding rms values. The power delivered to the network at any instant of time is given by

$$p(t) = v(t)i(t) = 2|V||I|\cos(\omega t + \alpha)\cos(\omega t + \beta)$$

$$= \left[|V||I|\cos(\alpha - \beta)\right] + \left[|V||I|\cos(2\omega t + \alpha + \beta)\right]$$

(3.5)

The last form is obtained by using trigonometric identities for the sum and difference of two angles. Whereas both the voltage and the current are sinusoidal, the instantaneous power contains a constant term (independent of time) in addition to a sinusoidal term. Furthermore, the frequency of the sinusoidal term is twice that of the voltage or current. Plots of *v, i,* and *p* are shown in Figure 3.24 for specific values of α and β. The power is sometimes positive, sometimes negative. This means that power is sometimes delivered to the terminals and sometimes extracted from them.

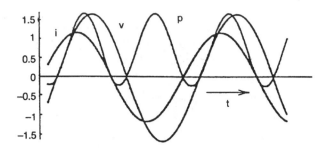

FIGURE 3.24 Instantaneous voltage, current, and power.

The energy that is transmitted into the network over some interval of time is found by integrating the power over this interval. If the area under the positive part of the power curve were the same as the area under the negative part, the net energy transmitted over one cycle would be zero. For the values of α and β used in the figure, however, the positive area is greater, so there is a net transmission of energy toward the network. The energy flows back from the network to the source over part of the cycle, but on average, more energy flows towards the network than away from it.

In Terms of RMS Values and Phase Difference

Consider the question from another point of view. The preceding equation shows the power to consist of a constant term and a sinusoid. The average value of a sinusoid is zero, so this term will contribute nothing to the net energy transmitted. Only the constant term will contribute. This constant term is the average value of the power, as can be seen either from the preceding figure or by integrating the preceding equation over one cycle. Denoting the average power by P and letting $\theta = \alpha - \beta$, which is the angle of the network impedance, the average power becomes

$$
\begin{aligned}
P &= |V||I| \cos \theta \\
&= |V||I| \mathrm{Re}\left(e^{j\theta}\right) = \mathrm{Re}\left[|V||I|e^{j(\alpha-\beta)}\right] \\
&= \mathrm{Re}\left[\left(|V|e^{j\alpha}\right)\left(|I|e^{-j\beta}\right)\right] \\
&= \mathrm{Re}\left(VI^*\right)
\end{aligned}
\tag{3.6}
$$

The third line is obtained by breaking up the exponential in the previous line by the law of exponents. The first factor between square brackets in this line is identified as the phasor voltage and the second factor as the conjugate of the phasor current. The last line then follows. It expresses the average power in terms of the voltage and current phasors and is sometimes more convenient to use.

Complex and Reactive Power

The average ac power is found to be the real part of a complex quantity VI^*, labeled S, that in rectangular form is

$$
\begin{aligned}
S = VI^* &= |V||I|e^{j\theta} = |V||I| \cos \theta + j|V||I| \sin \theta \\
&= P + jQ
\end{aligned}
\tag{3.7}
$$

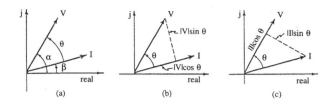

FIGURE 3.25 In-phase and quadrature components of *V* and *I*.

where

$$P = |V||I| \cos\theta \quad \text{(a)}$$
$$Q = |V||I| \sin\theta \quad \text{(b)} \qquad\qquad (3.8)$$
$$|S| = |V||I| \qquad\quad \text{(c)}$$

We already know P to be the average power. Since it is the real part of some complex quantity, it would be reasonable to call it the **real power**. The complex quantity S of which P is the real part is, therefore, called the *complex power*. Its magnitude is the product of the rms values of voltage and current: $|S| = |V||I|$. It is called the *apparent power* and its unit is the volt-ampere (VA). To be consistent, then, we should call Q the *imaginary power*. This is not usually done, however; instead, Q is called the **reactive power** and its unit is a VAR (volt-ampere reactive).

Phasor and Power Diagrams

An interpretation useful for clarifying and understanding the preceding relationships and for the calculation of power is a graphical approach. Figure 3.25(a) is a phasor diagram of V and I in a particular case. The phasor voltage can be resolved into two components, one parallel to the phasor current (or in phase with I) and another perpendicular to the current (or in quadrature with it). This is illustrated in Figure 3.25(b). Hence, the average power P is the magnitude of phasor I multiplied by the in-phase component of V; the reactive power Q is the magnitude of I multiplied by the quadrature component of V.

Alternatively, one can imagine resolving phasor I into two components, one in phase with V and one in quadrature with it, as illustrated in Figure 3.25(c). Then P is the product of the magnitude of V with the in-phase component of I, and Q is the product of the magnitude of V with the quadrature component of I. Real power is produced only by the in-phase components of V and I. The quadrature components contribute only to the reactive power.

The in-phase or quadrature components of V and I do not depend on the specific values of the angles of each, but on their phase difference. One can imagine the two phasors in the preceding diagram to be rigidly held together and rotated around the origin by any angle. As long as the angle θ is held fixed, all of the discussion of this section will still apply. It is common to take the current phasor as the reference for angle; that is, to choose $\beta = 0$ so that phasor I lies along the real axis. Then $\theta = \alpha$.

Power Factor

For any given circuit it is useful to know what part of the total complex power is real (average) power and what part is reactive power. This is usually expressed in terms of the **power factor** F_p, defined as the ratio of real power to apparent power:

$$\text{Power factor} \doteq F_P = \frac{P}{|S|} = \frac{P}{|V||I|} \qquad\qquad (3.9)$$

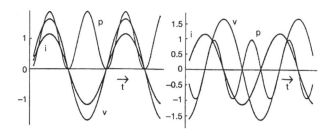

FIGURE 3.26   Power waveform for unity and zero power factors.

Not counting the right side, this is a general relationship although we developed it here for sinusoidal excitations. With $P = |V||I|\cos\theta$, we find that the power factor is simply $\cos\theta$. Because of this, θ itself is called the power factor angle.

Since the cosine is an even function [$\cos(-\theta) = \cos\theta$], specifying the power factor does not reveal the sign of θ. Remember that θ is the angle of the impedance. If θ is positive, this means that the current lags the voltage; we say that the power factor is a *lagging* power factor. However, if θ is negative, the current leads the voltage and we say this represents a *leading* power factor.

The power factor will reach its maximum value, unity, when the voltage and current are in phase. This will happen in a purely resistive circuit, of course. It will also happen in more general circuits for specific element values and a specific frequency.

We can now obtain a physical interpretation for the reactive power. When the power factor is unity, the voltage and current are in phase and $\sin\theta = 0$. Hence, the reactive power is zero. In this case, the instantaneous power is never negative. This case is illustrated by the current, voltage, and power waveforms in Figure 3.26; the power curve never dips below the axis, and there is no exchange of energy between the source and the circuit. At the other extreme, when the power factor is zero, the voltage and current are 90° out of phase and $\sin\theta = 1$. Now the reactive power is a maximum and the average power is zero. In this case, the instantaneous power is positive over half a cycle (of the voltage) and negative over the other half. All the energy delivered by the source over half a cycle is returned to the source by the circuit over the other half.

It is clear, then, that the reactive power is a measure of the exchange of energy between the source and the circuit without being used by the circuit. Although none of this exchanged energy is dissipated by or stored in the circuit, and it is returned unused to the source, nevertheless it is temporarily made available to the circuit by the source.[1]

Average Stored Energy

The average ac energy stored in an inductor or a capacitor can be established by using the expressions for the instantaneous stored energy for arbitrary time functions in Equation (3.2), specifying the time function to be sinusoidal, and taking the average value of the result:

$$W_L = \frac{1}{2}L|I|^2 \qquad W_C = \frac{1}{2}C|V|^2 \tag{3.10}$$

---

[1]Power companies charge their industrial customers not only for the average power they use but for the reactive power they return. There is a reason for this. Suppose a given power system is to deliver a fixed amount of average power at a constant voltage amplitude. Since $P = |V||I|\cos\theta$, the current will be inversely proportional to the power factor. If the reactive power is high, the power factor will be low and a high current will be required to deliver the given power. To carry a large current, the conductors carrying it to the customer must be correspondingly larger and better insulated, which means a larger capital investment in physical plant and facilities. It may be cost-effective for customers to try to reduce the reactive power they require, even if they have to buy additional equipment to do so.

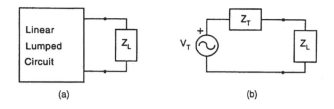

FIGURE 3.27   A linear circuit delivering power to a load in the steady state.

Application of Tellegen's Theorem to Complex Power

An example of two topologically equivalent networks was shown in Figure 3.23. Let us now specify that two such networks are linear, all sources are same-frequency sinusoids, they are operating in the steady state, and all variables are phasors. Furthermore, suppose the two networks are the same, except that the sources of network b have phasors that are the complex conjugates of those of network a. Then, if **V** and **I** denote the vectors of branch voltages and currents of network a, Tellegen's theorem in Equation (3.3) becomes

$$\sum_{\text{all } j} V_j^* I_j = \mathbf{V}^* \mathbf{I} = 0 \tag{3.11}$$

where $\mathbf{V}^*$ is the conjugate transpose of vector $\mathbf{V}$.

This result states that the sum of the complex power delivered to all branches of a linear circuit operating in the ac steady state is zero. Alternatively stated, the total complex power delivered to a network by its sources equals the sum of the complex power delivered to its nonsource branches. Again, this result is not surprising. Since, if a complex quantity is zero both the real and imaginary parts must be zero, the same result can be stated for the average power and for the reactive power.

Maximum Power Transfer Theorem

The diagram in Figure 3.27 illustrates a two-terminal linear circuit at whose terminals an impedance Z_L is connected. The circuit is assumed to be operating in the ac steady state. The problem to be addressed is this: given the two-terminal circuit, how can the impedance connected to it be adjusted so that the maximum possible average power is transferred from the circuit to the impedance?

The first step is to replace the circuit by its Thévenin equivalent, as shown in Figure 3.27(b). The current phasor in this circuit is $I = V_T/(Z_T + Z_L)$. The average power transferred by the circuit to the impedance is

$$P = |I|^2 \text{Re}(Z_L) = \frac{|V_T|^2 \text{Re}(Z_L)}{|Z_T + Z_L|^2} = \frac{|V_T|^2 R_L}{(R_T + R_L)^2 + (X_T + X_L)^2} \tag{3.12}$$

In this expression, only the load (that is, R_L and X_L) can be varied. The preceding equation, then, expresses a dependent variable (P) in terms of two independent ones (R_L and X_L).

What is required is to maximize P. For a function of more than one variable, this is done by setting the partial derivatives with respect to each of the independent variables equal to zero; that is, $\partial P/\partial R_L = 0$ and $\partial P/\partial X_L = 0$. Carrying out these differentiations leads to the result that maximum power will be transferred

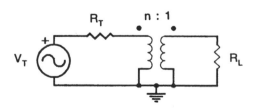

FIGURE 3.28   Matching with an ideal transformer.

when the load impedance is the conjugate of the Thévenin impedance of the circuit: $Z_L = Z_T^*$. If the Thévenin impedance is purely resistive, then the load resistance must equal the Thévenin resistance.

In some cases, both the load impedance and the Thévenin impedance of the source may be fixed. In such a case, the matching for maximum power transfer can be achieved by using a transformer, as illustrated in Figure 3.28, where the impedances are both resistive. The transformer is assumed to be ideal, with turns ratio n. Maximum power is transferred if $n^2 = R_T / R_L$.

Measuring AC Power and Energy

With ac steady-state average power given in the first line of Equation (3.6), measuring the average power requires measuring the rms values of voltage and current, as well as the power factor. This is accomplished by the arrangement shown in Figure 3.29, which includes a breakout of an electrodynamometer-type wattmeter. The current in the high-resistance pivoted coil is proportional to the voltage across the load. The current to the load and the pivoted coil together through the energizing coil of the electromagnet establishes a proportional magnetic field across the cylinder of rotation of the pivoted coil. The torque on the pivoted coil is proportional to the product of the magnetic field strength and the current in the pivoted coil. If the current in the pivoted coil is negligible compared to that in the load, then the torque becomes essentially proportional to the product of the voltage across the load (equal to that across the pivoted coil) and the current in the load (essentially equal to that through the energizing coil of the electromagnet). The dynamics of the pivoted coil together with the restraining spring, at ac power frequencies, ensures that the angular displacement of the pivoted coil becomes proportional to the average of the torque or, equivalently, the average power.

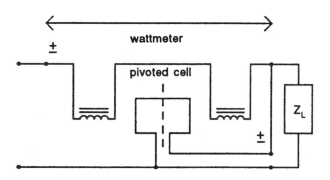

FIGURE 3.29   A wattmeter connected to a load.

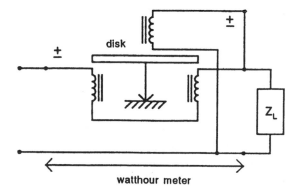

FIGURE 3.30   A watthour meter connected to a load.

One of the most ubiquitous of electrical instruments is the induction-type watthour meter, which measures the energy delivered to a load. Every customer of an electrical utility has one, for example. In this instance the pivoted coil is replaced by a rotating conducting (usually aluminum) disk, as shown in Figure 3.30. An induced eddy current in the disk replaces the pivoted coil current interaction with the load-current-established magnetic field. After compensating for the less-than-ideal nature of the electrical elements making up the meter as just described, the result is that the disk rotates at a rate proportional to the average power to the load and the rotational count is proportional to the energy delivered to the load.

At frequencies above the ac power frequencies and, in some instances, at the ac power frequencies, electronic instruments are available to measure power and energy. They are not a cost-effective substitute for these meters in the monitoring of power and energy delivered to most of the millions upon millions of homes and businesses.

Defining Terms

AC steady-state power: Consider an ac source connected at a pair of terminals to an otherwise isolated network. Let $\sqrt{2}\cdot|V|$ and $\sqrt{2}\cdot|I|$ denote the peak values, respectively, of the ac steady-state voltage and current at the terminals. Furthermore, let θ denote the phase angle by which the voltage leads the current. Then the average power delivered by the source to the network would be expressed as $P = |V|\cdot|I|\cos(\theta)$.

Power and energy: Consider an electrical source connected at a pair of terminals to an otherwise isolated network. Power, denoted by p, is the time rate of change in the energy delivered to the network by the source. This can be expressed as $p = vi$, where v, the voltage across the terminals, is the energy expended per unit charge in moving the charge between the pair of terminals and i, the current through the terminals, is the time rate of charge motion.

Power factor: Consider an ac source connected at a pair of terminals to an otherwise isolated network. The power factor, the ratio of the real power to the apparent power $|V|\cdot|I|$, is easily established to be $\cos(\theta)$, where θ is the power factor angle.

Reactive power: Consider an ac source connected at a pair of terminals to an otherwise isolated network. The reactive power is a measure of the energy exchanged between the source and the network without being dissipated in the network. The reactive power delivered would be expressed as $Q = |V|\cdot|I|\sin(\theta)$.

Real power: Consider an ac source connected at a pair of terminals to an otherwise isolated network. The real power, equal to the average power, is the power dissipated by the source in the network.

Tellegen's theorem: Two networks, here including all sources, are topologically equivalent if they are similar structurally, component by component. Tellegen's theorem states that the sum over all products of the product of the current of a component of one network, network a, and of the voltage of the corresponding component of the other network, network b, is zero. This would be expressed as $\Sigma_{\text{all }j}v_{bj}i_{aj} = 0$. From this general relationship it follows that in any electrical network, the sum of the power supplied by the sources is equal to the sum of the power delivered to all the nonsource components.

References

N. Balabanian, *Electric Circuits,* New York: McGraw-Hill, 1994.

J.D. Irwin, *Basic Engineering Circuit Analysis,* New York: Macmillan, 1995.

D.E. Johnson, J.L. Hilburn, and J.R. Johnson, *Basic Electric Circuit Analysis,* 3rd ed., Englewood Cliffs, NJ: Prentice-Hall, 1990.

R.C. Dorf, *Electric Circuits*, 6th ed., New York: Wiley, 2004.

3.5   Three-Phase Circuits

Norman Balabanian

Figure 3.31(a) represents the basic circuit for considering the flow of power from a single sinusoidal source to a load. The power can be thought to cross an imaginary boundary surface (represented by the dotted line in the figure) separating the source from the load. Suppose that

$$v(t) = \sqrt{2}\,|V|\cos(\omega t + \alpha)$$
$$i(t) = \sqrt{2}\,|I|\cos(\omega t + \beta) \tag{3.13}$$

Then the power to the load at any instant of time is

$$p(t) = |V||I|\,[\cos(\alpha - \beta) + \cos(2\omega t + \alpha + \beta)] \tag{3.14}$$

The instantaneous power has a constant term and a sinusoidal term at twice the frequency. The quantity in brackets fluctuates between a minimum value of $\cos(\alpha - \beta) - 1$ and a maximum value of $\cos(\alpha - \beta) + 1$. This fluctuation of power delivered to the load has certain disadvantages in some situations where the transmission of power is the purpose of a system. An electric motor, for example, operates by receiving electric power and transmitting mechanical (rotational) power at its shaft. If the electric power is delivered to the motor in spurts, the motor is likely to vibrate. In order to run satisfactorily, a physically larger motor will be needed, with a larger shaft and flywheel, to provide inertia than would be the case if the delivered power were constant.

This problem is overcome in practice by the use of what is called a *three-phase* system. This section will provide a brief discussion of three-phase systems.

Consider the circuit in Figure 3.31(b). This arrangement is similar to a combination of three of the simple circuits in Figure 3.31(a) connected in such a way that each one shares the return connection from O to N. The three sources can be viewed collectively as a single source and the three loads—which are assumed to be

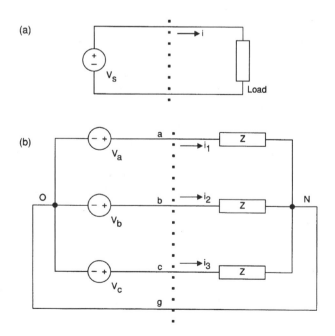

FIGURE 3.31   Flow of power from source to load.

identical—can be viewed collectively as a single load. Then, as before, the dotted line represents a surface separating the source from the load. Each of the individual sources and loads is referred to as one *phase* of the three-phase system.

The three sources are assumed to have the same frequency; they are said to be *synchronized*. It is also assumed that the three voltages have the same rms values and the phase difference between each pair of voltages is ±120° (2π/3 rad). Thus, they can be written:

$$v_a = \sqrt{2}\,|V|\cos(\omega t + \alpha_1) \quad \leftrightarrow \quad V_a = |V|\,e^{j0°}$$

$$v_b = \sqrt{2}\,|V|\cos(\omega t + \alpha_2) \quad \leftrightarrow \quad V_b = |V|\,e^{-j120°} \qquad (3.15)$$

$$v_c = \sqrt{2}\,|V|\cos(\omega t + \alpha_3) \quad \leftrightarrow \quad V_c = |V|\,e^{j120°}$$

The **phasors** representing the sinusoids have also been shown. For convenience, the angle of v_a has been chosen as the reference for angles; v_b lags v_a by 120° and v_c leads v_a by 120°.

Because the loads are identical, the rms values of the three currents shown in Figure 3.32 will also be the same and the phase difference between each pair of them will be ±120°. Thus, the currents can be written:

$$i_1 = \sqrt{2}\,|I|\cos(\omega t + \beta_1) \quad \leftrightarrow \quad I_1 = |I|\,e^{j\beta_1}$$

$$i_2 = \sqrt{2}\,|I|\cos(\omega t + \beta_2) \quad \leftrightarrow \quad I_2 = |I|\,e^{j(\beta_1 - 120°)} \qquad (3.16)$$

$$i_3 = \sqrt{2}\,|I|\cos(\omega t + \beta_3) \quad \leftrightarrow \quad I_3 = |I|\,e^{j(\beta_1 + 120°)}$$

Perhaps a better form for visualizing the voltages and currents is a graphical one. Phasor diagrams for the voltages separately and the currents separately are shown in Figure 3.32. The value of angle β_1 will depend on the load. An interesting result is clear from these diagrams. First, V_2 and V_3 are each other's conjugates. So if we add them, the imaginary parts cancel and the sum will be real, as illustrated by the construction in the voltage diagram. Furthermore, the construction shows that this real part is negative and equal in size to V_1. Hence, the sum of the three voltages is zero. The same is true of the sum of the three currents, as can be established graphically by a similar construction.

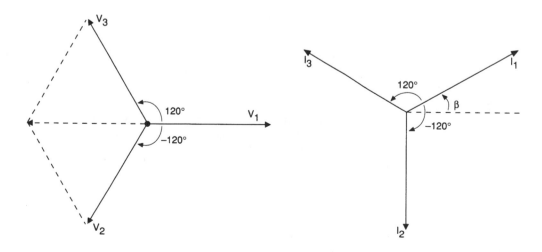

FIGURE 3.32   Voltage and current phasor diagrams.

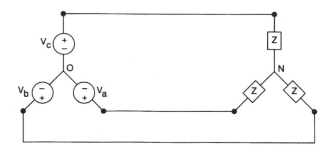

FIGURE 3.33 Wye-connected three-phase system.

By Kirchhoff's current law applied at node N in Figure 3.31(b), we find that the current in the return line is the sum of the three currents in Equation (3.16). However, since this sum was found to be zero, the return line carries no current. Hence, it can be removed entirely without affecting the operation of the system. The resulting circuit is redrawn in Figure 3.33. Because of its geometrical form, this connection of both the sources and the loads is said to be a wye (Y) connection.

The instantaneous power delivered by each of the sources has the form given in Equation (3.14), consisting of a constant term representing the average power and a double-frequency sinusoidal term. The latter, being sinusoidal, can be represented by a phasor also. The only caution is that a different frequency is involved here, so this power phasor should not be mixed with the voltage and current phasors in the same diagram or calculations. Let $|S| = |V||I|$ be the apparent power delivered by each of the three sources and let the three power phasors be S_a, S_b, and S_c, respectively. Then:

$$S_a = |S|e^{j(\alpha_1 + \beta_1)} = |S|e^{j\beta_1}$$

$$S_b = |S|e^{j(\alpha_2 + \beta_2)} = |S|e^{j(-120° + \beta_1 - 120°)} = |S|e^{j(\beta_1 + 120°)} \qquad (3.17)$$

$$S_c = |S|e^{j(\alpha_3 + \beta_3)} = |S|e^{j(+120° + \beta_1 + 120°)} = |S|e^{j(\beta_1 - 120°)}$$

It is evident that the phase relationships among these three phasors are the same as the ones among the voltages and the currents. That is, the second leads the first by 120° and the third lags the first by 120°. Hence, just like the voltages and the currents, the sum of these three phasors will also be zero. This is a very significant result. Although the instantaneous power delivered by each source has a constant component and a sinusoidal component, when the three powers are added, the sinusoidal components add to zero, leaving only the constants. Thus, the total power delivered to the three loads is constant.

To determine the value of this constant power, use Equation (3.14) as a model. The contribution of the kth source to the total (constant) power is $|S|\cos(\alpha_k - \beta_k)$. One can easily verify that $\alpha_k - \beta_k = \alpha_1 - \beta_1 = -\beta_1$. The first equality follows from the relationships among the α's from Equation (3.15) and among the β's from Equation (3.16). The choice of $\alpha_1 = 0$ leads to the last equality. Hence, the constant terms contributed to the power by each source are the same. If P is the total average power, then

$$P = P_a + P_b + P_c + = 3P_a = 3|V||I|\cos(\alpha_1 - \beta_1) \qquad (3.18)$$

Although the angle α_1 has been set equal to zero, for the sake of generality we have shown it explicitly in this equation.

What has just been described is a *balanced* three-phase three-wire power system. The three sources in practice are not three independent sources but consist of three different parts of the same generator. The same

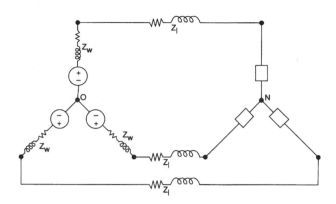

FIGURE 3.34 Three-phase circuit with nonzero winding and line impedances.

is true of the loads.[1] What has been described is ideal in a number of ways. First, the circuit can be *unbalanced*—for example, by the loads being somewhat unequal. Second, since the real devices whose ideal model is a voltage source are coils of wire, each source should be accompanied by a branch consisting of the coil inductance and resistance. Third, since the power station (or the distribution transformer at some intermediate point) may be at some distance from the load, the parameters of the physical line carrying the power (the line inductance and resistance) must also be inserted in series between the source and the load.

For an unbalanced system, the analysis of this section does not apply. An entirely new analytical technique is required to do full justice to such a system.[2] However, an understanding of balanced circuits is a prerequisite for tackling the unbalanced case.

The last two of the conditions that make the circuit less than ideal (line and source impedances) introduce algebraic complications, but nothing fundamental is changed in the preceding theory. If these two conditions are taken into account, the appropriate circuit takes the form shown in Figure 3.34. Here, the internal impedance of a source and the line impedance connecting that source to its load are both connected in series with the corresponding load. Thus, instead of the impedance in each phase being Z, it is $Z + Z_w + Z_l$, where w and l are subscripts standing for "winding" and "line," respectively. Hence, the rms value of each current is

$$|I| = \frac{|V|}{|Z + Z_w + Z_l|} \tag{3.19}$$

instead of $|V|/|Z|$. All other results we had arrived at remain unchanged, namely that the sum of the phase currents is zero and that the sum of the phase powers is a constant. The detailed calculations simply become a little more complicated.

One other point, illustrated for the loads in Figure 3.35, should be mentioned. Given wye-connected sources or loads, the wye and the **delta** can be made equivalent by proper selection of the arms of the delta. Thus, either the sources in Figure 3.33, or the loads, or both, can be replaced by a delta equivalent; thus, we can conceive of four different three-phase circuits; wye–wye, delta–wye, wye–delta, and delta–delta. Not only can we conceive of them, they are extensively used in practice.

---

[1]An ac power generator consists of (a) a rotor, which produces a magnetic field and which is rotated by a prime mover (say a turbine), and (b) a stator on which are wound one or more coils of wire. In three-phase systems, the number of coils is three. The rotating magnetic field induces a voltage in each of the coils. The 120° leading and lagging phase relationships among these voltages are obtained by distributing the conductors of the coils around the circumference of the stator so that they are separated geometrically by 120°. Thus, the three sources described in the text are in reality a single physical device, a single generator. Similarly, the three loads might be the three windings on a three-phase motor, again a single physical device.

[2]The technique for analyzing unbalanced circuits utilizes what are called *symmetrical components*.

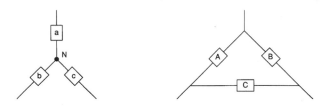

FIGURE 3.35 Wye connection and delta connection.

It is not worthwhile to carry out detailed calculations for these four cases. Once the basic properties described here are understood, one should be able to make the calculations. Observe, however, that in the delta structure there is no neutral connection, so the phase voltages cannot be measured. The only voltages that can be measured are the *line-to-line* or simply the *line* voltages. These are the differences of the phase voltages taken in pairs, as is evident from Figure 3.34.

Defining Terms

Delta connection: The sources or loads in a three-phase system connected end-to-end, forming a closed path, like the Greek letter Δ.
Phasor: A complex number representing a sinusoid; its magnitude and angle are the rms value and phase of the sinusoid, respectively.
Wye connection: The three sources or loads in a three-phase system connected to have one common point, like the letter Y.

References

V. del Toro, *Electric Power Systems,* Englewood Cliffs, NJ: Prentice-Hall, 1992.
R.C. Dorf, *Electric Circuits*, 6th ed., New York: Wiley, 2004.
P.Z. Peebles and T.A. Giuma, *Principles of Electrical Engineering,* New York: McGraw-Hill, 1991.
J.J. Grainger and W.D. Stevenson, Jr., *Power Systems Analysis,* New York: McGraw-Hill, 1994.
G.T. Heydt, *Electric Power Quality,* Stars in a Circle Publications, 1996.

3.6   Graph Theory[1]

Shu-Park Chan

Topology is a branch of mathematics; it may be described as "the study of those properties of geometric forms that remain invariant under certain transformations, as bending, stretching, etc."[2] Network topology (or network graph theory) is a study of (electrical) networks in connection with their nonmetric geometrical (namely topological) properties by investigating the interconnections between the branches and the nodes of the networks. Such a study will lead to important results in network theory such as algorithms for formulating network equations and the proofs of various basic network theorems (Seshu and Reed, 1961; Chan, 1969).

---

[1]Based on S.-Pchan, "Graph theory and some of its applications in electrical network," in *Mathematical Aspects of Electrical Network Analysis*, Vol. 3, *SIAM/AMS Proceedings*, American Mathematical Society, Providence, RI,1971. With permission.
[2]This brief description of topology is quoted directly from the *Random House Dictionary of the English Language*, Random House, New York, 1967.

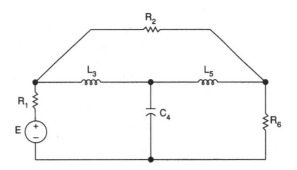

FIGURE 3.36 A passive network N with a voltage driver *E*.

The following are some basic definitions in network graph theory, which will be needed in the development of topological formulas in the analysis of linear networks and systems.

A **linear graph** (or simply a *graph*) is a set of line segments called *edges* and points called *vertices,* which are the endpoints of the edges, interconnected in such a way that the edges are connected to (or *incident* with) the vertices. The *degree* of a vertex of a graph is the number of edges incident with that vertex.

A subset G_i of the edges of a given graph G is called a **subgraph** of G. If G_i does not contain all of the edges of G, it is a **proper subgraph** of G. A **path** is a subgraph having all vertices of degree 2 except for the two endpoints, which are of degree 1 and are called the terminals of the path. The set of all edges in a path constitutes a **path-set**. If the two terminals of a path coincide, the path is a closed path and is called a **circuit** (or **loop**). The set of all edges contained in a circuit is called a **circuit-set** (or **loop-set**).

A graph or subgraph is said to be **connected** if there is at least one path between *every* pair of its vertices. A **tree** of a connected graph G is a connected subgraph which contains all the vertices of G but no circuits. The edges contained in a tree are called the **branches of the tree**. A 2-tree of a connected graph G is a (proper) subgraph of G consisting of two unconnected circuitless subgraphs, each subgraph itself being connected, which together contain all the vertices of G. Similarly, a *k*-tree is a subgraph of *k* unconnected circuitless subgraphs, each subgraph being connected, which together include all the vertices of G. The ***k*-tree admittance product of a *k*-tree** is the product of the admittances of all the branches of the *k*-tree.

Example 3.5

The graph G shown in Figure 3.37 is the graph of the network N of Figure 3.36. The edges of G are e_1, e_2, e_4, e_5, and e_6; the vertices of G are V_1, V_2, V_3, and V_4. A path of G is the subgraph G_1 consisting of edges e_2, e_3, and e_6 with vertices V_2 and V_4 as terminals. Thus, the set $\{e_2, e_3, e_6\}$ is a path-set. With edge e_4 added to G_1, we form another subgraph G_2, which is a circuit since as far as G_2 is concerned all its vertices are of degree 2. Hence the set $\{e_2, e_3, e_4, e_6\}$ is a circuit-set. Obviously, G is a connected graph since there exists a path between every pair of vertices of G. A tree of G may be the subgraph consisting of edges e_1, e_4, and e_6. Two other trees of G are $\{e_2, e_5, e_6\}$ and $\{e_3, e_4, e_5\}$. A 2-tree of G is $\{e_2, e_4\}$; another one is $\{e_3, e_6\}$; and still another one is $\{e_3, e_5\}$. Note that both $\{e_2, e_4\}$ and $\{e_3, e_6\}$ are subgraphs which obviously satisfy the definition of a 2-tree in the sense that each contains two disjoint circuitless connected subgraphs, both of which include all the four vertices of G. Thus, $\{e_3, e_5\}$ does not seem to be a 2-tree. However, if we agree to consider $\{e_3, e_5\}$ as a subgraph which contains edges e_3 and e_5 plus the isolated vertex V_4, we

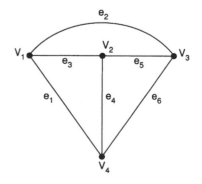

FIGURE 3.37 The graph G of the network N of Figure 3.33.

see that {e_3, e_5} will satisfy the definition of a 2-tree since it now has two circuitless connected subgraphs with e_3 and e_5 forming one of them and the vertex V_4 alone forming the other. Moreover, both subgraphs together indeed include all the four vertices of G. It is worth noting that a 2-tree is obtained from a tree by removing *any one* of the branches from the tree; in general, a k-tree is obtained from a $(k-1)$ tree by removing from it any one of its branches. Finally, the tree-admittance product of the tree {e_2, e_5, e_6} is 1/2 1/5 1/6; the 2-tree admittance product of the 2-tree {e_3, e_5} is 1/3 1/5 (with the admittance of a vertex defined to be 1).

The k-Tree Approach

The development of the analysis of passive electrical networks using topological concepts may be dated back to 1847 when Kirchhoff formulated his set of topological formulas in terms of resistances and the branch-current system of equations. In 1892, Maxwell developed another set of topological formulas based on the k-tree concept, which are the duals of Kirchhoff's. These two sets of formulas were supported mainly by heuristic reasoning and no formal proofs were then available.

In the following we shall discuss only Maxwell's topological formulas for linear networks without mutual inductances.

Consider a network N with n independent nodes, as shown in Figure 3.38. The node $1'$ is taken as the reference (datum) node. The voltages V_1, $V_2, \ldots$, V_n (which are functions of s) are the transforms of the node-pair voltages (or simply node voltages) v_1, $v_2, \ldots$, v_n (which are function s of t) between the n nodes and the reference node $1'$ with the plus polarity marks at the n nodes. It can be shown (Aitken, 1956) that the matrix equation for the n independent nodes of N is given by

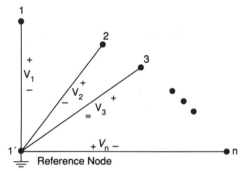

FIGURE 3.38 A network N with n independent nodes.

$$\begin{bmatrix} y_{11} & y_{12} & \cdots & y_{1n} \\ y_{21} & y_{22} & \cdots & y_{2n} \\ \vdots & \vdots & \vdots & \vdots \\ y_{n1} & y_{n2} & \cdots & y_{nn} \end{bmatrix} \begin{bmatrix} V_1 \\ V_2 \\ \vdots \\ V_n \end{bmatrix} = \begin{bmatrix} I_1 \\ I_2 \\ \vdots \\ I_n \end{bmatrix} \quad (3.20)$$

or in abbreviated matrix notation:

$$Y_n V_n = I_n \tag{3.21}$$

where Y_n is the node-admittance matrix, V_n the $n \times 1$ matrix of the node voltage transforms, and I_n the $n \times 1$ matrix of the transforms of the known current sources.

For a relaxed passive one-port (with zero initial conditions) shown in Figure 3.39, the driving-point impedance function $Z_d(s)$ and its reciprocal, namely driving-point admittance function $Y_d(s)$, are given by

$$Z_d(s) = V_1/I_1 = \Delta_{11}/\Delta$$

and

$$Y_d(s) = 1/Z_d(s) = \Delta/\Delta_{11}$$

respectively, where Δ is the determinant of the node-admittance matrix Y_n, and Δ_{11} is the (1,1)-cofactor of Δ.

Similarly, for a passive reciprocal *RLC* two-port (Figure 3.40), the open-circuit impedances and the short-circuit admittances are seen to be

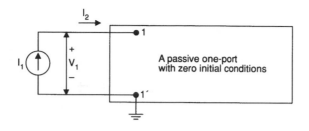

FIGURE 3.39   The network N driven by a single current source.

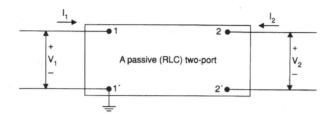

FIGURE 3.40   A passive two-port.

$$z_{11} = \Delta_{11}/\Delta \tag{3.22a}$$

$$z_{12} = z_{21} = (\Delta_{12} - \Delta_{12'})/\Delta \tag{3.22b}$$

$$z_{22} = (\Delta_{22} + \Delta_{2'2'} - 2\Delta_{22'})/\Delta \tag{3.22c}$$

and

$$y_{11} = (\Delta_{22} + \Delta_{2'2'} - 2\Delta_{22'})/(\Delta_{1122} + \Delta_{112'2'} - 2\Delta_{1122'}) \tag{3.23a}$$

$$y_{12} = y_{21} = \Delta_{12'} - \Delta_{12}/(\Delta_{1122} + \Delta_{112'2'} - 2\Delta_{1122'}) \tag{3.23b}$$

$$y_{22} = \Delta_{11}/(\Delta_{1122} + \Delta_{112'2'} - 2\Delta_{1122'}) \tag{3.23c}$$

respectively, where Δ_{ij} is the (i,j)-cofactor of Δ, and Δ_{ijkm} is the cofactor of Δ by deleting rows i and k and columns j and m from Δ (Aitken, 1956).

Expressions in terms of network determinants and cofactors for other network transfer functions are given by (Figure 3.41)

$$z_{12} = \frac{V_2}{I_1} = \frac{\Delta_{12} - \Delta_{12'}}{\Delta} \quad \text{(transfer-impedance function)} \tag{3.24a}$$

$$G_{12} = \frac{V_2}{V_1} = \frac{\Delta_{12} - \Delta_{12'}}{\Delta_{11}} \quad \text{(voltage-ratio transfer function)} \tag{3.24b}$$

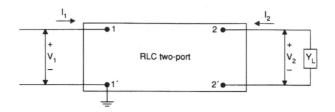

FIGURE 3.41   A loaded passive two-port.

$$Y_{12} = Y_L G_{12} = Y_L\left(\frac{\Delta_{12} - \Delta_{12'}}{\Delta_{11}}\right) \quad \text{(transfer-admittance function)} \qquad (3.24c)$$

$$\alpha_{12} = Y_1 Z_{12} = Y_L\left(\frac{\Delta_{12} - \Delta_{12'}}{\Delta}\right) \quad \text{(current-ratio transfer function)} \qquad (3.24d)$$

The topological formulas for the various network functions of a passive one-port or two-port are derived from the following theorems which are stated without proof (Chan, 1969).

Theorem 3.1.   Let N be a passive network without mutual inductances. The determinant Δ of the node admittance matrix Y_n is equal to the sum of all tree-admittances of N, where a tree-admittance product $t^{(i)}(y)$ is defined to be the product of the admittance of all the branches of the tree $T^{(i)}$. That is:

$$\Delta = \det Y_n = \sum_i T^{(i)}(y) \qquad (3.25)$$

Theorem 3.2.   Let Δ be the determinant of the node-admittance matrix Y_n of a passive network N with $n+1$ nodes and without mutual inductances. Also let the reference node be denoted by $1'$. Then the (j,j)-cofactor Δ_{jj} of Δ is equal to the sum of all the 2-tree-admittance products $T_{2j,1'}(y)$ of N, each of which contains node j in one part and node $1'$ as the reference node) and without mutual inductances is given by

$$\Delta_{jj} = \sum_k T_{2j,1'}^{(k)}(y) \qquad (3.26)$$

where the summation is taken over all the 2-tree-admittance products of the form $T_{2j,1'}(y)$.

Theorem 3.3.   The (i,j)-cofactor Δ_{ij} of Δ of a relaxed passive network N with n independent nodes (with node $1'$ as the reference node) and without mutual inductances is given by

$$\Delta_{ij} = \sum_k T_{2ij,1'}^{(k)}(y) \qquad (3.27)$$

where the summation is taken over all the 2-tree-admittance products of the form $T_{2ij,1'}(y)$ with each containing nodes i and j in one connected port and the reference node $1'$ in the other.

For example, the topological formulas for the driving-point function of a passive one-port can be readily obtained from Equations (3.25) and (3.26) in Theorems 3.1 and 3.2 as stated in the next theorem.

Theorem 3.4.   With the same notation as in Theorems 3.1 and 3.2, the driving-point admittance $Y_d(s)$ and the driving-point impedance $Z_d(s)$ of a passive one-port containing no mutual inductances at terminals 1 and $1'$ are given by

$$Y_d(s) = \frac{\Delta}{\Delta_{11}} = \frac{\sum_i T^{(i)}(y)}{\sum_k T^{(k)}_{2_{1,1}}(y)} \quad \text{and} \quad Z_d(s) = \frac{\Delta_{11}}{\Delta} = \frac{\sum_k T^{(k)}_{2_{1,1}}(y)}{\sum_i T^{(i)}(y)} \qquad (3.28)$$

respectively.

For convenience we define the following shorthand notation:

$$(a) V(Y) \equiv \sum_i T^{(i)}(y) = \text{sum of all tree-admittance products, and}$$

$$(b) W_{j,r}(y) \equiv \sum_k T_{2j,r}(y) = \text{sum of all 2-tree-admittance products with node } j \qquad (3.29)$$

and the reference node r contained in different parts.

Thus Equation (3.28) may be written as

$$Y_d(s) = V(Y)/W_{1,1'}(Y) \quad \text{and} \quad Z_d(s) = W_{1,1'}(Y)/V(Y) \qquad (3.30)$$

In a two-port network N, there are four nodes to be specified, namely, nodes 1 and $1'$ at the input port $(1,1')$ and nodes 2 and $2'$ at the output port $(2,2')$, as illustrated in Figure 3.41. However, for a 2-tree of the type $T_{2ij,1'}$, only three nodes have been used, thus leaving the fourth one unidentified.

With very little effort, it can be shown that, in general, the following relationship holds:

$$W_{ij,1'}(Y) = W_{ijk,1'}(Y) + W_{ij,k1'}(Y)$$

or simply:

$$W_{ij,1'} = W_{ijk,1'} + W_{ij,k1'} \qquad (3.31)$$

where i, j, k, and $1'$ are the four terminals of N with $1'$ denoting the datum (reference) node. The symbol $W_{ijk,1'}$ denotes the sum of all the 2-tree admittance products, each containing nodes i, j, and k in one connected part and the reference node, $1'$, in the other.

We now state the next theorem.

Theorem 3.5.   With the same hypothesis and notation as stated earlier in this section:

$$\Delta_{12} - \Delta_{12'} = W_{12,12'}(Y) - W_{12',1'2}(Y) \qquad (3.32)$$

It is interesting to note that Equation (3.32) is stated by Percival (1953) in the following descriptive fashion:

$$\Delta_{12} - \Delta_{12'} = W_{12,12'} - W_{12',1'2} = \begin{pmatrix} 1 \circ\!\!-\!\!\circ 2 \\ 1'\circ\!\!-\!\!\circ 2' \end{pmatrix} - \begin{pmatrix} 1 \circ \quad \circ 2 \\ \diagdown\!\!\diagup \\ 1'\circ \quad \circ 2' \end{pmatrix}$$

which illustrates the two types of 2-trees involved in the formula. Hence, we state the topological formulas for z_{11}, z_{12}, and z_{22} in the following theorem.

Theorem 3.6.   With the same hypothesis and notation as stated earlier in this section:

$$z_{11} = W_{1,1'}(Y)/V(Y) \qquad (3.33a)$$

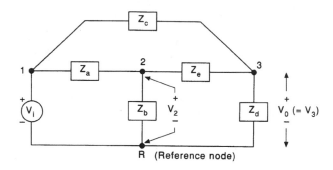

FIGURE 3.42   The network N of Example 3.7.

$$z_{12} = z_{21} = \{W_{12,1'2'}(Y) - W_{12',1'2}(Y)\}/V(Y) \tag{3.33b}$$

$$z_{22} = W_{2,2'}(Y)/V(Y) \tag{3.33c}$$

We shall now develop the topological expressions for the short-circuit admittance functions. Let us denote by $U_{a,b,c}(Y)$ the sum of all 3-tree-admittance products of the form $T_{3a,b,c}(Y)$ with identical subscripts in both symbols to represent the same specified distribution of vertices. Then, following arguments similar to those of Theorem 3.5, we readily see that

$$\Delta_{1122} = \sum_i T^{(i)}_{3_{1,2,1'}}(y) \equiv U_{1,2,1'}(Y) \tag{3.34a}$$

$$\Delta_{112'2'} = \sum_j T^{(j)}_{3_{1,2',1'}}(y) \equiv U_{1,2',1'}(Y) \tag{3.34b}$$

$$\Delta_{1122'} = \sum_k T^{(k)}_{3_{1,22',1'}}(y) \equiv U_{1,22',1'}(U) \tag{3.34c}$$

where $1,1',2,2'$ are the four terminals of the two-port with $1'$ denoting the reference node (Figure 3.42). However, we note that in Equations (3.34a) and (3.34b) only three of the four terminals have been specified. We can therefore further expand $U_{1,2,1'}$ and $U_{1,2',1'}$ to obtain the following:

$$\Delta_{1122} + \Delta_{112'2'} - 2\Delta_{1122'} = U_{12',2,1'} + U_{1,2,1'2'} + U_{12,2',1'} + U_{1,2',1'2} \tag{3.35}$$

For convenience, we shall use the shorthand notation ΣU to denote the sum of the right of Equation (3.35). Thus, we define

$$\Sigma U = U_{12',2,1'} + U_{1,2,1'2'} + U_{12,2',1'} + U_{1,2',1'2} \tag{3.36}$$

Hence, we obtain the topological formulas for the short-circuit admittances as stated in the following theorem.

Theorem 3.7.   The short-circuit admittance functions y_{11}, y_{12}, and y_{22} of a passive two-port network with no mutual inductances are given by

$$y_{11} = W_{2,2'}/\Sigma U \tag{3.37a}$$

$$y_{12} = y_{21} = (W_{12',1'2} - W_{12,1'2'})/\Sigma U \tag{3.37b}$$

$$y_{22} = W_{1,1'}/\Sigma U \tag{3.37c}$$

where ΣU is defined in Equation (3.36) above.

Finally, following similar developments, other network functions are stated in Theorem 3.8.

Theorem 3.8. With the same notation as before:

$$Z_{12}(s) = \frac{W_{12,1'2'} - W_{12',1'2}}{V} \tag{3.38a}$$

$$G_{12}(s) = \frac{W_{12,1'2'} - W_{12',1'2}}{W_{1,1'}} \tag{3.38b}$$

$$Y_{12}(s) = Y_{\mathrm{L}} \frac{W_{12,1'2'} - W_{12',1'2}}{W_{1,1'}} \tag{3.38c}$$

$$\alpha_{12}(s) = Y_{\mathrm{L}} \frac{W_{12,1'2'} - W_{12',1'2}}{V} \tag{3.38d}$$

The Flowgraph Approach

Mathematically speaking, a linear electrical network or, more generally, a linear system can be described by a set of simultaneous linear equations. Solutions to these equations can be obtained either by the method of successive substitutions (elimination theory), by the method of determinants (Cramer's rule), or by any of the topological techniques such as Maxwell's k-tree approach discussed in the preceding subsection and the flowgraph techniques represented by the works of Mason (1953, 1956) and Coates (1959).

Although the methods using algebraic manipulations can be amended and executed by a computer, they do not reveal the physical situations existing in the system. The flowgraph techniques, however, show intuitively the causal relationships between the variables of the system of interest and hence enable the network analyst to have an excellent physical insight into the problem.

In the following, two of the more well-known flowgraph techniques are discussed; namely, the **signal-flowgraph** technique devised by Mason and the method based on the flowgraph of Coates and recently modified by Chan and Bapna (1967).

A *signal-flowgraph* G_{m} of a system S of n independent linear (algebraic) equations in n unknowns:

$$\sum_{j=1}^{n} a_{ij}x_j = b_i \quad i = 1, 2, \ldots, n \tag{3.39}$$

is a graph with junction points called *nodes* that connected by directed line segments called *branches* with signals traveling along the branches only in the direction described by the arrows of the branches. A signal x_k traveling along a branch between x_k and x_j is multiplied by the gain of the branches g_{kj}, so that a signal of $g_{kj}x_k$

is delivered at node x_j. An *input node* (*source*) is a node which contains only outgoing branches; an *output node* (*sink*) is a node which has only incoming branches. A *path* is a continuous unidirectional succession of branches, all of which are traveling in the same direction; a *forward path* is a path from the input node to the output node along which all nodes are encountered exactly once; and a *feedback path* (*loop*) is a closed path which originates from and terminates at the same node, and along which all other nodes are encountered exactly once (the trivial case is a *self-loop* which contains exactly one node and one branch). A *path gain* is the product of all the branch gains of the path; similarly, a loop gain is the product of all the branch gains of the branches in a loop.

The procedure for obtaining the Mason graph from a system of linear algebraic equations may be described in the following steps:

1. Arrange all the equations of the system in such a way that the *j*th dependent (output) variable x_j in the *j*th equation is expressed explicitly in terms of the other variables. Thus, if the system under study is given by Equation (3.39), namely:

$$
\begin{aligned}
a_{11}x_1 + a_{12}x_2 + \cdots + a_{1n}x_n &= b_1 \\
a_{21}x_1 + a_{22}x_2 + \cdots + a_{2n}x_n &= b_2 \\
\vdots \qquad \vdots \qquad \vdots \qquad \vdots \qquad \vdots & \\
a_{n1}x_1 + a_{n2}x_2 + \cdots + a_{nn}x_n &= b_n
\end{aligned}
\tag{3.40}
$$

where $b_1, b_2, \ldots, b_n$ are inputs (sources) and $x_1, x_2, \ldots, x_n$ are outputs, the equations may be rewritten as

$$
\begin{aligned}
x_1 &= \frac{1}{a_{11}}b_1 - \frac{a_{12}}{a_{11}}x_2 - \frac{a_{13}}{a_{11}}x_3 - \cdots - \frac{a_{1n}}{a_{11}}x_n \\
x_2 &= \frac{1}{a_{22}}b_2 - \frac{a_{21}}{a_{22}}x_1 - \frac{a_{23}}{a_{22}}x_3 - \cdots - \frac{a_{2n}}{a_{22}}x_n \\
\vdots \quad \vdots \quad \vdots \quad \vdots \quad \vdots \quad \vdots & \\
x_n &= \frac{1}{a_{nn}}b_n - \frac{a_{n1}}{a_{nn}}x_1 - \frac{a_{n2}}{a_{nn}}x_2 - \cdots - \frac{a_{n-1,n-1}}{a_{nn}}x_{n-1}
\end{aligned}
\tag{3.41}
$$

2. The number of input nodes in the flowgraph is equal to the number of nonzero *b*'s. That is, each of the source nodes corresponds to a nonzero b_j.
3. To each of the output nodes is associated one of the dependent variables $x_1, x_2, \ldots, x_n$.
4. The value of the variable represented by a node is equal to the sum of all the incoming signals.
5. The value of the variable represented by any node is transmitted onto all branches leaving the node.

It is a simple matter to write the equations from the flowgraph since every node, except the source nodes of the graph, represents an equation, and the equation associated with node *k*, for example, is obtained by equating to x_k the sum of all incoming branch gains multiplied by the values of the variables from which these branches originate.

Mason's general gain formula is now stated in the following theorem.

Theorem 3.9. Let G be the overall graph gain and G_k be the gain of the *k*th forward path from the source to the sink. Then

$$
G = \frac{1}{\Delta}\sum_k G_k\Delta_k
\tag{3.42}
$$

where

$$\Delta = 1 \sum_m p_{m1} + \sum_m p_{m2} - \sum_m p_{m3} + \cdots + (-1)^j \sum_m p_{mj}$$

$p_{m1} =$ loop gain (the product of all the branch gains around a loop)
$p_{m2} =$ product of the loop gains of the mth set of two nontouching loops
$p_{m3} =$ product of the loop gains of the mth set of three nontouching loops, and in general
$p_{mj} =$ product of the loop gains of the mth set of j nontouching loops
$\Delta_k =$ the value of Δ for that subgraph of the graph obtained by removing the kth forward path along with those branches touching the path

Mason's signal-flowgraphs constitute a very useful graphical technique for the analysis of linear systems. This technique not only retains the intuitive character of the block diagrams but at the same time allows one to obtain the gain between an input node and an output node of a signal-flowgraph by inspection. However, the derivation of the gain formula (Equation (3.42)) is by no means simple, and, more importantly, if more than one input is present in the system, the gain cannot be obtained directly; that is, the principle of superposition must be applied to determine the gain due to the presence of more than one input. Thus, by slight modification of the conventions involved in Mason's signal-flowgraph, Coates (1959) was able to introduce the so-called "flowgraphs" which are suitable for direct calculation of gain.

Recently, Chan and Bapna (1967) further modified Coates's flowgraphs and developed a simpler gain formula based on the modified graphs. The definitions and the gain formula based on the modified Coates graphs are presented in the following discussion.

The **flowgraph G₁** (called the *modified Coates graph*) of a system S of n independent linear equations in n unknowns

$$\sum_{j=1}^n a_{ij}x_j = b_i \qquad i = 1, 2, \ldots, n$$

is an oriented graph such that the variable x_j in S is represented by a *node* (also denoted by x_j) in G_1, and the coefficient a_{ij} of the variable x_j in S by a *branch* with a branch gain a_{ij} connected between nodes x_i and x_j in G_1 and directed from x_j to x_i. Furthermore, a *source node* is included in G_1 such that for each constant b_k in S there is a node with gain b_k in G_1 from node 1 to node s_k. Graph G_{10} is the subgraph of G_1 obtained by deleting the source node l and all the branches connected to it. Graph G_{ij} is the subgraph of G_1 obtained by first removing all the outgoing branches from node x_j and then short-circuiting node l to node x_j. A *loop set* l is a subgraph of G_{10} that contains all the nodes of G_{10} with each node having exactly one incoming and one outgoing branch. The product p of the gains of all the branches in l is called a *loop-set product*. A 2-loop-set I_2 is a subgraph of G_{1j} containing all the nodes of G_{1j} with each node having exactly one incoming and one outgoing branch. The product p_2 of the gains of all the branches in l_2 is called a *2-loop-set product*.

The modified Coates gain formula is now stated in the following theorem.

Theorem 3.10. In a system of n independent linear equations in n unknowns

$$a_{ij}x_j = b_i \quad i = 1, 2, \ldots, n$$

the value of the variable x_j is given by

$$x_j = \sum_{(\text{all } p_2)} (-1)^{N_{l_2}} p_2 \Big/ \sum_{(\text{all } p_2)} (-1)^{N_l} p \qquad (3.43)$$

where N_{l_2} *is the number of loops in a 2-loop-set* l_2 *and* N_l *is the number of loops in a loop set* l.

Since both the Mason graph G_m and the modified Coates graph G_l are topological representations of a system of equations it is logical that certain interrelationships exist between the two graphs so that one can be transformed into the other. Such interrelationships have been noted (Chan, 1969), and the transformations are briefly stated as follows:

1. *Transformation of G_m into G_l.* Graph G_m can be transformed into an equivalent Coates graph G_l (representing an equivalent system of equations) by the following steps:

   a. Subtract 1 from the gain of each existing self-loop.
   b. Add a self-loop with a gain of -1 to each branch devoid of self-loop.
   c. Multiply by $-b_k$ the gain of the branch at the kth source node b_k ($k = 1, 2, \ldots, r$, r being the number of source nodes) and then combine all the (r) nodes into one source node (now denoted by 1).

2. *Transformation of G_l into G_m.* Graph G_l can be transformed into G_m by the following steps:

   a. Add 1 to the gain of each existing self-loop.
   b. Add a self-loop with a gain of 1 to each node devoid of self-loop except the source node l.
   c. Break the source node l into r source nodes (r being the number of branches connected to the source node l before breaking), and identify the r new sources nodes by $b_1, b_2, \ldots, b$, with the gain of the corresponding r branches multiplied by $-1/b_1, -1/b_2, \ldots, -1/b_r$, respectively, so that the new gains of these branches are all equal to l, keeping the edge orientations unchanged.

The gain formulas of Mason and Coates are the classical ones in the theory of flowgraphs. From the systems viewpoint, the Mason technique provides excellent physical insight as one can visualize the signal flow through the subgraphs (forward paths and feedback loops) of G_m. The graph-reduction technique based on the Mason graph enables one to obtain the gain expression using a step-by-step approach and at the same time observe the cause-and-effect relationships in each step. However, since the Mason formula computes the ratio of a specified output over *one* particular input, the principle of superposition must be used in order to obtain the overall gain of the system if more than one input is present. The Coates formula, however, computes the output directly regardless of the number of inputs present in the system, but because of such a direct computation of a given output, the graph reduction rules of Mason cannot be applied to a Coates graph since the Coates graph is *not* based on the same cause-and-effect formulation of equations as Mason's.

The *k*-Tree Approach versus the Flowgraph Approach

When a linear network is given, loop or node equations can be written from the network, and the analysis of the network can be accomplished by means of either Coates's or Mason's technique.

However, it has been shown (Chan, 1969) that if the Maxwell k-tree approach is employed in solving a linear network, the redundancy inherent either in the direct expansion of determinants or in the flowgraph techniques described above can be either completely eliminated for passive networks or greatly reduced for active networks. This point and others will be illustrated in the following example.

Example 3.7

Consider the network N as shown in Figure 3.42. Let us determine the voltage gain, $G_{12} = V_0/V_1$, using (1) Mason's method, (2) Coates's method, and (3) the k-tree method.

The two node equations for the network are given by

$$\text{for node 2:} \quad (Y_a + Y_b + Y_e)V_2 + (-Y_3)V_0 = Y_a V_i$$

$$\text{for node 3:} \quad (-Y_e)V_2 + (Y_c + Y_d + Y_e)V_0 = Y_c V_i$$

$$(3.44)$$

where

$$Y_a = 1/Z_a, Y_b = 1/Z_b, Y_c = 1/Z_c, Y_d = 1/Z_d \text{ and } Y_e = 1/Z_e$$

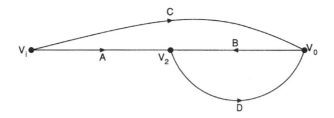

FIGURE 3.43 The Mason graph of N.

(1) *Mason's approach.* Rewrite the system of both parts of Equation (3.44) as follows:

$$V_2 = \left(\frac{Y_a}{Y_a + Y_b + Y_e}\right)V_i + \left(\frac{Y_e}{Y_a + Y_b + Y_e}\right)V_0$$

$$V_0 = \left(\frac{Y_c}{Y_c + Y_d + Y_e}\right)V_i + \left(\frac{Y_e}{Y_c + Y_d + Y_e}\right)V_2$$

(3.45)

or

$$V_2 = AV_i + BV_0 \quad V_0 = CV_i + DV_2$$

(3.46)

where

$$A = \frac{Y_a}{Y_a + Y_b + Y_e} \quad B = \frac{Y_e}{Y_a + Y_b + Y_e}$$

$$C = \frac{Y_c}{Y_c + Y_d + Y_e} \quad D = \frac{Y_e}{Y_c + Y_d + Y_c}$$

The Mason graph of system (Equation (3.46)) is shown in Figure 3.43, and according to the Mason graph formula (Equation (3.42)), we have

$$\Delta = 1 - BD$$

$$G_C = C \quad \Delta_C = 1$$

$$G_{AD} = AD \quad \Delta_{AD} = 1$$

and hence

$$G_{12} = \frac{V_0}{V_1} = \frac{1}{\Delta}\sum_k G_k \Delta_k = \frac{1}{1 - BD}(C + AD)$$

$$= \frac{Y_c/(Y_c + Y_d + Y_e) + Y_a/(Y_a + Y_b + Y_e)(Y_c + Y_d + Y_e)}{1 - Y_e^2/(Y_a + Y_b + Y_e)(Y_c + Y_d + Y_e)}$$

Upon cancellation and rearrangement of terms

$$G_{12} = \frac{Y_a Y_c + Y_a Y_e + Y_b Y_c + Y_c Y_e}{Y_a Y_c + Y_a Y_d + Y_a Y_e + Y_b Y_c + Y_b Y_d + Y_b Y_e + Y_c Y_e + Y_d Y_e}$$

(3.47)

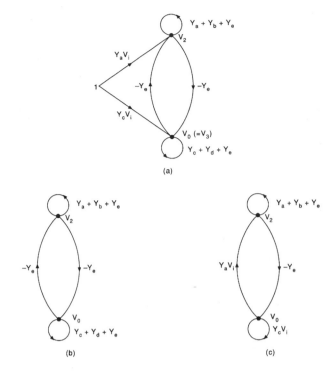

FIGURE 3.44   The Coates graphs: (a) G_I, (b) G_{I0}, and (c) G_{I3}.

(2) *Coates's approach.* From Equation (3.44) we obtain the Coates graphs G_1, G_{10}, and G_{13}, as shown in Figure 3.44(a), (b), and (c), respectively. The set of all loop-sets of G_{10} is shown in Figure 3.45, and the set of all 2-loop-sets of G_{13} is shown in Figure 3.46. Thus, by Equation (3.43):

$$V_0 = \frac{\sum\limits_{(\text{all } p_2)} (-1)^{N_{l_2}} p_2}{\sum\limits_{(\text{all } p)} (-1)^{N_l} p} = \frac{(-1)^1(-Y_e)(Y_a V_i) + (-1)^2(Y_a + Y_b + Y_e)(Y_c V_i)}{(-1)^1(-Y_e)(-Y_e) + (-1)^2(Y_a + Y_b + Y_e)(Y_c + Y_d + Y_e)}$$

Or, after simplification, we find

$$V_0 = \frac{(Y_a Y_c + Y_a Y_e + Y_b Y_c + Y_c Y_e)V_i}{Y_a Y_c + Y_a Y_d + Y_a Y_e + Y_b Y_c + Y_b Y_d + Y_b Y_e + Y_c Y_e + Y_d Y_e} \tag{3.48}$$

which gives the same ratio V_0/V_i as Equation (3.47).

(3) *The k-tree approach.* Recall that the gain formula for V_0/V_i using the *k*-tree approach is given (Chan, 1969) by

$$\frac{V_0}{V_i} = \frac{\Delta_{13}}{\Delta_{11}} = \frac{W_{13,\text{R}}}{W_{1,\text{R}}}$$

$$= \frac{\sum \begin{pmatrix} \text{all 2-tree admittance products with nodes 1 and 3 in one part} \\ \text{and the reference node R in the other part of each such 2-tree} \end{pmatrix}}{\sum \begin{pmatrix} \text{all 2-tree admittance products with nodes 1 in one part} \\ \text{and the reference node R in the other part of each such 2-tree} \end{pmatrix}} \tag{3.49}$$

FIGURE 3.45 The set of all loop-sets of G_{l0}.

FIGURE 3.46 The set of all 2-loop-sets of G_{l3}.

where Δ_{13} and Δ_{11} are cofactors of the determinant Δ of the node-admittance matrix of the network. Furthermore, it is noted that the 2-trees corresponding to Δ_{ii} may be obtained by finding all the trees of the modified graph G_i, which is obtained from the graph G of the network by short-circuiting node i (i being any node other than R) to the reference node R, and that the 2-trees corresponding to Δ_{ij} can be found by taking all those 2-trees, each of which is a tree of both G_i and G_j (Chan, 1969). Thus, for Δ_{11}, we first find G and G_1 (Figure 3.47), and then find the set S_1 of all trees of G_1 (Figure 3.48); then for Δ_{13}, we find G_3 (Figure 3.49) and the set S_3 of all trees of G_3 (Figure 3.50), and then from S_1 and S_3 we find all the terms common to both sets (which correspond to the set of all trees common to G_1 and G_3), as shown in Figure 3.51. Finally, we form the ratio of 2-tree admittance products according to Equation (3.49). Thus, from Figures 3.48 and 3.51, we find

$$\frac{V_0}{V_i} = \frac{Y_a Y_c + Y_a Y_e + Y_b Y_c + Y_c Y_e}{Y_a Y_c + Y_a Y_d + Y_a Y_e + Y_b Y_c + Y_b Y_d + Y_b Y_e + Y_c Y_e + Y_d Y_e}$$

which is identical to the results obtained by the flowgraph techniques.

From the above discussions and Example 3.7, we see that the Mason approach is the best from a systems viewpoint, especially when a single source is involved. It gives an excellent physical insight to the system and reveals the cause-and-effect relationships at various stages when graph reduction technique is employed. While the Coates approach enables one to compute the output directly regardless of the number of inputs involved in the system, thus overcoming one of the difficulties associated with Mason's approach, it does not

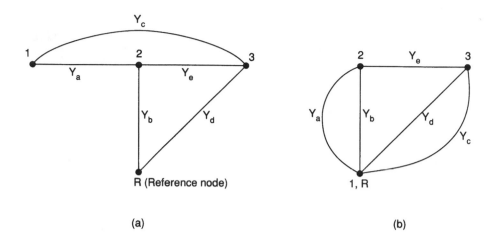

FIGURE 3.47   (a) Graph G and (b) the modified graph G_l of G.

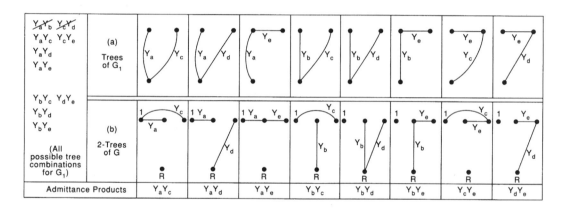

FIGURE 3.48   (a) The set of all trees of the modified graph G_l which corresponds to (b) the set of all 2-trees of G (with nodes l and R in separate parts in each of such 2-trees).

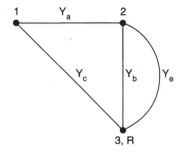

FIGURE 3.49   The modified graph G_3 of G.

allow one to reduce the graph step-by-step toward the final solution as Mason's does. However, it is interesting to note that in the modified Coates technique the introduction of the loop-sets (analogous to trees) and the 2-loop-sets (analogous to 2-trees) brings together the two different concepts—the flowgraph approach and the *k*-tree approach.

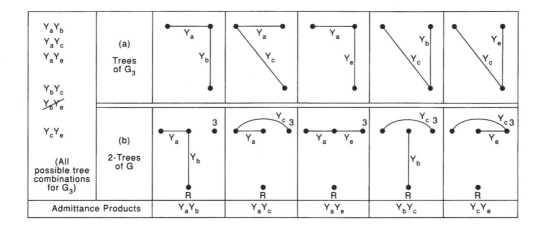

FIGURE 3.50 (a) The set of all trees of the modified graph G_3, which corresponds to (b) the set of all 2-trees of G (with nodes 3 and R in separate parts in each such 2-trees).

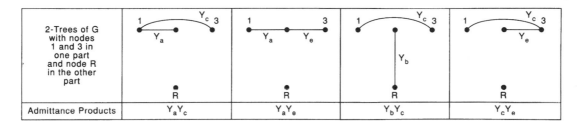

FIGURE 3.51 The set of all 2-trees of G (with nodes l and 3 in one part of the reference node R in the other part of each such 2-trees).

From the network's point of view, the Maxwell k-tree approach not only enables one to express the solution in terms of the topology (namely, the trees and 2-trees in Example 3.7) of the network but also avoids the cancellation problem inherent in all the flowgraph techniques since, as evident from Example 3.7, the trees and the 2-trees in the gain expression by the k-tree approach correspond (one-to-one) to the *uncanceled terms* in the final expressions of the gain by the flowgraph techniques. Finally, it should be obvious that the k-tree approach depends upon the knowledge of the graph of a given network. Thus, if in a network problem only the system of (loop or node) equations is given and the network is not known, or more generally, if a system is characterized by a block diagram or a system of equations, the k-tree approach cannot be applied and one must resort to the flowgraph techniques between the two approaches.

Some Topological Applications in Network Analysis and Design

In practice, a circuit designer often has to make approximations and analyze the same network structure many times with different sets of component values before the final network realization is obtained. Conventional analysis techniques which require the evaluation of high-order determinants are undesirable even on a digital computer because of the large amount of redundancy inherent in the determinant expansion process. The extra calculation in the evaluation (expansion of determinants) and simplification (cancelation of terms) is time consuming and costly and thereby contributes much to the undesirability of such methods.

The k-tree topological formulas presented in this section, however, eliminate completely the cancellation of terms. Also, they are particularly suited for digital computation when the size of the network is not exceedingly large. All of the terms involved in the formulas can be computed by means of a digital computation using a single "tree-finding" program (Chan, 1969). Thus, the application of topological formulas in analyzing a network with the aid of a digital computer can mean saving a considerable amount of time and cost to the circuit designer, especially true when it is necessary to repeat the same analysis procedure a large number of times. In a preliminary system design, the designer usually seeks one or more concepts that will meet the specifications, and in engineering practice each concept is generally subjected to some form of analysis. For linear systems, the signal flowgraph of Mason is widely used in this activity. The flowgraph analysis is popular because it depicts the relationships existing between system variables, and the graphical structure may be manipulated using Mason's formulas to obtain system transfer functions in symbolic or symbolic/numerical form. Although the preliminary design problems are usually of limited size (several variables), hand derivation of transfer functions is nonetheless difficult and often prone to error arising from the omission of terms. The recent introduction of remote, time-shared computers into modern design areas offers a means to handle such problems swiftly and effectively.

An efficient algorithm suitable for digital compution of transfer functions from the signal flowgraph description of a linear system has been developed (Dunn and Chan, 1969) which provides a powerful analytical tool in the conceptual phases of linear system design.

In the past several decades, graph theory has been widely used in electrical engineering, computer science, social science, and in the solution of economic problems (Swamy and Thulasiraman, 1981; Chen, 1990). Finally, the application of graph theory in conjunction with symbolic network analysis and computer-aided simulation of electronic circuits has been well recognized in recent years (Lin, 1991).

Defining Terms

Branches of a tree: The edges contained in a tree.

Circuit (or loop):   A closed path where all vertices are of degree 2, thus having *no* endpoints in the path.

Circuit-set (or loop-set):   The set of all edges contained in a circuit (loop).

Connectedness: A graph or subgraph is said to be connected if there is at least one path between *every* pair of its vertices.

Flowgraph G_l (or modified Coates graph G_l):   The flowgraph G_l (called *the modified Coates graph*) of a system S of n independent linear equations in n unknowns:

$$\sum_{j=1}^{n} a_{ij}x_j = b_i \quad i = 1, 2, \ldots, n$$

is an oriented graph such that the variable x_j in S is represented by a *node* (also denoted by x_j) in G_l, and the coefficient a_{ij} of the variable x_j in S by a *branch* with a branch gain a_{ij} connected between nodes x_i and x_j in G_l and directed from x_j to x_i. Furthermore, a *source node l* is included in G_l such that for each constant b_k in S there is a node with gain b_k in G_l from node l to node s_k. Graph G_{ij} is the subgraph of G_l obtained by first removing all the outgoing branches from node x_j and then short-circuiting node l to node x_j. A *loop set l* is a subgraph of G_{l0} that contains all the nodes of G_{l0} with each node having exactly one incoming and one outgoing branch. The product p of the gains of all the branches in l is called a *loop-set product*. A 2-loop-set l_2 is a subgraph of G_{lj} containing all the nodes of G_{lj} with each node having exactly one incoming and one outgoing branch. The product p_2 of the gains of all the branches in l_2 is called a *2-loop-set product*.

k-Tree admittance product of a k-tree: The product of the admittances of all the branches of the k-tree.

k-Tree of a connected graph G: A proper subgraph of G consisting of k unconnected circuitless subgraphs, each subgraph itself being connected, which together contain all the vertices of G.

Linear graph: A set of line segments called edges and points called vertices, which are the endpoints of the edges, interconnected in such a way that the edges are connected to (or incident with) the vertices. The degree of a vertex of a graph is the number of edges incident with that vertex.

Path: A subgraph having all vertices of degree 2 except for the two endpoints which are of degree 1 and are called the terminals of the path, where the degree of a vertex is the number of edges connected to the vertex in the subgraph.

Path-set: The set of all edges in a path.

Proper subgraph: A subgraph which does not contain all of the edges of the given graph.

Signal-flowgraph G_m (or Mason's graph G_m): A signal-flowgraph G_m of a system S of n independent linear (algebraic) equations in n unknowns:

$$\sum_{j=1}^{n} a_{ij}x_j = b_i \quad i = 1, 2, \ldots, n$$

is a graph with junction points called *nodes* which are connected by directed line segments called *branches* with signals traveling along the branches only in the direction described by the arrows of the branches. A signal x_k traveling along a branch between x_k and x_j is multiplied by the gain of the branches g_{kj}, so that a signal $g_{kj} x_k$ is delivered at node x_j. An *input node* (*source*) is a node which contains only outgoing branches; an *output node* (*sink*) is a node which has only incoming branches. A *path* is a continuous unidirectional succession of branches, all of which are traveling in the same direction; a *forward path* is a path from the input node to the output node along which all nodes are encountered exactly once; and a *feedback path* (*loop*) is a closed path which originates from and terminates at the same node, and along with all other nodes are encountered exactly once (the trivial case is a *self-loop* which contains exactly one node and one branch). A *path* gain is the product of all the branch gains of the branches in a loop.

Subgraph: A subset of the edges of a given graph.

Tree: A connected subgraph of a given connected graph G which contains all the vertices of G but no circuits.

References

A.C. Aitken, *Determinants and Matrices,* 9th ed., New York: Interscience, 1956.

S.P. Chan, *Introductory Topological Analysis of Electrical Networks,* New York: Holt, Rinehart and Winston, 1969.

S.P. Chan and B.H. Bapna, "A modification of the Coates gain formula for the analysis of linear systems," *Inst. J. Control,* vol. 5, pp. 483–495, 1967.

S.P. Chan and S.G. Chan, "Modifications of topological formulas," *IEEE Trans. Circuit Theory,* vol. CT15, pp. 84–86, 1968.

W.K. Chen, *Theory of Nets: Flows in Networks,* New York: Wiley Interscience, 1990.

C.L. Coates, "Flow-graph solutions of linear algebraic equations," *IRE Trans. Circuit Theory,* vol. CT6, pp. 170–187, 1959.

W.R. Dunn, Jr., and S.P. Chan, "Flowgraph analysis of linear systems using remote timeshared computation," *J. Franklin Inst.,* vol. 288, pp. 337–349, 1969.

G. Kirchhoff, "Über die Auflösung der Gleichungen, auf welche man bei der Untersuchung der linearen Vertheilung galvanischer Ströme, geführt wird," *Ann. Physik Chemie,* vol. 72, pp. 497–508, 1847; English transl., *IRE Trans. Circuit Theory,* vol. CT5, pp. 4–7, 1958.

P.M. Lin, *Symbolic Network Analysis,* New York: Elsevier, 1991.

S.J. Mason, "Feedback theory—Some properties of signal flow graphs," *Proc. IRE,* vol. 41, pp. 1144–1156, 1953.

S.J. Mason, "Feedback theory—Further properties of signal flow graphs," *Proc. IRE,* vol. 44, pp. 920–926, 1956.

J.C. Maxwell, *Electricity and Magnetism,* Oxford: Clarendon Press, 1892.

W.S. Percival, "Solution of passive electrical networks by means of mathematical trees," *Proc. IEE,* vol. 100, pp. 143–150, 1953.

S. Seshu and M.B. Reed, *Linear Graphs and Electrical Networks,* Reading, MA.: Addison-Wesley, 1961.
M.N.S. Swamy ad K. Thulasiraman, *Graphs, Networks, and Algorithms,* New York: Wiley, 1981.

Further Information

All defining terms used in this section can be found in S.P. Chan, *Introductory Topological Analysis of Electrical Networks,* Holt, Rinehart, and Winston, New York, 1969. Also an excellent reference for the applications of graph theory in electrical engineering (i.e., network analysis and design) is S. Seshu and M.B. Reed, *Linear Graphs and Electrical Networks,* Reading, MA: Addison-Wesley, 1961.

For applications of graph theory in computer science, see M.N.S. Swamy and K. Thulasiraman, *Graphs, Networks, and Algorithms,* New York: Wiley, 1981.

For flowgraph applications, see W.K. Chen, *Theory of Nets: Flows in Networks,* New York: Wiley Interscience, 1990.

For applications of graph theory in symbolic network analysis, see P.M. Lin, *Symbolic Network Analysis,* New York: Elsevier, 1991.

3.7   Two-Port Parameters and Transformations

Norman S. Nise

Many times we want to model the behavior of an electric network at only two terminals, as shown in Figure 3.52. Here, only V_1 and I_1, not voltages and currents internal to the circuit, need to be described. To produce the model for a linear circuit, we use **Thévenin's** or **Norton's theorem** to simplify the network as viewed from the selected terminals. We define the pair of terminals shown in Figure 3.52 as a **port**, where the current, I_1, entering one terminal equals the current leaving the other terminal.

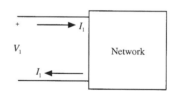

FIGURE 3.52   An electrical network port.

If we further restrict the network by stating that (1) all external connections to the circuit, such as sources and impedances, are made at the port and (2) the network can have internal **dependent sources**, but not **independent sources**, we can mathematically model the network at the port as

$$V_1 = ZI_1 \tag{3.50}$$

or

$$I_1 = YV_1 \tag{3.51}$$

where Z is the Thévenin impedance and Y is the Norton admittance at the terminals. Z and Y can be constant resistive terms, Laplace transforms $Z(s)$ or $Y(s)$, or sinusoidal steady-state functions $Z(j\omega)$ or $Y(j\omega)$.

Defining Two-Port Networks

Electrical networks can also be used to transfer signals from one port to another. Under this requirement, connections to the network are made in two places, the input and the output. For example, a transistor has an input between the base and emitter and an output between the collector and emitter. We can model such circuits as **two-port networks**, as shown in Figure 3.53. Here, we see the input port, represented by V_1 and I_1, and the output port, represented by V_2 and I_2. Currents are assumed positive if they flow as shown in

Figure 3.53. The same restrictions about external connections and internal sources mentioned above for the single port also apply.

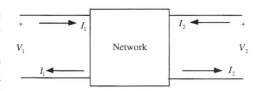

FIGURE 3.53 A two-port network.

Now that we have defined two-port networks, let us discuss how to create a mathematical model of the network by establishing relationships among all of the input and output voltages and currents. Many possibilities exist for modeling. In the next section we arbitrarily begin by introducing the z-parameter model to establish the technique. In subsequent sections we present alternative models and draw relationships among them.

Mathematical Modeling of Two-Port Networks via z Parameters

In order to produce a mathematical model of circuits represented by Figure 3.53, we must find relationships among V_1, I_1, V_2, and I_2. Let us visualize placing a current source at the input and a current source at the output. Thus, we have selected two of the variables, I_1 and I_2. We call these variables the independent variables. The remaining variables, V_1 and V_2, are dependent upon the selected applied currents. We call V_1 and V_2 the dependent variables. Using **superposition** we can write each dependent variable as a function of the independent variables as follows:

$$V_1 = z_{11}I_1 + z_{12}I_2 \tag{3.52a}$$

$$V_2 = z_{21}I_1 + z_{22}I_2 \tag{3.52b}$$

We call the coefficients z_{ij} in Equations (3.52) parameters of the two-port network or, simply, **two-port parameters**.

From Equations (3.52), the two-port parameters are evaluated as

$$z_{11} = \left.\frac{V_1}{I_1}\right|_{I_2=0} ; \quad z_{12} = \left.\frac{V_1}{I_2}\right|_{I_1=0}$$

$$z_{21} = \left.\frac{V_2}{I_1}\right|_{I_2=0} ; \quad z_{22} = \left.\frac{V_2}{I_2}\right|_{I_1=0} \tag{3.53}$$

Notice that each parameter can be measured by setting a port current, I_1 or I_2, equal to zero. Since the parameters are found by setting these currents equal to zero, this set of parameters is called **open-circuit parameters**. Also, since the definitions of the parameters as shown in Equations (3.53) are the ratio of voltages to currents, we alternatively refer to them as **impedance parameters**, or **z parameters**. The parameters themselves can be impedances represented as Laplace transforms, $Z(s)$, sinusoidal steady-state impedance functions, $Z(j\omega)$, or simply pure resistance values, R.

Evaluating Two-Port Network Characteristics in Terms of z Parameters

The two-port parameter model can be used to find the following characteristics of a two-port network when used in some cases with a source and load, as shown in Figure 3.54:

$$\text{input impedance} = Z_{\text{in}} = V_1/I_1 \tag{3.54a}$$

$$\text{output impedance} = Z_{\text{out}} = V_2/I_2|V_s = 0 \tag{3.54b}$$

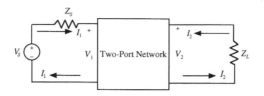

FIGURE 3.54 Terminated two-port network for finding two-port network characteristics.

$$\text{network voltage gain} = V_g = V_2/V_1 \tag{3.54c}$$

$$\text{total voltage gain} = V_{gt} = V_2/V_s \tag{3.54d}$$

$$\text{network current gain} = I_g = I_2/I_1 \tag{3.54e}$$

To find Z_{in} of Figure 3.54, determine V_1/I_1. From Figure 3.54, $V_2 = -I_2 Z_L$. Substituting this value in Equation (3.52b) and simplifying, Equations (3.52) become

$$V_1 = z_{11}I_1 + z_{12}I_2 \tag{3.55a}$$

$$0 = z_{21}I_1 + (z_{22} + Z_L)I_2 \tag{3.55b}$$

Solving simultaneously for I_1 and then forming $V_1/I_1 = Z_{in}$, we obtain

$$Z_{in} = \frac{V_1}{I_1} = z_{11} - \frac{z_{12}z_{21}}{(z_{22} + Z_L)} \tag{3.56}$$

To find Z_{out}, set $V_S = 0$ in Figure 3.54. This step terminates the input with Z_S. Next, determine V_2/I_2. From Figure 3.54 with V_S shorted, $V_1 = -I_1 Z_S$. By substituting this value into Equation (3.52a)) and simplifying, Equations (3.52) become

$$0 = (z_{11} + z_s)I_1 + z_{12}I_2 \tag{3.57a}$$

$$V_2 = z_{21}I_1 + z_{22}I_2 \tag{3.57b}$$

By solving simultaneously for I_2 and then forming $V_2/I_2 = Z_{out}$:

$$Z_{out} = \frac{V_2}{I_2}\bigg|_{V_S=0} = z_{22} - \frac{z_{12}z_{21}}{(z_{11} + z_s)} \tag{3.58}$$

To find V_g, we see from Figure 3.54 that $I_2 = -V_2/Z_L$. Substituting this value in Equations (3.52) and simplifying, we obtain

$$V_1 = z_{11}I_1 - \frac{z_{12}}{z_L}V_2 \tag{3.59a}$$

$$0 = z_{21}I_1 - \left(\frac{z_{22} + Z_L}{Z_L}\right)V_2 \tag{3.59b}$$

TABLE 3.1 Network Characteristics Developed from z-Parameter Defining Equations (3.52)

Network Characteristic Definition	From Figure 3.54	Substitute in Defining Equations (3.52) and Obtain	Solve for Network Characteristic
Input impedance $z_{in} = \dfrac{V_1}{I_1}$	$V_2 = -I_2 Z_L$	$V_1 = z_{11}I_1 + z_{12}I_2$ $0 = z_{21}I_1 + (z_{22}+Z_L)I_2$	$Z_{in} = z_{11} - \dfrac{z_{12}z_{21}}{z_{22}+z_L}$
Output impedance $Z_{out} = \dfrac{V_2}{I_2}\Big|_{V_s=0}$	$V_1 = V_s - I_1 Z_s$ $V_s = 0$	$0 = (z_{11}+z_s)I_1 + z_{12}I_2$ $V_2 = Z_{21}I_1 + z_{22}I_2$	$Z_{out} = z_{22} - \dfrac{z_{12}z_{21}}{z_{11}+Z_s}$
Network voltage gain $V_g = \dfrac{V_2}{V_1}$	$I_2 = -\dfrac{V_2}{Z_L}$	$V_1 = z_{11}I_1 - \dfrac{z_{12}V_2}{Z_L}$ $0 = z_{21}I_1 - \dfrac{(z_{22}+Z_L)V_2}{Z_L}$	$V_g = \dfrac{z_{21}Z_L}{z_{11}(z_{22}+Z_L) - z_{12}z_{21}}$
Total voltage gain $V_{gt} = \dfrac{V_2}{V_s}$	$V_1 = V_s - I_1 Z_s$ $I_2 = -\dfrac{V_2}{Z_L}$	$V_s = (z_{11}+Z_s)I_1 - \dfrac{z_{12}V_2}{Z_L}$ $0 = z_{21}I_1 - \dfrac{(z_{22}+Z_L)V_2}{Z_L}$	$V_{gt} = \dfrac{z_{21}Z_L}{(z_{11}+Z_s)(z_{22}+Z_L) - z_{12}z_{21}}$
Network current gain $I_g = \dfrac{I_2}{I_1}$	$V_2 = -I_2 Z_L$	$V_1 = z_{11}I_1 + z_{12}I_2$ $0 = z_{21}I_1 + (z_{22}+Z_L)I_2$	$I_g = -\dfrac{z_{21}}{z_{22}+Z_L}$

By solving simultaneously for V_2 and then forming $V_2/V_1 = V_g$:

$$V_g = \frac{V_2}{V_1} = \frac{z_{21}Z_L}{z_{11}(z_{22}+Z_L) - z_{12}z_{21}} \tag{3.60}$$

Similarly, other characteristics such as current gain and the total voltage gain from the source voltage to the load voltage can be found. Table 3.1 summarizes many of the network characteristics that can be found using z parameters as well as the process to arrive at the result.

To summarize the process of finding network characteristics:

1. Define the network characteristic.
2. Use appropriate relationships from Figure 3.54.
3. Substitute the relationships from Step 2 into Equations (3.52).
4. Solve the modified equations for the network characteristic.

An Example Finding z Parameters and Network Characteristics

To solve for two-port network characteristics we can first represent the network with its two-port parameters and then use these parameters to find the characteristics summarized in Table 3.1. To find the parameters, we terminate the network adhering to the definition of the parameter we are evaluating. Then we can use mesh or nodal analysis, current or voltage division, or equivalent impedance to solve for the parameters. The following example demonstates the technique.

Consider the network of Figure 3.55(a). The first step is to evaluate the z parameters. From their definition, z_{11} and z_{21} are found by open-circuiting the output and applying a voltage at the input, as shown in Figure 3.55(b). Thus, with $I_2 = 0$:

$$6I_1 - 4I_a = V_1 \tag{3.61a}$$

$$-4I_1 + 18I_a = 0 \tag{3.61b}$$

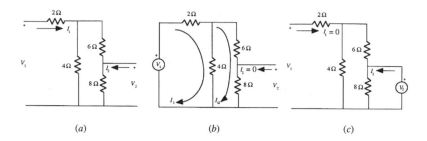

FIGURE 3.55   (a) Two-port network example; (b) two-port network modified to find z_{11} and z_{21}; (c) two-port network modified to find z_{22} and z_{12}.

Solving for I_1 yields

$$I_1 = \frac{\begin{vmatrix} V_1 & -4 \\ 0 & 18 \end{vmatrix}}{\begin{vmatrix} 6 & -4 \\ -4 & 18 \end{vmatrix}} = \frac{18V_1}{92} \tag{3.62}$$

from which

$$z_{11} = \frac{V_1}{I_1}\bigg|_{I_2=0} = \frac{46}{9} \tag{3.63}$$

We now find z_{21}. From Equation (3.61b):

$$\frac{I_a}{I_1} = \frac{2}{9} \tag{3.64}$$

But, from Figure 3.55(b), $I_a = V_2/8$. Thus

$$z_{21} = \frac{V_2}{I_2}\bigg|_{I_2=0} = \frac{16}{9} \tag{3.65}$$

Based on their definitions, z_{22} and z_{12} are found by placing a source at the output and open-circuiting the input as shown in Figure 3.55(c). The equivalent resistance, R_{2eq}, as seen at the output with $I_1 = 0$ is

$$R_{2eq} = \frac{8 \times 10}{8 + 10} = \frac{40}{9} \tag{3.66}$$

Therefore

$$z_{22} = \frac{V_2}{I_2}\bigg|_{I_1=0} = \frac{40}{9} \tag{3.67}$$

From Figure 3.55(c), using voltage division:

$$V_1 = (4/10)V_2 \tag{3.68}$$

But

$$V_2 = I_2 R_{2eq} = I_2(40/9) \tag{3.69}$$

Substituting Equation (3.69) into Equation (3.68) and simplifying yields

$$z_{12} = \left.\frac{V_1}{I_2}\right|_{I_1=0} = \frac{16}{9} \tag{3.70}$$

Using the z-parameter values found in Equation (3.63), Equation (3.65), Equation (3.67), and Equation (3.70) and substituting into the network characteristic relationships shown in the last column of Table 3.1, assuming $Z_S = 20\ \Omega$ and $Z_L = 10\ \Omega$, we obtain $Z_{in} = 4.89\ \Omega$, $Z_{out} = 4.32\ \Omega$, $V_g = 0.252$, $V_{gt} = 0.0494$, and $I_g = -0.123$.

Additional Two-Port Parameters and Conversions

We defined the z parameters by establishing I_1 and I_2 as the independent variables and V_1 and V_2 as the dependent variables. Other choices of independent and dependent variables lead to definitions of alternative two-port parameters. The total number of combinations one can make with the four variables, taking two at a time as independent variables, is six. Table 3.2 defines the six possibilities as well as the names and symbols given to the parameters.

The table also presents the expressions used to calculate directly the parameters of each set based upon their definition as we did with z parameters. For example, consider the y or **admittance parameters**. These parameters are seen to be **short-circuit parameters**, since their evaluation requires V_1 or V_2 to be zero. Thus, to find y_{22} we short-circuit the input and find the admittance looking back from the output. For Figure 3.55(a), $y_{22} = 23/88$. Any parameter in Table 3.2 is found either by open-circuiting or short-circuiting a terminal and then performing circuit analysis to find the defining ratio.

Another method of finding the parameters is to convert from one set to another. Using the "Definition" row in Table 3.2, we can convert the defining equations of one set to the defining equations of another set. For example, we have already found the z parameters. We can find the **h parameters** as follows.

Solve for I_2 using the second z-parameter equation (Equation (3.52b)) and obtain the second h-parameter equation as

$$I_2 = -\frac{z_{21}}{z_{22}}I_1 + \frac{I}{z_{22}}V_2 \tag{3.71}$$

which is of the form $I_2 = h_{21}I_1 + h_{22}V_2$, the second h-parameter equation. Now, substitute Equation (3.71) into the first z-parameter equation (Equation (3.52a)) rearrange, and obtain

$$V_1 = \frac{z_{11}z_{22} - z_{12}z_{21}}{z_{22}}I_1 + \frac{z_{12}}{z_{22}}V_2 \tag{3.72}$$

which is of the form $V_1 = h_{11}I_1 + h_{12}V_2$, the first h-parameter equation. Thus, for example, $h_{21} = -z_{21}/z_{22}$ from Equation (3.71). Other transformations are found through similar manipulations and are summarized in Table 3.2.

Finally, there are other parameter sets that are defined differently from the standard sets covered here. Specifically, they are scattering parameters used for microwave networks and image parameters used for filter design. A detailed discussion of these parameters is beyond the scope of this section. The interested reader should consult the bibliography in the "Further Information" section.

Two-Port Parameter Selection

The choice of parameters to use in a particular analysis or design problem is based on analytical convenience or the physics of the device or network at hand. For example, an ideal transformer cannot be represented with z parameters. I_1 and I_2 are not linearly independent variables, since they are related through the turns ratio. A similar argument applies to the **y-parameter** representation of a transformer. Here, V_1 and V_2 are not independent, since they too are related via the turns ratio. A possible choice for the transformer is the **transmission parameters**. For an ideal transformer, B and C would be zero. For a BJT transistor, there is effectively linear independence between the input current and the output voltage. Thus, the hybrid parameters are the parameters of choice for the transistor.

TABLE 3.2 Two-Port Parameter Definitions and Conversions

	Impedance Parameters (Open-Circuit Parameters) z	Admittance Parameters (Short-Circuit Parameters) y	Hybrid Parameters h
Definition	$V_1 = z_{11}I_1 + z_{12}I_2$ $V_2 = z_{21}I_1 + z_{22}I_2$	$I_1 = y_{11}V_1 + y_{12}V_2$ $I_2 = y_{21}V_1 + y_{22}V_2$	$V_1 = h_{11}I_1 + h_{12}V_2$ $I_2 = h_{21}I_1 + h_{22}V_2$
Parameters	$z_{11} = \dfrac{V_1}{I_1}\Big|_{I_2=0}$; $z_{12} = \dfrac{V_1}{I_2}\Big|_{I_1=0}$ $z_{21} = \dfrac{V_2}{I_1}\Big|_{I_2=0}$; $z_{22} = \dfrac{V_2}{I_2}\Big|_{I_1=0}$	$y_{11} = \dfrac{I_1}{V_1}\Big|_{V_2=0}$; $y_{12} = \dfrac{I_1}{V_1}\Big|_{V_1=0}$ $y_{21} = \dfrac{I_2}{V_1}\Big|_{V_2=0}$; $y_{22} = \dfrac{I_2}{V_1}\Big|_{V_1=0}$	$h_{11} = \dfrac{V_1}{I_1}\Big|_{V_2=0}$; $h_{12} = \dfrac{V_1}{V_2}\Big|_{I_1=0}$ $h_{21} = \dfrac{I_2}{I_1}\Big|_{V_2=0}$; $h_{22} = \dfrac{I_2}{V_2}\Big|_{I_1=0}$
Conversion to z parameters		$z_{11} = \dfrac{y_{22}}{\Delta_y}$; $z_{12} = \dfrac{-y_{12}}{\Delta_y}$ $z_{21} = \dfrac{-y_{21}}{\Delta_y}$; $z_{22} = \dfrac{y_{11}}{\Delta_y}$	$z_{11} = \dfrac{\Delta_h}{h_{22}}$; $z_{12} = \dfrac{h_{12}}{h_{22}}$ $z_{21} = \dfrac{-h_{21}}{h_{22}}$; $z_{22} = \dfrac{1}{h_{22}}$
Conversion to y parameters	$y_{11} = \dfrac{z_{22}}{\Delta_z}$; $y_{12} = \dfrac{-z_{12}}{\Delta_z}$ $y_{21} = \dfrac{-z_{21}}{\Delta_z}$; $y_{22} = \dfrac{z_{11}}{\Delta_z}$		$y_{11} = \dfrac{1}{h_{11}}$; $y_{12} = \dfrac{-h_{12}}{h_{11}}$ $y_{21} = \dfrac{h_{21}}{h_{11}}$; $y_{22} = \dfrac{\Delta_h}{h_{11}}$
Conversion to h parameters	$h_{11} = \dfrac{\Delta_z}{z_{22}}$; $h_{12} = \dfrac{z_{12}}{z_{22}}$ $h_{21} = \dfrac{-z_{21}}{z_{22}}$; $h_{22} = \dfrac{1}{z_{22}}$	$h_{11} = \dfrac{1}{y_{11}}$; $h_{12} = \dfrac{-y_{12}}{y_{11}}$ $h_{21} = \dfrac{y_{21}}{y_{11}}$; $h_{22} = \dfrac{\Delta_y}{y_{11}}$	
Conversion to g parameters	$g_{11} = \dfrac{1}{z_{11}}$; $g_{12} = \dfrac{-z_{12}}{z_{11}}$ $g_{21} = \dfrac{z_{21}}{z_{11}}$; $g_{22} = \dfrac{\Delta_z}{z_{11}}$	$g_{11} = \dfrac{\Delta_y}{y_{22}}$; $g_{12} = \dfrac{y_{12}}{y_{22}}$ $g_{21} = \dfrac{-y_{21}}{y_{22}}$; $g_{22} = \dfrac{1}{y_{22}}$	$g_{11} = \dfrac{h_{22}}{\Delta_h}$; $g_{12} = \dfrac{-h_{12}}{\Delta_h}$ $g_{21} = \dfrac{-h_{21}}{\Delta_h}$; $g_{22} = \dfrac{h_{11}}{\Delta_h}$
Conversion to T parameters	$A = \dfrac{z_{11}}{z_{21}}$; $B = \dfrac{\Delta_z}{z_{21}}$ $C = \dfrac{1}{z_{21}}$; $D = \dfrac{z_{22}}{z_{21}}$	$A = \dfrac{-y_{22}}{y_{21}}$; $B = \dfrac{-1}{y_{21}}$ $C = \dfrac{-\Delta_y}{y_{21}}$; $D = \dfrac{-y_{11}}{y_{21}}$	$A = \dfrac{-\Delta_h}{h_{21}}$; $B = \dfrac{-h_{11}}{h_{21}}$ $C = \dfrac{-h_{22}}{h_{21}}$; $D = \dfrac{-1}{h_{21}}$
Conversion to T' parameters	$A' = \dfrac{z_{22}}{z_{12}}$; $B' = \dfrac{\Delta_z}{z_{12}}$ $C' = \dfrac{1}{z_{12}}$; $D' = \dfrac{z_{11}}{z_{12}}$	$A' = \dfrac{-y_{11}}{y_{12}}$; $B' = \dfrac{-1}{y_{12}}$ $C' = \dfrac{-\Delta_y}{y_{12}}$; $D' = \dfrac{-y_{22}}{y_{12}}$	$A' = \dfrac{1}{h_{12}}$; $B' = \dfrac{h_{11}}{h_{12}}$ $C' = \dfrac{h_{22}}{h_{12}}$; $D' = \dfrac{\Delta_h}{h_{12}}$
Δ	$\Delta_z = z_{11}z_{22} - z_{12}z_{21}$	$\Delta_y = y_{11}y_{22} - y_{12}y_{21}$	$\Delta h = h_{11}h_{22} - h_{12}h_{21}$

TABLE 3.2 (Continued)

	Inverse Hybrid Parameters g	Transmission Parameters T	Inverse Transmission Par. T'
Definition	$I_1 = g_{11}V_1 + g_{12}I_2$ $V_2 = g_{21}V_1 + g_{22}I_2$	$V_1 = AV_2 - BI_2$ $I_1 = CV_2 - DI_2$	$V_2 = A'V_1 - B'I_1$ $I_2 = C'V_1 - D'I_1$
Parameters	$g_{11} = \left.\dfrac{I_1}{V_1}\right\|_{I_2=0}$; $\quad g_{12} = \left.\dfrac{I_1}{I_2}\right\|_{V_1=0}$ $g_{21} = \left.\dfrac{V_2}{V_1}\right\|_{I_2=0}$; $\quad g_{22} = \left.\dfrac{V_2}{I_2}\right\|_{V_1=0}$	$A = \left.\dfrac{V_1}{V_2}\right\|_{I_2=0}$; $\quad B = \left.\dfrac{-V_1}{I_2}\right\|_{V_2=0}$ $C = \left.\dfrac{I_1}{V_2}\right\|_{I_2=0}$; $\quad D = \left.\dfrac{-I_1}{I_2}\right\|_{V_2=0}$	$A' = \left.\dfrac{V_2}{V_1}\right\|_{I_1=0}$; $\quad B' = \left.\dfrac{-V_2}{I_1}\right\|_{V_1=0}$ $C' = \left.\dfrac{I_2}{V_1}\right\|_{I_1=0}$; $\quad D' = \left.\dfrac{-I_2}{I_1}\right\|_{V_1=0}$
Conversion to z parameters	$z_{11} = \dfrac{1}{g_{11}}$; $\quad z_{12} = \dfrac{-g_{12}}{g_{11}}$ $z_{21} = \dfrac{g_{21}}{g_{11}}$; $\quad z_{22} = \dfrac{\Delta_g}{g_{11}}$	$z_{11} = \dfrac{A}{C}$; $\quad z_{12} = \dfrac{\Delta_T}{C}$ $z_{21} = \dfrac{1}{C}$; $\quad z_{22} = \dfrac{D}{C}$	$z_{11} = \dfrac{D'}{C'}$; $\quad z_{12} = \dfrac{1}{C'}$ $z_{21} = \dfrac{\Delta_{T'}}{C'}$; $\quad z_{22} = \dfrac{A'}{C'}$
Conversion to y parameters	$y_{11} = \dfrac{\Delta_g}{g_{22}}$; $\quad y_{12} = \dfrac{g_{12}}{g_{22}}$ $y_{21} = \dfrac{-g_{21}}{g_{22}}$; $\quad y_{22} = \dfrac{1}{g_{22}}$	$y_{11} = \dfrac{D}{B}$; $\quad y_{12} = \dfrac{-\Delta_T}{B}$ $y_{21} = \dfrac{-1}{B}$; $\quad y_{22} = \dfrac{A}{B}$	$y_{11} = \dfrac{A'}{B'}$; $\quad y_{12} = \dfrac{-1}{B'}$ $y_{21} = \dfrac{-\Delta_{T'}}{B'}$; $\quad y_{22} = \dfrac{D'}{B'}$
Conversion to h parameters	$h_{11} = \dfrac{g_{22}}{\Delta_g}$; $\quad h_{12} = \dfrac{-g_{12}}{\Delta_g}$ $h_{21} = \dfrac{-g_{21}}{\Delta_g}$; $\quad h_{22} = \dfrac{g_{11}}{\Delta_g}$	$h_{11} = \dfrac{B}{D}$; $\quad h_{12} = \dfrac{\Delta_T}{D}$ $h_{21} = \dfrac{-1}{D}$; $\quad h_{22} = \dfrac{C}{D}$	$h_{11} = \dfrac{B'}{A'}$; $\quad h_{12} = \dfrac{1}{A'}$ $h_{21} = \dfrac{-\Delta_{T'}}{A'}$; $\quad h_{22} = \dfrac{C'}{A'}$
Conversion to g parameters		$g_{11} = \dfrac{C}{A}$; $\quad g_{12} = \dfrac{-\Delta_T}{A}$ $g_{21} = \dfrac{1}{A}$; $\quad g_{22} = \dfrac{B}{A}$	$g_{11} = \dfrac{C'}{D'}$; $\quad g_{12} = \dfrac{-1}{D'}$ $g_{21} = \dfrac{\Delta_{T'}}{D'}$; $\quad g_{22} = \dfrac{B'}{D'}$
Conversion to T parameters	$A = \dfrac{1}{g_{21}}$; $\quad B = \dfrac{g_{22}}{g_{21}}$ $C = \dfrac{g_{11}}{g_{21}}$; $\quad D = \dfrac{\Delta_g}{g_{21}}$		$A = \dfrac{D'}{\Delta_{T'}}$; $\quad B = \dfrac{B'}{\Delta_{T'}}$ $C = \dfrac{C'}{\Delta_{T'}}$; $\quad D = \dfrac{A'}{\Delta_{T'}}$
Conversion to T' parameters	$A' = \dfrac{-\Delta_g}{g_{12}}$; $\quad B' = \dfrac{-g_{22}}{g_{12}}$ $C' = \dfrac{-g_{11}}{g_{12}}$; $\quad D' = \dfrac{-1}{g_{12}}$	$A' = \dfrac{D}{\Delta_T}$; $\quad B' = \dfrac{B}{\Delta_T}$ $C' = \dfrac{C}{\Delta_T}$; $\quad D' = \dfrac{A}{\Delta_T}$	
Δ	$\Delta_g = g_{11}g_{22} - g_{12}g_{21}$	$\Delta_T = AD - BC$	$\Delta_{T'} = A'D' - B'C'$

Adapted from Van Valkenburg, M.E. 1974. *Network Analysis*, 3rd ed. Table 3.11–2, p. 337. Prentice-Hall, Englewood Cliffs, NJ. With permission.

The choice of parameters can be based also upon the ease of analysis. For example, Table 3.3 shows that T networks lend themselves to easy evaluation of the z parameters, while y parameters can be easily evaluated for Π networks. Table 3.3 summarizes other suggested uses and selections of network parameters for a few specific cases. When electric circuits are interconnected, a judicious choice of parameters can simplify the calculations to find the overall parameter description for the interconnected networks. For example, Table 3.3 shows that the z parameters for series-connected networks are simply the sum of the z parameters of the individual circuits [see Ruston et al. (1966) for derivations of the parameters for some of the interconnected networks]. The bold entries imply 2×2 matrices containing the four parameters. For example

$$\mathbf{h} = \begin{bmatrix} h_{11} & h_{12} \\ h_{21} & h_{22} \end{bmatrix} \tag{3.73}$$

TABLE 3.3   Two-Port Parameter Set Selection

	Common Circuit Applications	Interconnected Network Applications
Impedance parameters z	• T networks $z_{11}=Z_a+Z_c$; $z_{12}=z_{21}=Z_c$ $z_{22}=Z_b+Z_c$	• Series connected $\mathbf{z}=\mathbf{z_A}+\mathbf{z_B}$
Admittance parameters y	• Π networks $y_{11}=Y_a+Y_c$; $y_{12}=y_{21}=-Y_c$ $y_{22}=Y_b+Y_c$ • Field effect transistor equivalent circuit where typically: $1/y_{11}=\infty$, $y_{12}=0$, $y_{21}=gm$, $1/y_{22}=r_d$	• Parallel connected $\mathbf{y}=\mathbf{y_A}+\mathbf{y_B}$
Hybrid parameters h	• Transistor equivalent circuit where typically for common emitter: $h_{11}=h_{ie}$, $h_{12}=h_{re}$, $h_{21}=h_{fe}$, $h_{22}=h_{oe}$	• Series-parallel connected $\mathbf{h}=\mathbf{h_A}+\mathbf{h_B}$
Inverse hybrid parameters g		• Parallel-series connected $\mathbf{g}=\mathbf{g_A}+\mathbf{g_B}$

TABLE 3.3 (Continued)

	Common Circuit Applications	Interconnected Network Applications
Transmission parameters T	• Ideal transformer circuits	• Cascade connected

$$T = T_A T_B$$

		Interconnected Network Applications
Inverse transmission parameters T		• Cascade connected

$$T' = T'_B T'_A$$

Summary

In this section, we developed two-port parameter models for two-port electrical networks. The models define interrelationships among the input and output voltages and currents. A total of six models exists, depending upon which two variables are selected as independent variables. Any model can be used to find such network characteristics as input and output impedance and voltage and current gains. Once one model is found, other models can be obtained from transformation equations. The choice of parameter set is based upon physical reality and analytical convenience.

Defining Terms

Admittance parameters: That set of two-port parameters, such as y parameters, where all the parameters are defined to be the ratio of current to voltage. See Table 3.2 for the specific definition.

Dependent source: A voltage or current source whose value is related to another voltage or current in the network.

g Parameters: See hybrid parameters.

h Parameters: See hybrid parameters.

Hybrid (inverse hybrid) parameters: That set of two-port parameters, such as $h(g)$ parameters, where input current (voltage) and output voltage (current) are the independent variables. The parenthetical expressions refer to the inverse hybrid parameters. See Table 3.2 for specific definitions.

Impedance parameters: That set of two-port parameters, such as z parameters, where all the parameters are defined to be the ratio of voltage to current. See Table 3.2 for the specific definition.

Independent source: A voltage or current source whose value is not related to any other voltage or current in the network.

Norton's theorem: At a pair of terminals a linear electrical network can be replaced with a current source in parallel with an admittance. The current source is equal to the current that flows through the terminals when the terminals are short-circuited. The admittance is equal to the admittance at the terminals with all independent sources set equal to zero.

Open-circuit parameters: Two-port parameters, such as z parameters, evaluated by open-circuiting a port.

Port: Two terminals of a network where the current entering one terminal equals the current leaving the other terminal.

Short-circuit parameters: Two-port parameters, such as y parameters, evaluated by short-circuiting a port.

Superposition: In linear networks, a method of calculating the value of a dependent variable. First, the value of the dependent variable produced by each independent variable acting alone is calculated. Then, these values are summed to obtain the total value of the dependent variable.

Thévenin's theorem: At a pair of terminals a linear electrical network can be replaced with a voltage source in series with an impedance. The voltage source is equal to the voltage at the terminals when the terminals are open-circuited. The impedance is equal to the impedance at the terminals with all independent sources set equal to zero.

T parameters: See transmission parameters.

T' parameters: See transmission parameters.

Transmission (inverse transmission) parameters: That set of two-port parameters, such as the $T(T')$ parameters, where the dependent variables are the input (output) variables of the network and the independent variables are the output (input) variables. The parenthetical expressions refer to the inverse transmission parameters. See Table 3.2 for specific definitions.

Two-port networks: Networks that are modeled by specifying two ports, typically input and output ports.

Two-port parameters: A set of four constants, Laplace transforms, or sinusoidal steady-state functions used in the equations that describe a linear two-port network. Some examples are z, y, h, g, T, and T' parameters.

y Parameters: See admittance parameters.

z Parameters: See impedance parameters.

References

H. Ruston, and J. Bordogna, "Two-port networks," in *Electric Networks: Functions, Filters, Analysis,* New York: McGraw-Hill, 1966, chap. 4, pp. 244–266.

M.E. Van Valkenburg, "Two-port parameters," in *Network Analysis*, 3rd ed., Englewood Cliffs, NJ: Prentice-Hall, 1974, chap. 11, pp. 325–350.

Further Information

The following texts cover standard two-port parameters:

J.W. Nilsson, "Two-port circuits," in *Electric Circuits*, 4th ed., Reading, MA: Addison-Wesley, 1995, chap. 21, pp. 755–786.

H. Ruston, and J. Bordogna, "Two-port networks," in *Electric Networks: Functions, Filters, Analysis,* New York: McGraw-Hill, 1966, chap. 4, pp. 206–311.

The following texts have added coverage of scattering and image parameters:

H. Ruston and J. Bordogna, "Two-port networks," in *Electric networks: Functions, Filters, Analysis,* New York: McGraw-Hill, 1966, chap. 4, pp. 266–297.

S. Seshu and N. Balabanian, "Two-port networks" and "Image parameters and filter theory," in *Linear Network Analysis,* New York: Wiley, 1959, chaps. 8 and 11, pp. 291–342, 453–504.

The following texts show applications to electronic circuits:

F.H. Mitchell, Jr. and F. H. Mitchell, Sr., "Midrange AC amplifier design," in *Introduction to Electronics Design,* Englewood Cliffs, NJ: Prentice-Hall, 1992, chap. 7, pp. 335–384.

C.J. Savant, Jr., M.S. Roden, and G.L. Carpenter, "Bipolar transistors," "Design of bipolar junction transistor amplifiers," and "Field-effect transistor amplifiers," in *Electronic Design*, 2nd ed., Redwood City, CA: Benjamin/Cummings, 1991, chaps. 2, 3, and 4, pp. 69–212.

S.S. Sedra and K.C. Smith, "Frequency response" and "Feedback," in *Microelectronic Circuits*, 3rd ed., Philadelphia, PA: Saunders, 1991, chaps. 7 and 8, pp. 488–645.

4

Passive Signal Processing

William J. Kerwin
The University of Arizona

4.1   Introduction

This chapter will include detailed design information for passive RLC filters, including Butterworth, Thomson, and Chebyshev, both singly and doubly terminated. As the filter slope is increased in order to obtain greater rejection of frequencies beyond cut-off, the complexity and cost are increased and the response to a step input is worsened. In particular, the overshoot and the settling time are increased. The element values given are for normalized low-pass configurations to fifth order. All higher order doubly-terminated Butterworth filter element values can be obtained using Takahasi's equation, and an example is included. In order to use this information in a practical filter these element values must be scaled. Scaling rules to denormalize in frequency and impedance are given with examples. Since all data is for low-pass filters the transformation rules to change from low-pass to high-pass and to band-pass filters are included with examples.

Laplace Transform

We will use the Laplace operator, $s = \sigma + j\omega$. Steady-state impedance is thus Ls and $1/Cs$, respectively, for an inductor (L) and a capacitor (C), and admittance is $1/Ls$ and Cs. In steady state $\sigma = 0$ and therefore $s = j\omega$.

Transfer Functions

We will consider only lumped, linear, constant, and bilateral elements, and we will define the **transfer function** $T(s)$ as response over excitation:

$$T(s) = \frac{\text{signal output}}{\text{signal input}} = \frac{N(s)}{D(s)}$$

The roots of the numerator polynomial $N(s)$ are the zeros of the system, and the roots of the denominator $D(s)$ are the poles of the system (the points of infinite response). If we substitute $s = j\omega$ into $T(s)$ and separate

Adapted from *Instrumentation and Control: Fundamentals and Applications,* edited by Chester L. Nachtigal, pp. 487–497, copyright 1990, John Wiley & Sons, Inc. Reproduced by permission of John Wiley & Sons, Inc.

the result into real and imaginary parts (numerator and denominator) we obtain

$$T(j\omega) = \frac{A_1 + jB_1}{A_2 + jB_2} \tag{4.1}$$

Then the magnitude of the function, $|T(j\omega)|$, is

$$|T(j\omega)| = \left(\frac{A_1^2 + B_1^2}{A_2^2 + B_2^2}\right)^{\frac{1}{2}} \tag{4.2}$$

and the phase $\overline{T(j\omega)}$ is

$$\overline{T(j\omega)} = \tan^{-1}\frac{B_1}{A_1} - \tan^{-1}\frac{B_2}{A_2} \tag{4.3}$$

Analysis

Although mesh or nodal analysis can always be used, since we will consider only ladder networks we will use a method commonly called *linearity* or *working your way through*. The method starts at the output and assumes either 1 volt or 1 ampere as appropriate and uses Ohm's law and Kirchhoff's current law only.

Example 4.1.   Analysis of the circuit in Figure 4.1 for $V_o = 1$ V.

$$I_3 = \frac{3}{2}s; \quad V_1 = 1 + \left(\frac{3}{2}s\right)\left(\frac{4}{3}s\right) = 1 + 2s^2$$

$$I_2 = V_1\left(\frac{1}{2}s\right) = \frac{1}{2}s + s^3; \quad I_1 = I_2 + I_3$$

$$V_i = V_1 + I_1 = s^3 + 2s^2 + 2s + 1$$

$$T(s) = \frac{V_o}{V_i} = \frac{1}{s^3 + 2s^2 + 2s + 1}$$

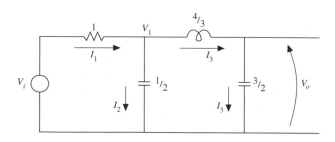

FIGURE 4.1   Singly terminated third-order low-pass filter (Ω, H, F).

Example 4.2. Determine the magnitude and phase of $T(s)$ in Example 4.1:

$$|T(s)| = \left| \frac{1}{s^3 + 2s^2 + 2s + 1} \right|_{s=j\omega}$$

$$|T(s)| = \sqrt{\frac{1}{(1 - 2\omega^2) + (2\omega - \omega^3)^2}} = \frac{1}{\sqrt{\omega^6 + 1}}$$

$$\overline{T(s)} = \tan^{-1} 0 - \tan^{-1} \frac{2\omega - \omega^3}{1 - 2\omega^2} = -\tan^{-1} \frac{2\omega - \omega^3}{1 - 2\omega^2}$$

The values used for the circuit of Figure 4.1 were normalized; that is, they are all near unity in ohms, henrys, and farads. These values simplify computation and, as we will see later, can easily be scaled to any desired set of actual element values. In addition, this circuit is low-pass because of the shunt capacitors and the series inductor. By low-pass we mean a circuit that passes the lower frequencies and attenuates higher frequencies. The cut-off frequency is the point at which the magnitude is 0.707 (-3 dB) of the dc level and is the dividing line between the **passband** and the **stopband**. In the above example we see that the magnitude of V_o/V_i at $\omega = 0$ (dc) is 1.00 and that at $\omega = 1$ rad/sec we have

$$|T(j\omega)| = \frac{1}{\sqrt{(\omega^6 + 1)}} \Bigg| = 0.707 \quad \omega = 1 \, \text{rad/sec} \tag{4.4}$$

and therefore this circuit has a cut-off frequency of 1 rad/sec.

Thus, we see that the normalized element values used here give us a cut-off frequency of 1 rad/sec.

4.2 Low-Pass Filter Functions[1]

The most common function in signal processing is the Butterworth. It is a function that has only poles (i.e., no finite zeros) and has the flattest magnitude possible in the passband. This function is also called **maximally flat magnitude (MFM)**. The derivation of this function is illustrated by taking a general all-pole function of third-order with a dc gain of 1 as follows:

$$T(s) = \frac{1}{as^3 + bs^2 + cs + 1} \tag{4.5}$$

The squared magnitude is

$$|T(j\omega)|^2 = \frac{1}{(1 - b\omega^2)^2 + (c\omega - a\omega^3)^2} \tag{4.6}$$

or

$$|T(j\omega)|^2 = \frac{1}{a^2\omega^6 + (b^2 - 2ac)\omega^4 + (c^2 - 2b)\omega^2 + 1} \tag{4.7}$$

MFM requires that the coefficients of the numerator and the denominator match term by term (or be in the same ratio) except for the highest power.

---

[1] Adapted from *Handbook of Measurement Science*, edited by Peter Sydenham, copyright 1982, John Wiley & Sons Limited. Reproduced by permission of John Wiley & Sons Limited.

Therefore

$$c^2 - 2b = 0; \quad b^2 - 2ac = 0 \tag{4.8}$$

We will also impose a normalized cut-off (-3 dB) at $\omega = 1$ rad/sec; that is

$$|T(j\omega)|_{\omega=1} = \frac{1}{\sqrt{(a^2+1)}} = 0.707 \tag{4.9}$$

Thus, we find $a = 1$, then $b = 2$, $c = 2$ are solutions to the flat magnitude conditions of Equation (4.8) and our third-order Butterworth function is

$$T(s) = \frac{1}{s^3 + 2s^2 + 2s + 1} \tag{4.10}$$

Table 4.1 gives the Butterworth denominator polynomials up to $n = 5$.

In general, for all Butterworth functions the normalized magnitude is

$$|T(j\omega)| = \frac{1}{\sqrt{(\omega^{2n}+1)}} \tag{4.11}$$

Note that this is down 3 dB at $\omega = 1$ rad/sec for all n.

This may, of course, be multiplied by any constant less than 1 for circuits whose dc gain is deliberately set to be less than 1.

Example 4.3. A low-pass Butterworth filter is required whose cut-off frequency (-3 dB) is 3 kHz and in which the response must be down 40 dB at 12 kHz. Normalizing to a cut-off frequency of 1 rad/sec, the -40-dB frequency is

$$\frac{12\,\text{kHz}}{3\,\text{kHz}} = 4\,\text{rad/s}$$

thus

$$-40 = 20 \quad \log\frac{1}{\sqrt{4^{2n}+1}}$$

TABLE 4.1   Butterworth Polynomials

$s+1$
$s^2 + 1.414s + 1$
$s^3 + 2s^2 + 2s + 1$
$s^4 + 2.6131s^3 + 3.4142s^2 + 2.6131s + 1$
$s^5 + 3.2361s^4 + 5.2361s^3 + 5.2361s^2 + 3.2361s + a$

Source: Handbook of Measurement Science, edited by Peter Sydenham, copyright 1982, John Wiley & Sons. Reproduced by permission of John Wiley & Sons.

therefore $n = 3.32$. Since n must be an integer, a fourth-order filter is required for this specification.

There is an extremely important difference between the singly terminated (dc gain = 1) and the doubly terminated filters (dc gain = 0.5). As was shown by John Orchard, the sensitivity in the passband (ideally at maximum output) to all L, C components in an L, C filter with *equal* terminations is *zero*. This is true regardless of the circuit.

This, of course, means component tolerances and temperature coefficients are of much less importance in the equally terminated case. For this type of Butterworth low-pass filter (normalized to equal 1-Ω terminations),

Takahasi has shown that the normalized element values are exactly given by

$$L, C = 2 \quad \sin\left(\frac{(2k-1)\pi}{2n}\right) \tag{4.12}$$

for any order n, where k is the L or C element from 1 to n.

Example 4.4. Design a normalized ($\omega_{-3\,\mathrm{dB}} = 1$ rad/sec) doubly terminated (i.e., source and load $= 1\ \Omega$) Butterworth low-pass filter of order 6; that is, $n = 6$.

The element values from Equation (4.12) are

$$L_1 = 2 \quad \sin\frac{(2-1)\pi}{12} = 0.5176 \quad \mathrm{H}$$

$$C_2 = 2 \quad \sin\frac{(4-1)\pi}{12} = 1.4141 \quad \mathrm{F}$$

$$L_3 = 2 \quad \sin\frac{(6-1)\pi}{12} = 1.9319 \quad \mathrm{H}$$

The values repeat for C_4, L_5, C_6 so that

$$C_4 = L_3, L_5 = C_2, C_6 = L_1$$

Thomson Functions

The Thomson function is one in which the time delay of the network is made maximally flat. This implies a linear phase characteristic since the steady-state time delay is the negative of the derivative of the phase. This function has excellent time-domain characteristics and is used wherever excellent step response is required. These functions have very little overshoot to a step input and have far superior settling times compared to the Butterworth functions. The slope near cut-off is more gradual than the Butterworth. Table 4.2 gives the Thomson denominator polynomials. The numerator is a constant equal to the dc gain of the circuit multiplied by the denominator constant. The cut-off frequencies are *not* all 1 rad/sec. They are given in Table 4.2.

TABLE 4.2 Thomson Polynomials

	$\omega_{-3\,\mathrm{dB}}$ (rad/sec)
$s + 1$	1.0000
$s^2 + 3s + 3$	1.3617
$s^3 + 6s^2 + 15s + 15$	1.7557
$s^4 + 10s^3 + 45s^2 + 105s + 105$	2.1139
$s^5 + 15s^4 + 105s^3 + 420s^2 + 945s + 945$	2.4274

Source: Handbook of Measurement Science, edited by Peter Sydenham, copyright 1982, John Wiley & Sons. Reproduced by permission of John Wiley & Sons.

Chebyshev Functions

A second function defined in terms of magnitude, the Chebyshev, has an **equal ripple** character within the passband. The ripple is determined by ϵ.

$$\epsilon = \sqrt{(10^{A/10} - 1)} \tag{4.13}$$

where $A = $ decibels of ripple; then for a given order n, we define v:

TABLE 4.3   Chebyshev Polynomials

$$s + \sinh v$$

$$s^2 + (\sqrt{2} \sinh v)s + \sinh^2 v + 1/2$$

$$(s + \sinh v)[s^2 + (\sinh v)s + \sinh^2 v + 3/4]$$

$$[s^2 + (0.75637 \sinh v)s + \sinh^2 v + 0.85355] \times [s^2 + (1.84776 \sinh v)s + \sinh^2 v + 0.14645]$$

$$(s + \sinh v)[s^2 + (0.61803 \sinh v)s + \sinh^2 v + 0.90451] \times [s^2 + (1.61803 \sinh v)s + \sinh^2 v + 0.34549]$$

Source: Handbook of Measurement Science, edited by Peter Sydenham, copyright 1982, John Wiley & Sons. Reproduced by permission of John Wiley & Sons.

$$v = \frac{1}{n} \sinh^{-1}\left(\frac{1}{\epsilon}\right) \tag{4.14}$$

Table 4.3 gives denominator polynomials for the Chebyshev functions. In all cases, the cut-off frequency (defined as the end of the ripple) is 1 rad/sec. The -3-dB frequency for the Chebyshev function is

$$\omega_{-3\text{dB}} = \cosh\left[\frac{\cosh^{-1}(1/\epsilon)}{n}\right] \tag{4.15}$$

The magnitude in the *stopband* ($\omega > 1$ rad/sec) for the normalized filter is

$$|T(j\omega)|^2 = \frac{1}{1 + \epsilon^2 \cosh^2(n \quad \cosh^{-1} \omega)} \tag{4.16}$$

for the singly terminated filter. For equal terminations the above magnitude is multiplied by 1/2 (1/4 in Equation (4.16)).

Example 4.5.   What order of singly terminated Chebyshev filter having 0.25-dB ripple (A) is required if the magnitude must be -60 dB at 15 kHz and the cut-off frequency (-0.25 dB) is to be 3 kHz? The normalized frequency for a magnitude of -60 dB is

$$\frac{15 \quad \text{kHz}}{3 \quad \text{kHz}} = 5 \text{ rad/sec}$$

Thus, for a ripple of $A = 0.25$ dB, we have from Equation (4.13)

$$\epsilon = \sqrt{(10^{A/10} - 1)} = 0.2434$$

and solving Equation (4.16) for n with $\omega = 5$ rad/sec and $|T(j\omega)| = -60$ dB, we obtain $n = 3.93$. Therefore, we must use $n = 4$ to meet these specifications.

4.3   Low-Pass Filters[1]

Introduction

Normalized element values are given here for both singly and doubly terminated filters. The source and load resistors are normalized to 1 Ω. Scaling rules will be given in Section 4.4 that will allow these values to be

---

[1]Adapted from *Handbook of Measurement Science*, edited by Peter Sydenham, copyright 1982, John Wiley & Sons. Reproduced by permission of John Wiley & Sons.

modified to any specified impedance value and to any cut-off frequency desired. In addition, we will cover the **transformation** of these **low-pass filters** to **high-pass** or **bandpass filters**.

Butterworth Filters

For $n = 2, 3, 4$, or 5, Figure 4.2 gives the element values for the singly terminated filters and Figure 4.3 gives the element values for the doubly terminated filters. All cut-off frequencies (-3 dB) are 1 rad/sec.

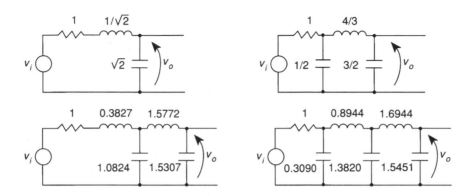

FIGURE 4.2 Singly terminated Butterworth filter element values (in Ω, H, F). *(Source: Handbook of Measurement Science, edited by Peter Sydenham, copyright 1982, John Wiley & Sons. Reproduced by permission of John Wiley & Sons.)*

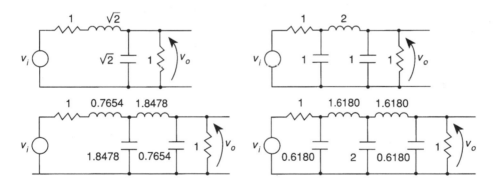

FIGURE 4.3 Doubly terminated Butterworth filter element values (in Ω, H, F). *(Source: Handbook of Measurement Science, edited by Peter Sydenham, copyright 1982, John Wiley & Sons. Reproduced by permission of John Wiley & Sons.)*

Thomson Filters

Singly and doubly terminated Thomson filters of order $n = 2, 3, 4$, and 5 are shown in Figures 4.4 and 4.5. All time delays are 1 sec. The cut-off frequencies are given in Table 4.2.

Chebyshev Filters

The amount of ripple can be specified as desired so that only a selective sample can be given here. We will use 0.1, 0.25, and 0.5 dB. All cut-off frequencies (end of ripple for the Chebyshev function) are at 1 rad/sec. Since the maximum power transfer condition precludes the existence of an equally terminated even-order

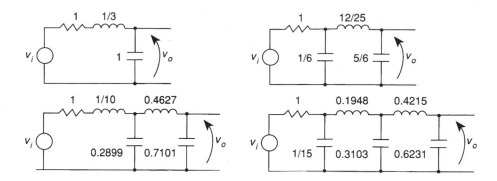

FIGURE 4.4   Singly terminated Thomson filter element values (in Ω, H, F). (*Source: Handbook of Measurement Science,* edited by Peter Sydenham, copyright 1982, John Wiley & Sons. Reproduced by permission of John Wiley & Sons.)

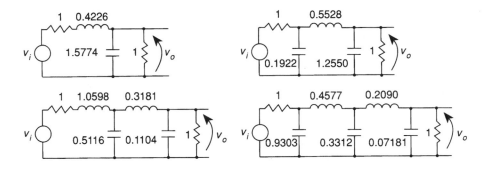

FIGURE 4.5   Doubly terminated Thomson filter element values (in Ω, H, F). (*Source: Handbook of Measurement Science,* edited by Peter Sydenham, copyright 1982, John Wiley & Sons. Reproduced by permission of John Wiley & Sons.)

filter, only odd orders are given for the doubly terminated case. Figure 4.6 gives the singly terminated Chebyshev filters for $n = 2$, 3, 4, and 5 and Figure 4.7 gives the doubly terminated Chebyshev filters for $n = 3$ and $n = 5$.

4.4   Filter Design

We now consider the steps necessary to convert normalized filters into actual filters by scaling both in frequency and in impedance. In addition, we will cover the transformation laws that convert low-pass filters to high-pass filters and low-pass to bandpass filters.

Scaling Laws and a Design Example

Since all data previously given are for normalized filters, it is necessary to use the scaling rules to design a low-pass filter for a specific signal processing application:

Rule 1.   All impedances may be multiplied by any constant without affecting the transfer-voltage ratio.
Rule 2.   To modify the cut-off frequency, divide all inductors and capacitors by the ratio of the desired frequency to the normalized frequency.

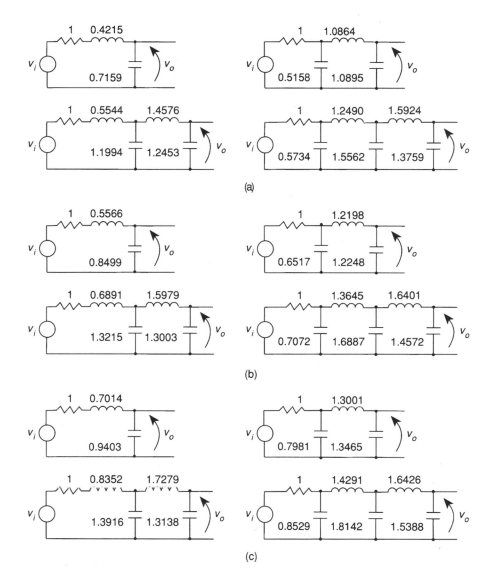

FIGURE 4.6 Singly terminated Chebyshev filter element values (in Ω, H, F): (a) 0.1-dB ripple; (b) 0.25-dB ripple; (c) 0.50-dB ripple. (*Source: Handbook of Measurement Science,* edited by Peter Sydenham, copyright 1982, John Wiley & Sons. Reproduced by permission of John Wiley & Sons.)

Example 4.6. Design a low-pass filter of MFM type (Butterworth) to operate from a 600-Ω source into a 600-Ω load, with a cut-off frequency of 500 Hz. The filter must be at least 36 dB below the dc level at 2 kHz, that is, -42 dB (dc level is -6 dB).

Since 2 kHz is four times 500 Hz, it corresponds to $\omega = 4$ rad/sec in the normalized filter. Thus at $\omega = 4$ rad/sec we have

$$-42 \text{ dB} = 20 \log \frac{1}{2}\left[\frac{1}{\sqrt{4^{2n}+1}}\right]$$

therefore, $n = 2.99$, so $n = 3$ must be chosen. The 1/2 is present because this is a doubly terminated (equal values) filter so that the dc gain is 1/2.

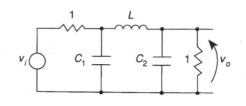

Ripple (dB)	C_1	L	C_2
0.10	1.0316	1.1474	1.0316
0.25	1.3034	1.1463	1.3034
0.50	1.5963	1.0967	1.5963

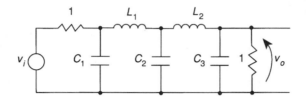

Ripple (dB)	C_1	L_1	C_2	L_2	C_2
0.10	1.1468	1.3712	1.9750	1.3712	1.1468
0.25	1.3824	1.3264	2.2091	1.3264	1.3824
0.50	1.7058	1.2296	2.5408	1.2296	1.7058

FIGURE 4.7   Doubly terminated Chebyshev filter element values (in Ω, H, F).

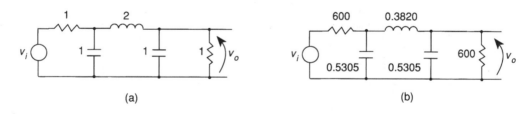

FIGURE 4.8   Third-order Butterworth low-pass filter: (a) normalized (in Ω, H, F); (b) scaled (in Ω, H, μF).

Thus a third-order, doubly terminated Butterworth filter is required. From Figure 4.3 we obtain the normalized network shown in Figure 4.8(a).

The **impedance scaling** factor is $600/1 = 600$ and the **frequency scaling** factor is $2\pi500/1 = 2\pi500$: that is, the ratio of the desired radian cut-off frequency to the normalized cut-off frequency (1 rad/sec). Note that the impedance scaling factor increases the size of the resistors and inductors, but reduces the size of the capacitors. The result is shown in Figure 4.8(b).

Transformation Rules, Passive Circuits

All information given so far applies only to low-pass filters, yet we frequently need high-pass or bandpass filters in signal processing.

Low-Pass to High-Pass Transformation

To transform a low-pass filter to high-pass, we first scale it to a cut-off frequency of 1 rad/sec if it is not already at 1 rad/sec. This allows a simple frequency rotation about 1 rad/sec of $s \rightarrow 1/s$. All L's become C's, all C's become L's, and all values reciprocate. The cut-off frequency does not change.

Example 4.7. Design a third-order, high-pass Butterworth filter to operate from a 600-Ω source to a 600-Ω load with a cut-off frequency of 500 Hz.

Starting with the normalized third-order low-pass filter of Figure 4.3 for which $\omega_{-3} = 1$ rad/sec, we reciprocate all elements and all values to obtain the filter shown in Figure 4.9(a) for which $\omega_{-3} = 1$ rad/sec.

Now we apply the scaling rules to raise all impedances to 600 Ω and the radian cut-off frequency to $2\pi500$ rad/sec as shown in Figure 4.9(b).

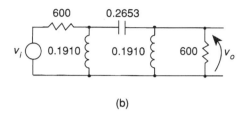

FIGURE 4.9 Third-order Butterworth high-pass filter: (a) normalized (in Ω, H, F); (b) scaled (in Ω, H, μF).

Low-Pass to Bandpass Transformation

To transform a low-pass filter to a bandpass filter we must first scale the low-pass filter so that the cut-off frequency is equal to the bandwidth of the normalized bandpass filter. The normalized center frequency of the bandpass filter is $\omega_0 = 1$ rad/sec. Then we apply the transformation $s \rightarrow s + 1/s$. For an inductor:

$$Z = Ls \text{ transforms to } Z = L\left(s + \frac{1}{s}\right)$$

For a capacitor:

$$Y = Cs \text{ transforms to } Y = C\left(s + \frac{1}{s}\right)$$

The first step is then to determine the Q of the bandpass filter where

$$Q = \frac{f_0}{B} = \frac{\omega_0}{B_r}$$

(f_0 is the center frequency in Hz and B is the 3-dB bandwidth in Hz). Now we scale the low-pass filter to a cut-off frequency of $1/Q$ rad/sec, then series tune every inductor, L, with a capacitor of value $1/L$ and parallel tune every capacitor, C, with an inductor of value $1/C$.

Example 4.8. Design a bandpass filter centered at 100 kHz having a 3-dB bandwidth of 10 kHz starting with a third-order Butterworth low-pass filter. The source and load resistors are each to be 600 Ω.

The Q required is

$$Q = \frac{100\,\text{kHz}}{10\,\text{kHz}} = 10, \text{ or } \frac{1}{Q} = 0.1$$

Scaling the normalized third-order low-pass filter of Figure 4.10(a) to $\omega_{-3\,\text{dB}} = 1/Q = 0.1$ rad/sec, we obtain the filter of Figure 4.10(b).

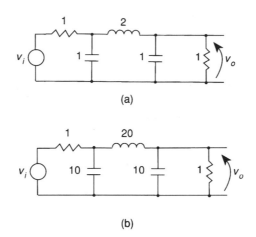

FIGURE 4.10   Third-order Butterworth low-pass filter: (a) normalized (in Ω, H, F); (b) scaled in (in Ω, H, F).

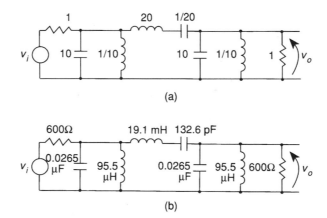

FIGURE 4.11   Sixth-order Butterworth bandpass filter ($Q = 10$): (a) normalized, $\omega_0 = 1$ rad/sec (in Ω, H, F); (b) scaled.

Now converting to bandpass with $\omega_0 = 1$ rad/sec, we obtain the normalized bandpass filter of Figure 4.11(a). Next, scaling to an impedance of 600 Ω and to a center frequency of $f_0 = 100$ kHz ($\omega_0 = 2\pi 100$ k rad/sec), we obtain the filter of Figure 4.11(b).

Defining Terms

Bandpass filter: A filter whose passband extends from a finite lower cut-off frequency to a finite upper cut-off frequency.

Equal ripple: A frequency response function whose magnitude has equal maxima and equal minima in the passband.

Frequency scaling: The process of modifying a filter to change from a normalized set of element values to other usually more practical values by dividing all L, C elements by a constant equal to the ratio of the scaled (cut-off) frequency desired to the normalized cut-off frequency.

High-pass filter: A filter whose band extends from some finite cut-off frequency to infinity.

Impedance scaling: Modifying a filter circuit to change from a normalized set of element values to other usually more practical element values by multiplying all impedances by a constant equal to the ratio of the desired (scaled) impedance to the normalized impedance.

Low-pass filter: A filter whose passband extends from dc to some finite cut-off frequency.

Maximally flat magnitude (MFM) filter: A filter having a magnitude that is as flat as possible versus frequency while maintaining a monotonic characteristic.

Passband: A frequency region of signal transmission usually within 3 dB of the maximum transmission.

Stopband: The frequency response region in which the signal is attenuated, usually by more than 3 dB from the maximum transmission.

Transfer function: The Laplace transform of the response (output voltage) divided by the Laplace transform of the excitation (input voltage).

Transformation: The modification of a low-pass filter to convert it to an equivalent high-pass or bandpass filter.

References

A. Budak, *Passive and Active Network Analysis and Synthesis,* Boston: Houghton Mifflin, 1974.

C. Nachtigal, Ed., *Instrumentation and Control: Fundamentals and Applications,* New York: John Wiley, 1990.

H.-J. Orchard, "Inductorless filters," *Electron. Lett.,* vol. 2, pp. 224–225, 1966.

P. Sydenham, Ed., *Handbook of Measurement Science,* Chichester, UK: John Wiley, 1982.

W.E. Thomson, "Maximally flat delay networks," *IRE Transactions,* vol. CT-6, p. 235, 1959.

L. Weinberg, *Network Analysis and Synthesis,* New York: McGraw-Hill, 1962.

L. Weinberg and P. Slepian, "Takahasi's results on Tchebycheff and Butterworth ladder networks," *IRE Transactions, Professional Group on Circuit Theory,* vol. CT-7, no. 2, pp. 88–101, 1960.

5

Nonlinear Circuits

Jerry L. Hudgins
University of Nebraska

Theodore F. Bogart, Jr.
University of Southern Missisppi

Taan El Ali
Benedict College

Mahamudunnabi Basunia
University of Dayton

Kartikeya Mayaram
Oregon State University

5.1 Diodes and Rectifiers

Jerry L. Hudgins

A diode generally refers to a two-terminal solid-state semiconductor device that presents a low impedance to current flow in one direction and a high impedance to current flow in the opposite direction. These properties allow the diode to be used as a one-way current valve in electronic circuits. *Rectifiers* are a class of circuits whose purpose is to convert ac waveforms (usually sinusoidal and with zero average value) into a waveform that has a significant nonzero average value (dc component). Simply stated, rectifiers are ac-to-dc energy converter circuits. Most rectifier circuits employ diodes as the principal elements in the energy conversion process; thus the almost inseparable notions of diodes and rectifiers. The general electrical characteristics of common diodes and some simple rectifier topologies incorporating diodes are discussed.

Diodes

Most diodes are made from a host crystal of silicon (Si) with appropriate impurity elements introduced to modify, in a controlled manner, the electrical characteristics of the device. These diodes are the typical **pn-junction** (or **bipolar**) devices used in electronic circuits. Another type is the **Schottky diode** (unipolar), produced by placing a metal layer directly onto the semiconductor (Mott, 1938; Schottky, 1938;). The metal–semiconductor interface serves the same function as the *pn*-junction in the common diode structure. Other semiconductor materials such as gallium-arsenide (GaAs) and silicon-carbide (SiC) are also in use for new and specialized applications of diodes. Detailed discussion of diode structures and the physics of their operation can be found in later paragraphs of this section.

The electrical circuit symbol for a bipolar diode is shown in Figure 5.1. The polarities associated with the forward voltage drop for forward current flow are also included. Current or voltage opposite to the polarities indicated in Figure 5.1 are considered to be negative values with respect to the diode conventions shown.

The characteristic curve shown in Figure 5.2 is representative of the current–voltage dependencies of typical diodes. The diode conducts forward current with a small forward voltage drop across the device, simulating a closed switch. The relationship between the forward current and forward voltage is approximately given by the Shockley diode equation (Shockley, 1949):

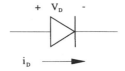

FIGURE 5.1 Circuit symbol for a bipolar diode indicating the polarity associated with the forward voltage and current directions.

$$i_\mathrm{D} = I_\mathrm{s}\left[\exp\!\left(\frac{qV_D}{nkT}\right) - 1\right] \qquad (5.1)$$

where I_s is the leakage current through the diode, q is the electronic charge, n is a correction factor, k is Boltzmann's constant, and T is the temperature of the semiconductor. Around the knee of the curve in Figure 5.2 is a positive voltage that is termed the turn-on or sometimes the threshold voltage for the diode. This value is an approximate voltage above which the diode is considered turned on and can be modeled to first degree as a closed switch with constant forward drop. Below the threshold voltage value the diode is considered weakly conducting and approximated as an open switch. The exponential relationship shown in Equation (5.1) means that the diode forward current can change by orders of magnitude before there is a large change in diode voltage, thus providing the simple circuit model during conduction. The nonlinear relationship of Equation (5.1) also provides a means of frequency mixing for applications in modulation circuits.

Reverse voltage applied to the diode causes a small leakage current (negative according to the sign convention) to flow that is typically orders of magnitude lower than current in the forward direction. The diode can withstand reverse voltages up to a limit determined by its physical construction and the semiconductor material used. Beyond this value the reverse voltage imparts enough energy to the charge carriers to cause large increases in current. The mechanisms by which this current increase occurs are impact ionization (avalanche) (McKay, 1954) and a tunneling phenomenon (Zener breakdown) (Moll, 1964). Avalanche break-down results in large power dissipation in the diode, is generally destructive, and should be avoided at all times. Both breakdown regions are superimposed in Figure 5.2 for comparison of their effects on the shape of the diode characteristic curve. Avalanche breakdown occurs for reverse applied voltages in the range of volts to kilovolts depending on the exact design of the diode. Zener breakdown occurs at much lower voltages than the avalanche mechanism. Diodes specifically designed to operate in the Zener breakdown mode are used extensively as voltage regulators in regulator integrated circuits and as discrete components in large regulated power supplies.

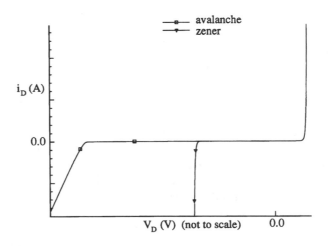

FIGURE 5.2   A typical diode dc characteristic curve showing the current dependence on voltage.

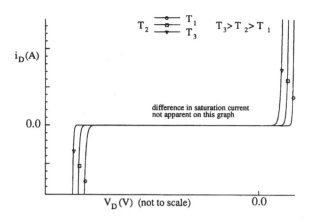

FIGURE 5.3 The effects of temperature variations on the forward voltage drop and the avalanche breakdown voltage in a bipolar diode.

During forward conduction the power loss in the diode can become excessive for large current flow. Schottky diodes have an inherently lower turn-on voltage than pn-junction diodes and are therefore more desirable in applications where the energy losses in the diodes are significant (such as output rectifiers in switching power supplies). Other considerations such as recovery characteristics from forward conduction to reverse blocking may also make one diode type more desirable than another. Schottky diodes conduct current with one type of charge carrier and are therefore inherently faster to turn off than bipolar diodes. However, one of the limitations of Schottky diodes is their excessive forward voltage drop when designed to support reverse biases above about 200 V. Therefore, high-voltage diodes are the pn-junction type.

The effects due to an increase in the temperature in a bipolar diode are many. The forward voltage drop during conduction will decrease over a large current range, the reverse leakage current will increase, and the reverse avalanche breakdown voltage (V_{BD}) will increase as the device temperature climbs. A family of static characteristic curves highlighting these effects is shown in Figure 5.3 where $T_3 > T_2 > T_1$. In addition, a major effect on the switching characteristic is the increase in the reverse recovery time during turn-off. Some of the key parameters to be aware of when choosing a diode are its repetitive peak inverse voltage rating, V_{RRM} (relates to the avalanche breakdown value), the peak forward surge current rating, I_{FSM} (relates to the maximum allowable transient heating in the device), the average or rms current rating, I_O (relates to the steady-state heating in the device), and the reverse recovery time, t_{rr} (relates to the switching speed of the device).

Rectifiers

This section discusses some simple **uncontrolled rectifier** circuits that are commonly encountered. The term *uncontrolled* refers to the absence of any control signal necessary to operate the primary switching elements (diodes) in the rectifier circuit. The discussion of controlled rectifier circuits, and the controlled switches themselves, is more appropriate in the context of power electronics applications (Hoft, 1986). Rectifiers are the fundamental building block in dc power supplies of all types and in dc power transmission used by some electric utilities.

A single-phase full-wave rectifier circuit with the accompanying input and output voltage waveforms is shown in Figure 5.4. This topology makes use of a center-tapped transformer with each diode conducting on opposite half-cycles of the input voltage. The forward drop across the diodes is ignored on the output graph, which is a valid approximation if the peak voltages of the input and output are large compared to 1 V. The circuit changes a sinusoidal waveform with no dc component (zero average value) to one with a dc component of $2V_{peak}/\pi$. The rms value of the output is $0.707V_{peak}$.

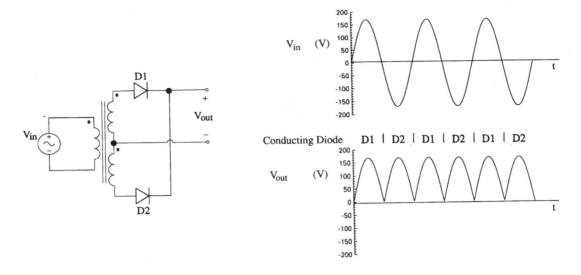

FIGURE 5.4   A single-phase full-wave rectifier circuit using a center-tapped transformer with the associated input and output waveforms.

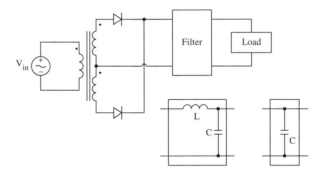

FIGURE 5.5   A single-phase full-wave rectifier with the addition of an output filter.

The dc value can be increased further by adding a low-pass filter in cascade with the output. The usual form of this filter is a shunt capacitor or an LC filter as shown in Figure 5.5. The resonant frequency of the LC filter should be lower than the fundamental frequency of the rectifier output for effective performance. The ac portion of the output signal is reduced while the dc and rms values are increased by adding the filter. The remaining ac portion of the output is called the **ripple**. Though somewhat confusing, the transformer, diodes, and filter are often collectively called the rectifier circuit.

Another circuit topology commonly encountered is the bridge rectifier. Figure 5.6 illustrates single- and three-phase versions of the circuit. In the single-phase circuit diodes D1 and D4 conduct on the positive half-cycle of the input while D2 and D3 conduct on the negative half-cycle of the input. Alternate pairs of diodes conduct in the three-phase circuit depending on the relative amplitude of the source signals.

The three-phase inputs with the associated rectifier output voltage are shown in Figure 5.7 as they would appear without the low-pass filter section. The three-phase bridge rectifier has a reduced ripple content of 4% as compared to a ripple content of 47% in the single-phase bridge rectifier (Milnes, 1980). The corresponding diodes that conduct are also shown at the top of the figure. This output waveform assumes a purely resistive load connected as shown in Figure 5.6. Most loads (motors, transformers, etc.) and many sources (power grid)

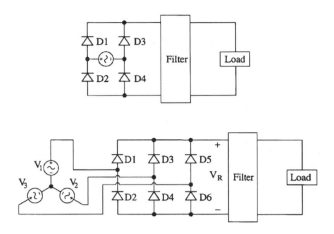

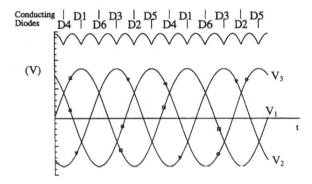

FIGURE 5.6   Single- and three-phase bridge rectifier circuits.

FIGURE 5.7   Three-phase rectifier output compared to the input signals. The input signals as well as the conducting diode labels are those referenced to Figure 5.6.

include some inductance, and in fact may be dominated by inductive properties. This causes phase shifts between the input and output waveforms. The rectifier output may thus vary in shape and phase considerably from that shown in Figure 5.7 (Kassakian et al., 1991). When other types of switches are used in these circuits the inductive elements can induce large voltages that may damage sensitive or expensive components. Diodes are used regularly in such circuits to shunt current- and clamp-induced voltages at low levels to protect expensive components such as electronic switches.

One variation of the typical rectifier is the Cockroft-Walton circuit used to obtain high voltages without the necessity of providing a high-voltage transformer. The circuit in Figure 5.8 multiplies the peak secondary voltage by a factor of six. The steady-state voltage level at each filter capacitor node is shown in the figure. Adding additional stages increases the load voltage further. As in other rectifier circuits, the value of the capacitors will determine the amount of ripple in the output waveform for given load-resistance values. In general, the capacitors in a lower voltage stage should be larger than in the next highest voltage stage.

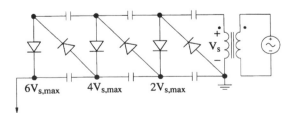

FIGURE 5.8   Cockroft-Walton circuit used for voltage multiplication.

Defining Terms

Bipolar device: Semiconductor electronic device that uses positive and negative charge carriers to conduct electric current.

Diode: Two-terminal solid-state semiconductor device that presents a low impedance to current flow in one direction and a high impedance to current flow in the opposite direction.

***pn*-junction:**   Metallurgical interface of two regions in a semiconductor where one region contains impurity elements that create equivalent positive charge carriers (p-type) and the other semiconductor region contains impurities that create negative charge carriers (n-type).

Ripple: The ac (time-varying) portion of the output signal from a rectifier circuit.

Schottky diode: A diode formed by placing a metal layer directly onto a unipolar semiconductor substrate.

Uncontrolled rectifier: A rectifier circuit employing switches that do not require control signals to operate them in their on or off states.

References

R.G. Hoft, *Semiconductor Power Electronics,* New York: Van Nostrand Reinhold, 1986.

J.G. Kassakian, M.F. Schlecht, and G.C. Verghese, *Principles of Power Electronics,* Reading, MA: Addison-Wesley, 1991.

K.G. McKay, "Avalanche breakdown in silicon," *Physical Review,* vol. 94, p. 877, 1954.

A.G. Milnes, *Semiconductor Devices and Integrated Electronics,* New York: Van Nostrand Reinhold, 1980.

J.L. Moll, *Physics of Semiconductors,* New York: McGraw-Hill, 1964.

N.F. Mott, "Note on the contact between a metal and an insulator or semiconductor," *Proc. Cambridge Philos. Soc.,* vol. 34, p. 568, 1938.

W. Schottky, "Halbleitertheorie der Sperrschicht," *Naturwissenschaften,* vol. 26, p. 843, 1938.

W. Shockley, "The theory of p-n junctions in semiconductors and p-n junction transistors," *Bell System Tech. J.,* vol. 28, p. 435, 1949.

Further Information

A good introduction to solid-state electronic devices with a minimum of mathematics and physics is *Solid State Electronic Devices,* 3rd edition, by B.G. Streetman, Prentice-Hall, 1989. A rigorous and more detailed discussion is provided in *Physics of Semiconductor Devices,* 2nd edition, by S.M. Sze, John Wiley & Sons, 1981. Both of these books discuss many specialized diode structures as well as other semiconductor devices. Advanced material on the most recent developments in semiconductor devices, including diodes, can be found in technical journals such as the *IEEE Transactions on Electron Devices, Solid State Electronics,* and *Journal of Applied Physics.* A good summary of advanced rectifier topologies and characteristics is given in *Basic Principles of Power Electronics* by K. Heumann, Springer-Verlag, 1986. Advanced material on rectifier designs as well as other power electronics circuits can be found in *IEEE Transactions on Power Electronics, IEEE Transactions on Industry Applications,* and the *EPE Journal.* Two good industry magazines that cover power devices such as diodes and power converter circuitry are *Power Control and Intelligent Motion* (PCIM) and *Power Technics.*

5.2   Limiter (Clipper)

Theodore F. Bogart, Jr., Taan El Ali, and Mahamudunnabi Basunia

A limiter is a device that can keep the voltage excursions at its output to a prescribed level. This is also called clipping because of the circuit ability to clip both alternations of the input signal. The simplest type consists simply of diodes (including Zener diodes) and resistors. To improve performance and add precision, operational amplifiers are usually used. Limiters are used in a wide variety of electronic systems. They are generally used to perform one of two functions: (1) altering the shape of a waveform or (2) circuit transient protection.

Limiter Operator and Circuits

Figure 5.9 shows the general transfer characteristic of the limiter operator. As indicated in the figure, the limiter acts as an amplifier of gain k, which can be either positive or negative, for inputs in a certain range; $a_2/k <= v_{in} <= a_1/k$. If input v_{in} exceeds the upper threshold (a_1/k), the output voltage limits or clamps to the upper limiting level, a_1, and holds this voltage.

However, if v_{in} is reduced below the lower limiting threshold (a_2/k), the output voltage v_o will be limited to lower limiting level a_2 and it will hold this voltage.

The general transfer characteristic of Figure 5.9 describes a *double limiter*, which works for both positive and negative peaks of input waveform with different limiting levels (a_1 and a_2). Here, a_1 and a_2 might have equal or different absolute values.

Figure 5.9 shows the characteristic of a hard limiter (where linear region to saturation region changes rapidly). For a soft limiter, there is a smooth transition between linear region and saturation region of its characteristic curve, as shown in Figure 5.10. The slope in the saturation region should be greater than zero.

If a clipping circuit follows the transfer characteristic of Figure 5.9, then for the sine wave input shown in Figure 5.11, the output will be as shown in Figure 5.12.

In Figure 5.13, all circuits output are assumed to have unity gain (e.g., $k = 1$); as a result, the slope in the linear region is 1. Clipping circuits rely on the fact that diodes have very low impedances when they are forward biased and are essentially open circuits when reverse-biased.

Figures 5.13(a) and (b) are positive limiting circuits. They are identical except for the additional biased dc voltage source in Figure 5.13(b). Feeding a sinusoid to Figure 5.13(a) results in a 0.7-V output for positive cycle and −10 V for negative cycle (same as negative input cycle). In Figure 5.13(b), for 4 V fixed voltage in the

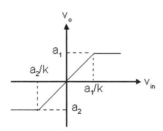

FIGURE 5.9 General transfer characteristic of the limiter.

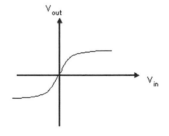

FIGURE 5.10 Soft limiter characteristic curve.

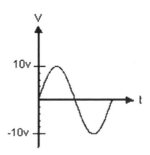

FIGURE 5.11 A sine wave with amplitude 10 V (this is the input v_{in} of all circuits in this chapter).

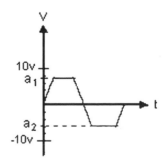

FIGURE 5.12 Applying a sine wave to a limiter can result in clipping off its two peaks.

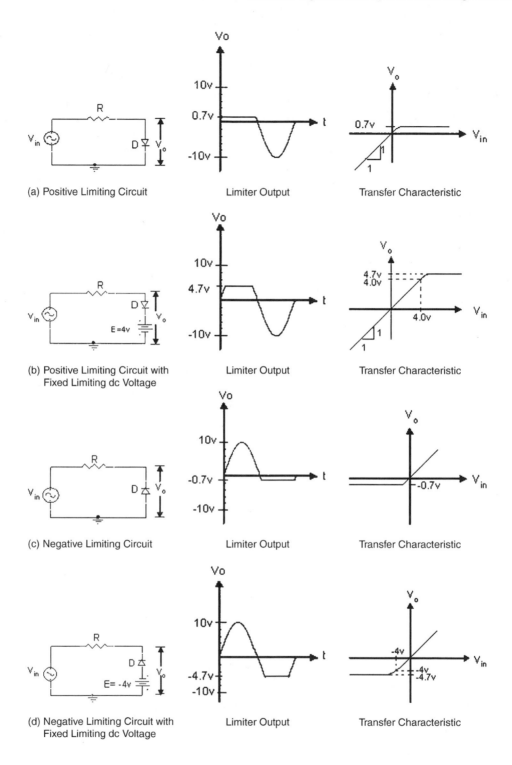

FIGURE 5.13 Different limiter circuits.

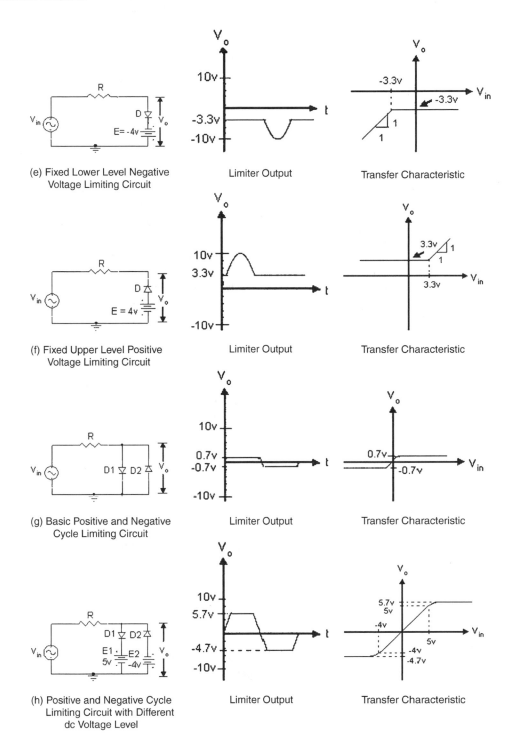

(e) Fixed Lower Level Negative Voltage Limiting Circuit

Limiter Output

Transfer Characteristic

(f) Fixed Upper Level Positive Voltage Limiting Circuit

Limiter Output

Transfer Characteristic

(g) Basic Positive and Negative Cycle Limiting Circuit

Limiter Output

Transfer Characteristic

(h) Positive and Negative Cycle Limiting Circuit with Different dc Voltage Level

Limiter Output

Transfer Characteristic

FIGURE 5.13 (Continued).

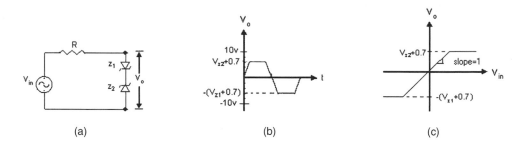

FIGURE 5.14 (a) Zener shunt clipping circuit, (b) limiter output, and (c) transfer characteristic.

circuit, output is $4 + 0.7 = 4.7$ V peak for positive cycle and -10 V for negative cycle (in this case the diode works as an open circuit, output will follow the input cycle).

Figure 5.13(c) and (d) are negative limiting circuits. They are identical except for the additional fixed negative voltage in Figure 5.13(d). Feeding a sinusoid to Figure 5.13(c) results in a 10-V peak voltage for positive cycle and -0.7 V for negative cycle. For positive cycle, the diode acts as an open circuit, so the output will follow the input cycle. For negative cycle, the diode is forward biased and works as a short circuit. So the output is the forward biasing voltage of -0.7 V.

Note that the type of clipping we showed in Figure 5.13(e and f) occurs when the fixed bias voltage tends to forward bias the diode and the clipping will occur only when the fixed bias voltage tends to reverse bias the diode.

Figure 5.13(g and h) are double-ended limiting circuits using diodes, where two opposite-polarity diodes are put in parallel. Figure 5.13(g) results in a crude approximation of a square wave, with about 1.4 V peak-to-peak amplitude. Figure 5.13(h) with two dc-biased voltages in series with diodes made a different level clipping circuit.

In Figure 5.13 the output equals the dc source voltage (if we consider diodes are ideal, so forward biasing voltage $V_F = 0$) when the input reaches the value necessary to forward bias the diode. When the diode is reverse biased by the input signal, it is like an open circuit that disconnects the dc source, and the output follows the input. These circuits are called parallel clippers because the biased diode is parallel to the output.

Figure 5.13 illustrates a different kind of limiting action where the output follows the input when the signal is above or below a certain level.

Figure 5.14, shows a zener shunt clipping circuit, which is used to clip both alternations of the input signal. The zener shunt clipper uses both the forward and reverse operation characteristics of the zener diode. Feeding a sinusoid to this circuit results in a crude approximation of a square wave, with the approximately 1.4-V peak-to-peak amplitude.

When input is positive, Z_1 is forward biased (0.7 V) and Z_2 is reverse biased (V_{Z_2})(assuming that the value of V_{in} is sufficiently high to turn both diodes on).

In this circuit, limiting occurs in the positive direction at a voltage of $V_{Z_2} + 0.7$, where 0.7 V represents the voltage drop across zener diode Z_1 when conducting in the forward direction. For negative inputs, the opposite condition will exist. Z_1 acts as a zener, while Z_2 conducts in the forward direction. So, the output voltage will be $-(V_{Z_2} + 0.7)$. It should be mentioned that pairs of zener diodes connected in series are available commercially for applications of this type under the name double-anode zener.

In most symmetrical zener shunt clippers, the V_Z ratings of the two diodes are equal. V_{Z_1} and V_{Z_2} have different values in rare practical situations. The symmetrical zener shunt clipper is used primarily for circuit protection.

Operational Amplifier Limiting Circuits

Figure 5.15(a) shows a biased diode connected in the feedback path of an operational amplifier. It looks like a clipping circuit. Since inverting terminal (2) is at virtual ground, the output voltage v_o is the same voltage across R_f.

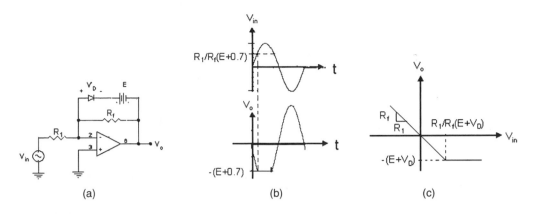

(a)   (b)   (c)

FIGURE 5.15 (a) An operational amplifier limiting circuit, (b) output clamps at $E + V_D$ volts when input reaches $R_1/R_f (E + 0.7)$, and (c) transfer characteristic.

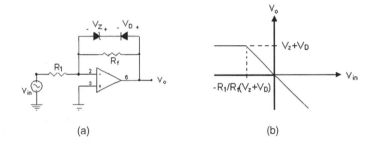

(a)   (b)

FIGURE 5.16 (a) Positive limiting circuit and (b) transfer characteristic.

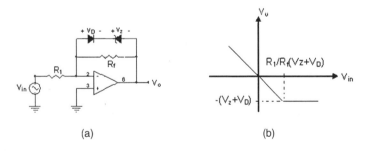

(a)   (b)

FIGURE 5.17 (a) Negative limiter circuit and (b) transfer characteristic.

In Figure 5.15(b), where input is the sinusoidal voltage, the output is the bias voltage $-(E + 0.7)$, the output is held at $-(E + 0.7)$ V. Notice that output clipping occurs at input voltage $(R_1/R_f)(E + 0.7)$ because the amplifier inverts and has closed-loop gain magnitude R_f/R_1. A characteristic curve for this is shown in Figure 5.15(c). This circuit is a limiting circuit because it limits the output to the dc level clamped by the diode.

In practice, the fixed voltage source is replaced by a zener diode. So, in Figure 5.16 and Figure 5.17, the zener diode is in series with a conventional diode. The zener diode works as a conventional diode when it is forward biased. In reverse bias, the zener is in a breakdown region, which exhibits a voltage drop (V_Z) that is almost constant and independent of the current through the diode. Every zener diode has a specified value for its breakdown voltage, also called the zener voltage (V_Z).

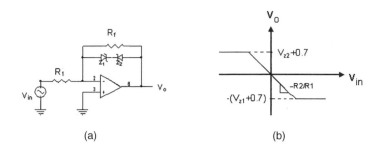

FIGURE 5.18   (a) Double-ended limiting circuit and (b) transfer characteristic.

These are useful and practical circuits that function as a comparator if the feedback resistance R_f is not included and as a limiter when R_f is included. The output voltage v_0 equals forward diode voltage plus zener voltage V_Z.

Figure 5.18 shows double-ended limiting circuits, where zeners are in a back-to-back position. Here, both positive and negative peaks of the output waveform are clipped. In both positive and negative peaks, one zener diode is forward biased and another one is reverse biased.

References

A.S. Sedra, K.C. Smith, *Microelectronic Circuits*, New York: Saunders College Publishing, 1987.
T.F. Bogart, Jr., *Electronic Devices and Circuits*, 6th ed., Columbus, OH: Macmillan/Merrill, 2004.
R.T. Paynter, *Introductory Electronic Devices and Circuits*, Englewood Cliffs, NJ: Prentice-Hall, 1991.
S. Franco, *Electric Circuits Fundamentals*, Orlando, FL: Saunders College Publishing, 1995.

5.3   Distortion

Kartikeya Mayaram

The diode was introduced in the previous sections as a nonlinear device that is used in rectifiers and limiters. These are applications that depend on the nonlinear nature of the diode. Typical electronic systems are composed not only of diodes, but also of other nonlinear devices such as transistors. In analog applications, transistors are used to amplify weak signals (amplifiers) and to drive large loads (output stages). For such situations it is desirable that the output be an amplified true reproduction of the input signal; therefore, the transistors must operate as linear devices. However, the inherent nonlinearity of transistors results in an output that is a "distorted" version of the input.

The distortion due to a nonlinear device is illustrated in Figure 5.19. For an input X, the output is $Y = F(X)$, where F denotes the nonlinear transfer characteristics of the device; the dc operating point is given by X_0. Sinusoidal input signals of two different amplitudes are applied and the output responses corresponding to these inputs are also shown.

For an input signal of small amplitude, the output faithfully follows the input; whereas for large-amplitude signals, the output is distorted; a flattening occurs at the negative peak value. The distortion in amplitude results in the output having frequency components that are integer multiples of the input frequency, *harmonics*, and this type of distortion is referred to as **harmonic distortion**.

The distortion level places a restriction on the amplitude of the input signal that can be applied to an electronic system. Therefore, it is essential to characterize the distortion in a circuit. In this section different types of distortion are defined and techniques for distortion calculation are presented. These techniques are applicable to simple circuit configurations. For larger circuits, a circuit simulation program is invaluable.

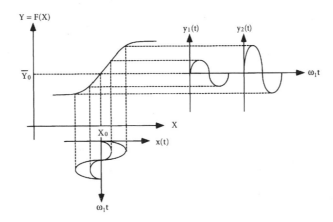

FIGURE 5.19 The dc transfer characteristics of a nonlinear circuit and the input and output waveforms. For a large input amplitude the output is distorted.

Harmonic Distortion

When a sinusoidal signal of a single frequency is applied at the input of a nonlinear device or circuit, the resulting output contains frequency components that are integer multiples of the input signal. These harmonics are generated by the nonlinearity of the circuit and the *harmonic distortion* is measured by comparing the magnitudes of the harmonics with the fundamental component (input frequency) of the output.

Consider the input signal to be of the form

$$x(t) = X_1 \cos \omega_1 t \tag{5.2}$$

where $f_1 = \omega_1/2\pi$ is the frequency and X_1 is the amplitude of the input signal. Let the output of the nonlinear circuit be

$$y(t) = Y_0 + Y_1 \cos \omega_1 t + Y_2 \cos 2\omega_1 t + Y_3 \cos 3 \omega_1 t + \cdots \tag{5.3}$$

where Y_0 is the dc component of the output, Y_1 is the amplitude of the fundamental component, and Y_2, Y_3 are the amplitudes of the second and third harmonic components, respectively. The *second harmonic distortion factor* (HD$_2$), the *third harmonic distortion factor* (HD$_3$), and the *nth harmonic distortion factor* (HD$_n$) are defined as

$$HD_2 = \frac{|Y_2|}{|Y_1|} \tag{5.4}$$

$$HD_3 = \frac{|Y_3|}{|Y_1|} \tag{5.5}$$

$$HD_n = \frac{|Y_n|}{|Y_1|} \tag{5.6}$$

The **total harmonic distortion** (THD) of a waveform is defined to be the ratio of the rms (root-mean-square) value of the harmonics to the amplitude of the fundamental component.

$$\mathrm{THD} = \frac{\sqrt{Y_2^2 + Y_3^2 + \cdots + Y_n^2}}{|Y_1|} \tag{5.7}$$

THD can be expressed in terms of the individual **harmonic distortion factors**:

$$\mathrm{THD} = \sqrt{\mathrm{HD}_2^2 + \mathrm{HD}_3^2 + \cdots + \mathrm{HD}_n^2} \tag{5.8}$$

Various methods for computing the harmonic distortion factors are described next.

Power-Series Method

In this method a truncated power-series expansion of the dc transfer characteristics of a nonlinear circuit is used. Therefore, the method is suitable only when energy storage effects in the nonlinear circuit are negligible and the input signal is small. In general, the input and output signals comprise both dc and time-varying components. For distortion calculation we are interested in the time-varying or incremental components around a quiescent[1] operating point. For the transfer characteristic of Figure 5.19, denote the quiescent operating conditions by X_0 and $\bar{Y}_0$ and the incremental variables by $x(t)$ and $y(t)$ at the input and output, respectively. The output can be expressed as a function of the input using a series expansion:

$$\bar{Y}_0 + y = F(X_0 + x) = a_0 + a_1 x + a_2 x^2 + a_3 x^3 + \cdots \tag{5.9}$$

where $a_0 = \bar{Y}_0 = F(X_0)$ is the output at the dc operating point. The incremental output is

$$y = a_1 x + a_2 x^2 + a_3 x^3 + \cdots \tag{5.10}$$

Depending on the amplitude of the input signal, the series can be truncated at an appropriate term. Typically only the first few terms are used, which makes this technique applicable only to small input signals. For a pure sinusoidal input (Equation (5.2)), the distortion in the output can be estimated by substituting for x in Equation (5.10) and by use of trigonometric identities one can arrive at the form given by Equation (5.3). For a series expansion that is truncated after the cubic term

$$
\begin{aligned}
Y_0 &= \frac{a_2 X_1^2}{2} \\[4pt]
Y_1 &= a_1 X_1 + \frac{3 a_3 X_1^3}{4} \cong a_1 X_1 \\[4pt]
Y_2 &= \frac{a_2 X_1^2}{2} \\[4pt]
Y_3 &= \frac{a_3 X_1^3}{4}
\end{aligned}
\tag{5.11}
$$

Notice that a dc term Y_0 is present in the output (produced by the even-powered terms) that results in a shift of the operating point of the circuit due to distortion. In addition, depending on the sign of a_3 there can be an *expansion* or *compression* of the fundamental component. The harmonic distortion factors (assuming $Y_1 = a_1 X_1$) are

$$\mathrm{HD}_2 = \frac{|Y_2|}{|Y_1|} = \frac{1}{2}\left|\frac{a_2}{a_1} X_1\right|$$

$$\mathrm{HD}_3 = \frac{|Y_3|}{|Y_1|} = \frac{1}{4}\left|\frac{a_3}{a_1} X_1^2\right| \tag{5.12}$$

As an example, choose as the transfer function $Y = F(X) = \exp(X)$; then $a_1 = 1$, $a_2 = 1/2$, $a_3 = 1/6$. For an input signal amplitude of 0.1, $\mathrm{HD}_2 = 2.5\%$ and $\mathrm{HD}_3 = 0.04\%$.

---

[1]Defined as the operating condition when the input has no time-varying component.

Differential-Error Method

This technique is also applicable to nonlinear circuits in which energy storage effects can be neglected. The method is valuable for circuits that have small distortion levels and relies on one's ability to calculate the small-signal gain of the nonlinear function at the quiescent operating point and at the maximum and minimum excursions of the input signal. Again the power-series expansion provides the basis for developing this technique. The small-signal gain[1] at the quiescent state ($x = 0$) is a_1. At the extreme values of the input signal X_1 (positive peak) and $-X_1$ (negative peak) let the small-signal gains be a^+ and a^-, respectively. By defining two new parameters, the differential errors, E^+ and E^-, as

$$E^+ = \frac{a^+ - a_1}{a_1} \quad E^- = \frac{a^- - a_1}{a_1} \tag{5.13}$$

the distortion factors are given by

$$HD_2 = \frac{E^+ - E^-}{8}$$

$$HD_3 = \frac{E^+ + E^-}{24} \tag{5.14}$$

The advantage of this method is that the transfer characteristics of a nonlinear circuit can be directly used; an explicit power-series expansion is not required. Both the power-series and the differential-error techniques cannot be applied when only the output waveform is known. In such a situation the distortion factors are calculated from the output signal waveform by a simplified Fourier analysis as described in the next section.

Three-Point Method

The three-point method is a simplified analysis applicable to small levels of distortion and can only be used to calculate HD_2. The output is written directly as a Fourier cosine series as in Equation (5.3) where only terms up to the second harmonic are retained. The dc component includes the quiescent state and the contribution due to distortion that results in a shift of the dc operating point. The output waveform values at $\omega_1 t = 0$ (F_0), $\omega_1 t = \pi/2$ ($F_{\pi/2}$), $\omega_1 t = \pi$ (F_π), as shown in Figure 5.20, are used to calculate Y_0, Y_1, and Y_2:

$$Y_0 = \frac{F_0 + 2F_{\pi/2} + F_\pi}{4}$$

$$Y_1 = \frac{F_0 - F_\pi}{2} \tag{5.15}$$

$$Y_2 = \frac{F_0 - 2F_{\pi/2} + F_\pi}{4}$$

The second harmonic distortion is calculated from the definition. From Figure 5.20, $F_0 = 5$, $F_{\pi/2} = 3.2$, $F_\pi = 1$, $Y_0 = 3.1$, $Y_1 = 2.0$, $Y_2 = -0.1$, and $HD_2 = 5.0\%$.

Five-Point Method

The five-point method is an extension of the above technique and allows calculation of third and fourth harmonic distortion factors. For distortion calculation the output is expressed as a Fourier cosine series with terms up to the fourth harmonic where the dc component includes the quiescent state and the shift due to distortion. The output waveform values at $\omega_1 t = 0$ (F_0), $\omega_1 t = \pi/3$ ($F_{\pi/3}$), $\omega_1 t = \pi/2$ ($F_{\pi/2}$), $\omega_1 t = 2\pi/3$ ($F_{2\pi/3}$), $\omega_1 t = \pi$ ($F\pi$), as shown in Figure 5.20, are used to calculate Y_0, Y_1, Y_2, Y_3, and Y_4:

---

[1]Small-signal gain $= dy/dx = a_1 + 2a_2 x + 3a_3 x^2 + \cdots$.

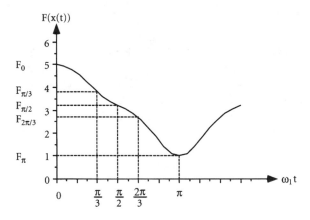

FIGURE 5.20 Output waveform from a nonlinear circuit.

$$Y_0 = \frac{F_0 + 2F_{\pi/3} + 2F_{2\pi/3} + F_\pi}{6}$$

$$Y_1 = \frac{F_0 + F_{\pi/3} - F_{2\pi/3} - F_\pi}{3}$$

$$Y_2 = \frac{F_0 - 2F_{\pi/2} + F_\pi}{4} \qquad\qquad (5.16)$$

$$Y_3 = \frac{F_0 - 2F_{\pi/3} + 2F_{2\pi/3} - F_\pi}{6}$$

$$Y_4 = \frac{F_0 - 4F_{\pi/3} + 6F_{\pi/2} - 4F_{2\pi/3} + F_\pi}{12}$$

For $F_0 = 5$, $F_{\pi/3} = 3.8$, $F_{\pi/2} = 3.2$, $F_{2\pi/3} = 2.7$, $F_\pi = 1$, $Y_0 = 3.17$, $Y_1 = 1.7$, $Y_2 = -0.1$, $Y_3 = 0.3$, $Y_4 = -0.07$, and $HD_2 = 5.9\%$, $HD_3 = 17.6\%$. This particular method allows calculation of HD_3 and also gives a better estimate of HD_2. To obtain higher-order harmonics a detailed Fourier series analysis is required and for such applications a circuit simulator such as SPICE should be used.

Intermodulation Distortion

The previous sections have examined the effect of nonlinear device characteristics when a single-frequency sinusoidal signal is applied at the input. However, if there are two or more sinusoidal inputs, then the nonlinearity results in not only the fundamental and harmonics but also additional frequencies called the *beat frequencies* at the output. The distortion due to the components at the beat frequencies is called **intermodulation distortion.** To characterize this type of distortion consider the incremental output given by Equation (5.10) and the input signal to be

$$x(t) = X_1 \cos \omega_1 t + X_2 \cos \omega_2 t \qquad\qquad (5.17)$$

where $f_1 = \omega_1/2\pi$ and $f_2 = \omega_2/2\pi$ are the two input frequencies. The output frequency spectrum due to the quadratic term is shown in Table 5.1.

TABLE 5.1 Output Frequency Spectrum due to the Quadratic Term

Frequency	0	$2f_1$	$2f_2$	$f_1 \pm f_2$
Amplitude	$\frac{a_2}{2}[X_1^2 + X_2^2]$	$\frac{a_2}{2}X_1^2$	$\frac{a_2}{2}X_2^2$	$a_2X_1X_2$

TABLE 5.2 Output Frequency Spectrum due to the Cubic Term

Frequency	f_1	f_2	$2f_1 \pm f_2$	$2f_2 \pm f_1$	$3f_1$	$3f_2$
Amplitude	$\frac{3a_3}{4}[X_1^3 + X_1X_2^2]$	$\frac{3a_3}{4}[X_2^3 + X_1^2X_2]$	$\frac{3}{4}a_3X_1^2X_2$	$\frac{3}{4}a_3X_1X_2^2$	$\frac{1}{4}a_3X_1^3$	$\frac{1}{4}a_3X_2^3$

In addition to the dc term and the second harmonics of the two frequencies, there are additional terms at the sum and difference frequencies, $f_1 + f_2, f_1 - f_2$, which are the beat frequencies. The *second-order intermodulation distortion* (IM$_2$) is defined as the ratio of the amplitude at a beat frequency to the amplitude of the fundamental component,

$$\text{IM}_2 = \left| \frac{a_2X_1X_2}{a_1X_1} \right| = \left| \frac{a_2X_2}{a_1} \right| \tag{5.18}$$

where it has been assumed that the contribution to second-order intermodulation by higher-order terms is negligible. In defining IM$_2$ the input signals are assumed to be of equal amplitude and for this particular condition, IM$_2$ = 2 HD$_2$ (Equation (5.12)).

The cubic term of the series expansion for the nonlinear circuit gives rise to components at frequencies $2f_1 + f_2, 2f_2 + f_1, 2f_1 - f_2, 2f_2 - f_1$, and these terms result in *third-order intermodulation distortion* (IM$_3$). The frequency spectrum obtained from the cubic term is shown in Table 5.2.

For definition purposes the two input signals are assumed to be of equal amplitude and IM$_3$ is given by (assuming negligible contribution to the fundamental by the cubic term)

$$\text{IM}_3 = \frac{3}{4} \left| \frac{a_3X_1^3}{a_1X_1} \right| = \frac{3}{4} \left| \frac{a_3X_1^2}{a_1} \right| \tag{5.19}$$

Under these conditions IM$_3$ = 3 HD$_3$ (Equation (5.12)). When f_1 and f_2 are close to one another, then the third-order intermodulation components, $2f_1 - f_2, 2f_2 - f_1$, are close to the fundamental and are difficult to filter out.

Triple-Beat Distortion

When three sinusoidal signals are applied at the input, then the output consists of components at the triple-beat frequencies. The cubic term in the nonlinearity results in the triple-beat terms:

$$\frac{3}{2}a_3X_1X_2X_2 \cos[\omega_1 \pm \omega_2 \pm \omega_3]t \tag{5.20}$$

and the *triple-beat distortion factor* (TB) is defined for equal-amplitude input signals:

$$\text{TB} = \frac{3}{2} \left| \frac{a_3X_1^2}{a_1} \right| \tag{5.21}$$

From the above definition, TB = 2 IM$_3$. If all of the frequencies are close to one another, the triple beats will be close to the fundamental and cannot be easily removed.

Cross Modulation

Another form of distortion that occurs in amplitude-modulated (AM) systems due to the circuit nonlinearity is **cross modulation**. The modulation from an unwanted AM signal is transferred to the signal of interest and results in distortion. Consider an AM signal:

$$x(t) = X_1 \cos \omega_1 t + X_2[1 + m \cos \omega_m t]\cos \omega_2 t \qquad (5.22)$$

where $m < 1$ is the modulation index. Because of the cubic term of the nonlinearity, the modulation from the second signal is transferred to the first and the modulated component corresponding to the fundamental is

$$a_1 X_1 \left[1 + \frac{3a_3 X_2^2 m}{a_1} \cos \omega_m t \right] \cos \omega_1 t \qquad (5.23)$$

The *cross-modulation factor* (CM) is defined as the ratio of the transferred modulation index to the original modulation:

$$\mathrm{CM} = 3\left| \frac{a_3 X_2^2}{a_1} \right| \qquad (5.24)$$

The cross modulation for equal amplitude input signals is a factor of four larger than IM_3 and 12 times as large as HD_3.

Compression and Intercept Points

For high-frequency circuits, distortion is specified in terms of compression and intercept points. These quantities are derived from extrapolated small-signal output power levels. The 1-dB *compression point* is defined as the value of the fundamental output power for which the power is 1 dB below the extrapolated small-signal value.

The *nth-order intercept point* (IP_n), $n \geq 2$, is the output power at which the extrapolated small-signal power of the fundamental and the nth intermodulation term intersect. The third-order intercept (TOI or IP_3) is the point at which the extrapolated small-signal power of the fundamental and the third-order intermodulation term are identical. IP_3 is an important specification for narrow-band communication systems.

Crossover Distortion

This type of distortion occurs in circuits that use devices operating in a "push–pull" manner. The devices are used in pairs and each device operates only for half a cycle of the input signal (Class AB operation). One advantage of such an arrangement is the cancellation of even harmonic terms resulting in smaller total harmonic distortion. However, if the circuit is not designed to achieve a smooth crossover or transition from one device to another, then there is a region of the transfer characteristics when the output is zero. The resulting distortion is called **crossover distortion**.

Failure-to-Follow Distortion

When a properly designed peak detector circuit is used for AM demodulation, the output follows the envelope of the input signal whereby the original modulation signal is recovered. A simple peak detector is a diode in series with a low-pass RC filter. The critical component of such a circuit is a linear element, the filter capacitance C. If C is large, then the output fails to follow the envelope of the input signal, resulting in **failure-to-follow distortion**.

Frequency Distortion

Ideally, an amplifier circuit should provide the same amplification for all input frequencies. However, due to the presence of energy storage elements, the gain of the amplifier is frequency dependent. Consequently, different frequency components have different amplifications resulting in **frequency distortion.** The distortion is specified by a frequency response curve in which the amplifier output is plotted as a function of frequency. An ideal amplifier has a flat frequency response over the frequency range of interest.

Phase Distortion

When the phase shift (θ) in the output signal of an amplifier is not proportional to the frequency, the output does not preserve the form of the input signal, resulting in **phase distortion.** If the phase shift is proportional to frequency, different frequency components have a constant delay time (θ/ω) and no distortion is observed. In TV applications phase distortion can result in a smeared picture.

Computer Simulation of Distortion Components

Distortion characterization is important for nonlinear circuits. However, the techniques presented for distortion calculation can only be used for simple circuit configurations and at best to determine the second and third harmonic distortion factors. In order to determine the distortion generation in actual circuits one must fabricate the circuit and then use a harmonic analyzer for sine curve inputs to determine the harmonics present in the output. An attractive alternative is the use of circuit simulation programs that allow one to investigate circuit performance before fabricating the circuit. In this section a brief overview of the techniques used in circuit simulators for distortion characterization is provided.

The simplest approach is to simulate the time-domain output for a circuit with a specified sinusoidal input signal and then perform a Fourier analysis of the output waveform. The simulation program SPICE2 provides a capability for computing the Fourier components of any waveform using a *.FOUR* command and specifying the voltage or current for which the analysis has to be performed. A simple diode circuit, the SPICE input file, and transient voltage waveforms for an input signal frequency of 1 MHz and amplitudes of 10 and 100 mV are shown in Figure 5.21. The Fourier components of the resistor voltage are shown in Figure 5.22; only the fundamental and first two significant harmonics are shown (SPICE provides information to the ninth harmonic).

In this particular example the input signal frequency is 1 MHz, and this is the frequency at which the Fourier analysis is requested. Since there are no energy storage elements in the circuit another frequency would

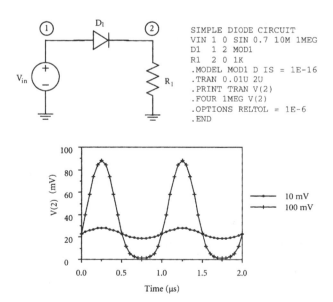

FIGURE 5.21 Simple diode circuit, SPICE input file, and output voltage waveforms.

```
FOURIER COMPONENTS OF TRANSIENT RESPONSE V(2)                    VIN = 10mV
DC COMPONENT =     2.330D-02
   HARMONIC    FREQUENCY      FOURIER       NORMALIZED     PHASE      NORMALIZED
     NO          (HZ)        COMPONENT      COMPONENT      (DEG)     PHASE (DEG)
      1        1.000D+06     4.695D-03      1.000000      -0.001        0.000
      2        2.000D+06     1.242D-04      0.026462     -89.989      -89.988
      3        3.000D+06     1.705D-06      0.000363      -3.241       -3.239

   TOTAL HARMONIC DISTORTION =      2.646409 PERCENT


FOURIER COMPONENTS OF TRANSIENT RESPONSE V(2)                    VIN = 100mV
DC COMPONENT =     3.445D-02
   HARMONIC    FREQUENCY      FOURIER       NORMALIZED     PHASE      NORMALIZED
     NO          (HZ)        COMPONENT      COMPONENT      (DEG)     PHASE (DEG)
      1        1.000D+06     4.402D-02      1.000000      -0.011        0.000
      2        2.000D+06     1.059D-02      0.240634     -89.993      -89.983
      3        3.000D+06     1.658D-04      0.015127      -0.686       -0.675

   TOTAL HARMONIC DISTORTION =     24.132679 PERCENT
```

FIGURE 5.22 Fourier components of the resistor voltage for input amplitudes of 10 and 100 mV, respectively.

have given identical results. To determine the Fourier components accurately a small value of the parameter RELTOL is used and a sufficient number of points for transient analysis are specified. From the output voltage waveforms and the Fourier analysis it is seen that the harmonic distortion increases significantly when the input voltage amplitude is increased from 10 to 100 mV.

The transient approach can be computationally expensive for circuits that reach their periodic steady state after a long simulation time. Results from the Fourier analysis are meaningful only in the periodic steady state, and although this approach works well for large levels of distortion it is inaccurate for small distortion levels.

For small distortion levels accurate distortion analysis can be performed by use of the Volterra series method. This technique is a generalization of the power-series method and is useful for analyzing harmonic and intermodulation distortion due to frequency-dependent nonlinearities. The SPICE3 program supports this analysis technique (in addition to the Fourier analysis of SPICE2) whereby the second and third harmonic and intermodulation components can be efficiently obtained by three small-signal analyses of the circuit.

An approach based on the *harmonic balance* technique is applicable to both large and small levels of distortion. The periodic steady state of a circuit with sinusoidal input signal can be determined using this technique. The unknowns are the magnitudes of the circuit variables at the fundamental frequency and at all the significant harmonics of the fundamental. The distortion levels can be simply calculated by taking the ratios of the magnitudes of the appropriate harmonics to the fundamental.

Defining Terms

Compression and intercept points: Characterize distortion in high-frequency circuits. These quantities are derived from extrapolated small-signal output power levels.

Cross modulation: Occurs in amplitude-modulated systems when the modulation of one signal is transferred to another by the nonlinearity of the system.

Crossover distortion: Present in circuits that use devices operating in a push–pull arrangement such that one device conducts when the other is off. Crossover distortion results if the transition or crossover from one device to the other is not smooth.

Failure-to-follow distortion: Can occur during demodulation of an amplitude-modulated signal by a peak detector circuit. If the capacitance of the low-pass RC filter of the peak detector is large, then the output fails to follow the envelope of the input signal, resulting in failure-to-follow distortion.

Frequency distortion: Caused by the presence of energy storage elements in an amplifier circuit. Different frequency components have different amplifications, resulting in frequency distortion and the distortion is specified by a frequency–response curve.

Harmonic distortion: Caused by the nonlinear transfer characteristics of a device or circuit. When a sinusoidal signal of a single frequency (the *fundamental* frequency) is applied at the input of a nonlinear circuit, the output contains frequency components that are integer multiples of the fundamental frequency (*harmonics*). The resulting distortion is called harmonic distortion.

Harmonic distortion factors: A measure of the harmonic content of the output. The *nth harmonic distortion factor* is the ratio of the amplitude of the *n*th harmonic to the amplitude of the fundamental component of the output.

Intermodulation distortion: Distortion caused by the mixing or beating of two or more sinusoidal inputs due to the nonlinearity of a device. The output contains terms at the sum and difference frequencies called the *beat frequencies*.

Phase distortion: Occurs when the phase shift in the output signal of an amplifier is not proportional to the frequency.

Total harmonic distortion: The ratio of the root-mean-square value of the harmonics to the amplitude of the fundamental component of a waveform.

References

K.K. Clarke and D.T. Hess, *Communication Circuits: Analysis and Design*, Reading, MA: Addison-Wesley, 1971.

P.R. Gray, P.J. Hurst, S.H. Lewis, and R.G. Meyer, *Analysis and Design of Analog Integrated Circuits*, New York: Wiley, 2001.

K.S. Kundert, *Spectre User's Guide: A Frequency Domain Simulator for Nonlinear Circuits*, Berkeley, CA: EECS Industrial Liaison Program Office, University of California, 1987.

K.S. Kundert, *The Designer's Guide to SPICE and SPECTRE*, Boston, MA: Kluwer Academic Publishers, 1995.

L.W. Nagel, *SPICE2: A Computer Program to Simulate Semiconductor Circuits*, Memo No. ERL-M520, Berkeley, CA: Electronics Research Laboratory, University of California, 1975.

D.O. Pederson and K. Mayaram, *Analog Integrated Circuits for Communication: Principles, Simulation and Design*, Boston, MA: Kluwer Academic Publishers, 1991.

T.L. Quarles, *SPICE3C.1 User's Guide*, Berkeley, CA: EECS Industrial Liaison Program Office, University of California, 1989.

J.S. Roychowdhury, *SPICE 3 Distortion Analysis*, Memo No. UCB/ERL M89/48, Berkeley, CA: Electronics Research Laboratory, University of California, 1989.

P. Wambacq and W. Sansen, *Distortion Analysis of Analog Integrated Circuits*, Boston, MA: Kluwer Academic Publishers, 1998.

D.D. Weiner and J.F. Spina, *Sinusoidal Analysis and Modeling of Weakly Nonlinear Circuits*, New York: Van Nostrand Reinhold Company, 1980.

Further Information

Characterization and simulation of distortion in a wide variety of electronic circuits (with and without feedback) is presented in detail in Pederson and Mayaram (1991). Also derivations for the simple analysis techniques are provided and verified using SPICE2 simulations. Algorithms for computer-aided analysis of distortion are available in Weiner and Spina (1980), Nagel (1975), Roychowdhury (1989), Kundert (1987), and Wambacq and Sansen (1998). Chapter 5 of Kundert (1995) gives valuable information on use of Fourier analysis in SPICE for distortion calculation in circuits. The software packages SPICE2, SPICE3, and SPECTRE are available from EECS Industrial Liaison Program Office, University of California, Berkeley, CA 94720.

6

Laplace Transform

Richard C. Dorf
University of California, Davis

David E. Johnson
Birmingham-Southern College

6.1 Definitions and Properties

Richard C. Dorf

The **Laplace transform** is a useful analytical tool for converting time-domain signal descriptions into functions of a complex variable. This *complex domain* description of a signal provides new insight into the analysis of signals and systems. In addition, the Laplace transform method often simplifies the calculations involved in obtaining system response signals.

Laplace Transform Integral

The Laplace transform completely characterizes the exponential response of a time-invariant linear function. This transformation is formally generated through the process of multiplying the linear characteristic signal $x(t)$ by the signal e^{-st} and then integrating that product over the time interval $(-\infty, +\infty)$. This systematic procedure is more generally known as *taking the Laplace transform* of the signal $x(t)$.

Definition: The Laplace transform of the continuous-time signal $x(t)$ is

$$X(s) = \int_{-\infty}^{+\infty} x(t)e^{-st}\,dt$$

The variable s that appears in this integrand exponential is generally complex valued and is therefore often expressed in terms of its rectangular coordinates:

$$s = \sigma + j\omega$$

where $\sigma = Re(s)$ and $\omega = Im(s)$ are referred to as the *real* and *imaginary* components of s, respectively.

The signal $x(t)$ and its associated Laplace transform $X(s)$ are said to form a *Laplace transform pair*. This reflects a form of equivalency between the two apparently different entities $x(t)$ and $X(s)$. We may symbolize this interrelationship in the following suggestive manner:

$$X(s) = \mathscr{L}[x(t)]$$

where the operator notation $\mathscr{L}$ means to multiply the signal $x(t)$ being operated upon by the complex exponential e^{-st} and then to integrate that product over the time interval $(-\infty, +\infty)$.

Region of Absolute Convergence

In evaluating the Laplace transform integral that corresponds to a given signal, it is generally found that this integral will exist (that is, the integral has finite magnitude) for only a restricted set of s values.

The definition of **region of absolute convergence** is as follows. The set of complex numbers s for which the magnitude of the Laplace transform integral is finite is said to constitute the region of absolute convergence for that integral transform. This region of convergence is always expressible as

$$\sigma_+ < \text{Re}(s) < \sigma_-$$

where σ_+ and σ_- denote real parameters that are related to the causal and anticausal components, respectively, of the signal whose Laplace transform is being sought.

Laplace Transform Pair Tables

It is convenient to display the Laplace transforms of standard signals in one table. Table 6.1 displays the time signal $x(t)$ and its corresponding Laplace transform and region of absolute convergence and is sufficient for our needs.

Example

To find the Laplace transform of the first-order causal exponential signal:

$$x_1(t) = e^{-at}u(t)$$

where the constant a can in general be a complex number.

The Laplace transform of this general exponential signal is determined upon evaluating the associated Laplace transform integral:

$$X_1(s) = \int_{-\infty}^{+\infty} e^{-at}u(t)e^{-st}dt = \int_{0}^{+\infty} e^{-(s+a)t}dt$$

$$= \frac{e^{-(s+a)t}}{-(s+a)}\Bigg|_{0}^{+\infty} \tag{6.1}$$

In order for $X_1(s)$ to exist, it must follow that the real part of the exponential argument be positive, that is:

$$\text{Re}(s+a) = \text{Re}(s) + \text{Re}(a) > 0$$

If this were not the case, the evaluation of expression (Equation (6.1)) at the upper limit $t = +\infty$ would either be unbounded if $\text{Re}(s) + \text{Re}(a) < 0$ or undefined when $\text{Re}(s) + \text{Re}(a) = 0$. However, the upper limit evaluation is zero when $\text{Re}(s) + \text{Re}(a) > 0$, as is already apparent. The lower limit evaluation at $t = 0$ is equal to $1/(s+a)$ for all choices of the variable s.

TABLE 6.1 Laplace Transform Pairs

	Time Signal $x(t)$	Laplace Transform $X(s)$	Region of Absolute Convergence
1.	$e^{-at}u(t)$	$\dfrac{1}{s+a}$	$\text{Re}(s) > -\text{Re}(a)$
2.	$t^k e^{-at}u(-t)$	$\dfrac{k!}{(s+a)^{k+1}}$	$\text{Re}(s) > -\text{Re}(a)$
3.	$-e^{-at}u(-t)$	$\dfrac{1}{s+a}$	$\text{Re}(s) < -\text{Re}(a)$
4.	$(-t)^k e^{-at}u(-t)$	$\dfrac{k!}{(s+a)^{k+1}}$	$\text{Re}(s) < -\text{Re}(a)$
5.	$U(t)$	$\dfrac{1}{s}$	$\text{Re}(s) > 0$
6.	$\delta(t)$	1	All s
7.	$\dfrac{d^k\delta(t)}{dt^k}$	s^k	All s
8.	$t^k u(t)$	$\dfrac{k!}{s^{k+1}}$	$\text{Re}(s) > 0$
9.	$\text{sgn}\,t = \begin{cases}1, t \geq 0 \\ -1, t < 0\end{cases}$	$\dfrac{2}{s}$	$\text{Re}(s) = 0$
10.	$\sin \omega_0 t\, u(t)$	$\dfrac{\omega_0}{s^2 + \omega_0^2}$	$\text{Re}(s) > 0$
11.	$\cos \omega_0 t\, u(t)$	$\dfrac{s}{s^2 + \omega_0^2}$	$\text{Re}(s) > 0$
12.	$e^{-at}\sin \omega_0 t\, u(t)$	$\dfrac{\omega}{(s+a)^2 + \omega_0^2}$	$\text{Re}(s) > -\text{Re}(a)$
13.	$e^{-at}\cos \omega_0 t\, u(t)$	$\dfrac{s+a}{(s+a)^2 + \omega_0^2}$	$\text{Re}(s) > -\text{Re}(a)$

The Laplace transform of exponential signal $e^{-at}u(t)$ has therefore been found and is given by

$$L[e^{-ut}u(t)] = \frac{1}{s+a} \text{ for } \text{Re}(s) > -\text{Re}(a)$$

Properties of Laplace Transform
Linearity

Let us obtain the Laplace transform of a signal, $x(t)$, that is composed of a linear combination of two other signals:

$$x(t) = \alpha_1 x_1(t) + \alpha_2 x_2(t)$$

where α_1 and α_2 are constants.

The linearity property indicates that

$$\mathscr{L}[\alpha_1 x_1(t) + \alpha_2 x_2(t)] = \alpha_1 X_1(s) + \alpha_2 X_2(s)$$

and the region of absolute convergence is *at least as large* as that given by the expression

$$\max(\sigma_+^1; \sigma_+^2) < \text{Re}(s) < \min(\sigma_-^1; \sigma_-^2)$$

where the pairs $(\sigma_+^1; \sigma_+^2) < \text{Re}(s)\min(\sigma_-^1; \sigma_-^2)$ identify the regions of convergence for the Laplace transforms $X_1(s)$ and $X_2(s)$, respectively.

Time-Domain Differentiation

The operation of time-domain differentiation has then been found to correspond to a multiplication by s in the Laplace variable s domain.

The Laplace transform of differentiated signal $dx(t)/dt$ is

$$\mathscr{L}\left[\frac{dx(t)}{dt}\right] = sX(s)$$

Furthermore, it is clear that the region of absolute convergence of $dx(t)/dt$ is at least as large as that of $x(t)$. This property may be envisioned as shown in Figure 6.1.

Time Shift

The signal $x(t - t_0)$ is said to be a version of the signal $x(t)$ right shifted (or delayed) by t_0 sec. Right shifting (delaying) a signal by a t_0 second duration in the time domain is seen to correspond to a multiplication by e^{-st_0} in the Laplace transform domain. The desired Laplace transform relationship is

$$\mathscr{L}[x(t - t_0)] = e^{-st_0}X(s)$$

where $X(s)$ denotes the Laplace transform of the unshifted signal $x(t)$. As a general rule, any time a term of the form e^{-st_0} appears in $X(s)$, this implies some form of time shift in the time domain. This most important property is depicted in Figure 6.2. It should be further noted that the regions of absolute convergence for the signals $x(t)$ and $x(t - t_0)$ are identical.

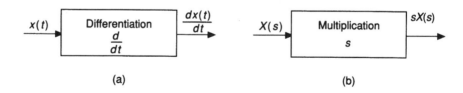

(a)   **(b)**

FIGURE 6.1 Equivalent operations in the (a) time-domain operation and (b) Laplace transform-domain operation. (*Source:* J.A. Cadzow and H.F. Van Landingham, *Signals, Systems, and Transforms,* Englewood Cliffs, NJ: Prentice-Hall, 1985, p. 138. With permission.)

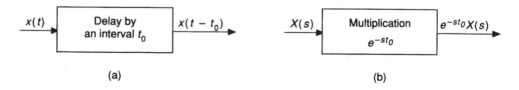

(a)   **(b)**

FIGURE 6.2 Equivalent operations in (a) the time domain and (b) the Laplace transform domain. (*Source:* J.A. Cadzow and H.F. Van Landingham, *Signals, Systems, and Transforms,* Englewood Cliffs, NJ: Prentice-Hall, 1985, p. 140. With permission.)

Time-Convolution Property

The convolution integral signal $y(t)$ can be expressed as

$$y(t) = \int_{-\infty}^{\infty} h(\tau)x(t-\tau)d\tau$$

where $x(t)$ denotes the input signal, the $h(t)$ characteristic signal identifying the operation process.
The Laplace transform of the response signal is simply given by

$$Y(s) = H(s)X(s)$$

where $H(s) = \mathscr{L}[h(t)]$ and $X(s) = \mathscr{L}[x(t)]$. Thus, the convolution of two time-domain signals is seen to correspond to the multiplication of their respective Laplace transforms in the s-domain. This property may be envisioned as shown in Figure 6.3.

Time-Correlation Property

The operation of correlating two signals $x(t)$ and $y(t)$ is formally defined by the integral relationship

$$\phi_{xy}(\tau) = \int_{-\infty}^{\infty} x(t)y(t+\tau)dt$$

The Laplace transform property of the correlation function $\phi_{xy}(\tau)$ is

$$\Phi_{xy}(s) = X(-s)Y(s)$$

in which the region of absolute convergence is given by

$$\max(-\sigma_{x^-}, \sigma_{y^+}) < \mathrm{Re}(s) < \min(-\sigma_{x^+}, \sigma_{y^-})$$

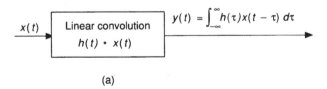

(a)

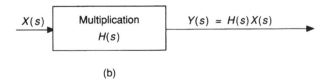

(b)

FIGURE 6.3 Representation of a time-invariant linear operator in (a) the time domain and (b) the s-domain. *(Source:* J.A. Cadzow and H.F. Van Landingham, *Signals, Systems, and Transforms,* Englewood Cliffs, NJ: Prentice-Hall, 1985, p. 144. With permission.)

Autocorrelation Function

The autocorrelation function of the signal $x(t)$ is formally defined by

$$\phi_{xx}(\tau) = \int_{-\infty}^{\infty} x(t)x(t + \tau)dt$$

The Laplace transform of the autocorrelation function is

$$\Phi_{xx}(s) = X(-s)X(s)$$

and the corresponding region of absolute convergence is

$$\max(-\sigma_{x^-}, \sigma_{y^+}) < \text{Re}(s) < \min(-\sigma_{x^+}, \sigma_{y^-})$$

Other Properties

A number of properties that characterize the Laplace transform are listed in Table 6.2. Application of these properties often enables one to efficiently determine the Laplace transform of seemingly complex time functions.

TABLE 6.2　Laplace Transform Properties

Property	Signal $x(t)$ Time Domain	Laplace Transform $X(s)$ s Domain	Region of Convergence of $X(s)$ $\sigma_+ < \text{Re}(s) < \sigma_-$
Linearity	$\alpha_1 x_1(t) + \alpha_2 x_2(t)$	$\alpha_1 X_1(s) + \alpha_2 X_2(s)$	At least the intersection of the region of convergence of $X_1(s)$ and $X_2(s)$
Time differentiation	$\dfrac{dx(t)}{dt}$	$sX(s)$	At least $\sigma_+ < \text{Re}(s)$ and $X_2(s)$
Time shift	$x(t - t_0)$	$e^{-st_0}X(s)$	$\sigma_+ < \text{Re}(s) < \sigma_-$
Time convolution	$\displaystyle\int_{-\infty}^{\infty} h(\tau)x(t - \tau)dt$	$H(s)X(s)$	At least the intersection of the region of convergence of $H(s)$ and $X(s)$
Time scaling	$x(at)$	$\dfrac{1}{\|a\|}X\!\left(\dfrac{s}{a}\right)$	$\sigma_+ < \text{Re}\!\left(\dfrac{s}{a}\right) < \sigma_-$
Frequency shift	$e^{-at}x(t)$	$X(s + a)$	$\sigma_+ - \text{Re}(a) < \text{Re}(s) < \sigma_- - \text{Re}(a)$
Multiplication (frequency convolution)	$x_1(t)x_2(t)$	$\dfrac{1}{2\pi j}\displaystyle\int_{c-j\infty}^{c+j\infty} X_1(u)X_2(s - u)d$	$\sigma_+^{(1)} + \sigma_+^{(2)} < \text{Re}(s) < \sigma_-^{(1)} + \sigma_-^{(2)}$
Time integration	$\displaystyle\int_{-\infty}^{1} x(\tau)d\tau$	$\dfrac{1}{s}X(s)$ for $X(0)$	At least $\sigma_+ < \text{Re}(s) < \sigma_-$
Frequency differentiation	$(-t)^k x(t)$	$\dfrac{d^k X(s)}{ds^k}$	At least $\sigma_+ < \text{Re}(s) < \sigma_-$
Time correlation	$\displaystyle\int_{-\infty}^{+\infty} x(t)y(t + z)dt$	$X(-s)Y(s)$	$\max(-\sigma_{x-}, \sigma_{y+}) < \text{Re}(s) < \min(-\sigma_{x+}, \sigma_{y-})$
Autocorrelation function	$\displaystyle\int_{-\infty}^{+\infty} x(t)x(t + z)dt$	$X(-s)X(s)$	$\max(-\sigma_{x-}, \sigma_{x+}) < \text{Re}(s) < \min(-\sigma_{x+}, \sigma_{x-})$

Source: J.A. Cadzow and H.F. Van Landingham, *Signals, Systems, and Transforms*, Englewood Cliffs, NJ: Prentice-Hall, 1985. With permission.

Inverse Laplace Transform

Given a transform function $X(s)$ and its region of convergence, the procedure for finding the signal $x(t)$ that generated that transform is called *finding the inverse Laplace transform* and is symbolically denoted as

$$x(t) = \mathscr{L}^{\pm 1}[X(s)]$$

The signal $x(t)$ can be recovered by means of the relationship

$$x(t) = \frac{1}{2\pi j} \int_{c-j\infty}^{c+j\infty} X(s)e^{st}ds$$

In this integral, the real number c is to be selected so that the complex number $c + j\omega$ lies entirely within the region of convergence of $X(s)$ for all values of the imaginary component ω. For the important class of rational Laplace transform functions, there exists an effective alternate procedure that does not necessitate directly evaluating this integral. This procedure is generally known as the *partial-fraction expansion method*.

Partial-Fraction Expansion Method

As just indicated, the partial fraction expansion method provides a convenient technique for reacquiring the signal that generates a given rational Laplace transform. Recall that a transform function is said to be rational if it is expressible as a ratio of polynomial in s, that is:

$$X(s) = \frac{B(s)}{A(s)} = \frac{b_m s^m + b_{m-1} s^{m-1} + \cdots + b_1 s + b_0}{s^n + a_{n-1} s^{n-1} + \cdots + a_1 s + a_0}$$

The partial-fraction expansion method is based on the appealing notion of equivalently expressing this rational transform as a sum of n elementary transforms whose corresponding inverse Laplace transforms (i.e., generating signals) are readily found in standard Laplace transform pair tables. This method entails the simple five-step process as outlined in Table 6.3. A description of each of these steps and their implementation is now given.

TABLE 6.3 Partial-Fraction Expansion Method for Determining the Inverse Laplace Transform

I.	Put rational transform into proper form whereby the degree of the numerator polynomial is less than or equal to that of the denominator polynomial.
II.	Factor the denominator polynomial.
III.	Perform a partial fraction expansion.
IV.	Separate partial fraction expansion terms into causal and anticausal components using the associated region of absolute convergence for this purpose.
V.	Using a Laplace transform pair table, obtain the inverse Laplace transform.

Source: J.A. Cadzow and H.F. Van Landingham, *Signals, Systems, and Transforms,* Englewood Cliffs, NJ: Prentice-Hall, 1985, p. 153. With permission.

I. Proper Form for Rational Transform. This division process yields an expression in the proper form as given by

$$X(s) = \frac{B(s)}{A(s)}$$
$$= Q(s) + \frac{R(s)}{A(s)}$$

in which $Q(s)$ and $R(s)$ are the quotient and remainder polynomials, respectively, with the division made so that the degree of $R(s)$ is less than or equal to that of $A(s)$.

II. Factorization of Denominator Polynomial. The next step of the partial-fraction expansion method entails the factorizing of the nth-order denominator polynomial $A(s)$ into a product of n first-order factors. This factorization is always possible and results in the equivalent representation of $A(s)$ as given by

$$A(s) = (s - p_1)(s - p_2)\ldots(s - p_n)$$

The terms $p_1, p_2, \ldots, p_n$ constituting this factorization are called the roots of polynomial $A(s)$, or the *poles of $X(s)$*.

III. Partial-Fraction Expansion. With this factorization of the denominator polynomial accomplished, the rational Laplace transform $X(s)$ can be expressed as

$$X(s) = \frac{B(s)}{A(s)} = \frac{b_n s^n + b_{n-1} s^{n-1} + \ldots + b_0}{(s - p_1)(s - p_2)\ldots(s - p_n)} \tag{6.2}$$

We shall now *equivalently represent* this transform function as a linear combination of elementary transform functions.

Case 1: A(s) Has Distinct Roots.

$$X(s) = \alpha_0 + \frac{\alpha_1}{s - p_1} + \frac{\alpha_2}{s - p_2} + \ldots + \frac{\alpha_n}{s - p_n}$$

where the α_k are constants that identify the expansion and must be properly chosen for a valid representation:

$$\alpha_k = (s - p_k)X(s)\big|_{s=p_k} \quad \text{for } k = 1, 2, \ldots, n$$

and

$$\alpha_0 = b_n$$

The expression for parameter α_0 is obtained by letting s become unbounded (i.e., $s = +\infty$) in expansion (Equation (6.2)).

Case 2: A(s) Has Multiple Roots.

$$X(s) = \frac{B(s)}{A(s)} = \frac{B(s)}{(s - p_1)^q A_1(s)}$$

The appropriate partial fraction expansion of this rational function is then given by

$$X(s) = \alpha_0 + \frac{\alpha_1}{(s-p_1)^1} + \cdots + \frac{\alpha_q}{(s-p_1)^q} + (n-q)$$

$$\text{other elementary}$$
$$\text{terms due to the}$$
$$\text{roots of } A_1(s)$$

The coefficient α_0 may be expediently evaluated by letting s approach infinity, whereby each term on the right side goes to zero except α_0. Thus:

$$\alpha_0 = \lim_{s \to +\infty} X(s) = 0$$

The α_q coefficient is given by the convenient expression:

$$\alpha_q = (s-p_1)^q X(s)\Big|_{s=p_1}$$
$$= \frac{B(p_1)}{A_1(p_1)} \tag{6.3}$$

The remaining coefficients $\alpha_1, \alpha_2, \ldots, \alpha_{q-1}$ associated with the multiple root p_1 may be evaluated by solving Equation (6.3) by setting s to a specific value.

IV. Causal and Anticausal Components.
In a partial-fraction expansion of a rational Laplace transform $X(s)$ whose region of absolute convergence is given by

$$\sigma_+ < \text{Re}(s) < \sigma_-$$

it is possible to decompose the expansion's elementary transform functions into causal and anticausal functions (and possibly impulse-generated terms). Any elementary function is interpreted as being (1) *causal* if the real component of its pole is less than or equal to σ_+ and (2) *anticausal* if the real component of its pole is greater than or equal to σ_-.

The poles of the rational transform that lie to the left (right) of the associated region of absolute convergence correspond to the causal (anticausal) component of that transform. Figure 6.4 shows the location of causal and anticausal poles of rational transform.

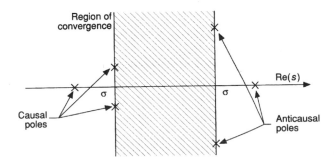

FIGURE 6.4 Location of causal and anticausal poles of a rational transform. (*Source:* J.A. Cadzow and H.F. Van Landingham, *Signals, Systems, and Transforms*, Englewood Cliffs, NJ: Prentice-Hall, 1985, p. 161. With permission.)

V. Table Look-Up of Inverse Laplace Transform. To complete the inverse Laplace transform procedure, one need simply refer to a standard Laplace transform function table to determine the time signals that generate each of the elementary transform functions. The required time signal is then equal to the same linear combination of the inverse Laplace transforms of these elementary transform functions.

Defining Terms

Laplace transform: A transformation of a function $f(t)$ from the time domain into the complex frequency domain yielding $F(s)$:

$$F(s) = \int_{-\infty}^{\infty} f(t)e^{-st}dt$$

where $s = \sigma + j\omega$.

Region of absolute convergence: The set of complex numbers s for which the magnitude of the Laplace transform integral is finite. The region can be expressed as

$$\sigma_+ < \text{Re}(s) < \sigma_-$$

where σ_+ and σ_- denote real parameters that are related to the causal and anticausal components, respectively, of the signal whose Laplace transform is being sought.

References

R.C. Dorf, *Engineering Handbook*, 2nd ed., Boca Raton, Fla.: CRC Press, 2004.
B.P. Lathi, *Signals and Systems*, 2nd ed., Carmichael, CA: Berkeley-Cambridge Press, 2001.

6.2   Applications[1]

David E. Johnson

In applications such as electric circuits, we start counting time at $t = 0$, so that a typical function $f(t)$ has the property $f(t) = 0$, $t < 0$. Its transform is given therefore by

$$F(s) = \int_0^{\infty} f(t)e^{-st}dt$$

which is sometimes called the *one-sided Laplace transform*. Since $f(t)$ is like $x(t)u(t)$, we may still use Table 6.1 of the previous section to look up the transforms, but for simplicity we will omit the factor $u(t)$, which is understood to be present.

Differentiation Theorems

Time-Domain Differentiation

If we replace $f(t)$ in the one-sided transform by its derivative $f(t)$ and integrate by parts, we have the transform of the derivative

$$\mathscr{L}[f'(t)] = sF(s) - f(0) \qquad (6.4)$$

---

[1]Based on D.E. Johnson, J.R. Johnson, and J.L. Hilburn, *Electric Circuit Analysis*, 2nd ed., Englewood Cliffs, NJ: Prentice-Hall, 1992, chapters 19 and 20. With permission.

We may formally replace f by f' to obtain

$$\mathscr{L}[f''(t)] = s\,\mathscr{L}[f'(t)] - f'(0)$$

or by (Equation (6.4)),

$$\mathscr{L}[f''(t)] = s^2 F(s) - sf(0) - f'(0) \tag{6.5}$$

We may replace f by f' again in (Equation (6.5)) to obtain $\mathscr{L}[f'''(t)]$, and so forth, obtaining the general result:

$$\mathscr{L}[f^{(n)}(t)] = s^n F(s) - s^{n-1}f(0) - s^{n-2}f'(0) - \cdots - f^{(n-1)}(0) \tag{6.6}$$

where $f^{(n)}$ is the nth derivative. The functions $f, f', \ldots, f^{(n-1)}$ are assumed to be continuous on $(0,\infty)$, and $f^{(n)}$ is continuous except possibly for a finite number of finite discontinuities.

Example 6.1

As an example, let $f(t) = t^n$, for n a nonnegative integer. Then $f^{(n)}(t) = n!$ and $f(0) = f'(0) = \cdots = f^{(n-1)}(0) = 0$. Therefore, we have

$$\mathscr{L}[n!] = s^n \mathscr{L}[t^n]$$

or

$$\mathscr{L}[t^n] = \frac{1}{s^n}\mathscr{L}[n!] = \frac{n!}{s^{n+1}}; \quad n = 0, 1, 2, \ldots \tag{6.7}$$

∎

Example 6.2

As another example, let us invert the transform

$$F(s) = \frac{8}{s^3(s+2)}$$

which has the partial fraction expansion

$$F(s) = \frac{A}{s^3} + \frac{B}{s^2} + \frac{C}{s} + \frac{D}{s+2}$$

where

$$A = s^3 F(s)|_{s=0} = 4$$

and

$$D = (s+2)F(s)|_{s=-2} = -1$$

To obtain B and C, we clear $F(s)$ of fractions, resulting in

$$8 = 4(s+2) + Bs(s+2) + Cs^2(s+2) - s^3$$

Equating coefficients of s^3 yields $C = 1$, and equating those of s^2 yields $B = -2$. The transform is therefore

$$F(s) = 2\frac{2!}{s^3} - 2\frac{1!}{s^2} + \frac{1}{s} - \frac{1}{s+2}$$

so that

$$f(t) = 2t^2 - 2t + 1 - e^{-2t}$$ ∎

Frequency-Domain Differentiation

Frequency-domain differentiation formulas may be obtained by differentiating the Laplace transform with respect to s. That is, if $F(s) = \mathcal{L}[f(t)]$:

$$\frac{dF(s)}{ds} = \frac{d}{ds}\int_0^\infty f(t)e^{-st}dt$$

Assuming that the operations of differentiation and integration may be interchanged, we have

$$\frac{dF(s)}{ds} = \frac{d}{ds}\int_0^\infty \frac{d}{ds}[f(t)e^{-st}]dt$$

$$= \int_0^\infty [-tf(t)]e^{-st}dt$$

From the last integral it follows by definition of the transform that

$$\mathcal{L}[tf(t)] = -\frac{dF(s)}{ds} \tag{6.8}$$

Example 6.3

As an example, if $f(t) = \cos kt$, then $F(s) = s/(s^2 + k^2)$, and we have

$$\mathcal{L}[t \cos kt] = -\frac{d}{ds}\left(\frac{s}{s^2+k^2}\right) = \frac{s^2 - k^2}{(s^2+k^2)^2}$$ ∎

We may repeatedly differentiate the transform to obtain the general case:

$$\frac{d^n F(s)}{ds^n} = \int_0^\infty [(-t)^n f(t)]e^{-st}dt$$

from which we conclude that

$$\mathcal{L}[t^n f(t)] = (-1)^n \frac{d^n F(s)}{ds^n}; \quad n = 0, 1, 2, \dots \tag{6.9}$$

Properties of the Laplace transform obtained in this and the previous section are listed in Table 6.4.

TABLE 6.4   One-Sided Laplace Transform Properties

	$f(t)$	$F(s)$
1.	$cf(t)$	$cF(s)$
2.	$f_1(t) + f_2(t)$	$F_1(s) + F_2(s)$
3.	$\dfrac{df(t)}{dt}$	$sF(s) - f(0)$
4.	$\dfrac{d^n f(t)}{dt^n}$	$s^n F(s) - s^{n-1} f(0) - s^{n-2} f'(0)$ $-s^{n-1} f''(0) - \cdots - f^{n-1}(0)$
5.	$\displaystyle\int_0^t f(\tau)d\tau$	$\dfrac{F(s)}{s}$
6.	$e^{-at}f(t)$	$F(s + a)$
7.	$f(t - \tau)u(t - \tau)$	$e^{-s\tau}F(s)$
8.	$f * g = \displaystyle\int_0^1 f(\tau)g(t - \tau)d\tau$	$F(s)G(s)$
9.	$f(ct),\ c > 0$	$\dfrac{1}{c}F\left(\dfrac{s}{c}\right)$
10.	$t^n f(t),\ n = 0,1,2,\ldots$	$(-1)^n F^{(n)}(s)$

Applications to Integrodifferential Equations

If we transform both members of a linear differential equation with constant coefficients, the result will be an algebraic equation in the transform of the unknown variable. This follows from Equation (6.6), which also shows that the initial conditions are automatically taken into account. The transformed equation may then be solved for the transform of the unknown and inverted to obtain the time-domain answer.

Thus, if the differential equation is

$$a_n x^{(n)} + a_{n-1} x^{(n-1)} + \ldots + a_0 x = f(t)$$

the transformed equation is

$$a_n\left[s^n X(s) - s^{n-1} x(0) - 4 .. - x^{(n-1)}(0)\right]$$
$$+a_{n-1}\left[s^{n-1} X(s) - s^{n-2} x(0) - \ldots - x^{(n-2)}(0)\right]$$
$$+ \ldots + a_0 X(s) = F(s)$$

The transform $X(s)$ may then be found and inverted to give $x(t)$.

Example 6.4

As an example, let us find the solution $x(t)$, for $t > 0$, of the system of equations:

$$x'' + 4x' + 3x = e^{-2t} \qquad x(0) = 1, \quad x'(0) = 2$$

Transforming, we have

$$s^2 X(s) - s - 2 + 4[sX(s) - 1] + 3X(s) = \frac{1}{s + 2}$$

from which

$$X(s) = \frac{s^2 + 8s + 13}{(s + 1)(s + 2)(s + 3)}$$

The partial-fraction expansion is

$$X(s) = \frac{3}{s+1} - \frac{1}{s+2} - \frac{1}{s+3}$$

from which

$$x(t) = 3e^{-t} - e^{-2t} - e^{-3t}$$

Integration Property

Certain integrodifferential equations may be transformed directly without first differentiating to remove the integrals. We need only transform the integrals by means of

$$\mathscr{L}\left[\int_0^t f(\tau)d\tau\right] = \frac{F(s)}{s}$$

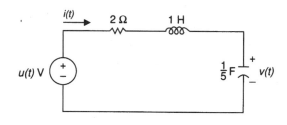

FIGURE 6.5   An *RLC* circuit.

Example 6.5

As an example, the current $i(t)$ in Figure 6.5, with no initial stored energy, satisfies the system of equations:

$$\frac{di}{dt} + 2i + 5\int_0^t i\,dt = u(t) \quad i$$

$$(0) = 0$$

Transforming yields

$$sI(s) + 2I(s) + \frac{5}{s}I(s) = \frac{1}{s}$$

or

$$I(s) = \frac{1}{s^2 + 2s + 5} = \frac{1}{2}\left[\frac{2}{(s+1)^2 + 4}\right]$$

Therefore the current is

$$i(t) = 0.5e^{-t}\sin 2t \; A$$

Applications to Electric Circuits

As the foregoing example shows, the Laplace transform method is an elegant procedure than can be used for solving electric circuits by transforming their describing integrodifferential equations into algebraic equations and applying the rules of algebra. If there is more than one loop or nodal equation, their transformed equations are solved simultaneously for the desired circuit current or voltage transforms, which are then inverted to obtain the time-domain answers. Superposition is not necessary because the various source functions appearing in the equations are simply transformed into algebraic quantities.

The Transformed Circuit

Instead of writing the describing circuit equations, transforming the results, and solving for the transform of the circuit current or voltage, we may go directly to a **transformed circuit**, which is the original circuit with the currents, voltages, sources, and passive elements replaced by transformed equivalents. The current or voltage transforms are then found using ordinary circuit theory and the results inverted to the time-domain answers.

Voltage Law Transformation

First, let us note that if we transform Kirchhoff's voltage law:

$$v_1(t) + v_2(t) + \ldots + v_n(t) = 0$$

we have

$$V_1(s) + V_2(s) + \ldots + V_n(s) = 0$$

where $V_i(s)$ is the transform of $v_i(t)$. The transformed voltages thus satisfy Kirchhoff's voltage law. A similar procedure will show that transformed currents satisfy Kirchhoff's current law as well. Next, let us consider the passive elements. For a resistance R, with current i_R and voltage v_R, for which

$$v_R = Ri_R$$

the transformed equation is

$$V_R(s) = RI_R(s) \tag{6.10}$$

This result may be represented by the transformed resistor element of Figure 6.6(a).

Inductor Transformation

For an inductance L, the voltage is

$$v_L = L\, di_L/dt$$

Transforming, we have

$$V_L(s) = sLI_L(s) - Li_L(0) \tag{6.11}$$

which may be represented by an inductor with impedance sL in series with a source, $Li_L(0)$, with the

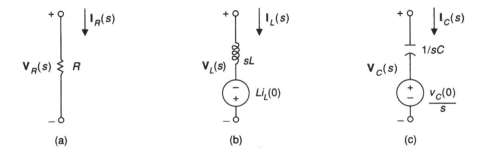

FIGURE 6.6 Transformed circuit elements.

proper polarity, as shown in Figure 6.6(b). The included voltage source takes into account the initial condition $i_L(0)$.

Capacitor Transformation

In the case of a capacitance C we have

$$v_c = \frac{1}{C}\int_0^t i_C\, dt + v_C(0)$$

which transforms to

$$V_c(s) = \frac{1}{sC}I_C(s) + \frac{1}{s}v_C(0) \tag{6.12}$$

This is represented in Figure 6.6(c) as a capacitor with impedance $1/sC$ in series with a source, $v_C(0)/s$, accounting for the initial condition.

We may solve Equation (6.10), Equation (6.11), and Equation (6.12) for the transformed currents and use the results to obtain alternate transformed elements useful for nodal analysis, as opposed to those of Figure 6.6, which are ideal for loop analysis. The alternate elements are shown in Figure 6.7.

Source Transformation

Independent sources are simply labeled with their transforms in the transformed circuit. Dependent sources are transformed in the same way as passive elements. For example, a controlled voltage source defined by

$$v_1(t) = Kv_2(t)$$

transforms to

$$V_1(s) = KV_2(s)$$

which in the transformed circuit is the transformed source controlled by a transformed variable. Since Kirchhoff's laws hold and the rules for impedance hold, the transformed circuit may be analyzed exactly as we would an ordinary resistive circuit.

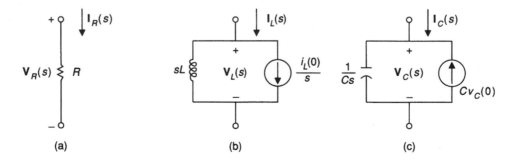

FIGURE 6.7   Transformed elements useful for nodal analysis.

Example 6.6

To illustrate, let us find $i(t)$ in Figure 6.8(a), given that $i(0) = 4$ A and $v(0) = 8$ V. The transformed circuit is shown in Figure 6.8(b), from which we have

$$I(s) = \frac{[2/(s+3)] + 4 - (8/s)}{3 + s + (2/s)}$$

This may be written

$$I(s) = -\frac{13}{s+1} + \frac{20}{s+2} - \frac{3}{s+3}$$

so that

$$i(t) = -13e^{-t} + 20e^{-2t} - 3e^{-3t} \text{ A} \qquad \blacksquare$$

Thévenin's and Norton's Theorems

Since the procedure using transformed circuits is identical to that using the phasor equivalent circuits in the ac steady-state case, we may obtain transformed Thévenin and Norton equivalent circuits exactly as in the phasor case. That is, the Thévenin impedance will be $Z_{th}(s)$ seen at the terminals of the transformed circuit with the sources made zero, and the open-circuit voltage and the short-circuit current will be $V_{oc}(s)$ and $I_{sc}(s)$, respectively, at the circuit terminals. The procedure is exactly like that for resistive circuits, except that in the transformed circuit the quantities involved are functions of s. Also, as in the resistor and phasor cases, the open-circuit voltage and short-circuit current are related by

$$V_{oc}(s) = Z_{th}(s)I_{sc}(s) \qquad (6.13)$$

Example 6.7

As an example, let us consider the circuit of Figure 6.9(a) with the transformed circuit shown in Figure 6.9(b). The initial conditions are $i(0) = 1$ A and $v(0) = 4$ V. Let us find $v(t)$ for $t > 0$ by replacing everything to the right of the 4-ω resistor in Figure 6.9(b) by its Thévenin equivalent circuit. We may find $Z_{th}(s)$ directly from Figure 6.9(b) as the impedance to the right of the resistor with the two current sources

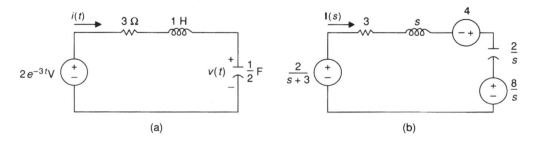

(a) (b)

FIGURE 6.8 (a) A circuit and (b) its transformed counterpart.

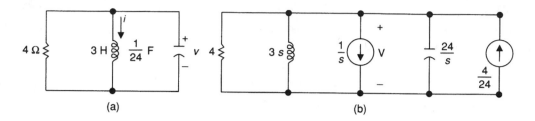

FIGURE 6.9   (a) An *RLC* parallel circuit and (b) its transformed circuit.

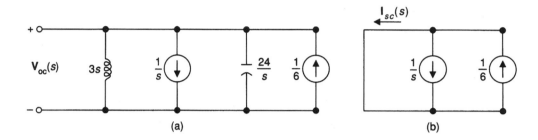

FIGURE 6.10   Circuit for obtaining (a) $V_{oc}(s)$ and (b) $I_{sc}(s)$.

made zero (open circuited). For illustrative purposes we choose, however, to find the open-circuit voltage and short-circuit current shown in Figure 6.10(a) and (b), respectively, and use Equation (6.13) to get the Thévenin impedance.

The nodal equation in Figure 6.10(a) is

$$\frac{V_{oc}(s)}{3s} + \frac{1}{s} + \frac{s}{24}\,V_{oc}(s) = \frac{1}{6}$$

from which we have

$$V_{oc}(s) = \frac{4(s-6)}{s^2+8}$$

From Figure 6.10(b)

$$I_{sc}(s) = \frac{s-6}{6s}$$

The Thévenin impedance is therefore

$$Z_{th}(s) = \frac{V_{oc}(s)}{I_{sc}(s)} = \frac{\left[\dfrac{4(s-6)}{s^2+8}\right]}{\left[\dfrac{s-6}{6s}\right]} = \frac{24s}{s^2+8}$$

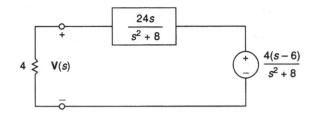

FIGURE 6.11   Thévenin equivalent circuit terminated in a resistor.

and the Thévenin equivalent circuit, with the 4 ω connected, is shown in Figure 6.11. From this circuit we find the transform

$$V(s) = \frac{4(s-6)}{(s+2)(s+4)} = \frac{-16}{s+2} + \frac{20}{s+4}$$

from which

$$v(t) = -16e^{-2t} + 20e^{-4t} \text{ V}$$

Network Functions

A **network function** or **transfer function** is the ratio $H(s)$ of the Laplace transform of the output function, say $v_o(t)$, to the Laplace transform of the input, say $v_i(t)$, assuming that there is only one input. (If there are multiple inputs, the transfer function is based on one of them with the others made zero.) Suppose that in the general case the input and output are related by the differential equation

$$a_n \frac{d^n v_o}{dt^n} + a_{n-1} \frac{d^{n-1} v_o}{dt^{n-1}} + \cdots + a_1 \frac{dv_o}{dt} + a_0 v_o$$
$$= b_m \frac{d^m v_i}{dt^m} + b_{m-1} \frac{d^{m-1} v_i}{dt^{m-1}} + \cdots + b_1 \frac{dv_i}{dt} + b_0 v_i$$

and that the initial conditions are all zero; that is:

$$v_o(0) = \frac{dv_o(0)}{dt} = \cdots = \frac{d^{n-1} v_o(0)}{dt^{n-1}} = v_i(0) = \frac{dv_i(0)}{dt} = \cdots = \frac{d^{m-1} v_i(0)}{dt^{m-1}} = 0$$

Then, transforming the differential equation results in

$$(a_n s^n + a_{n-1} s^{n-1} + \cdots + a_1 s + a_0) V_o(s)$$
$$= (b_m s^m + b_{m-1} s^{m-1} + \cdots + b_1 s + b_0) V_i(s)$$

from which the network function, or transfer function, is given by

$$H(s) = \frac{V_o(s)}{V_i(s)} = \frac{b_m s^m + b_{m-1} s^{m-1} + \cdots + b_1 s + b_0}{a_n s^n + a_{n-1} s^{n-1} + \cdots + a_1 s + a_0} \qquad (6.14)$$

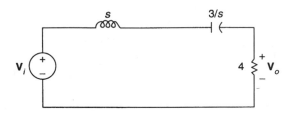

FIGURE 6.12   An *RLC* circuit.

Example 6.8

As an example, let us find the transfer function for the transformed circuit of Figure 6.12, where the transfer function is $V_o(s)/V_i(s)$. By voltage division we have

$$H(s) = \frac{V_o(s)}{V_i(s)} = \frac{4}{s + 4 + (3/s)} = \frac{4s}{(s+1)(s+3)} \tag{6.15}$$

■

Step and Impulse Responses

In general, if $Y(s)$ and $X(s)$ are the transformed output and input, respectively, then the network function is $H(s) = Y(s)/X(s)$ and the output is

$$Y(s) = H(s)X(s) \tag{6.16}$$

The **step response $r(t)$** is the output of a circuit when the input is the unit step function $u(t)$, with transform $1/s$. Therefore, the transform of the step response $R(s)$ is given by

$$R(s) = H(s)/s \tag{6.17}$$

The **impulse response $h(t)$** is the output when the input is the unit impulse $\delta(t)$. Since $\mathscr{L}[\delta(t)] = 1$, we have from Equation (6.16):

$$h(t) = \mathscr{L}^{-1}[H(s)/1] = \mathscr{L}^{-1}[H(s)] \tag{6.18}$$

Example 6.9

As an example, for the circuit of Figure 6.12, $H(s)$, given in Equation (6.15), has the partial-fraction expansion:

$$H(s) = \frac{-2}{s+1} + \frac{6}{s+3}$$

so that

$$h(t) = -2e^{-t} + 6e^{-3t} \text{ V}$$

■

If we know the impulse response, we can find the transfer function:

$$H(s) = \mathscr{L}[h(t)]$$

from which we can find the response to *any* input. In the case of the step and impulse responses, it is understood that there are no other inputs except the step or the impulse. Otherwise, the transfer function would not be defined.

Stability

An important concern in circuit theory is whether the output signal remains bounded or increases indefinitely following the application of an input signal. An unbounded output could damage or even destroy the circuit, and thus it is important to know before applying the input if the circuit can accommodate the expected output. This question can be answered by determining the *stability* of the circuit.

A circuit is defined to have **bounded input–bounded output** (BIBO) stability if any bounded input results in a bounded output. The circuit in this case is said to be **absolutely stable** or *unconditionally stable*. BIBO stability can be determined by examining the *poles* of the network function (Equation (6.14)).

If the denominator of $H(s)$ in Equation (6.14) contains a factor $(s-p)^n$, then p is said to be a pole of $H(s)$ of *order n*. The output $V_o(s)$ would also contain this factor, and its partial fraction expansion would contain the term $K/(s-p)^n$. Thus, the inverse transform $v_o(t)$ is of the form:

$$v_o(t) = A_n t^{n-1} e^{pt} + A_{n-1} t^{n-2} e^{pt} + \cdots + A_1 e^{pt} + v_1(t) \tag{6.19}$$

where $v_1(t)$ results from other poles of $V_o(s)$. If p is a real positive number or a complex number with a positive real part, $v_o(t)$ is unbounded because e^{pt} is a growing exponential. Therefore, for absolute stability there can be no pole of $V_o(s)$ that is positive or has a positive real part. This is equivalent to saying that $V_o(s)$ has no poles in the right half of the s-plane. Since $v_i(t)$ is bounded, $V_i(s)$ has no poles in the right half-plane. Therefore, since the only poles of $V_o(s)$ are those of $H(s)$ and $V_i(s)$, no pole of $H(s)$ for an absolutely stable circuit can be in the right-half of the s-plane.

From Equation (6.19) we see that $v_i(t)$ is bounded, as far as pole p is concerned, if p is a *simple* pole (of order 1) and is purely imaginary. That is, $p = j\omega$, for which

$$e^{pt} = \cos \omega t + j \sin \omega t$$

which has a bounded magnitude. Unless $V_i(s)$ contributes an identical pole $j\omega$, $v_o(t)$ is bounded. Thus, $v_o(t)$ is bounded on the *condition* that any $j\omega$ pole of $H(s)$ is simple.

In summary, a network is *absolutely stable* if its network function $H(s)$ has only left half-plane poles. It is **conditionally stable** if $H(s)$ has only simple $j\omega$-axis poles and possibly left half-plane poles. It is *unstable* otherwise (right half-plane or multiple $j\omega$-axis poles).

Example 6.10

As an example, the circuit of Figure 6.12 is absolutely stable, since from Equation (6.15) the only poles of its transfer function are $s = -1, -3$, which are both in the left half-plane. There are countless examples of conditionally stable circuits that are extremely useful, for example, a network consisting of a single capacitor with $C = 1$ F with input current $I(s)$ and output voltage $V(s)$. The transfer function is $H(s) = Z(s) = 1/Cs = 1/s$, which has the simple pole $s = 0$ on the $j\omega$-axis. Figure 6.13 illustrates a circuit that is

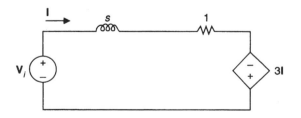

FIGURE 6.13 Unstable circuit.

unstable. The transfer function is

$$H(s) = I(s)/V_i(s) = 1/(s - 2)$$

which has the right half-plane pole $s = 2$.

Defining Terms

Absolute stability: When the network function $H(s)$ has only left half-plane poles.

Bounded input–bounded output stability: When any bounded input results in a bounded output.

Conditional stability: When the network function $H(s)$ has only simple $j\omega$-axis poles and possibly left half-plane poles.

Impulse response, $h(t)$: The output when the input is the unit impulse $\delta(t)$.

Network or transfer function: The ratio $H(s)$ of the Laplace transform of the output function to the Laplace transform of the input function.

Step response, $r(t)$: The output of a circuit when the input is the unit step function $u(t)$, with transform $1/s$.

Transformed circuit: An original circuit with the currents, voltages, sources, and passive elements replaced by transformed equivalents.

References

R.C. Dorf, *Introduction to Electric Circuits,* 2nd ed., New York: John Wiley, 1993.

J.D. Irwin, *Basic Engineering Circuit Analysis,* 3rd ed., New York: Macmillan, 1989.

D.E. Johnson, J.R. Johnson, J.L. Hilburn, and P.D. Scott, *Electric Circuit Analysis,* 3rd ed., Englewood Cliffs, NJ: Prentice-Hall, 1997.

J.W. Nilsson, *Electric Circuits,* 5th ed., Reading, MA: Addison-Wesley, 1996.

7

State Variables: Concept and Formulation

Wai-Kai Chen
University of Illinois

7.1 Introduction

An electrical network is describable by a system of algebraic and differential equations known as the primary system of equations obtained by applying the Kirchhoff's current and voltage laws and the element v–i relations. In the case of linear networks, these equations can be transformed into a system of linear algebraic equations by means of the Laplace transformation, which is relatively simple to manipulate. The main drawback is that it contains a large number equations. To reduce this number, three secondary systems of equations are available: the nodal system, the cutset system, and the loop system. If a network has n nodes, b branches, and c components, there are $n - c$ linearly independent equations in nodal or cutset analysis and $b - n + c$ linearly independent equations in loop analysis. These equations can then be solved to yield the Laplace transformed solution. To obtain the final time-domain solution, we must take the inverse Laplace transformation. For most practical networks, the procedure is usually long and complicated and requires an excessive amount of computer time.

As an alternative we can formulate the network equations in the time domain as a system of first-order differential equations, which describe the dynamic behavior of the network. Some advantages of representing the network equations in this form are the following. First, such a system has been widely studied in mathematics, and its solution, both analytic and numerical, is known and readily available. Second, the representation can easily and naturally be extended to time-varying and nonlinear networks. In fact, computer-aided solution of time-varying, nonlinear network problems is almost always accomplished using the state-variable approach. Finally, the first-order differential equations can easily be programmed for a digital computer or simulated on an analog computer. Even if it were not for the above reasons, the approach provides an alternative view of the physical behavior of the network.

The term *state* is an abstract concept that may be represented in many ways. If we call the set of instantaneous values of all the branch currents and voltages as the *state* of the network, then the knowledge of the instantaneous values of all these variables determines this instantaneous state. Not all of these instantaneous values are required in order to determine the instantaneous state, however, because some can be calculated from the others. A set of data qualifies to be called the *state* of a system if it fulfills the following two requirements:

1. The state of any time, say t_0, and the input to the system from t_0 on determine uniquely the state at any time $t > t_0$.
2. The state at time t and the inputs together with some of their derivatives at time t determine uniquely the value of any system variable at the time t.

The state may be regarded as a vector, the components of which are *state variables*. Network variables that are candidates for the state variables are the branch currents and voltages. Our problem is to choose state variables in order to formulate the *state equations*. Like the nodal, cutset, or loop system of equations, the state equations are formulated from the primary system of equations. For our purposes, we shall focus our attention on how to obtain state equations for linear systems.

7.2   State Equations in Normal Form

For a linear network containing k energy storage elements and h independent sources, our objective is to write a system of k first-order differential equations from the primary system of equations, as follows:

$$\dot{x}_i(t) = \sum_{j=1}^{k} a_{ij}x_j(t) + \sum_{j=1}^{h} b_{ij}u_j(t), \quad (i = 1, 2, \ldots, k) \tag{7.1}$$

In matrix notation, Equation (7.1) becomes

$$
\begin{bmatrix} \dot{x}_1(t) \\ \dot{x}(t) \\ \cdot \\ \cdot \\ \cdot \\ \dot{x}_k(t) \end{bmatrix}
=
\begin{bmatrix}
a_{11} & a_{12} & \cdots & a_{1k} \\
a_{21} & a_{22} & \cdots & a_{2k} \\
\cdot & \cdot & \cdots & \cdot \\
\cdot & \cdot & \cdots & \cdot \\
a_{k1} & a_{k2} & \cdots & a_{kk}
\end{bmatrix}
\begin{bmatrix} x_1(t) \\ x_2(t) \\ \cdot \\ \cdot \\ \cdot \\ x_k(t) \end{bmatrix}
$$

$$
+
\begin{bmatrix}
b_{11} & b_{12} & \cdots & b_{1h} \\
b_{21} & b_{22} & \cdots & b_{2h} \\
\cdot & \cdot & \cdots & \cdot \\
\cdot & \cdot & \cdots & \cdot \\
b_{k1} & b_{k2} & \cdots & b_{kh}
\end{bmatrix}
\begin{bmatrix} u_1(t) \\ u_2(t) \\ \cdot \\ \cdot \\ \cdot \\ u_h(t) \end{bmatrix}
\tag{7.2}
$$

or, more compactly:

$$\dot{x}(t) = \mathbf{A}\mathbf{x}(t) + \mathbf{B}\mathbf{u}(t) \tag{7.3}$$

The real functions $x_1(t)$, $x_2(t)$,..., $x_k(t)$ of the time t are called the **state variables**, and the k-vector $\mathbf{x}(t)$ formed by the state variables is known as the **state vector**. The h-vector $\mathbf{u}(t)$ formed by the h known forcing functions or excitations $u_j(t)$ is referred to as the **input vector**. matrices $\mathbf{A}$ and $\mathbf{B}$, depending only upon the network parameters, are of orders $k \times k$ and $k \times h$, respectively. Equation (7.3) is usually called the **state equation in normal form**.

The state variables x_j may or may not be the desired output variables. We therefore must express the desired output variables in terms of the state variables and excitations. In general, if there are q output variables $y_j(t)$ ($j = 1, 2, ..., q$) and h input excitations, the **output vector** $\mathbf{y}(t)$ formed by the q output variables $y_j(t)$ can be expressed in terms of the state vector $\mathbf{x}(t)$ and the input vector $\mathbf{u}(t)$ by the matrix equation:

$$\mathbf{y}(t) = \mathbf{C}\mathbf{x}(t) + \mathbf{D}\mathbf{u}(t) \tag{7.4}$$

where the known coefficient matrices $\mathbf{C}$ and $\mathbf{D}$, depending only on the network parameters, are of orders $q \times k$ and $q \times h$, respectively. Equation (7.4) is called the **output equation**. The state equation, Equation (7.3), and the output equation, Equation (7.4), together are known as the **state equations**.

7.3   The Concept of State and State Variables and Normal Tree

Our immediate problem is to choose the network variables as the state variables in order to formulate the state equations. If we call the set of instantaneous values of all the branch currents and voltages the *state* of the network, then the knowledge of the instantaneous values of all these variables determines this instantaneous state. Not all of these instantaneous values are required in order to determine the instantaneous state, however, because some can be calculated from the others. For example, the instantaneous voltage of a resistor can be obtained from its instantaneous current through Ohm's law. The question arises as to the minimum number of instantaneous values of branch voltages and currents that are sufficient to determine completely the instantaneous state of the network.

In a given network, a minimal set of its branch variables is said to be a **complete set of state variables** if their instantaneous values are sufficient to determine completely the instantaneous values of all the branch variables. For a linear time-invariant nondegenerate network, it is convenient to choose the capacitor voltages and inductor currents as the state variables. A **nondegenerate network** is one that contains neither a circuit composed only of capacitors and/or independent or dependent voltage sources nor a cutset composed only of inductors and/or independent or dependent current sources, where a cutset is a minimal subnetwork, the removal of which cuts the original network into two connected pieces. Thus, not all the capacitor voltages and inductor currents of a degenerate network can be state variables. To help systematically select the state variables, we introduce the notion of normal tree.

A **tree** of a connected network is a connected subnetwork that contains all the nodes but does not contain any circuit. A **normal tree** of a connected network is a tree that contains all the independent voltage sources, the maximum number of capacitors, the minimum number of inductors, and none of the independent current sources. This definition excludes the possibility of having unconnected networks. In the case of unconnected networks, we can consider the normal trees of the individual components. We remark that the representation of the state of a network is generally not unique, but the state of a network itself is.

7.4   Systematic Procedure in Writing State Equations

In the following we present a systematic step-by-step procedure for writing the state equation for a network. They are a systematic way to eliminate the unwanted variables in the primary system of equations:

1. In a given network N, assign the voltage and current references of its branches.
2. In N select a normal tree T and choose as the state variables the capacitor voltages of T and the inductor currents of the **cotree** $\bar{T}$, the complement of T in N.
3. Assign each branch of T a voltage symbol, and assign each element of $\bar{T}$, called the **link**, a current symbol.

4. Using Kirchhoff's current law, express each tree-branch current as a sum of cotree-link currents, and indicate it in N if necessary.
5. Using Kirchhoff's voltage law, express each cotree-link voltage as a sum of tree-branch voltages, and indicate it in N if necessary.
6. Write the element v–i equations for the passive elements and separate these equations into two groups:
   a. Those element v–i equations for the tree-branch capacitors and the cotree-link inductors
   b. Those element v–i equations for all other passive elements
7. Eliminate the nonstate variables among the equations obtained in the preceding step. **Nonstate variables** are defined as those variables that are neither state variables nor known independent sources.
8. Rearrange the terms and write the resulting equations in normal form.

We illustrate the preceding steps by the following examples.

Example 1

We write the state equations for the network N of Figure 7.1 by following the eight steps outlined above.

Step 1

The voltage and current references of the branches of the active network N are as indicated in Figure 7.1.

Step 2

Select a normal tree T consisting of the branches R_1, C_3, and v_g. The subnetwork $C_3 i_5 v_g$ is another example of a normal tree.

Step 3

The tree branches R_1, C_3, and v_g are assigned the voltage symbols v_1, v_3, and v_g; and the cotree-links R_2, L_4, i_5, and i_g are assigned the current symbols i_2, i_4, i_3, and i_g, respectively. The controlled current source i_5 is given the current symbol i_3 because its current is controlled by the current of the branch C_3, which is i_3.

Step 4

Applying Kirchhoff's current law, the branch currents i_1, i_3, and i_7 can each be expressed as the sums of cotree-link currents:

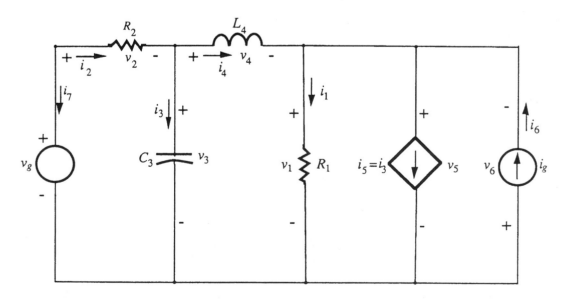

FIGURE 7.1   An active network used to illustrate the procedure for writing the state equations in normal form.

$$i_1 = i_4 + i_g - i_3 \tag{7.5a}$$

$$i_3 = i_2 - i_4 \tag{7.5b}$$

$$i_7 = -i_2 \tag{7.5c}$$

Step 5

Applying Kirchhoff's voltage law, the cotree-link voltages v_2, v_4, v_5, and v_6 can each be expressed as the sums of tree-branch voltages:

$$v_2 = v_g - v_3 \tag{7.6a}$$

$$v_4 = v_3 - v_1 \tag{7.6b}$$

$$v_5 = v_1 \tag{7.6c}$$

$$v_6 = -v_1 \tag{7.6d}$$

Step 6

The element v–i equations for the tree-branch capacitor and the cotree-link inductor are found to be

$$C_3 \dot{v}_3 = i_3 = i_2 - i_4 \tag{7.7a}$$

$$L_4 \dot{i}_4 = v_4 = v_3 - v_1 \tag{7.7b}$$

Likewise, the element v–i equations for other passive elements are obtained as

$$v_1 = R_1 i_1 = R_1(i_4 + i_g - i_3) \tag{7.8a}$$

$$i_2 = \frac{v_2}{R_2} = \frac{v_g - v_3}{R_2} \tag{7.8b}$$

variables

Step 7

The state variables are the capacitor voltage v_3 and inductor current i_4, and the known independent sources are i_g and v_g. To obtain the state equation, we must eliminate the nonstate variables v_1 and i_2 in Equation (7.7). From Equations (7.5b) and Equation (7.8) we express v_1 and i_2 in terms of the state variables and obtain

$$v_1 = R_1 \left(2i_4 + i_g + \frac{v_3}{R_2} - \frac{v_g}{R_2} \right) \tag{7.9a}$$

$$i_2 = \frac{v_g - v_3}{R_2} \tag{7.9b}$$

Substituting these in Equation (7.7) yields

$$C_3 \dot{v}_3 = \frac{v_g - v_3}{R_2} - i_4 \tag{7.10a}$$

$$L_4 \dot{i}_4 = \left(1 - \frac{R_1}{R_2}\right) v_3 - 2R_1 i_4 - R_1 i_g + \frac{R_1 v_g}{R_2} \tag{7.10b}$$

Step 8

Equations (7.10a) and (7.10b) are written in matrix form as

$$\begin{bmatrix} \dot{v}_3 \\ \dot{i}_4 \end{bmatrix} = \begin{bmatrix} -\dfrac{1}{R_2 C_3} & -\dfrac{1}{C_3} \\ \dfrac{1}{L_4} - \dfrac{R_1}{R_2 L_4} & -\dfrac{2R_1}{L_4} \end{bmatrix} \begin{bmatrix} v_3 \\ i_4 \end{bmatrix} + \begin{bmatrix} \dfrac{1}{R_2 C_3} & 0 \\ \dfrac{R_1}{R_2 L_4} & -\dfrac{R_1}{L_4} \end{bmatrix} \begin{bmatrix} v_g \\ i_g \end{bmatrix} \tag{7.11}$$

This is the state equation in normal form for the active network N of Figure 7.1.

Suppose that resistor voltage v_1 and capacitor current i_3 are the output variables. Then from Equation (7.5b) and Equation (7.9) we obtain

$$v_1 = \frac{R_1}{R_2} v_3 + 2R_1 i_4 + R_1 \left(i_g - \frac{v_g}{R_2}\right) \tag{7.12a}$$

$$i_3 = -\frac{v_3}{R_2} - i_4 + \frac{v_g}{R_2} \tag{7.12b}$$

In matrix form, the output equation of the network becomes

$$\begin{bmatrix} v_1 \\ i_3 \end{bmatrix} \begin{bmatrix} \dfrac{R_1}{R_2} & 2R_1 \\ -\dfrac{1}{R_2} & -1 \end{bmatrix} \begin{bmatrix} v_3 \\ i_4 \end{bmatrix} + \begin{bmatrix} -\dfrac{R_1}{R_2} & R_1 \\ \dfrac{1}{R_2} & 0 \end{bmatrix} \begin{bmatrix} v_g \\ i_g \end{bmatrix} \tag{7.13}$$

Equation (7.11) and Equation (7.13) together are the state equations of the active network of Figure 7.1.

7.5   State Equations for Networks Described by Scalar Differential Equations

In many situations we are faced with networks that are described by scalar differential equations of order higher than one. Our purpose here is to show that these networks can also be represented by the state equations in normal.

Consider a network that can be described by the nth-order linear differential equation:

$$\frac{d^n y}{dt^n} + a_1 \frac{d^{n-1} y}{dt^{n-1}} + a_2 \frac{d^{n-2} y}{dt^{n-2}} + \cdots + a_{n-1} \frac{dy}{dt} + a_n y = bu \tag{7.14}$$

Then its state equation can be obtained by defining

$$x_1 = y$$
$$x_2 = \dot{x}_1$$
$$\cdot$$
$$\cdot \qquad\qquad\qquad\qquad\qquad\qquad (7.15)$$
$$\cdot$$
$$x_n = \dot{x}_{n-1}$$

showing that the nth-order linear differential Equation (7.14) is equivalent to

$$\dot{x}_1 = x_2$$
$$\dot{x}_2 = x_3$$
$$\cdot$$
$$\cdot \qquad\qquad\qquad\qquad\qquad\qquad (7.16)$$
$$\cdot$$
$$\dot{x}_{n-1} = x_n$$
$$\dot{x}_n = -a_n x_1 - a_{n-1} x_2 - \cdots - a_2 x_{n-1} - a_1 x_n + bu$$

or, in matrix form:

$$
\begin{bmatrix} \dot{x}_1 \\ \dot{x}_2 \\ \cdot \\ \cdot \\ \cdot \\ \dot{x}_{n-1} \\ \dot{x}_n \end{bmatrix}
=
\begin{bmatrix} 0 & 1 & 0 & \cdots & 0 \\ 0 & 0 & 1 & \cdots & 0 \\ \cdot & \cdot & \cdot & \cdots & \cdot \\ \cdot & \cdot & \cdot & \cdots & \cdot \\ \cdot & \cdot & \cdot & \cdots & \cdot \\ 0 & 0 & 0 & \cdots & 1 \\ -a_n & -a_{n-1} & -a_{n-2} & \cdots & -a_1 \end{bmatrix}
\begin{bmatrix} x_1 \\ x_2 \\ \cdot \\ \cdot \\ \cdot \\ x_{n-1} \\ x_n \end{bmatrix}
+
\begin{bmatrix} 0 \\ 0 \\ \cdot \\ \cdot \\ \cdot \\ 0 \\ b \end{bmatrix} [u] \qquad (7.17)
$$

More compactly, Equation (7.17) can be written as

$$\dot{x}(t) = Ax(t) + Bu(t) \qquad\qquad\qquad\qquad (7.18)$$

The coefficient matrix **A** is called the **companion matrix** of Equation (7.14), and Equation (7.17) is the state-equation representation of the network describable by the linear differential Equation (7.14).

Let us now consider the more general situation where the right-hand side of (Equation (7.14)) includes derivatives of the input excitation u. In this case, the different equation takes the general form:

$$\frac{d^n y}{dt^n} + a_1 \frac{d^{n-1} y}{dt^{n-1}} + a_2 \frac{d^{n-2} y}{dt^{n-2}} + \cdots + a_{n-1} \frac{dy}{dt} + a_n y$$
$$= b_0 \frac{d^n u}{dt^n} + b_1 \frac{d^{n-1} u}{dt^{n-1}} + \cdots + b_{n-1} \frac{du}{dt} + b_n u \qquad (7.19)$$

Its state equation can be obtained by defining

$$
\begin{aligned}
x_1 &= y - c_0 u \\
x_2 &= \dot{x}_1 - c_1 u \\
&\;\;\vdots \\
x_n &= \dot{x}_{n-1} u
\end{aligned}
\tag{7.20}
$$

The general state equation becomes

$$
\begin{bmatrix} \dot{x}_1 \\ \dot{x}_2 \\ \vdots \\ \dot{x}_{n-1} \\ \dot{x}_n \end{bmatrix}
=
\begin{bmatrix}
0 & 1 & 0 & \cdots & 0 \\
0 & 0 & 1 & \cdots & 0 \\
\vdots & \vdots & \vdots & \vdots & \vdots \\
0 & 0 & 0 & \cdots & 1 \\
-a_n & -a_{n-1} & -a_{n-2} & \cdots & -a_1
\end{bmatrix}
\begin{bmatrix} x_1 \\ x_2 \\ \vdots \\ x_{n-1} \\ x_n \end{bmatrix}
+
\begin{bmatrix} c_1 \\ c_2 \\ \vdots \\ c_{n-1} \\ c_n \end{bmatrix}
[u]
\tag{7.21}
$$

where $n > 1$:

$$
\begin{aligned}
c_1 &= b_1 - a_1 b_0 \\
c_2 &= (b_2 - a_2 b_0) - a_1 c_1 \\
c_3 &= (b_3 - a_3 b_0) - a_2 c_1 - a_1 c_2 \\
&\;\;\vdots \\
c_n &= (b_n - a_3 b_0) - a_{n-1} c_1 - a_{n-2} c_2 - \cdots - a_2 c_{n-2} - a_1 c_{n-1}
\end{aligned}
\tag{7.22}
$$

and

$$
x_1 = y - b_0 u
\tag{7.23}
$$

Finally, if y is the output variable, the output equation becomes

$$
y(t) = [1 \; 0 \; 0 \; \cdots \; 0]
\begin{bmatrix} x_1 \\ x_2 \\ \vdots \\ x_n \end{bmatrix}
+ [b_0][u]
\tag{7.24}
$$

7.6   Extension to Time-Varying and Nonlinear Networks

A great advantage in the state-variable approach to network analysis is that it can easily be extended to time-varying and nonlinear networks, which are often not readily amenable to the conventional methods of analysis. In these cases, it is more convenient to choose the capacitor charges and inductor flux as the the state variables instead of capacitor voltages and inductor currents.

In the case of a linear time-varying network, its state equations can be written the same as before except that now the coefficient matrices are time-dependent:

$$
\dot{\mathbf{x}}(t) = \mathbf{A}(t)\mathbf{x}(t) + \mathbf{B}(t)\mathbf{u}(t)
\tag{7.25a}
$$

$$
\mathbf{y}(t) = \mathbf{C}(t)\mathbf{x}(t) + \mathbf{D}(t)\mathbf{u}(t)
\tag{7.25b}
$$

Thus, with the state-variable approach, it is no more difficult to write the governing equations for a linear time-varying network than it is for a linear time-invariant network. Their solutions are, of course, a different matter.

For a nonlinear network, its state equation in normal form is describable by a coupled set of first-order differential equations:

$$\dot{\mathbf{x}} = \mathbf{f}(\mathbf{x}, \mathbf{u}, t) \tag{7.26}$$

If the function $\mathbf{f}$ satisfies the familiar Lipshitz condition with respect to $\mathbf{x}$ in a given domain, then for every set of initial conditions $\mathbf{x}_0(t_0)$ and every input $\mathbf{u}$ there exists a unique solution $\mathbf{x}(t)$, the components of which are the state variables of the network.

Defining Terms

Companion matrix: The coefficient matrix in the state-equation representation of the network describable by a linear differential equation.

Complete set of state variables: A minimal set of network variables, the instantaneous values of which are sufficient to determine completely the instantaneous values of all the network variables.

Cotree: The complement of a tree in a network.

Cutset: A minimal subnetwork, the removal of which cuts the original network into two connected pieces.

Cutset system: A secondary system of equations using cutset voltages as variables.

Input vector: A vector formed by the input variables to a network.

Link: An element of a cotree.

Loop system: A secondary system of equations using loop currents as variables.

Nodal system: A secondary system of equations using nodal voltages as variables.

Nondegenerate network: A network that contains neither a circuit composed only of capacitors and/or independent or dependent voltage sources nor a cutset composed only of inductors and/or independent or dependent current sources.

Nonstate variables: Network variables that are neither state variables nor known independent sources.

Normal tree: A tree that contains all the independent voltage sources, the maximum number of capacitors, the minimum number of inductors, and none of the independent current sources.

Output equation: An equation expressing the output vector in terms of the state vector and the input vector.

Output vector: A vector formed by the output variables of a network.

Primary system of equations: A system of algebraic and differential equations obtained by applying the Kirchhoff's current and voltage laws and the element v–i relations.

Secondary system of equations: A system of algebraic and differential equations obtained from the primary system of equations by transformation of network variables.

State: A set of data, the values of which at any time t, together with the input to the system at the time, determine uniquely the value of any network variable at the time t.

State equation in normal form: A system of first-order differential equations that describes the dynamic behavior of a network and that is put into a standard form.

State equations: Equations formed by the state equation and the output equation.

State variables: Network variables used to describe the state.

State vector: A vector formed by the state variables.

Tree: A connected subnetwork that contains all the nodes of the original network but does not contain any circuit.

References

W.K. Chen, *Linear Networks and Systems: Algorithms and Computer-Aided Implementations,* Singapore: World Scientific Publishing, 1990.

W.K. Chen, *Active Network Analysis,* Singapore: World Scientific Publishing, 1991.

L.O. Chua and P.M. Lin, *Computer-Aided Analysis of Electronics Circuits: Algorithms & Computational Techniques,* Englewood Cliffs, NJ: Prentice-Hall, 1975.

E.S. Kuh and R.A. Rohrer, "State-variables approach to network analysis," *Proc. IEEE,* vol. 53, pp. 672–686, July 1965.

Further Information

An expository paper on the application of the state-variables technique to network analysis was originally written by E.S. Kuh and R.A. Rohrer ("State-variables approach to network analysis," *Proc. IEEE,* vol. 53, pp. 672–686, July 1965). A computer-aided network analysis based on state-variables approach is extensively discussed in the book by Wai-Kai Chen, *Linear Networks and Systems: Algorithms and Computer-Aided Implementations* (World Scientific Publishing Co., Singapore, 1990). The use of state variables in the analysis of electronics circuits and nonlinear networks is treated in the book by L.O. Chua and P.M. Lin, *Computer-Aided Analysis of Electronics Circuits: Algorithms & Computational Techniques* (Prentice-Hall, Englewood Cliffs, NJ, 1975). The application of state-variables technique to active network analysis is contained in the book by Wai-Kai Chen, *Active Network Analysis* (World Scientific Publishing Co., Singapore, 1991).

8

The z-Transform

Richard C. Dorf
University of California, Davis

8.1 Introduction

Discrete-time signals can be represented as sequences of numbers. Thus, if x is a discrete-time signal, its values can, in general, be indexed by n as follows:

$$x = \{\ldots, x(-2), x(-1), x(0), x(1), x(2), \ldots, x(n), \ldots\}$$

In order to work within a transform domain for discrete-time signals, we define the z-transform as follows. The z-transform of the sequence x in the previous equation is

$$Z\{x(n)\} = X(z) = \sum_{n=-\infty}^{\infty} x(n)z^{-n}$$

in which the variable z can be interpreted as being either a time-position marker or a complex-valued variable, and the script Z is the z-transform operator. If the former interpretation is employed, the number multiplying the marker z^{-n} is identified as being the nth element of the x sequence, i.e., $x(n)$. It will be generally beneficial to take z to be a complex-valued variable.

The z-transforms of some useful sequences are listed in Table 8.1.

8.2 Properties of the z-Transform

Linearity

Both the direct and inverse z-transform obey the property of linearity. Thus, if $Z\{f(n)\}$ and $Z\{g(n)\}$ are denoted by $F(z)$ and $G(z)$, respectively, then:

$$Z\{af(n) + bg(n)\} = aF(z) + bG(z)$$

where a and b are constant multipliers.

TABLE 8.1   Partial-Fraction Equivalents Listing Causal and Anticausal z-Transform Pairs

z-Domain: $F(z)$	Sequence Domain: $f(n)$
1a. $\dfrac{1}{z-a}$, for $\lvert z \rvert > \lvert a \rvert$	$a^{n-1}u(n-1) = \left\{0,\, 1,\, a,\, a^2,\, \ldots\right\}$
1b. $\dfrac{1}{z-a}$, for $\lvert z \rvert < \lvert a \rvert$	$-a^{n-1}u(-n) = \left\{\cdots,\, \dfrac{-1}{a^3},\, \dfrac{-1}{a^2},\, \dfrac{-1}{a}\right\}$
2a. $\dfrac{1}{(z-a)^2}$, for $\lvert z \rvert > \lvert a \rvert$	$(n-1)a^{n-2}u(n-1) = \left\{0,\, 1,\, 2a,\, 3a^2,\, \ldots\right\}$
2b. $\dfrac{1}{(z-a)^2}$, for $\lvert z \rvert < \lvert a \rvert$	$-(n-1)a^{n-2}u(-n) = \left\{\ldots,\, \dfrac{3}{a^4},\, \dfrac{2}{a^3},\, \dfrac{1}{a^2}\right\}$
3a. $\dfrac{1}{(z-a)^3}$, for $\lvert z \rvert > \lvert a \rvert$	$\dfrac{1}{2}(n-1)(n-2)a^{n-3}u(n-1) = \left\{0,\, 0,\, 1,\, 3a,\, 6a^2,\, \ldots\right\}$
3b. $\dfrac{1}{(z-a)^3}$, for $\lvert z \rvert < \lvert a \rvert$	$\dfrac{-1}{2}(n-1)(n-2)a^{n-3}u(-n) = \left\{\ldots,\, \dfrac{-6}{a^5},\, \dfrac{-3}{a^4},\, \dfrac{-1}{a^3}\right\}$
4a. $\dfrac{1}{(z-a)^m}$, for $\lvert z \rvert > \lvert a \rvert$	$\dfrac{1}{(m-1)!}\displaystyle\prod_{k=1}^{m-1}(n-k)a^{n-m}u(n-1)$
4b. $\dfrac{1}{(z-a)^m}$, for $\lvert z \rvert < \lvert a \rvert$	$\dfrac{-1}{(m-1)!}\displaystyle\prod_{k=1}^{m-1}(n-k)a^{n-m}u(-n)$
5a. z^{-m}, for $z \neq 0$, $m \geqslant 0$	$\delta(n-m) = \left\{\ldots,\, 0,\, 0,\, \ldots,\, 1,\, 0,\, \ldots,\, 0,\, \ldots\right\}$
5b. z^{+m}, for $\lvert z \rvert < \infty$, $m \geqslant 0$	$\delta(n+m) = \left\{\ldots,\, 0,\, 0,\, \ldots,\, 1,\, \ldots 0,\, \ldots,\, 0,\, \ldots\right\}$

Source: J.A. Cadzow and H.F. Van Landingham, *Signals, Systems and Transforms,* Englewood Cliffs, NJ: Prentice-Hall, 1985, p.191. With permission.

Translation

An important property when transforming terms of a difference equation is the *z-transform* of a sequence shifted in time. For a constant shift, we have

$$Z\{f(n+k)\} = z^k F(z)$$

for positive or negative integer k. The region of convergence of $z^k F(z)$ is the same as for $F(z)$ for positive k; only the point $z = 0$ need be eliminated from the convergence region of $F(z)$ for negative k.

Convolution

In the z-domain, the time-domain convolution operation becomes a simple product of the corresponding transforms, that is:

$$Z\{f(n) * g(n)\} = F(z)G(z)$$

Multiplication by a^n

This operation corresponds to a rescaling of the z-plane. For $a > 0$:

$$Z\{a^n f(n)\} = F\left(\frac{z}{a}\right) \quad \text{for } aR_1 < |z| < aR_2$$

where $F(z)$ is defined for $R_1 < |z| < R_2$.

Time Reversal

$$Z\{f(-n)\} = F(z^{-1}) \quad \text{for } R_2^{-1}|z|R_1^{-1}$$

where $F(z)$ is defined for $R_1 < |z| < R_2$

8.3   Unilateral z-Transform

The unilateral z-transform is defined as

$$Z_+\{x(n)\} = X(z) = \sum_{n=0}^{\infty} x(n)z^{-n} \quad \text{for } |z| > R$$

where it is called single-sided since $n \geq 0$, just as if the sequence $x(n)$ was in fact single-sided. If there is no ambiguity in the sequel, the subscript plus is omitted and we use the expression *z-transform* to mean either the double- or the single-sided transform. It is usually clear from the context which is meant. By restricting signals to be single-sided, the following useful properties can be proved.

Time Advance

For a single-sided signal $f(n)$:

$$Z_+\{f(n+1)\} = zF(z) - zf(0)$$

More generally:

$$Z_+\{f(n+k)\} = z^k F(z) - z^k f(0) - z^{k-1}f(1) - \cdots - zf(k-1)$$

This result can be used to solve linear constant-coefficient difference equations. Occasionally, it is desirable to calculate the initial or final value of a single-sided sequence without a complete inversion. The following two properties present these results.

Initial Signal Value

If $f(n) = 0$ for $n < 0$:

$$f(0) = \lim_{z \Rightarrow \infty} F(z)$$

where $F(z) = Z\{f(n)\}$ for $|z| > R$.

Final Value

If $f(n) = 0$ for $n < 0$ and $Z\{f(n)\} = F(z)$ is a rational function with all its denominator roots (poles) strictly inside the unit circle except possibly for a first-order pole at $z = 1$:

$$f(\infty) = \lim_{n \Rightarrow \infty} f(n) = \lim_{z \Rightarrow \infty} (1 - z^{-1})F(z)$$

8.4   z-Transform Inversion

We operationally denote the inverse transform of $F(z)$ in the form:

$$f(n) = Z^{-1}\{F(z)\}$$

There are three useful methods for inverting a transformed signal. They are:

1. Expansion into a series of terms in the variables z and z^{-1}
2. Complex integration by the method of residues
3. Partial-fraction expansion and table look-up

We discuss two of these methods in turn.

Method 1

For the expansion of $F(z)$ into a series, the theory of functions of a complex variable provides a practical basis for developing our inverse transform techniques. As we have seen, the general region of convergence for a transform function $F(z)$ is of the form $a < |z| < b$, i.e., an annulus centered at the origin of the z-plane. This first method is to obtain a series expression of the form:

$$F(z) = \sum_{n=-\infty}^{\infty} c_n z^{-n}$$

which is valid in the annulus of convergence. When $F(z)$ has been expanded as in the previous equation, that is, when the coefficients c_n, $n = 0, \pm 1, \pm 2, \dots$ have been found, the corresponding sequence is specified by $f(n) = c_n$ *by uniqueness of the transform.*

Method 2

We evaluate the inverse transform of $F(z)$ by the method of residues. The method involves the calculation of residues of a function both inside and outside of a simple closed path that lies inside the region of convergence. A number of key concepts are necessary in order to describe the required procedure.

A complex-valued function $G(z)$ has a pole of order k at $z = z_0$ if it can be expressed as

$$G(z) = \frac{G_1(z_0)}{(z - z_0)^k}$$

where $G_1(z_0)$ is finite.

The residue of a complex function $G(z)$ at a pole of order k at $z = z_0$ is defined by

$$\text{Res}[G(z)]\Big|_{z=z_0} = \frac{1}{(k-1)!} \frac{d^{k-1}}{dz^{k-1}} [(z - z_0)^k G(z)]\Big|_{z=z_0}$$

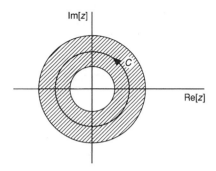

FIGURE 8.1 Typical convergence region for a transformed discrete-time signal (*Source:* J.A. Cadzow and H.F. Van Landingham, *Signals, Systems and Transforms*, Englewood Cliffs, NJ: Prentice-Hall, 1985, p. 191. With permission.)

Inverse Transform Formula (Method 2)

If $F(z)$ is convergent in the annulus $0 < a < |z| < b$ as shown in Figure 8.1 and C is the closed path shown (the path C must lie entirely within the annulus of convergence), then

$$f(n) \begin{cases} \text{sum of residues of } F(z)z^{n-1} \text{at poles of } F(z) \text{ inside } C, \ m \geq 0 \\ -(\text{sum of residues of } F(z)^{n-1} \text{at poles of } F(z) \text{ outside } C), \ m < 0 \end{cases}$$

where m is the least power of z in the numerator of $F(z)z^{n-1}$, e.g., m might equal $n-1$. Figure 8.1 illustrates the previous equation.

8.5   Sampled Data

Data obtained for a signal only at discrete intervals (sampling period) is called sampled data. One advantage of working with sampled data is the ability to represent sequences as combinations of sampled time signals. Table 8.2 provides some key z-transform pairs. So that the table can serve a multiple purpose, there are three items per line: the first is an indicated sampled continuous-time signal, the second is the Laplace transform of the continuous-time signal, and the third is the z-transform of the uniformly sampled continous-time signal. To illustrate the interrelation of these entries, consider Figure 8.2. For simplicity, only single-sided signals have been used in Table 8.2. Consequently, the convergence regions are understood in this context to be $\text{Re}[s] < \sigma_0$ and $|z| > \rho_0$ for the Laplace and z-transforms, respectively. The parameters σ_0 and ρ_0 depend on the actual transformed functions; in factor z, the inverse sequence would begin at $n = 0$. Thus, we use a modified partial-fraction expansion whose terms have this extra z-factor.

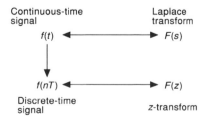

FIGURE 8.2 Signal and transform relationships for Table 8.2.

TABLE 8.2   *z*-Transforms for Sampled Data

$f(t)$, $t = nT$, $n = 0, 1, 2, \ldots$	$F(s)$, $\text{Re}[s] > \sigma_0$	$F(z)$, $\lvert z \rvert > \rho_0$
1. (unit step)	$\dfrac{1}{s}$	$\dfrac{z}{z-1}$
2. t (unit ramp)	$\dfrac{1}{s^2}$	$\dfrac{Tz}{(z-1)^2}$
3. t^2	$\dfrac{2}{s^3}$	$\dfrac{T^2 z(z+1)}{(z-1)^3}$
4. e^{-at}	$\dfrac{1}{s+a}$	$\dfrac{z}{z-e^{-aT}}$
5. te^{-at}	$\dfrac{1}{(s+a)^2}$	$\dfrac{Tze^{-aT}}{(z-e^{-aT})^2}$
6. $\sin \omega t$	$\dfrac{\omega}{s^2+\omega^2}$	$\dfrac{z \sin \omega T}{z^2 - 2z \cos \omega T + 1}$
7. $\cos \omega t$	$\dfrac{s}{s^2+\omega^2}$	$\dfrac{z(z - \cos \omega T)}{z^2 - 2z \cos \omega T + 1}$
8. $e^{-at} \sin \omega t$	$\dfrac{\omega}{(s+a)^2+\omega^2}$	$\dfrac{ze^{-aT} \sin \omega T}{z^2 - 2ze^{-aT} \cos \omega T + e^{-2aT}}$
9. $e^{-at} \cos \omega t$	$\dfrac{s+a}{(s+a)^2+\omega^2}$	$\dfrac{z(z - e^{-aT} \cos \omega T)}{z^2 - 2ze^{-aT} \cos \omega T + e^{-2aT}}$

Source: J.A. Cadzow and H.F. Landingham, *Signals, Systems and Transforms*, Englewood Cliffs, NJ: Prentice-Hall, 1985, p.191. With permission.

Defining Terms

Sampled data: Data obtained for a variable only at discrete intervals. Data are obtained once every sampling period.

Sampling period: The period for which the sampled variable is held constant.

z-Transform: A transform from the *s*-domain to the *z*-domain by $z = e^{sT}$.

References

R.C. Dorf, *Modern Control Systems*, 10th ed. Reading, MA: Addison-Wesley, 2004.

Further Information

IEEE Transactions on Education
IEEE Transactions on Automatic Control
IEEE Transactions on Signal Processing
Contact IEEE, Piscataway, NJ 08855-1313

9

T–Π Equivalent Networks

Richard C. Dorf
University of California, Davis

9.1   Introduction

Two very important two-ports are the T and Π networks shown in Figure 9.1. Because we encounter these two geometrical forms often in two-port analyses, it is useful to determine the conditions under which these two networks are equivalent. In order to determine the equivalence relationship, we will examine Z-parameter equations for the T network and the Y-parameter equations for the Π network.

For the T network the equations are

$$\mathbf{V}_1 = (\mathbf{Z}_1 + \mathbf{Z}_3)\mathbf{I}_1 + \mathbf{Z}_3\mathbf{I}_2$$

$$\mathbf{V}_2 = \mathbf{Z}_3\mathbf{I}_1 + (\mathbf{Z}_2 + \mathbf{Z}_3)\mathbf{I}_2$$

and for the Π network the equations are

$$\mathbf{I}_1 = (\mathbf{Y}_a + \mathbf{Y}_b)\mathbf{V}_1 - \mathbf{Y}_b\mathbf{V}_2$$

$$\mathbf{I}_2 = -\mathbf{Y}_b\mathbf{V}_1 + (\mathbf{Y}_b + \mathbf{Y}_c)\mathbf{V}_2$$

Solving the equations for the T network in terms of $\mathbf{I}_1$ and $\mathbf{I}_2$, we obtain

$$\mathbf{I}_1 = \left(\frac{\mathbf{Z}_2 + \mathbf{Z}_3}{\mathbf{D}_1}\right)\mathbf{V}_1 - \frac{\mathbf{Z}_3\mathbf{V}_2}{\mathbf{D}_1}$$

$$\mathbf{I}_2 = -\frac{\mathbf{Z}_3\mathbf{V}_1}{\mathbf{D}_1} + \left(\frac{\mathbf{Z}_1 + \mathbf{Z}_3}{\mathbf{D}_1}\right)\mathbf{V}_2$$

where $\mathbf{D}_1 = \mathbf{Z}_1\mathbf{Z}_2 + \mathbf{Z}_2\mathbf{Z}_3 + \mathbf{Z}_1\mathbf{Z}_3$. Comparing these equations with those for the Π network, we find that

$$\mathbf{Y}_a = \frac{\mathbf{Z}_2}{\mathbf{D}_1}$$

$$\mathbf{Y}_b = \frac{\mathbf{Z}_3}{\mathbf{D}_1}$$

$$\mathbf{Y}_c = \frac{\mathbf{Z}_1}{\mathbf{D}_1}$$

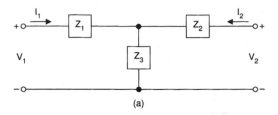

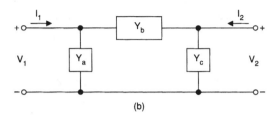

FIGURE 9.1　T and Π two-port networks.

or in terms of the impedances of the Π network:

$$Z_a = \frac{D_1}{Z_2}$$

$$Z_b = \frac{D_1}{Z_3}$$

$$Z_c = \frac{D_1}{Z_1}$$

If we reverse this procedure and solve the equations for the Π network in terms of V_1 and V_2 and then compare the resultant equations with those for the T network, we find that

$$Z_1 = \frac{Y_c}{D_2}$$

$$Z_2 = \frac{Y_a}{D_2} \tag{9.1}$$

$$Z_3 = \frac{Y_b}{D_2}$$

where $D_2 = Y_a Y_b + Y_b Y_c + Y_a Y_c$. Equation (9.1) can also be written in the form:

$$Z_1 = \frac{Z_a Z_b}{Z_a + Z_b + Z_c}$$

$$Z_2 = \frac{Z_b Z_c}{Z_a + Z_b + Z_c}$$

$$Z_3 = \frac{Z_a Z_c}{Z_a + Z_b + Z_c}$$

The T is a wye-connected network and the Π is a delta-connected network, as we discuss in the next section.

9.2 Three-Phase Connections

By far the most important polyphase voltage source is the balanced three-phase source. This source, as illustrated by Figure 9.2, has the following properties. The phase voltages, that is, the voltage from each line a, b, and c to the neutral *n*, are given by

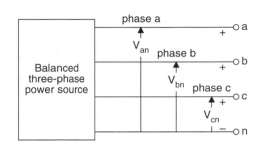

FIGURE 9.2   Balanced three-phase voltage source.

$$\mathbf{V}_{an} = \mathbf{V}_p \angle 0°$$

$$\mathbf{V}_{bn} = \mathbf{V}_p \angle -120° \qquad (9.2)$$

$$\mathbf{V}_{cn} = \mathbf{V}_p \angle +120°$$

An important property of the balanced voltage set is that

$$\mathbf{V}_{an} + \mathbf{V}_{bn} + \mathbf{V}_{cn} = 0 \qquad (9.3)$$

From the standpoint of the user who connects a load to the balanced three-phase voltage source, it is not important how the voltages are generated. It is important to note, however, that if the load currents generated by connecting a load to the power source shown in Figure 9.2 are also *balanced*, there are two possible equivalent configurations for the load. The equivalent load can be considered as being connected in either a wye (Y) or a delta (Δ) configuration. The balanced wye configuration is shown in Figure 9.3. The delta configuration is shown in Figure 9.4. Note that in the case of the delta connection, there is no neutral line. The actual function of the neutral line in the wye connection will be examined and it will be shown that in a balanced system the neutral line carries no current and therefore may be omitted.

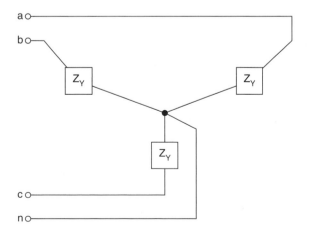

FIGURE 9.3   Wye (Y)-connected loads.

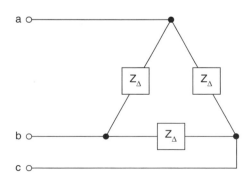

FIGURE 9.4   Delta (Δ)-connected loads.

9.3   Wye⇔Delta Transformations

For a balanced system, the equivalent load configuration may be either wye or delta. If both of these configurations are connected at only three terminals, it would be very advantageous if an equivalence could be established between them. It is, in fact, possible characteristics are the same. Consider, for example, the two networks shown in Figure 9.5.

TABLE 9.1   Current–Voltage Relationships for the Wye and Delta Load Configurations

Parameter	Wye Configuration	Delta Configuration
Voltage	$\mathbf{V}_{\text{line to line}} = \sqrt{3}\mathbf{V}_\gamma$	$\mathbf{V}_{\text{line to line}} = \mathbf{V}_\Delta$
Current	$\mathbf{I}_{\text{line}} = \mathbf{I}_\gamma$	$\mathbf{I}_{\text{line}} = \sqrt{3}\mathbf{I}_\Delta$

For these two networks to be equivalent at each corresponding pair of terminals it is necessary that the input impedances at the corresponding terminals be equal, for example, if at terminals a and b, with c open-circuited, the impedance is the same for both configurations. Equating the impedances at each port yields

$$\mathbf{Z}_{ab} = \mathbf{Z}_a + \mathbf{Z}_b = \frac{\mathbf{Z}_1(\mathbf{Z}_2 + \mathbf{Z}_3)}{\mathbf{Z}_1 + \mathbf{Z}_2 + \mathbf{Z}_3}$$

$$\mathbf{Z}_{bc} = \mathbf{Z}_b + \mathbf{Z}_c = \frac{\mathbf{Z}_3(\mathbf{Z}_1 + \mathbf{Z}_2)}{\mathbf{Z}_1 + \mathbf{Z}_2 + \mathbf{Z}_3} \tag{9.4}$$

$$\mathbf{Z}_{ca} = \mathbf{Z}_c + \mathbf{Z}_a = \frac{\mathbf{Z}_2(\mathbf{Z}_1 + \mathbf{Z}_3)}{\mathbf{Z}_1 + \mathbf{Z}_2 + \mathbf{Z}_3}$$

Solving this set of equations for $\mathbf{Z}_a$, $\mathbf{Z}_b$, and $\mathbf{Z}_c$ yields

$$\mathbf{Z}_a = \frac{\mathbf{Z}_1\mathbf{Z}_2}{\mathbf{Z}_1 + \mathbf{Z}_2 + \mathbf{Z}_3}$$

$$\mathbf{Z}_b = \frac{\mathbf{Z}_1\mathbf{Z}_3}{\mathbf{Z}_1 + \mathbf{Z}_2 + \mathbf{Z}_3} \tag{9.5}$$

$$\mathbf{Z}_c = \frac{\mathbf{Z}_2\mathbf{Z}_3}{\mathbf{Z}_1 + \mathbf{Z}_2 + \mathbf{Z}_3}$$

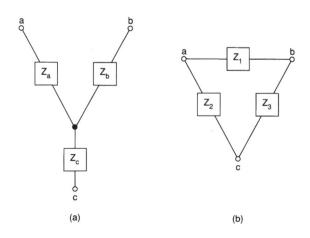

FIGURE 9.5   General wye- and delta-connected loads.

Similary, if we solve Equation (9.4) for $\mathbf{Z}_1$, $\mathbf{Z}_2$, and $\mathbf{Z}_3$, we obtain

$$\mathbf{Z}_1 = \frac{\mathbf{Z}_a\mathbf{Z}_b + \mathbf{Z}_b\mathbf{Z}_c + \mathbf{Z}_c\mathbf{Z}_a}{\mathbf{Z}_c}$$

$$\mathbf{Z}_2 = \frac{\mathbf{Z}_a\mathbf{Z}_b + \mathbf{Z}_b\mathbf{Z}_c + \mathbf{Z}_c\mathbf{Z}_a}{\mathbf{Z}_b} \tag{9.6}$$

$$\mathbf{Z}_3 = \frac{\mathbf{Z}_a\mathbf{Z}_b + \mathbf{Z}_b\mathbf{Z}_c + \mathbf{Z}_c\mathbf{Z}_a}{\mathbf{Z}_a}$$

Equations (9.5) and (9.6) are general relationships and apply to any set of impedances connected in a wye or delta configuration. For the balanced case where $\mathbf{Z}_a = \mathbf{Z}_b = \mathbf{Z}_c$ and $\mathbf{Z}_1 = \mathbf{Z}_2 = \mathbf{Z}_3$, the equations above reduce to

$$\mathbf{Z}_y = \frac{1}{3}\mathbf{Z} \tag{9.7}$$

and

$$\mathbf{Z}_\Delta = 3\mathbf{Z}_y \tag{9.8}$$

Defining Terms

Balanced voltages of the three-phase connection: The three voltages satisfy

$$\mathbf{V}_{an} + \mathbf{V}_{bn} + \mathbf{V}_{cn} = 0$$

where

$$\mathbf{V}_{an} = \mathbf{V}_p\angle 0°$$

$$\mathbf{V}_{bn} = \mathbf{V}_p\angle -120°$$

$$\mathbf{V}_{cn} = \mathbf{V}_p\angle +120°$$

T network: The equations of the T network are

$$\mathbf{V}_1 = (\mathbf{Z}_1 + \mathbf{Z}_3)\mathbf{I}_1 + \mathbf{Z}_3\mathbf{I}_2$$

$$\mathbf{V}_2 = \mathbf{Z}_3\mathbf{I}_1 + (\mathbf{Z}_2 + \mathbf{Z}_3)\mathbf{I}_2$$

Π network: The equations of Π network are

$$\mathbf{I}_1 = (\mathbf{Y}_a + \mathbf{Y}_b)\mathbf{V}_1 - \mathbf{Y}_b\mathbf{V}_2$$

$$\mathbf{I}_2 = -\mathbf{Y}_b\mathbf{V}_1 + (\mathbf{Y}_b + \mathbf{Y}_c)\mathbf{V}_2$$

T and Π can be transferred to each other.

References

J.D. Irwin, *Basic Engineering Circuit Analysis,* 7th ed., New York: MacMillan, 2003.
R.C. Dorf, *Introduction to Electric Circuits,* 6th ed., New York: John Wiley & Sons, 2004.

Further Information

IEEE Transactions on Power Systems
IEEE Transactions on Circuits and Systems, Part II: Analog and Digital Signal Processing

10

Transfer Functions of Filters

Richard C. Dorf

University of California, Davis

10.1 Introduction

Filters are widely used to pass signals at selected frequencies and reject signals at other frequencies. An *electrical filter* is a circuit that is designed to introduce gain or loss over a prescribed range of frequencies. In this section, we will describe ideal filters and then a selected set of practical filters.

10.2 Ideal Filters

An **ideal filter** is a system that completely rejects sinusoidal inputs of the form $x(t) = A \cos \omega t$, $-\infty < t < \infty$, for ω in certain frequency ranges and does not attenuate sinusoidal inputs whose frequencies are outside these ranges. There are four basic types of ideal filters: low-pass, high-pass, bandpass, and bandstop. The magnitude functions of these four types of filters are displayed in Figure 10.1. Mathematical expressions for these magnitude functions are as follows:

$$\text{ideal low-pass: } |H(\omega)| = \begin{cases} 1, & -B \leq \omega \leq B \\ 0, & |\omega| > B \end{cases} \tag{10.1}$$

$$\text{ideal high-pass: } |H(\omega)| = \begin{cases} 0, & -B < \omega < B \\ 1, & |\omega| \geq B \end{cases} \tag{10.2}$$

$$\text{ideal bandpass: } |H(\omega)| = \begin{cases} 1, & B_1 \leq |\omega| \leq B_2 \\ 0, & \text{all other } \omega \end{cases} \tag{10.3}$$

$$\text{ideal bandstop: } |H(\omega)| = \begin{cases} 0, & B_1 \leq |\omega| \leq B_2 \\ 1, & \text{all other } \omega \end{cases} \tag{10.4}$$

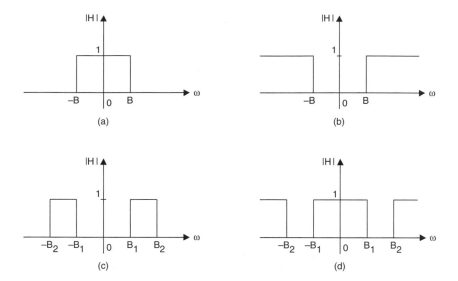

FIGURE 10.1   Magnitude functions of ideal filters: (a) low-pass; (b) high-pass; (c) bandpass; (d) bandstop.

The **stopband** of an ideal filter is defined to be the set of all frequencies ω for which the filter completely stops the sinusoidal input $x(t) = A \cos \omega t, -\infty < t < \infty$. The **passband** of the filter is the set of all frequencies ω for which the input $x(t)$ is passed without attenuation.

More complicated examples of ideal filters can be constructed by cascading ideal low-pass, high-pass, bandpass, and bandstop filters. For instance, by cascading bandstop filters with different values of B_1 and B_2, we can construct an ideal comb filter, whose magnitude function is illustrated in Figure 10.2.

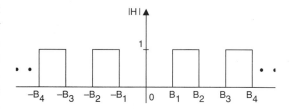

FIGURE 10.2   Magnitude function of an ideal comb filter.

10.3   The Ideal Linear-Phase Low-Pass Filter

Consider the ideal low-pass filter with the frequency function:

$$H(\omega) = \begin{cases} e^{-j\omega t_d}, & -B \leq \omega \leq B \\ 0, & \omega < -B, \omega > B \end{cases} \qquad (10.5)$$

where t_d is a positive real number. Equation (10.5) is the polar-form representation of $H(\omega)$. From Equation (10.5) we have

$$|H(\omega)| = \begin{cases} 1, & -B \leq \omega \leq B \\ 0, & \omega < -B, \omega > B \end{cases}$$

and

$$\angle H(\omega) = \begin{cases} -\omega t_d, & -B \leq \omega \leq B \\ 0, & \omega < -B, \omega > B \end{cases}$$

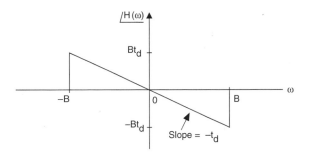

FIGURE 10.3 Phase function of ideal low-pass filter defined by Equation (10.5).

The phase function $\angle H(\omega)$ of the filter is plotted in Figure 10.3. Note that over the frequency range 0 to B, the phase function of the system is linear with slope equal to $-t_d$.

The impulse response of the low-pass filter defined by Equation (10.5) can be computed by taking the inverse Fourier transform of the frequency function $H(\omega)$. The impulse response of the ideal lowpass filter is

$$h(t) = \frac{B}{\pi} Sa[B(t - t_d)], \quad -\infty < t < \infty \tag{10.6}$$

where $Sa(x) = (\sin x)/x$. The impulse response $h(t)$ of the ideal low-pass filter is not zero for $t < 0$. Thus, the filter has a response before the impulse at $t=0$ and is said to be noncausal. As a result, it is not possible to build an ideal low-pass filter.

10.4   Ideal Linear-Phase Bandpass Filters

One can extend the analysis to ideal linear-phase bandpass filters. The frequency function of an ideal linear-phase bandpass filter is given by

$$H(\omega) = \begin{cases} e^{-j\omega t_d}, & B_1 \leq |\omega| \leq B_2 \\ 0, & \text{all other } \omega \end{cases}$$

where t_d, B_1, and B_2 are positive real numbers. The magnitude function is plotted in Figure 10.1(c) and the phase function is plotted in Figure 10.4. The passband of the filter is from B_1 to B_2. The filter will pass the signal within the band with no distortion, although there will be a time delay of t_d sec.

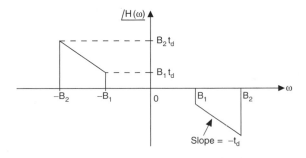

FIGURE 10.4 Phase function of ideal linear-phase bandpass filter.

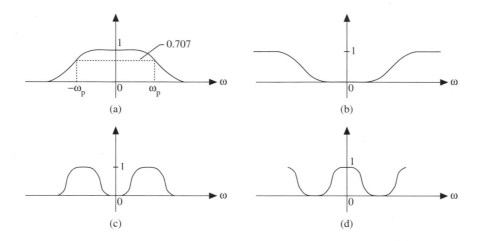

FIGURE 10.5   Causal filter magnitude functions: (a) low-pass; (b) high-pass; (c) bandpass; (d) bandstop.

10.5   Causal Filters

As observed in the preceding section, ideal filters cannot be utilized in real-time filtering applications since they are noncausal. In such applications, one must use **causal filters**, which are necessarily nonideal; that is, the transition from the passband to the stopband (and vice versa) is gradual. In particular, the magnitude functions of causal versions of low-pass, high-pass, bandpass, and bandstop filters have gradual transitions from the passband to the stopband. Examples of magnitude functions for the basic filter types are shown in Figure 10.5.

For a causal filter with frequency function $H(\omega)$, the passband is defined as the set of all frequencies ω for which

$$|H(\omega)| \geq \frac{1}{\sqrt{2}}\left|H(\omega_{\mathrm{p}})\right| \simeq 0.707\left|H(\omega_{\mathrm{p}})\right| \tag{10.7}$$

where ω_{p} is the value of ω for which $|H(\omega)|$ is maximum. Note that Equation (10.7) is equivalent to the condition that $|H(\omega)|_{\mathrm{dB}}$ is less than 3 dB down from the peak value $|H(\omega_{\mathrm{p}})|_{\mathrm{dB}}$. For low-pass or bandpass filters, the width of the passband is called the **3-dB bandwidth**.

A stopband in a causal filter is a set of frequencies ω for which $|H(\omega)|_{\mathrm{dB}}$ is down some desired amount (e.g., 40 or 50 dB) from the peak value $|H(\omega_{\mathrm{p}})|_{\mathrm{dB}}$. The range of frequencies between a passband and a stopband is called a **transition region**. In causal filter design, a key objective is to have the transition regions be suitably small in extent.

10.6   Butterworth Filters

The transfer function of the two-pole Butterworth filter is

$$H(s) = \frac{\omega_n^2}{s^2 + \sqrt{2}\,\omega_n s + \omega_n^2}$$

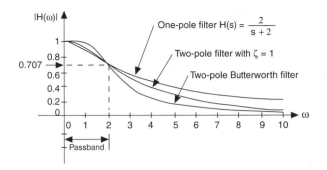

FIGURE 10.6   Magnitude curves of one- and two-pole low-pass filters.

Factoring the denominator of $H(s)$, we see that the poles are located at

$$s = -\frac{\omega_n}{\sqrt{2}} \pm j\frac{\omega_n}{\sqrt{2}}$$

Note that the magnitude of each of the poles is equal to ω_n.

Setting $s = j\omega$ in $H(s)$, we have that the magnitude function of the two-pole Butterworth filter is

$$|H(\omega)| = \frac{1}{\sqrt{1 + (\omega/\omega_n)^4}} \tag{10.8}$$

From Equation (10.8) we see that the 3-dB bandwidth of the Butterworth filter is equal to ω_n. For the case $\omega_n = 2$ rad/sec, the frequency response curves of the Butterworth filter are plotted in Figure 10.6. Also displayed are the frequency response curves for the one-pole low-pass filter with transfer function $H(s) = 2/(s+2)$, and the two-pole low-pass filter with $\zeta = 1$ and with 3-dB bandwidth equal to 2 rad/sec. Note that the Butterworth filter has the sharpest cutoff of all three filters.

10.7   Chebyshev Filters

The magnitude function of the n-pole Butterworth filter has a monotone characteristic in both the passband and stopband of the filter. Here *monotone* means that the magnitude curve is gradually decreasing over the passband and stopband. In contrast to the Butterworth filter, the magnitude function of a type 1 Chebyshev filter has ripple in the passband and is monotone decreasing in the stopband (a type 2 Chebyshev filter has the opposite characteristic). By allowing ripple in the passband or stopband, we are able to achieve a sharper transition between the passband and stopband in comparison with the Butterworth filter.

The n-pole type 1 Chebyshev filter is given by the frequency function:

$$|H(\omega)| = \frac{1}{\sqrt{1 + \epsilon^2 T_n^2(\omega/\omega_1)}} \tag{10.9}$$

where $T_n(\omega/\omega_1)$ is the nth-order Chebyshev polynomial. Note that ϵ is a numerical parameter related to the level of ripple in the passband. The Chebyshev polynomials can be generated from the recursion

$$T_n(x) = 2xT_{n-1}(x) - T_{n-2}(x)$$

where $T_0(x) = 1$ and $T_1(x) = x$. The polynomials for $n = 2, 3, 4,$ and 5 are

$$
\begin{aligned}
T_2(x) &= 2x(x) - 1 = 2x^2 - 1 \\
T_3(x) &= 2x(2x^2 - 1) - x = 4x^3 - 3x \\
T_4(x) &= 2x(4x^3 - 3x) - (2x^2 - 1) = 8x^4 - 8x^2 + 1 \\
T_5(x) &= 2x(8x^4 - 8x^2 + 1) - (4x^3 - 3x) = 16x^5 - 20x^3 + 5x
\end{aligned}
\tag{10.10}
$$

Using Equation (10.10), the two-pole type 1 Chebyshev filter has the following frequency function:

$$
|H(\omega)| = \frac{1}{\sqrt{1 + \epsilon^2[2(\omega/\omega_1)^2 - 1]^2}}
$$

For the case of a 3-dB ripple ($\epsilon = 1$), the transfer functions of the two-pole and three-pole type 1 Chebyshev filters are

$$
H(s) = \frac{0.50\omega_c^2}{s^2 + 0.645\omega_c s + 0.708\omega_c^2}
$$

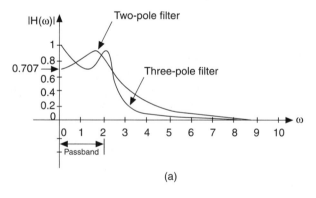

(a)

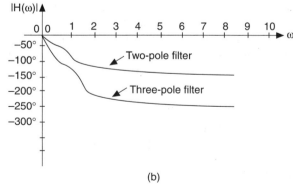

(b)

FIGURE 10.7 Frequency curves of two- and three-pole Chebyshev filters with $\omega_c = 2.5$ rad/sec: (a) magnitude curves; (b) phase curves.

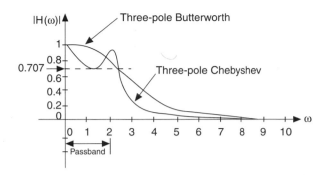

FIGURE 10.8 Magnitude curves of three-pole Butterworth and three-pole Chebyshev filters with 3-dB bandwidth equal to 2.5 rad/sec.

$$H(s) = \frac{0.251\omega_c^3}{s^2 + 0.597\omega_c s^2 + 0.928\omega_c^2 s + 0.251\omega_c^3}$$

where $\omega_c = $ 3-dB bandwidth. The frequency curves for these two filters are plotted in Figure 10.7 for the case $\omega_c = 2.5$ rad.

The magnitude response functions of the three-pole Butterworth filter and the three-pole type 1 Chebyshev filter are compared in Figure 10.8 with the 3-dB bandwidth of both filters equal to 2 rad. Note that the transition from passband to stopband is sharper in the Chebyshev filter; however, the Chebyshev filter does have the 3-dB ripple over the passband.

Defining Terms

Causal filter: A filter of which the transition from the passband to the stopband is gradual, not ideal. This filter is realizable.

3-dB bandwidth: For a causal low-pass or bandpass filter with a frequency function $H(j\omega)$: the frequency at which $|H(\omega)|_{dB}$ is less than 3 dB down from the peak value $|H(\omega_p)|_{dB}$.

Ideal filter: An ideal filter is a system that completely rejects sinusoidal inputs of the form $x(t) = A \cos \omega t, -\infty < t < \infty$, for ω within a certain frequency range, and does not attenuate sinusoidal inputs whose frequencies are outside this range. There are four basic types of ideal filters: low-pass, high-pass, bandpass, and bandstop.

Passband: Range of frequencies ω for which the input is passed without attenuation.

Stopband: Range of frequencies ω for which the filter completely stops the input signal.

Transition region: The range of frequencies of a filter between a passband and a stopband.

References

R.C. Dorf, *Introduction to Electrical Circuits*, 6th ed., New York: Wiley, 2004.
E.W. Kamen, *Introduction to Signals and Systems*, 2nd ed., New York: Macmillan, 1990.

Further Information

IEEE Transactions on Circuits and Systems, Part I: Fundamental Theory and Applications.
IEEE Transactions on Circuits and Systems, Part II: Analog and Digital Signal Processing.

11

Frequency Response

Paul Neudorfer
Seattle University

11.1   Introduction

The Institute of Electrical and Electronics Engineers defines **frequency response** in stable, linear systems as "the frequency-dependent relation in both gain and phase difference between steady-state sinusoidal inputs and the resultant steady-state sinusoidal outputs" (IEEE, 1988). The frequency-response characteristics of a system can be found analytically from its transfer function. They are also commonly measured in laboratory or field tests. A single-input/single-output linear time-invariant system is shown in Figure 11.1.

For dynamic linear systems with no time delay, the transfer function $H(s)$ is in the form of a ratio of polynomials in the complex frequency s:

$$H(s) = K \frac{N(s)}{D(s)}$$

where K is a frequency-independent constant. For a system in the sinusoidal steady state, s is replaced by the sinusoidal frequency $j\omega$ ($j = \sqrt{-1}$) and the system function becomes

$$H(j\omega) = K \frac{N(j\omega)}{D(j\omega)} = |H(j\omega)| e^{j\angle H(j\omega)}$$

$H(j\omega)$ is a complex quantity. Its magnitude $|H(j\omega)|$ and its angle or argument $\angle H(j\omega)$ relate, respectively, the amplitudes and phase angles of sinusoidal steady-state input and output signals. Referring to Figure 11.1, if the input and output signals are

$$x(t) = X \cos(\omega t + \theta_x)$$

$$y(t) = Y \cos(\omega t + \theta_y)$$

then the output's magnitude Y and phase angle θ_y are related to those of the input by the two equations:

$$Y = |H(j\omega)| X$$

$$\theta_y = \angle H(j\omega) + \theta_x$$

The phrase *frequency-response characteristics* usually implies a complete description of a system's sinusoidal steady-state behavior as a function of frequency. Because $H(j\omega)$ is complex, frequency-response characteristics cannot be graphically displayed as a single curve plotted with respect to frequency. Instead, the magnitude and angle of $H(j\omega)$ can be separately plotted as functions of frequency. It is often advantageous to plot

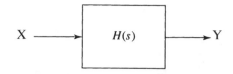

FIGURE 11.1   A single-input/single-output linear system.

frequency-response curves on other than linearly scaled Cartesian coordinates. **Bode diagrams** (developed in the 1930s by H.W. Bode of Bell Labs) use a logarithmic scale for frequency and a decibel measure for magnitude. In **Nyquist plots** (from Harry Nyquist, also of Bell Labs), $H(j\omega)$ is displayed in Argand (polar) diagram form on the complex number plane, $\mathrm{Re}[H(j\omega)]$ being on the horizontal axis and $\mathrm{Im}[H(j\omega)]$ on the vertical. Frequency is a parameter of such curves. It is sometimes numerically identified at selected points of the curve and sometimes omitted. The **Nichols chart** (developed by N.B. Nichols) graphs magnitude versus phase for the system function, frequency again being a parameter of the curve.

Frequency-response techniques are most obviously applicable to topics such as communications and electrical filters in which the frequency-response behaviors of systems are central to an understanding of their operations. It is, however, in the area of control systems that frequency-response techniques are most fully developed as analytical and design tools. The Nichols chart, for instance, is used exclusively in the analysis and design of classical feedback control systems.

The remaining sections of this chapter describe several frequency-response plotting methods. Applications of the methods can be found in other chapters throughout the handbook.

11.2   Frequency-Response Plotting

Frequency-response plots are prepared by computing the magnitude and angle of $H(j\omega)$.

Linear Plots

In linear plots $|H(j\omega)|$ and $\angle H(j\omega)$ are shown in separate diagrams as functions of frequency (either f or ω). Cartesian coordinates are used and all scales are linear.

Example 11.1

Consider the transfer function

$$H(s) = \frac{160{,}000}{s^2 + 220s + 160{,}000}$$

The complex frequency variable s is replaced by the sinusoidal frequency $j\omega$ and the magnitude and angle are found.

$$H(j\omega) = \frac{160{,}000}{(j\omega)^2 + 220(j\omega) + 160{,}000}$$

$$|H(j\omega)| = \frac{160{,}000}{\sqrt{(160{,}000 - \omega^2)^2 + (220\omega)^2}}$$

$$\angle H(j\omega) = -\tan^{-1}\frac{220\omega}{160{,}000 - \omega^2}$$

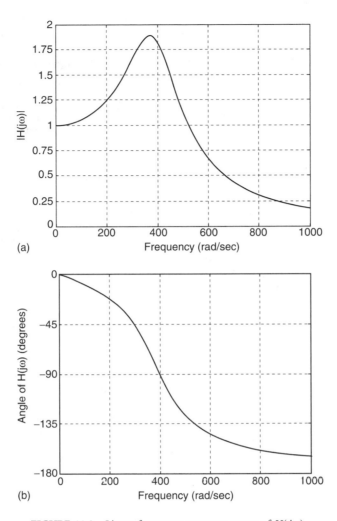

(a)

(b)

FIGURE 11.2   Linear frequency-response curves of $H(j\omega)$.

The plots of magnitude and angle are shown in Figure 11.2. Linear plots are most useful when the frequency range of interest is small. Such plots give a straightforward representation of system-response characteristics.

Bode Diagrams

A Bode diagram consists of plots of the gain and angle of a transfer function, each with respect to logarithmically scaled frequency axes. In addition, the gain of the transfer function is scaled in **decibels (dB)** according to the definition

$$H_{dB} = 20 \log_{10} |H(j\omega)|$$

Bode diagrams have the advantage of clearly identifying system features even if they occur over wide ranges of frequency and dynamic response. Before constructing a Bode diagram, the transfer function is normalized so that each pole or zero term (except those at $s = 0$) has a dc gain of one. For instance:

$$H(s) = K \frac{s + \omega_z}{s(s + \omega_p)} = \frac{K\omega_z}{\omega_p} \frac{s/\omega_z + 1}{s(s/\omega_p + 1)} = K' \frac{s\tau_z + 1}{s(s\tau_p + 1)}$$

It is common to draw Bode diagrams directly from $H(s)$ without making the formal substitution $s = j\omega$.

When drawn by hand, Bode magnitude and angle curves are developed by adding the individual contributions of the factored terms of the transfer function's numerator and denominator polynomials. In general, these factored terms may include (1) a constant K; (2) a simple s term corresponding to either a zero at the origin (if in the numerator) or a pole at the origin (if in the denominator); (3) a term such as $(s\tau + 1)$ corresponding to a real-valued (nonzero) pole or zero; and (4) a quadratic term with a possible standard form of $[(s/\omega_n)^2 + 2\zeta(s/\omega_n) + 1]$ corresponding to a pair of complex conjugate poles or zeros and for which $0 < \zeta < 1$. With the exception of quadratic terms having small ζ (**damping ratio**), Bode magnitude, and angle curves can be reasonably approximated by a series of straight line segments. Detailed procedures for drawing Bode diagrams are described in many references.

The Bode magnitude and angle curves for the factored terms listed above are shown in Figure 11.3 to Figure 11.5. Note that both decibel magnitude and angle are plotted semi-logarithmically. The frequency axis is logarithmically scaled so that every tenfold, or **decade**, change in frequency occurs over an equal distance. The magnitude axis is given in decibels. Customarily this axis is marked in 20 dB increments. Positive decibel magnitudes correspond to amplifications between input and output that are greater than one (output amplitude larger than input). Negative decibel gains correspond to attenuation between input and output (output amplitude smaller than input).

Figure 11.3 shows three separate magnitude functions. Curve 1 is trivial: the Bode magnitude of a constant K is simply the decibel-scaled constant $20\log_{10}K$, shown in the figure for an arbitrary value of $K = 5\,(20\log_{10}5 = 13.98)$. The angle is not shown. However, a constant of $K > 0$ has an angle of $0°$ for all frequencies. For $K < 0$, the angle would be $\pm180°$. Curve 2 shows the magnitude frequency-response curve for a pole at the origin $(1/s)$. It is a straight line with slope of -20 dB/decade. The line passes through 0 dB at $\omega = 1$ rad/sec. The angle associated with a single pole at the origin is $-90°$, independent of frequency. The effect of a zero at the origin is shown in Curve 3. It is again a straight line that passes through 0 dB at $\omega = 1$ rad/sec; however, the slope is $+20$ dB/decade. The angle associated with a single zero at $s = 0$ is $+90°$, independent of frequency.

Note from Figure 11.3 and the foregoing discussion that in Bode diagrams the effect of a pole term at a given location is simply the negative of that of a zero at the same location. This is true of both magnitude and angle curves.

Figure 11.4 shows the magnitude and angle curves for a zero term of the form $(s/\omega_z + 1)$ and pole term of the form $1/(s/\omega_p + 1)$. Exact plots of the curves are shown as solid lines. Straight-line approximations are shown as dotted lines. Notice that at low and high frequencies the straight-line approximations are virtually identical to the exact curves. The straight-line approximations differ most significantly from the

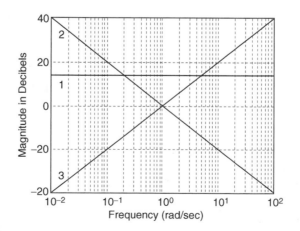

FIGURE 11.3 Bode (decibel) magnitude curves for (1) a constant of $K = 5$, (2) a pole at the origin $(1/s)$, and (3) a zero at the origin (s).

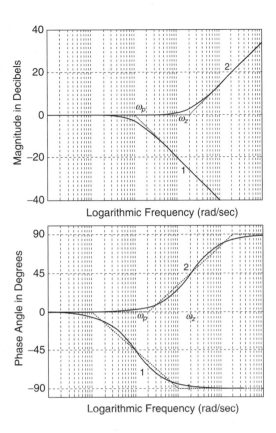

FIGURE 11.4   Bode curves for (1) a pole at $s = -\omega_p$ and (2) a zero at $s = -\omega_z$.

exact curves at points where the approximations change slope. In magnitude curves these are called **breakpoints**. At breakpoints the straight-line approximation and the exact curve differ by 3 dB for each pole or zero.

In Bode analysis complex conjugate poles or zeros are always treated as pairs in the corresponding quadratic form $[(s/\omega_n)^2 + 2\zeta(s/\omega_n) + 1]$.[1] For quadratic terms in stable, minimum phase systems, the damping ratio ζ (Greek letter zeta) is within the range $0 < \zeta < 1$. Quadratic terms cannot always be adequately represented by straight-line approximations. This is especially true for lightly damped systems (small ζ).

The traditional approach was to draw a preliminary representation consisting of a straight line of 0 dB from dc to the breakpoint at ω_n followed by a straight line of slope ± 40 dB/decade beyond the breakpoint, depending on whether the quadratic term related to conjugate poles or conjugate zeros. Then, referring to families of curves such as shown in Figure 11.5, the preliminary representation could be improved based on the value of ζ.

Bode diagrams are easily constructed by hand because, with the exception of lightly damped quadratic terms, each contribution can be reasonably approximated with straight lines. Also, the overall frequency response curve is found by adding the individual contributions. Today, many commercially available mathematical analysis software packages have built-in utilities for creating Bode diagrams. These have rendered the plotting of Bode diagrams by hand almost obsolete. Still, there is benefit in understanding the traditional methods because they give insight into the meanings of Bode diagram features. Breakpoints, for instance, can be related to the locations of poles and/or zeros and slopes can be related to the numbers of poles and/or zeros. Two examples of Bode diagram construction follow.

---

[1]Several such standard forms are used. This is the one most commonly encountered in controls engineering.

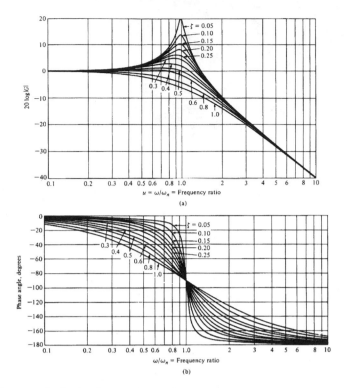

FIGURE 11.5   Bode diagram of $1/[(s/\omega_n)^2 + 2\zeta(s/\omega_n) + 1]$. (*Source*: R.C. Dorf, *Modern Control Systems*, 4th ed., Reading, MA: Addison-Wesley, 1986, p. 258. With permission.)

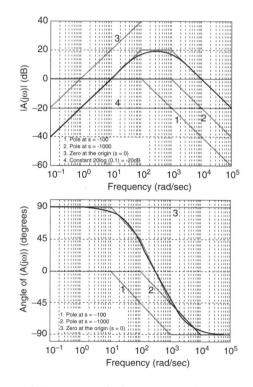

FIGURE 11.6   Bode diagram curves for $A(j\omega)$.

Example 11.2

$$A(s) = \frac{10^4 s}{s^2 + 1100s + 10^5} = \frac{10^4 s}{(s + 100)(s + 1000)} = 10^{-1} \frac{s}{(s/100 + 1)(s/1000 + 1)}$$

In Figure 11.6 the individual contributions of the four factored terms of $H(s)$ are shown as dotted lines. The overall straight-line approximations for gain and angle are shown with dashed lines. The exact curves are plotted with solid lines.

Example 11.3

$$G(s) = \frac{1000(s + 500)}{s^2 + 70s + 10000} = \frac{50(s/500 + 1)}{(s/100)^2 + 2(0.35)(s/100) + 1}$$

Note that for the quadratic term in the denominator the damping ratio is 0.35, an indication of resonance. For small damping ratios the straight-line approximations of Bode magnitude and phase plots can vary significantly from the exact curves. For improved accuracy the approximations would have to be adjusted near the frequency of $\omega = 100$ rad/sec. This is not a consideration when a computer is used to generate a Bode diagram. Figure 11.7 shows the exact gain and angle frequency response curves for $G(s)$.

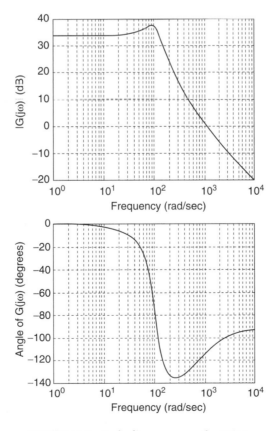

FIGURE 11.7 Bode diagram curves for $G(j\omega)$.

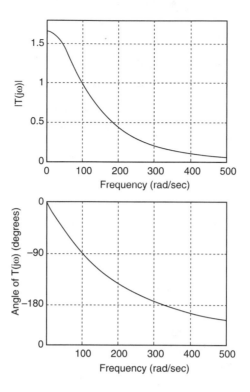

FIGURE 11.8   Linear frequency-response curves for $T(j\omega)$.

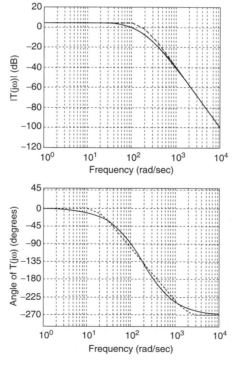

FIGURE 11.9   Bode diagram curves for $T(j\omega)$.

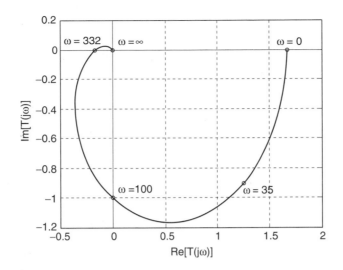

FIGURE 11.10 Nyquist plot of $T(j\omega)$.

11.3 A Comparison of Methods

This chapter ends with the frequency response of a simple system plotted in three different ways.

Example 11.4

$$T(s) = \frac{10^7}{(s+100)(s+200)(s+300)}$$

Figure 11.8 shows linear frequency-response curves for $T(s)$. Corresponding Bode and Nyquist diagrams are shown in Figure 11.9 and Figure 11.10, respectively. The information contained in the three sets of diagrams is the same.

Defining Terms

Bode diagram: A frequency response plot of 20-log gain and phase angle on a log-frequency base.

Breakpoint: A point of abrupt change in slope in the straight-line approximation of a Bode magnitude curve.

Damping ratio: The ratio between a system's damping factor (measure of rate of decay of response) and the damping factor when the system is critically damped.

Decade: Power of ten. In context, a tenfold change in frequency.

Decibel: A measure of relative size. The decibel gain between voltages V_{out} and V_{in} is $20\log_{10}(V_{out}/V_{in})$. The decibel ratio between two powers is $10\log_{10}(P_{out}/P_{in})$

Frequency response: The frequency-dependent relation in both gain and phase difference between steady-state sinusoidal inputs and the resultant steady-state sinusoidal outputs.

Nichols chart: A plot showing magnitude contours and phase contours of the closed-loop transfer function referred to ordinates of logarithmic loop gain and abscissas of loop phase.

Nyquist plot: A parametric frequency response plot with the real part of the transfer function on the abscissa and the imaginary part of the transfer function on the ordinate.

Resonance: The enhancement of the response of a physical system to the steady-state sinusoidal input when the excitation frequency is near a natural frequency of the system.

References

R.C. Dorf, *Modern Control Systems*, 4th ed., Reading, MA: Addison-Wesley, 1986.

G.F. Franklin, J.D. Powell, and A. Emani-Naeini, *Feedback Control of Dynamic Systems*, 3rd ed., Reading, MA: Addison-Wesley, 1994.

J. Golten and A. Verwer, *Control System Design and Simulation*, New York: McGraw-Hill, 1991.

IEEE, *IEEE Standard Dictionary of Electrical and Electronics Terms*, 4th ed., Piscataway, NJ: Institute of Electrical and Electronics Engineers, 1988.

P.O. Neudorfer and M. Hassul, *Introduction to Circuit Analysis*, Englewood Cliffs, NJ: Prentice-Hall, 1990.

W.J. Palm III, *Control Systems Engineering*, New York: Wiley, 1986.

Further Information

Good coverage of frequency response theory and techniques can be found in many undergraduate-level electrical engineering textbooks. Refer especially to classical automatic controls or circuit analysis books.

Useful information can also be found in books on active filter design.

Examples of the application of frequency response methods abound in journal articles ranging over such diverse topics as controls, acoustics, electronics, and communications.

12
Stability Analysis

Ferenc Szidarovszky
University of Arizona

A. Terry Bahill
*University of Arizona
and BAE systems*

12.1 Introduction

In this chapter, which is based on Szidarovszky and Bahill (1998), we discuss **stability**. We start by discussing interior stability, where the stability of the state trajectory or equilibrium state is examined, and then we discuss exterior stability, in which we guarantee that a bounded input always evokes a bounded output. We present four techniques for examining **interior stability**: (1) Lyapunov functions, (2) checking the boundedness or limit of the fundamental matrix, (3) finding the location of the eigenvalues for state-space notation, and (4) finding the location of the poles in the complex frequency plane of the closed-loop transfer function. We present two techniques for examining **exterior** (or bounded-input/bounded-output [BIBO]) **stability** (1) use of the weighting pattern of the system and (2) finding the location of the eigenvalues for state-space notation.

Proving stability with Lyapunov functions is very general: it even works for nonlinear and time-varying systems. It is also good for doing proofs. However, proving the stability of a system with Lyapunov functions is difficult, and failure to find a Lyapunov function that proves a system is stable does not prove that the system is unstable. The next technique we present, finding the fundamental matrix, requires the solution of systems of differential equations, or in the time-invariant case, the computation of the eigenvalues. Determining the eigenvalues or the poles of the transfer function is sometimes difficult because it requires factoring high-order polynomials. However, many commercial software packages are available for this task. We think most engineers would benefit by having one of these computer programs. Jamshidi et al. (1992) and advertisements in technical publications such as the *IEEE Control Systems Magazine* and *IEEE Spectrum* describe many appropriate software packages. The last concept we present, BIBO stability, is very general.

Let us begin our discussion of stability and instability of systems informally. In an *unstable system,* the state can have large variations, and small inputs or small changes in the initial state may produce large variations in the output. A common example of an unstable system is illustrated by someone pointing the microphone of a public address (PA) system at a speaker; a loud high-pitched tone results. Often instabilities are caused by too much gain; so to quiet the PA system, decrease the gain by pointing the microphone away from the speaker. Discrete systems can also be unstable. A friend of ours once provided an example. She was sitting in a chair reading and she got cold. So she went over and turned up the thermostat on the heater. The house warmed up. She got hot, so she got up and turned down the thermostat. The house cooled off. She got cold and turned up the thermostat. This process continued until someone finally suggested that she put on a sweater (reducing the gain of her heat loss system). She did, and was much more comfortable. We called this a discrete system, because she seemed to sample the environment and produce outputs at discrete intervals about 15 minutes apart.

12.2   Using the State of the System to Determine Stability

The stability of a system can be defined with respect to a given equilibrium point in state space. If the initial state x_0 is selected at an equilibrium state $\bar{x}$ of the system, then the state will remain at $\bar{x}$ for all future time. When the initial state is selected close to an equilibrium state, the system might remain close to the equilibrium state or it might move away. In this section, we introduce conditions that guarantee that whenever the system starts near an equilibrium state, it remains near it, perhaps even converging to the equilibrium state as time increases. For simplicity, only time-invariant systems are considered in this section. Time-variant systems are discussed in Section "BIBO Stability."

Continuous, time-invariant systems have the form:

$$\dot{x}(t) = f(x(t)) \tag{12.1}$$

and discrete, time-invariant systems are modeled by the difference equation:

$$x(t + 1) = f(x(t)) \tag{12.2}$$

Here we assume that $f : X \rightarrow R^n$, where $X \subset R^n$ is the state space. We also assume that function f is continuous; furthermore, for arbitrary initial state $x_0 \in X$, there is a unique solution of the corresponding initial value problem $x(t_0) = x_0$, and the entire trajectory $x(t)$ is in X. Assume furthermore that t_0 denotes the initial time period of the system. It is also known that the vector $\bar{x} \in X$ is an equilibrium state of the continuous system (Equation (12.1)) if and only if $f(x) = 0$, and it is an equilibrium state of the discrete system (Equation (12.2)) if and only if $\bar{x} = f(\bar{x})$. In this chapter the equilibrium of a system will always mean the equilibrium *state*, if it is not specified otherwise. In analyzing the dependence of the state trajectory $x(t)$ on the selection of the initial state x_0 nearby the equilibrium, the following stability types are considered.

Definition 12.1

1. An equilibrium state $\bar{x}$ is *stable* if there is an $\varepsilon_0 > 0$ with the following property For all ε_1, $0 < \varepsilon_1 < \varepsilon_0$, there is an $\varepsilon > 0$ such that if $\| \bar{x} - x_0 \| < \varepsilon$, then $\| \bar{x} - x(t) \| < \varepsilon_1$ for all $t > t_0$.
2. An equilibrium state $\bar{x}$ is *asymptotically stable* if it is stable and there is an $\varepsilon > 0$ such that whenever $\| \bar{x} - x_0 \| < \varepsilon$, then $x(t) \rightarrow \bar{x}$ as $t \rightarrow \infty$.
3. An equilibrium state $\bar{x}$ is *globally asymptotically stable* if it is stable and with arbitrary initial state $x_0 \in X, x(t) \rightarrow \bar{x}$ as $t \rightarrow \infty$.

The first definition says an equilibrium state $\bar{x}$ is stable if the entire trajectory $x(t)$ is closer to the equilibrium state than any small ε_1, if the initial state x_0 is selected close enough to the equilibrium state. For asymptotic stability, in addition $x(t)$ converges to the equilibrium state as $t \rightarrow \infty$. If the equilibrium state is globally asymptotically stable, then $x(t)$ converges to the equilibrium state regardless of how the initial state x_0 is selected.

These stability concepts are called *internal*, because they represent properties of the state of the system. They are illustrated in Figure 12.1, where the block dots are the initial states and $\bar{x}$ is the origin. In the electrical engineering literature, sometimes our stability definition is called marginal stability and our asymptotic stability is called stability.

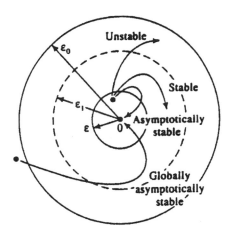

FIGURE 12.1 Stability concepts. (*Source*: F. Szidarovszky and A.T. Bahill, *Linear Systems Theory*, Boca Raton, FL: CRC Press, 1998, p. 199. With permission.)

12.3 Lyapunov Stability Theory

Assume that $\bar{x}$ is an equilibrium state of a continuous or discrete system, and let Ω denote a subset of the state space X such that $\bar{x} \in \Omega$.

Definition 12.2

A real-valued function V defined on Ω is called a Lyapunov function if:

1. V is continuous.
2. V has a unique global minimum at $\bar{x}$ with respect to all other points in Ω.
3. For any state trajectory $x(t)$ contained in Ω, $V(x(t))$ is nonincreasing in t.

The Lyapunov function can be interpreted as the generalization of the energy function in electrical systems. The first requirement simply means that the graph of V has no breaks. The second requirement means that the graph of V has its lowest point at the equilibrium, and the third requirement generalizes the well-known fact of electrical systems that the energy in a free electrical system with resistance always decreases unless the system is at rest.

Theorem 12.1

Assume that there exists a Lyapunov function V on the spherical region:

$$\Omega = \{x \mid \|x - \bar{x}\| < \varepsilon_0\} \tag{12.3}$$

where $\varepsilon_0 > 0$ is given; furthermore, $\Omega \subseteq X$. Then the equilibrium state is stable.

Theorem 12.2

Assume that in addition to the conditions of Theorem 12.1, the Lyapunov function $V(x(t))$ is strictly decreasing in t unless $x(t) = \bar{x}$. Then the equilibrium state is asymptotically stable.

Theorem 12.3

Assume that the Lyapunov function defined on the entire state space X, $V(x(t))$ is strictly decreasing in t unless $x(t) = \bar{x}$; furthermore, $V(x)$ tends to infinity as any component of x gets arbitrarily large in magnitude. Then the equilibrium state is globally asymptotically stable.

Example 12.1

Consider the differential equation:

$$\dot{x} = \begin{pmatrix} 0 & \omega \\ -\omega & 0 \end{pmatrix} x + \begin{pmatrix} 0 \\ 1 \end{pmatrix}$$

which describes a harmonic oscillator.

The stability of the equilibrium state $(1/\omega, 0)^T$ can be verified directly by using Theorem 12.1 without computing the solution. Select the Lyapunov function:

$$V(x) = (x - \bar{x})^T (x - \bar{x}) = \| x - \bar{x} \|_2^2$$

where the Euclidean norm is used.

This is continuous in x; furthermore, it has its minimal (zero) value at $x = \bar{x}$. Therefore, to establish the stability of the equilibrium state we have to show that $V(x(t))$ is decreasing. Simple differentiation shows that

$$\frac{d}{dt} V(x(t)) = 2(x - \bar{x})^T \dot{x} = 2(x - \bar{x})^T (Ax + b)$$

with

$$A = \begin{pmatrix} 0 & \omega \\ -\omega & 0 \end{pmatrix} \quad \text{and} \quad b = \begin{pmatrix} 0 \\ 1 \end{pmatrix}$$

That is, with $x = (x_1, x_2)^T$,

$$\frac{d}{dt} V(x(t)) = 2\left(x_1 - \frac{1}{\omega}, x_2\right)\begin{pmatrix} \omega x_2 \\ -\omega x_1 + 1 \end{pmatrix}$$
$$= 2(\omega x_1 x_2 - x_2 - \omega x_1 x_2 + x_2) = 0$$

Therefore, function $V(x(t))$ is a constant, which is a nonincreasing function. That is, all conditions of Theorem 12.1 are satisfied, which implies the stability of the equilibrium state.

Theorem 12.1, Theorem 12.2, and Theorem 12.3 guarantee, respectively, the stability, asymptotic stability, and global asymptotic stability of the equilibrium state if a Lyapunov function is found. Failure to find such a Lyapunov function does not mean that the system is unstable or that the stability is not asymptotic or globally asymptotic. It only means that you were not clever enough to find a Lyapunov function that proved stability.

12.4 Stability of Time-Invariant Linear Systems

This section is divided into two subsections. In the first subsection, the stability of linear time-invariant systems given in state-space notation is analyzed. In the second subsection, methods based on transfer functions are discussed.

Stability Analysis with State-Space Notation

Consider the time-invariant continuous linear system:

$$\dot{x} = Ax + b \tag{12.4}$$

and the time-invariant discrete linear system:

$$\mathbf{x}(t+1) = \mathbf{A}\mathbf{x}(t) + \mathbf{b} \tag{12.5}$$

Assume that $\bar{x}$ is an equilibrium state, and let $\boldsymbol{\phi}(t, t_0)$ denote the fundamental matrix.

Theorem 12.4

1. The equilibrium state $\bar{x}$ is stable if and only if $\boldsymbol{\phi}(t, t_0)$ is bounded for $t \geqslant t_0$.
2. The equilibrium state $\bar{x}$ is asymptotically stable if and only if $\boldsymbol{\phi}(t, t_0)$ is bounded and tends to zero as $t \to \infty$.

In the case of linear systems, asymptotic stability and global asymptotic stability are equivalent.

We use the symbol s to denote complex frequency, i.e., $s = \sigma + j\omega$. For specific values of s, such as eigenvalues and poles, we use the symbol λ.

Theorem 12.5

1. Assume that for all eigenvalues λ_i of A, Re $\lambda_i \leqslant 0$ in the continuous case (or $|\lambda_i| \leqslant 1$ in the discrete case), and all eigenvalues with the property Re $\lambda_i = 0$ (or $|\lambda_i| = 1$) have single multiplicity; then the equilibrium state is stable.
2. The stability is asymptotic if and only if for all i, Re $\lambda_i < 0$ (or $|\lambda_i| < 1$).
3. If for at least one eigenvalue of A, Re $\lambda_i > 0$ (or $|\lambda_i| > 1$) then the equilibrium is unstable.

Remark 1. Note that Part 1 gives only sufficient conditions for the stability of the equilibrium state. As the following examples show, these conditions are not necessary.

If there is at least one multiple eigenvalue with zero-real part (unit absolute value) then we cannot decide the stability of the equilibrium based on only the eigenvalues. In such cases the boundedness of the fundamental matrix has to be checked.

Example 12.2

Consider first the continuous system $\dot{x} = \mathbf{O}x$, where $\mathbf{O}$ is the zero matrix. Note that all constant functions $x(t) \equiv \bar{x}$ are solutions and also equilibrium states. Since

$$\boldsymbol{\phi}(t, t_0) = e^{\mathbf{O}(t-t_0)} = \mathbf{I}$$

is bounded (being independent of t), all equilibrium states are stable, but $\mathbf{O}$ has only one eigenvalue $\lambda_1 = 0$ with zero real part and multiplicity n, where n is the order of the system.

Consider next the discrete systems $x(t+1) = \mathbf{I}x(t)$, when all constant functions $x(t) \equiv \bar{x}$ are also solutions and equilibrium states. Furthermore:

$$\boldsymbol{\phi}(t, t_0) = \mathbf{A}^{t-t_0} = \mathbf{I}^{t-t_0} = \mathbf{I}$$

which is obviously bounded. Therefore, all equilibrium states are stable, but the condition of Part 1 of the theorem is violated again.

Remark 2. The following extension of Theorem 12.5 can be proven. The equilibrium state is stable if and only if for all eigenvalues of A, Re $\lambda_i \leqslant 0$ (or $|\lambda_i| \leqslant 1$), and if λ_i is a repeated eigenvalue of A such that Re $\lambda_i = 0$ (or $|\lambda_i| = 1$), then the size of each block containing λ_i in the Jordan canonical form of A is 1×1.

Remark 3. The equilibrium states of inhomogeneous equations are stable or asymptotically stable if and only if the same holds for the equilibrium states of the corresponding homogeneous equations.

Example 12.3

Consider again the continuous system:

$$\dot{x} = \begin{pmatrix} 0 & \omega \\ -\omega & 0 \end{pmatrix} x + \begin{pmatrix} 0 \\ 1 \end{pmatrix}$$

the stability of which was analyzed earlier in Example 12.1 by using the Lyapunov function method. The characteristic polynomial of the coefficient matrix is

$$\varphi(s) = \det\begin{pmatrix} -s & \omega \\ -\omega & -s \end{pmatrix} = s^2 + \omega^2$$

therefore the eigenvalues are $\lambda_1 = j\omega$ and $\lambda_2 = -j\omega$. Both eigenvalues have single multiplicities, and $\operatorname{Re} \lambda_1 = \operatorname{Re} \lambda_2 = 0$. Hence, the conditions of Part 1 are satisfied, and therefore the equilibrium state is stable. The conditions of Part 2 do not hold. Consequently, the system is not asymptotically stable.

If a time-invariant system is nonlinear, then the Lyapunov method is the most popular choice for stability analysis. If the system is linear, then the direct application of Theorem 12.5 is more attractive, since the eigenvalues of the coefficient matrix A can be obtained by standard methods. In addition, several conditions are known from the literature that guarantee the asymptotic stability of time-invariant discrete and continuous systems even without computing the eigenvalues. For examining asymptotic stability, linearization is an alternative approach to the Lyapunov method as is shown here. Consider the time-invariant continuous and discrete systems

$$\dot{x}(t) = f(x(t))$$

and

$$x(t + 1) = f(x(t))$$

Let $J(x)$ denote the Jacobian of $f(x)$, and let $\bar{x}$ be an equilibrium state of the system. It is known that the method of linearization around the equilibrium state results in the time-invariant linear systems

$$\dot{x}_\delta(t) = J(\bar{x})x_\delta(t)$$

and

$$x_\delta(t + 1) = J(\bar{x})x_\delta(t)$$

where $x_\delta(t) = x(t) - \bar{x}$. It is also known from the theory of difference and ordinary differential equations that the asymptotic stability of the zero vector in the linearized system implies the asymptotic stability of the equilibrium state $\bar{x}$ in the original nonlinear system. The asymptotic stability of the linearized system can be examined by the methodology being discussed above.

For continuous systems, the following results have special importance.

Theorem 12.6

The equilibrium state of a continuous system (Equation (12.4)) is asymptotically stable if and only if equation

$$A^T Q + QA = -M \tag{12.6}$$

has positive definite solution Q with some positive definite matrix M. We note that in practical applications the identity matrix is usually selected for M.

Theorem 12.7

Let $\varphi(\lambda) = \lambda_n + p_{n-1}\lambda^{n-1} + \cdots + p_1\lambda + p_0$ be the characteristic polynomial of matrix **A**. Assume that all eigenvalues of matrix **A** have negative real parts. Then $p_i > 0$ $(i = 0, 1, \ldots, n-1)$.

Corollary. If any of the coefficients p_i is negative or zero, the equilibrium state of the system with coefficient matrix A cannot be asymptotically stable. This result can be used as an initial stability test. However, the conditions of the theorem do not imply that the eigenvalues of A have negative real parts, the corresponding sufficient and necessary conclusions are known as the Routh–Hurwitz stability criteria (see, for example, Szidarovszky and Bahill, 1998).

Example 12.4

For matrix

$$A = \begin{pmatrix} 0 & \omega \\ -\omega & 0 \end{pmatrix}$$

the characteristic polynomial is $\varphi(s) = s^2 + \omega^2$. Since the coefficient of s^1 is zero, this system is not asymptotically stable.

The Transfer Function Approach

The transfer function of the continuous system

$$\dot{x} = \mathbf{Ax} + \mathbf{Bu}$$
$$\mathbf{y} = \mathbf{Cx} \tag{12.7}$$

and that of the discrete system

$$x(t+1) = \mathbf{Ax}(t) + \mathbf{Bu}(t)$$
$$y(t) = \mathbf{Cx}(t) \tag{12.8}$$

have the common form

$$\mathbf{TF}(s) = \mathbf{C}(s\mathbf{I} - \mathbf{A})^{-1}\mathbf{B}$$

If both the input and output are single, then

$$TF(s) = \frac{Y(s)}{U(s)}$$

or in the familiar electrical engineering notation:

$$TF(s) = \frac{KG(s)}{1 + KG(s)H(s)} \tag{12.9}$$

where K is the gain term in the forward loop, $G(s)$ represents the dynamics of the forward loop or the plant, and $H(s)$ models the dynamics in the feedback loop. We note that in the case of continuous systems, s is the variable of the transfer function, and for discrete systems, the variable is denoted by z.

After World War II, systems and control theory flourished. The transfer function representation was the most popular representation for systems. To determine the stability of a system, we merely had to factor the denominator of the transfer function (Equation (12.9)) and see if the poles were in the left half of the complex frequency plane. However, with manual techniques, factoring polynomials of large order was difficult. So, engineers, being naturally lazy people, developed several ways to determine the stability of a system without factoring the polynomials (Dorf, 1992). First, we have methods of Routh and Hurwitz, developed a century ago, that looked at the coefficients of the characteristic polynomial. These methods showed whether the system was stable or not, but they did not show how close the system was to being stable.

What we want to know is, for what value of gain, **K**, and at what frequency, ω, will the denominator of the transfer function (Equation (12.9)) become zero. Or, when $KG(s)H(s) = -1$, meaning when the magnitude of **KGH** equals 1 with a phase angle of $-180°$. These parameters can be determined easily with a Bode diagram. Construct a Bode diagram for **KHG** of the system, look at the frequency where the phase angle equals $-180°$, and look up at the magnitude plot. If it is smaller than 1.0, then the system is stable. If it is larger than 1.0, then the system is unstable.

The quantity $KG(s)H(s)$ is called the open-loop transfer function of the system, because it is the effect that would be encountered by a signal making one loop around the system if the feedback loop were artificially open (Bahill, 1981).

To gain some intuition, think of a closed-loop negative feedback system. Apply a small sinusoid at frequency ω to the input. Assume that the gain around the loop, **KGH**, is 1 or more, and that the phase angle is $-180°$. The summing junction will flip over the feedback signal and add it to the original signal. The result is a signal that is bigger than what came in. This signal will circulate around this loop, getting bigger and bigger on every loop until the real system no longer matches the model. This is what we call instability.

The question of stability can also be answered with Nyquist diagrams. They are related to Bode diagrams, but they give more information. A simple way to construct a Nyquist diagram is to make a polar plot on the complex frequency plane of the Bode diagram. Simply stated, if this contour encircles the −1 point in the complex frequency plane, then the system is unstable (see Figure 12.2).

The two advantages of the Nyquist technique are: (1) in addition to the information of Bode diagrams, there are about a dozen rules that can be used to help construct Nyquist diagrams, and (2) Nyquist diagrams handle bizarre systems better, as is shown in the following rigorous statement of the Nyquist stability criterion. The number of clockwise encirclements minus the number of counter clockwise encirclements of the point $s = -1+j0$ by the Nyquist plot of $KG(s)H(s)$ is equal to the number of poles of $Y(s)/U(s)$ minus the number of poles of $KG(s)H(s)$ in the right half of the s-plane.

The root-locus technique was another popular technique for assessing stability. It furthermore allowed the engineer to see the effects of small changes in the gain, **K**, on the stability of the system. The root-locus diagram shows the location in the s-plane of the poles of the closed-loop transfer function, $Y(s)/U(s)$. All branches of the root-locus diagram start on poles of open-loop transfer function, **KGH**, and end either on

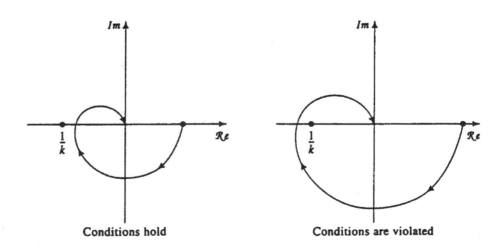

FIGURE 12.2 Illustration of Nyquist stability criteria. (*Source:* F. Szidarovszky and A.T. Bahill, *Linear Systems Theory*, Boca Raton, FL: CRC Press, 1998, p. 219. With permission.)

zeros of the open-loop transfer function, **KGH**, or at infinity. There are about a dozen rules to help draw these trajectories.

We consider all these techniques to be old-fashioned. They were developed to help answer the question of stability without factoring the characteristic polynomial. However, many computer programs are currently available that factor polynomials. We recommend that engineers merely buy one of these computer packages and find the roots of the closed-loop transfer function to assess the stability of a system.

The poles of a system are defined as all values of s such that $s\mathbf{I}\text{-}\mathbf{A}$ is singular. The poles of the closed-loop transfer function are the same as the eigenvalues of the system: engineers prefer the term *poles* and the symbol s, and mathematicians prefer the term *eigenvalues* and the symbol λ for specific values of s. We will use s for complex frequency and λ for specific values of s.

Sometimes, some poles could be canceled in the rational function form of $\mathbf{TF}(s)$ so that they would not be explicitly shown. However, even if some poles could be canceled by zeros, we would still have to consider all the poles in the following criteria. The equilibrium state of a continuous system (Equation (12.7)) with constant input is stable if all poles of $\mathbf{TF}(s)$ have nonpositive real parts and all poles with zero real parts are single. The equilibrium state is asymptotically stable if and only if all poles of $\mathbf{TF}(s)$ have negative real parts; that is, all poles are in the left half of the s-plane. Similarly, the equilibrium state of a discrete system (Equation (12.8)) with constant input is stable if all poles of $\mathbf{TF}(z)$ have absolute values less than or equal to one and all poles with unit absolute values are single. The equilibrium state is asymptotically stable if and only if all poles of $\mathbf{TF}(z)$ have absolute values less than one; that is, the poles are all inside the unit circle of the z-plane.

Example 12.5

Consider again the system

$$\dot{x} = \begin{pmatrix} 0 & \omega \\ -\omega & 0 \end{pmatrix} x + \begin{pmatrix} 0 \\ 1 \end{pmatrix}$$

that was discussed earlier. Assume that the output equation has the form:

$$y = (1, 1)x$$

then

$$TF(s) = \frac{s + \omega}{s^2 + \omega^2}$$

The poles are $+j\omega$ and $-j\omega$, which have zero real parts; that is, they are on the imaginary axis of the s-plane. Consequently, the equilibrium state is stable but not asymptotically stable. A system such as this would produce constant amplitude sinusoids at frequency ω. So it seems natural to assume that such systems would be used to build sinusoidal signal generators and to model oscillating systems. However, this is not the case, because (1) zero resistance circuits are hard to make and therefore, most function generators use other techniques to produce sinusoids; and (2) most real-world oscillating systems (i.e., biological systems) have energy dissipation elements in them.

More generally, real-world function generators are seldom made from closed-loop feedback control systems with $-180°$ of phase shift because (1) it would be difficult to get a broad range of frequencies and several waveforms from such systems, (2) precise frequency selection would require expensive high-precision components, and (3) it would be difficult to maintain constant frequency in such circuits in the face of changing temperatures and power supply variations. Likewise, closed-loop feedback control systems with $-180°$ of phase shift are not good models for oscillating biological systems because most biological systems oscillate because of nonlinear network properties.

A special stability criterion for single-input, single-output time-invariant, continuous systems will be introduced next. Consider the system

$$\dot{x} = \mathbf{A}x + \mathbf{b}u \text{ and } y = \mathbf{c}^\mathsf{T}x \tag{12.10}$$

where **A** is an $n \times n$ constant matrix, and **b** and **c** are constant n-dimensional vectors. The transfer function of this system is

$$\mathbf{TF}_1(s) = \mathbf{c}^\mathsf{T}(s\mathbf{I} - \mathbf{A})^{-1}\mathbf{b}$$

which is obviously a rational function of s. Now let us add negative feedback around this system so that $u = ky$ where k is a constant. The resulting system can be described by the differential equation

$$\dot{x} = \mathbf{A}x + k\mathbf{b}\mathbf{c}^\mathsf{T}x = (\mathbf{A} + k\mathbf{b}\mathbf{c}^T)x \tag{12.11}$$

The transfer function of this feedback system is

$$TF(s) = \frac{TF_1(s)}{1 - kTF_1(s)} \tag{12.12}$$

To help show the connection between the asymptotic stability of system (Equation (12.10)) and system (Equation (12.11)), we introduce the following definition.

Definition 12.3

Let $r(s)$ be a rational function of s. Then the locus of points

$$L(r) = \left\{a + jb \middle| a = \mathrm{Re}(r(jv)), \ b = \mathrm{Im}(r(jv)), \ v \in \mathbf{R}\right\}$$

is called the *response diagram of r*. Note that $L(r)$ is the image of the imaginary line $\mathrm{Re}(s) = 0$ under the mapping r. We shall assume that $L(r)$ is bounded, which is the case if and only if the degree of the denominator is not less than that of the numerator and r has no poles on the line $\mathrm{Re}(s) = 0$.

Theorem 12.8

The Nyquist stability criterion. Assume that TF_1 has a bounded response diagram $L(TF_1)$. If TF_1 has v poles in the right half of the s-plane, where $\mathrm{Re}(s) > 0$, then TF has $\rho + v$ poles in the right half of the s-plane if the point $1/k + j \cdot 0$ is not on $L(\mathrm{TF}_1)$, and $L(\mathrm{TF}_1)$ encircles $1/k + j \cdot 0$ ρ times in the clockwise sense.

Corollary. Assume that system (Equation (12.10)) is asymptotically stable with constant input and that $L(TF_1)$ is bounded and traversed in the direction of increasing v and has the point $1/k + j \cdot 0$ on its left. Then the feedback system (Equation (12.11)) is also asymptotically stable.

This result has many applications since feedback systems have a crucial role in constructing stabilizers, observers, and filters for given systems. Figure 12.2 illustrates the conditions of the corollary. The application of this result is especially convenient, if system (Equation (12.10)) is given and only appropriate values k of the feedback are to be determined. In such cases, the locus $L(TF_1)$ has to be computed first, and then the region of the appropriate k values can be determined easily from the graph of $L(TF_1)$.

This analysis has dealt with the closed-loop transfer function, whereas the techniques of Bode, root-locus, etc., use the open-loop transfer function. This should cause little confusion as long as the distinction is kept in mind.

12.5   BIBO Stability

In the previous sections, internal stability of time-invariant systems was examined, i.e., the stability of the state was investigated. In this section, the **external stability** of systems is discussed; this is usually called the **BIBO** (*bounded-input, bounded-output*) **stability**. Here, we drop the simplifying assumption of the previous section that the system is time-invariant; we will include time-variant systems in the following analysis.

Definition 12.4

A system is called BIBO stable if for zero initial conditions, a bounded input always evokes a bounded output.

   For continuous systems, a necessary and sufficient condition for BIBO stability can be formulated as follows.

Theorem 12.9

Let $T(t, \tau) = (t_{ij}(t, \tau))$ be the weighing pattern, $C(t)\phi(t, \tau)B(\tau)$, of the system. Then the continuous time-variant linear system is BIBO stable if and only if the integral

$$\int_{t_0}^{t} \left| t_{ij}(t, \tau) \right| d\tau \tag{12.13}$$

is bounded for all $t \geq t_0$, i and j.

   Corollary.   Integrals (Equation (12.13)) are all bounded if and only if

$$I(t) = \int_{t_0}^{t} \sum_{i} \sum_{j} \left| t_{ij}(t, \tau) \right| d\tau \tag{12.14}$$

is bounded for $t \geq t_0$. Therefore, it is sufficient to show the boundedness of only one integral in order to establish BIBO stability.

   The discrete counterpart of this theorem can be given in the following way.

Theorem 12.10

Let $T(t, \tau) = (t_{ij}(t, \tau))$ be the weighing pattern of the discrete linear system. Then it is BIBO stable if and only if the sum

$$I(t) = \sum_{\tau=t_0}^{t-1} \left| t_{ij}(t, \tau) \right| \tag{12.15}$$

is bounded for all $t \geq t_0$, i and j.

   Corollary.   The sums (Equation (12.15)) are all bounded if and only if

$$\sum_{\tau=t_0}^{t-1} \sum_{i} \sum_{j} \left| t_{ij}(t, \tau) \right| \tag{12.16}$$

is bounded. Therefore, it is sufficient to verify the boundedness of only one sum in order to establish BIBO stability.

Consider next the time-invariant case, when $\mathbf{A}(t) \equiv \mathbf{A}, \mathbf{B}(t) \equiv \mathbf{B}, \mathbf{C}(t) \equiv \mathbf{C}$. From the foregoing theorems and the definition $\mathbf{T}(t, \tau)$ we have immediately the following sufficient condition.

Theorem 12.11

Assume that for all eigenvalues λ_i of $\mathbf{A}$, $\mathrm{Re}\, \lambda_i < 0$ (or $|\lambda_i| < 1$). Then the time-invariant linear continuous (or discrete) system is BIBO stable.

BIBO stability is different from stability in the sense of Definition 12.1. For example, a system with a zero eigenvalue might not be BIBO stable; however, if the eigenvalue with zero real part is single, then the system still might be stable in the sense of Definition 12.1.

Finally, we note that BIBO stability is not implied by an observation that a certain bounded input generates bounded output. All bounded inputs must generate bounded outputs in order to guarantee BIBO stability.

Adaptive-control systems are time-varying systems. Therefore, it is usually difficult to prove that they are stable. Szidarovszky et al. (1990), however, show a technique for doing this. This result gives a necessary and sufficient condition for the existence of an asymptotically stable model-following adaptive-control system, and in the case of the existence of such systems, they present an algorithm for finding the appropriate feedback parameters.

12.6   Bifurcations

The asymptotic properties of the equilibrium state of any dynamic system depend on the particular values of the model parameters. For certain values, the system might be asymptotically stable and with a sudden change of one or more parameter values this stability disappears. Such sudden change in asymptotical behavior is called bifurcation. As a simple illustration, consider the following extension of Example 12.1.

Example 12.6

Consider the system

$$\dot{x} = \begin{pmatrix} \delta & \omega \\ -\omega & \delta \end{pmatrix} x + \begin{pmatrix} 0 \\ 1 \end{pmatrix}$$

in which $\omega > 0$ and δ is a real parameter. The characteristic polynomial of the system can be written as

$$\lambda^2 - 2\lambda\delta + (\delta^2 + \omega^2) = 0$$

so the eigenvalues are

$$\lambda_1 = \delta + j\omega \text{ and } \lambda_2 = \delta - j\omega$$

If $\delta < 0$, then both eigenvalues have negative real parts implying asymptotical stability. If $\delta > 0$, then the real part of the eigenvalues becomes positive, so the equilibrium becomes unstable. So at $\delta = 0$ bifurcation occurs with an eigenvalues with zero-real part.

Parameter δ is called the **bifurcation parameter**, since we examine the change of stability behavior as a function of the change in its value. The eigenvalues also depend on the value of the bifurcation parameter. If the real part of the derivative of the pure complex eigenvalue with respect to the bifurcation parameter is nonzero, then *Hopf-bifurcation* occurs, which guarantees the birth of limit cycles around the equilibrium. Other bifurcation types and their conditions with applications are discussed, for example, in Guckenheimer and Holmes (1983).

12.7   Physical Examples

In this section, we show some examples of stability analysis of physical systems.

1. Consider a simple *harmonic oscillator* constructed of a mass and an ideal spring. Its dynamic response is summarized by

$$\dot{x} = \begin{pmatrix} 0 & \omega \\ -\omega & 0 \end{pmatrix} x + \begin{pmatrix} 0 \\ 1 \end{pmatrix} u$$

In Example 12.3, we showed that this system is stable but not asymptotically stable. This means that if we leave it alone in its equilibrium state, it will remain stationary, but if we jerk on the mass it will oscillate forever. There is no damping term to remove the energy, so the energy will be transferred back and forth between potential energy in the spring and kinetic energy in the moving mass. A good approximation of such a harmonic oscillator is the pendulum clock. The more expensive it is (i.e., the smaller the damping), the less often we have to wind it (i.e., add energy).

2. A *linear second-order electrical system* composed of a series connection of an input voltage source, an inductor, a resistor, and a capacitor, with the output defined as the voltage across the capacitor, can be characterized by the second-order equation

$$\frac{V_{\text{out}}}{V_{\text{in}}} = \frac{1}{LCs^2 + RCs + 1}$$

For convenience, let us define

$$\omega_n = \sqrt{\frac{1}{LC}} \text{ and } \zeta = \frac{R}{2}\sqrt{\frac{C}{L}}$$

With these parameters the transfer function becomes

$$\frac{V_{\text{out}}}{V_{\text{in}}} = \frac{\omega_n^2}{s^2 + 2\zeta\omega_n s + \omega_n^2}.$$

Is this system stable? The roots of the characteristic equation are

$$\lambda_{1,2} = -\zeta\omega_n \pm j\omega_n\sqrt{1 - \zeta^2}$$

If $\zeta > 0$, the poles are in the left half of the *s*-plane, and therefore the systems is asymptotically stable. If $\zeta = 0$, as in the previous example, the poles are on the imaginary axis; therefore, the system is stable but not asymptotically stable. If $\zeta < 0$, the poles are in the right half of the *s*-plane and the system is unstable.

3. An *electrical system* is shown in Figure 12.3. Simple calculation shows that by introducing the variables

$$x_1 = i_L, x_2 = v_c, \quad \text{and} \quad u = v_s$$

the system can be described by the differential equations

$$\dot{x}_1 = -\frac{R_1}{L}x_1 - \frac{1}{L}x_2 + \frac{1}{L}u$$

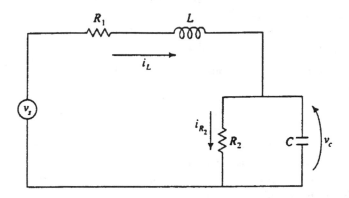

FIGURE 12.3 A simple electrical system. (*Source:* F. Szidarovszky and A.T. Bahill, *Linear Systems Theory*, Boca Raton, FL: CRC Press, 1998, p. 158. With permission.)

$$\dot{x}_2 = \frac{1}{C}x_1 - \frac{1}{CR_2}x_2$$

The characteristic equation has the form

$$\left(-s - \frac{R_1}{L}\right)\left(-s - \frac{1}{CR_2}\right) + \frac{1}{LC} = 0$$

which simplifies as

$$s^2 + s\left(\frac{R_1}{L} + \frac{1}{CR_2}\right) + \left(\frac{R_1}{LCR_2} + \frac{1}{LC}\right) = 0$$

Since R_1, R_2, L, and C are positive numbers, the coefficients of this equation are all positive. The constant term equals $\lambda_1\lambda_2$, and the coefficient of s^1 is $-(\lambda_1 + \lambda_2)$. Therefore

$$\lambda_1 + \lambda_2 < 0 \quad \text{and} \quad \lambda_1\lambda_2 > 0$$

If the eigenvalues are real, then these relations hold if and only if both eigenvalues are negative. If they were positive, then $\lambda_1 + \lambda_2 > 0$. If they had different signs, then $\lambda_1\lambda_2 < 0$. Furthermore, if at least one eigenvalue is zero, then $\lambda_1\lambda_2 = 0$. Assume next that the eigenvalues are complex:

$$\lambda_{1,2} = \operatorname{Re} s \pm j \operatorname{Im} s$$

Then

$$\lambda_1 + \lambda_2 = 2\operatorname{Re} s$$

and

$$\lambda_1\lambda_2 = (\operatorname{Re} s)^2 \pm (\operatorname{Im} s)^2$$

Hence, $\lambda_1 + \lambda_2 < 0$ implies that $\operatorname{Re} s < 0$.

In summary, the system is asymptotically stable, since in both the real and complex cases the eigenvalues have negative values and negative real parts, respectively.

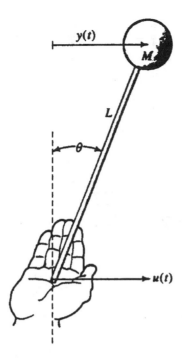

FIGURE 12.4 Stick balancing. (*Source:* F. Szidarovszky and A.T. Bahill, *Linear Systems Theory*, Boca Raton, FL: CRC Press, 1998, p. 165. With permission.)

4. The classical *stick-balancing* problem shown in Figure 12.4. Simple analysis shows that $y(t)$ satisfies the second-order equation

$$\ddot{y} = \frac{g}{L}(y - u)$$

If one selects $L = 1$, then the characteristic equation has the form

$$s^2 - g = 0$$

So the eigenvalues are

$$\lambda_{1,2} = \pm\sqrt{g}$$

One is in the right half of the s-plane and the other is in the left half of the s-plane, so the system is unstable. This instability is understandable, since without an intelligent input to control the system, if the stick is not upright with zero velocity, it will fall over.

5. A simple transistor circuit can be modeled as shown in Figure 12.5. The state variables are related to the input and output of the circuit: The base current, i_b, is x_1 and the output voltage, v_{out}, is x_2. Therefore:

$$\dot{x} = \begin{pmatrix} -\dfrac{h_{ie}}{L} & 0 \\ \dfrac{h_{fe}}{C} & 0 \end{pmatrix} x + \begin{pmatrix} \dfrac{1}{L} \\ 0 \end{pmatrix} e_s \quad \text{and} \quad c^{\mathrm{T}} = (0, 1)$$

The **A** matrix looks strange with a column of all zeros, and indeed the circuit does exhibit odd behavior. For example, as we will show, there is no equilibrium state for a unit step input of e_s. This is reasonable, however, because the model is for mid-frequencies, and a unit step does not qualify. In response to a unit

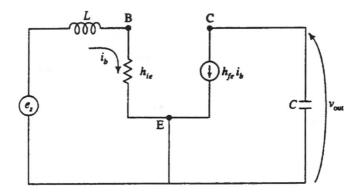

FIGURE 12.5 A model for a simple transistor circuit. (*Source:* F. Szidarovszky and A.T. Bahill, *Linear Systems Theory*, Boca Raton, FL: CRC Press, 1998, p. 160. With permission.)

step the output voltage will increase linearly until the model is no longer valid. If e_s is considered the input, the system is

$$\dot{x} = \begin{pmatrix} -\dfrac{h_{ie}}{L} & 0 \\ \dfrac{h_{fe}}{C} & 0 \end{pmatrix} x + \begin{pmatrix} \dfrac{1}{L} \\ 0 \end{pmatrix} u$$

If $u(t) \equiv 1$, then at the equilibrium state

$$\begin{pmatrix} -\dfrac{h_{ie}}{L} & 0 \\ \dfrac{h_{fe}}{C} & 0 \end{pmatrix} \begin{pmatrix} \bar{x}_1 \\ \bar{x}_2 \end{pmatrix} + \begin{pmatrix} \dfrac{1}{L} \\ 0 \end{pmatrix} = \begin{pmatrix} 0 \\ 0 \end{pmatrix}$$

That is:

$$-\frac{h_{ie}}{L}\bar{x}_1 + \frac{1}{L} = 0$$

$$\frac{h_{fe}}{C}\bar{x}_1 = 0$$

Since $h_{fe}/C \neq 0$, the second equation implies that $\bar{x}_1 = 0$, and by substituting this value into the first equation we get the obvious contradiction $1/L = 0$. Hence, with nonzero constant input *no* equilibrium state exists.

Let us now investigate the stability of this system. First let $\tilde{x}(t)$ denote a fixed trajectory of this system, and let $x(t)$ be an arbitrary solution. Then the difference $\delta x(t) = x(t) - \tilde{x}(t)$ satisfies the homogenous equation

$$\delta \dot{x} = \begin{pmatrix} -\dfrac{h_{ie}}{L} & 0 \\ \dfrac{h_{fe}}{C} & 0 \end{pmatrix} \delta x$$

This system has an equilibrium $\boldsymbol{\delta x}(t) = 0$. Next, the stability of this equilibrium is examined by solving for the poles of the transfer function. The characteristic equation is

$$\det \begin{pmatrix} -\dfrac{h_{ie}}{L} - s & 0 \\ \dfrac{h_{fe}}{C} & -s \end{pmatrix} = 0$$

which can be simplified as

$$s^2 + s\frac{h_{ie}}{L} + 0 = 0$$

The roots are

$$\lambda_1 = 0 \text{ and } \lambda_2 = -\frac{h_{ie}}{L}$$

Therefore, the system is stable but not asymptotically stable. This stability means that for small changes in the initial state, the entire trajectory $\boldsymbol{x}(t)$ remains close to $\tilde{\boldsymbol{x}}(t)$.

Defining Terms

Asymptotic stability: An equilibrium state $\bar{x}$ of a system is asymptotically stable if, in addition to being stable, there is an $\varepsilon > 0$ such that whenever $\|\bar{x} - x_0\| < \varepsilon$, then $x(t) \rightarrow \bar{x}$ as $t \rightarrow \infty$. A system is asymptotically stable if all the poles in a closed-loop transfer function are in the left half of the s-plane (inside the unit circle of the z-plane for discrete systems).

BIBO stability: A system is BIBO stable if for zero initial conditions any bounded input always evokes a bounded output.

Bifurcation: If a sudden change of a model parameter value results in a change of the asymptotic behavior of the system. This model parameter is called the *bifurcation parameter*, and its specific value where the change occurs, is called the *critical value.*

External stability: Stability concepts related to the input–ouput behavior of the system.

Global asymptotic stability: An equilibrium state $\bar{x}$ of a system is globally asymptotically stable if it is stable and with arbitrary initial state $x_0 \in X, x(t) \rightarrow \bar{x}$ as $t \rightarrow \infty$.

Instability: An equilibrium state of a system is unstable if it is not stable. A system is unstable if at least one pole of the closed-loop transfer is in the right half of the s-plane (outside the unit circle of the z-plane for discrete systems). A system might be unstable if poles with zero-real parts (with unit absolute values) are multiple.

Internal stability: Stability concepts related to the state of the system.

Stability: An equilibrium state $\bar{x}$ of a system is stable if there is an $\varepsilon_0 > 0$ with the following property: for all $\varepsilon_1, 0 < \varepsilon_1 < \varepsilon_0$, there is an $\varepsilon > 0$ such that if $\|\bar{x} - x_0\| < \varepsilon$, then $\|\bar{x} - x(t)\| < \varepsilon_1$ for all $t > t_0$. A system is stable if the poles of its closed-loop transfer function are (1) in the left half of the complex frequency plane, called the s-plane (inside the unit circle of the z-plane for discrete systems), or (2) on the imaginary axis, and all of the poles on the imaginary axis are single (on the unit circle and such poles are single for discrete systems). Stability for a system with repeated poles on the $j\omega$ axis (the unit circle) is complicated and is examined in the discussion after Theorem 12.5. In the electrical engineering literature, this definition of stability is sometimes called *marginal stability* and sometimes *stability in the sense of Lyapunov.*

References

A.T. Bahill, *Bioengineering: Biomedical, Medical and Clinical Engineering*, Englewood Cliffs, NJ: Prentice-Hall, 1981, pp. 214–215, 250–252.

R.C. Dorf, *Modern Control Systems*, 6th ed., Reading, MA: Addison-Wesley, 1992.

J. Guckenheimer and P. Holmes, *Nonlinear Oscillations, Dynamic Systems and Bifurcation of Vector Fields*, Berlin/Heidelberg/New York: Springer-Verlag, 1983.

M. Jamshidi, M. Tarokh, and B. Shafai, *Computer-Aided Analysis and Design of Linear Control Systems*, Englewood Cliffs, NJ: Prentice-Hall, 1992.

A.M. Lyapunov, "Problème général de la stabilité du mouvement" (in French), *Ann. Fac. Sci. Toulouse*, vol. 9, 1907, pp. 203–474. Reprinted in *Ann. Math. Study*, no. 17, 1949, Princeton University Press.

F. Szidarovszky and A.T. Bahill, *Linear Systems Theory*, 2nd ed., Boca Raton, FL: CRC Press, 1998.

F. Szidarovszky, A.T. Bahill and S. Molnar, "On stable adaptive control systems," *Pure Math. and Appl.*, vol. 1, ser. B, no. 2–3, pp. 115–121, 1990.

Further Information

For further information, consult the textbooks *Modern Control Systems* by Dorf (1992) or *Linear Systems Theory* by Szidarovszky and Bahill (1998).

13

Computer Software for Circuit Analysis and Design[1]

J. Gregory Rollins
Technology Modeling Associates Inc.

Sina Balkir
University of Nebraska–Lincoln

Peter Bendix
LSI Logic Corp.

13.1   Analog Circuit Simulation

J. Gregory Rollins (revised by Sina Balkir)

Introduction

Computer-aided simulation is a powerful aid during the design or analysis of electronic circuits and semiconductor devices. The first part of this chapter focuses on analog circuit simulation. The second part covers simulations of semiconductor processing and devices. While the main emphasis is on analog circuits, the same simulation techniques may, of course, be applied to digital circuits (which are, after all, composed of analog circuits). The main limitation will be the size of these circuits because the techniques presented here provide a very detailed analysis of the circuit in question and, therefore, would be too costly in terms of computer resources to analyze a large digital system.

The most widely known and used circuit simulation program is SPICE (simulation program with integrated circuit emphasis). This program was first written at the University of California at Berkeley by Laurence Nagel in 1975. Research in the area of circuit simulation is ongoing at many universities and industrial sites. Commercial versions of SPICE or related programs are available on a wide variety of computing platforms, from small personal computers to large mainframes. A list of some commercial simulator vendors can be found in the Appendix to this section.

It is possible to simulate virtually any type of circuit using a program like SPICE. The programs have built-in elements for resistors, capacitors, inductors, dependent and independent voltage and current sources, diodes, MOSFETs, JFETs, BJTs, transmission lines, transformers, and even transformers with saturating cores in some versions. Libraries of standard components which have all the necessary parameters prefitted to typical specifications are found in commercial versions. These libraries include items such as discrete transistors, op amps, phase-locked loops, voltage regulators, logic integrated circuits (ICs), and saturating transformer cores.

---

[1]The material in this chapter was previously published by CRC Press in *The Circuits and Filters Handbook*, Wai-Kai Chen, Ed., 1995.

Computer-aided circuit simulation is now considered an essential step in the design of integrated circuits, because without simulation the number of "trial runs" necessary to produce a working IC would greatly increase the development cost of the IC. Currently, silicon foundries provide very accurate analog simulation models of the devices they fabricate. This facilitates the design of robust circuits that exhibit tolerance to process induced device variations during fabrication.

Simulation provides other advantages, however:

- The ability to measure "inaccessible" voltages and currents. Because a mathematical model is used all voltages and currents are available. No loading problems are associated with placing a voltmeter or oscilloscope in the middle of the circuit, with measuring difficult one-shot wave forms, or probing a microscopic die.
- Mathematically ideal elements are available. Creating an ideal voltage or current source is trivial with a simulator, but impossible in the laboratory. In addition, all component values are exact and no parasitic elements exist, making it easier to test an initial design idea quickly using ideal models.
- It is easy to change the values of components or the configuration of the circuit. Unsoldering leads or redesigning IC masks are unnecessary.

Unfortunately, computer-aided simulation has its own problems:

- Real circuits are distributed systems, not the "lumped element models" which are assumed by simulators. Real circuits, therefore, have resistive, capacitive, and inductive parasitic elements present besides the intended components. In high-speed circuits these parasitic elements are often the dominant performance-limiting elements in the circuit, and must be painstakingly modeled. In addition, this modeling effort requires accompanying parasitic extractor software for the circuit under development. The results of parasitic extraction then need to be back-annotated to the original design for further verification and/or fine-tuning, rendering the overall design flow complicated.
- Suitable predefined numerical models have not yet been developed for certain types of devices or electrical phenomena. The software user may be required, therefore, to create his or her own models out of other models which are available in the simulator. (An example is the solid-state thyristor which may be created from a NPN and PNP bipolar transistor.)
- The numerical methods used may place constraints on the form of the model equations used.

The three primary analog simulation modes are dc, ac, and transient analyses. Other simulation modes are based on the primary ones and can be listed as: ac/dc sensitivity, dc sweep, distortion analysis, model parameter sweep, Monte Carlo analysis, Fourier analysis, transfer function, pole-zero, noise, and worst-case analysis. State-of-the-art simulation packages typically contain all these analysis modes and display results through advanced graphical user interfaces. The following sections consider the three primary simulation modes. In each section an overview is given of the numerical techniques used. Some examples are then given, followed by a brief discussion of common pitfalls.

DC (Steady-State) Analysis

DC analysis calculates the state of a circuit with fixed (non-time varying) inputs after an infinite period of time. DC analysis is useful to determine the operating point (Q-point) of a circuit, power consumption, regulation and output voltage of power supplies, transfer functions, noise margin and fan-out in logic gates, and many other types of analysis. In addition dc analysis is used to find the starting point for ac and transient analysis. To perform the analysis the simulator performs the following steps:

1. All capacitors are removed from the circuit (replaced with opens).
2. All inductors are replaced with shorts.
3. Modified nodal analysis is used to construct the nonlinear circuit equations. This results in one equation for each circuit node plus one equation for each voltage source. Modified nodal analysis is used rather than standard nodal analysis because an ideal voltage source or inductance cannot be

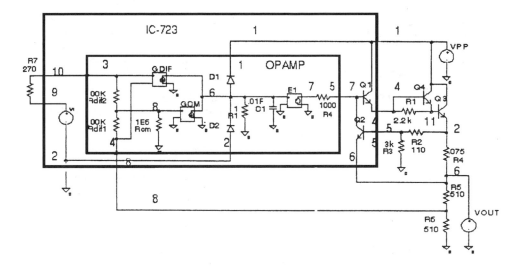

FIGURE 13.1 Regulator circuit to be used for dc analysis, created using PSPICE.

represented using normal nodal analysis. To represent the voltage sources, loop equations (one for each voltage source or inductor), are included as well as the standard node equations. The node voltages and voltage source currents then represent the quantities which are solved for. These form a vector **x**. The circuit equations can also be represented as a vector $\mathbf{F(x)} = \mathbf{0}$.

4. Because the equations are nonlinear, Newton's method (or a variant thereof) is then used to solve the equations.

Example 13.1. Simulation of a Voltage Regulator

We shall now consider simulation of the type 723 voltage regulator IC, shown in Figure 13.1. We wish to simulate the IC and calculate the sensitivity of the output *I–V* characteristic and verify that the output current follows a "fold-back" type characteristic under overload conditions.

The IC itself contains a voltage reference source and operational amplifier. Simple models for these elements are used here rather than representing them in their full form, using transistors, to illustrate model development. The use of simplified models can also greatly reduce the simulation effort. (For example, the simple op amp used here requires only eight nodes and ten components, yet realizes many advanced features.)

Note in Figure 13.1 that the numbers next to the wires represent the circuit nodes. These numbers are used to describe the circuit to the simulator. In most SPICE-type simulators the nodes are represented by numbers, with the ground node being node zero. Referring to Figure 13.2, the 723 regulator and its internal op amp are represented by subcircuits. Each subcircuit has its own set of nodes and components. Subcircuits are useful for encapsulating sections of a circuit or when a certain section needs to be used repeatedly (see next section).

The following properties are modeled in the op amp:

1. Common mode gain
2. Differential mode gain
3. Input impedance
4. Output impedance
5. Dominant pole
6. Output voltage clipping

The input terminals of the op amp connect to a "T" resistance network, which sets the common and differential mode input resistance. Therefore, the common mode resistance is RCM+RDIF = 1.1E6 and the differential mode resistance is RDIF1 + RDIF2 = 2.0E5.

```
Regulator circuit.                          .subckt ic723 1 2  4 5 6 7 8 9 10
* Complete circuit *                        * Type 723 voltage regulator *
* Load source*                              x1 1 2 10 8 7 opamp
vout 6 0                                    * Internal voltage reference *
* Power input *                             vr 9 2 2.5
vpp 1 0 11                                  q1 3 7 4 mm
x1 1 0 4 5 6 7 8 9 10 ic723                 q2 7 5 6 mm
* Series Pass transistors *                 .model mm npn (is=1e-12 bf=100
q3 1 4 11 mq3                               + br=5)
q4 1 11 2 mq4                               .ends ic723
r1 4 11 2.2k                                * Ideal opamp with limiting
r2 5 2 110                                  .subckt opamp 1  2  3  4  5
r3 5 0 3k                                   *            vcc vee +in -in out
r4 2 6 0.075                                rdif1 3 8 1e5
r5 6 8 510                                  rdif2 4 8 1e5
r6 8 0 510                                  rcm  8 0 1e6
r7 9 10 270                                 * Common mode gain *
* Control cards *                           gcm 6 0 8 0 1e-1
.op                                         * Differential mode gain *
.model mq3 npn(is=1e-9 bf=30                gdif 6 0 4 3 100
+ br=5 ikf=50m)                             r1 6 0 1
.model mq4 npn(is=1e-6 bf=30                * Single pole response *
+ br=5 ikf=10)                              c1 6 0 0.01
.dc vout 1 5.5 .01                          d1 6 1 ideal
.plot dc i(vout)                            d2 2 6 ideal
.probe                                      e1 7 0 6 0 1
                                            rout 5 7 1e3
                                            .model ideal d (is=1e-6 n=.01)
                                            .ends opamp
```

FIGURE 13.2   SPICE input listing of regulator circuit shown in Figure 13.1.

Dependent current sources are used to create the main gain elements. Because these sources force current into a 1-Ω resistor, the voltage gain is Gm*R at low frequency. In the differential mode this gives (GDIF*R1 = 100). In the common mode this gives (GCM*R1*(RCM/(RDIF1 + RCM)) = 0.0909). The two diodes D1 and D2 implement clipping by preventing the voltage at node 6 from exceeding VCC or going below VEE. The diodes are made "ideal" by reducing the ideality factor n. Note that the diode current is $I_d = I_s[\exp(V_d/(nV_t)) - 1]$, where V_t is the thermal voltage (0.026 V). Thus, reducing n makes the diode turn on at a lower voltage.

A single pole is created by placing a capacitor ($C1$) in parallel with resistor $R1$. The pole frequency is therefore given by $1.0/(2^*\pi^*R1^*C1)$. Finally, the output is driven by the voltage-controlled voltage source $E1$ (which has a voltage gain of unity), through the output resistor $R4$. The output resistance of the op amp is therefore equal to $R4$.

To observe the output voltage as a function of resistance, the regulator is loaded with a voltage source (VOUT) and the voltage source is swept from 0.05 to 6.0 V. A plot of output voltage vs. resistance can then be obtained by plotting VOUT vs. VOUT/I(VOUT) (using PROBE in this case; see Figure 13.3). Note that for this circuit, even though a current source would seem a more natural choice, a voltage source must be used as a load rather than a current source because the output characteristic curve is multivalued in current. If a current source were used it would not be possible to easily simulate the entire curve. Of course, many other interesting quantities can be plotted; for example, the power dissipated in the pass transistor can be approximated by plotting IC($Q3$)*VC($Q3$).

For these simulations PSPICE was used running on an IBM PC. The simulation took <1 min of CPU time.

Pitfalls.   Convergence problems are sometimes experienced if "difficult" bias conditions are created. An example of such a condition is if a diode is placed in the circuit backwards, resulting in a large forward bias voltage, SPICE will have trouble resolving the current. Another difficult case is if a current source is used

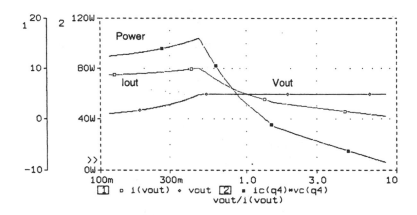

FIGURE 13.3   Output characteristics of regulator circuit using PSPICE.

instead of a voltage to bias the output in the previous example. If the user then tries to increase the output current above 10 A, SPICE would not be able to converge because the regulator will not allow such a large current.

AC Analysis

AC analysis uses phasor analysis to calculate the frequency response of a circuit. The analysis is useful for calculating the gain, 3 dB frequency input, and output impedance, and noise of a circuit as a function of frequency, bias conditions, temperature, etc.

Numerical Method

1. A DC solution is performed to calculate the Q-point for the circuit.
2. A linearized circuit is constructed at the Q-point. To do this, all nonlinear elements are replaced by their linearized equivalents. For example, a nonlinear current source $I = aV_1^2 + bV_2^3$ would be replaced by a linear voltage controlled current source $I = V_1(2aV_{1q}) + V_2(3bV_{2q}^2)$.
3. All inductors and capacitors are replaced by complex impedances, and conductances evaluated at the frequency of interest.
4. Nodal analysis is now used to reduce the circuit to a linear algebraic complex matrix. The ac node voltages may now be found by applying an excitation vector (which represents the independent voltage and current sources) and using Gaussian elimination (with complex arithmetic) to calculate the node voltages.

AC analysis does have limitations and the following types of nonlinear or large signal problems cannot be modeled:

1. Distortion due to nonlinearities such as clipping, etc.
2. Slew rate-limiting effects
3. Analog mixers
4. Oscillators

Noise analysis is performed by including noise sources in the models. Typical noise sources include thermal noise in resistors $I_n^2 = 4kT \, \Delta f/R$, and shot $I_n^2 = 2qI_d \, \Delta f$, and flicker noise in semiconductor devices. Here, T is temperature in Kelvin, k is Boltzmann's constant, and Δf is the bandwidth of the circuit. These noise sources are inserted as independent current sources $In_j(f)$ into the ac model. The resulting current due to the noise source is then calculated at a user-specified summation node(s) by multiplying by the gain function between the noise source and the summation node $A_{js}(f)$. This procedure is repeated for each noise source and then the contributions at the reference node are root-mean-square (rms) summed to give the total noise at the reference node. The equivalent input noise is then easily calculated from the transfer function between the circuit input and the reference node $A_{is}(f)$. The equation describing the input noise is therefore:

$$I_i = \frac{1}{A_{is}(f)} \sqrt{\sum_j [A_{js}(f)In_j(f)]^2}$$

Example 13.2.   Cascode Amplifier with Macro Models

Here, we find the gain, bandwidth, input impedance, and output noise of a cascode amplifier. The circuit for the amplifier is shown in Figure 13.5. The circuit is assumed to be fabricated in a monolithic IC process, so it will be necessary to consider some of the parasitics of the IC process. A cross-section of a typical IC bipolar transistor is shown in Figure 13.4 along with some of the parasitic elements. These parasitic elements are easily included in the amplifier by creating a "macro model" for each transistor. The macro model is then implemented in SPICE form using subcircuits.

The input to the circuit is a voltage source (VIN), applied differentially to the amplifier. The output will be taken differentially across the collectors of the two upper transistors at nodes 2 and 3. The input impedance of the amplifier can be calculated as VIN/I(VIN) or because VIN = 1.0 just as 1/I(VIN). These quantities are

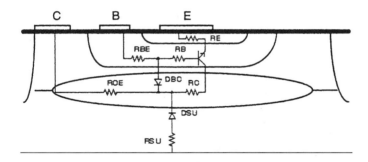

FIGURE 13.4   BJT cross-section with macro model elements.

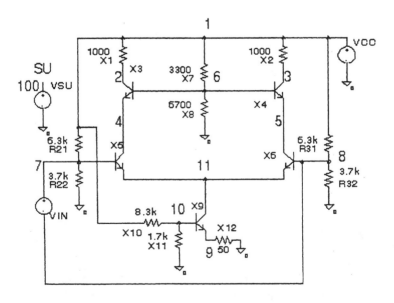

FIGURE 13.5   Cascode amplifier for ac analysis, created using PSPICE.

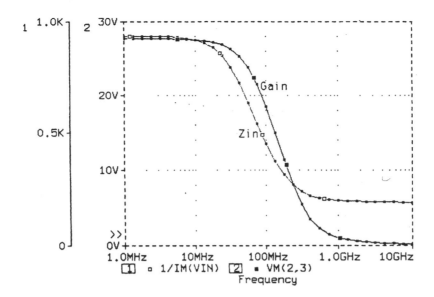

FIGURE 13.6 Gain and input impedance of cascode amplifier.

shown plotted using PROBE in Figure 13.6. It can be seen that the gain of the amplifier falls off at high frequency as expected. The input impedance also drops because parasitic capacitances shunt the input. This example took <1 min on an IBM PC.

Pitfalls. Many novice users will forget that ac analysis is a linear analysis. They will, for example, apply a 1-V signal to an amplifier with 5-V power supplies and a gain of 1000 and be surprised when SPICE tells them that the output voltage is 1000 V. Of course, the voltage generated in a simple amplifier must be less than the power supply voltage, but to examine such clipping effects, transient analysis must be used. Likewise, selection of a proper Q-point is important. If the amplifier is biased in a saturated portion of its response and ac analysis is performed, the gain reported will be much smaller than the actual large signal gain.

Transient Analysis

Transient analysis is the most powerful analysis capability of a simulator, because the transient response is so hard to calculate analytically. Transient analysis can be used for many types of analysis, such as switching speed, distortion, basic operation of certain circuits like switching power supplies. Transient analysis is also the most CPU intensive and can require 100 or 1000 times the CPU time as a dc or ac analysis.

Numerical Method

In a transient analysis time is discretized into intervals called time steps. Typically the time steps are of unequal length, with the smallest steps being taken during portions of the analysis when the circuit voltages and currents are changing most rapidly. The capacitors and inductors in the circuit are then replaced by voltage and current sources based on the following procedure.

The current in a capacitor is given by $I_c = C \, dV_c/dt$. The time derivative can be approximated by a difference equation:

$$I_c^k + I_c^{k-1} = 2C \frac{V_c^k - V_c^{k-1}}{t^k - t^{k-1}}$$

In this equation the superscript k represents the number of the time step. Here, k is the time step we are presently solving for and $(k-1)$ is the previous time step. This equation can be solved to give the capacitor current at the present time step:

$$I_c^k = V_c^k(2C/\Delta t) - V_c^{k-1}(2C/\Delta t) - I_c^{k-1}$$

Here, $\Delta t = t^k - t^{k-1}$, or the length of the time step. As time steps are advanced, $V_c^{k-1} \rightarrow V_c^k$; $I_c^{k-1} \rightarrow I_c^k$. Note that the second two terms on the right-hand side of the above equation are dependent only on the capacitor voltage and current from the previous time step, and are therefore fixed constants as far as the present step is concerned. The first term is effectively a conductance ($g = 2C/\Delta t$) multiplied by the capacitor voltage, and the second two terms could be represented by an independent current source. The entire transient model for the capacitor therefore consists of a conductance in parallel with two current sources (the numerical values of these are, of course, different at each time step). Once the capacitors and inductors have been replaced as indicated, the normal method of dc analysis is used. One complete dc analysis must be performed for each time point. This is the reason that transient analysis is so CPU intensive. The method outlined here is the trapezoidal time integration method and is used as the default in SPICE. Moreover, Gear's method of numerical integration is also available in SPICE for stiff systems with largely varying time constants.

Example 13.3. Phase-Locked Loop Circuit

Figure 13.7 shows the phase-locked loop circuit. The phase detector and voltage-controlled oscillator are modeled in separate subcircuits. Examine the VCO subcircuit and note the PULSE-type current source ISTART connected across the capacitor. The source gives a current pulse 03.E-6 s wide at the start of the simulation to start the VCO running. To start a transient simulation SPICE first computes a dc operating point (to find the initial voltages V_c^{k-1} on the capacitors). As this dc point is a valid, although not necessarily stable, solution, an oscillator will remain at this point indefinitely unless some perturbation is applied to start the oscillations. Remember, this is an ideal mathematical model and no noise sources or asymmetries exist that would start a real oscillator—it must be done manually. The capacitor C1 would have to be placed off-chip, and bond pad capacitances (CPAD1 and CPAD2) have been included at the capacitor nodes. Including the pad capacitances is very important if a small capacitor C1 is used for high-frequency operation.

In this example, the PLL is to be used as an FM detector circuit and the FM signal is applied to the input using a single frequency FM voltage source. The carrier frequency is 600 kHz and the modulation frequency is 60 kHz. Figure 13.8 shows the input voltage and the output voltage of the PLL at the VCO output and at the phase detector output. It can be seen that, after a brief starting transient, the PLL locks onto the input signal and that the phase detector output has a strong 60-kHz component. This example took 251 sec on a Sun SPARC workstation (3046 time steps, with an average of 5 Newton iterations per time step).

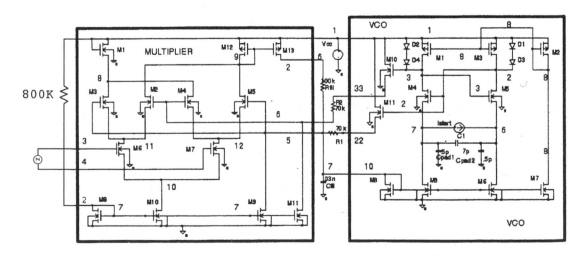

FIGURE 13.7   Phase-locked loop circuit for transient analysis, created with PSPICE.

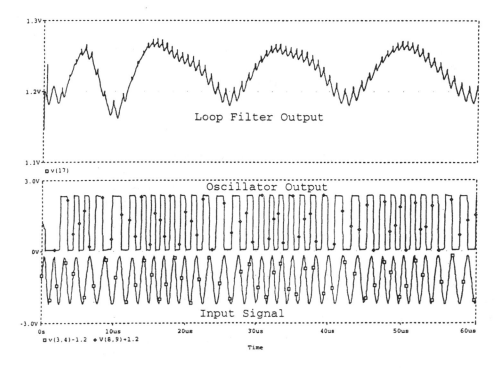

FIGURE 13.8   Transient analysis results of PLL circuit, created using PSPICE.

Pitfalls.   Occasionally SPICE will fail and give the message "Timestep too small in transient analysis," which means that the process of Newton iterations at certain time steps could not be made to converge. One of the most common causes of this is the specification of a capacitor with a value that is much too large, for example, specifying a 1-F capacitor instead of a 1 pF capacitor (an easy mistake to make by not adding the "p" in the value specification). Unfortunately, we usually have no way of telling which capacitor is at fault from the type of failure generated other than to manually search the input deck.

Other transient failures are caused by MOSFET models. Some models contain discontinuous capacitances (with respect to voltage) and others do not conserve charge. These models can vary from version to version so it is best to check the user's guide.

Process and Device Simulation

Process and devices simulation are the steps that precede analog circuit simulation in the overall simulation flow (see Figure 13.9). The simulators are also different in that they are not measurement driven as are analog circuit simulators. The input to a process simulator is the sequence of process steps performed (times, temperatures, gas concentrations) as well as the mask dimensions. The output from the process simulator is a

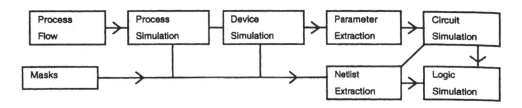

FIGURE 13.9   Data flow for complete process–device–circuit modeling.

detailed description of the solid-state device (doping profiles, oxide thickness, junction depths, etc.). The input to the device simulator is the detailed description generated by the process simulator (or via measurement). The output of the device simulator is the electrical characteristics of the device (IV curves, capacitances, switching transient curves).

Process and device simulation are becoming increasingly important and widely used during the integrated circuit design process. A number of reasons exist for this:

- As device dimensions shrink, second-order effects can become dominant. Modeling of these effects is difficult using analytical models.
- Computers have greatly improved, allowing time-consuming calculations to be performed in a reasonable amount of time.
- Simulation allows access to impossible to measure physical characteristics.
- Analytic models are not available for certain devices, for example, thyristors, heterojunction devices, and IGBTS.
- Analytic models have not been developed for certain physical phenomena, for example, single event upset, hot electron aging effects, latchup, and snap-back.
- Simulation runs can be used to replace split lot runs. As the cost to fabricate test devices increases, this advantage becomes more important.
- Simulation can be used to help device, process, and circuit designers understand how their devices and processes work.

Clearly, process and device simulation is a topic which can be and has been the topic of entire texts. The following sections attempt to provide an introduction to this type of simulation, give several examples showing what the simulations can accomplish, and provide references to additional sources of information.

Process Simulation

Integrated circuit processing involves a number of steps which are designed to deposit (deposition, ion implantation), remove (etching), redistribute (diffusion), or transform (oxidation) the material of which the IC is made. Most process simulation work has been in the areas of diffusion, oxidation, and ion implantation; however, programs are available that can simulate the exposure and development of photo-resist, the associated optical systems, as well as gas and liquid phase deposition and etch.

In the following section a very brief discussion of the governing equations used in SUPREM (from Stanford University, California) will be given along with the results of an example simulation showing the power of the simulator.

Diffusion

The main equation governing the movement of electrically charged impurities (acceptors in this case) in the crystal is the diffusion equation:

$$\frac{\partial C}{\partial t} = \nabla \cdot \left(D\nabla C - \frac{DqC_a}{kT} E \right)$$

Here, C is the concentration (#/cm^3) of impurities, C_a is the number of electrically active impurities (#/cm^3), q is the electron charge, k is Boltzmann's constant, T is temperature in degrees Kelvin, D is the diffusion constant, and E is the built-in electric field. The built-in electric field E in (V/cm) can be found from

$$E = \frac{kT}{q}\frac{1}{n}\nabla n$$

In this equation n is the electron concentration (#/cm^3), which in turn can be calculated from the number of electrically active impurities (C_a). The diffusion constant (D) is dependent on many factors. In silicon the following expression is commonly used:

$$D = F_{IV}\left(D_x + D_+\frac{n_i}{n_i} + D_-\frac{n}{n_i} + D_=\left(\frac{n}{n_i}\right)^2\right)$$

The four D components represent the different possible charge states for the impurity: (x) neutral, $(+)$ positive, $(-)$ negative, $(=)$ doubly negatively charged. n_i is the intrinsic carrier concentration, which depends only on temperature. Each D component is in turn given by an expression of the type:

$$D = A \exp\left(-\frac{B}{kT}\right)$$

Here, A and B are experimentally determined constants, different for each type of impurity $(x, +, -, =)$. B is the activation energy for the process. This expression derives from the Maxwellian distribution of particle energies and will be seen many times in process simulation. It is easily seen that the diffusion process is strongly influenced by temperature. The term F_{IV} is an enhancement factor which is dependent on the concentration of interstitials and vacancies within the crystal lattice (an interstitial is an extra silicon atom which is not located on a regular lattice site; a vacancy is a missing silicon atom which results in an empty lattice site) $F_{IV} \propto C_I + C_v$. The concentration of vacancies, C_v, and interstitials, C_I, are in turn determined by their own diffusion equation:

$$\frac{\partial C_v}{\partial t} = +\nabla \cdot D_v \cdot \nabla C_v - R + G$$

In this equation D_V is another diffusion constant of the form $A \exp(-B/kT)$. R and G represent the recombination and generation of vacancies and interstitials. Note that an interstitial and a vacancy may recombine and in the process destroy each other, or an interstitial and a vacancy pair may be simultaneously generated by knocking a silicon atom off its lattice site. Recombination can occur anywhere in the device via a bulk recombination process $R = A(C_V C_I)\exp(-B/kT)$. Generation occurs where there is damage to the crystal structure, in particular at interfaces where oxide is being grown or in regions where ion implantation has occurred, as the high-energy ions can knock silicon atoms off their lattice sites.

Oxidation

Oxidation is a process whereby silicon reacts with oxygen (or with water) to form new silicon dioxide. Conservation of the oxidant requires the following equation:

$$\frac{dy}{dt} = \frac{F}{N}$$

Here, F is the flux of oxidant (#/cm^2/sec), N is the number of oxidant atoms required to make up a cubic centimeter of oxide, and dy/dt is the velocity with which the Si–SiO$_2$ interface moves into the silicon. In general the greater the concentration of oxidant (C_0), the faster the growth of the oxide and the greater the flux of oxidant needed at the Si–SiO$_2$ interface. Thus, $F = k_s C_0$. The flux of oxidant into the oxide from the gaseous environment is given by

$$F = h(HP_{ox} - C_0)$$

Here H is a constant, P is the partial pressure of oxygen in the gas, and C_0 is the concentration of oxidant in the oxide at the surface and h is of the form $A \exp(-B/kT)$. Finally, the movement of the oxidant within the already existing oxide is governed by diffusion: $\mathbf{F} = D_0 \nabla C$. When all these equations are combined, it is found that (in the one-dimensional case) oxides grow linearly $dy/dt \propto t$ when the oxide is thin and the oxidant can move easily ∝ through the existing oxide. As the oxide grows thicker $dy/dt \propto \sqrt{t}$ because the movement of the oxidant through the existing oxide becomes the rate-limiting step.

Modeling two-dimensional oxidation is a challenging task. The newly created oxide must "flow" away from the interface where it is begin generated. This flow of oxide is similar to the flow of a very thick or viscous liquid and can be modeled by a creeping flow equation:

$$\nabla_2 V \propto \nabla P$$

$$\nabla \cdot V = 0$$

V is the velocity at which the oxide is moving and P is the hydrostatic pressure. The second equation results from the incompressibility of the oxide. The varying pressure P within the oxide leads to mechanical stress, and the oxidant diffusion constant D_0 and the oxide growth rate constant k_s are both dependent on this stress. The oxidant flow and the oxide flow are therefore coupled because the oxide flow depends on the rate at which oxide is generated at the interface and the rate at which the new oxide is generated depends on the availability of oxidant, which is controlled by the mechanical stress.

Ion Implantation

Ion implantation is normally modeled in one of two ways. The first involves tables of moments of the final distribution of the ions which are typically generated by experiment. These tables are dependent on the energy and the type of ion being implanted. The second method involves Monte-Carlo simulation of the implantation process. In Monte-Carlo simulation the trajectories of individual ions are followed as they interact with (bounce off) the silicon atoms in the lattice. The trajectories of the ions, and the recoiling Si atoms (which can strike more Si atoms) are followed until all come to rest within the lattice. Typically several thousand trajectories are simulated (each will be different due to the random probabilities used in the Monte-Carlo method) to build up the final distribution of implanted ions.

Process simulation is always done in the transient mode using time steps as was done with transient circuit simulation. Because partial differential equations are involved, rather than ordinary differential equations, spatial discretization is needed as well. To numerically solve the problem, the differential equations are discretized on a grid. Either rectangular or triangular grids in one, two, or three dimensions are commonly used. This discretization process results in the conversion of the partial differential equations into a set of nonlinear algebraic equations. The nonlinear equations are then solved using a Newton method in a way very similar to the method used for the circuit equations in SPICE.

Example 13.4.   NMOS Transistor

In this example the process steps used to fabricate a typical NMOS transistor will be simulated using SUPREM-IV. These steps are:

1. Grow initial oxide (30 min at 1000 K).
2. Deposit nitride layer (a nitride layer will prevent oxidation of the underlying silicon).
3. Etch holes in nitride layer.
4. Implant $P+$ channel stop (boron dose = 5e12, energy = 50 keV).
5. Grow the field oxide (180 min at 1000 K wet O_2).
6. Remove all nitride.
7. Perform P channel implant (boron dose = 1e11, energy = 40 keV).
8. Deposit and etch polysilicon for gate.
9. Oxidize the polysilicon (30 min at 1000 K, dry O_2).
10. Implant the light doped drain (arsenic dose = 5e13 energy = 50 keV).
11. Deposit sidewall space oxide.
12. Implant source and drain (arsenic, dose = 1e15, energy = 200 keV).
13. Deposit oxide layer and etch contact holes.
14. Deposit and etch metal.

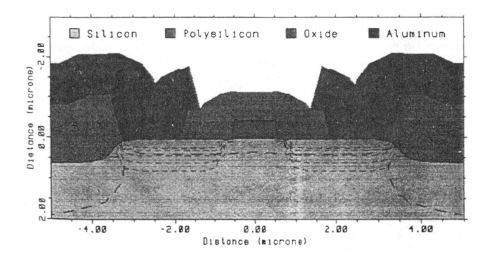

FIGURE 13.10   Complete NMOS transistor cross-section generated by process simulation, created with TMA SUPREM-IV.

The top 4 μm of the completed structure, as generated by SUPREM-IV, is shown in Figure 13.10. The actual simulation structure used is 200 μm deep to allow correct modeling of the diffusion of the vacancies and interstitials. The gate is at the center of the device. Notice how the edges of the gate have lifted up due to the diffusion of oxidant under the edges of the polysilicon (the polysilicon, as deposited in step 8, is flat). The dashed contours show the concentration of dopants in both the oxide and silicon layers. The short dashes indicate N-type material, while the longer dashes indicate P-type material. This entire simulation requires about 30 min on a Sun SPARC-2 workstation.

Device Simulation

Device simulation uses a different approach from that of conventional lumped circuit models to determine the electrical device characteristics. Whereas with analytic or empirical models all characteristics are determined by fitting a set of adjustable parameters to measured data, device simulators determine the electrical behavior by numerically solving the underlying set of differential equations. The first is the Poisson equation, which describes the electrostatic potential within the device:

$$\nabla \cdot \varepsilon \cdot \nabla \Psi = q(N_a^- - N_d^+ - p + n - Q_f)$$

N_d and N_a are the concentration of donors and acceptors, i.e., the N- and P-type dopants. Q_f is the concentration of fixed charge due, for example, to traps or interface charge. The electron and hole concentrations are given by n and p, respectively, and Ψ is the electrostatic potential.

A set of continuity equations describes the conservation of electrons and holes:

$$\frac{\partial n}{\partial t} = \left(\frac{1}{q} \nabla \cdot J_n - R + G \right)$$

$$\frac{\partial p}{\partial t} = \left(-\frac{1}{q} \nabla \cdot J_p - R + G \right)$$

In these equations R and G describe the recombination and generation rates for the electrons and holes. The recombination process is influenced by factors such as the number of electrons and holes present as well as the doping and temperature. The generation rate is also dependent upon the carrier concentrations, but is most strongly influenced by the electric field, with increasing electric fields giving larger generation rates. Because this generation process is included, device simulators are capable of modeling the breakdown of devices at high voltage. J_n and J_p are the electron and hole current densities (in amperes per square centimeter). These current densities are given by another set of equations

$$J_n = q\mu\left(-n\nabla\Psi + \frac{kT_n}{q}\nabla_n\right)$$

$$J_p = q\mu\left(-p\nabla\Psi - \frac{kT_p}{q}\nabla_p\right)$$

In this equation k is Boltzmann's constant, μ is the carrier mobility, which is actually a complex function of the doping, n, p, electric field, temperature, and other factors. In silicon the electron mobility will range between 50 and 1000 and the hole mobility will normally be a factor of 2 smaller. In other semiconductors such as gallium arsenide the electron mobility can be as high as 5000. T_n and T_p are the electron and hole mean temperatures, which describe the average carrier energy. In many models these default to the device temperature (300 K). In the first term the current is proportional to the electric field ($\nabla\Psi$), and this term represents the drift of carriers with the electric field. In the second term the current is proportional to the gradient of the carrier concentration (∇n), so this term represents the diffusion of carriers from regions of high concentration to those of low concentration. The model is therefore called the drift-diffusion model.

In devices in which self-heating effects are important, a lattice heat equation can also be solved to give the internal device temperature:

$$\sigma(T)\frac{\partial T}{\partial t} = H + \nabla\cdot\lambda(T)\cdot\nabla T$$

$$H = -(J_n + J_p)\cdot\nabla\Psi + H_R$$

where H is the heat generation term, which includes resistive (Joule) heating as well as recombination heating, H_u. The terms $\sigma(T)$, $\lambda(T)$ represent the specific heat and the thermal conductivity of the material (both temperature dependent). Inclusion of the heat equation is essential in many power device problems.

As with process simulation partial differential equations are involved, therefore, a spatial discretization is required. As with circuit simulation problems, various types of analysis are available:

- Steady state (DC), used to calculate characteristic curves of MOSFETs, BJTs diodes, etc.
- AC analysis, used to calculate capacitances, Y-parameters, small signal gains, and S-parameters
- Transient analysis used for calculation of switching and large signal behavior, and special types of analysis such as radiation effects

Example 13.5. NMOS IV Curves

The structure generated in the previous SUPREM-IV simulation is now passed into the device simulator and bias voltages are applied to the gate and drain. Models were included with account for Auger and Shockley Reed Hall recombination, doping and electric field-dependent mobility, and impact ionization. The set of drain characteristics obtained is shown in Figure 13.11. Observe how the curves bend upward at high V_{ds} as the device breaks down. The $V_g = 1$ curve has a negative slope at $I_d = 1.5\text{E-4A}$ as the device enters snap-back. It is possible to model this type of behavior because impact ionization is included in the model.

Figure 13.12 shows the internal behavior of the device with $V_{gs} = 3$ V and $I_d = 3\text{E-4A}$. The filled contours indicate impact ionization, with the highest rate being near the edge of the drain right beneath the gate. This is to be expected because this is the region in which the electric field is largest due to the drain depletion region. The dark lines indicate current flow from the source to the drain. Some current also flows from the drain to the substrate. This substrate current consists of holes generated by the impact ionization. The triangular grid

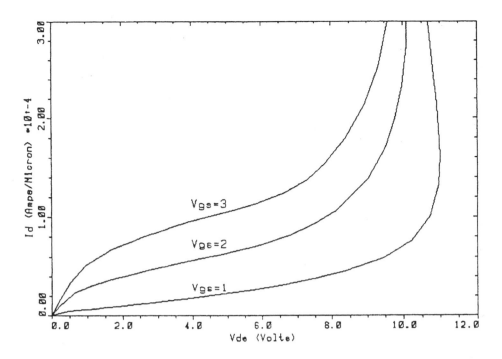

FIGURE 13.11 I_d vs. V_{ds} curves generated by device simulation, created with TMA MEDICI.

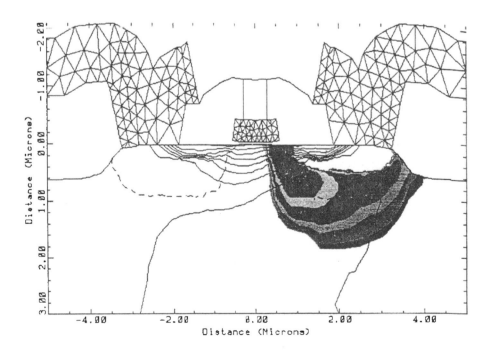

FIGURE 13.12 Internal behavior of MOSFET under bias, created with TMA MEDICI.

used in the simulation can be seen in the source, drain, and gate electrodes. A similar grid was used in the oxide and silicon regions.

Appendix

Circuit Analysis Software

> SPICE2, SPICE3: University of California, Berkeley, CA
> PSPICE: Part of OrCAD product line, Cadence Design Systems, San Jose, CA (used in this chapter)
> HSPICE: Synopsys, Mountain View, CA
> ICAP/4: Intusoft, Gardena, CA
> SPECTRE: Cadence Design Systems, San Jose, CA
> SABER: Synopsys, Mountain View, CA
> MULTISIM: Electronics Workbench, Toronto, Ontario

Process and Device Simulators

> SUPREM-IV, PISCES: Stanford University, Palo Alto, CA
> MINIMOS: Technical University, Vienna, Austria
> TSUPREM-4, MEDICI, DAVINCI: Synopsys, Mountain View, CA (used in this chapter)
> SEMICAD: Dawn Technologies, Sunnyvale, CA

References

P. Antognetti and G. Massobrio, *Semiconductor Device Modeling with SPICE*, New York: McGraw-Hill, 1988.

P.W. Tuinenga, SPICE, *A Guide to Circuit Simulation and Analysis Using PSPICE*, Englewood Cliffs, NJ: Prentice-Hall, 1988.

J.A. Connelly and P. Choi, *Macromodeling with SPICE*, Englewood Cliffs, NJ: Prentice-Hall, 1992.

S. Selberherr, *Analysis and Simulation of Semiconductor Devices*, Berlin: Springer-Verlag, 1984.

R. Dutton and Z. Yu, *Technology CAD, Computer Simulation of IC Process and Devices*, Boston, MA: Kluwer Academic, 1993.

13.2   Parameter Extraction for Analog Circuit Simulation

Peter Bendix

Introduction

Definition of Device Modeling

We use various terms such as device characterization, parameter extraction, optimization, and model fitting to address an important engineering task. In all of these, we start with a mathematical model that describes the transistor behavior. The model has a number of parameters which are varied or adjusted to match the IV (current–voltage) characteristics of a particular transistor or set of transistors. The act of determining the appropriate set of model parameters is what we call device modeling. We then use the model with the particular set of parameters that represent our transistors in a circuit simulator such as SPICE[1] to simulate how circuits with our kinds of transistors will behave. Usually the models are supplied by the circuit simulator we chose. Occasionally we may want to modify these models or construct our own models. In this case we need access to the circuit simulator model subroutines as well as the program that performs the device characterization.

---

[1]SPICE is a circuit simulation program from the Department of Electrical Engineering and Computer Science at the University of California at Berkeley.

Steps Involved in Device Characterization

Device characterization begins with a test chip. Without the proper test chip structures, proper device modeling cannot be done from measured data. A good test chip for MOS technology would include transistors of varying geometries, gate oxide capacitance structures, junction diode capacitance structures, and overlap capacitance structures. This would be a minimal test chip. Additional structures might include ring oscillators and other circuits for checking the ac performance of the models obtained. It is very important that the transistors be well designed and their geometries be chosen appropriate for the technology as well as the desired device model. Although a complete test chip description is beyond the scope of this book, be aware that even perfect device models cannot correct for a poor test chip.

Next we need data that represent the behavior of a transistor or set of transistors of different sizes. These data can come from direct measurement or they can be produced by a device simulator such as PISCES.[1]

It is also possible to use a combination of a process simulator like SUPREM-IV[2] coupled to a device simulator, to provide the simulated results. The benefits of using simulation over measurement are that no expensive measurement equipment or fabricated wafers are necessary. This can be very helpful when trying to predict the device characteristics of a new fabrication process before any wafers have been produced.

Once the measured (or simulated) data are available, parameter extraction software is used to find the best set of model parameter values to fit the data.

Least-Squares Curve Fitting (Analytical)

We begin this section by showing how to do least-squares curve fitting by analytical solutions, using a simple example to illustrate the method. We then mention least-squares curve fitting using numerical solutions in the next section. We can only find analytical solutions to simple problems. The more complex ones must rely on numerical techniques.

Assume a collection of measured data, $m_1, \ldots, m_n$. For simplicity, let these measured data values be functions of a single variable, v, which was varied from v_1 through v_n, measuring each m_i data point at each variable value v_i, i running from 1 to n. For example, the m_i data points might be drain current of an MOS transistor, and the v_i might be the corresponding values of gate voltage. Assume that we have a model for calculating simulated values of the measured data points, and let these simulated values be denoted by $s_1, \ldots, s_n$. We define the least-squares, rms error as

$$\text{error}_{\text{rms}} = \left[\frac{\sum\limits_{i=1}^{n} \{\text{weight}_i(s_i - m_i)\}^2}{\sum\limits_{i=1}^{n} \{\text{weight}_i m_i\}^2} \right]^{1/2} \tag{13.1}$$

where a weighting term is included for each data point. The goal is to have the simulated data match the measured data as closely as possible, which means we want to minimize the rms error. Actually, what we have called the rms error is really the relative rms error, but the two terms are used synonymously. There is another way of expressing the error, called the absolute rms error, defined as follows:

$$\text{error}_{\text{rms}} = \left[\frac{\sum\limits_{i=1}^{n} \{\text{weight}_i(s_i - m_i)\}^2}{\sum\limits_{i=1}^{n} \{\text{weight}_i m_{\text{min}}\}^2} \right]^{1/2} \tag{13.2}$$

---

[1]PISCES is a process simulation program from the Department of Electrical Engineering at Stanford University, Stanford, CA.

[2]SUPREM-IV is a process simulation program from the Department of Electrical Engineering at Stanford University, Stanford, CA.

where we have used the term m_{min} in the denominator to represent some minimum value of the measured data. The absolute rms error is usually used when the measured values approach zero to avoid problems with small or zero denominators in Equation (13.1). For everything that follows, we consider only the relative rms error. The best result is obtained by combining the relative rms formula with the absolute rms formula by taking the maximum of the denominator from Equation (13.1) or Equation (13.2).

We have a simple expression for calculating the simulated data points, s_i, in terms of the input variable, v, and a number of model parameters, $p_1, \ldots, p_m$. That is:

$$s_i = f(v_i, p_1, \ldots, p_m) \tag{13.3}$$

where f is some function. Minimizing the rms error function is equivalent to minimizing its square. Also, we can ignore the term in the denominator of Equation (13.1) as concerns minimizing, because it is a normalization term. In this spirit, we can define a new error term:

$$\text{error} = (\text{error}_{rms})^2 \left[\sum_{i=1}^{n} \{\text{weight}_i m_i\}^2 \right] \tag{13.4}$$

and claim that minimizing error is equivalent to minimizing error_{rms}. To minimize error, we set all partial derivatives of it with respect to each model parameter equal to zero; that is, write

$$\frac{\partial(\text{error})}{\partial p_j} = 0, \quad \text{for } j = 1, \ldots, m \tag{13.5}$$

Then solve the above equations for the value of p_j.

Least-Square Curve Fitting (Numerical)

For almost all practical applications we are forced to do least-squares curve fitting numerically because the analytic solutions as previously discussed are not obtainable in closed form. What we are calling least-squares curve fitting is more generally known as nonlinear optimization. Many fine references on this topic are available. We refer the reader to Gill et al. (1981) for details.

Extraction (as Opposed to Optimization)

The terms "extraction" and "optimization" are, unfortunately, used interchangeably in the semiconductor industry; however, strictly speaking, they are not the same. By optimization, we mean using generalized least-squares curve fitting methods such as the Levenberg–Marquardt algorithm (Gill et al., 1981) to find a set of model parameters. By extraction, we mean any technique that does not use general least-squares fitting methods. This is a somewhat loose interpretation of the term extraction. The main point is that we write the equations we want and then solve them by whatever approximations we choose, as long as these approximations allow us to get the extracted results in closed form. This is parameter extraction.

Extraction vs. Optimization

Extraction has the advantage of being much faster than optimization, but it is not always as accurate. It is also much harder to supply extraction routines for models that are being developed. Each time you make a change in the model, you must make suitable changes in the corresponding extraction routine. For optimization, however, no changes are necessary other than the change in the model itself, because least-squares curve fitting routines are completely general. Also, if anything goes wrong in the extraction algorithm (and no access to the source code is available), almost nothing can be done to correct the problem. With optimization, one can always change the range of data, weighting, upper and lower bounds, etc. A least-squares curve fitting program can be steered toward a correct solution.

Novices at device characterization find least-squares curve fitting somewhat frustrating, because a certain amount of user intervention and intuition is necessary to obtain the correct results. These beginners prefer extraction methods because they do not have to do anything. However, after being burned by extraction routines that do not work, a more experienced user will usually prefer the flexibility, control, and accuracy that optimization provides.

Commercial software is available that provides both extraction and optimization together. The idea here is to first use extraction techniques to make reasonable initial guesses and then use these results as a starting point for optimization, because optimization can give very poor results if poor initial guesses for the parameters are used. Nothing is wrong with using extraction techniques to provide initial guesses for optimization, but for an experienced user this is rarely necessary, assuming that the least-squares curve fitting routine is robust (converges well) and the experienced user has some knowledge of the process under characterization. Software that relies heavily on extraction may do so because of the nonrobustness of its optimizer.

These comments apply when an experienced user is doing optimization locally, not globally. For global optimization (a technique we do not recommend), the above comparisons between extraction and optimization are not valid. The following section contains more detail about local vs. global optimization.

Strategies: General Discussion

The most naive way of using an optimization program would be to take all the measured data for all devices, put them into one big file, and fit to all these data with all model parameters simultaneously. Even for a very high quality, robust optimization program the chances of this method converging are slight. Even if the program does converge, it is almost certain that the values of the parameters will be very unphysical. This kind of approach is an extreme case of global optimization. We call any optimization technique that tries to fit with parameters to data outside their region of applicability a global approach. That is, if we try to fit to saturation region data with linear region parameters such as threshold voltage, mobility, etc., we are using a global approach. In general, we advise avoiding global approaches, although in the strategies described later, sometimes the rules are bent a little.

Our recommended approach is to fit subsets of relevant parameters to corresponding subsets of relevant data in a way that makes physical sense. For example, in the MOS level 3 model, VT0 is defined as the threshold voltage of a long, wide transistor at zero back-bias. It does not make sense to use this parameter to fit to a short channel transistor, or to fit at nonzero back-bias values, or to fit to anywhere outside the linear region. In addition, subsets of parameters should be obtained in the proper order so that those obtained at a later step do not affect those obtained at earlier steps. That is, we would not obtain saturation region parameters before we have obtained linear region parameters because the values of the linear region parameters would influence the saturation region fits; we would have to go back and reoptimize the saturation region parameters after obtaining the linear region parameters. Finally, never use optimization to obtain a parameter value when the parameter can be measured directly. For example, the MOS oxide thickness, TOX, is a model parameter, but we would never use optimization to find it. Always measure its value directly on a large oxide capacitor provided on the test chip. The recommended procedure for proper device characterization follows:

1. Have all the appropriate structures necessary on your test chip. Without this, the job cannot be performed properly.
2. Always measure whatever parameters are directly measurable. Never use otpimization for these.
3. Fit the subset of parameters to corresponding subsets of data, and do so in physically meaningful ways.
4. Fit parameters in the proper order so that those obtained later do not affect those obtained previously. If this is not possible, iteration may be necessary.

Naturally, a good strategy cannot be mounted if one is not intimately familiar with the model used. There is no substitute for learning as much about the model as possible. Without this knowledge, one must rely on strategies provided by software vendors, and these vary widely in quality.

Finally, no one can provide a completely general strategy applicable to all models and all process technologies. At some point the strategy must be tailored to suit the available technology and circuit performance requirements. This not only requires familiarity with the available device models, but also information from the circuit designers and process architects.

MOS DC Models

Available MOS Models

A number of MOS models have been provided over time with the original circuit simulation program, SPICE. In addition, some commercially available circuit simulation programs have introduced their own proprietary models, most notably HSPICE.[1] This section is concentrated on the standard MOS models provided by UC Berkeley's SPICE, not only because they have become the standard models used by all circuit simulation programs, but also because the proprietary models provided by commercial vendors are not well documented and no source code is available for these models to investigate them thoroughly.

MOS Levels 1, 2, and 3.   Originally, SPICE came with three MOS models known as level 1, level 2, and level 3. The level 1 MOS model is a very crude first-order model that is rarely used. The level 2 and level 3 MOS models are extensions of the level 1 model and have been used extensively in the past and present (Vladimirescu and Liu, 1980). These two models contain about 15 dc parameters each and are usually considered useful for digital circuit simulation down to 1 μm channel length technologies. They can fit the drain current for wide transistors of varying length with reasonable accuracy (about 5% rms error), but have very little advanced fitting capability for analog application. They have only one parameter for fitting the subthreshold region, and no parameters for fitting the derivative of drain current with respect to drain voltage, G_{ds} (usually considered critical for analog applications). They also have no ability to vary the mobility degradation with back-bias, so the fits to I_{ds} in the saturation region at high back-bias are not very good. Finally, these models do not interpolate well over device geometry; e.g., if a fit it made to a wide-long device and a wide-short device, and then one observes how the models track for lengths between these two extremes, they usually do not perform well. For narrow devices they can be quite poor as well. Level 3 has very little practical advantage over level 2, although the level 2 model is proclaimed to be more physically based, whereas the level 3 model is called semiempirical. If only one can be used, perhaps level 3 is slightly better because it runs somewhat faster and does not have quite such an annoying kink in the transition region from linear to saturation as does level 2.

Berkeley Short-Channel Igfet Model (BSIM).   To overcome the many shortcomings of level 2 and level 3, the BSIM and BSIM2 models were introduced. The most fundamental difference between these and the level 2 and 3 models is that BSIM and BSIM2 use a different approach to incorporate the geometry dependence (Jeng et al., 1987; Ouster et al., 1988). In level 2 and 3 the geometry dependence is built directly into the model equations. In BSIM and BSIM2 each parameter (except for a very few) is written as a sum of three terms:

$$\text{parameter} - \text{Par}_0 + \frac{\text{Par}_L}{L_{\text{eff}}} + \frac{\text{Par}_w}{W_{\text{eff}}}, \tag{13.6}$$

where Par_0 is the zero-order term, Par_L accounts for the length dependence of the parameter, Par_W accounts for the width dependence, and L_{eff} and W_{eff} are the effective channel width and length, respectively. This approach has a large influence on the device characterization strategy, as discussed later. Because of this tripling of the number of parameters and for other reasons as well, the BSIM model has about 54 DC parameters and the BSIM2 model has over 100.

The original goal of the BSIM model was to fit better than the level 2 and 3 models for submicron channel lengths, over a wider range of geometries, in the subthreshold region, and for nonzero back-bias. Without

---

[1]HSPICE is a commercially available, SPICE-like circuit simulation program from Meta Software, Campbell, CA.

question, BSIM can fit individual devices better than level 2 and level 3. It also fits the subthreshold region better and it fits better for nonzero back-biases. However, its greatest shortcoming is its inability to fit over a large geometry variation. This occurs because Equation (13.6) is a truncated Taylor series in $1/L_{eff}$ and $1/W_{eff}$ terms, and in order to fit better over varying geometries, higher power terms in $1/L_{eff}$ and $1/W_{eff}$ are needed. In addition, no provision was put into the BSIM model for fitting G_{ds}, so its usefulness for analog applications is questionable. Many of the BSIM model parameters are unphysical, so it is very hard to understand the significance of these model parameters. This has profound implications for generating skew models (fast and slow models to represent the process corners) and for incorporating temperature dependence. Another flaw of the BSIM model is its wild behavior for certain values of the model parameters. If model parameters are not specified for level 2 or 3, they will default to values that will at least force the model to behave well. For BSIM, not specifying certain model parameters, setting them to zero, or various combinations of values can cause the model to become very ill-behaved.

BSIM2.   The BSIM2 model was developed to address the shortcomings of the BSIM model. This was basically an extension of the BSIM model, removing certain parameters that had very little effect, fixing fundamental problems such as currents varying the wrong way as a function of certain parameters, adding more unphysical fitting parameters, and adding parameters to allow fitting G_{ds}. BSIM2 does fit better than BSIM, but with more than twice as many parameters as BSIM, it should. However, it does not address the crucial problem of fitting large geometry variations. Its major strengths over BSIM are fitting the subthreshold region better, and fitting G_{ds} better. Most of the other shortcomings of BSIM are also present in BSIM2, and the large number of parameters in BSIM2 makes it a real chore to use in device characterization.

BSIM3.   Realizing the shortcomings of BSIM2, UC Berkeley recently introduced the BSIM3 model. This is an unfortunate choice of name because it implies BSIM3 is related to BSIM and BSIM2. In reality, BSIM3 is an entirely new model that in some sense is related more to level 2 and 3 than BSIM or BSIM2. The BSIM3 model abandons the length and width dependence approach of BSIM and BSIM2, preferring to go back to incorporating the geometry dependence directly into the model equations, as do level 2 and 3. In addition, BSIM3 is a more physically based model, with about 30 fitting parameters (the model has many more parameters, but the majority of these can be left untouched for fitting), making it more manageable, and it has abundant parameters for fitting G_{ds}, making it a strong candidate for analog applications.

It is an evolving model, so perhaps it is unfair to criticize it at this early stage. Its greatest shortcoming is, again, the inability to fit well over a wide range of geometries. It is hoped that future modifications will address this problem. In all fairness, however, it is a large order to ask a model to be physically based, have not too many parameters, be well behaved for all default values of the parameters, fit well over temperature, fit G_{ds}, fit over a wide range of geometries, and still fit individual geometries as well as a model with over 100 parameters, such as BSIM2. Some of these features were compromised in developing BSIM3.

Proprietary Models.   A number of other models are available from commercial circuit simulator vendors, the literature, etc. Some circuit simulators also offer the ability to add a researcher's own models. In general, we caution against using proprietary models, especially those which are supplied without source code and complete documentation. Without an intimate knowledge of the model equations, it is very difficult to develop a good device characterization strategy. Also, incorporating such models into device characterization software is almost impossible. To circumvent this problem, many characterization programs have the ability to call the entire circuit simulator as a subroutine in order to exercise the proprietary model subroutines. This can slow program execution by a factor of 20 or more, seriously impacting the time required to characterize a technology. Also, if proprietary models are used without source code, the circuit simulator results can never be checked against other circuit simulators. Therefore, we want to stress the importance of using standard models. If these do not meet the individual requirements, the next best approach is to incorporate a proprietary model whose source code one has access to. This requires being able to add the individual model not only to circuit simulators, but also to device characterization programs; it can become a very large task.

MOS Level 3 Extraction Strategy in Detail

The strategy discussed here is one that we consider to be a good one, in the spirit of our earlier comments. Note, however, that this is not the only possible strategy for the level 3 model. The idea here is to illustrate basic concepts so that this strategy can be refined to meet particular individual requirements.

In order to do a dc characterization, the minimum requirement is one each of the wide-long, wide-short, and narrow-long devices. We list the steps of the procedure and then discuss them in more detail.

STEP 1. Fit the wide-long device in the linear region at zero back-bias at V_{gs} values above the subthreshold region, with parameters VT0 (threshold voltage), U0 (mobility), and THETA (mobility degradation with V_{gs}).

STEP 2. Fit the wide-short device in the linear region at zero back-bias, at V_{gs} values above the subthreshold region, with parameters VT0, LD (length encroachment), and THETA. When finished with this step, replace VT0 and THETA with the values from step 1, but keep the value of LD.

STEP 3. Fit the narrow-long device in the linear region at zero back-bias, at V_{gs} values above the subthreshold region, with parameters VT0, DW (width encroachment), and THETA. When finished with this step, replace VT0 and THETA with the values from step 1, but keep the value of DW.

STEP 4. Fit the wide-short device in the linear region at zero back-bias, at V_{gs} values above the subthreshold region, with parameters RS and RD (source and drain series resistance).

STEP 5. Fit the wide-long device in the linear region at all back-biases, at V_{gs} values above the subthreshold region, with parameter NSUB (channel doping affects long channel variation of threshold voltage with backbias).

STEP 6. Fit the wide-short device in the linear region at zero back-bias, at V_{gs} values above the subthreshold region, with parameter XJ (erroneously called the junction depth; affects short-channel variation of threshold voltage with back-bias).

STEP 7. Fit the narrow-long device in the linear region at zero back-bias, at V_{gs} values above the subthreshold region, with parameter DELTA (narrow channel correction to threshold voltage).

STEP 8. Fit the wide-short device in the saturation region at zero back-bias (or all back-biases) with parameters VMAX (velocity saturation), KAPPA (saturation region slope fitting parameter), and ETA (V_{ds} dependence of threshold voltage).

STEP 9. Fit the wide-short device in the subthreshold region at whatever back-bias and drain voltage is appropriate (usually zero back-bias and low V_{ds}) with parameter NES (subthreshold slope fitting parameter). One may need to fit with VT0 also and then VT0 is replaced after this step with the value of VT0 obtained from step 1.

This completes the dc characterization steps for the MOS level 3 model. One would then go on to do the junction and overlap capacitance terms (discussed later). Note that this model has no parameters for fitting over temperature, although temperature dependence is built into the model that the user cannot control.

In Step 1 VT0, U0, and THETA are defined in the model for a wide-long device at zero back-bias. They are zero-order fundamental parameters without any short or narrow channel corrections. We therefore fit them to a wide-long device. It is absolutely necessary that such a device be on the test chip. Without it, one cannot obtain these parameters properly. The subthreshold region must be avoided also because these parameters do not control the model behavior in subthreshold.

In Step 2 we use LD to fit the slope of the linear region curve, holding U0 fixed from step 1. We also fit with VT0 and THETA because without them the fitting will not work. However, we want only the value of LD that fits the slope, so we throw away VT0 and THETA, replacing them with the values from Step 1.

Step 3 is the same as Step 2, except that we are getting the width encroachment instead of the length.

In Step 1 the value of THETA that fits the high V_{gs} portion of the wide-long device linear region curve was found. Because the channel length of a long transistor is very large, the source and drain series resistances have almost no effect here, but for a short-channel device, the series resistance will also affect the high V_{gs} portion of the linear region curve. Therefore, in Step 4 we fix THETA from Step 1 and use RS and RD to fit the wide-short device in the linear region, high V_{gs} portion of the curve.

In Step 5 we fit with NSUB to get the variation of threshold voltage with back-bias. We will get better results if we restrict ourselves to lower values of V_{gs} (but still above subthreshold) because no mobility degradation adjustment exists with back-bias, and therefore the fit may not be very good at higher V_{gs} values for the nonzero back-bias curves.

Step 6 is just like Step 5, except we are fitting the short-channel device. Some people think that the value of XJ should be the true junction depth. This is not true. The parameter XJ is loosely related to the junction depth, but XJ is really the short-channel correction to NSUB. Do not be surprised if XJ is not equal to the true junction depth.

Step 7 uses DELTA to make the narrow channel correction to the threshold voltage. This step is quite straightforward.

Step 8 is the only step that fits in the saturation region. The use of parameters VMAX and KAPPA is obvious, but one may question using ETA to fit in the saturation region. The parameter ETA adjusts the threshold voltage with respect to V_{ds}, and as such one could argue that ETA should be used to fit measurements of I_{ds} sweeping V_{gs} and stepping V_{ds} to high values. In doing so, one will corrupt the fit in the saturation region, and usually we want to fit the saturation region better at the expense of the linear region.

Step 9 uses NFS to fit the slope of the $\log(I_{ds})$ vs. V_{gs} curve. Often the value of VT0 obtained from Step 1 will prevent one from obtaining a good fit in the subthreshold region. If this happens, try fitting with VT0 and NFS, but replacing the final value of VT0 with that from Step 1 at the end, keeping only NFS from this final step.

The above steps illustrate the concepts of fitting relevant subsets of parameters to relevant subsets of data to obtain physical values of the parameters, as well as fitting parameters in the proper order so that those obtained in the later steps will affect those obtained in earlier steps minimally. Please refer to Figure 13.13 and Figure 13.14 for how the resulting fits typically appear (all graphs showing model fits are provided by the device modeling software package Aurora, from Technology Modeling Associates, Inc., Palo Alto, CA).

An experienced person may notice that we have neglected some parameters. For example, we did not use parameters KP and GAMMA. This means KP will be calculated from U0, and GAMMA will be calculated

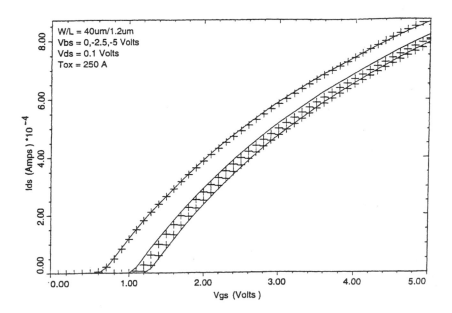

FIGURE 13.13 Typical MOS level 3 linear region measured and simulated plots at various V_{bs} values for a wide-short device.

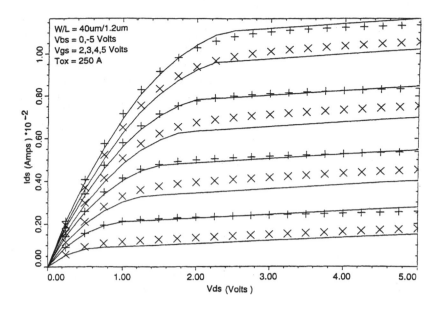

FIGURE 13.14 Typical MOS level 3 saturation region measured and simulated plots at various V_{gs} and V_{bs} values for a wide-short device.

from NSUB. In a sense U0 and NSUB are more fundamental parameters than KP and GAMMA. For example, KP depends on U0 and TOX; GAMMA depends on NSUB and TOX. If one is trying to obtain skew models, it is much more advantageous to analyze statistical distributions of parameters that depend on a single effect than those that depend on multiple effects. KP will depend on mobility and oxide thickness; U0 is therefore a more fundamental parameter. We also did not obtain parameter PHI, so it will be calculated from NSUB. The level 3 model is very insensitive to PHI, so using it for curve fitting is pointless. This illustrates the importance of being very familiar with the model equations. The kind of judgments described here cannot be made without such knowledge.

Test Chip Warnings. The following hints will greatly assist in properly performing device characterization.

1. Include a wide-long device; without this, the results will not be physically correct.
2. All MOS transistors with the same width should be drawn with their sources and drains identical. No difference should be seen in the number of source/drain contacts, contact spacing, source/drain contact overlap, poly gate to contact spacing, etc.
3. Draw devices in pairs. That is, if the wide-long device is W/L = 20/20, make the wide-short device the same width as the wide-long device; e.g., make the short device 20/1, not 19/1. If the narrow-long device is 2/20, make the narrow-short device of the same width; i.e., make is 2/1, not 3/1, and similarly for the lengths. (Make the wide-short and the narrow-short devices have the same length.)

BSIM Extraction Strategy in Detail

All MOS model strategies have basic features in common; namely, fit the linear region at zero back-bias to get the basic zero-order parameters, fit the linear region at nonzero back-bias, fit the saturation region at zero back-bias, fit the saturation region at nonzero back-bias, and then fit the subthreshold region. It is possible to extend the type of strategy we covered for level 3 to the BSIM model, but that is not the way BSIM was intended to be used.

The triplet sets of parameters for incorporating geometry dependence into the BSIM model, (Equation (13.6)), allow an alternate strategy. We obtain sets of parameters without geometry dependence by fitting to individual devices without using the Par_L and Par_W terms. We do this for each device size individually. This produces sets of parameters relevant to each individual device. So, for device number 1 of width W(1) and length L(1) we would have a value for the parameter VFB which we will call VFB(1); for device number n of width W(n) and length L(n) we will have VFB(n). To get the Par_0, Par_L, and VFB_W we fit to the "data points" VFB(1), ..., VFB(n) with parameters VFB_0, VFB_L, and VFB_W using Equation (13.6) where L_{eff} and W_{eff} are different for each index, 1 through n.

Note that as L and W become very large, the parameters must approach Par_0. This suggests that we use the parameter values for the wide-long device as the Par_0 terms and only fit the other geometry sizes to get the Par_L and Par_W terms. For example, if we have obtained VFB(1) for our first device which is our wide-long device, we would set $VFB_0 = VFB(1)$, and then fit to VFB(2), ..., VBF(n) with parameters VFB_L and VFB_W, and similarly for all the other triplets of parameters. In order to use a general least-squares optimization program in this way the software must be capable of specifying parameters as targets, as well as measured data points.

We now list a basic strategy for the BSIM model:

STEP 1. Fit the wide-long device in the linear region at zero back-bias, at V_{gs} values above the subthreshold region, with parameters VFB (flatband voltage), MUZ (mobility), and U0 (mobility degradation), with DL (length encroachment) and DW (width encroachment) set to zero.

STEP 2. Fit the wide-short device in the linear region at zero back-bias, at V_{gs} values above the subthreshold region, with parameters VFB, U0, and DL.

STEP 3. Fit the narrow-long device in the linear region at zero back-bias, at V_{gs} values above the subthreshold region, with parameters VFB, U0, and DW.

STEP 4. Refit the wide-long device in the linear region at zero back-bias, at V_{gs} values above the subthreshold region, with parameters VFB, MUZ, and U0, now that DL and DW are known.

STEP 5. Fit the wide-short device in the linear region at zero back-bias, at V_{gs} values above the subthreshold region, with parameters VFB, RS, and RD. When finished, replace the value of VFB with the value found in Step 4.

STEP 6. Fit the wide-long device in the linear region at all back-biases, at V_{gs} values above the subthreshold region, with parameters K1 (first-order body effect), K2 (second-order body effect), U0, and X2U0 (V_{bs} dependence of U0).

STEP 7. Fit the wide-long device in the saturation region at zero back-bias with parameters U0, ETA (V_{ds} dependence of threshold voltage), MUS (mobility in saturation), U1 (V_{ds} dependence of mobility), and X3MS (V_{ds} dependence of MUS).

STEP 8. Fit the wide-long device in the saturation region at all back-biases with parameter X2MS (V_{bs} dependence of MUS).

STEP 9. Fit the wide-long device in the subthreshold region at zero back-bias and low V_{ds} value with parameter N0; then fit the subthreshold region nonzero back-bias low V_{ds} data with parameter NB; and finally fit the subthreshold region data at higher V_{ds} values with parameter ND. Or, fit all the subthreshold data simultaneously with parameters N0, NB, and ND.

Repeat steps 6 through 10 for all the other geometries, with the result of sets of geometry-independent parameters for each different size device. Then follow the procedure described previously for obtaining the geometry-dependent terms Par_0, Par_L, and Par_W.

In the above strategy we have omitted various parameters either because they have minimal effect or because they have the wrong effect and were modified in the BSIM2 model. Because of the higher complexity of the BSIM model over the level 3 model, many more strategies are possible than the one just listed. One may be able to find variations of the above strategy that suit the individual technology better. Whatever modifications are made, the general spirit of the above strategy probably will remain.

Some prefer to use a more global approach with BSIM, fitting to measured data with Par_L and Par_W terms directly. Although this is certainly possible, it is definitely not a recommended approach. It represents the worst form of blind curve fitting, with no regard for physical correctness or understanding. The BSIM model was originally developed with the idea of obtaining the model parameters via extraction as opposed to optimization. In fact, UC Berkeley provides software for obtaining BSIM parameters using extraction algorithms, with no optimization at all. As stated previously, this has the advantage of being relatively fast and easy. Unfortunately, it does not always work. One of the major drawbacks of the BSIM model is that certain values of the parameters can cause the model to produce negative values of G_{ds} in saturation. This is highly undesirable, not only from a modeling standpoint, but also because of the convergence problems it can cause in circuit simulators. If an extraction strategy is used that does not guarantee non-negative G_{ds}, very little can be done to fix the problem when G_{ds} becomes negative. Of course, the extraction algorithms can be modified, but this is difficult and time consuming. With optimization strategies, one can weight the fitting for G_{ds} more heavily and thus force the model to produce non-negative G_{ds}. We, therefore, do not favor extraction strategies for BSIM, or anything else. As with most things in life, minimal effort provides minimal rewards.

BSIM2 Extraction Strategy

We do not cover the BSIM2 strategy in complete detail because it is very similar to the BSIM strategy, except more parameters are involved. The major difference in the two models is the inclusion of extra terms in BSIM2 for fitting G_{ds} (refer to Figure 13.15, which shows how badly BSIM typically fits $1/G_{ds}$ vs. V_{ds}). Basically, the BSIM2 strategy follows the BSIM strategy for the extraction of parameters not related to G_{ds}. Once these have been obtained, the last part of the strategy includes steps for fitting to G_{ds} with parameters that account for channel length modulation and hot electron effects. The way this proceeds in BSIM2 is to fit I_{ds} first, and then parameters MU2, MU3, and MU4 are used to fit to $1/G_{ds}$ vs. V_{ds} curves for families of V_{gs} and V_{bs}. This can be a very time-consuming and frustrating experience, because fitting to $1/G_{ds}$ is quite difficult. Also, the equations describing how G_{ds} is modeled with MU2, MU3, and MU4 are very unphysical and the interplay between the parameters makes fitting awkward. The reader is referred to Figure 13.16, which shows how BSIM2 typically fits $1/G_{ds}$ vs. V_{ds}. BSIM2 is certainly better than BSIM but it has its own problems fitting $1/G_{ds}$.

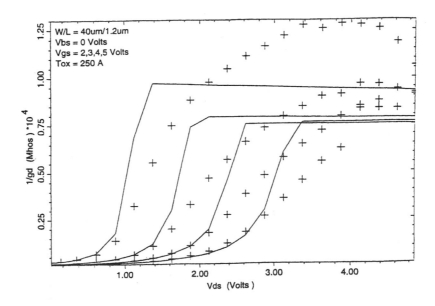

FIGURE 13.15 Typical BSIM $1/G_d$ vs. V_{ds} measured and simulated plots at various V_{gs} values for a wide-short device.

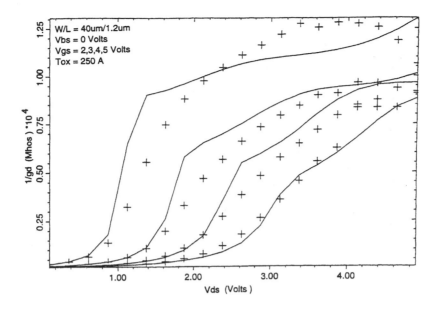

FIGURE 13.16 Typical BSIM2 $1/G_d$ vs. V_{ds} measured and simulated plots at various V_{gs} values for a wide-short device.

BSIM3 Comments

The BSIM3 model is very new and will undoubtedly change in the future (Huang et al., 1993). We will not list a BSIM3 strategy here, but focus instead on the features of the model that make it appealing for analog modeling.

BSIM3 has terms for fitting G_{ds} that relate to channel length modulation, drain-induced barrier lowering, and hot electron effects. They are incorporated completely differently from the G_{ds} fitting parameters of BSIM2. In BSIM3 these parameters enter through a generalized Early voltage relation, with the drain current in saturation written as

$$I_{ds} = I_{dsat}\left[1 + \frac{(V_{ds} - V_{dsat})}{V_A}\right] \tag{13.7}$$

where V_A is a generalized Early voltage made up of three terms as

$$\frac{1}{V_A} = \frac{1}{V_{ACLM}} + \frac{1}{V_{ADIBL}} + \frac{1}{V_{AHCE}} \tag{13.8}$$

with the terms in Equation (13.8) representing generalized early voltages for channel length modulation (CLM), drain-induced barrier lowering (DIBL), and hot carrier effects (HCE). This formulation is more physically appealing than the one used in BSIM2, making it easier to fit $1/G_{ds}$ vs. V_{ds} curves with BSIM2. Figure 13.17 and Figure 13.18 show how BSIM3 typically fits I_{ds} vs. V_{ds} and $1/G_{ds}$ vs. V_{ds}.

Most of the model parameters for BSIM3 have physical significance so they are obtained in the spirit of the parameters for the level 2 and 3 models. The incorporation of temperature dependence is also easier in BSIM3 because the parameters are more physical. All this, coupled with the fact that about 30 parameters exist for BSIM3 as compared to over 100 for BSIM2, makes BSIM3 a logical choice for analog design. However, BSIM3 is evolving, and shortcomings to the model may still exist that may be corrected in later revisions.

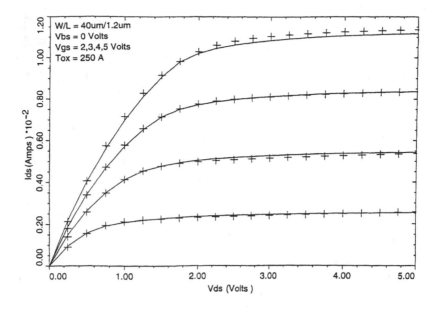

FIGURE 13.17   Typical BSIM3 saturation region measured and simulated plots at various V_{gs} values for a wide-short device.

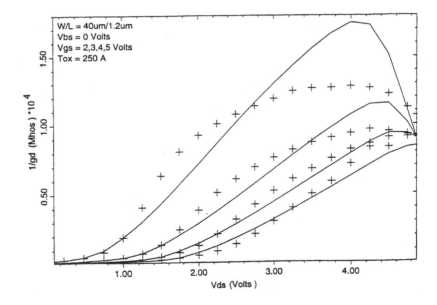

FIGURE 13.18   Typical BSIM3 $1/G_d$ vs. V_{ds} measured and simulated plots at various V_{gs} values for a wide-short device.

Which MOS Model to Use?

Many MOS models are available in circuit simulators, and the novice is bewildered as to which model is appropriate. No single answer exists, but some questions must be asked before making a choice:

1. What kind of technology am I characterizing?
2. How accurate a model do I need?
3. Do I want to understand the technology?

4. How important are the skew model files (fast and slow parameter files)?
5. How experienced am I? Do I have the expertise to handle a more complicated model?
6. How much time can I spend doing device characterization?
7. Do I need to use this model in more than one circuit simulator?
8. Is the subthreshold region important?
9. Is fitting G_{ds} important?

Let us approach each question with regard to the models available. If the technology is not submicron, perhaps a simpler model such as level 3 is capable of doing everything needed. If the technology is deep submicron, then use a more complicated model such as BSIM, BSIM2, or BSIM3. If high accuracy is required, then the best choice is BSIM3, mainly because it is more physical than all the other models and is capable of fitting better.

For a good physical understanding of the process being characterized. BSIM and BSIM2 are not good choices. These are the least physically based of all the models. The level 2 and 3 models have good physical interpretation for most of the parameters, although they are relatively simple models. BSIM3 is also more physically based, with many more parameters than level 2 or 3, so it is probably the best choice.

If meaningful skew models need to be generated, then BSIM and BSIM2 are very difficult to use, again, because of their unphysical parameter sets. Usually, the simplest physically based model is the best for skew model generation. A more complicated physically based model such as BSIM3 may also be difficult to use for skew model generation.

If the user is inexperienced, none of the BSIM models should be used until the user's expertise improves. Our advice is to practice using simpler models before tackling the harder ones.

If time is critical, the simpler models will definitely be much faster for use in characterization. The more complicated models require more measurements over wider ranges of voltages as well as wider ranges of geometries. This, coupled with the larger number of parameters, means they will take some time with which to work. The BSIM2 model will take longer than all the rest, especially if the G_{ds} fitting parameters are to be used.

The characterization results may need to be used in more than one circuit simulator. For example, if a foundry must supply models to various customers, they may be using different circuit simulators. In this case proprietary models applicable to a single circuit simulator should not be used. Also, circuit designers may want to check the circuit simulation results on more than one circuit simulator. It is better to use standard Berkeley models (level 2, level 3, BSIM, BSIM2, and BSIM3) in such cases.

If the subthreshold region is important, then level 2 or level 3 cannot be used, and probably not even BSIM; BSIM2 or BSIM3 must be used instead. These two models have enough parameters for fitting the subthreshold region.

If fitting G_{ds} is important, BSIM2 and BSIM3 are, again, the only choices. None of the other models has enough parameters for fitting G_{ds}.

Finally, if a very unusual technology is to be characterized, none of the standard models may be appropriate. In this case commercially available specialized models or the user's own models must be used. This will be a large task, so the goals must justify the effort.

Skew Parameter Files

This chapter discussed obtaining model parameters for a single wafer, usually one that has been chosen to represent a typical wafer for the technology being characterized. The parameter values obtained from this wafer correspond to a typical case. Circuit designers also want to simulate circuits with parameter values representing the extremes of process variation, the so-called fast and slow corners, or skew parameter files. These represent the best and worst case of the process variation over time.

Skew parameter values are obtained usually by tracking a few key parameters, measuring many wafers over a long period of time. The standard deviation of these key parameters is found and added to or subtracted from the typical parameter values to obtain the skew models. This method is extremely crude and will not normally produce a realistic skew model. It will almost always overestimate the process spread, because the various model parameters are not independent—they are correlated.

Obtaining realistic skew parameter values, taking into account all the subtle correlations between parameters, is more difficult. In fact, skew model generation is often more an art than a science. Many attempts have been made to utilize techniques from a branch of statistics called multivariate analysis (Dillon and Goldstein, 1984). In this approach principal component or factor analysis is used to find parameters that are linear combinations of the original parameters. Only the first few of these new parameters will be kept; the others will be discarded because they have less significance. This new set will have fewer parameters than the original set and therefore will be more manageable in terms of finding their skews. The user sometimes must make many choices in the way the common factors are utilized, resulting in different users obtaining different results.

Unfortunately, a great deal of physical intuition is often required to use this approach effectively. To date, we have only seen it applied to the simpler MOS models such as level 3. It is not known if this is a viable approach for a much more complicated model such as BSIM2 (Power et al., 1993).

References

W.R. Dillon and M. Goldstein, *Multivariate Analysis Methods and Applications*, New York: John Wiley & Sons, 1984.

P.E. Gill, W. Murray, and M. Wright, *Practical Optimization*, Orlando, FL: Academic Press, 1981.

J.S. Duster, J.-C. Jeng, P.K. Ko, and C. Hu, *User's Guide for BSIM2 Parameter Extraction Program and The SPICE3 with BSIM Implementation*, Electronic Research Laboratory, Berkeley, CA: University of California, 1988.

J.-H. Huang, Z.H. Liu, M.-C. Jeng, P.K. Ko, and C. Hu, *BSIM3 Manual*, Berkeley, CA: University of California, 1993.

M.-C. Jeng, P.M. Lee, M.M. Kuo, P.K. Ko, and C. Hu, *Theory, Algorithms, and User's Guide for BSIM and SCALP*, Version 2.0, Electronic Research Laboratory, Berkeley, CA: University of California, 1987.

J.A. Power, A. Mathewson, and W.A. Lane, "An Approach for Relating Model Parameter Variabilities to Process Fluctuations," *Proc. IEEE Int. Conf. Microelectronic Test Struct.*, vol. 6, Mar. 1993.

W.H. Press, B.P. Flannery, S.A. Teukolsky, and W.T. Vetterling, *Numerical Recipes in C*, Cambridge, UK: Cambridge University Press, 1988.

B.J. Sheu, D.L. Scharfetter, P.K. Ko, and M.-C. Jeng, "BSIM: Berkeley Short-Channel IGFET Model for MOS Transistors," *IEEE J. Solid-State Circuits*, vol. SC-22, no. 4, Aug. 1987.

A. Vladimirescu and S. Liu, *The Simulation of MOS Integrated Circuits Using SPICE2*, memorandum no. UCB/ERL M80/7, Berkeley, CA: University of California, 1980.

Further Information

Other recommended publications that are useful in device characterization are:

L.W. Nagel, *SPICE2: A Computer Program to Simulate Semiconductor Circuits*, memorandum no. ERL-M520, Berkeley, CA: University of California, 1975.

G. Massobrio and P. Antognetti, *Semiconductor Device Modeling with SPICE*, New York: McGraw-Hill, 1993.

II

Signal Processing

14

Digital Signal Processing

W. Kenneth Jenkins
The Pennsylvania State University

Alexander D. Poularikas
University of Alabama

Bruce W. Bomar
*University of Tennessee Space
Institute*

L. Montgomery Smith
*University of Tennessee Space
Institute*

James A. Cadzow
Vanderbilt University

Dean J. Krusienski
Wadsworth Center

14.1 Fourier Transforms

W. Kenneth Jenkins

The Fourier transform is a mathematical tool that is used to expand signals into a spectrum of sinusoidal components to facilitate signal representation and the analysis of system performance. In certain applications the Fourier transform is used for spectral analysis, while in others it is used for spectrum shaping that adjusts

TABLE 14.1   Continuous Time Fourier Transform Pairs

Signal	Fourier transform	Fourier series coefficients (if periodic)
$\sum_{k=-\infty}^{+\infty} a_k e^{jk\omega_0 t}$	$2\pi \sum_{k=-\infty}^{+\infty} a_k \delta(\omega - k\omega_0)$	a_k
$e^{j\omega_0 t}$	$2\pi\delta(\omega - \omega_0)$	$a_1 = 1$ $a_k = 0,$ otherwise
$\cos \omega_0 t$	$\pi[\delta(\omega - \omega_0) + \delta(\omega + \omega_0)]$	$a_1 = a_{-1} = \dfrac{1}{2}$ $a_k = 0,$ otherwise
$\sin \omega_0 t$	$\dfrac{\pi}{j}[\delta(\omega - \omega_0) - \delta(\omega + \omega_0)]$	$a_1 = -a_{-1} = \dfrac{1}{2j}$ $a_k = 0,$ otherwise
$x(t) = 1$	$2\pi\delta(\omega)$	$a_0 = 1, \quad a_k = 0, k \neq 0$ (has this Fourier series representation for any choice of $T_0 > 0$)
Periodic square wave $x(t) = \begin{cases} 1, & \|t\| < T_1 \\ 0, & T_1 < \|t\| \le \dfrac{T_0}{2} \end{cases}$ and $x(t + T_0) = x(t)$	$\sum_{k=-\infty}^{+\infty} \dfrac{2 \sin k\omega_0 T_1}{k} \delta(\omega - k\omega_0)$	$\dfrac{\omega_0 T_1}{\pi} \mathrm{sinc}\left(\dfrac{k\omega_0 T_1}{\pi}\right) = \dfrac{\sin k\omega_0 T_1}{k\pi}$
$\sum_{n=-\infty}^{+\infty} \delta(t - nT)$	$\dfrac{2\pi}{T} \sum_{k=-\infty}^{+\infty} \delta\left(\omega - \dfrac{2\pi k}{T}\right)$	$a_k = \dfrac{1}{T}$ for all k
$x(t) = \begin{cases} 1, & \|t\| < T_1 \\ 0, & \|t\| > T_1 \end{cases}$	$2T_1 \mathrm{sinc}\left(\dfrac{\omega T_1}{\pi}\right) = \dfrac{2 \sin \omega T_1}{\omega}$	—
$\dfrac{W}{\pi} \mathrm{sinc}\left(\dfrac{Wt}{\pi}\right) = \dfrac{\sin Wt}{\pi t}$	$X(\omega) = \begin{cases} 1, & \|\omega\| < W \\ 0, & \|\omega\| > W \end{cases}$	—
$\delta(t)$	1	—
$u(t)$	$\dfrac{1}{j\omega} + \pi\delta(\omega)$	—
$\delta(t - t_0)$	$e^{-j\omega t_0}$	—
$e^{-at}u(t), \ \mathcal{R}e\{a\} > 0$	$\dfrac{1}{a + j\omega}$	—
$te^{-at}u(t), \ \mathcal{R}e\{a\} > 0$	$\dfrac{1}{(a + j\omega)^2}$	—
$\dfrac{t^{n-1}}{(n-1)!} e^{-at}u(t), \ \mathcal{R}e\{a\} > 0$	$\dfrac{1}{(a + j\omega)^n}$	—

Source: A.V. Oppenheim, A.S. Willsky, and I.T. Young, *Signals and Systems*, Englewood Cliffs, NJ: Prentice-Hall, 1983. With permission.

the relative contributions of different frequency components in the filtered result. In certain applications the Fourier transform is used for its ability to decompose the input signal into uncorrelated components, so that signal processing can be more effectively implemented on the individual spectral components. Different forms of the Fourier transform, such as the continuous-time Fourier series, the continuous-time Fourier transform, the discrete-time Fourier transform (DTFT), the discrete Fourier Transform (DFT), and the fast Fourier transform are applicable in different circumstances. One goal of this section is to clearly define the various Fourier transforms, to discuss their properties, and to illustrate how each form is related to the others in the context of a family tree of Fourier signal processing methods.

Classical Fourier methods such as the Fourier series and the Fourier integral are used for continuous-time (CT) signals and systems, i.e., systems in which the signals are defined at all values of t on the continuum $-\infty < t < \infty$. A more recently developed set of discrete Fourier methods, including the DTFT and the DFT, are extensions of basic Fourier concepts for discrete-time (DT) signals and systems. A DT signal is defined only for integer values of n in the range $-\infty < n < \infty$. The class of DT Fourier methods is particularly useful as a basis for digital signal processing (DSP) because it extends the theory of classical Fourier analysis to DT signals and leads to many effective algorithms that can be directly implemented on general computers or special purpose DSP devices.

The Classical Fourier Transform for Continuous-Time Signals

A continuous time signal $s(t)$ and its Fourier transform $S(j\omega)$ form a transform pair that are related by Equation 14.1 for any $s(t)$ for which the integral Equation (14.1a) converges:

$$S(j\omega) = \int_{-\infty}^{\infty} s(t)e^{-j\omega t}dt \tag{14.1a}$$

$$s(t) = \frac{1}{2\Pi} \int_{-\infty}^{\infty} S(j\omega)e^{j\omega t}d\omega \tag{14.1b}$$

In most literature, Equation (14.1a) is simply called the Fourier transform, whereas Equation (14.1b) is called the Fourier integral. The relationship $S(j\omega) = F\{s(t)\}$ denotes the Fourier transformation of $s(t)$, where $F\{\cdot\}$ is a symbolic notation for the integral operator, and where ω is the continuous frequency variable expressed in rad/sec. A transform pair $s(t) \leftrightarrow S(j\omega)$ represents a one-to-one invertible mapping as long as $s(t)$ satisfies conditions which guarantee that the Fourier transform converges.

In the following discussion the symbol $\delta(t)$ is used to denote a **CT impulse function** that is defined to be zero for all $t \neq 0$, undefined for $t = 0$, and has unit area when integrated over the range $-\infty < t < \infty$. From Equation (14.1a) it is found that $F\{\delta(t - t_o)\} = e^{-j\omega t_o}$ due to the well-known sifting property of $\delta(t)$. Similarly, from Equation (14.1b) we find that $F^{-1}\{2\pi\delta(\omega - \omega_o)\} = e^{j\omega_o t}$, so that $\delta(t - t_o) \leftrightarrow e^{j\omega t_o}$ and $e^{j\omega_o t} \leftrightarrow 2\pi\delta(\omega - \omega_o)$ are Fourier transform pairs. Using these relationships it is easy to establish the Fourier transforms of $\cos(\omega_o t)$ and $\sin(\omega_o t)$, as well as many other useful waveforms, many of which are listed in Table 14.1.

The CT Fourier transform is useful in the analysis and design of CT systems, i.e., systems that process CT signals. Fourier analysis is particularly applicable to the design of CT filters which are characterized by Fourier magnitude and phase spectra, i.e., by $|H(j\omega)|$ and arg $H(j\omega)$, where $H(j\omega)$ is commonly called the frequency response of the filter.

Properties of the Continuous Time (CT) Fourier Transform

The CT Fourier transform has many properties that make it useful for the analysis and design of linear CT systems. Some of the more useful properties are summarized in this section, while a more complete list of the CT Fourier transform properties is given in Table 14.2. Proofs of these properties are found in Oppenheim et al. (1983) and Bracewell (1986). Note that $F\{.\}$ denotes the Fourier transform operation, $F^{-1}\{\cdot\}$ denotes the inverse Fourier transform operation, and "*" denotes the convolution operation defined as:

$$f_1(t)\,{}^*f_2(t) = \int_{-\infty}^{\infty} f_1(t - \tau)f_2(\tau)d\tau$$

Linearity (superposition), a and b complex constants	$F\{af_1(t) + bf_2(t)\} = aF\{f(t)\} + bF\{f_2(t)\}$
Time-shifting	$F\{f(t - t_0)\} = e^{-j\omega t_0}F\{f(t)\}$
Frequency-shifting	$e^{j\omega_0 t}F^{-1}\{F\{j(\omega-\omega_0)\}$
Time-domain convolution	$F\{f_1(t) * f_2(t)\} = F\{f_1(t)\} \cdot F\{f_2(t)\}$
Frequency-domain convolution	$F\{f_1(t) \cdot f_2(t)\} = \frac{1}{2\Pi}F\{f_1(t)\} * F\{f_2(t)\}$
Time-differentiation	$-j\omega F(j\omega) = F\{d(f(t))/dt\}$
Time-integration	$F\{\int_{-\infty}^{t} f(\tau)d\tau\} = \frac{1}{j\omega}F(j\omega) + \pi F(0)\delta(\omega)$

TABLE 14.2   Properties of the Continuous Time Fourier Transform

Name:	If $\mathscr{F}f(t) = F(j\omega)$, then:
Definition	$F(j\omega) = \int_{-\infty}^{\infty} f(t)e^{-j\omega t}\,dt$
	$f(t) = \frac{1}{2\pi}\int_{-\infty}^{\infty} F(j\omega)e^{j\omega t}\,d\omega$
Superposition	$\mathscr{F}[af_1(t) + bf_2(t)] = aF_1(j\omega) + bF_2(j\omega)$
Simplification if: (a) $f(t)$ is even	$F(j\omega) = 2\int_{0}^{\infty} f(t)\cos\omega t\,dt$
(b) $f(t)$ is odd	$F(j\omega) = 2j\int_{0}^{\infty} f(t)\sin\omega t\,dt$
Negative t	$\mathscr{F}f(-t) = F^{*}(j\omega)$
Scaling: (a) time	$\mathscr{F}f(at) = \frac{1}{|a|}F\left(\frac{j\omega}{a}\right)$
(b) magnitude	$\mathscr{F}af(t) = aF(j\omega)$
Differentiation	$\mathscr{F}\left[\frac{d^{n}}{dt^{n}}f(t)\right] = (j\omega)^{n}F(j\omega)$
Integration	$\mathscr{F}\left[\int_{-\infty}^{t} f(x)\,dx\right] = \frac{1}{j\omega}F(j\omega) + \pi F(0)\delta(\omega)$
Time Shifting	$\mathscr{F}f(t-a) = F(j\omega)e^{-j\omega a}$
Modulation	$\mathscr{F}f(t)e^{j\omega_0 t} = F[j(\omega - \omega_0)]$
	$\mathscr{F}f(t)\cos\omega_0 t = \frac{1}{2}\{F[j(\omega - \omega_0)] + F[j(\omega + \omega_0)]\}$
	$\mathscr{F}f(t)\sin\omega_0 t = \frac{1}{2}j\{F[j(\omega - \omega_0)] - F[j(\omega + \omega_0)]\}$
Time Convolution	$\mathscr{F}^{-1}[F_1(j\omega)F_2(j\omega)] = \int_{-\infty}^{\infty} f_1(\tau)f_2(t - \tau)\,d\tau$
Frequency Convolution	$\mathscr{F}[f_1(t)f_2(t)] = \frac{1}{2\pi}\int_{-\infty}^{\infty} F_1(j\lambda)F_2[j(\omega - \lambda)]\,d\lambda$

Source: M.E. VanValkenburg, *Network Analysis*, Englewood Cliffs, NJ: Prentice-Hall, 1974. With permission.

The above properties are particularly useful in CT system analysis and design, especially when the system characteristics are easily specified in the frequency domain, as in linear filtering. Note that properties (1), (6), and (7) are useful for solving differential or integral equations. Property (4) (time-domain convolution) provides the basis for many signal processing algorithms, since many systems can be specified directly by their impulse or frequency response. Property (3) (frequency-shifting) is useful for analyzing the performance of communication systems where different modulation formats are commonly used to shift spectral energy among different frequency bands.

Fourier Spectrum of a Continuous Time Sampled Signal

The operation of uniformly sampling a continuous time signal $s(t)$ at every T sec is characterized by Equation 14.2, where $\delta(t)$ is the CT time impulse function defined earlier:

$$s_{a}(t) = \sum_{n=-\infty}^{\infty} s_a(t)\delta(t - nT) = \sum_{n=-\infty}^{\infty} s_a(nT)\delta(t - nT) \qquad (14.2)$$

Since $s_a(t)$ is in fact a CT signal, it is appropriate to apply the CT Fourier transform to obtain an expression for the spectrum of the sampled signal:

$$F\{s_a(t)\} = F\left\{ \sum_{n=-\infty}^{\infty} s_a(nT)\delta(t - nT) \right\} = \sum_{n=-\infty}^{\infty} s_a(nT)[e^{j\omega T}]^{-n} \qquad (14.3)$$

Since the expression on the right-hand side of Equation (14.3) is a function of $e^{j\omega T}$ it is customary to express the transform as $F(e^{j\omega T}) = F\{s_a(t)\}$. If ω is replaced with a normalized frequency $\omega' = \omega/T$, so that $-\pi < \omega' < \pi$, then the right side of Equation (14.3) becomes identical to the discrete time Fourier transform that is defined directly for the sequence $s[n] = s_a(nT)$ (to be discussed further in a later section).

Fourier Series Representation of Continuous Time Periodic Signals

The classical Fourier series representation of a periodic time domain signal $s(t)$ involves an expansion of $s(t)$ into an infinite series of terms that consist of sinusoidal basis functions, each weighted by a complex constant (Fourier coefficient) that provides the proper contribution of that frequency component to the complete waveform. The conditions under which a periodic signal $s(t)$ can be expanded in a Fourier series are known as the **Dirichlet conditions**. They require that in each period $s(t)$ has a finite number of discontinuities, a finite number of maxima and minima, and that $s(t)$ satisfies the absolute convergence criterion of Equation (14.4) (VanValkenburg, 1974):

$$\int_{-T/2}^{T/2} |s(t)| \mathrm{d}t < \infty \qquad (14.4)$$

It is assumed throughout the following discussion that the Dirichlet conditions are satisfied by all functions that will be represented by a Fourier series.

The Exponential Fourier Series

If $s(t)$ is a CT periodic signal with period T, then the exponential Fourier series expansion of $s(t)$ is given by

$$s(t) = \sum_{n=-\infty}^{\infty} a_n e^{jn\omega_0 t} \qquad (14.5a)$$

where $\omega_o = 2\pi/T$, and the $a_n's$ are the complex Fourier coefficients given by

$$a_n = \frac{1}{T} \int_{\frac{-T}{2}}^{\frac{T}{2}} s(t) e^{-jn\omega_0 t} \mathrm{d}t \quad -\infty < n < \infty \qquad (14.5b)$$

For every value of t where $s(t)$ is continuous, the right side of Equation (14.5a) converges to $s(t)$. At values of t where $s(t)$ has a finite jump discontinuity, the right side of Equation (14.5a) converges to the average of $s(t^-)$ and $s(t^+)$, where $s(t^-) = \lim_{\varepsilon \to 0}(t - \varepsilon)$ and $s(t^+) = \lim_{\varepsilon \to 0}(t + \varepsilon)$.

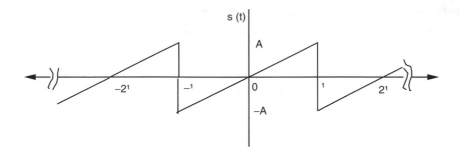

FIGURE 14.1 Periodic continuous time signal used in Fourier series example.

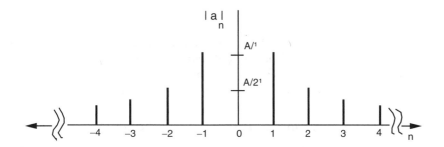

FIGURE 14.2 Magnitude of the Fourier coefficients for the example in Figure 14.1.

For example, the Fourier series expansion of the sawtooth waveform illustrated in Figure 14.1 is characterized by $T = 2\pi$, $\omega_o = 1$, $a_0 = 0$, and $a_n = a_{-n} = A\cos(n\pi)/(jn\pi)$ for $n = 1, 2, \ldots$. The coefficients of the exponential Fourier series given by Equation (14.5b) can be interpreted as a spectral representation of $s(t)$, since the a_nth coefficient represents the contribution of the $(n\omega_o)$th frequency component to the complete waveform. Since the a_n's are complex valued, the Fourier domain (spectral) representation has both magnitude and phase spectra. For example, the magnitude of the a_n's is plotted in Figure 14.2 for the saw tooth waveform of Figure 14.1. The fact that the a_n's constitute a discrete set is consistent with the fact that a periodic signal has a spectrum that contains only integer multiples of the fundamental frequency ω_o. The equation pair given by Equations (14.5a) and (14.5b) can be interpreted as a transform pair that is similar to the CT Fourier transform for periodic signals. This leads to the observation that the classical Fourier series can be interpreted as a special transform that provides a one-to-one invertible mapping between the discrete-spectral domain and the continuous-time domain.

Trigonometric Fourier Series

Although the complex form of the Fourier series expansion is useful for complex periodic signals, the Fourier series can be more easily expressed in terms of real-valued sine and cosine functions for real-valued periodic signals. In the following discussion it is assumed that the signal $s(t)$ is real-valued for the sake of simplifying the discussion. When $s(t)$ is periodic and real-valued it is convenient to replace the complex exponential form of the Fourier series with a **trigonometric expansion** that contains $\sin(\omega_o t)$ and $\cos(\omega_o t)$ terms with corresponding real-valued coefficients (VanValkenburg, 1974). The trigonometric form of the Fourier series for a real-valued signal $s(t)$ is given by

$$s(t) = \sum_{n=0}^{\infty} b_n\cos(n\omega_0) + \sum_{n=1}^{\infty} c_n\sin(n\omega_0) \tag{14.6a}$$

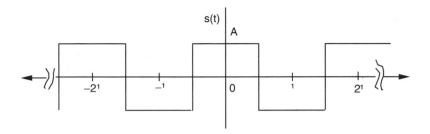

FIGURE 14.3 Periodic CT signal used in Fourier series example.

where $\omega_o = 2\pi/T$. The b_n's and c_n's are real-valued Fourier coefficients determined by

$$b_0 = \frac{1}{T}\int_{-T/2}^{T/2} s(t)\mathrm{d}t \tag{14.6b}$$

$$b_n = \frac{2}{T}\int_{-T/2}^{T/2} s(t)\cos(n\omega_0 t)\mathrm{d}t \quad n = 1, 2, \dots$$

and

$$c_n = \frac{2}{T}\int_{-T/2}^{T/2} s(t)\sin(n\omega_0 t)\mathrm{d}t \quad n = 1, 2, \dots$$

An arbitrary real-valued signal $s(t)$ can be expressed as a sum of even and odd components, $s(t) = s_{\text{even}}(t) + s_{\text{odd}}(t)$, where $s_{\text{even}}(t) = s_{\text{even}}(-t)$ and $s_{\text{odd}}(t) = -s_{\text{odd}}(-t)$, and where $s_{\text{even}}(t) = [s(t) + s(-t)]/2$ and $s_{\text{odd}}(t) = [s(t) - s(-t)]/2$. For the trigonometric Fourier series, it can be shown that $s_{\text{even}}(t)$ is represented by the (even) cosine terms in the infinite series, $s_{\text{odd}}(t)$ is represented by the (odd) sine terms, and b_0 is the dc level of the signal. Therefore, if it can be determined by inspection that a signal has a dc level, or if it is even or odd, then the correct form of the trigonometric series can be chosen to simplify the analysis. For example, it is easily seen that the signal shown in Figure 14.3 is an even signal with a zero dc level, and therefore can be accurately represented by the cosine series with $b_n = 2A\sin(\pi n/2)/(\pi n/2)$, $n = 1, 2, \dots$, as shown in Figure 14.4. In contrast, note that the sawtooth waveform used in the previous example is an odd signal with zero DC level, so that it can be completely specified by the sine terms of the trigonometric series. This result can be demonstrated by pairing each positive frequency component from the exponential series with its

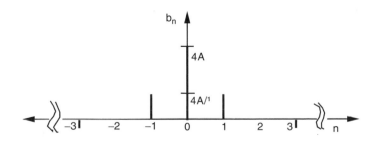

FIGURE 14.4 Fourier coefficients for example of Figure 14.4.

conjugate partner, i.e., $c_n = \sin(n\omega_o t) = a_{-n}e^{jn\omega_o t} + a_{-n}e^{-jn\omega_o t}$, whereby it is found that $c_n = 2A\cos(n\pi)/(n\pi)$ for this example. In general it is found that $a_n = (b_n - jc_n)/2$ for $n = 1, 2, \ldots$, $a_0 = b_0$, and $a_{-n} = a_n^*$. The trigonometric Fourier series is common in the signal processing literature because it replaces complex coefficients with real ones and often results in a simpler and more intuitive interpretation of the results.

Convergence of the Fourier Series

The Fourier series representation of a periodic signal is an approximation that exhibits mean squared convergence to the true signal. If $s(t)$ is a periodic signal of period T and $s'(t)$ denotes the Fourier series approximation of $s(t)$, then $s(t)$ and $s'(t)$ are equal in the mean square sense if

$$\text{mse} = \int_{-T/2}^{T/2} |s(t) - s'(t)|^2 dt = 0 \qquad (14.7)$$

Even with Equation (14.7) satisfied, **mean square error** convergence does not guarantee that $s(t) = s'(t)$ at every value of t. In particular, it is known that at values of t where $s(t)$ is discontinuous, the Fourier series converges to the average of the limiting values to the left and right of the discontinuity. For example, if t_0 is a point of discontinuity, then $s'(t_0) = [s(t_0^-) + s(t_0^+)]/2$, where $s(t_0^-)$ and $s(t_0^+)$ were defined previously (note that at points of continuity, this condition is also satisfied by the very definition of continuity). Since the Dirichlet conditions require that $s(t)$ have at most a finite number of points of discontinuity in one period, the set S_t such that $s(t) \neq s'(t)$ within one period contains a finite number of points, and S_t is a set of measure zero in the formal mathematical sense. Therefore $s(t)$ and its Fourier series expansion $s'(t)$ are *equal almost everywhere*, and $s(t)$ can be considered identical to $s'(t)$ for analysis in most practical engineering problems.

The condition of convergence described above is satisfied almost everywhere only in the limit as an infinite number of terms are included in the Fourier series expansion. If the infinite series expansion of the Fourier series is truncated to a finite number of terms, as it must always be in practical applications, then the approximation will exhibit an oscillatory behavior around the discontinuity, known as the **Gibbs phenomenon** (VanValkenburg, 1974). Let $s'_N(t)$ denote a truncated Fourier series approximation of $s(t)$, where only the terms in Equation (14.5a) from $n = -N$ to $n = N$ are included if the complex Fourier series representation is used, or where only the terms in Equation (14.6a) from $n = 0$ to $n = N$ are included if the trigonometric form of the Fourier series is used. It is well known that in the vicinity of a discontinuity at t_0 the Gibbs phenomenon causes $s'_N(t)$ to be a poor approximation to $s(t)$. The peak magnitude of the Gibbs oscillation is 13% of the size of the jump discontinuity $s(t_0^-) - s(t_0^+)$ regardless of the number of terms used in the approximation. As N increases, the region which contains the oscillation becomes more concentrated in the neighborhood of the discontinuity, until, in the limit as N approaches infinity, the Gibbs oscillation is squeezed into a single point of mismatch at t_0. The Gibbs phenomenon is illustrated in Figure 14.5 where an ideal lowpass frequency response is approximated by impulse response function that has been limited to having only N nonzero coefficients, and hence the Fourier series expansion contains only a finite number of terms.

An important property of the Fourier series is that the exponential basis functions $e^{jn\omega_o t}$ (or $\sin(n\omega_o t)$ and $\cos(n\omega_o t)$ for the trigonometric form) for $n = 0, \pm 1, \pm 2, \ldots$ (or $n = 0, 1, 2, \ldots$ for the trigonometric form) constitute an **orthonormal set**, i.e., $t_{nk} = 1$ for $n = k$, and $t_{nk} = 0$ for $n \neq k$, where

$$t_{nk} = \frac{1}{T} \int_{-T/2}^{T/2} (e^{-jn\omega_o t})(e^{jk\omega_o t}) dt$$

As terms are added to the Fourier series expansion, the orthogonality of the basis functions guarantees

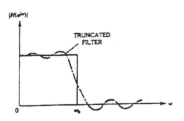

FIGURE 14.5 Gibbs phenomenon in a lowpass digital filter caused by truncating the impulse response to N terms.

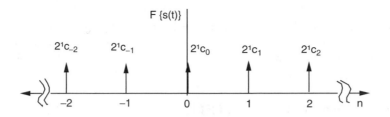

FIGURE 14.6   Spectrum of the Fourier representation of a periodic signal.

that the approximation error decreases in the mean square sense, i.e., that mse_N monotonically decreases as N is increased, where

$$\text{mse}_N = \int_{-T/2}^{T/2} |s(\mathbf{t}) - s_N'(t)|^2 dt$$

Therefore, when applying Fourier series analysis, including more terms always improves the accuracy of the signal representation.

Fourier Transform of Periodic Continuous Time Signals

For a periodic signal $s(t)$ the CT Fourier transform can then be applied to the Fourier series expansion of $s(t)$ to produce a mathematical expression for the "line spectrum" that is characteristic of periodic signals:

$$F\{s(t)\} = \left\{F \sum_{n=-\infty}^{\infty} a_n e^{jn\omega_0 t}\right\} = 2\pi \sum_{n=-\infty}^{\infty} a_n \delta(\omega - \omega_0) \tag{14.8}$$

The spectrum is shown in Figure 14.6. Note the similarity between the spectral representation of Figure 14.6 and the plot of the Fourier coefficients in Figure 14.2, which was heuristically interpreted as a line spectrum. Figure 14.2 and Figure 14.6 are different but equivalent representations of the Fourier line spectrum that is characteristic of periodic signals.

Generalized Complex Fourier Transform

The CT Fourier transform characterized by Equation (14.1) can be generalized by considering the variable $j\omega$ to be the special case of $u = \sigma + j\omega$ with $\sigma = 0$, writing Equation (14.1) in terms of u, and interpreting u as a complex frequency variable. The resulting complex Fourier transform pair is given by Equation (14.9a) and Equation (14.9b):

$$s(t) = \frac{1}{2\Pi j} \int_{\sigma - j\infty}^{\sigma + j\infty} S(u) e^{jut} du \tag{14.9a}$$

$$S(u) = \int_{-\infty}^{\infty} s(t) e^{-jut} dt \tag{14.9b}$$

The set of all values of u for which the integral of Equation (14.9b) converges is called the region of convergence, denoted ROC. Since the transform $S(u)$ is defined only for values of u within the ROC, the path of integration in Equation (14.9a) must be defined so the entire path lies within the ROC. In some literature this transform pair is called the *bilateral Laplace transform* because it is the same result obtained by including both the negative and positive portions of the time axis in the classical Laplace transform integral.

The complex Fourier transform (bilateral Laplace transform) is not often used in solving practical problems, but its significance lies in the fact that it is the most general form that represents the place where Fourier and Laplace transform concepts merge together. Identifying this connection reinforces the observation that Fourier and Laplace transform concepts share common properties because they are derived by placing different constraints on the same parent form.

Discrete-Time Fourier Transform (DTFT)

The DTFT is obtained directly in terms of the sequence samples $s[n]$ by taking the relationship obtained in Equation (14.3) to be the definition of the DTFT. Letting $T = 1$ so that the sampling period is removed from the equations and the frequency variable is replaced with a normalized frequency $\omega' = \omega T$, the DTFT pair is defined by Equation (14.10). In order to simplify notation it is not customary to distinguish between ω and ω', but rather to rely on the context of the discussion to determine whether ω refers to the normalized ($T = 1$) or the unnormalized ($T \neq 1$) frequency variable:

$$S(e^{j\omega'}) = \sum_{n=-\infty}^{\infty} s[n]e^{-j\omega'n} \tag{14.10a}$$

$$s[n] = \frac{1}{2\Pi} \int_{-\Pi}^{\Pi} S(e^{j\omega'})e^{jn\omega'}\,d\omega' \tag{14.10b}$$

The spectrum $S(e^{j\omega'})$ is periodic in ω' with period 2π. The fundamental period in the range $-\pi < \omega' \leq \pi$, sometimes referred to as the baseband, is the useful frequency range of the DT system because frequency components in this range can be represented unambiguously in sampled form (without aliasing error). In much of the signal processing literature the explicit primed notation is omitted from the frequency variable. However, the explicit primed notation will be used throughout this section because there is a potential for confusion when so many related Fourier concepts are discussed within the same framework.

By comparing Equation (14.3) and Equation (14.10a), and noting that $\omega' = \omega T$, we see that

$$F\{s_a(t) = \text{DTFT}\{s[n]\}$$

where $s[n] = s_a(t)|_{t=nT}$. This demonstrates that the spectrum of $s_a(t)$ as calculated by the CT Fourier transform is identical to the spectrum of $s[n]$ as calculated by the DTFT. Therefore although $s_a(t)$ and $s[n]$ are quite different sampling models, they are equivalent in the sense that they have the same Fourier domain representation. A list of common DTFT pairs is presented in Table 14.3. Just as the CT Fourier transform is useful in CT signal system analysis and design, the DTFT is equally useful for DT system analysis and design.

In the same way that the CT Fourier transform was found to be a special case of the complex Fourier transform (or bilateral Laplace transform), the DTFT is a special case of the *bilateral z-transform* with $z = e^{j\omega't}$. The more general bilateral z-transform is given by:

$$S(z) = \sum_{n=-\infty}^{\infty} s[n]z^{-n} \tag{14.11a}$$

$$s[n] = \frac{1}{2\pi j} \oint_C S(z)z^{n-1}\,dz \tag{14.11b}$$

where C is a counter-clockwise contour of integration which is a closed path completely contained within the region of convergence of $S(z)$. Recall that the DTFT was obtained by taking the CT Fourier transform of the CT sampling model $s_a(t)$. Similarly, the bilateral z-transform results by taking the bilateral Laplace transform of $s_a(t)$.

TABLE 14.3   Some Basic Discrete Time Fourier Transform Pairs

Sequence	Fourier Transform
1. $\delta[n]$	1
2. $\delta[n - n_0]$	$e^{-j\omega n_0}$
3. $1 \quad (-\infty < n < \infty)$	$\sum\limits_{k=-\infty}^{\infty} 2\pi\delta(\omega + 2\pi k)$
4. $a^n u[n] \quad (|a| < 1)$	$\dfrac{1}{1 - ae^{-j\omega}}$
5. $u[n]$	$\dfrac{1}{1 - e^{-j\omega}} + \sum\limits_{k=-\infty}^{\infty} \pi\delta(\omega + 2\pi k)$
6. $(n + 1)a^n u[n] \quad (|a| < 1)$	$\dfrac{1}{(1 - ae^{-j\omega})^2}$
7. $\dfrac{r^n \sin \omega_p(n + 1)}{\sin \omega_p} u[n] \quad (|r| < 1)$	$\dfrac{1}{1 - 2r \cos \omega_p e^{-j\omega} + r^2 e^{-j2\omega}}$
8. $\dfrac{\sin \omega_c n}{\pi n}$	$X(e^{j\omega}) = \begin{cases} 1, & |\omega| < \omega_c, \\ 0, & \omega_c < |\omega| \le \pi \end{cases}$
9. $x[n] = \begin{cases} 1, & 0 \le n \le M \\ 0, & \text{otherwise} \end{cases}$	$\dfrac{\sin[\omega(M + 1)/2]}{\sin(\omega/2)} e^{-j\omega M/2}$
10. $e^{j\omega_0 n}$	$\sum\limits_{k=-\infty}^{\infty} 2\pi\delta(\omega - \omega_0 + 2\pi k)$
11. $\cos(\omega_0 n + \phi)$	$\pi \sum\limits_{k=-\infty}^{\infty} [e^{j\phi}\delta(\omega - \omega_0 + 2\pi k) + e^{-j\phi}\delta(\omega + \omega_0 + 2\pi k)]$

Source: A.V. Oppenheim and R.W. Schafer, 1989. *Discrete-Time Signal Processing*, Englewood Cliffs, NJ: Prentice-Hall. With permission.

If the lower limit on the summation of Equation (14.11a) is taken to be $n = 0$, then Equations (14.11a) and (14.11b) become the one-sided z-transform, which is the DT equivalent of the one-sided Laplace transform for CT signals.

Properties of the Discrete-Time Fourier Transform (DTFT)

Since the DTFT is a close relative of the classical CT Fourier transform, it should come as no surprise that many properties of the DTFT are similar to those of the CT Fourier transform. In fact, for many of the properties presented earlier there is an analogous property for the DTFT. The following list parallels the list that was presented in the previous section for the CT Fourier transform, to the extent that the same property exists. A more complete list of DTFT pairs is given in Table 14.4.

Note that the time–differentiation and time–integration properties of the CT Fourier transform do not have analogous counterparts in the DTFT because time-domain differentiation and integration are not defined for DT signals. When working with DT systems practitioners must often manipulate difference equations in the frequency domain. For this purpose the properties of linearity and index-shifting are very important. As with the CT Fourier transform, time-domain convolution is also important for DT systems because it allows engineers to work with the frequency response of the system in order to achieve proper shaping of the input spectrum, or to achieve frequency selective filtering for noise reduction or signal detection.

Linearity (superposition), a and b complex constants	$\mathrm{DTFT}\{a f_1[n] + b f_2[n]\} = a \cdot \mathrm{DTFT}\{f_1[n]\} + b \cdot \mathrm{DTFT}\{f_2[n]\}$
Index-shifting	$\mathrm{DTFT}\{f[n - n_0]\} = e^{-j\omega n_0}\mathrm{DTFT}\{f[n]\}$
Frequency-shifting	$e^{j\omega_0 n}f[n] = \mathrm{DTFT}^{-1}\{F(j(\omega - \omega_0))\}$
Time-domain convolution	$\mathrm{DTFT}\{f_1[n] * f_2[n]\} = F\{f_1[n]\} \cdot F\{f_2[n]\}$
Frequency-domain convolution	$DTFT\{f_1[n] \cdot f_2[n]\} = \dfrac{1}{2\Pi}\mathrm{DTFT}\{f_1[n]\} * \mathrm{DTFT}\{f_2[n]\}$
Frequency-differentiation	$nf[n] = \mathrm{DTFT}^{-1}\{dF(j\omega)/d\omega\}$

TABLE 14.4   Properties of the Discrete Time Fourier Transform

Sequence $x[n]$ $y[n]$	Fourier Transform $X(e^{j\omega})$ $Y(e^{j\omega})$
1. $ax[n] + by[n]$	$aX(e^{j\omega}) + bY(e^{j\omega})$
2. $x[n - n_d]$ (n_d an integer)	$e^{-j\omega n_d}X(e^{j\omega})$
3. $e^{j\omega_0 n}x[n]$	$X(e^{j(\omega - \omega_0)})$
4. $x[-n]$	$X(e^{-j\omega})$ if $x[n]$ real $X^*(e^{j\omega})$
5. $nx[n]$	$j\dfrac{dX(e^{j\omega})}{d\omega}$
6. $x[n] * y[n]$	$X(e^{j\omega})Y(e^{j\omega})$
7. $x[n]y[n]$	$\dfrac{1}{2\pi}\displaystyle\int_{-\pi}^{\pi} X(e^{j\theta})Y(e^{j(\omega - \theta)})d\theta$

Parseval's Theorem

8. $\displaystyle\sum_{n=-\infty}^{\infty} |x[n]|^2 = \frac{1}{2\pi}\int_{-\pi}^{\pi} |X(e^{j\omega})|^2 \, d\omega$

9. $\displaystyle\sum_{n=-\infty}^{\infty} x[n]y^*[n] = \frac{1}{2\pi}\int_{-\pi}^{\pi} X(e^{j\omega})Y^*(e^{j\omega}) \, d\omega$

Source: A.V. Oppenheim and R.W. Schafer, *Discrete-Time Signal Processing*, Englewood Cliffs, NJ: Prentice-Hall, 1989. With permission.

Relationship between the CT and DT Spectra

Since DT signals often originate by sampling a CT signal, it is important to develop the relationship between the original spectrum of the CT signal and the spectrum of the DT signal that results. First, the CT Fourier transform is applied to the CT sampling model, and the properties are used to produce the following result:

$$F\{s_a(t)\} = F\left\{s_a(t) \sum_{n=-\infty}^{\infty} \delta(t - nT)\right\} = \frac{1}{2\pi}S_a(j\omega)F\left\{\sum_{n=-\infty}^{\infty} \delta(t - nT)\right\} \qquad (14.12)$$

Since the sampling function (summation of shifted impulses) on the right-hand side of Equation (14.12) is periodic with period T it can be replaced with a CT Fourier series expansion and the frequency-domain convolution property of the CT Fourier transform can be applied to yield two equivalent expressions for the DT spectrum:

$$S(e^{j\omega T}) = \frac{1}{T} \sum_{n=-\infty}^{\infty} S_a(j[\omega - n\omega_s]) \quad \text{or} \quad S(e^{j\omega'}) = \frac{1}{T} \sum_{n=-\infty}^{\infty} S_a(j[\omega' - n2\pi/T]) \qquad (14.13)$$

In Equation (14.13) $\omega_s = (2\pi/T)$ is the sampling frequency and $\omega' = \omega T$ is the normalized DT frequency axis expressed in radians. Note that $S(e^{j\omega T}) = S(e^{j\omega'})$ consists of an infinite number of replicas of the CT spectrum $S(j\omega)$, positioned at intervals of $(2\pi/T)$ on the ω axis (or at intervals of 2π on the ω' axis), as illustrated in Figure 14.7. Note that if $S(j\omega)$ is band-limited with a bandwidth ω_c, and if T is chosen sufficiently small so that $\omega_s > 2\omega_c$, then the DT spectrum is a copy of $S(j\omega)$ (scaled by $1/T$) in the baseband. The limiting case of $\omega_s = 2\omega_c$ is called the **Nyquist sampling frequency**. Whenever a CT signal is sampled at or above the Nyquist rate, no aliasing distortion occurs (i.e., the baseband spectrum does not overlap with the higher order replicas) and the CT signal can be exactly recovered from its samples by extracting the baseband spectrum of $S(e^{j\omega'})$

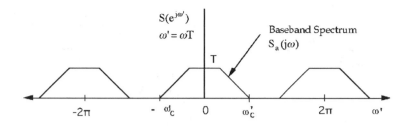

FIGURE 14.7 Relationship between the CT and DT spectra.

with an ideal low-pass filter that recovers the original CT spectrum by removing all spectral replicas outside the baseband and scaling the baseband by a factor of T.

Discrete Fourier Transform (DFT)

To obtain the DFT the continuous-frequency domain of the DTFT is sampled at N points uniformly spaced around the unit circle in the z-plane, i.e., at the points $\omega_k = (2\pi k/N)$, $k = 0, 1, \ldots, N-1$. The result is the DFT transform pair defined by Equations (14.14a) and (14.14b). The signal $s[n]$ is either a finite length sequence of length N, or it is a periodic sequence with period N:

$$S[k] = \sum_{n=0}^{N-1} s[n] e^{-j\frac{2\pi kn}{N}}, \quad k = 0, 1, \ldots, N-1 \tag{14.14a}$$

$$s[k] = \frac{1}{N} \sum_{k=0}^{N-1} S[k] e^{j\frac{2\pi kn}{N}}, \quad n = 0, 1, \ldots, N-1 \tag{14.14b}$$

Regardless of whether $s[n]$ is a finite length or periodic sequence, the DFT treats the N samples of $s[n]$ as though they are one period of a periodic sequence. This is a peculiar feature of the DFT, and one that must be handled properly in signal processing to prevent the introduction of artifacts.

Properties of the DFT

Important properties of the DFT are summarized in Table 14.5. The notation $([k])_N$ denotes k modulo N, and $R_N[n]$ is a rectangular window such that $R_N[n] = 1$ for $n = 0, \ldots, N-1$, and $R_N[n] = 0$ for $n < 0$ and $n \geq N$. The transform relationship given by Equations (14.14a) and (14.14b) is also valid when $s[n]$ and $S[k]$ are periodic sequences, each of period N. In this case n and k are permitted to range over the complete set of real integers, and $S[k]$ is referred to as the discrete Fourier series (DFS). The DFS is developed by some authors as a distinct transform pair in its own right (Oppenheim, 1975). Whether or not the DFT and the DFS are considered identical or distinct is not very important in this discussion. The important point to be emphasized here is that the DFT treats $s[n]$ as though it were a single period of a periodic sequence, and all signal processing done with the DFT will inherit the consequences of this assumed periodicity.

Most of the properties listed in Table 14.5 for the DFT are similar to those of the Z-transform and the DTFT, although there are some important differences. For example, Property 5 (time-shifting property) holds for circular shifts of the finite length sequence $s[n]$, which is consistent with the notion that the DFT treats $s[n]$ as one period of a periodic sequence. Also, the multiplication of two DFTs results in the **circular convolution** of the corresponding DT sequences, as specified by Property 7. This latter property is quite different from the linear convolution property of the DTFT. Circular convolution is simply a linear convolution of the periodic extensions of the finite sequences being convolved, where each of the finite sequences of length N defines the structure of one period of the periodic extensions.

TABLE 14.5   Properties of the Discrete Fourier Transform (DFT)

Finite-Length Sequence (Length N)	N-Point DFT (Length N)
1. $x[n]$	$X[k]$
2. $x_1[n], x_2[n]$	$X_1[k], X_2[k]$
3. $ax_1[n] + bx_2[n]$	$aX_1[k] + bX_2[k]$
4. $X[n]$	$Nx[((-k))_N]$
5. $x[((n-m))_N]$	$W_N^{km} X[k]$
6. $W_N^{-\ell n}x[n]$	$X[((k-\ell))_N]$
7. $\displaystyle\sum_{m=0}^{N-1} x_1(m)x_2[((n-m))_N]$	$X_1[k]X_2[k]$
8. $x_1[n]x_2[n]$	$\dfrac{1}{N}\displaystyle\sum_{\ell=0}^{N-1} X_1(\ell)X_2[((k-\ell))_N]$
9. $x^*[n]$	$X^*[((-k))_N]$
10. $x^*[((-n))_N]$	$X^*[k]$
11. $\mathcal{R}e\{x[n]\}$	$X_{ep}[k] = \frac{1}{2}\{X[((k))_N] + X^*[((-k))_N]\}$
12. $j\mathcal{I}m\{x[n]\}$	$X_{op}[k] = \frac{1}{2}\{X[((k))_N] - X^*[((-k))_N]\}$
13. $x_{ep}[n] = \frac{1}{2}\{x[n] + x^*[((-n))_N]\}$	$\mathcal{R}e\{X[k]\}$
14. $x_{op}[n] = \frac{1}{2}\{x[n] - x^*[((-n))_N]\}$	$j\mathcal{I}m\{X[k]\}$
Properties 15–17 apply only when $x[n]$ is real.	
15. Symmetry properties	$\begin{cases} X[k] = X^*[((-k))_N] \\ \mathcal{R}e\{X[k]\} = \mathcal{R}e\{X[((-k))_N]\} \\ \mathcal{I}m\{X[k]\} = -\mathcal{I}m\{X[((-k))_N]\} \\ |X[k]| = |X[((-k))_N]| \\ \sphericalangle\{X[k]\} = -\sphericalangle\{X[((-k))_N]\} \end{cases}$
16. $x_{ep}[n] = \frac{1}{2}\{x[n] + x[((-n))_N]\}$	$\mathcal{R}e\{X[k]\}$
17. $x_{op}[n] = \frac{1}{2}\{x[n] - x[((-n))_N]\}$	$j\mathcal{I}m\{X[k]\}$

Source: A.V. Oppenheim and R.W. Schafer, *Discrete-Time Signal Processing*, Englewood Cliffs, NJ: Prentice-Hall, 1989. With permission.

Figure 14.8 illustrates the functional relationships among the various forms of CT and DT Fourier transforms that have been discussed in the previous sections. The family of CT Fourier transforms is shown on the left side of Figure 14.8, whereas the right side of the figure shows the hierarchy of DT Fourier transforms. The complex Fourier transform is identical to the bilateral Laplace transform, and it is at this level that the classical Laplace transform techniques and Fourier transform techniques become identical.

Walsh-Hadamard Transform

The Walsh-Hadamard Transform (WHT) is a computationally attractive orthogonal transform that is structurally related to the DFT, and which can be implemented in practical applications without multiplication, and with a computational complexity for addition that is of the same order of complexity as that of an FFT. The t_{mk}th element of the WHT matrix $\mathbf{T}_{\text{WHT}}$ is given by

$$t_{\text{mk}} = \frac{1}{\sqrt{N}} \prod_{\ell=0}^{p-1} (-1)^{b_\ell(m)b_{p-1-\ell}(k)} \quad m \quad \text{and} \quad k = 0,\ldots,N-1$$

where $b_\ell(m)$ is the ℓth order bit in the binary representation of m, and $N = 2^p$. The WHT is defined only when N is a power-of-2. Note that the columns of $\mathbf{T}_{\text{WHT}}$ form a set of orthogonal basis vectors whose elements are all 1's or -1's, so that the calculation of the matrix-vector product $\mathbf{T}_{\text{WHT}}\mathbf{X}$ can be accomplished with only additions and subtractions. It is well known that $\mathbf{T}_{\text{WHT}}$ of dimension $(N \times N)$ for N a power-of-2, can be

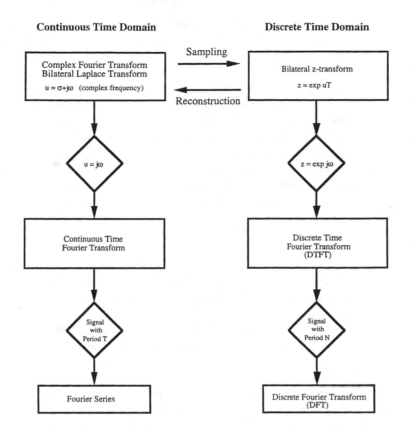

FIGURE 14.8   Functional relationships among various Fourier transforms.

computed recursively according to:

$$\mathbf{T}_k = \begin{bmatrix} \mathbf{T}_{k/2} & \mathbf{T}_{k/2} \\ \mathbf{T}_{k/2} & -\mathbf{T}_{k/2} \end{bmatrix} \quad \text{for} \quad K = 4, \dots, N(\text{even}), \text{ and} \quad \mathbf{T}_2 = \begin{bmatrix} 1 & 1 \\ 1 & -1 \end{bmatrix}$$

The above relationship provides a convenient way of quickly constructing the Walsh-Hadamard matrix for any arbitrary (even) size N.

Because of structural similarities between the DFT and the FFT matrices, the WHT transform can be implemented using a modified FFT algorithm. The core of any FFT program is a butterfly calculation that is characterized by a pair of coupled equations that have the following form:

$$X_{i+1}(\ell, m) = X_i(\ell, m) + e^{j\theta(\ell, m, k, s)} X_i(k, s)$$
$$X_{i+1}(\ell, m) = X_i(\ell, m) - e^{j\theta(\ell, m, k, s)} X_i(k, s)$$

If the exponential factor in the butterfly calculation is replaced by a "1," so the "modified butterfly" calculation becomes

$$X_{i+1}(\ell, m) = X_i(\ell, m) + X_i(k, s)$$
$$X_{i+1}(\ell, m) = X_i(\ell, m) - X_i(k, s)$$

the modified FFT program will in fact perform a WHT on the input vector. This property not only provides a quick and convenient way to implement the WHT, but it also establishes clearly that in addition to the WHT requiring no multiplication, the number of additions required has an order of complexity of $(N/2)\log_2 N$, i.e., the same as that of the FFT.

The WHT is used in many signal processing applications that require signals to be decomposed in real time into a set of orthogonal components. A typical application in which the WHT has been used in this manner is in CDMA wireless communication systems. A CDMA system requires spreading of each user's signal spectrum using a PN sequence. In addition to the PN spreading codes, a set of length-64 mutually orthogonal codes, called the Walsh codes, are used to ensure orthogonality among the signals for users received from the same base station. The length $N = 64$ Walsh codes can be thought of as the orthogonal column vectors from a (64×64) Walsh-Hadamard matrix, and the process of demodulation in the receiver can be interpreted as performing a WHT on the complex input signal containing all the modulated user's signals so they can be separated for accurate detection.

DFT (FFT) Spectral Analysis

A FFT program is often used to perform spectral analysis on signals that are sampled and recorded as part of laboratory experiments, or in certain types of data acquisition systems. There are several issues to be addressed when spectral analysis is performed on (sampled) analog waveforms that are observed over a finite interval of time.

Windowing: The FFT treats the block of data as though it were one period of a periodic sequence. If the underlying waveform is not periodic, then harmonic distortion may occur because the periodic waveform created by the FFT may have sharp discontinuities at the boundaries of the blocks. This effect is minimized by removing the mean of the data (it can always be reinserted) and by windowing the data so the ends of the block are smoothly tapered to zero. A good rule of thumb is to taper 10% of the data on each end of the block using either a cosine taper or one of the other common windows. An alternate interpretation of this phenomenon is that the finite length observation has already windowed the true waveform with a rectangular window that has large spectral sidelobes. Hence, applying an additional window results in a more desirable window that minimizes frequency-domain distortion.

Zero-Padding: An improved spectral analysis is achieved if the block length of the FFT is increased. This can be done by: (1) taking more samples within the observation interval, (2) increasing the length of the observation interval, or (3) augmenting the original dataset with zeros. First, it must be understood that the finite observation interval results in a fundamental limit on the spectral resolution, even before the signals are sampled. The CT rectangular window has a $(\sin x)/x$ spectrum, which is convolved with the true spectrum of the analog signal. Therefore, the frequency resolution is limited by the width of the mainlobe in the $(\sin x)/x$ spectrum, which is inversely proportional to the length of the observation interval. Sampling causes a certain degree of aliasing, although this effect can be minimized by sampling at a high enough rate. Therefore, lengthening the observation interval increases the fundamental resolution limit, while taking more samples within the observation interval minimizes aliasing distortion and provides a better definition (more sample points) on the underlying spectrum.

Padding the data with zeros and computing a longer FFT does give more frequency domain points (improved spectral resolution), but it does not improve the fundamental limit, nor does it alter the effects of aliasing error. The resolution limits are established by the observation interval and the sampling rate. No amount of zero padding can improve these basic limits. However, zero padding is a useful tool for providing more spectral definition, i.e., it enables one to get a better look at the (distorted) spectrum that results once the observation and sampling effects have occurred.

Leakage and the Picket-Fence Effect: An FFT with block length N can accurately resolve only frequencies $\omega_k = (2\pi/N)k$, $k = 0, \ldots, N - 1$ that are integer multiples of the fundamental $\omega_1 = (2\pi/N)$. An analog waveform that is sampled and subjected to spectral analysis may have frequency components between the harmonics. For example, a component at frequency $\omega_{k+1/2} = (2\pi/N)(k+1/2)$ will appear scattered throughout the spectrum. The effect is illustrated in Figure 14.9 for a sinusoid that is observed through a rectangular window and then sampled at N points. The "picket-fence effect" means that not all frequencies can be seen by

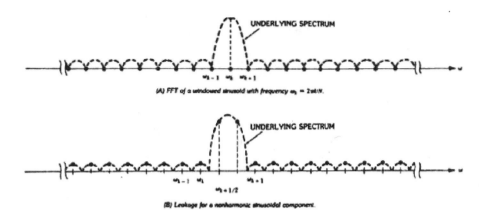

FIGURE 14.9 Illustration of leakage and the picket-fence effect.

the FFT. Harmonic components are seen accurately, but other components "slip through the picket fence" while their energy is "leaked" into the harmonics. These effects produce artifacts in the spectral domain that must be carefully monitored to ensure that an accurate spectrum is obtained from FFT processing.

Defining Terms

Continuous time (CT) impulse function: A generalized function $\delta(t)$ defined to be zero for all $t \neq 0$, undefined at $t = 0$, and having the special property that $\int_{-\infty}^{\infty} \delta(t)\mathrm{d}t = 1$.

Circular convolution: A convolution of finite length sequences in which the shifting operation is performed circularly within the finite support interval. Alternately called *periodic convolution*.

Dirichlet conditions: Conditions that must be satisfied in order to expand a periodic signal $s(t)$ in a Fourier series: each period of $s(t)$ must have a finite number of discontinuities, a finite number of maxima and minima, and $\int_{-T/2}^{T/2} |s(t)|\mathrm{d}t < \infty$ must be satisfied, where T is the period.

DFT (FFT) leakage: An effect that occurs when a signal containing a frequency component that falls between frequencies is sampled by the DFT and the power from this component incorrectly appears in adjacent frequency channels.

Gibbs phenomenon: Oscillatory behavior of Fourier series approximations in the vicinity of finite jump discontinuities.

Line spectrum: A common term for Fourier transforms of periodic **signals** for which the spectrum has nonzero components only at integer multiples of the fundamental frequency.

Mean squared error (mse): A measure of "closeness" between two functions given by $\mathrm{mse} = \frac{1}{T} \int_{-T/2}^{T/2} |f_1(t) - f_2(t)|^2 \mathrm{d}t$, where T is the period.

Nyquist sampling frequency: Minimum sampling frequency for which a CT signal $s(t)$ can be perfectly reconstructed from a set of uniformly spaced samples $s(nT)$.

Orthonormal set: A countable set of functions for which every pair in the set is mathematically orthogonal according to a valid norm, and for which each element of the set has unit length according to the same norm. The Fourier basis functions form an orthonormal set according to the mean squared error norm.

Trigonometric expansion: A Fourier series expansion for a real valued signal in which the basis functions are chosen to be $\sin(n\omega_0 t)$ and $\cos(n\omega_0 t)$.

References

R.N. Bracewell, *The Fourier Transform*, 2nd ed., New York: McGraw-Hill, 1986.

W.K. Jenkins, "Fourier series, Fourier transforms, and the discrete Fourier transform," in *The Circuits and Filters Handbook*, 2nd ed., Wai-Kai Chen, Ed., Boca Raton, FL: CRC Press, 2002a.

W.K. Jenkins, "Discrete-time signal processing," in *Reference Data for Engineers: Radio, Electronics, Computers, and Communications*, 9th ed., Wendy M. Middleton, Editor-in-Chief. Newnes (Butterworth-Heinemann), 2002b.

A.V. Oppenheim, A.S. Willsky, and I.T. Young, *Signals and Systems*, Englewood Cliffs, NJ: Prentice-Hall, 1983.

A.V. Oppenheim and R.W. Schafer, *Discrete-Time Signal Processing*, Englewood Cliffs, NJ: Prentice-Hall, 1989.

A.V. Oppenheim and R.W. Schafer, *Digital Signal Processing*, Englewood Cliffs, NJ: Prentice-Hall, 1975.

M.E. VanValkenburg, *Network Analysis*, Englewood Cliffs, NJ: Prentice-Hall, 1974.

For Further Information

A more thorough treatment of the complete family of CT and DT Fourier transform concepts is given in Jenkins (2002a). This article emphasizes the parallels between CT and DT Fourier-based signal processing. Digital filtering design and implementation is discussed in more detail in Jenkins (2002b).

An excellent treatment of Fourier Waveform Analysis is given by D. C. Munson, Jr. in Chapter 7 of *Reference Data for Engineers: Radio, Electronics, Computers, and Communications*, Ninth Edition, Wendy M. Middleton, Editor-in-Chief, Newnes (Butterworth-Heinemann), 2002.

A classic reference on the CT Fourier transform is Bracewell (1986).

14.2   Fourier Transforms and the Fast Fourier Transform

Alexander D. Poularikas

The Discrete-Time Fourier Transform (DTFT)

The discrete-time Fourier transform (DTFT) of a signal $\{f(n)\}$ is defined by

$$\mathscr{F}_{dt}\{f(n)\} \equiv F(\omega) \equiv F(e^{j\omega}) = \sum_{n=-\infty}^{\infty} f(n)e^{-j\omega n} \tag{14.15}$$

and its inverse discrete-time Fourier transform (IDTFT) is given by

$$f(n) = \frac{1}{2\pi} \int_{-\pi}^{\pi} F(\omega)e^{j\omega n} d\omega \tag{14.16}$$

The amplitude and phase spectra are periodic with a period of 2π and thus the frequency range of any discrete signal is limited to the range $(-\pi, \pi]$ or $(0, 2\pi]$.

Example 14.1

Find the DTFT of the sequence $f(n) = 0.8^n$ for $n = 0, 1, 2, 3\ldots$

Solution

From Equation (14.15), we write:

$$F(\omega) = \sum_{n=0}^{\infty} 0.8^n e^{-j\omega n} = \sum_{n=0}^{\infty} (0.8e^{-j\omega})^n = \frac{1}{1 - 0.8e^{-j\omega}} \tag{14.17}$$

TABLE 14.6 The DTFT Properties of Discrete-Time Sequences

Property	Time Domain	Frequency Domain
Linearity	$af_1(n) + bf_2(n)$	$aF_1(\omega) + bF_2(\omega)$
Time Shifting	$f(n - n_0)$	$e^{-j\omega n_0}F(\omega)$
Time Reversal	$f(-n)$	$F(-\omega)$
Convolution	$f_1(n)*f_2(n)$	$F_1(\omega)F_2(\omega)$
Frequency Shifting	$e^{j\omega_0 n}f(n)$	$F(\omega - \omega_0)$
Time Multiplication	$nf(n)$	$-z\dfrac{dF(z)}{dz}\Big\|z = e^{j\omega}$
Modulation	$f(n)\cos\omega_0 n$	$\dfrac{1}{2}F(\omega - \omega_0) + \dfrac{1}{2}F(\omega + \omega_0)$
Correlation	$f_1(n)\cdot f_2(n)$	$F_1(\omega)F_2(-\omega)$
Parseval's Formula	$\sum\limits_{n=-\infty}^{\infty}\|f\|(n)\|^2 = \dfrac{1}{2\pi}\int_{-\pi}^{\pi}\|F\|(\omega)\|^2 d\omega$	

$$|F(\omega)| = \frac{1}{\sqrt{1.64 - 1.6\cos\omega}}\,; \text{Arg}F(\omega)\tan^{-1}\left(\frac{0.8\sin\omega}{1 - 0.8\cos\omega}\right) \qquad (14.18)$$

If we set $\omega = -\omega$ in the last two equations we find that the amplitude is an even function and the argument is an odd function.

Relationship to the Z-Transform

$$F(z)\Big|_{z=e^{j\omega}} = \sum_{n=-\infty}^{\infty} f(n)z^{-n}\Big|_{z=e^{j\omega}}$$

Properties

Table 14.6 tabulates the DTFT properties of discrete time sequences.

Fourier Transforms of Finite-Time Sequences

The trancated Fourier transform of a sequence is given by

$$F_N(\omega) = \sum_{n=0}^{N-1} f(n)e^{-j\omega n} = \sum_{n=-\infty}^{\infty} f(n)w(n)e^{-j\omega n} = \frac{1}{2\pi}F(\omega)*W(\omega) \qquad (14.19)$$

where $w(n)$ is a window function that extends from $n = 0$ to $n = N - 1$. If the value of the sequence is unity for all n's, the window is known as the rectangular one. From Equation (14.19) we observe that the truncation of a sequence results in a smoother version of the exact spectrum.

Frequency Response of LTI Discrete Systems

A first-order LTI discrete system is described by the difference equation

$$y(n) + a_1 y(n - 1) = b_0 x(n) + b_1 x(n - 1)$$

The DTFT of the above equation is given by

$$Y(\omega) + a_1 e^{-j\omega}Y(\omega) = b_0 X(\omega) + b_1 e^{-j\omega}X(\omega)$$

from which we write the system function

$$H(\omega) = \frac{Y(\omega)}{X(\omega)} = \frac{b_0 + b_1 e^{-j\omega}}{1 + a_1 e^{-j\omega}}$$

To approximate the continuous time Fourier transform using the DTFT we follow the following steps:

1. Select the time interval T such that $F(\omega_c) \approx 0$ for all $\omega_c > \pi/T$. ω_c designates the frequency of a continuous time function.
2. Sample $f(t)$ at times nT to obtain $f(nT)$.
3. Compute the DFT using the sequence $\{Tf(nT)\}$.
4. The resulting approximation is then $F(\omega_c) \approx F(\omega)$ for $-\pi/T < \omega_c < \pi/T$.

The Discrete Fourier Transform

One of the methods, and one that is used extensively, calls for replacing continuous Fourier transforms by an equivalent *discrete Fourier transform* (DFT) and then evaluating the DFT using the discrete data. However, evaluating a DFT with 512 samples (a small number in most cases) requires more than 1.5×10^6 mathematical operations. It was the development of the *fast Fourier transform* (FFT), a computational technique that reduces the number of mathematical operations in the evaluation of the DFT to $N \log_2(N)$ (approximately 2.5×10^4 operations for the 512-point case mentioned above), that makes DFT an extremely useful tool in most all fields of science and engineering.

A data sequence is available only with a finite time window from $n = 0$ to $n = N - 1$. The transform is discretized for N values by taking samples at the frequencies $2\pi/NT$, where T is the time interval between sample points. Hence, we define the DFT of a sequence of N samples for $0 \leqslant k \leqslant N - 1$ by the relation

$$F(k\Omega) \doteq \mathcal{F}_d\{f(nT)\} = T \sum_{n=0}^{N-1} f(nT) e^{-j2\pi nkT/NT}$$

$$= T \sum_{n=0}^{N-1} f(nT) e^{-j\Omega Tnk} \quad n = 0, 1, \ldots, N - 1 \tag{14.20}$$

where N = number of sample values, T = sampling time interval, $(N - 1)T$ = signal length, $f(nT)$ = sampled form of $f(t)$ at points nT, $\Omega = (2\pi/T)\,1/N = \omega_s/N$ = frequency sampling interval, $e^{-i\Omega T}$ = Nth principal root of unity, and $j = \sqrt{-1}$. The inverse DFT is given by

$$f(nT) \doteq \mathcal{F}_d^{-1}\{F(k\Omega)\} = \frac{1}{NT} \sum_{k=0}^{N-1} F(k\Omega) e^{j2\pi nkT/NT}$$

$$= \frac{1}{NT} \sum_{k=0}^{N-1} F(k\Omega) e^{i\Omega Tnk} \tag{14.21}$$

The sequence $f(nT)$ can be viewed as representing N consecutive samples $f(n)$ of the continuous signal, while the sequence $F(k\Omega)$ can be considered as representing N consecutive samples $F(k)$ in the frequency domain. Therefore, Equations (14.20) and (14.21) take the compact form:

$$F(k) \doteq \mathcal{F}_d\{f(n)\} = \sum_{n=0}^{N-1} f(n) e^{-j2\pi nk/N}$$

$$= \sum_{n=0}^{N-1} f(n) W_N^{nk} \quad k = 0, \ldots, N - 1 \tag{14.22}$$

$$f(n) \doteq \mathcal{F}_d^{-1}\{F(k)\} = \frac{1}{N}\sum_{k=0}^{N-1} F(k)e^{j2\pi nk/N}$$

$$= \sum_{k=0}^{N-1} F(k)W_N^{-nk} \quad k = 0, \ldots, N-1 \qquad (14.23)$$

where

$$W_N = e^{-j2\pi/N} \quad j = \sqrt{-1}$$

An important property of the DFT is that $f(n)$ and $F(k)$ are uniquely related by the transform pair, Equations (14.22) and (14.23).

We observe that the functions W^{kn} are N-periodic; that is:

$$W_N^{kn} = W_N^{k(n+N)} \quad k, n = 1, \pm 1, \pm 2, \ldots \qquad (14.24)$$

As a consequence, the sequences $f(n)$ and $F(k)$ as defined by Equations (14.22) and (14.23) are also N-periodic.

It is generally convenient to adopt the convention

$$\{f(n)\} \leftrightarrow \{F(k)\} \qquad (14.25)$$

to represent the transform pair, Equations (14.22) and (14.23).

Properties of the DFT

A detailed discussion of the properties of DFT can be found in the cited references at the end of this section. In what follows we consider a few of these properties that are of value for the development of the FFT.

1. *Linearity*:

$$\{af(n) + by(n)\} \leftrightarrow \{aF(k)\} + \{bY(k)\} \qquad (14.26)$$

2. *Complex conjugate*: If $f(n)$ is real, $N/2$ is an integer and $\{f(n)\} \leftrightarrow \{F(k)\}$, then

$$F\left(\frac{N}{2}+l\right) = F*\left(\frac{N}{2}-l\right) \quad l = 0, 1, \ldots, \frac{N}{2} \qquad (14.27)$$

where $F^*(k)$ denotes the complex conjugate of $F(k)$. The preceding identity shows the folding property of the DFT.

3. *Reversal*:

$$\{f(-n)\} \leftrightarrow \{F(-k)\} \qquad (14.28)$$

4. *Time shifting*:

$$\{f(n+l)\} \leftrightarrow \{W^{-lk}F(k)\} \qquad (14.29)$$

5. *Convolution of real sequences*: If

$$y(n) = \frac{1}{N}\sum_{l=0}^{N-1} f(l)h(n-l) \quad n = 0, 1, \ldots, N-1 \qquad (14.30)$$

then

$$\{y(n)\} \leftrightarrow \{F(k)\,H(k)\} \tag{14.31}$$

6. *Correlation of real sequences*: If

$$y(n) = \frac{1}{N}\sum_{l=0}^{N-1} f(l)h(n+l) \quad n = 0, 1, \ldots, N-1 \tag{14.32}$$

then

$$\{y(n)\} \leftrightarrow \{F(r)\,H^*(k)\} \tag{14.33}$$

7. *Symmetry*:

$$\left\{\frac{1}{N}F(n)\right\} \leftrightarrow \{f(-k)\} \tag{14.34}$$

8. *Parseval's theorem*:

$$\sum_{n=0}^{N-1} |f(n)|^2 = \frac{1}{N}\sum_{k=0}^{N-1} \left|F(k)\right|^2 \tag{14.35}$$

where $|F(k)| = F(k)F^*(k)$.

Example 14.2

Verify Parseval's theorem for the sequence $\{f(n)\} = \{1, 2, -1, 3\}$.

Solution. With the help of Equation (14.22) we obtain

$$\begin{aligned}
F(k)\Big|_{k=0} &= F(0) = \sum_{n=0}^{3} f(n)\mathrm{e}^{-j(2\pi/4)kn}\Big|_{k=0} \\
&= (1\mathrm{e}^{-j(\pi/2)0.0} + 2\mathrm{e}^{-j(\pi/2)0.1} - \mathrm{e}^{-j(\pi/2)0.2} + 3\mathrm{e}^{-j(\pi/2)0.3}) \\
&= 5
\end{aligned}$$

Similarly, we find

$$F(1) = 2 + j \quad F(2) = -5 \quad F(3) = 2 - j$$

Introducing these values in Equation (14.35) we obtain

$$1^2 + 2^2 + (-1)^2 + 3^2 = 1/4[5^2 + (2+j)(2-j) + 5^2 + (2-j)(2+j)] \quad \text{or } 15 = 60/4$$

which is an identity, as it should have been.

Relation between DFT and Fourier Transform

The sampled form of a continuous function $f(t)$ can be represented by N equally spaced sampled values $f(n)$ such that

$$f(n) = f(nT) \quad n = 0, 1, \ldots, N - 1 \tag{14.36}$$

where T is the sampling interval. The length of the continuous function is $L = NT$, where $f(N) = f(0)$.

We denote the sampled version of $f(t)$ by $f_s(t)$, which may be represented by a sequence of impulses. Mathematically it is represented by the expression

$$f_s(t) = \sum_{n=0}^{N-1} [Tf(n)]\delta(t - nT) \tag{14.37}$$

where $\delta(t)$ is the Dirac or impulse function.

Taking the Fourier transform of $f_s(t)$ in Equation (14.37) we obtain

$$\begin{aligned}
F_s(\omega) &= T \int_{\infty}^{\infty} \sum_{n=0}^{N-1} f(n)\delta(t - nT)e^{-j\omega t}\mathrm{d}t \\
&= T \sum_{n=0}^{N-1} f(n) \int_{\infty}^{\infty} \delta(t - nT)e^{-j\omega t}\mathrm{d}t \\
&= T \sum_{n=0}^{N-1} f(n)e^{-j\omega nT}
\end{aligned} \tag{14.38}$$

Equation (14.38) yields $F_s(\omega)$ for all values of ω. However, if we are only interested in the values of $F_s(\omega)$ at a set of discrete equidistant points, then Equation (14.38) is expressed in the form (see also Equation (14.20)):

$$F_s(k\Omega) = T \sum_{n=0}^{N-1} f(n)e^{-jkn\Omega t} \quad k = 0, \pm 1, \pm 2, \ldots, \pm N/2 \tag{14.39}$$

where $\Omega = 2\pi/L = 2\pi/NT$. Therefore, comparing Equation (14.22) and Equation (14.39) we observe that we can find $F(\omega)$ from $F_s(\omega)$ using the relation

$$F(k) = F_s(\omega)|_{\omega = k\Omega} \tag{14.40}$$

Power, Amplitude, and Phase Spectra

If $f(t)$ represents voltage or current waveform supplying a load of 1 Ω, the left-hand side of Parseval's theorem, Equation (14.35) represents the power dissipated in the 1-Ω resistor. Therefore, the right-hand side represents the power contributed by each harmonic of the spectrum. Thus the DFT **power spectrum** is defined as

$$P(k) = F(k)F^\star(k) = |F(k)|^2 \quad k = 0, 1, \ldots, N - 1 \tag{14.41}$$

For real $f(n)$ there are only $(N/2+1)$ independent DFT spectral points as the complex conjugate property shows (Equation (14.27)). Hence we write

$$P(k) = |F(k)|^2 \quad k = 0, 1, \ldots, N/2 \tag{14.42}$$

The *amplitude spectrum* is readily found from that of a power spectrum, and it is defined as

$$A(k) = |F(k)| \quad k = 0, 1, \ldots, N - 1 \tag{14.43}$$

The power and amplitude spectra are invariant with respect to shifts of the data sequence $\{f(n)\}$.

The **phase spectrum** of a sequence $\{f(n)\}$ is defined as

$$\phi_f(k) = \tan^{-1} \frac{\text{Im}\{F(k)\}}{\text{Re}\{F(k)\}} \quad k = 0, 1, \ldots, N - 1 \tag{14.44}$$

As in the case of the power spectrum, only $(N/2 + 1)$ of the DFT phase spectral points are independent for real $\{f(n)\}$. For a real sequence $\{f(n)\}$ the power spectrum is an *even function* about the point $k = N/2$ and the phase spectrum is an *odd function* about the point $k = N/2$.

Observations

1. The frequency spacing $\Delta\omega$ between coefficients is

$$\Delta\omega = \Omega = \frac{2\pi}{NT} = \frac{\omega_s}{N} \quad \text{or} \quad \Delta f = \frac{1}{NT} = \frac{f_s}{N} = \frac{1}{T_0} \tag{14.45}$$

2. The reciprocal of the record length defines the frequency resolution.
3. If the number of samples N is fixed and the sampling time is increased, the record length and the precision of frequency resolution is increased. When the sampling time is decreased, the opposite is true.
4. If the record length is fixed and the sampling time is decreased (N increases), the resolution stays the same and the computed accuracy of $F(n\Omega)$ increases.
5. If the record length is fixed and the sampling time is increased (N decreases), the resolution stays the same and the computed accuracy of $F(n\Omega)$ decreases.

Data Windowing

To produce more accurate frequency spectra it is recommended that the data are weighted by a **window** function. Hence, the new dataset will be of the form $\{f(n)\, w(n)\}$. The following are the most commonly used windows:

1. Triangle (Fejer, Bartlet) window:

$$w(n) = \begin{cases} \dfrac{n}{N/2} & n = 0, 1, \ldots, \dfrac{N}{2} \\ w(N - n) & n = \dfrac{N}{2}, \ldots, N - 1 \end{cases} \tag{14.46}$$

2. $\text{Cos}^\alpha(x)$ windows:

$$w(n) = \sin^2\left(\frac{n}{N}\pi\right)$$

$$= 0.5\left[1 - \cos\left(\frac{2n}{N}\pi\right)\right] \quad n = 0, 1 \ldots, N - 1 \quad \alpha = 2 \tag{14.47}$$

This window is also called the raised cosine or Hamming window.

TABLE 14.7

No. of Terms in (14.29)	Maximum Sidelobe, dB	Parameter Values			
		a_0	a_1	a_2	a_3
3	−70.83	0.42323	0.49755	0.07922	
3	−62.05	0.44959	0.49364	0.05677	
4	−92	0.35875	0.48829	0.14128	0.01168
4	−74.39	0.40217	0.49703	0.09892	0.00188

3. Hamming window:

$$w(n) = 0.54 - 0.46\cos\left(\frac{2\pi}{N}n\right) \quad n = 0, 1, \ldots, N-1 \tag{14.48}$$

4. Blackman window:

$$w(n) = \sum_{m=0}^{k} (-1)^m a_m \cos\left(2\pi m\frac{n}{N}\right) \quad n = 0, 1, \ldots, N-1 \quad K \leq \frac{N}{2} \tag{14.49}$$

for $K = 2$, $a_0 = 0.42$, $a_1 = 0.50$, and $a_2 = 0.08$.
5. Blackman–Harris window. Harris used a gradient search technique to find three- and four-term expansion of Equation (14.49) that either minimized the maximum sidelobe level for fixed mainlobe width, or traded mainlobe width versus minimum sidelobe level (see Table 14.7).
6. Centered Gaussian window:

$$w(n) = \exp\left[-\frac{1}{2}\alpha\left(\frac{n}{N/2}\right)^2\right] \quad 0 \leq |n| \leq \frac{N}{2} \quad \alpha = 2, 3, \ldots \tag{14.50}$$

As α increases, the mainlobe of the frequency spectrum becomes broader and the sidelobe peaks become lower.
7. Centered Kaiser–Bessel window:

$$w(n) = \frac{I_0\left[\pi a\sqrt{1.0 - \left(\frac{n}{N/2}\right)^2}\right]}{I_0(\pi\alpha)} \quad 0 \leq |n| \leq \frac{N}{2} \tag{14.51}$$

where

$$I_0(x) = \text{zero-order modified Bessel function}$$

$$= \sum_{k=0}^{\infty} \left(\frac{(x/2)^k}{k!}\right)^2 \tag{14.52}$$

$$k! = 1 \times 2 \times 3 \times \cdots \times k$$

$$\alpha = 2, 2.5, 3 \quad \text{(typical values)}$$

Fast Fourier Transform

One of the approaches to speed the computation of the DFT of a sequence is the *decimation-in-time* method. This approach is one of breaking the N-point transform into two $(N/2)$-point transforms, breaking each $(N/2)$-point transform into two $(N/4)$-point transforms, and continuing the above process until we obtain the two-point transform. We start with the DFT expression and factor it into two DFTs of length $N/2$:

$$F(k) = \sum_{n=0}^{N-2} f(n) W_N^{kn} \qquad n \text{ even}$$

$$+ \sum_{n=1}^{N-1} f(n) W_N^{kn} \quad n \text{ odd} \tag{14.53}$$

Letting $n = 2m$ in the first sum and $n = 2m+1$ in the second, Equation (14.53) becomes

$$F(k) = \sum_{m=0}^{(N/2)-1} f(2m) W_N^{2mk} + \sum_{m=0}^{(N/2)-1} f(2m+1) W_N^{(2m+1)k} \tag{14.54}$$

However, because of the identities

$$W_N^{2mk} = \left(W_N^2 \right)^{mk} = e^{-j(2\pi/N)2mk} = e^{-j(4\pi mk/N)} = W_{N/2}^{mk} \tag{14.55}$$

and the substitution $f(2m) = f_1(m)$ and $f(2m+1) = f_2(m)$, $m = 0, 1,\ldots, N/2-1$, takes the form

$$F(k) = \sum_{m=0}^{(N/2)-1} f_1(m) W_{N/2}^{mk} \qquad \frac{N}{2} - \text{point DFT of even-indexed sequence}$$

$$+ W_N^k \sum_{m=0}^{(N/2)-1} f_2(m) W_{N/2}^{mk} \qquad \frac{N}{2} - \text{point DFT of odd-indexed sequence} \tag{14.56}$$

$$k = 0, \ldots, N/2 - 1$$

We can also write Equation (14.56) in the form

$$F(k) = F_1(k) + W_N^k F_2(k) \qquad k = 0, 1, \ldots, N/2 - 1$$

$$F\left(k + \frac{N}{2}\right) = F_1(k) + W_N^{k+N/2} F_2(k)$$

$$= F_1(k) - W_N^k F_2(k) \qquad k = 0, 1, \ldots, N/2 - 1 \tag{14.57}$$

where $W_N^{k+N/2} = -W_N^k$ and $W_{N/2}^{m(k+N/2)} = W_{N/2}^{mk}$. Since the DFT is periodic, $F_1(k) = F_1(k + N/2)$ and $F_2(k) = F_2(k + N/2)$.

We next apply the same procedure to each $N/2$ samples, where $f_{11}(m) = f_1(2m)$ and $f_{21}(m) = f_2(2m + 1)$, $m = 0,1,\ldots, (N/4) - 1$. Hence:

$$F_1(k) = \sum_{m=0}^{(N/4)-1} f_{11}(m) W_{N/4}^{mk} + W_N^{2k} \sum_{m=0}^{(N/4)-1} f_{21}(m) W_{N/4}^{mk} \tag{14.58}$$

$$k = 0, 1, \ldots, N/4 - 1$$

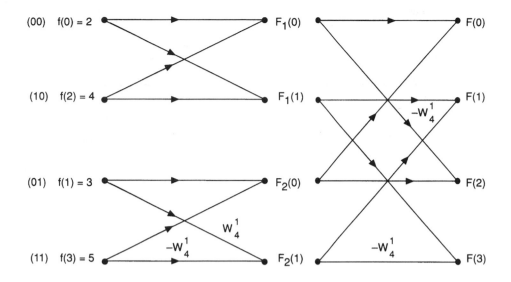

FIGURE 14.10 Illustration of Example 14.2.

or

$$F_1(k) = F_{11}(k) + W_N^{2k}F_{21}(k)$$

$$F_1\left(k + \frac{N}{4}\right) = F_{11}(k) - W_N^{2k}F_{21}(k) \quad k = 0, 1, \ldots, N/4 - 1 \tag{14.59}$$

Therefore, each one of the sequences f_1 and f_2 has been split into two DFTs of length $N/4$.

Example 14.3

To find the FFT of the sequence $\{2, 3, 4, 5\}$ we first bit reverse the position of the elements from their priority $\{00, 01, 10, 11\}$ to $\{00, 10, 01, 11\}$ position. The new sequence is $\{2, 4, 3, 5\}$ (see also Figure 14.10). Using Equations (14.56) and (14.57) we obtain

$$F_1(0) = \sum_{m=0}^{1} f_1(m) W_2^{m0} = f_1(0) W_2^0 + f_1(1) W_2^0 = f(0) \cdot 1 + f(2) \cdot 1$$

$$F_1(1) = \sum_{m=0}^{1} f_1(m) W_2^{m \cdot 1} = f_1(0) W_2^{0 \cdot 1} + f_1(1) W_2^1 = f(0) + f(2)(-j)$$

$$F_2(0) = W_4^0 \sum_{m=0}^{1} f_2(m) W_2^{m \cdot 0} = f_2(0) W_2^0 + f_2(1) W_2^0 = f(1) + f(3)$$

$$F_2(1) = W_4^1 \sum_{m=0}^{1} f_2(m) W_2^{m \cdot 1} = W_4^1 \Big[f(1) W_2^0 + f(3) W_2^1 \Big] = W_4^1 f(1) - W_4^1 f(3)$$

TABLE 14.8　　FFT Subroutine

```
SUBROUTINE FOUR1 (DATA, NN, ISIGN)
    Replaces DATA by its discrete Fourier transform, if SIGN is input as 1; or replaces DATA by NN times its inverse
    discrete Fourier transform, if ISIGN is input as−1. DATA is a complex array of length NN or, equivalently,
    a real array of length 2*NN. NN must be an integer power of 2.
    REAL*8 WR, WI, WPR, WPI, WTEMP, THETA               Double precision for the trigonometric recurrences.
    DIMENSION DATA (2*NN)
    N=2*NN
    J=1
    DO 11 I=1, N, 2                                      This is the bit-reversal section of the routine.
        IF (J.GT.I) THEN                                 Exchange the two complex numbers.
            TEMPR=DATA(J)
            TEMPI=DATA(J+1)
            DATA(J)=DATA(I)
            DATA(J+1)=DATA(I+1)
            DATA(I)=TEMPR
            DATA(I+1)=TEMPI
        ENDIF
        M=N/2
1       IF ((M.GE.2).AND. (J.GT.M)) THEN
            J=J−M
            M=M/2
            GO TO 1
        ENDIF
        J=J+M
11  CONTINUE
    MMAX=2                                              Here begins the Danielson−Lanczos section of the routine.
2   IF (N.GT.MMAX) THEN                                 Outer loop executed log₂ NN times.
        ISTEP=2*MMAX
        THETA=6.28318530717959D0/(ISIGN*MMAX)          Initialize for the trigonometric recurrence.
        WPR=-2.D0*DSIN(0.5D0*THETA)**2
        WPI=DSIN(THETA)
        WR=1.D0
        WI=0.D0
        DO 13 M=1,MMAX,2                               Here are the two nested inner loops.
            DO 12 I=M,N,ISTEP
            J=I+MMAX                                    This is the Danielson−Lanczos formula:
            TEMPR=SNGL(WR)*DATA(J)-SNGL(WI)*DATA(J+1)
            TEMPI=SNGL(WR)*DATA(J+1)+SNGL(WI)*DATA(J)
            DATA(J)=DATA(I)−TEMPR
            DATA(J+1)=DATA(I+1)−TEMPI
            DATA(I)=DATA(I)+TEMPR
            DATA(I+1)=DATA(I+1)+TEMPI
12      CONTINUE
        WTEMP=WR                                        Trigonometric recurrence.
        WR=WR*WPR-WI*WPI+WR
        WI=WI*WPR+WTEMP*WPI+WI
13  CONTINUE
    MMAX=STEP
    GO TO 2
    ENDIF
    RETURN
    END
```

Source: © 1986 Numerical Recipes Software. From *Numerical Recipes: The Art of Scientific Computing,* published by Cambridge University Press. Used by permission.

From Equation (14.57) the output is

$$F(0) = F_1(0) + W_4^0 F_2(0)$$

$$F(1) = F_1(1) + W_4^1 F_2(1)$$

$$F(2) = F_1(0) - W_4^0 F_2(0)$$

$$F(3) = F_1(1) - W_4^1 F_2(1)$$

Computation of the Inverse DFT

To find the inverse FFT using an FFT algorithm, we use the relation

$$f(n) = \frac{[\text{FFT}(F^*(k))]^*}{N} \qquad (14.60)$$

For other transforms and their fast algorithms the reader should consult the references given at the end of this section.

Table 14.8 gives the FFT subroutine for fast implementation of the DFT of a finite sequence.

Defining Terms

FFT: A computational technique that reduces the number of mathematical operations in the evaluation of the discrete Fourier transform (DFT) to $N \log_2 N$.

Phase spectrum: All phases associated with the spectrum harmonics.

Power spectrum: A power contributed by each harmonic of the spectrum.

Window: Any appropriate function that multiplies the data with the intent to minimize the distortions of the Fourier spectra.

References

A. Ahmed and K.R. Rao, *Orthogonal Transforms for Digital Signal Processing*, New York: Springer-Verlag, 1975.

E.R. Blahut, *Fast Algorithms for Digital Signal Processing*, Reading, MA: Addison-Wesley, 1987.

E.O. Bringham, *The Fast Fourier Transform*, Englewood Cliffs, NJ: Prentice-Hall, 1974.

F.D. Elliot, *Fast Transforms, Algorithms, Analysis, Applications*, New York: Academic Press, 1982.

H.J. Nussbaumer, *Fast Fourier Transform and Convolution Algorithms*, New York: Springer-Verlag, 1982.

A.D. Poularikas and S. Seely, *Signals and System*. 2nd ed., Melbourne, Fla.: Krieger Publishing, 1995.

Further Information

A historical overview of the fast Fourier transform can be found in J.W. Cooley, P.A.W. Lewis, and P.D. Welch, "Historical notes on the fast Fourier transform," *IEEE Trans. Audio Electroacoust.*, vol. AV-15, pp. 76–79, June 1967.

Fast algorithms appear frequently in the monthly magazine *Signal Processing*, published by The Institute of Electrical and Electronics Engineers.

14.3   Design and Implementation of Digital Filters

Bruce W. Bomar and L. Montgomery Smith

A *digital filter* is a linear, shift-invariant system for computing a **discrete output sequence** from a **discrete input sequence**. The input/output relationship is defined by the *convolution sum*:

$$y(n) = \sum_{m=-\infty}^{\infty} h(m)x(n-m)$$

where $x(n)$ is the input sequence, $y(n)$ is the output sequence, and $h(n)$ is the *impulse response* of the filter. The filter is often conveniently described in terms of its frequency characteristics that are given by the *transfer function* $H(e^{j\omega})$. The impulse response and transfer function are a Fourier transform pair:

$$H(e^{j\omega}) = \sum_{n=-\infty}^{\infty} h(n)e^{-j\omega n} \qquad -\pi \leqslant \omega \leqslant \pi$$

$$h(n) = \frac{1}{2\pi}\int_{-\pi}^{\pi} H(e^{j\omega})e^{j\omega n}\,d\omega \qquad -\infty \leqslant n \leqslant \infty$$

Closely related to the Fourier transform of $h(n)$ is the *z*-transform defined by

$$H(z) = \sum_{n=-\infty}^{\infty} h(n)z^{-n}$$

The Fourier transform is then the *z*-transform evaluated on the unit circle in the *z*-plane ($z = e^{j\omega}$). An important property of the *z*-transform is that $z^{-1}\,H(z)$ corresponds to $h(n-1)$, so z^{-1} represents a one-sample delay, termed a *unit delay*.

In this section, attention will be restricted to *frequency-selective* filters. These filters are intended to pass frequency components of the input sequence in a given band of the spectrum while blocking the rest. Typical frequency-selective filter types are *low-pass, high-pass, bandpass,* and *band-reject*. Other special-purpose filters exist, but their design is an advanced topic that will not be addressed here. In addition, special attention is given to *causal* filters, that is, those for which the impulse response is identically zero for negative *n* and thus can be realized in real time. Digital filters are further separated into two classes depending on whether the impulse response contains a finite or infinite number of nonzero terms.

Finite Impulse Response Filter Design

The objective of **finite impulse response (FIR) filter** design is to determine $N + 1$ coefficients:

$$h(0), h(1), \ldots, h(N)$$

so that the transfer function $H(e^{j\omega})$ approximates a desired frequency characteristic $H_d(e^{j\omega})$. All other impulse response coefficients are zero. An important property of FIR filters for practical applications is that they can be designed to be *linear phase*; that is, the transfer function has the form:

$$H(e^{j\omega}) = A(e^{j\omega})e^{-j\omega N/2}$$

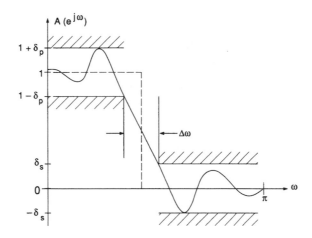

FIGURE 14.11 Amplitude frequency characteristics of a FIR low-pass filter showing definitions of passband ripple δ_p, stopband attenuation δ_s, and transition bandwidth $\Delta\omega$.

where the amplitude $A(e^{j\omega})$ is a real function of frequency. The desired transfer function can be similarly written

$$H_d(e^{j\omega}) = A_d(e^{j\omega})e^{-j\omega N/2}$$

where $A_d(e^{j\omega})$ describes the amplitude of the desired frequency-selective characteristics. For example, the amplitude frequency characteristics of an ideal low-pass filter are given by

$$A_d(e^{j\omega}) = \begin{cases} 1 & \text{for } |\omega| \leqslant \omega_c \\ 0 & \text{otherwise} \end{cases}$$

where ω_c is the *cutoff frequency* of the filter.

A linear phase characteristic ensures that a filter has a constant group delay independent of frequency. Thus, all frequency components in the signal are delayed by the same amount, and the only signal distortion introduced is that imposed by the filter's frequency-selective characteristics. Since a FIR filter can only approximate a desired frequency-selective characteristic, some measures of the accuracy of approximation are needed to describe the quality of the design. These are the *passband ripple* δ_p, the *stopband attenuation* δ_s, and the *transition bandwidth* $\Delta\omega$. These quantities are illustrated in Figure 14.11 for a prototype low-pass filter. The passband ripple gives the maximum deviation from the desired amplitude (typically unity) in the region where the input signal spectral components are desired to be passed unattenuated. The stopband attenuation gives the maximum deviation from zero in the region where the input signal spectral components are desired to be blocked. The transition bandwidth gives the width of the spectral region in which the frequency characteristics of the transfer function change from the passband to the stopband values. Often, the passband ripple and stopband attenuation are specified in decibels, in which case their values are related to the quantities δ_p and δ_s by

$$\text{passband ripple (dB)} = P = -20\log_{10}(1-\delta_p)$$

$$\text{stopband attenuation (dB)} = S = -20\log_{10}\delta_s$$

FIR Filter Design by Windowing

The windowing design method is a computationally efficient technique for producing nonoptimal filters. Filters designed in this manner have equal passband ripple and stopband attenuation:

$$\delta_p = \delta_s = \delta$$

The method begins by finding the impulse response of the desired filter from

$$h_d(n) = \frac{1}{2\pi} \int_{-\pi}^{\pi} A_d(e^{j\omega}) e^{j\omega(n-N/2)} d\omega$$

For ideal low-pass, high-pass, bandpass, and band-reject frequency-selective filters, the integral can be solved in closed form. The impulse response of the filter is then found by multiplying this ideal impulse response with a window $w(n)$ that is identically zero for $n < 0$ and for $n > N$:

$$h(n) = h_d(n)w(n) \quad n = 0, 1, \ldots, N$$

Some commonly used windows are defined as follows:

1. Rectangular (truncation)

$$w(n) = \begin{cases} 1 & \text{for } 0 \leq n \leq N \\ 0 & \text{otherwise} \end{cases}$$

2. Hamming

$$w(n) = \begin{cases} \left(0.54 - 0.46 \cos \dfrac{2\pi n}{N}\right) & \text{for } 0 \leq n \leq N \\ 0 & \text{otherwise} \end{cases}$$

3. Kaiser

$$w(n) = \begin{cases} \dfrac{I_0\left(\beta\sqrt{1 - [(2n-N)/N]^2}\right)}{I_0(\beta)} & \text{for } 0 \leq n \leq N \\ 0 & \text{otherwise} \end{cases}$$

In general, windows that slowly taper the impulse response to zero result in lower passband ripple and a wider transition bandwidth. Other windows (e.g., Hamming, Blackman) are also sometimes used but not as often as those shown above.

Of particular note is the Kaiser window where $I_0(.)$ is the 0th-order modified Bessel function of the first kind and β is a shape parameter. The proper choice of N and β allows the designer to meet given passband ripple/stopband attenuation and transition bandwidth specifications. Specifically, using S, the stopband attenuation in dB, the filter order must satisfy

$$N = \frac{S - 8}{2.285 \Delta \omega}$$

Then, the required value of the shape parameter is given by

$$\beta = \begin{cases} 0 & \text{for } S < 21 \\ 0.5842(S - 21)^{0.4} + 0.07886(S - 21) & \text{for } 21 \leq S \leq 50 \\ 0.1102(S - 8.7) & \text{for } S > 50 \end{cases}$$

As an example of this design technique, consider a low-pass filter with a cutoff frequency of $\omega_c = 0.4\pi$. The ideal impulse response for this filter is given by

$$h_d(n) = \frac{\sin[0.4\pi(n - N/2)]}{\pi(n - N/2)}$$

Choosing $N = 8$ and a Kaiser window with a shape parameter of $\beta = 0.5$ yields the following impulse response coefficients:

$$h(0) = h(8) = -0.07568267$$

$$h(1) = h(7) = -0.06236596$$

$$h(2) = h(6) = 0.09354892$$

$$h(3) = h(5) = 0.30273070$$

$$h(4) = 0.40000000$$

Design of Optimal FIR Filters

The accepted standard criterion for the design of optimal FIR filters is to minimize the maximum value of the error function:

$$E(e^{j\omega}) = W_d(e^{j\omega})_d(e^{j\omega}) - A(e^{j\omega})|$$

over the full range of $-\pi \le \omega \le \pi$. $W_d(e^{j\omega})$ is a desired weighting function used to emphasize specifications in a given frequency band. The ratio of the deviation in any two bands is inversely proportional to the ratio of their respective weighting.

A consequence of this optimization criterion is that the frequency characteristics of optimal filters are *equiripple*. although the maximum deviation from the desired characteristic is minimized, it is reached several times in each band. Thus, the passband and stopband deviations oscillate about the desired values with equal amplitude in each band. Such approximations are frequently referred to as *minimax* or *Chebyshev* approximations. In contrast, the maximum deviations occur near the band edges for filters designed by windowing.

Equiripple FIR filters are usually designed using the *Parks–McClellan* computer program (Parks and Burrus, 1987), which uses the *Remez exchange algorithm* to determine iteratively the *extremal frequencies* at which the maximum deviations in the error function occur. A listing of this program along with a detailed description of its use is available in several references including Parks and Burrus (1987) and DSP Committee (1979). The program is executed by specifying as inputs the desired band edges, gain for each band (usually 0 or 1), band weighting, and FIR length. If the resulting filter has too much ripple in some bands, those bands can be weighted more heavily and the filter redesigned. Details on this design procedure are discussed in Rabiner (1973), along with approximate design relationships which aid in selecting the filter length needed to meet a given set of specifications.

Although we have focused attention on the design of frequency-selective filters, other types of FIR filters exist. For example, the Parks–McClellan program will also design linear-phase FIR filters for differentiating broadband signals and for approximating the Hilbert transform of such signals. A simple modification to this program permits arbitrary magnitude responses to be approximated with linear-phase filters. Other design techniques are available that permit the design of FIR filters which approximate an arbitrary complex response (Chen and Parks, 1987; Parks and Burrus, 1987), and, in cases where a nonlinear phase response is acceptable, design techniques are available that give a shorter impulse response length than would be required by a linear-phase design (Goldberg et al., 1981).

As an example of an equiripple filter design, an 8th-order low-pass filter with a passband $0 \leq \omega \leq 0.3\pi$, a stopband $0.5\pi \leq \omega \leq \pi$, and equal weighting for each band was designed. The impulse response coefficients generated by the Parks–McClellan program were as follows:

$$h(0) = h(8) = -0.06367859$$

$$h(1) = h(7) = -0.06912276$$

$$h(2) = h(6) = 0.10104360$$

$$h(3) = h(5) = 0.28574990$$

$$h(4) = 0.41073000$$

These values can be compared to those for the similarly specified filter designed in the previous subsection using the windowing method.

Infinite Impulse Response Filter Design

An **infinite impulse response (IIR) digital filter** requires less computation to implement than a FIR digital filter with a corresponding frequency response. However, IIR filters cannot generally achieve a perfect linear-phase response and are more susceptible to **finite wordlength effects.**

Techniques for the design of IIR analog filters are well established. For this reason, the most important class of IIR digital filter design techniques is based on forcing a digital filter to behave like a reference analog filter. This can be done in several different ways. For example, if the analog filter impulse response is $h_a(t)$ and the digital filter impulse response is $h(n)$, then it is possible to make $h(n) = h_a(nT)$, where T is the sample spacing of the digital filter. Such designs are referred to as *impulse-invariant* (Parks and Burrus, 1987). Likewise, if $g_a(t)$ is the unit step response of the analog filter and $g(n)$ is the unit step response of the digital filter, it is possible to make $g(n) = g_a(nT)$, which gives a *step-invariant* design (Parks and Burrus, 1987).

The step-invariant and impulse-invariant techniques perform a time domain matching of the analog and digital filters but can produce aliasing in the frequency domain. For frequency-selective filters it is better to attempt matching frequency responses. This task is complicated by the fact that the analog filter response is defined for an infinite range of frequencies ($\Omega = 0$ to ∞), while the digital filter response is defined for a finite range of frequencies ($\omega = 0$ to π). Therefore, a method for mapping the infinite range of analog frequencies Ω into the finite range from $\omega = 0$ to π, termed the *bilinear transform*, is employed.

Bilinear Transform Design of IIR Filters

Let $H_a(s)$ be the Laplace transform transfer function of an analog filter with frequency response $H_a(j\Omega)$. The bilinear transform method obtains the digital filter transfer function $H(z)$ from $H_a(s)$ using the substitution:

$$s = \frac{2(1 - z^{-1})}{T(1 + z^{-1})}$$

That is:

$$H(z) = H_a(s)|s = \frac{2}{T} \frac{1 - z^{-1}}{1 + z^{-1}}$$

This maps analog frequency Ω to digital frequency ω according to

$$\omega = 2 \tan^{-1} \frac{\Omega T}{2}$$

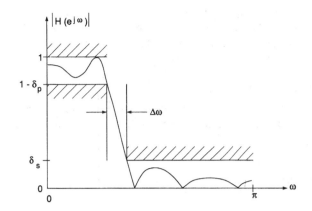

FIGURE 14.12   Frequency characteristics of an IIR digital low-pass filter showing definitions of passband ripple δ_p, stopband attenuation δ_s, and transition bandwith $\Delta\omega$.

thereby warping the frequency response $H_a(j\Omega)$ and forcing it to lie between 0 and π for $H(e^{j\omega})$. Therefore, to obtain a digital filter with a cutoff frequency of ω_c it is necessary to design an analog filter with cutoff frequency

$$\Omega_c = \frac{2}{T} \tan \frac{\omega_c}{2}$$

This process is referred to as *prewarping* the analog filter frequency response to compensate for the warping of the bilinear transform. Applying the bilinear transform substitution to this analog filter will then give a digital filter that has the desired cutoff frequency.

Analog filters and hence IIR digital filters are typically specified in a slightly different fashion than FIR filters. Figure 14.12 illustrates how analog and IIR digital filters are usually specified. Notice by comparison to Figure 14.11 that the passband ripple in this case never goes above unity, whereas in the FIR case the passband ripple is specified about unity.

Four basic types of analog filters are generally used to design digital filters: (1) Butterworth filters that are maximally flat in the passband and decrease monotonically outside the passband, (2) Chebyshev filters that are equiripple in the passband and decrease monotonically outside the passband, (3) inverse Chebyshev filters that are flat in the passband and equiripple in the stopband, and (4) elliptic filters that are equiripple in both the passband and stopband. Techniques for designing these analog filters are covered elsewhere (see, for example, Van Valkenberg, 1982) and will not be considered here.

To illustrate the design of an IIR digital filter using the bilinear transform, consider the design of a second-order Chebyshev low-pass filter with 0.5 dB of passband ripple and a cutoff frequency of $\omega_c = 0.4\,\pi$. The sample rate of the digital filter is to be 5 Hz, giving $T = 0.2$ sec. To design this filter we first design an analog Chebyshev low-pass filter with a cutoff frequency of

$$\Omega_c = \frac{2}{0.2} \tan 0.2\pi = 7.2654 \text{ rad/sec}$$

This filter has a transfer function:

$$H(s) = \frac{0.9441}{1 + 0.1249s + 0.01249s^2}$$

Substituting

$$s = \frac{2}{0.2} \frac{z-1}{z+1}$$

gives

$$H(z) = \frac{0.2665(z+1)^2}{z^2 - 0.1406z + 0.2695}$$

Computer programs are available that accept specifications on a digital filter and carry out all steps required to design the filter, including prewarping frequencies, designing the analog filter, and performing the bilinear transform. Two such programs are given in Parks and Burrus (1987) Antoniou (1979).

Design of Other IIR Filters

For frequency-selective filters, the bilinear transformation of an elliptic analog filter provides an optimal equiripple design. However, if a design other than standard low-pass, high-pass, bandpass, or bandstop is needed or if it is desired to approximate an arbitrary magnitude or group delay characteristic, some other design technique is needed. Unlike the FIR case, there is no standard IIR design program for obtaining optimal approximations to an arbitrary response.

Four techniques that have been used for designing optimal equiripple IIR digital filters are (Parks and Burrus, 1987) (1) minimizing the L_p norm of the weighted difference between the desired and actual responses, (2) linear programming, (3) iteratively using the Remez exchange algorithm on the numerator and denominator of the transfer function, and (4) the differential correction algorithm. A computer program for implementing the first method is available in DSP Committee (1979).

Finite Impulse Response Filter Implementation

For FIR filters, the convolution sum represents a computable process, and so filters can be implemented by directly programming the arithmetic operations. Nevertheless, some options are available that may be preferable for a given processor architecture, and means for reducing computational loads exist. This section outlines some of these methods and presents schemes for FIR filter realization.

Direct Convolution Methods

The most obvious method for the implementation of FIR filters is to directly evaluate the sum of products in the convolution sum:

$$y(n) = h(0)\,x(n) + h(1)\,x(n-1) + \ldots + h(N)\,x(n-N)$$

The block diagram for this is shown in Figure 14.13. This method involves storing the present and previous N values of the input, multiplying each sample by the corresponding impulse response coefficient, and summing the products to compute the output. This method is referred to as a *tapped delay line* structure.

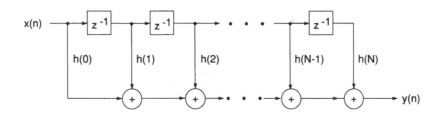

FIGURE 14.13   A direct-form implementation of a FIR filter.

A modification to this approach is suggested by writing the convolution as

$$y(n) = h(0)\,x(n) + \sum_{m=1}^{N} h(m)\,x(n-m)$$

In this approach, the output is computed by adding the product of $h(0)$ with the present input sample to a previously computed sum of products and updating a set of N sums of products with the present input sample value. The signal flow graph for this method is shown in Figure 14.14.

FIR filters designed to have linear phase are usually obtained by enforcing the symmetry constraint:

$$h(n) = h(N - n)$$

For these filters, the convolution sum can be written:

$$y(n) = \begin{cases} \displaystyle\sum_{m=0}^{N/2-1} h(m)[x(n-m) + x(n+m-N)] + h\!\left(\dfrac{N}{2}\right)\!x\!\left(n - \dfrac{N}{2}\right) & N \text{ even} \\[1em] \displaystyle\sum_{m=0}^{(N-1)/2} h(m)[x(n-m) + x(n+m-N)] & N \text{ odd} \end{cases}$$

Implementation of the filter according to these formulas reduces the number of multiplications by approximately a factor of 2 over direct-form methods. The block diagrams for these filter structures are shown in Figure 14.15 and Figure 14.16.

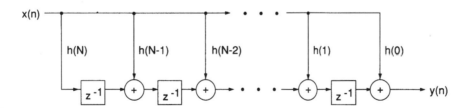

FIGURE 14.14 Another direct-form implementation of a FIR filter.

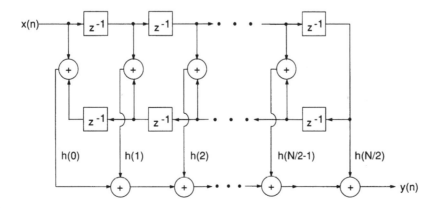

FIGURE 14.15 Implementation of a linear-phase FIR filter for even N.

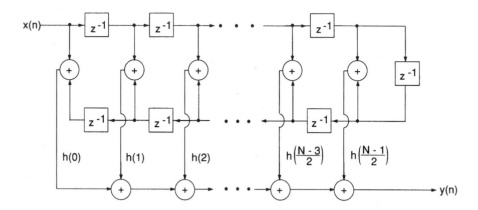

FIGURE 14.16   Implementation of a linear-phase FIR filter for odd N.

Implementation of FIR Filters Using the Discrete Fourier Transform

A method for implementing FIR filters that can have computational advantages over direct-form convolution involves processing the input data in blocks using the discrete Fourier transform (DFT) via the *overlap-save* method. The computational advantage arises primarily from use of the fast Fourier transform (FFT) algorithm (discussed in Section 14.2) to compute the DFTs of the individual data blocks. In this method, the input data sequence $\{x(n); -\infty < n < \infty\}$ is divided into L-point blocks:

$$x_i(n) \quad 0 \leqslant n \leqslant L - 1 \quad -\infty < i < \infty$$

where $L > N + 1$, the length of the FIR filter. The L-point DFT of the impulse response is precomputed from

$$H[k] = \sum_{n=0}^{L-1} h(n) e^{-j2\pi kn/L} \quad k = 0, 1, \ldots, L - 1$$

where square brackets are used to distinguish the DFT from the continuous-frequency transfer function of the filter $H(e^{j\omega})$. Then, the DFT of each data block is computed according to

$$X_i[k] = \sum_{n=0}^{L-1} x_i(n) e^{-j2\pi kn/L} \quad k = 0, 1, \ldots, L - 1$$

These two complex sequences are multiplied together term by term to form the DFT of the output data block:

$$Y_i[k] = H[k]X_i[k] \quad k = 0, 1, \ldots, L - 1$$

and the output data block is computed by the inverse DFT:

$$y_i(n) = \frac{1}{L} \sum_{k=0}^{L-1} Y_i[k] e^{j2\pi kn/L} \quad n = 0, 1, \ldots, L - 1$$

However, the output data block computed in this manner is the *circular convolution* of the impulse response of the filter and the input data block given by

$$y_i(n) = \sum_{m=0}^{N} h(m)x_i((n - m) \text{ modulo } L)$$

Thus, only the output samples from $n = N$ to $n = L - 1$ are the same as those that would result from the convolution of the impulse response with the infinite-length data sequence $x(n)$. The first N data points are corrupted and must therefore be discarded. So that the output data sequence does not have N-point "gaps" in it, it is therefore necessary to *overlap* the data in adjacent input data blocks. In carrying out the processing, samples from block to block are *saved* so that the last N points of the ith data block $x_i(n)$ are the same as the first N points of the following data block $x_{i+1}(n)$. Each processed L-point data block thus produces $L-N$ output samples.

Another technique of block processing of data using DFTs is the *overlap-add* method in which $L-N$-point blocks of input data are zero-padded to L points, the resulting output blocks are overlapped by N points, and corresponding samples added together. This method requires more computation than the overlap-save method and is somewhat more difficult to program. Therefore, its usage is not as widespread as the overlap-save method.

Infinite Impulse Response Filter Implementation

Direct-Form Realizations

For an IIR filter the convolution sum does not represent a computable process. Therefore, it is necessary to examine the general transfer function, which is given by

$$H(z) = \frac{Y(z)}{X(z)} = \frac{\gamma_0 + \gamma_1 z^{-1} + \gamma_2 z^{-2} + + \gamma_M z^{-M}}{1 + \beta_1 z^{-1} + \beta_2 z^{-2} + + \beta_N z^{-N}}$$

where $Y(z)$ is the z-transform of the filter output $y(n)$ and $X(z)$ is the z-transform of the filter input $x(n)$. The unit-delay characteristic of z^{-1} then gives the following *difference equation* for implementing the filter:

$$y(n) = \gamma_0 x(n) + \gamma_1 x(n-1) + \ldots + \gamma_M x(n-M) - \beta_1 y(n-1) - \ldots - \beta_N y(n-N)$$

When calculating $y(0)$, the values of $y(-1), y(-2), \ldots, y(-N)$ represent initial conditions on the filter. If the filter is started in an initially relaxed state, then these initial conditions are zero.

Figure 14.17 gives a block diagram realizing the filter's difference equation. This structure is referred to as the *direct-form I* realization. Notice that this block diagram can be separated into two parts, giving two cascaded networks, one of which realizes the filter zeros and the other the filter poles. The order of these networks can be reversed without changing the transfer function. This results in a structure where the two strings of delays are storing the same values, so a single string of delays of length $\max(M, N)$ is sufficient,

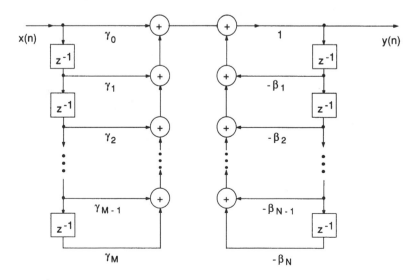

FIGURE 14.17 Direct-form I realization.

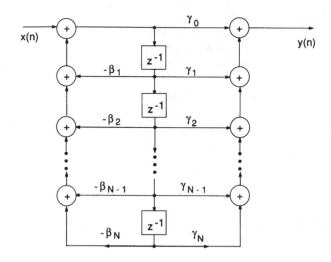

FIGURE 14.18 Direct-form II realization.

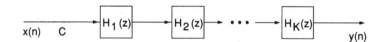

FIGURE 14.19 Cascade realization of an IIR filter.

as shown in Figure 14.18. The realization of Figure 14.18 requires the minimum number of z^{-1} delay operations and is referred to as the *direct-form II* realization.

Cascade and Parallel Realizations

The transfer function of an IIR filter can always be factored into the product of second-order transfer functions as

$$H(z) = C\prod_{k=1}^{K}\frac{1 + a_{1k}z^{-1} + a_{2k}z^{-2}}{1 + b_{1k}z^{-1} + b_{2k}z^{-2}} = C\prod_{k=1}^{K}H_k(z)$$

where we have assumed $M = N$ in the original transfer function and where K is the largest integer contained in $(N+1)/2$. If N is odd, the values of a_{2k} and b_{2k} in one term are zero. The realization corresponding to this transfer function factorization is shown in Figure 14.19. Each second-order $H_k(z)$ term in this realization is referred to as a *biquad*. The digital filter design programs in Parks and Burrus (1987) and Antoniou (1979) give the filter transfer function in factored form.

If the transfer function of an IIR filter is written as a partial-fraction expansion and first-order sections with complex-conjugate poles are combined, $H(z)$ can be expressed in the form:

$$H(z) = D + \sum_{k=1}^{K}\frac{\alpha_{0k} + \alpha_{1k}z^{-1}}{1 + b_{1k}z^{-1} + b_{2k}z^{-2}} = D + \sum_{k=1}^{K}G_k(z)$$

This results in the parallel realization of Figure 14.20.

Finite Wordlength Effects in IIR Filters

Since practical digital filters must be implemented with limited-precision arithmetic, four types of finite wordlength effects result: (1) roundoff noise, (2) coefficient quantization error, (3) overflow oscillations, and

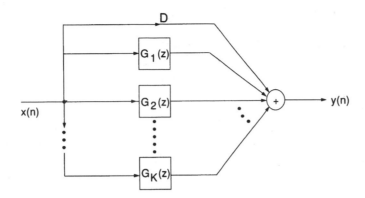

FIGURE 14.20   Parallel realization of an IIR filter.

(4) limit cycles. *Round-off noise* is that error in the filter output which results from rounding (or truncating) calculations within the filter. This error appears as low-level noise at the filter output. *Coefficient quantization error* refers to the deviation of a practical filter's frequency response from the ideal due to the filter's coefficients being represented with finite precision. The term *overflow oscillation*, sometimes also referred to as *adder overflow limit cycle*, refers to a high-level oscillation that can exist in an otherwise stable filter due to the nonlinearity associated with the overflow of internal filter calculations. A *limit cycle*, sometimes referred to as a *multiplier round-off limit cycle*, is a low-level oscillation that can exist in an otherwise stable filter as a result of the nonlinearity associated with rounding (or truncating) internal filter calculations. Overflow oscillations and limit cycles require recursion to exist and do not occur in nonrecursive FIR filters.

The direct-form I and direct-form II IIR filter realizations generally have very poor performance in terms of all finite wordlength effects. Therefore, alternative realizations are usually employed. The most common alternatives are the cascade and parallel realizations where the direct-form II realization is used for each second-order section. By simply factoring or expanding the original transfer function, round-off noise and coefficient quantization error are significantly reduced. A further improvement is possible by implementing the cascade or parallel sections using *state-space* realizations (Roberts and Mullis, 1987). The price paid for using state-space realizations is an increase in the computation required to implement each section. Another realization that has been used to reduce round-off noise and coefficient quantization error is the *lattice realization* (Parks and Burrus, 1987), which is usually formed directly from the unfactored and unexpanded transfer function.

Overflow oscillations can be prevented in several different ways. One technique is to employ floating-point arithmetic that renders overflow virtually impossible due to the large dynamic range which can be represented. In fixed-point arithmetic implementations it is possible to *scale* the calculations so that overflow is impossible (Roberts and Mullis, 1987), to use saturation arithmetic (Ritzerfeld, 1989), or to choose a realization for which overflow transients are guaranteed to decay to zero (Roberts and Mullis, 1987).

Limit cycles can exist in both fixed-point and floating-point digital filter implementations. Many techniques have been proposed for testing a realization for limit cycles and for bounding their amplitude when they do exist. In fixed-point realizations it is possible to prevent limit cycles by choosing a state-space realization for which any internal transient is guaranteed to decay to zero and then using magnitude truncation of internal calculations in place of rounding (Diniz and Antoniou, 1986).

Defining Terms

Discrete sequence: A set of values constituting a signal whose values are known only at distinct sampled points. Also called a digital signal.

Filter design: The process of determining the coefficients of a difference equation to meet a given frequency or time response characteristic.

Filter implementation: The numerical method or algorithm by which the output sequence is computed from the input sequence.

Finite impulse response (FIR) filter: A filter whose output in response to a unit impulse function is identically zero after a given bounded number of samples. A FIR filter is defined by a linear constant-coefficient difference equation in which the output depends only on the present and previous sample values of the input.

Finite wordlength effects: Perturbations of a digital filter output due to the use of finite precision arithmetic in implementing the filter calculations. Also called quantization effects.

Infinite impulse response (IIR) filter: A filter whose output in response to a unit impulse function remains nonzero for indefinitely many samples. An IIR filter is defined by a linear constant-coefficient difference equation in which the output depends on the present and previous samples of the input and on previously computed samples of the output.

References

A. Antoniou, *Digital Filters: Analysis, Design, and Applications*, 2nd ed., New York: McGraw-Hill, 1993.

X. Chen and T.W. Parks, "Design of FIR filters in the complex domain," *IEEE Trans. Acoust., Speech, Signal Process.*, vol. ASSP-35, pp. 144–153, 1987.

P.S.R. Diniz and A. Antoniou, "More economical state-space digital filter structures which are free of constant-input limit cycles," *IEEE Trans. Acoust., Speech, Signal Process.*, vol. ASSP-34, pp. 807–815, 1986.

DSP Committee, IEEE ASSP (Eds.), *Programs for Digital Signal Processing*, New York: IEEE Press, 1979.

E. Goldberg, R. Kurshan, and D. Malah, "Design of finite impulse response digital filters with nonlinear phase response," *IEEE Trans. Acoust., Speech, Signal Process.*, vol. ASSP-29, pp. 1003–1010, 1981.

T.W. Parks and C.S. Burrus, *Digital Filter Design*, New York: Wiley, 1987.

L.R. Rabiner, "Approximate design relationships for low-pass FIR digital filters," *IEEE Trans. Audio Electroacoust.*, vol. AU-21, pp. 456–460, 1973.

J.H.F. Ritzerfeld, "A condition for the overflow stability of second-order digital filters that is satisfied by all scaled state-space structures using saturation," *IEEE Trans. Circuits Syst.*, vol. CAS-36, pp. 1049–1057, 1989.

R.A. Roberts and C.T. Mullis, *Digital Signal Processing*, Reading, MA: Addison-Wesley, 1987.

M.E. Van Valkenberg, *Analog Filter Design*, New York: Holt, Rinehart and Winston, 1982.

Further Information

The monthly journal *IEEE Transactions on Circuits and Systems II* routinely publishes articles on the design and implementation of digital filters. Finite wordlength effects are discussed in articles published in the April 1988 issue (pp. 365–374) and in the February 1992 issue (pp. 90–98).

Another journal containing articles on digital filters is *IEEE Transactions on Signal Processing*. Overflow oscillations and limit cycles are discussed in the August 1978 issue (pp. 334–338).

The bimonthly journal *IEEE Transactions on Instrumentation and Measurement* also contains related information. The use of digital filters for integration and differentiation is discussed in the December 1990 issue (pp. 923–927).

14.4   Minimum ℓ_1, ℓ_2, and ℓ_∞ Norm Approximate Solutions to an Overdetermined System of Linear Equations

James A. Cadzow

Many practical problems encountered in digital signal processing and other quantitative oriented disciplines entail finding a best approximate solution to an overdetermined system of linear equations. Invariably, the least squares error approximate solution (i.e., minimum ℓ_2 norm) is chosen for this task due primarily to the existence of a convenient closed expression for its determination. It should be noted, however, that in many applications a minimum ℓ_1 or ℓ_∞ norm approximate solution is preferable. For example, in cases where the data being analyzed contains a few data outliers a minimum ℓ_1 approximate solution is preferable, since it tends to ignore the bad data points. In other applications one may wish to determine an approximate solution whose largest error magnitude is the smallest possible (i.e., a minimum ℓ_∞ norm approximate solution). Unfortunately, there do not exist convenient closed form expressions for either the minimum ℓ_1, or minimum ℓ_∞ norm approximate solution and one must resort to nonlinear programming methods for their determination. Effective algorithms for determining these two solutions are presented here.

Introduction

Many practical problems encountered in digital signal processing and other quantitative oriented disciplines entail finding a best approximate solution to an overdetermined system of linear equations. It is therefore not surprising that data analysts have extensively made use of many of the theoretical concepts of linear algebra for identifying the salient features of a system of linear equations. A general system of M real valued linear equations in N real unknowns can be compactly represented as

$$\mathbf{y} = A\mathbf{x} \tag{14.61}$$

where matrix $A \in R^{M \times N}$ and vector $\mathbf{y} \in R^{M \times 1}$ are given while the vector $\mathbf{x} \in R^{N \times 1}$ is unknown.[1] If there exists at least one choice for the vector $\mathbf{x}$ which satisfies this linear relationship then the system of equations is said to be *consistent*. However, if no such vector exists then this system of linear equations is said to be *inconsistent*.

In many data-processing applications, the system of linear equations under consideration is inconsistent and it is desired to find a best approximation solution. For such situations, the traditional approach is to find a selection of vector $\mathbf{x}$ so that the ℓ_2 norm (sum of squared errors criterion) of the *residual error vector*:

$$\mathbf{r}(\mathbf{x}) = \mathbf{y} - A\mathbf{x} \tag{14.62}$$

is minimized. One of the main benefits accrued in employing a minimum ℓ_2 norm criterion is the existence of a closed form solution to this approximation problem. In particular, it is well known that a minimum ℓ_2 norm approximation solution is any solution of the consistent linear system of *normal equations*:

$$A^T A \mathbf{x}^o = A^T \mathbf{y} \tag{14.63}$$

In many applications, however, it is more preferable to determine an approximate solution which minimizes either the ℓ_1 norm or the ℓ_∞ norm of the residual error vector. Unfortunately, a closed form solution in these two norm cases is not available and one must resort to nonlinear programming techniques for finding such approximate solutions. In this section, efficient algorithms are developed for finding optimal minimum ℓ_1 and ℓ_∞ norm approximate solutions to an inconsistent system of linear equations.

To illustrate how linear equations arise in practical applications let us consider the classical problem of modeling a set of real valued *empirical data* $y(1), y(2), \ldots, y(M)$ that forms the components of the *empirical data vector* $\mathbf{y} \in R^{M \times 1}$. It is now hypothesized that the primary component of the empirical data vector is composed of a linear combination of N given *signal vectors* $\mathbf{a}_1, \mathbf{a}_2, \ldots, \mathbf{a}_N \in R^{M \times 1}$. The $\mathbf{a}_n$ component signal vectors are typically selected by the data analyst to represent information believed (or conjectured) to be present in the empirical data. The quality of approximating the empirical data vector as a linear combination

---

[1]The vector spaces $R^{M \times 1}$ and $R^{M \times N}$ designate the set of all real valued $M \times 1$ vectors and the set of all real valued $M \times N$ matrices, respectively.

of the signal vectors results in the *residual error vector*:

$$\mathbf{r}(\mathbf{x}) = \mathbf{y} - \sum_{n=1}^{N} x(n)\mathbf{a}_n$$
$$= \mathbf{y} - A\mathbf{x}$$
(14.64)

In this expression the $\mathbf{a}_n$ signal vectors form the columns of the matrix $A \in R^{M \times N}$ while the *coefficient vector* $\mathbf{x} \in R^{N \times 1}$ has as its elements the $x(n)$ linear combination coefficients.

If the coefficient vector $\mathbf{x}$ can be chosen so that the corresponding residual error vector $\mathbf{r}(\mathbf{x}) = \mathbf{y} - A\mathbf{x}$ is made "small" in size then the data vector $\mathbf{y}$ is said to be richly endowed with the signals represented by the composite feature vector $A\mathbf{x}$. Moreover, each $x(n)$ coefficient provides a quantitative measure of the amount of the hypothesized signal $\mathbf{a}_n$ that is present in the data vector $\mathbf{y}$. However, if the residual error cannot be made sufficiently small then one might conclude that there is little if any of the hypothesized signals present in the empirical data vector.

Vector Approximation Problem

The primary objective of this section is that of finding a best approximate solution to an inconsistent system of M linear equations in N unknowns as represented by $A\mathbf{x} = \mathbf{y}$. To begin this investigation, let us consider the simplest case of M linear equations in one unknown $N = 1$. The residual error vector in this case is specified by

$$\mathbf{r}(x) = \mathbf{y} - x\mathbf{a}$$
(14.65)

where $\mathbf{a}$ and $\mathbf{y}$ are given nonzero vectors contained in $R^{M \times 1}$ and $x \in R$ is a real parameter to be chosen to give the best approximation. In particular, this parameter is to be selected so as to minimize the norm-induced functional:

$$f(x) = \|\mathbf{y} - x\mathbf{a}\|$$
(14.66)

where $\| \cdot \|$ designates a norm defined on vector space $R^{M \times 1}$. Using the properties defining a norm it is shown that $f(x)$ is a continuous and convex function of the real variable x. The convexity property is important since it ensures that any local minimum of $f(x)$ is also a global minimum. A characterization of a minimizing solution of function $f(x)$ for three commonly employed norms is now described.

Minimum Sum of Squared Errors Approximation

The most commonly employed measure in finding an approximate solution to a linear system of equations is the sum of squared errors measure. For the case of a linear system of M equations in one variable function as specified by Equation (14.65) this measure corresponds to the quadratic function in x[1]

$$f_2(x) = \sum_{m=1}^{M} |y(m) - xa(m)|^2 = [\mathbf{y} - x\mathbf{a}]^T [\mathbf{y} - x\mathbf{a}]$$
$$= \mathbf{y}^T\mathbf{y} - 2x\mathbf{a}^T\mathbf{y} + x^2\mathbf{a}^T\mathbf{a}$$
(14.67)

A necessary condition for an optimal selection of the real variable x is obtained by setting to zero this function's derivative with respect to the variable x. This partial derivative is specified by $df(x)/dx = -2\mathbf{a}^T\mathbf{y} + 2x\mathbf{a}^T\mathbf{a}$ thereby giving the unique minimum sum of squared errors solution:

---

[1]This criterion is equal to the square of the ℓ_2 norm of the residual error vector as specified by $\|\mathbf{y} - a x\|_2^2$.

$$x^{\mathrm{o}} = \frac{\mathbf{a}^{\mathrm{T}}\mathbf{y}}{\mathbf{a}^{\mathrm{T}}\mathbf{a}} \tag{14.68}$$

This selection corresponds to a global minimum since the second derivative of $f_2(x)$ is the positive number $2\mathbf{a}^{\mathrm{T}}\mathbf{a}$.

Minimum Sum of Error Magnitudes Approximation

Another popularly used measure of approximation fidelity is the sum of error magnitudes. This particular measure is appropriate when it is suspected that a small minority of the data points being analyzed are unreliable (i.e., data outliers). The sum of error magnitudes associated with residual error vector Equation (14.65) corresponds to the ℓ_1 norm of the residual error vector, that is:

$$f_1(x) = \|\mathbf{y} - x\mathbf{a}\|_1 = \sum_{m=1}^{M} |y(m) - xa(m)| \tag{14.69}$$

To illustrate why this measure is less sensitive to a bad data point(s) let the element $y(p)$ denote a single data outlier. The impact of this bad data point on the sum of error magnitude squared criterion is given by $|y(p) - xa(p)|^2$ while its impact on the sum of error magnitudes is $|y(p) - xa(p)|$. Clearly, the squaring of a large residual error is more deleterious thereby suggesting that the ℓ_1 norm is less susceptible to bad data points.

Upon examination of the sum of error magnitudes criterion, it is first noted that if the term $a(m) = 0$, it follows that the mth term of the residual error vector is not a function of x and therefore has no influence in minimizing $f_1(x)$. A plot of $f_1(x)$ versus x reveals it to be a continuous convex linear function of x where the transition from one slope to the next slope occurs at the roots $x_m = y(m)/a(m)$ with $a(m) \neq 0$. This claim is made apparent by observing that the derivative of $f_1(x)$ for all values of x not equal to the roots:

$$x_m = \frac{y(m)}{a(m)} \quad \text{for } a(m) \neq 0 \tag{14.70}$$

is given by

$$\frac{\mathrm{d}f_1(x)}{\mathrm{d}x} = \sum_{x_m < x} a(m) - \sum_{x_m > x} a(m) = -\sum_{m=1}^{M} |a(m)| + 2\sum_{x_m < x} |a(m)| \tag{14.71}$$

Use of the indentity $|y(m) - xa(m)| = [y(m) - xa(m)]\,\mathrm{sgn}[y(m) - xa(m)]$ has been made in arriving at this derivative expression where the sgn function is defined by $\mathrm{sgn}(x) = 1$ if $x > 0$, $\mathrm{sgn}(x) = -1$ if $x < 0$ and $\mathrm{sgn}(0) = 0$. Thus, the derivative is a piecewise constant function of x which has a jump of value $a(m)0 \neq$ at the root $x_m = y(m)/a(m)$.

The value of the piecewise constant derivative function at $x = -\infty$ is seen to be equal to the negative quantity $-\sum_{m=1}^{M} |a(m)|$ and it increases in a piecewise constant fashion to $\sum_{m=1}^{M} |a(m)|$ at $x = \infty$. An optimal choice of x corresponds to: (i) a root where the derivative changes from a negative number to a positive number, or (ii) for any x lying in the interval $x_k \leqslant x \leqslant x_{k+1}$ where $\mathrm{d}f_1(x_k)/\mathrm{d}x = 0$. In the first case, the optimum value of x is unique, while in the second case the optimum is not unique and can be any point contained in the interval where the derivative is zero. Whichever case applies, an optimum selection of x is always equal to one of the roots, that is:

$$\|\mathbf{y} - x^{\mathrm{o}}\mathbf{a}\|_1 = \min_{\substack{x_m = y(m)/a(m) \\ a(m) \neq 0}} \|\mathbf{y} - x_m\mathbf{a}\|_1 \tag{14.72}$$

with the solution x^{o} being unique if there exists only one root x_m giving rise to this minimum. If two or more different roots give rise to the minimum then any point lying within the interval whose end points are the two most separated of such roots will also be a minimum.

A more efficient method for determining an optimal x^o than the direct evaluation procedure (Equation (14.73)) is possible (see [3]) and requires the rearrangement of the roots (Equation (14.71)) in the monotonically increasing fashion so that $x_{m_1} < x_{m_2} < \ldots < x_{m_q}$ where $q \leq M$ in which the $a(m)$ coefficient associated with root x_{m_k} is designated by a_{m_k}. It is readily shown that an optimum solution $x^o = x_{m_k}$ corresponds to the integer n for which the inequality

$$-\sum_{m=1}^{M} |a(m)| + 2\sum_{k=1}^{n} |a(m_k)| \geq 0 \tag{14.73}$$

is first satisfied. In both the direct evaluation method (Equation (14.72)) and this alternative procedure it is necessary to compute all the roots (Equation (14.70)). Only a partial number of functional evaluations are required in the alternative method as reflected by expression (Equation (14.73)), however, as opposed to the evaluation for all roots needed in the direct evaluation method. This advantage is partially offset by the rearrange of roots operation required in the more efficient method.

Minimum Largest Error Magnitude Approximation

In many applications, it is desired to minimize the size of the largest error magnitude incurred in an inconsistent system of linear equations. The minimum largest error magnitude of the residual error vector (Equation (14.61)) corresponds to the ℓ_∞ norm of that vector, that is:

$$f_\infty(x) = \|\mathbf{y} - x\mathbf{a}\|_\infty = \max_{1 \leq m \leq M} |y(m) - xa(m)| \tag{14.74}$$

Upon examination of this expression it is observed that if a term $a(m) = 0$ then the choice of x has no influence on the error element $y(m) - xa(m) = y(m)$. With this in mind, any row of the vectors $\mathbf{a}$ and $\mathbf{y}$ for which $a(m) = 0$ is removed so that this preprocessing operation results in a vector $\mathbf{a}$ which has no zeros. As in the ℓ_1 norm case, $f_\infty(x)$ is found to be a piecewise linear convex function of x. Furthermore, it is now shown that a solution to this problem can be confined to values of x where two residual error vector components have the same magnitude, that is:

$$\|\mathbf{y} - x^o\mathbf{a}\|_\infty = \min_{\substack{|y(m) - xa(m)| = |y(n) - xa(n)| \\ m \neq n}} \|\mathbf{y} - x\mathbf{a}\|_\infty \tag{14.75}$$

To prove this conjecture, let x^o be a solution for which only one of the components of the residual error vector as designated by $|y(m^o) - x^o a(m^o)|$ has maximum magnitude where $a(m^o) \neq 0$. Let this solution be perturbed from x^o to $x^o + \epsilon$ where ϵ is a scalar chosen small enough in magnitude so as to preserve the sign requirement $\text{sgn}\{y(m^o) - (x^o + \epsilon)a(m^o)\} = \text{sgn}\{y(m^o) - x^o a(m^o)\}$. Under this sign preservation requirement it is seen that $|y(m^o) - (x^o + \epsilon)a(m^o)| = |y(m^o) - x^o a(m^o)| - \epsilon a(m^o)\text{sgn}\{y(m^o) - x^o a(m^o)\}$. Clearly, it is always possible to either gradually increase or decrease ε from zero in a manner such that $|y(m^o) - (x^o + \epsilon)a(m^o)| < |y(m^o) - x^o a(m^o)|$ which contradicts the assumption that x^o is a solution. It therefore follows that two or more components of the optimum residual error vector $\mathbf{y} - x^o\mathbf{a}$ must have the same maximum magnitude.

Example 14.4

Let it be desired to find the best ℓ_1, ℓ_2, and ℓ_∞ norm approximations of the vector $\mathbf{y} = [2\ 1]^T$ by a scalar multiple of the vector $\mathbf{a} = [1\ 1]^T$. The residual error vector associated with these vector choices is then:

$$\mathbf{r}(x) = \mathbf{y} - x\mathbf{a} = \begin{bmatrix} 2 \\ 1 \end{bmatrix} - \begin{bmatrix} 1 \\ 1 \end{bmatrix} x = \begin{bmatrix} 2 - x \\ 1 - x \end{bmatrix}$$

where x is real valued scalar. It is readily shown that the ℓ_1 norm of this residual error vector as given by $\|\mathbf{r}(x)\|_1 = |2 - x| + |1 - x|$ has the convex function representation:

$$\|\mathbf{r}(x)\|_1 = \begin{cases} 3 - 2x & \text{for } x < 1 \\ 1 & \text{for } 1 \leqslant x \leqslant 2 \\ 2x - 3 & \text{for } x > 2 \end{cases}$$

Clearly, any selection of x_1^o in the interval $1 \leqslant x^o \leqslant 2$ results in a minimum ℓ_1 solution with $\|\mathbf{r}(x_1^o)\|_1 = 1$. Only the two end points of this interval $x_1^o = 1$ and $x_1^o = 2$ result in a residual error vector with one zero component. However, the unique minimum ℓ_2 as specified by expression (Equation (14.68)) is given by $x_2^o = (\mathbf{y}^T a)/(a^T a) = 3/2$ while the unique minimum ℓ_∞ as specified by expression (Equation (14.75)) is found to be $x_\infty^o = 3/2$. In summary, these three approximations are specified by

$$x_1^o = x \begin{bmatrix} 1 \\ 1 \end{bmatrix} \quad \text{for all } 1 \leqslant x \leqslant 2, \quad \text{and} \quad x_2^o = x_\infty^o = \frac{3}{2} \begin{bmatrix} 1 \\ 1 \end{bmatrix}$$

Although the best ℓ_1, ℓ_2, and ℓ_∞ norm approximations are different in general problems, it is seen in this example that the unique ℓ_2 and ℓ_∞ approximate solutions are identical. However, the ℓ_1 approximate solution is not unique but it does equal to $x_2^o = x_\infty^o$ for the scalar selection $x = 3/2$.

General Minimum Norm Approximation Problem

We now address the more general case of M inconsistent linear equations in N unknowns which generates the residual error vector $\mathbf{r}(\mathbf{x}) = \mathbf{y} - A\mathbf{x}$. As indicated in the introduction section, a problem of interdisciplinary importance is that of selecting a vector $x \in R^{N \times 1}$ so as to make the residual error vector as small as possible in some sense. Although there exist several metrics that measure the size of a vector, we shall appeal to a norm induced function as specified by

$$f(x) = \|y - Ax\| \tag{14.76}$$

where $\| \cdot \|$ designates any suitable chosen norm defined on vector space $R^{M \times 1}$.[1] The approximation problem then corresponds to finding a vector $\mathbf{x}^o$ which minimizes this norm function. A minimizing selection then generates the best approximation $A\mathbf{x}^o$ of vector y in this norm sense. The minimization problem to be solved then takes one of the two equivalent forms:

$$\min_{\mathbf{x} \in R^{N \times 1}} \|\mathbf{y} - A\mathbf{x}\| = \min_{\tilde{\mathbf{y}} \in R(A)} \|\mathbf{y} - \tilde{\mathbf{y}}\| \tag{14.77}$$

In the second equivalent optimization problem, $R(A) = \{\tilde{\mathbf{y}} \in R^{M \times 1} : \tilde{\mathbf{y}} = A\mathbf{x} \text{ for } \mathbf{x} \in R^{N \times 1}\}$ designates the *range space* of matrix A and this problem is formulated as that of finding a vector $\tilde{\mathbf{y}}^o \in R(A)$ which most closely approximates $\mathbf{y}$ in the prescribed norm fashion.

There are numerous selections for the norm used in the relationship in Equation (14.76), of which the most commonly employed are the class of ℓ_p norms. Independent of which norm is used, however, the following theorem shows that this norm induced function is a continuous as well as a convex function of x. As such, it follows that any vector x^o which is a local minimum of $f(x)$ is also a global minimum and the set of global minima is a convex set [18]. This property is of particular importance when applying nonlinear programming algorithms for finding a numerical solution to the problem of minimizing $f(x)$. If the algorithm employed has assured convergence to a local minimum then we are guaranteed that the local minimum is also a global minimum.

---

[1]The mapping $\|$ of a vector space X over the field of scalars F is said to be a norm if it satisfies the three axioms:

(i) $\|x\| \geqslant 0$ for all $x \in X$ and $\|x\| = 0$ if and only if $x \neq 0$

(ii) $\|x + y\| \leqslant \|x\| + \|y\|$ for all $x, y \in X$

(iii) $\|\alpha x\| = |\alpha| \, \|x\|$ for all $x \in X$ and scalars α

Theorem 14.1. Consider the norm induced function $f(\mathbf{x}) = \|\mathbf{y} - A\mathbf{x}\|$ of the vector $\mathbf{x} \in R^{N \times 1}$ in which the vector $\mathbf{y} \in R^{M \times 1}$ and matrix $A \in R^{M \times N}$ are given and $\| \cdot \|$ designates any norm defined on vector space $R^{M \times 1}$. It then follows that $f(\mathbf{x})$ is a convex and continuous function of $\mathbf{x}$.

The properties of continuity and convexity of function $f(\mathbf{x})$ with respect to vector $\mathbf{x}$ are now established. Since function $f(\mathbf{x})$ is norm induced it follows that the three axioms associated with a norm are applicable. This being the case, it directly follows that for any choice of the scalar λ satisfying $0 \leqslant \lambda \leqslant 1$ and arbitrary vectors $\mathbf{x}_1, \mathbf{x}_2 \in R^{N \times 1}$, we have

$$f(\lambda \mathbf{x}_1 + (1 - \lambda)\mathbf{x}_2) = \|\mathbf{y} - A[\lambda \mathbf{x}_1 + (1 - \lambda)\mathbf{x}_2]\| = \|\lambda(\mathbf{y} - A\mathbf{x}_1) + (1 - \lambda)(\mathbf{y} - A\mathbf{x}_2)\|$$
$$\leqslant \|\lambda(\mathbf{y} - A\mathbf{x}_1)\| + \|(1 - \lambda)(\mathbf{y} - A\mathbf{x}_2)\| = \lambda f(\mathbf{x}_1) + (1 - \lambda)f(\mathbf{x}_2)$$

In going from line one to line two the triangular inequality axiom of a norm was employed while in going to the last equality on line two the scalar multiplier axiom of a norm was used. Convexity of $f(\mathbf{x})$ with respect to vector $\mathbf{x}$ is therefore established. Continuity with respect to vector $\mathbf{x}$ is shown in a similar fashion by using the triangular inequality of a norm to give

$$f(\mathbf{x} + \Delta) = \|\mathbf{y} - A[\mathbf{x} + \Delta]\| \leqslant \|\mathbf{y} - A\mathbf{x}\| + \|A\Delta\| \leqslant \|\mathbf{y} - A\mathbf{x}\| + \|A\| \cdot \|\Delta\| = f(\mathbf{x}) + \|A\| \cdot \|\Delta\|$$

In a similar fashion it can be shown that $f(\mathbf{x}) \leqslant f(\mathbf{x} + \Delta) + \|A\| \cdot \|\Delta\|$. It then follows that for any value of the positive scalar ε, no matter how small, the requirement for continuity of $f(\mathbf{x})$ relative to vector $\mathbf{x}$ as specified by

$$\|\Delta\| \leqslant \frac{\varepsilon}{\|A\|} \Rightarrow |f(\mathbf{x} + \Delta) - f(\mathbf{x})| \leqslant \varepsilon$$

is satisfied. The following lemma can be proven in a similar fashion and its proof is left as an exercise to the reader.

Lemma 14.1. The norm induced function $g(\mathbf{x}, \mathbf{y}, A) = \|\mathbf{y} - A\mathbf{x}\|$ is a convex and continuous function of the vectors $\mathbf{x} \in R^{N \times 1}$, $\mathbf{y} \in R^{M \times 1}$ and matrix $A \in R^{M \times N}$

ℓ_p Norm-Induced Functions

As indicated above, the measurement of goodness of approximation is defined to be the norm of the residual error vector $\mathbf{r}(\mathbf{x}) = \mathbf{y} - A\mathbf{x}$. The particular norm used is of critical importance and should be selected to reflect the particular objectives of the approximation problem at hand. For the purposes of this chapter, the norm induced function to be minimized is taken as the ℓ_p norm of the residual error vector, that is:

$$f_{\mathrm{p}}(\mathbf{x}) = \|\mathbf{y} - A\mathbf{x}\|_{\mathrm{p}} = \left[\sum_{m=1}^{M} |y(m) - \mathbf{e}_m^{\mathrm{T}} A\mathbf{x}|^{\mathrm{p}} \right]^{1/\mathrm{p}} \tag{14.78}$$

In this expression $\mathbf{e}_m \in R^{M \times 1}$ denotes the standard basis vector whose components are all zero except for its mth component which is 1. Thus the $1 \times N$ vector $\mathbf{e}_m^{\mathrm{T}} A$ corresponds to the mth row of matrix A while the entity $\mathbf{e}_m^{\mathrm{T}} A\mathbf{x}$ is the mth component of the vector $A\mathbf{x}$. This residual error vector measure corresponds to a norm provided that the parameter p is any real number greater than or equal to one. As Theorem 14.1 indicates, this norm function $f_{\mathrm{p}}(\mathbf{x})$ is a convex and continuous function of $\mathbf{x}$.

Unfortunately, a closed form solution for an optimal coefficient vector $\mathbf{x} \in R^{N \times 1}$ which minimizes norm function $f_{\mathrm{p}}(\mathbf{x})$ does not exist except for the widely used sum-squared errors selection $p = 2$. With this in mind, algorithms are herein developed for finding solutions for the important special cases: (i) $p = 1$ associated with the minimization of the *sum of error magnitude* criterion:

$$f_1(\mathbf{x}) = \sum_{m=1}^{M} \left| y(m) - \mathbf{e}_m^{\mathrm{T}} A\mathbf{x} \right| \tag{14.79}$$

and (ii) $P = \infty$ corresponding to the minimization of the *maximum error magnitude* (Chebyshev) norm as specified by

$$f_\infty(\mathbf{x}) = \max \left\{ \left| y(1) - \mathbf{e}_1^{\mathrm{T}} A\mathbf{x} \right|, \left| y(2) - \mathbf{e}_2^{\mathrm{T}} A\mathbf{x} \right|, \ldots, \left| y(M) - \mathbf{e}_M^{\mathrm{T}} A\mathbf{x} \right| \right\} \tag{14.80}$$

The sum of error magnitudes criterion $f_1(\mathbf{x})$ is of particular use in those applications where the data vector $\mathbf{y}$ contains a small number of *data outliers* (i.e., unrepresentative or bad data points). In such cases, the sum of squared errors criterion $f_2(\mathbf{x})$ is unduly influenced by these data outliers thereby often leading to a poor selection of the coefficient vector $\mathbf{x}$. The sum of error magnitudes criterion, however, tends to ignore the data outliers provided that they are relatively few in number. The maximum error magnitude criterion is used in those applications where the cost of making large errors is of central importance.

Minimum ℓ_2 Norm Approximate Solution

It has been previously stated that the problem of finding a vector $\mathbf{x}$ that minimizes the ℓ_2 norm of the residual error vector $\mathbf{r}(\mathbf{x}) = \mathbf{y} - A\mathbf{x}$ has a convenient closed form solution. Furthermore, it is shortly shown that this minimum ℓ_2 norm solution can be used for finding solutions to the associated problems of finding vectors $\mathbf{x}$ that minimize the ℓ_1 and ℓ_∞ norms of the residual error vector. With this importance established, let us now direct our attention to the problem of minimizing the sum of squared residual errors criterion:

$$f_2(\mathbf{x}) = \|\mathbf{y} - A\mathbf{x}\|_2 = \sqrt{[\mathbf{y} - A\mathbf{x}]^{\mathrm{T}}[\mathbf{y} - A\mathbf{x}]}$$
$$= \sqrt{\mathbf{y}^{\mathrm{T}}\mathbf{y} - 2\mathbf{y}^{\mathrm{T}} A\mathbf{x} + \mathbf{x}^{\mathrm{T}} A^{\mathrm{T}} A\mathbf{x}} \tag{14.81}$$

A necessary condition for a vector $\mathbf{x}^\circ$ to minimize function $f_2(\mathbf{x})$ is that the gradient of this function when evaluated at $\mathbf{x}^\circ$ must be equal to the zero vector. Using standard differentiation it is found that the gradient of function $f_2(\mathbf{x})$ is given by

$$\nabla_x f_2(\mathbf{x}) = \frac{1}{\|\mathbf{y} - A\mathbf{x}\|_2} (A^{\mathrm{T}} A\mathbf{x} - A^{\mathrm{T}}\mathbf{y}) \tag{14.82}$$

Upon setting this gradient equal to the zero vector, it is clear that a necessary condition for a least squared residual error selection for $\mathbf{x}^\circ$ is one that satisfies the consistent *normal system of equations* $A^{\mathrm{T}} A\mathbf{x}^\circ = A^{\mathrm{T}}\mathbf{y}$ which corresponds to N linear equations in the N unknowns $\mathbf{x}^\circ$. This fundamental result is now formally stated in theorem format.

Theorem 14.2. For any vector $\mathbf{y} \in R^{M \times 1}$ and matrix $A \in R^{M \times N}$, a necessary and sufficient condition that a vector $\mathbf{x}^\circ$ minimizes the sum of squared residual errors function $f_2(\mathbf{x}) = \|\mathbf{y} - A\mathbf{x}\|_2$ is that it satisfies the consistent linear system of *normal equations*:

$$A^{\mathrm{T}} A\mathbf{x}^\circ = A^{\mathrm{T}}\mathbf{y} \tag{14.83}$$

Furthermore, all solutions to these normal equations result in the same associated residual error vector $\mathbf{r}(\mathbf{x}^\circ) = \mathbf{y} - A\mathbf{x}^\circ$, which is orthogonal to the row vectors of matrix A; that is:

$$A^{\mathrm{T}}\mathbf{r}(\mathbf{x}^\circ) = \mathbf{o} \tag{14.84}$$

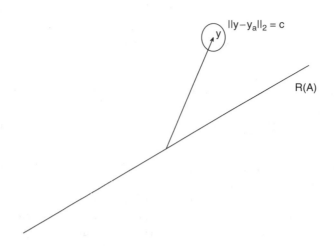

FIGURE 14.21   Approximate solution of $A\mathbf{x} = \mathbf{y}$ using the ℓ_2 norm.

If the matrix A has full column rank N, then there exists a unique solution to the normal equations as given by

$$\mathbf{x}^{\text{o}} = [A^{\text{T}}A]^{-1}A^{\text{T}}\mathbf{y} \tag{14.85}$$

Proof

The fact that all solutions to the problem of minimizing $f_2(\mathbf{x})$ satisfy the normal equations has already been proven. To establish the validity of orthogonality condition $A^{\text{T}}\mathbf{r}(\mathbf{x}^{\text{o}}) = \mathbf{o}$, one simply left multiplies the residual error $r(x^{\text{o}}) = y - Ax^0$ by matrix A^{T} to give $A^{\text{T}}\mathbf{r}(\mathbf{x}^{\text{o}}) = A^{\text{T}}\mathbf{y} - A^{\text{T}}Ax^{\text{o}}$, which is equal to the zero vector since $\mathbf{x}^{\text{o}}$ satisfies the normal equations. Finally, if matrix A has rank N, then the $N \times N$ matrix product $A^{\text{T}}A$ has rank N and is therefore invertible, which leads to the unique solution (Equation (14.85)).

An illustration of a minimum ℓ_2 norm approximation is shown in Figure 14.21 where $\mathcal{R}(A)$ designates the range space of matrix A (i.e., all vectors of the form $A\mathbf{x}$ with $x \in R^{N \times 1}$) whose dimension is equal to *rank*(A). The boundary of the hypersphere with center at $\mathbf{y}$ and radius c corresponds to all those vectors in $R^{M \times 1}$ that are located a distance c from $\mathbf{y}$. The minimum ℓ_2 approximate solution then corresponds to that smallest selection of radius c for which this hypersphere just touches the subspace $\mathcal{R}(A)$. Clearly, the point of contact is unique provided that the dimension of $\mathcal{R}(A)$ is N. Moreover, the residual error vector $\mathbf{r}_2^{\text{o}}$ is orthogonal to $\mathcal{R}(A)$, as shown by expression (Equation (14.84)).

It is useful to note that the results expressed in this theorem can be extended to the case in which the vectors x and $\mathbf{y}$ and matrix A are complex valued. The optimal solution in this complex data case are readily shown to give rise to the normal system of equations $A^{*}Ax^{0} = A^{*}y$, while the optimum residual error orthogonality condition becomes $A^{*}\mathbf{r}(\mathbf{x}_2^{\text{o}}) = \mathbf{o}$, the asterisk symbol $*$ designates the complex transpose operator. It is to be noted that the real valued data case Equation 14.83 is a special case of this complex data result in which the complex transpose operator $*$ is replaced by the transpose operator T.

Fundamental Properties of Minimum ℓ_1 Norm Approximation

The use of an ℓ_1 norm criterion is appropriate when it is suspected that a small portion of the data being analyzed is unreliable (i.e., contains data outliers). The ℓ_1 norm criterion has the capability of effectively ignoring a few bad data points while emphasizing the majority of data points that more properly reflect the true nature of the data. An insightful and useful characterization of a solution to the problem of minimizing the sum of residual error magnitude criterion as specified by

$$f_1(\mathbf{x}) = \|\mathbf{y} - A\mathbf{x}\|_1 = \sum_{m=1}^{M} |y(m) - \mathbf{e}_m^T A\mathbf{x}| \tag{14.86}$$

is now provided. To aid in our analysis, this summation for $1 \leq m \leq M$ may be decomposed over three disjoint integer sets where residual error components $r_m(\mathbf{x}) = y(m) - \mathbf{e}_m^T A\mathbf{x}$ are positive, negative, or zero. Under this decomposition the sum of residual error magnitude function can be expressed as

$$f_1(\mathbf{x}) = \sum_{y(m) > \mathbf{e}_m^T A\mathbf{x}} \left[y(m) - \mathbf{e}_m^T A\mathbf{x} \right] - \sum_{y(m) < \mathbf{e}_m^T A\mathbf{x}} \left[y(m) - \mathbf{e}_m^T A\mathbf{x} \right]$$

where the first-right side summation is taken over all summation index integers for which $r_m(\mathbf{x}) = y(m) - \mathbf{e}_m^T A\mathbf{x} > 0$ and the second summation is taken over all summation index integers for which $r_m(\mathbf{x}) = y(m) - \mathbf{e}_m^T A\mathbf{x} < 0$.

It is now shown that there exists a vector $\mathbf{x}^\circ$ which minimizes the sum of residual error magnitudes function $f_1(\mathbf{x})$, in which at least N components of the associated residual error vector $r(\mathbf{x}^\circ) = \mathbf{y} - A\mathbf{x}^\circ$ are zero. To prove this conjecture, let $\mathbf{x}^\circ$ be an optimum approximation solution for which only $N_0 \leq N - 1$ components of the residual error vector are zero. Let the vector $\mathbf{x}^\circ$ be perturbed to $\mathbf{x}^\circ + \epsilon\Delta$, where ϵ is a real valued *step size scalar* and Δ is a *perturbation direction vector*. The perturbation direction vector is chosen so that the original zeros in the unperturbed residual error vector $r(\mathbf{x}^\circ) = \mathbf{y} - A\mathbf{x}^\circ$ are maintained at zero in the perturbed residual error vector $\mathbf{r}(\mathbf{x}^\circ + \epsilon\Delta) = \mathbf{y} - A\mathbf{x}^\circ - \epsilon A\Delta$. This is always possible by choosing the nonzero perturbation direction vector Δ to be orthogonal to each of the $N_0 < N$ row vectors of matrix A associated with the zero components of $\mathbf{r}(\mathbf{x}^\circ) = \mathbf{y} - A\mathbf{x}^\circ$. The scalar ϵ is restricted to be small enough in magnitude so that the signs of the nonzero components in the unperturbed residual error vector $y - A\mathbf{x}^0$ and the perturbed residual error vector $r(x^0 + \epsilon\Delta) = y - Ax^0 - \epsilon A\Delta$ are maintained. Under this restriction the corresponding sum of residual error magnitude criterion at the perturbed point $\mathbf{x}^\circ + \epsilon\Delta$ is given by

$$f_1(\mathbf{x}^\circ + \epsilon\Delta) = \sum_{y(m) > \mathbf{e}_m^T A\mathbf{x}} \left[y_m - \mathbf{e}_m^T A(\mathbf{x}^\circ + \epsilon\Delta) \right] - \sum_{y(m) < \mathbf{e}_m^T A\mathbf{x}} [y_m - \mathbf{e}_m^T A(\mathbf{x}^\circ + \epsilon\Delta)]$$

$$= f_1(\mathbf{x}^\circ) - \epsilon \underbrace{\left[\sum_{y(m) > \mathbf{e}_m^T A\mathbf{x}} \mathbf{e}_m^T - \sum_{y(m) < \mathbf{e}_m^T A\mathbf{x}} \mathbf{e}_m^T \right]}_{\mathbf{p}^T} A\Delta = f_1(\mathbf{x}^\circ) - \epsilon \mathbf{p}^T A\Delta$$

Clearly, the vector $\mathbf{p}$ appearing in this expression has components which are exclusively one, minus one, or zero. It is further observed that if vector $\mathbf{p}$ is not orthogonal to vector $A\Delta$, then ϵ can be chosen so that $f_1(\mathbf{x}^\circ + \epsilon\Delta) < f_1(\mathbf{x}^\circ)$ which contradicts the fact that $\mathbf{x}^\circ$ is an optimal solution. Thus, the vector $\mathbf{p}$ must be orthogonal to $A\Delta$ and $f_1(\mathbf{x}^\circ + \epsilon\Delta) = f_1(\mathbf{x}^\circ)$. The step size scalar ϵ is now gradually increased or decreased from zero until a formally nonzero element of the unperturbed residual error vector $r(\mathbf{x}^\circ) = \mathbf{y} - A\mathbf{x}^\circ$ is first driven to zero. The new vector $\tilde{\mathbf{x}}^\circ = \mathbf{x}^\circ + \epsilon^\circ\Delta$ causes the perturbed residual error vector $\mathbf{y} - A\tilde{\mathbf{x}}^\circ$ to have one additional zero component while maintaining the functional's value. This perturbation procedure is continued in this manner until the perturbed residual vector eventually has N zero components while maintaining the functional value. The following fundamental theorem related to minimum ℓ_1 norm approximation solutions has therefore been proven.

Theorem 14.3. For any vector $\mathbf{y} \in R^{M \times 1}$ and matrix $A \in R^{M \times N}$, there exists a real $N \times 1$ vector $\mathbf{x}^\circ$ which minimizes the sum of residual error magnitudes criterion $f_1(\mathbf{x})$ such that the associated residual error vector

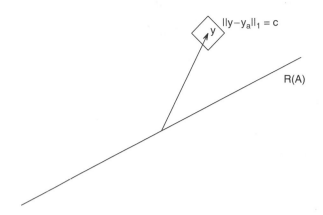

FIGURE 14.22 Approximate solution of $A\mathbf{x} = \mathbf{y}$ using a ℓ_1 norm.

$$\mathbf{r}(\mathbf{x}_1^o) = \mathbf{y} - A\mathbf{x}^o \qquad (14.87)$$

has at least N zero components. Furthermore, if the row vectors comprising the augmented matrix $[A \,\vdots\, \mathbf{y}] \in R^{M \times (N+1)}$ satisfy the Haar condition then there exists a real $N \times 1$ vector $\mathbf{x}^o$ which minimizes criterion $f_1(\mathbf{x})$ such that the residual vector has exactly N zero components.[1]

An illustration of the minimum ℓ_1 norm approximation is depicted in Figure 14.22 where the range space $\mathcal{R}(A)$ (i.e., all vectors of the form $A\mathbf{x}$) is a subspace of dimension rank(A). The boundary of the rotated by $45°$ hypercube with center at $\mathbf{y}$ and width c corresponds all those vectors in $R^{M \times 1}$ which are located a distance c from $\mathbf{y}$ in the ℓ_1 norm sense. The minimum ℓ_1 approximate solution corresponds to that smallest selection of width c for which this rotated hypercube just touches the subspace $\mathcal{R}(A)$. The point of contact may or may not be unique but one of the points of contact has an associated residual error vector r_1^o which has at least N zeros as indicated in the above theorem. Example 14.1 demonstrates that a minimum ℓ_1 norm solution need not be unique although it typically is in many applications.

In an algorithm to be shortly presented for determining a vector $\mathbf{x}$ which minimizes the ℓ_1 induced functional (Equation (14.61)), a basic requirement is that of finding a best ℓ_1 norm approximation of given vector $\mathbf{y} \in R^{M \times 1}$ by a scalar multiple of another given vector $\mathbf{a} \in R^{M \times 1}$. This is a special application of the above theorem in which $A = \mathbf{a}$ is a $M \times 1$ matrix so that $N = 1$. It therefore follows that there exists a minimum ℓ_1 norm approximation solution for which $N = 1$ components of the residual error vector are zero. This characterization is now formally described and agrees with the characterization (Equation (14.68)) previously established using a different approach.

Theorem 14.4. Let $\mathbf{a}$ and $\mathbf{y}$ be any two nonzero vectors contained in $R^{M \times 1}$ and consider the set of roots

$$x(m) = \frac{y(m)}{a(m)} \quad \text{for } 1 \leq m \leq M \text{ in which } a(m) \neq 0 \qquad (14.88)$$

which corresponds to those values of x for which the mth component of the residual error vector $\mathbf{e} = \mathbf{y} - x\mathbf{a}$ is zero. At least one root of these roots is a global minimum of the norm functional $f_1(x) = |\mathbf{y} - x\mathbf{a}|_1$, that is:

$$\min_{x \in R} f_1(x) = \min_{\substack{x(m) = y(m)/a(m) \\ \text{for all } a(m) \neq 0}} f_1(x) \qquad (14.89)$$

[1] Let there be given the set of vectors $\{a_1, a_2, \ldots, a_Q\}$ contained in vector space $R^{N \times 1}$ (or $R^{N \times 1}$) in which $Q \geq N$. This set of vectors is said to satisfy the *Harr condition* if every subset of N of these vectors forms a linearly independent set.

Furthermore, this global minimum is unique provided that only one of the roots achieves this global minimum. If two distinct roots achieve this global minimum, then any value of x contained in the closed interval with these roots serving as end points is also a global minimum.

Brute Force Method for Finding a Minimum ℓ_1 Approximate Solution

A brute force method of finding a minimum ℓ_1 approximate solution to a general system of M linear equations $\mathbf{y} = A\mathbf{x}$ in N unknowns is apparent from Theorem 14.3. This method makes use of the fact that an optimal selection of $\mathbf{x}$ exists for which at least N components of the associated error vector are zero. In this approach, one determines solutions to all subsets of N equations of the original M system of linear equations $\mathbf{y} = A\mathbf{x}$ and evaluates the ℓ_1 norm of the associated residual error vector $\mathbf{y} - A\mathbf{x}$ corresponding to each of these solutions. An optimal solution then corresponds to a residual error vector(s) that possesses the smallest ℓ_1 norm. Unfortunately, this brute force procedure is inefficient since the number of subsets of N equations in a system of M equations is given by the combination of M things taken N at a time, that is,

$$c_N^M = \frac{M!}{N!(M-N)!} \tag{14.90}$$

This number can be large, which often precludes the brute force method from being of use in real-time applications.

The algorithmic approach to be now described is predicated on systematically proceeding from one such subset of N linear equations to another subset of N linear equations so that the ℓ_1 norm functional $f_1(\mathbf{x})$ is decreased at each transition. The efficiency of this algorithm is measured by the number of such subsets (iterations) that need be evaluated until an optimal solution is reached. It has been empirically found that this number is significantly smaller than the brute force number (Equation (14.90)) as well as many proposed linear programming based methods.

Improving ℓ_1 Norm Perturbation Procedure

We shall now incorporate the characterization specified in Theorem 14.3 to develop an effective algorithm for numerically generating a minimum ℓ_1 norm approximate solution to the inconsistent system of linear equations $A\mathbf{x} = \mathbf{y}$ for a given vector $\mathbf{y} \in R^{M \times 1}$ and matrix $A \in R^{M \times N}$. Since this linear system of equations is hypothesized as being inconsistent, it follows that $\mathbf{y} \notin \mathcal{R}(A)$. Thus, associated with each $\mathbf{x} \in R^{N \times 1}$ there is generated a nonzero residual error vector as specified by

$$\mathbf{r}(\mathbf{x}) = \mathbf{y} - A\mathbf{x} \tag{14.91}$$

It is now desired to select $\mathbf{x}$ so that the transformed vector $A\mathbf{x}$ best approximates $\mathbf{y}$ in the sense of minimizing the ℓ_1 norm-induced functional:

$$f_1(\mathbf{x}) = \|\mathbf{y} - A\mathbf{x}\|_1 \tag{14.92}$$

It has been previously shown that functional $f_1(\mathbf{x})$ is a continuous and convex function of $\mathbf{x}$. Moreover, if one were to make a plot of the surface $f_1(\mathbf{x})$ vs. $\mathbf{x}$ in $R^{(N+1) \times 1}$, it would be a convex polyhedron. A nonunique minimum ℓ_1 norm solution corresponds to those choices of the pair $(A, \mathbf{y})$ for which this polyhedron has a flat bottom face.

Many of the algorithms for determining a minimum ℓ_1 norm solution are based on fundamental Theorem 14.3, which establishes the existence of a solution $\mathbf{x}^o$ for which at least N components of the residual error vector $\mathbf{r}(\mathbf{x}^o)$ are zero. The majority of these algorithms are based on an exchange principle whereby one of the N equations associated with the prevailing N zero residual components being considered for an optimum choice is exchanged for another equation in the remaining set of $M-N$ equations in such a manner that the new set of N equations results in a smaller ℓ_1 residual error vector norm. A primary goal of this and the next section is to develop a more effective solution algorithm that converges at a faster rate than this class of exchange algorithms.

In our development, a vector $\mathbf{x}$ is said to be an *extreme point* of a system of linear equations if the associated residual error vector $\mathbf{r}(\mathbf{x}) = A\mathbf{x} - \mathbf{y}$ has at least N zero components. Fundamental Theorem 14.3 indicates that at least one of the extreme points is a minimum ℓ_1 approximate solution. If more than N components of the residual error vector are zero, then the extreme point is said to be *degenerate*. A degenerate extreme point can arise if and only if the row vectors of the augmented matrix $[A\vdots\mathbf{y}]$ do not satisfy the Haar condition.

Perturbation Analysis

In the algorithm to be shortly proposed, the first step entails proceeding from an arbitrary nonextreme point $\mathbf{x}$ to an extreme point $\mathbf{x}_e$ in such a manner that the ℓ_1 norm of the residual error vector is either decreased or at worst maintained (i.e., $f_1(\mathbf{x}_e) \leqslant f_1(\mathbf{x})$). To achieve such a procedure, a perturbation analysis of the functional $f_1(\mathbf{x})$ is made whereby it is assumed that N_0 components of the residual error vector are zero where $0 \leqslant N_0 < N$. It is convenient to decompose the residual error vector $\mathbf{r}(\mathbf{x}) = \mathbf{y} - A\mathbf{x}$ into the two component vectors:

$$\mathbf{r}_1(\mathbf{x}) = \mathbf{y}_1 - A_1\mathbf{x} = \mathbf{0} \quad \text{and} \quad \mathbf{r}_2(\mathbf{x}) = \mathbf{y}_2 - A_2\mathbf{x} \tag{14.93}$$

where the entities $\mathbf{y}_1 \in R^{N_o \times 1}$ and $A_1 \in R^{N_o \times N}$ correspond to that subset of N_0 equations that have zero residuals and $\mathbf{y}_2 \in R^{(M-N_o) \times 1}$ and $A_2 \in R^{(M-N_o) \times N}$ correspond to the entities of the remaining subset of $M - N_o$ equations that have nonzero residuals. It therefore follows that the ℓ_1 norm of the residual error vector $\mathbf{r}(\mathbf{x}) = \mathbf{y} - A\mathbf{x}$ simplifies to

$$f_1(\mathbf{x}) = \|\mathbf{y}_2 - A_2\mathbf{x}\|_1$$

The prevailing vector $\mathbf{x}$ can now be perturbed to $\mathbf{x} + \epsilon\Delta$, where Δ is a nonzero *perturbation direction vector* and ε is a *step size* scalar. The perturbation vector Δ is now selected to lie in the null space of matrix A_1, that is,

$$A_1\Delta = \mathbf{0}$$

There always exists such a nonzero vector because the null space of the $N_0 \times N$ matrix A_1 has a dimension greater than or equal to $N - N_o > 0$. It therefore follows that for this perturbation vector selection, the perturbed residual error vector has its component vectors specified by

$$\mathbf{r}_1(\mathbf{x} + \epsilon\Delta) = \mathbf{0} \quad \text{and} \quad \mathbf{r}_2(\mathbf{x} + \epsilon\Delta) = \mathbf{r}_2(\mathbf{x}) - \epsilon A_2\Delta$$

In arriving at this result, it is seen that $\mathbf{r}_1(\mathbf{x} + \epsilon\Delta) = \mathbf{y}_1 - A_1(\mathbf{x} + \epsilon\Delta) = \mathbf{r}_1(\mathbf{x}) - \epsilon A_1\Delta = 0$ and $\mathbf{r}_2(\mathbf{x} + \epsilon\Delta) = \mathbf{y}_2 - A_2(\mathbf{x} + \epsilon\Delta) = \mathbf{r}_2(\mathbf{x}) - \epsilon A_2\Delta$. Since $\mathbf{r}_1(\mathbf{x} + \epsilon\Delta) = \mathbf{0}$ it follows that the ℓ_1 norm of the perturbed residual error vector as specified by $\|\mathbf{r}(\mathbf{x} + \epsilon\Delta)\|_1 = \|\mathbf{r}_2(\mathbf{x} + \epsilon\Delta)\|_1$. The step size scalar ε is now chosen to minimize the ℓ_1 norm of the perturbed residual error vector component $\mathbf{r}_2(\mathbf{x} + \epsilon\Delta) = \mathbf{r}_2(\mathbf{x}) - \epsilon A_2\Delta$. In accordance with Theorem 14.4, an optimal step size selection is obtained by employing relationship (14.93) in which $\mathbf{a} = A_2\Delta$ and $\mathbf{y} = \mathbf{r}_2(\mathbf{x})$, namely:

$$\|\mathbf{r}_2(\mathbf{x}) - \epsilon^\circ A_2\Delta\|_1 = \min_{\epsilon \in R} \|\mathbf{r}_2(\mathbf{x}) - \epsilon A_2\Delta\|_1 = \min_{\substack{\epsilon_m = \mathbf{e}_m^T \mathbf{r}_2(\mathbf{x})/\mathbf{e}_m^T A_2\Delta \\ \text{for all } \mathbf{e}_m^T A_2\Delta \neq 0}} \|\mathbf{r}_2(\mathbf{x}) - \epsilon A_2\Delta\|_1$$

This choice of $\epsilon = \epsilon^\circ$ causes at least one of the components of $\mathbf{r}_2(\mathbf{x} + \epsilon^\circ \Delta)$ to be zero. Based on this perturbation analysis the following fundamental theorem characterizing minimum ℓ_1 approximate solutions has been proven.

Theorem 14.5. Let there be given a vector $\mathbf{y} \in R^{M \times 1}$ and a matrix $A \in R^{M \times N}$. For any vector $\mathbf{x} \in R^N$ let the associated residual error vector $\mathbf{r}(\mathbf{x}) = \mathbf{y} - A\mathbf{x}$ have N_o zeros where $0 \leqslant N_o < N$ and let the residual error vector be decomposed as specified by relationship (14.93). The vector $\mathbf{x}$ is now perturbed to

$$\mathbf{x} + \epsilon^o \Delta \tag{14.94}$$

in which the nonzero perturbation direction vector Δ is chosen to lie in the null space of matrix A_1, that is:

$$A_1 \Delta = \mathbf{o} \tag{14.95}$$

The step size scalar ε^o is specified by

$$\| \mathbf{r}_2(\mathbf{x}) - \epsilon^o A_2 \Delta \|_1 = \min_{\substack{\epsilon_m = \mathbf{e}_m^T \mathbf{r}_2(\mathbf{x}) / \mathbf{e}_m^T A_2 \Delta \\ \text{for all } \mathbf{e}_m^T A_2 \Delta \neq 0}} \| \mathbf{r}_2(\mathbf{x}) - \epsilon A_2 \Delta \|_1 \tag{14.96}$$

The perturbed selection $\mathbf{x} + \epsilon^o \Delta$ causes the ℓ_1 norm criterion is be improved or at worst maintained so that $f_1(\mathbf{x} + \epsilon^o \Delta) \leqslant f_1(\mathbf{x})$ and the perturbed residual error vector $\mathbf{r}(\mathbf{x} + \epsilon^o \Delta)$ has at least $N_o + 1$ zero components in which N_o of these zero components are the same as those of the unperturbed residual error vector $\mathbf{r}(\mathbf{x})$.

It is to be noted that when the vector $\mathbf{x}$ appearing in this theorem is not a minimum ℓ_1 norm solution, it has been empirically found that this procedure for selecting the perturbation vector Δ has generally led to an improvement in the ℓ_1 norm's value so that $f_1(\mathbf{x} + \epsilon \Delta) < f_1(\mathbf{x})$. Moreover, by continuing this procedure in an iterative fashion an extreme point $\mathbf{x}_e$ is eventually arrived at in which the associated error vector $\mathbf{r}(\mathbf{x}_e) = \mathbf{y} - A\mathbf{x}_e$ has at least N zeros. The ℓ_1 norm of the residual error vector is either decreased or at worst maintained during the iterative process (i.e., $f_1(\mathbf{x}_e) \leqslant f_1(\mathbf{x})$). Once an extreme point has been obtained, it is necessary to determine whether that extreme point is a required minimum ℓ_1 norm solution. The following theorem provides a mechanism for making this determination.

Theorem 14.6. (Bloomfield and Steiger [3]) For a given vector $\mathbf{y} \in R^{M \times 1}$ and matrix $A \in R^{M \times N}$, let the vector $\mathbf{x} \in R^{N \times 1}$ be a nondegenerate extreme point of the residual error vector $\mathbf{r}(\mathbf{x}) = \mathbf{y} - A\mathbf{x}$. Furthermore, let the submatrix $A_1 \in R^{N \times N}$ designate the N rows of matrix A associated with the N zero elements of the residual error vector. If matrix A_1 is invertible then the nondegenerate extreme point $\mathbf{x}$ is a minimum ℓ_1 solution if and only if all the components of the vector:

$$\mathbf{c} = \left[A_1^T \right]^{-1} A_2^T \text{sgn}\{\mathbf{y}_2 - A_2 \mathbf{x}\} \tag{14.97}$$

have magnitudes less than or equal to one (i.e., $\|\mathbf{c}\|_\infty \leqslant 1$).[1] Moreover, this solution is unique if and only if all the components of $\mathbf{c}$ have magnitudes strictly less than one (i.e., $\|\mathbf{c}\|_\infty < 1$). In the case $\|\mathbf{c}\|_\infty > 1$, let $c(n_1)$

---

[1] The sgn function as applied to a real $N \times 1$ vector x is defined by

$$\text{sgn}(x) = \begin{bmatrix} \text{sgn}(x(1)) \\ \text{sgn}(x(2)) \\ \cdot \\ \cdot \\ \cdot \\ \text{sgn}(x(N)) \end{bmatrix}$$

where it is recalled that the sign of a real number α as specified by $\text{sgn}(\alpha) = 1$ if $\alpha > 0$, $= -1$ if $\alpha < 0$ and is 0 when $\alpha = 0$.

denote any component of $\mathbf{c}$ whose magnitude is larger than one. It then follows that a perturbation vector $\epsilon\Delta$ which renders the improvement $f_1(\mathbf{x} + \epsilon\Delta) < f_1(\mathbf{x})$ is given by

$$\epsilon\Delta = \alpha^o A_1^{-1}\mathbf{e}_{n_1} \tag{14.98}$$

where α^o is any value of the scalar α for which the vector $\alpha A_1^{-1}\mathbf{e}_{n_1}$ best approximates the residual error vector $\mathbf{r}(\mathbf{x})$ in the minimum ℓ_1 norm sense.

A proof of this theorem is found in Bloomfield and Steiger [3]. An alternative proof is now given that is both more concise and utilizes procedures which are subsequently employed in the minimum ℓ_1 norm algorithm to be presented in the next section. This proof makes use of the fact that the column vectors of A_1^{-1} form a basis of $R^{N\times 1}$ so that any perturbation vector can be uniquely represented as $\epsilon\Delta = A_1^{-1}\mathbf{b}$ for an appropriate choice of vector $\mathbf{b} \in R^{N\times 1}$. It is now assumed that the size of the perturbation vector $A_1^{-1}\mathbf{b}$ is restricted so that the sign preservation of the residual error vector $\text{sgn}\{\mathbf{y}_2 - A_2(\mathbf{x} + [A_1]^{-1}\mathbf{b})\} = \text{sgn}\{\mathbf{y}_2 - A_2\mathbf{x}\}$ is maintained. Under this restriction it follows that the perturbed residual error vectors are given by $\mathbf{r}_1(\mathbf{x} + A_1^{-1}\mathbf{b}) = -\mathbf{b}$ and $\mathbf{r}_2(\mathbf{x} + +A_1^{-1}\mathbf{b}) = \mathbf{y}_2 - A_2\mathbf{x} - A_2A_1^{-1}\mathbf{b}$. Thus:

$$\begin{aligned}
f_1(\mathbf{x} + A_1^{-1}\mathbf{b}) &= \|\mathbf{b}\|_1 + \left[\mathbf{y}_2 - A_2\mathbf{x} - A_2A_1^{-1}\mathbf{b})\right]^{\mathrm{T}}\text{sgn}\{\mathbf{y}_2 - A_2\mathbf{x}\} \\
&= \|\mathbf{b}\|_1 + f_1(\mathbf{x}) - \mathbf{b}^{\mathrm{T}}\left[A_1^{\mathrm{T}}\right]^{-1}A_2^{\mathrm{T}}\text{sgn}\{\mathbf{y}_2 - A_2\mathbf{x}\} \\
&= f_1(\mathbf{x}) + \|\mathbf{b}\|_1 - \mathbf{b}^{\mathrm{T}}\mathbf{c} = f_1(\mathbf{x}) + \sum_{n=1}^{N}|b(n)|[1 - \text{sgn}\{b(n)\}c(n)]
\end{aligned}$$

The fundamental issue to be now explored is that of determining conditions under which the vector $\mathbf{b}$ can be chosen so that the improvement condition $f_1(\mathbf{x} + A_1^{-1}\mathbf{b}) < f_1(\mathbf{x})$ is realized. Clearly, for the given vector $\mathbf{c}$ specified in Equation (14.97), the signs of the $b(n)$ coefficients should be chosen so that $sgn(b(n)) = sgn(c(n))$, in order that the summand terms $|b(n)|[1 - \text{sgn}\{b(n)\}c(n)]$ appearing in the above summation be made as small as possible. Under this choice, it follows that

$$f_1\left(\mathbf{x} + A_1^{-1}\mathbf{b}\right) = f_1(\mathbf{x}) + \sum_{n=1}^{N}|b(n)|[1 - |c(n)|]$$

From this expression it is apparent that if the $c(n)$ elements all have magnitude less or equal to 1 then this summation is always nonpositive and an improvement in the ℓ_1 norm functional $f_1(\mathbf{x})$ cannot be made. However, if $c(n_1)$ designates any element of $\mathbf{c}$ whose magnitude is greater than one then the choice $b(n) = 0$ for all n except for $b(n_1)$ which is set equal to a positive scalar β (i.e., $\mathbf{b} = \beta\mathbf{e}_{n_1}$). For this selection of $\mathbf{b}$ we have

$$f_1\left(\mathbf{x} + A_1^{-1}\beta\mathbf{e}_{n_1}\right) = f_1(\mathbf{x}) + \beta[1 - |c(n_1)|]$$

which results in the desired improvement $f_1(\mathbf{x} + \epsilon\Delta) < f_1(\mathbf{x})$. The positive scalar β must be chosen sufficiently small to achieve the aforementioned sign preservation of the residual error vector. When more than one component of $\mathbf{c}$ has a magnitude greater than 1, then a variety of different improving perturbation directions can be devised.

The validity of perturbation selection (Equation (14.98)) is established by appealing to the fact that an improving vector perturbation lies in the one-dimensional space spanned by $A_1^{-1}\mathbf{e}_{n_1}$. Application of Theorem 14.4 then generates the scalar β for which vector $\beta A_1^{-1}\mathbf{e}_{n_1}$ best approximates the residual error vector $\mathbf{r}(\mathbf{x})$ in the ℓ_1 norm sense. The above results have been predicated on the assumption that an extreme point is nondegenerate. When a degenerate extreme point is encountered, the following theorem addresses the issue of whether that point is optimum or not.

Theorem 14.7. Let the full column rank matrix A satisfy the Haar condition and let $\mathbf{x} \in R^N$ be a degenerate extreme point so that its associated residual error vector $\mathbf{r}(\mathbf{x}) = \mathbf{y} - A\mathbf{x}$ has $N_o > N$ zero elements. Furthermore, let A_1 designate the $N_o \times N$ matrix whose rows correspond to the row vectors of matrix A associated with these N_o zero residual error elements. This degenerate extreme point is a minimum ℓ_1 norm solution if and only if none of the vectors

$$\mathbf{c}^{(k)} = \left[A_1^{(k)T} \right]^{-1} A_2^T \mathrm{sgn}\{\mathbf{y}_2 - A_2\mathbf{x}\} \quad \text{for } 1 \leq k \leq \frac{N_o!}{N!(N_o - N)!}$$

has a $\|\mathbf{c}^{(k)}\|_\infty > 1$ where the $\{A_1^{(k)}\}$ designates the set of all $N \times N$ submatrices of matrix A_1.

Algorithm for Computing a Minimum ℓ_1 Norm Approximate Solution

A straightforward procedure for finding a minimum ℓ_1 approximate solution to the general system of linear equations $\mathbf{y} = A\mathbf{x}$ is obtained by appealing to the results spelled out in Theorem 14.3 indicating that an optimal selection of $\mathbf{x}$ exists for which at least N components of the associated residual error vector are zero. This algorithmic approach is predicated on systematically proceeding from one subset of N linear equations being set to zero to another subset of N linear equations being set to zero in a manner that the ℓ_1 norm functional $f_1(\mathbf{x})$ is decreased at each transition. The efficiency of this algorithm is measured by the number of such subsets (iterations) that need be evaluated until an optimal solution is reached. It has been empirically found that this number is significantly smaller than the brute force number (Equation (14.91)) as well as many so-called row exchange algorithms. The steps of this algorithm are outlined in Table 14.9.

Since the vector $\mathbf{x}$ arrived at in going to Step 5 has an associated residual error vector with N zero elements, it is a candidate for an optimal solution in accordance with Theorem 14.3. Furthermore, if an improvement in vector $\mathbf{x}$ cannot be made, the algorithm must have converged to an optimal solution due to the convexity of functional $f_1(\mathbf{x})$. Many of most widely employed algorithms for solving the minimum ℓ_1 norm approximation problem make use of linear programming methods. Three of the most popular of these linear programming algorithms share a common trait [1–3]. Specifically, once the condition that the residual error vector $r = \mathbf{y} - A\mathbf{x}$ has N zeros has been reached, an improving perturbation vector is then determined so that the perturbed residual error vector $\tilde{\mathbf{r}} = \mathbf{y} - A(\mathbf{x} + \Delta)$ maintains $N - 1$ of the unperturbed zero residual error elements at their zero value while setting to zero a previously nonzero residual error element. The proposed algorithm differs from these approaches in that once the state has been reached where the residual error vector has N zeros, an improving perturbation vector is then determined (from Steps 2 to 5) in which no restriction of maintaining $N - 1$ of the previous zero residual error elements is imposed. This property has the potential

TABLE 14.9 Minimum ℓ_1 Approximate Solution Algorithm

Step 1.	Using an arbitrary initial selection of x (e.g., the zero vector or the minimum ℓ_2 approximate solution) generate the associated residual error vector $r(x) = \mathbf{y} - A\mathbf{x}$.
Step 2.	If the residual error vector $r(\mathbf{x})$ has $N_o < N$ zero elements go to Step 3. Otherwise go to Step 5.
Step 3.	Set the new trial solution x equal to $x + \epsilon^o\Delta$ where the nonzero perturbation vector Δ and step size scalar ϵ are chosen according to Equations (14.95) and (14.96). The new residual error vector $\mathbf{r}(\mathbf{x} + \epsilon^o\Delta)$ has at least $N_o + 1$ zero elements with $f_1(\mathbf{x} + \epsilon^o\Delta) \leq f_1(\mathbf{x})$.
Step 4.	If the residual error vector associated with the new trial solution x vector has fewer than N zero elements go to Step 3. Otherwise go to Step 5.
Step 5.	Determine whether the vector $c = [A_1^T]^{-1}A_2^T\mathrm{sgn}\{y_2 - A_2x\}$ has at least one component whose magnitude is greater than one (i.e., $\|c\|_\infty > 1$). If $\|c\|_\infty > 1$ then make the improvement in the ℓ_1 norm functional $f_1(x)$ in accordance with Equation (14.98). Set the new trial. Solution x equal to $x + \epsilon\Delta$ and go to Step 2. If $\|c\|_\infty \leq 1$ then x is an optimal solution.

of making the proposed algorithm be more rapidly convergent than these linear programming based methods. As the next example indicates, empirical experience has shown that it takes far fewer than c_N^M iterations to reach convergence indicating a very efficient algorithm.

Example 14.5

To illustrate the efficiency of the proposed algorithm, an example of 24 overdetermined linear equations in four unknowns was considered in which the elements of the 24×4 matrix A and the 24×1 vector y where independent samples of: (i) a Gaussian random variable with zero mean and unit variance, or (ii) a uniform random variable taking on values in the interval $[0, 1]$. The proposed algorithm was found to have converged to the correct minimum ℓ_1 norm solution as verified by the brute force method in each of the 1000 trial runs made on the Gaussian and uniform random samples. Furthermore, the mean and variance of the number of iterations (i.e., algorithmic Steps 2 to 5) required to reach convergence in these 1000 trial runs was found to be 2.6880 and 1.0177, respectively, for the 1000 Gaussian random variable trial runs and 3.4210 and 1.3673, respectively, for the 1000 uniform random variable trial runs. Since there are a total of $c_4^{24} = 10,626$ possible trial solutions which render a residual error vector with four zeros, the fact that the proposed algorithm converges while evaluating on an average less than four of these possibilities indicates that the proposed algorithm is very efficient. This conclusion is further reinforced by the fact that in the 1000 trial runs made, a maximum of nine iterations was required for convergence on only one of the trial runs.

Solution in Degenerate Extreme Point Case

The algorithm described above operates in an efficient and effective manner provided that the row vectors of A satisfy the Haar condition. If this is not the case, it is possible that the matrix A_1 arising at Step 5 of the Algorithm given in (14.99) may not be invertible. Similarly, if more than N residual error vector components are zero then the requirements of Theorem 14.6 are not valid and the algorithm becomes stuck at Step 5. This condition can occur only if the row vectors of the $[A \vdots \mathbf{y}] \in R^{M \times (N+1)}$ do not satisfy the Haar condition. To overcome these two potential drawbacks, use is made of the fact that the functional

$$f_p(\mathbf{x}) = \|\mathbf{y} - A\mathbf{x}\|_p$$

is a continuous function of the vector $\mathbf{y}$ and matrix A for any p satisfying $1 \leqslant p \leqslant \infty$.

With these ideas in mind, let us consider the perturbed linear system of equations ℓ_1 based functional (i.e., $p = 1$) as specified by

$$\tilde{f}_1(\mathbf{x}, \alpha) = \|\mathbf{y} + \alpha\mathbf{y}_p - (A + \alpha A_p)\mathbf{x}\|_1 \qquad (14.99)$$

where $\mathbf{y}_p$ is a $M \times 1$ perturbation vector, A_p is a $M \times N$ perturbation matrix and α is a scalar. These vector and matrix perturbations are chosen so that no extreme degenerate points exist for this perturbed functional for nonzero selections of the scalar α. A little thought should convince oneself that this condition may be met with probability one by selecting the elements of this vector and matrix to be independent samples of a random variable. Once these these perturbation entities have been generated, use is made of the observation that

$$\lim_{\alpha \to 0} \tilde{f}_1(\mathbf{x}, \alpha) = f_1(\mathbf{x})$$

In particular, one solves the nondegenerate minimization of functional (14.99) for a sequence of α values which converge to zero (e.g., $\alpha = (0.5)^k$ for $k = 1, 2, \ldots$). The sequence of solutions will be such that exactly N residual error elements are zero in each case. One terminates this sequential process once the vector and matrix perturbations are sufficiently small. At this termination point, the N zero residual error elements are identified.

The required solution in minimizing $f_1(\mathbf{x})$ is then obtained by setting to zero the same residual error elements in the unperturbed residual error equation $\mathbf{y} - A\mathbf{x}$. Continuity of functional $f_1(\mathbf{x})$ ensures that this will be the desired solution.

Let us first consider the case in which row vectors of matrix A satisfy the Haar condition, but $N_o > N$ components of the residual error vector $\mathbf{y} - A\mathbf{x}$ obtained at Step 5 of the algorithm are zero. One then slightly perturbs $N_o - N$ of the components of vector $f_p(\mathbf{x}^o + \epsilon\Delta)$ associated with the zero components of the residual error vector so that the perturbed residual error vector then has exactly N zero components. These perturbations must be made sufficiently small in magnitude. We are then able to carry out Step 5 of the algorithm on the perturbed system of linear equations. The algorithm is then implemented in the normal fashion on the perturbed system until either convergence is obtained or when more than N components of the residual error vector are zero at Step 5. In this latter case, the perturbed vector $\mathbf{y} + \Delta_y$ is again perturbed in the same fashion described earlier in the paragraph to cause the reperturbed residual error vector to have exactly N zeros. Using this perturbation procedure, eventually convergence is achieved in which exactly N components of the residual error vector are zero. One then determines the unique vector $\mathbf{x}$ so that these same N components in the original residual error vector $r = \mathbf{y} - A\mathbf{x}$ are set to zero. Continuity of $f_1(\mathbf{x})$ relative to $\mathbf{y}$ indicates that this will be an optimum solution provided that the perturbations Δ_y that led to this result were made sufficiently small.

If the $N \times N$ matrix A_1 obtained at Step 5 of the algorithm fails to be invertible, the Haar assumption on matrix A is not valid. In this case, one slightly perturbs the rows of matrix A that are associated with A_z so that the perturbed A_z is invertible. One continues in a fashion similar to that taken for the perturbation of $\mathbf{y}$ described in the previous paragraph until convergence at Step 5 on the perturbed system of equations is achieved. One then selects that unique vector $\mathbf{x}$ so that the same N components in the original residual error vector $r(\mathbf{x}) = \mathbf{y} - A\mathbf{x}$ are zero. Continuity of $f_1(\mathbf{x})$ relative to A indicates that this will be an optimum solution provided that the perturbations on A that led to this result were made sufficiently small.

Best Line Fit to Two-Dimensional Data

In investigations involving a linear system of equations, one of the primary reasons for preferring an ℓ_1 approximate solution to the more widely employed ℓ_2 approximate solution lies in the ability of the ℓ_1 norm to ignore unrepresentative data (i.e., data outliers). To illustrate this predisposition, let us consider the task of the finding a best line fit to a set of two-dimensional data as specified by

$$(a(m), y(m)) \quad \text{for } 1 \leq m \leq M \tag{14.100}$$

A plot of this data in the (a, y) plane (i.e., a scatter diagram) provides a visual mechanism for determining the basic nature of the interrelationship between these two variables. It often happens that these data points cluster about a line in the (x, y) plane. In such situations, it is then desirable to employ a linear model of the form:

$$\hat{y}(m) = x_1 a(m) + x_2 \tag{14.101}$$

to approximate the $y(m)$ elements. The associated modeling residual errors $r(m) = y(m) - \hat{y}(m)$ for $1 \leq m \leq M$ may then be represented by the *model residual error vector*:

$$\begin{bmatrix} r(1) \\ r(2) \\ \vdots \\ r(M) \end{bmatrix} = \begin{bmatrix} y(1) \\ y(2) \\ \vdots \\ y(M) \end{bmatrix} - \begin{bmatrix} a(1) & 1 \\ a(2) & 1 \\ \vdots & \vdots \\ a(M) & 1 \end{bmatrix} \begin{bmatrix} x(1) \\ x(2) \end{bmatrix} \tag{14.102}$$

or more compactly:

$$\mathbf{r} = \mathbf{y} - A\mathbf{x} \tag{14.103}$$

The linear modeling problem then corresponds to selecting the *model parameter vector* $\mathbf{x} \in R^{2 \times 1}$ so as to minimize the model residual error vector's size.

As indicated previously, the ℓ_2 norm provides the most widely employed means for measuring the model residual error vector's size. It then follows from relationship (14.104) that the unique minimum ℓ_2 norm selection of the model parameter vector is specified by $\mathbf{x}^{\mathrm{o}} = [A^T A]^{-1} A^T \mathbf{y}$. Upon substitution of the vector $\mathbf{y}$ and matrix A appearing in relationship (14.103) into this expression for $\mathbf{x}^{\mathrm{o}}$ it is found that the parameters of the optimal least squares fit to the given two-dimensional data are given by

$$x_1^{\mathrm{o}} = \frac{M(\mathbf{a}^T\mathbf{y}) - (\mathbf{a}^T 1)(\mathbf{y}^T 1)}{M(\mathbf{a}^T\mathbf{a}) - (\mathbf{a}^T 1)^2} \quad \text{and} \quad x_2^{\mathrm{o}} = \frac{(\mathbf{a}^T\mathbf{a})(\mathbf{y}^T 1) - (\mathbf{a}^T 1)(\mathbf{y}^T\mathbf{a})}{M(\mathbf{a}^T\mathbf{a}) - (\mathbf{a}^T 1)^2} \tag{14.104}$$

in which the vectors $\mathbf{a}$ and $1 \in R^M$ have as their components $a(m)$ and 1, respectively, for $1 \leqslant m \leqslant M$. As is shown in the following example, this least squares residual error based model is particularly susceptible to data outliers while the ℓ_1 norm-based model is relatively immune to a few data outliers.

Example 14.6

Let us consider the case in which the data being analyzed is specified by

$$y(n) = -2 + 3n + w(n) \quad \text{for } 1 \leqslant n \leqslant 24$$

where $w(n)$ is Gaussian white noise of zero mean and unit variance. Furthermore, the data points at $n = 5$ and $n = 18$ are altered to $y(5) = 62$ and $y(18) = -12$ so as to represent two data outliers. These data points are plotted in Figure 14.23 and are represented by the asterisks. An examination of these data points suggests that, with the exception of the two data points $y(5)$ and $y(18)$, the points approximately lie on a line. With this in

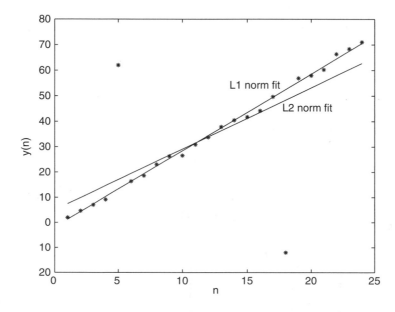

FIGURE 14.23 Minimum L1 and L2 linear fit to noise contaminated linear data with two data outliers.

mind, a linear fit to the data is next made where the 18×2 matrix A associated with this approximation problem is specified by

$$A = \begin{bmatrix} 1 & 1 \\ 2 & 1 \\ 3 & 1 \\ \vdots & \vdots \\ 18 & 1 \end{bmatrix}$$

with the true values of the parameter vector being $x(1) = 3$ and $x(2) = -2$. The optimum lines fit obtained by the closed form minimum ℓ_2 norm solution (14.104) and the ℓ_1 solution obtained by the proposed algorithm are also displayed in this plot. It is apparent that the minimum ℓ_1 norm line fit conforms more accurately to the given data points and tends to ignore two data outliers while the minimum ℓ_2 norm fit is deleteriously impacted by the data outliers. The proposed ℓ_1 algorithm only took two iterations before convergence and gave the parameter estimates $x_1(1)° = 3.0071$ and $x_1(2)° = -1.6582$ with the $x_1^{(1)}$ approximation seen to be very close to its true value of 3. However, a least squares residual error fit led to the less accurate estimates $x_2(1)° = 2.4011$ and $x_2(2)° = 5.0993$.

Example 14.7

We next consider the example treated by Hawley and Gallagher [13] whereby a best ℓ_1 linear fit $\hat{y}(m) = x(1)a(m) + x(2)$ to the data sequence:

$$y = \begin{bmatrix} 3 & 1 & 5 & 7 & 5 & 9 & 11 & 15 & 13 & 6 \end{bmatrix}^{\mathrm{T}}$$

is to be obtained. Linear model (Equation (14.102)) and this section's ℓ_1 algorithm described in section "Algorithm for Computing a Minimum ℓ_1 Norm Approximate Solution" were used to provide an optimum ℓ_1 norm linear fit in which the initial parameter vector was set at $\mathbf{x} = \mathbf{0}$. The algorithm converged in two iterations to an ℓ_1 optimum selection of the model parameters $x_1(1)° = 4/3$ and $x_1(2)° = 1$. This selection caused the third, sixth, and ninth components of the residual error vector to be zero and is in agreement with the solution obtained using the approach of Hawley and Gallagher.

Minimum ℓ_∞ Norm Approximate Solution

In certain applications, it is desired to find an approximate solution to an inconsistent system of linear equations $\mathbf{y} = A\mathbf{x}$ in which the largest error magnitude is to be minimized. The ℓ_∞ norm of the associated residual error vector provides this measure as formally specified by

$$f_\infty(\mathbf{x}) = \|\mathbf{y} - A\mathbf{x}\|_\infty = \max \left\{ |y_1 - \mathbf{e}_1^{\mathrm{T}} A\mathbf{x}|, |y_2 - \mathbf{e}_2^{\mathrm{T}} A\mathbf{x}|, \cdots, |y_M - \mathbf{e}_M^{\mathrm{T}} A\mathbf{x}| \right\} \qquad (14.105)$$

It is well known that a solution $\mathbf{x}°$ to this problem exists in which at least $N + 1$ components of the associated residual error vector $r° = \mathbf{y} - A\mathbf{x}°$ are equal to the maximum residual error magnitude $\|\mathbf{y} - A\mathbf{x}°\|_\infty$. We shall now prove this characterization using techniques that are subsequently employed in developing an effective algorithm for numerically finding a minimum ℓ_∞ approximate solution.

To begin this characterization, let $\mathbf{x}°$ designate any solution to the problem of minimizing the maximum residual error magnitude criterion $f_\infty(\mathbf{x})$. Furthermore, let the associated residual error vector $r° = \mathbf{y} - A\mathbf{x}°$ have N_m components whose magnitudes are equal to the maximum residual error magnitude criterion $f_\infty(\mathbf{x}°) = \|\mathbf{y} - A\mathbf{x}°\|_\infty$ where $1 \leq N_m \leq N$. Let us now isolate those equations which are associated with the N_m largest residual error magnitudes, that is:

$$\mathbf{r}_1(\mathbf{x}°) = \mathbf{y}_1 - A_1\mathbf{x}°$$

where the vectors $r_1(\mathbf{x}^o)$, $\mathbf{y}_1 \in R^{N_m \times 1}$ and matrix $A_1 \in R^{N_m \times N}$ are the components of the vectors $r(\mathbf{x}^o)$, $\mathbf{y} \in R^{M \times 1}$ and matrix $A \in R^{M \times N}$ associated with the N_m largest residual error magnitudes.

Let us now perturb $\mathbf{x}^o$ to $\mathbf{x}^o + \epsilon \Delta$ where ϵ is a small positive scalar multiplier and the perturbation vector Δ is any solution to the system of underdetermined equations:

$$A_1 \Delta = \mathrm{sgn}\{\mathbf{r}_1(\mathbf{x}^o)\}$$

This system of equations always has a solution provided that matrix $A_1 \in R^{N_m \times N}$ has full row rank N_m where it is recalled that $N_m \lesssim N$. This full rank condition is always assured provided that the row vectors of matrix A satisfy the Haar condition. Under this assumption, it follows that the $N_m \times 1$ residual error vector associated with the subset of equations corresponding to the largest residual error magnitude components becomes

$$\mathbf{y}_1 - A_1(\mathbf{x}^o + \varepsilon \Delta) = \mathbf{r}_1(\mathbf{x}^o) - \varepsilon \, \mathrm{sgn}\{\mathbf{r}_1(\mathbf{x}^o)\}$$

It therefore follows that the ℓ_∞ norm of this perturbed residual error vector is given by $\|\mathbf{y}_1 - A_1[\mathbf{x}^o + \epsilon\Delta]\|_\infty = \|\mathbf{y}_1 - A_1\mathbf{x}^o\|_\infty - \epsilon$ provided that ϵ is confined to the interval $0 \leqslant \epsilon \leqslant \|\mathbf{r}_1(\mathbf{x}^o)\|_\infty$. Furthermore, each component of the perturbed residual error vector $\mathbf{y}_1 - A_1[\mathbf{x}^o + \epsilon\Delta]$ has its magnitude decreased by ϵ. With this in mind, let ϵ^o be the smallest positive value for which one previously nonmaximum magnitude component has a ℓ_∞ norm equal to $\|\mathbf{y}_1 - A_1\mathbf{x}^o\|_\infty - \epsilon^o$. It then follows that the perturbed vector $\mathbf{x}^o + \epsilon^o\Delta$ renders a smaller ℓ_∞ norm than $f_\infty(\mathbf{x}^o) = \|\mathbf{y} - A\mathbf{x}^o\|_\infty$, which is contrary to the assumption that $\mathbf{x}^o$ was a minimum ℓ_∞ solution. It is therefore concluded that $N_m > N$. This result is now formalized as a theorem.

Theorem 14.8. For any vector $\mathbf{y} \in R^{M \times 1}$ and matrix $A \in R^{M \times N}$ whose row vectors satisfy the Haar condition there exists a vector $\mathbf{x}^o \in R^{N \times 1}$ which minimizes the maximum of residual error magnitude criterion $f_\infty(\mathbf{x}) = \|\mathbf{y} - A\mathbf{x}\|_\infty$ in which the associated residual error vector

$$\mathbf{r}(\mathbf{x}^o) = \mathbf{y} - A\mathbf{x}^o \tag{14.106}$$

has at least $N + 1$ components equal to the maximum residual error magnitude $\|\mathbf{r}(\mathbf{x}^o)\|_\infty$. Furthermore, this solution is unique if and only if the row vectors of the augmented matrix $[A \vdots \mathbf{y}]$ satisfy the Haar condition.

An illustration of the minimum ℓ_∞ norm approximation is illustrated in Figure 14.24, where the range space $R(A)$ has dimension $\mathrm{rank}(A)$ (i.e., of all vectors of the form $A\mathbf{x}$). The boundary of the hypercube with center at $\mathbf{y}$ and width $2c$ corresponds all those vectors in $R^{M \times 1}$ which are located a distance c from $\mathbf{y}$ in the ℓ_∞ norm sense. The minimum ℓ_∞ approximate solution corresponds to that smallest selection of width parameter c for which this hypercube just touches the range space $\mathcal{R}(A)$. In most applications, this point of

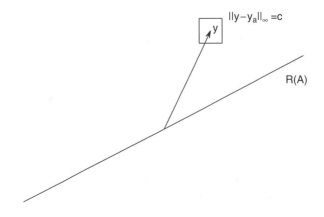

FIGURE 14.24 Approximate solution of $A\mathbf{x} = \mathbf{y}$ using a ℓ_∞ norm.

contact is unique and is such that the residual error vector $r(\mathbf{x}^o)$ has $N + 1$ components with the maximum magnitude. The following example, however, illustrates the fact that a minimum ℓ_∞ norm solution need not be unique although it typically is in most applications.

Example 14.8

Let it be desired to find the best ℓ_∞ approximation of the vector $\mathbf{y} = [2 \quad 1]^T$ by a scalar multiple of the vector $\mathbf{a} = [1 \quad 0]^T$. The residual error vector associated with these vector choices is then

$$\mathbf{r}(x) = \mathbf{y} - \mathbf{a}x = \begin{bmatrix} 2 \\ 1 \end{bmatrix} - \begin{bmatrix} 1 \\ 0 \end{bmatrix} x = \begin{bmatrix} 2 - x \\ 1 \end{bmatrix}$$

where x is real valued scalar. The ℓ_∞ norm of this residual error vector is specified by $\|\mathbf{r}(x)\|_\infty = \max\{|2 - x|, 1\}$ and it is readily shown that $\|\mathbf{r}(x)\|_\infty$ has the convex function representation:

$$\|\mathbf{r}(x)\|_\infty = \begin{cases} 2 - x & \text{for } x < 1 \\ 1 & \text{for } 1 \leqslant x \leqslant 3 \\ x - 2 & \text{for } x > 3 \end{cases}$$

Clearly, any selection of x^o in the interval $1 \leqslant x^o \leqslant 3$ results in a minimum ℓ_∞ solution with $\|\mathbf{r}(x^o)\|_\infty = 1$. Only the two end points of this interval $x^o = 1$ and $x^o = 3$ result in an residual error vector whose $N + 1 = 2$ largest component magnitudes are equal to $\|\mathbf{r}(x)\|_\infty = 1$.

Brute Force Method for Finding a Minimum ℓ_∞ Approximate Solution

As in the minimum ℓ_1 norm approximate problem, it is possible to apply a brute force method for finding a minimum ℓ_∞ norm approximate solution to the general system of linear equations by appealing to the results spelled out in Theorem 14.1. Specifically, use is made of the fact that an optimal selection of $\mathbf{x}$ exists for which at least $N + 1$ components of the associated residual error vector are equal to the maximum residual error magnitude $\|\mathbf{r}(x_m^o)\|_m$. In this brute force procedure, one finds a minimum ℓ_∞ approximate solution of all subsets of $N + 1$ linear equations of the system of linear equations $\mathbf{y} = Ax$ and then evaluates the ℓ_∞ norm of the associated residual error vector $\mathbf{y} - Ax$ corresponding to each of these solutions. A required optimal solution then corresponds to that solution(s) which possesses the largest ℓ_∞ norm. Unfortunately, this brute force procedure can be a computationally intensive process since the number of subsets of $N + 1$ equations from the original M equations as given by

$$c_{N+1}^M = \frac{M!}{(N + 1)!(M - N - 1)!} \tag{14.107}$$

can be a rather large number. Nonetheless, this direct approach will generate a desired minimum ℓ_∞ approximate solution.

In order to implement this brute force method, it is necessary to develop an effective method for finding a minimum ℓ_∞ approximate solution to a system of $N + 1$ linear equations in N unknowns. The following theorem provides the mechanism for determining this solution [9].

Theorem 14.9. Let there be given a vector $\mathbf{y} \in R^{(N+1)\times 1}$ and matrix $A \in R^{(N+1)\times N}$ of rank N. The residual error vector associated with the generally inconsistent system of $N + 1$ linear equations in N unknowns as specified by $Ax = \mathbf{y}$ is given by $\mathbf{r}(x) = \mathbf{y} - Ax$. The unique vector that minimizes the ℓ_2 norm of this residual error vector is given by $x_2^o = [A^T A]^{-1} A^T \mathbf{y}$ in which the residual error vector $\mathbf{r}(x_2^o) = \mathbf{y} - A[A^T A]^{-1} A^T \mathbf{y}$ satisfies the orthogonality property $A^T \mathbf{r}(x_2^o) = \mathbf{o}$. The vector x_∞^o that minimizes the ℓ_∞ norm of the residual error vector $\mathbf{r}(x)$ is equal to the unique solution of the consistent linear system of equations:

$$Ax = \mathbf{y} - \frac{\mathbf{r}(x_2^o)^{\mathrm{T}}\mathbf{y}}{\|\mathbf{r}(x_2^o)\|_1}\,\mathrm{sgn}\{r(\mathbf{x}_2^o)\} \tag{14.108}$$

with this solution being given by

$$x_\infty^o = x_2^o - \frac{\mathbf{r}(x_2^o)^{\mathrm{T}}\mathbf{y}}{\|\mathbf{r}(x_2^o)\|_1}[A^{\mathrm{T}}A]^{-1}A^{\mathrm{T}}\mathrm{sgn}\{\mathbf{r}(x_2^o)\} \tag{14.109}$$

Moreover, the associated ℓ_∞ norm of the minimum residual error vector is

$$\|\mathbf{r}(x_\infty^o)\|_\infty = \|\mathbf{y} - Ax_\infty^o\|_\infty = \frac{\mathbf{r}(x_2^o)^{\mathrm{T}}\mathbf{y}}{\|\mathbf{r}(x_2^o)\|_1} = \frac{\mathbf{r}(x_2^o)^{\mathrm{T}}\mathbf{r}(x_2^o)}{\|\mathbf{r}(x_2^o)\|_1} \tag{14.110}$$

Proof

It has been previously established that the vector $\mathbf{x}$ which minimizes the ℓ_2 norm of the residual error vector $\mathbf{r}(\mathbf{x}) = \mathbf{y} - A\mathbf{x}$ is given by $\mathbf{x}_2^o = [A^{\mathrm{T}}A]^{-1}A^{\mathrm{T}}\mathbf{y}$ where the subscript 2 on $\mathbf{x}_2^o$ is used to explicitly denote the fact that this is the unique minimum ℓ_2 norm approximate solution. If $\mathbf{x}_2^o$ satisfies the system of linear equations then it is also the minimum l_∞ norm approximation solution. We now treat the more interesting problem in which $\mathrm{r}(X_2^0)$ is a nonzero vector. Upon left multiplying each side of the associated minimum residual error vector $\mathrm{r}(X_2^0) = y - A\mathbf{x}_2^0$ by A^{T} it is seen that $A^{\mathrm{T}}\mathrm{r}(X_2^0) = o$ which indicates that the residual error vector $\mathrm{r}(X_2^0)$ is orthogonal to each column vector of matrix A. It is now shown that the vector appearing on the left side of Equation (14.108) is orthogonal to $\mathrm{r}(X_2^o)$. This conjecture is true since

$$r(\mathbf{x}_2^o)^{\mathrm{T}}\left[y - \frac{r(\mathbf{x}_2^o)^{\mathrm{T}}y}{\|r(\mathbf{x}_2^o)\|_1}\,\mathrm{sgn}\{r(\mathbf{x}_2^o)\}\right] = r(\mathbf{x}_2^o)^{\mathrm{T}}y - \frac{r(\mathbf{x}_2^o)^{\mathrm{T}}y}{\|r(\mathbf{x}_2^o)\|_1}r(\mathbf{x}_2^o)^{\mathrm{T}}\mathrm{sgn}\{r(\mathbf{x}_2^o)\} = 0$$

where use of the identity $\|\mathbf{r}(\mathbf{x}_2^o)^{\mathrm{T}}\|_1 = \mathbf{r}(\mathbf{x}_2^o)^{\mathrm{T}}\mathrm{sgn}\{\mathbf{r}(\mathbf{x}_2^o)\}$ has been made. This orthogonality condition implies that the vector in brackets must be equal to a linear combination of the column vectors of matrix A thereby proving that the system of linear equations (14.108) is consistent. The unique solution to this consistent system of equations is obtained by appealing to the relationship in which vector $\mathbf{y}$ is replaced by the vector appearing on the right side of Equation (14.108), that is:

$$\mathbf{x}_\infty^o = [A^{\mathrm{T}}A]^{-1}A^{\mathrm{T}}\left(\mathbf{y} - \frac{\mathbf{r}(\mathbf{x}_2^o)^{\mathrm{T}}\mathbf{y}}{\|\mathbf{r}(\mathbf{x}_2^o)\|_1}\,\mathrm{sgn}\{\mathbf{r}(\mathbf{x}_2^o)\}\right) = \mathbf{x}_2^o - \frac{\mathbf{r}(\mathbf{x}_2^o)^{\mathrm{T}}\mathbf{y}}{\|\mathbf{r}(\mathbf{x}_2^o)\|_1}[A^{\mathrm{T}}A]^{-1}A^{\mathrm{T}}\mathrm{sgn}\{\mathbf{r}(\mathbf{x}_2^o)\}$$

By employing the Holder inequality that states that for any vectors $\mathbf{u}, \mathbf{w} \in R^{M\times1}$ that $|\mathbf{u}^{\mathrm{T}}\mathbf{w}| \leq \|\mathbf{u}\|_\infty\|\mathbf{w}\|_1$, it follows that $\mathbf{x}_\infty^o$ is the minimum ℓ_∞ norm approximate solution since for all $\mathbf{x} \in R^{N\times1}$

$$\|\mathbf{y} - A\mathbf{x}\|_\infty = \sup_{\mathbf{w}\neq o}\frac{|(\mathbf{y} - A\mathbf{x})^{\mathrm{T}}\mathbf{w}|}{\|\mathbf{w}\|_1} \geq \frac{|(\mathbf{y} - A\mathbf{x})^{\mathrm{T}}\mathbf{r}(\mathbf{x}_2^o)|}{\|\mathbf{r}(\mathbf{x}_2^o)\|_1} = \frac{\mathbf{y}^{\mathrm{T}}\mathbf{r}(\mathbf{x}_2^o)}{\|\mathbf{r}(\mathbf{x}_2^o)\|_1} = \|\mathbf{r}(\mathbf{x}_2^o)\|_\infty$$

Algorithmic for Computing a Minimum ℓ_∞ Norm Approximate Solution

Similar to the procedure taken in solving the minimum ℓ_1 approximation problem, an effective algorithm for solving the minimum ℓ_∞ approximate solution is now developed. The basic nature of this algorithm is predicated on Theorem 14.8 which states that a minimum ℓ_∞ approximate solution x^o exists that minimizes the ℓ_∞ norm of the residual error vector $\mathbf{r}(\mathbf{x}) = \mathbf{y} - A\mathbf{x}$ in which $N+1$ components of the residual error vector $\mathbf{r}(\mathbf{x}_\infty^o)$ have magnitudes equal to $\|\mathbf{r}(\mathbf{x}_\infty^o)\|_\infty$. In this algorithmic approach, one first generates a vector $\mathbf{x}$ in a systematic manner for which the associated residual error vector has $N+1$ of it components equal to the residual error vector's ℓ_∞ norm. This residual error vector is then tested to determine if it is optimal. If it is found not to be optimal, then $\mathbf{x}$ is perturbed to $\mathbf{x} + \epsilon\Delta$ in an appropriate manner, so that the improvement condition $f_\infty(\mathbf{x} + \epsilon\Delta) < f_\infty(\mathbf{x})$ is met. The resultant perturbed residual error vector $\mathbf{r}(\mathbf{x} + \epsilon\Delta)$, however, typically has fewer that $N+1$ maximum magnitude elements. The vector $\mathbf{x} + \epsilon\Delta$ is then perturbed in a systematic manner until $N+1$ components of the perturbed residual vector all have the same maximum magnitude while maintaining or improving the ℓ_∞ norm criterion. These basic steps are repeated until a point $\mathbf{x}_\infty^o$ is found that satisfies the optimal condition. To explain the steps of the algorithm, a perturbation analysis of the residual error vector $\mathbf{r}(\mathbf{x}) = \mathbf{y} - A\mathbf{x}$ is now made.

The basic steps of this algorithm are implemented in the fashion now described. For any vector $\mathbf{x} \in R^{N \times 1}$, let the residual error vector $\mathbf{r}(\mathbf{x}) = \mathbf{y} - A\mathbf{x}$ have $1 \leq N_m \leq N$ components that have the same maximum magnitudes equal to $\|\mathbf{y} - A\mathbf{x}\|_\infty$. The residual error vector is now decomposed into two separate residual error vector components:

$$\mathbf{r}_1(\mathbf{x}) = \mathbf{y}_1 - A_1\mathbf{x} \quad \text{and} \quad \mathbf{r}_2(\mathbf{x}) = \mathbf{y}_2 - A_2\mathbf{x} \tag{14.111}$$

where vector $\mathbf{y}_1 \in R^{N_m \times 1}$ and matrix $A_1 \in R^{N_m \times N}$ correspond to those N_m equations of the residual error vector $\mathbf{r}(\mathbf{x}) = \mathbf{y} - A\mathbf{x}$ that have the same largest equal magnitudes while vector $\mathbf{y}_2$ and matrix $A_2 \in R^{(M-N_m) \times N}$ correspond to the remaining $M - N_m$ equations. The vector $\mathbf{x}$ is then perturbed to $\mathbf{x} + \epsilon\Delta$ where ϵ is a small positive step size scalar while the perturbation direction vector $\Delta \in R^{N \times 1}$ is specified by

$$\Delta = A_1^T[A_1 A_1^T]^{-1}\mathbf{r}_1(\mathbf{x}) \tag{14.112}$$

in which it is assumed that the matrix A_1 has rank N_m. The first and second component vectors of the perturbed residual error vector are then specified by

$$\mathbf{r}_1(\mathbf{x} + \varepsilon\Delta) = \mathbf{y}_1 - A_1(\mathbf{x} + \varepsilon\Delta) = (1 - \varepsilon)\mathbf{r}_1(\mathbf{x}) \quad \text{and}$$

$$\mathbf{r}_2(\mathbf{x} + \varepsilon\Delta) = \mathbf{r}_2(\mathbf{x}) - \varepsilon A_2 A_1^T[A_1 A_1^T]^{-1}\mathbf{r}_1(\mathbf{x})$$

Upon examination of first component vector $\mathbf{r}_1(\mathbf{x} + \epsilon\Delta)$, it is clear that as the scalar ϵ is gradually increased from zero, that the magnitude of each element of the residual error vector $\mathbf{r}_1(\mathbf{x} + \epsilon\Delta)$ is decreased by $(1 - \epsilon)\|\mathbf{r}_1(\mathbf{x})\|_\infty$ until they all become zero at $\epsilon = 1$. However, the majority of the elements of the second component vector $\mathbf{r}_2(\mathbf{x} + \epsilon\Delta)$ behave in a more irregular fashion. The parameter ε now increased from zero until the magnitude of a component of $\mathbf{r}_2(\mathbf{x} + \epsilon\Delta)$ is first equal to $(1 - \epsilon)\|\mathbf{r}_1(\mathbf{x})\|_\infty$. Such a positive value of ϵ must exist since otherwise the residual error vector could be driven to zero, indicating that the original system of equations had a solution. Thus, the required value for the step size is equal to that smallest positive step size ϵ for which the equality $\|\mathbf{r}_2(\mathbf{x} + \epsilon\Delta)\|_\infty = (1 - \epsilon)\|\mathbf{r}_1(\mathbf{x})\|_\infty$ is first satisfied. This smallest positive value ϵ^o is readily shown to be the smallest positive number in the set of roots:

$$\epsilon_k = \frac{\|\mathbf{r}_1(\mathbf{x})\|_\infty \pm e_k^T\mathbf{r}_2(\mathbf{x})}{\|\mathbf{r}_1(\mathbf{x})\|_\infty \pm e_k^T A_2 A_1^T[A_1 A_1^T]^{-1}\mathbf{r}_1(\mathbf{x})} \quad \text{for } 1 \leq k \leq M - N_m \tag{14.113}$$

At this choice of ϵ, the residual error vector $\mathbf{r}(\mathbf{x} + \epsilon^o\Delta)$ has at least N_m+1 or more components that have the same smaller maximum magnitude $(1 - \epsilon^o)\|\mathbf{r}_1(\mathbf{x})\|_\infty$.

Using this approach in an iterative manner, the state whereby $N_m \geq N + 1$ components of the residual error vector have the same maximum magnitude is eventually reached. It is then necessary to test whether the vector $\mathbf{x}$ producing this condition is optimal or not. This determination is readily made by examining the behavior of the residual error vector component associated with the N_m largest magnitudes when the prevailing vector $\mathbf{x}$ is perturbed to $\mathbf{x} + \epsilon\Delta$, that is:

$$\mathbf{r}_1(\mathbf{x} + \epsilon\Delta) = \mathbf{y}_1 - A_1\mathbf{x} - \epsilon A_1\Delta = \mathbf{r}_1(\mathbf{x}) - \epsilon A_1\Delta$$

To determine whether $\mathbf{x}$ is optimum or not, the perturbation direction vector Δ is now chosen so as to minimize the criterion $\|\mathbf{r}_1(\mathbf{x}) - \epsilon A_1\Delta\|_\infty$. In accordance with Theorem 14.9 this minimizing solution is given by

$$\Delta_\infty^\text{o} = \Delta_2^\text{o} - \frac{\mathbf{r}_1\left(\Delta_2^\text{o}\right)^{\mathrm{T}}\mathbf{r}_1(\mathbf{x})}{\|\mathbf{r}_1(\Delta_2^\text{o})\|_1}[A_1^{\mathrm{T}}A_1]^{-1}A_1^{\mathrm{T}}\mathrm{sgn}\{\mathbf{r}_1(\Delta_2^\text{o})\} \tag{14.114}$$

where

$$\Delta_2^\text{o} = [A_1^{\mathrm{T}}A_1]^{-1}A_1^{\mathrm{T}}\mathbf{r}_1(\mathbf{x}) \quad \text{and} \quad \mathbf{r}_1(\Delta_2^\text{o}) = \mathbf{r}_1(\mathbf{x}) - A_1\Delta_2^\text{o} \tag{14.115}$$

If $\Delta_\infty^\text{o} = \mathbf{o}$ then $\mathbf{x}$ is a local minimum solution and since $f_\infty(\mathbf{x}) = \|\mathbf{y} - A\mathbf{x}\|_\infty$ is a convex function it follows that this local minimum must also be a global minimum.

For $\Delta_\infty^\text{o} \neq \mathbf{o}$ an improving trial solution is obtained by gradually increasing the step size scalar ϵ from zero until the equality $\|\mathbf{r}_1(\mathbf{x}) - \epsilon A_1\Delta_\infty^\text{o}\|_\infty = \|\mathbf{r}_2(\mathbf{x}) - \epsilon A_2\Delta_\infty^\text{o}\|_\infty$ is first met. A little thought indicates that those components of the vector $A_1\Delta_\infty^\text{o}$ that have the largest and smallest magnitudes are the only candidates for the entity $\|\mathbf{r}_1(\mathbf{x}) - \epsilon^\text{o}A_1\Delta_\infty^\text{o}\|_\infty$, that is:

$$\left|e_{m_{\min}}^{\mathrm{T}}A_1\Delta_\infty^\text{o}\right| = \min_{1\leqslant k\leqslant N+1}\left\{\left|e_k^{\mathrm{T}}A_1\Delta_\infty^\text{o}\right|\right\} \quad \text{and} \quad \left|e_{m_{\max}}^{\mathrm{T}}A_1\Delta_\infty^\text{o}\right| = \max_{1\leqslant k\leqslant N+1}\left\{\left|e_k^{\mathrm{T}}A_1\Delta_\infty^\text{o}\right|\right\}$$

where $1 \leqslant m_{\min}, m_{\max} \leqslant N_m$. The optimum value of the step size scalar ε^o is that smallest positive number of the set of roots:

$$\varepsilon_{k,m} = \frac{e_m^{\mathrm{T}}r_1(x) \pm e_k^{\mathrm{T}}r_2(x)}{e_m^{\mathrm{T}}A_1\Delta_\infty^\text{o} \pm e_k^{\mathrm{T}}A_2\Delta_\infty^\text{o}} \quad \text{for } 1 \leqslant k \leqslant M - N - 1 \tag{14.116}$$

Using this perturbation analysis, it follows that the algorithm listed in Table 14.10 provides a systematic procedure for computing an optimum ℓ_∞ norm approximate solution to an inconsistent system of linear equations. It has the potential of being an efficient algorithm since when the state where $N_m \geqslant N + 1$ components of the residual error vector having the same magnitude has been reached (Step 2), the trial

TABLE 14.10 Minimum ℓ_∞ Approximate Norm Solution Algorithm

Step 1.	Using an arbitrary initial selection of $\mathbf{x}$ (e.g., the zero vector or the minimum ℓ_2 approximate solution) generate the associated residual error vector $\mathbf{r}(\mathbf{x}) = \mathbf{y} - A\mathbf{x}$.
Step 2.	If the residual error vector $r(\mathbf{x})$ has N_m components equal to $\|\mathbf{y} - A\mathbf{x}\|_\infty$ where $1 \leqslant N_m \leqslant N$ then go to Step 3. Otherwise go to Step 5.
Step 3.	Perturb the prevailing trial solution $\mathbf{x}$ to $\mathbf{x} + \varepsilon\Delta$ where the perturbation direction vector Δ is chosen according to relationship (14.112) and ε is selected to be the smallest positive scalar in the set (14.113). This results in $f_\infty(\mathbf{x} + \varepsilon^\text{o}\Delta) < f_\infty(\mathbf{x})$ and the perturbed residual error vector $\mathbf{r}(\mathbf{x} + \varepsilon^\text{o}\Delta) = \mathbf{y} - A(\mathbf{x} + \varepsilon^\text{o}\Delta)$ has the same N_m maximum magnitude elements and at least one additional maximum amplitude. Set the new trial solution $\mathbf{x}$ equal to $\mathbf{x} + \varepsilon\Delta$.
Step 4.	If the new perturbed residual error vector $\mathbf{y} - A\mathbf{x}$ has fewer than $N + 1$ maximum magnitude elements than go to Step 3. Otherwise go to Step 5.
Step 5.	Compute vector Δ_∞^o using Equation (14.113). If $\Delta_\infty^\text{o} = \mathbf{o}$ then $\mathbf{x}$ is an optimum solution. If $\Delta_\infty^\text{o} \neq \mathbf{o}$ then set the new improving trial solution x equal to $\mathbf{x} + \varepsilon^\text{o}\Delta_\infty^\text{o}$ where ε^o is the smallest positive number in the set of roots (14.116). Go to Step 2.

solution **x** is checked to determine if it is optimal or not optimal. If it is not optimal then an improving trial solution is generated resulting in a perturbed residual error vector which generally has fewer than $N + 1$ maximum magnitude components. Thus, many of the steps of algorithms based on the replacement of one of the $N + 1$ equations approach are bypassed. This has been found to result in a rapidly convergent algorithm.

Sequential Algorithm

The previous sections have been primarily directed to the problem of finding a best approximate solution to a system of inconsistent linear equations $A\mathbf{x} = \mathbf{y}$ in which a specific norm (i.e., the ℓ_p norm with $p = 1, 2,$ and ∞) is used to define the notion of optimal. Once the norm has been selected, salient properties associated with that specific optimum solution were then used in developing an algorithmic solution. For example, when an ℓ_1 norm was employed, it is known that an optimal vector selection $\mathbf{x}^o$ exists in which at least N components of the residual error vector $\mathbf{r}(\mathbf{x}) = \mathbf{y} - A\mathbf{x}$ are zero. This property was then used to develop an algorithm for finding an optimal ℓ_1 approximate solution.

In this section, a sequential algorithm is developed which finds a vector $\mathbf{x} \in R^{N \times 1}$ which provides a small value to the norm induced function:

$$f(\mathbf{x}) = \|\mathbf{y} - A\mathbf{x}\| \tag{14.117}$$

for a given vector $\mathbf{y} \in R^{M \times 1}$ and matrix $A \in R^{M \times N}$. The norm used here is general in nature and need not be an ℓ_p norm. It is to be emphasized that this algorithm does not have guaranteed convergence to a minimum norm solution. It does, however, typically converge to a good approximate solution at a reasonable computation expense. This sequential algorithm is not explicitly dependent on the particular vector norm used, but it is assumed that one has the ability to solve the basic problem of determining a scalar x^o that solves the following fundamental vector approximation problem:

$$\|\mathbf{y} - x^o\mathbf{a}\| = \min_{x \in R} \|\mathbf{y} - x\mathbf{a}\| \tag{14.118}$$

for given vectors $\mathbf{a}, \mathbf{y} \subset R^{N \times 1}$. For example, if the vector norm employed is the ℓ_1, ℓ_2, or ℓ_∞ norm, this fundamental vector approximation problem has the solution given by Equations (14.118), (14.114), and (14.121), respectively.

Sequential Algorithm

To initiate the sequential algorithm, an initial value of the vector **x** is assigned. For example, the least sum of errors squared selection given in Equation (14.115) often constitutes a good initial selection. Let the trial solution at the end of the kth iteration be designated by $\mathbf{x}^k$ so that the associated residual error vector is specified by

$$\mathbf{r}(\mathbf{x}^k) = \mathbf{y} - A\mathbf{x}^k = \mathbf{y} - \sum_{n=1}^{N} x^k(n)\mathbf{a}_n \tag{14.119}$$

in which $\mathbf{a}_n$ denotes the nth column of matrix A (i.e., $\mathbf{a}_n = A\mathbf{e}_n$). In the first step of the $k + 1$st iteration, the components $x^k(n)$ for $2 \leqslant n \leqslant N$ are fixed at their prevailing values while the first component (i.e., $x^k(1)$) takes on the variable value x. The residual error vector then is a function of this real variable x as expressed by

$$\tilde{\mathbf{r}}(x) = \left[\mathbf{y} - \sum_{n=2}^{N} x^k(n)\mathbf{a}_n \right] - x\mathbf{a}_1 \tag{14.120}$$

where the vector component enclosed within the rectangular braces is given. A selection of the variable x is now determined which minimizes the norm of this residual error vector by employing the solution (Equation (14.118)) in which **y** is replaced by the term within the braces and $\mathbf{a} = \mathbf{a}_1$. This leads to a selection designated by $x^{k+1}(1)$ which either improves upon $x^{k+1}(1)$ or at worst maintains the residual error norm value.

This process of extracting one component at a time and considering it as a variable x while maintaining the remaining $N - 1$ components at their prevailing values is continued on in a systematic fashion. For example, at the nth step of the $k + 1$st iteration, the residual error vector becomes

$$\tilde{\mathbf{r}}(x) = \left[\mathbf{y} - \sum_{m=1}^{n-1} x^{k+1}(m)\mathbf{a}_m - \sum_{m=n+1}^{N} x^k(m)\mathbf{a}_m \right] - x\mathbf{a}_n$$

Equation (14.118) in then used to obtain the minimizing value of **x** as designated by $x^{k+1}(n)$ in which **y** is replaced by the term within the braces and $\mathbf{a} = \mathbf{a}_n$. Application of this sequential process is continued for $1 \leq n \leq N$ to complete the $k + 1^{st}$ iteration. This iterative procedure is continued until no significant change is made in the trial solution from one iteration to the next. A convenient stopping condition is one in which the following inequality is first satisfied:

$$\rho_k = \frac{\|\mathbf{x}^k - \mathbf{x}^{k-1}\|}{\|\mathbf{x}^{k-1}\|} < \epsilon \tag{14.121}$$

for $k = 1, 2, 3, \cdots$ where ϵ is a suitably chosen small positive scalar (e.g., $\epsilon = 10^{-20}$).

Nonnorm Approach to Data Outlier Removal

When seeking the best approximate solution to a system of M linear equations in N unknowns as designated by the residual error expression $\mathbf{r}(\mathbf{x}) = \mathbf{y} - A\mathbf{x}$, an ℓ_1 norm criterion for measuring goodness of approximation is often useful in those cases in which a small number data outliers are present in the vector **y** or the matrix A. An ℓ_1 norm criterion tends to ignore such data outliers in contrast to the widely employed ℓ_2 norm criterion. The reason for this behavior is made apparent by examining the function to be minimized when seeking a best approximation as specified by

$$f_p(\mathbf{x}) = \sum_{m=1}^{M} |y_m - \mathbf{e}_m^{\mathrm{T}} A\mathbf{x}|^p \tag{14.122}$$

Clearly, the parameter selection $p = 1$ is seen to give less weight to residual errors whose magnitudes are relatively large in comparison to the parameter selection $p = 2$. As such, the selection of a minimizing **x** is less influenced by data outliers for the norm choice $p = 1$.

If data outlier removal is an important factor in a given data processing application, it is seen that a selection of parameter p lying in the interval $0 < p < 1$ would seem to provide an additional immunization to large data outliers. Unfortunately, the function $f_p(\mathbf{x})$ mapping $R^{N \times 1}$ into R is no longer convex for $0 < p < 1$ and as such a local minimum need not be a global minimum. The loss of this important convexity property, however, might be offset by possible improvements in data outlier immunity. The approximation error function (14.122) for selections $0 < p < 1$ is now shown to share many of the same properties as for the convex case $p = 1$. A determination of some characteristic properties of function $f_p(\mathbf{x})$ is obtained by decomposing the summation on the interval $1 \leq m \leq M$ into three disjoint integer sets over which the residual error components $r_m(\mathbf{x}) = y(m) - \mathbf{e}_m^{\mathrm{T}} A\mathbf{x}$ are positive, negative, or zero. This decomposition takes the form:

$$f_p(\mathbf{x}) = \sum_{y_m > \mathbf{e}_m^{\mathrm{T}} A\mathbf{x}} [y_m - \mathbf{e}_m^{\mathrm{T}} A\mathbf{x}]^p + \sum_{y_m < \mathbf{e}_m^{\mathrm{T}} A\mathbf{x}} [\mathbf{e}_m^{\mathrm{T}} A\mathbf{x} - y_m]^p$$

In this decomposition the first summation is taken over all integers in the integer interval $1 \leq m \leq M$ for which $r_m(\mathbf{x}) = y(m) - \mathbf{e}_m^T A\mathbf{x} > 0$ and the second summation is taken over all integers in the integer interval $1 \leq m \leq M$ for which $r_m(\mathbf{x}) = y(m) - \mathbf{e}_m^T A\mathbf{x} < 0$. The remaining residual error vector components are zero.

It is now shown that a vector $\mathbf{x}$ exists which minimizes $f_p(\mathbf{x})$ and causes at least N components of the associated residual error vector $r(\mathbf{x}^o) = \mathbf{y} - A\mathbf{x}^o$ to be zero. Specifically, let $\mathbf{x}^o$ minimize function $f_p(\mathbf{x})$ in which only N_o components of the residual error vector are zero where $0 \leq N_o \leq N - 1$. Let us now perturb $\mathbf{x}^o$ to $\mathbf{x}^o + \epsilon\Delta$ where ϵ is a real valued step size scalar and Δ is a perturbation direction vector. The perturbation direction vector is chosen so that the original zeros in the unperturbed residual error vector $r(\mathbf{x}^o) = \mathbf{y} - A\mathbf{x}^o$ are maintained at zero in the perturbed residual error vector $\mathbf{r}(\mathbf{x}^o + \epsilon\Delta) = \mathbf{y} - A\mathbf{x}^o - \epsilon A\Delta$. This is always possible since the perturbation direction vector Δ can be chosen to be any nonzero vector which is orthogonal to each of the $N_o < N$ row vectors of matrix A associated with the zero components of $\mathbf{r}(\mathbf{x}^o) = \mathbf{y} - A\mathbf{x}^o$.

The scalar ϵ is now chosen small enough in magnitude so that the signs of the nonzero components in the perturbed residual error vector $\mathbf{r}(\mathbf{x}^o + \epsilon\Delta) = \mathbf{y} - A\mathbf{x}^o - \epsilon A\Delta$ remain the same as the signs of the nonzero components in the unperturbed residual error vector $\mathbf{y} - A\mathbf{x}^o$. Under this restriction the corresponding sum of residual error magnitude criterion at the perturbed point $\mathbf{x}^o + \epsilon\Delta$ is given by

$$
\begin{aligned}
f_p(\mathbf{x}^o + \varepsilon\Delta) &= \sum_{y_m > \mathbf{e}_m^T A\mathbf{x}} \left[y_m - \mathbf{e}_m^T A(\mathbf{x}^o + \varepsilon\Delta) \right]^p + \sum_{y_m < \mathbf{e}_m^T A\mathbf{x}} \left[\mathbf{e}_m^T A(\mathbf{x}^o + \varepsilon\Delta) - y_m \right]^p \\
&= \sum_{y_m > \mathbf{e}_m^T A\mathbf{x}} [y_m - \mathbf{e}_m^T A\mathbf{x}^o]^p \left[1 - \frac{\varepsilon \mathbf{e}_m^T A\Delta}{y_m - \mathbf{e}_m^T A\mathbf{x}^o} \right]^p \\
&\quad + \sum_{y_m > \mathbf{e}_m^T A\mathbf{x}} \left[\mathbf{e}_m^T A\mathbf{x}^o - y_m \right]^p \left[1 + \frac{\varepsilon \mathbf{e}_m^T A\Delta}{\mathbf{e}_m^T A\mathbf{x}^o - y_m} \right]^p
\end{aligned}
$$

We now examine the behavior of this perturbed function for a small magnitude selection of the step size parameter ε. This behavior makes use of the following lemma.

Lemma 16. The function $[1 + \theta]^p$ has the Taylor series expansion about $\theta = 0$:

$$
[1 + \theta]^p = 1 + \sum_{k=1}^{\infty} p(p-1)\dots(p-k+1)\theta^k
$$

where $p \neq 1$. Moreover, for values of θ such that $|\theta| << 1$ this function can be approximated by the linear term

$$
[1 + \theta]^p \approx 1 + p\theta \tag{14.123}
$$

For small values of the step size parameter ϵ, use of approximation (14.123) in the perturbed function $f_p(\mathbf{x}^o + \epsilon\Delta)$ then results in

$$
f_p(\mathbf{x}^o + \epsilon\Delta) \approx f_p(\mathbf{x}^o) + \epsilon p \left[\sum_{y_m < \mathbf{e}_m^T A\mathbf{x}} \mathbf{e}_m^T A\Delta \, [y_m - \mathbf{e}_m^T A\mathbf{x}^o]^{p-1} - \sum_{y_m > \mathbf{e}_m^T A\mathbf{x}} \mathbf{e}_m^T A\Delta \, [y_m - \mathbf{e}_m^T A\mathbf{x}^o]^{p-1} \right]
$$

The term within the braces must be zero for otherwise a small magnitude selection of ε can be made in which $f_p(\mathbf{x}^o + \epsilon\Delta) < f_p(\mathbf{x}^o)$ which is contrary to the assumption that $\mathbf{x}^o$ is optimum. Let the parameter ε be gradually increased or decreased from zero until a formally nonzero element of the unperturbed residual error vector $r(\mathbf{x}^o) = \mathbf{y} - A\mathbf{x}^o$ is first driven to zero while maintaining $f_1(\mathbf{x}^o + \epsilon\Delta) = f_1(\mathbf{x}^o)$. This process is continued until eventually N components of the perturbed residual error vector are zero. This result is now formally stated in theorem format.

Theorem 14.10.   Let there be given a vector $\mathbf{y} \in R^{M \times 1}$ and matrix $A \in R^{M \times N}$ of rank N. It then follows that for any $0 < p \leq 1$ there exists a real $N \times 1$ vector $\mathbf{x}^{\circ}$ which minimizes the function:

$$f_p(\mathbf{x}) = \sum_{m=1}^{M} |y_m - \mathbf{e}_m^T A \mathbf{x}|^p$$

such that the associated residual error vector $\mathbf{r}(\mathbf{x}^{\circ}) = \mathbf{y} - A\mathbf{x}^{\circ}$ has at least N zero components.

   As in the ℓ_1 case, we can employ the computational costly brute force method for finding an optimum solution which entails determining solutions to all subsets of N linear equations of the system of linear equations $\mathbf{y} = A\mathbf{x}$ and then evaluating $f_p(\mathbf{x})$ at each of these solutions. A required optimal solution then corresponds to that solution(s) for which $f_p(\mathbf{x})$ takes on its smallest value. In order to ease the computational burden of the brute force method, one might be tempted to develop algorithms similar to the primal or duals algorithms for the case $p = 1$. Unfortunately, for $0 < p < 1$ the function $f_p(\mathbf{x})$ is not convex and therefore generally has several local minimum which are not a global minimum. As such, algorithms based on small perturbations are typically ineffective for the case $0 < p < 1$. This being the case, there is much to be said about the brute force method when $0 < p < 1$. Moreover, it is to be noted that on several simulated data examples, the optimal solutions for the case $p = 1$ also turned out to be optimal solutions for choices of $p = 0.25$ and $p = 0.5$. This provides further evidence of the utility of a minimum ℓ_1 criterion for removing data outliers.

Acknowledgment

This section first appeared in "Digital Signal Processing: A Review Journal", Academic Press, vol. 12, no. 4, 2002, pp. 524–559.

References

1.   I. Barrodale and F.D.K. Roberts, "An improved algorithm for discrete *L*1 linear approximation," *SIAM J. Numer. Anal.*, vol. 10, pp. 839–848, 1973.

2.   R.H. Bartels, A.R. Conn, and J.W. Sinclair, "Minimization techniques for piecewise differentiable functions: the L_1 solution to an overdetermined system," *SIAM J. Numer. Anal.*, vol. 15, pp. 224–241, 1978.

3.   P. Bloomfield and W.L. Steiger, "Least absolute deviations curve-fitting," *SIAM J. Sci. Stat. Comput.*, vol. 1, pp. 290–301, 1980.

4.   P. Bloomfield and W.L. Steiger, *Least Absolute Deviations, Theory, Applications and Algorithms*, Boston, MA: Birkhauser, 1983.

5.   J.A. Cadzow, *Digital Signal Processing: A Vector Space Approach*, a manuscript being considered for publication.

6.   J.F. Claerbout and F. Muir, "Robust modeling with erratic data," *Geophysics*, vol. 38, no. 5, pp. 826–844, 1973.

7.   E.W. Cheney, *Introduction to Approximation Theory*, New York: McGraw-Hill, 1966.

8.   E. Denoel and J-P. Solvay, "Linear prediction of speech with a least absolute value error criterion," *IEEE Trans. Acoust. Speech Signal Process.*, vol. ASSP-33, no. 6, pp. 1397–1403, 1985.

9.   C.S. Duris and V.P. Sreedharan, "Chebyshev and ℓ_1 solutions of linear equations using least squares solutions," *SIAM J. Numer. Anal.*, September, 491–505, 1968.

10.   F.Y. Edgeworth, "A new method of reducing observations relating to several quantities," *Phil. Mag.*, vol. 25, no. 5, pp. 184–191, 1888.

11.   C. Eisenhart, "Bascovitch and the combination of observations," in *Roger Joseph Boscovitch*, L.L. Whyte, Ed., New York: Fordham University Press, 1961.

12. T.E. Gentle, "Least absolute values estimation: an introduction," *Commun. Stat.*, vol. B6, pp. 313–328, 1977.

13. R.W. Hawley and N.C. Gallagher, "On Edgeworth's method for minimum absolute error linear regression," *IEEE Trans. Signal Process.*, vol. 42, no. 8, pp. 2045–2054, 1994.

14. R.V. Hogg, "An introduction to robust procedures," *Commun. Statist.-Theor. Meth.*, vol. A6, no. 9, pp. 789–794, 1977.

15. A. Kijko, "Seismological outliers: L_1 or adaptive L_p norm application," *Bull. Seismol. Soc. Am.*, vol. 84, no. 2, pp. 473–477, 1994.

16. P. Lancaster and M. Tismenetsky, *The Theory of Matrices with Applications*, 2nd ed., London: Academic Press, Inc, 1985.

17. Legendre, *Principle of Least Squares*.

18. D.G. Luenberger, *Optimization by Vector Space Methods*, New York: Wiley, 1969.

19. E.Ya. Remez, *General Computation Methods for Chebyshev Approximation, The Problems with Real Parameters*, Ukrainian S.S.R., Kiev: Academy of Sciences, 1957.

14.5   Adaptive Signal Processing

W. Kenneth Jenkins and Dean J. Krusienski

Introduction

Adaptive digital signal processing methodologies have become increasingly important in recent years due to demands for improved performance in high data rate digital communication systems and in wideband image/video processing systems. Adaptive system identification, adaptive noise cancellation, adaptive signal compression, and adaptive channel equalization are just a few of the important application areas that have seen significant applications in state-of-the-art systems. Much of the recent success in adaptive signal processing has been facilitated by improvements in VLSI digital signal processor (DSP) integrated circuit technology, which now enables large amounts of digital signal processing power in convenient and reliable forms. Since further advancements in integrated circuit technology are sure to develop in the future, it is expected that adaptive techniques will assume even more importance in high-performance electronic systems of the future.

Linear FIR Adaptive Filters

A finite impulse response (FIR) adaptive filter consists of a digital tapped delay line with variable multiplier coefficients that are adjusted by an adaptive algorithm. The adaptive algorithm attempts to minimize a cost function that is designed to provide an instantaneous on-line estimate of how closely the adaptive filter achieves a prescribed optimum condition. The cost function most frequently used is an approximation to the expected value of the squared error, $E\{|e|^2(n)\}$, where $e(n)=d(n) - y(n)$ is the difference between a training signal $d(n)$ (desired response) and the filter output $y(n)$, and $E\{\cdot\}$ denotes the statistical expected value. The acquisition of the training signal $d(n)$ is obtained by different means in different applications, and is often a limiting factor in the use of adaptive methodologies (Diniz, 2002; Haykin, 2002; Widrow, 1985).

The input vector and the coefficient weight vector of the adaptive filter at the nth iteration are defined as $\mathbf{X}(n) = [x(n), x(n - 1),\ldots, x(n - N + 1)]^{\mathrm{T}}$ and $\mathbf{W}(n) = [w_0(n), w_1(n),\ldots, w_{n - 1}(n)]^{\mathrm{T}}$, where the superscript T denotes vector transpose. The filter output at iteration n is given by

$$y(n) = \sum_{k=0}^{N-1} w_k x(n - k) = \mathbf{W}^{\mathrm{T}}(n)\mathbf{X}(n) \qquad (14.124)$$

In the following discussion, the training signal $d(n)$ and the input signal $x(n)$ are assumed to be stationary and ergodic. An adaptive filter uses an iterative method by which the tap weights $\mathbf{W}(n)$ are made to converge to the optimal solution $\mathbf{W}^*$ that minimizes the cost function. It is well known that $\mathbf{W}^*$, known in the literature as the Wiener solution, is given by

$$\mathbf{W}^* = \mathbf{R}_x^{-1}\mathbf{p}_{xd} \qquad (14.125)$$

where $\mathbf{R}_x = E\{\mathbf{X}(n)\mathbf{X}^T(n)\}$ is the autocorrelation matrix of the input and $\mathbf{p}_{xd} = E\{\mathbf{X}(n)d(n)\}$ is the cross correlation vector between the input and the desired response. The most common approach is to update each tap weight according to a steepest descent strategy; i.e., the tap weight vector is incremented in proportion to the gradient $\mathbf{Vw}$ according to

$$\mathbf{W}(n+1) = \mathbf{W}(n) - \mu\nabla_w \qquad (14.126)$$

where μ is the step size, $\mathbf{V_w} = [\nabla_{w0}(n),\ldots,\nabla_{wN-1}(n)]^T$, and $\nabla_{wk}(n) = \delta E\{|e|^2(n)\}/\delta w_k$ is the partial derivative of the cost function with respect to $w_k(n)$, $k = 0,\ldots, N-1$. In typical applications the precise value of the cost function is not known, nor is the gradient known explicitly, so it is necessary to make some simplifying assumptions that allow gradient estimates to be computed on-line. Different approaches to estimating the cost function (or the gradient) lead to different adaptive algorithms, such as the well known least mean squares (LMS) and the recursive least squares (RLS) algorithms. Including the Hessian matrix in Equation (14.126) to accelerate the steepest descent optimization strategy leads to the family of quasi-Newton algorithms that are characterized by rapid convergence at the expense of greater computational complexities.

The LMS algorithm (Widrow, 1985) makes the simplifying assumption that the expected value of the squared error is approximated by the squared error itself, i.e., $E\{|e(n)|^2\} \sim |e(n)|^2$. In deriving the algorithm, the error squared is differentiated with respect to $\mathbf{W}$ to approximate the true gradient. In vector notation the LMS update relation becomes

$$\mathbf{W}(n+1) = \mathbf{W}(n) + 2\mu e(n)\mathbf{X}(n) \qquad (14.127)$$

The value of μ determines both the convergence rate of the adaptive process and the minimum mean squared error after convergence. To ensure stability and guarantee convergence of both the tap weight vector and the mean squared error estimate, μ must satisfy the condition $0 < \mu < 1/NE\{|x(n)|^2\}$, where $E\{x^2(n)\} = (1/N)\mathrm{tr}[\mathbf{R_x}]$ is the average input signal power which can be calculated from $\mathbf{R_x}$ if the input autocorrelation matrix is known. When μ is properly chosen, the weight vector converges to a neighborhood of the optimal (Wiener) solution.

Convergence Properties of the LMS Adaptive Filter

It is well known that the convergence behavior of the LMS algorithm, as applied to a direct form FIR filter structure, is controlled by the autocorrelation matrix $\mathbf{R_x}$ of the input process, where

$$\mathbf{R_x} \equiv E[\mathbf{x}^*(n)\mathbf{x}^T(n)] \qquad (14.128)$$

and where * denotes complex conjugate to account for the general case of complex input signals. The autocorrelation matrix $\mathbf{R_x}$ is usually positive definite, which is one of the conditions necessary to guarantee convergence to the Wiener solution. Another necessary condition for convergence is $0 < \mu < 1/\lambda_{max}$, where λ_{max} is the largest eigenvalue of $\mathbf{R_x}$. The convergence of this algorithm is directly related to the condition number of $\mathbf{R_x}$, defined as $\kappa = \lambda_{max}/\lambda_{min}$, where λ_{min} is the minimum eigenvalue of $\mathbf{R_x}$ Ideal conditioning occurs when $\kappa = 1$ (white noise); as this ratio increases, slower convergence results. The condition number depends on the spectral distribution of the input signal, and can be shown to be related to the maximum and minimum values of the input power spectrum. From this line of reasoning it becomes clear why white noise is the ideal input signal for rapidly training an LMS adaptive filter. The adaptive process becomes slower and requires more computation for input signals that are more severely colored.

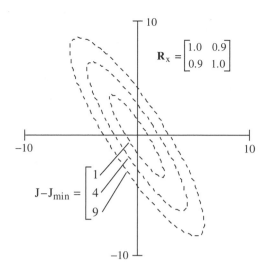

FIGURE 14.25   2-D error surface with a colored input signal.

Convergence properties are reflected in the geometry of the MSE surface, which is simply the mean squared output error $E[|e(n)|^2]$ expressed as a function of the N adaptive filter coefficients in $(N+1)$-space. An expression for the error surface of the direct form filter is

$$J(z) \equiv E[|e(n)|^2] = J_{\min} + \mathbf{z}^{*\mathrm{T}}\mathbf{R}_{\mathbf{x}}\mathbf{z} \tag{14.129}$$

with $\mathbf{R}_{\mathbf{x}}$ defined in Equation (14.128) and $\mathbf{z} \equiv \mathbf{w} - \mathbf{w}_{\mathrm{opt}}$, where $\mathbf{w}_{\mathrm{opt}}$ is the optimum filter coefficient vector in the sense of minimizing the mean squared error ($\mathbf{w}_{\mathrm{opt}}$ is the Wiener solution). An error surface for a simple two-tap filter is shown in Figure 14.25. In this example $x(n)$ was specified to be a colored noise input signal with an autocorrelation matrix:

$$\mathbf{R}_{\mathbf{x}} = \begin{bmatrix} 1.0 & 0.9 \\ 0.9 & 1.0 \end{bmatrix}$$

Figure 14.25 shows three equal-error contours on the three dimensional surface. The term $\mathbf{z}^{*\mathrm{T}}\mathbf{R}_{\mathbf{x}}\mathbf{z}$ in Equation (14.129) is a quadratic form that describes the bowl shape of the FIR error surface. When $\mathbf{R}_{\mathbf{x}}$ is positive definite, the equal-error contours of the surface are hyperellipses (N-dimensional ellipses) centered at the origin of the coefficient parameter space. Furthermore, the principal axes of these hyperellipses are the eigenvectors of R_x, and their lengths are proportional to the eigenvalues of R_x. Since the convergence rate of the LMS algorithm is inversely related to the ratio of the maximum to the minimum eigenvalues of R_x, large eccentricity of the equal-error contours implies slow convergence of the adaptive system. In the case of an ideal white noise input, R_x has a single eigenvalue of multiplicity N, so that the equal-error contours are hyperspheres.

Common Applications of Adaptive Methods

An adaptive filter is said to be used in the *system identification configuration* when both the adaptive filter and an unknown system are excited by the same input signal $x(n)$, the system ouput signals are compared to form the error signal $e(n) = d(n) - y(n)$, and the parameters of the adaptive filter are iteratively adjusted to minimize some specified function of the error $e(n)$. In the system identification configuration shown in Figure 14.26, the desired signal is produced as the output of an unknown plant whose input is accessible

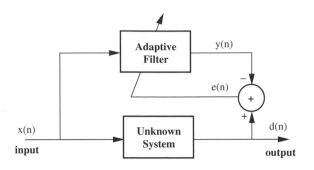

FIGURE 14.26   System identification configuration.

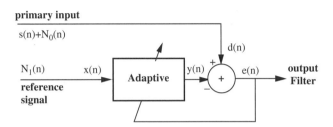

FIGURE 14.27   Noise canceling configuration.

for excitation. When the minimum of the cost function is achieved and the adaptive filter parameters have converged to stable values, the adaptive filter provides a model of the unknown system in the sense that the adaptive process has formed the best approximation in the MSE sense using the structure imposed by the adaptive system. The converged coefficients provide good estimates of the model parameters.

In order for the adaptive system to form a good model of the unknown system at all frequencies, it is important that the input signal have sufficiently rich spectral content. For example, if the adaptive filter is an FIR filter structure with N adjustable coefficients, the input signal must contain at least N distinct frequency components in order to uniquely determine the set of coefficients that minimizes the MSE. A white noise input signal is ideal because it excites all frequencies with equal power. A broadband colored noise input will also provide a good excitation signal in the sense of driving the adaptive filter to the minimum MSE solution, although in general the convergence rate of the learning process will be slower than for white noise inputs because the frequencies that are excited with small power levels will converge slowly. Many adaptive algorithms attempt to normalize (or whiten) the input power spectrum in order to improve the convergence rate of the learning process.

The system identification configuration is a fundamental adaptive filtering concept that underlies many applications of adaptive filters. The major attraction of the system identification configuration is that the training signal is automatically generated as the output of the unknown system. The disadvantage is that the input of the unknown system must be accessible to be excited by an externally applied input noise signal. In some applications obtaining a model of the unknown system is the desired result, and the accuracy of the adaptive coefficients is a primary concern. In other applications it is not necessary that the unknown system be identified explicitly, but rather that the adaptive filter is required to model the unknown system only to generate accurate estimates of its output signal.

A block diagram for an *adaptive noise cancelling configuration* is shown in Figure 14.27, where the unknown system in this configuration is neither shown explicitly nor identified in a direct way. The primary signal is assumed to be the sum of an information bearing signal $s(n)$ and an additive noise component $N_0(n)$, which is

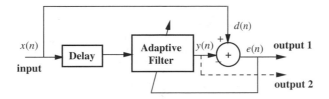

FIGURE 14.28 Linear prediction configuration.

uncorrelated with $s(n)$. The primary signal is used to train the adaptive noise canceler, so that $d(n) = s(n) + N_0(n)$ and the error signal becomes $e(n) = d(n) - y(n) = s(n) + N_0(n) - y(n)$. The reference signal, which is used as the input to the adaptive filter, should be a reference noise $N_1(n)$, which is uncorrelated to $s(n)$, but which is correlated in an unknown way with $N_0(n)$. The adaptive filter forms an estimate of $N_0(n)$ and subtracts this estimate from the primary input signal, thereby forming a good estimate of the information signal at the output of the noise canceled. Note that:

$$E[|e(n)|^2] = E[|s(n)|^2] + E[|N_0(n) - (y(n)|^2] \tag{14.130}$$

so that minimizing $E[|e(n)|^2]$ will also minimize $E[|N_0(n) - (y(n)|^2]$ because the first term in Equation (14.130) is dependent only on the information signal $s(n)$ and its mean squared value cannot be affected by the adaptive filter as long as $s(n)$ and $N_1(n)$ are uncorrelated. After the adaptive filter converges, $y(n)$ becomes the best estimate of $N_0(n)$ according to the MSE criterion.

Since the unknown system in the adaptive noise canceling configuration is implicit, there is no need for access to its input in this configuration. However, it is necessary to find a suitable reference signal that does not contain significant amount of the information signal $s(n)$. If the reference contains even small levels of $s(n)$, then some part of the primary signal $s(n)$ will be canceled and the overall signal-to-noise ratio will degrade.

Adaptive linear prediction is a very important and well-developed subject that spans many different areas of engineering. A block diagram of this configuration is shown in Figure 14.28. In this configuration the input vector is delayed, usually by one time sample, and the delayed input vector $\mathbf{x}(n - 1) = [\mathbf{x}(n - 1), \mathbf{x}(n - 2), \ldots, x(n - N)]^T$ is then used to predict $x(n)$, the current value of the input. The *prediction error* is given by $e(n) = d(n) - y(n) = x(n) - y(n)$.

Sometimes the entire system of Figure 14.28 from the input $x(n)$ to output 1 is considered to be a complete system, in which case it is referred to as a *prediction error filter*. Whenever the mean squared prediction error is minimized, $e(n)$ will become uncorrelated with $x(n)$, while $y(n)$ remains highly correlated with $x(n)$. Therefore, since $d(n) = y(n) + e(n)$, the prediction filter decomposes the input signal into two components, one that is uncorrelated to the input and one that is highly correlated to the input. In this sense the linear predictor is a type of correlation filter.

Two distinct outputs, output 1 and output 2, in Figure 14.28 provide access to both the correlated and uncorrelated components. Output 1 is used in applications such as adaptive linear predictive coding (LPC) for speech analysis and synthesis, and in adaptive differential pulse code modulation (ADPCM) for speech (and image) waveform compression. Since the prediction error is a difference between the actual value of $x(n)$ and its predicted value $y(n)$, the dynamic range needed for accurately encoding $e(n)$ is usually much smaller than $x(n)$ itself. This is the fundamental mechanism by which a linear prediction filter is able to compress waveforms. Alternately, output 2 produces a filtered version of $x(n)$ with the uncorrelated noise component removed. When used in this mode, the adaptive linear predictor becomes a *line enhancer*, which is capable of removing broadband noise from a narrow band information signal, a function frequently needed in communication systems.

The fourth adaptive filtering configuration is the *inverse system configuration* shown in Figure 14.29. In this case the adaptive filter is placed in series with an unknown system and the output $y(n)$ is driven by the adaptive

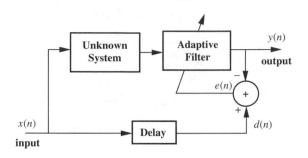

FIGURE 14.29 Inverse system configuration.

algorithm to form the best MSE approximation to a delayed copy of the input signal. When the adaptive filter reaches convergence, the series combination of the unknown and adaptive systems forms an overall frequency response that approximates a pure delay, i.e., the overall system approximates a flat magnitude response and a linear phase characteristic across the usable bandwidth of the excited spectrum. In this case the adaptive filter estimates $H^{-1}(j\omega)$, where $H(j\omega)$ is the frequency response of the unknown system. The inverse system configuration is the basis for adaptive equalization, in which nonideal communication channels are equalized in order to reduce dispersion and to eliminate inter-symbol interference in high-speed digital communications. The adaptive equalizer forms an essential component of most state-of-the-art modems today, where the equalization function is required to maintain acceptable bit error rates when binary information is transmitted across narrowband (4 kHz) telephone channels. Equalizers have also been used to equalize the dispersive channel that a computer faces when transferring high-speed digital data to a magnetic recording medium (disk or tape). It has been shown that properly designed equalizers will permit symbols to be more densely written on the magnetic recording medium due to the reduction in inter-symbol interference.

The training of an adaptive equalizer in the inverse system configuration raises a number of problems that are unique to this configuration. Note that by the nature of the configuration, the input to the adaptive filter has been filtered by the unknown system. Hence, in most situations, the input to the equalizer cannot be a white noise signal, and depending on the severity of the channel imperfections, the equalizer may experience trouble converging quickly. In a communication system, the transmitter and the receiver are typically located at separate physical locations, so it may not be a simple matter to provide a training signal that is an exact delayed copy of the transmitted waveform. For this reason, channel equalizers are often trained during prescribed "hand shaking" intervals, during which time a pseudorandom binary sequence with known spectral characteristics is transmitted. Once the equalizer has converged to equalize the present characteristics of the unknown channel, the parameters of the equalizer are frozen and held at their converged values during the data transfers that follow.

Because of the difficulty in obtaining a suitable training reference, there has been a great deal of interest in combining certain blind equalization schemes with decision feedback equalizers. In these cases the blind equalization technique is used to bring the equalizer into the neighborhood of proper convergence, at which point the scheme is switched over to a decision feedback algorithm that performs well as long as the equalizer remains in the neighborhood of its optimum solution.

Linear IIR Adaptive Filters

The mean squared error approximation that led to the conventional LMS algorithm for FIR filters has also been applied to infinite impulse response (IIR) filters (Jenkins, 1996; Farhang-Boroujeny, 1999). An IIR digital filter is characterized by a difference equation:

$$y(n) = \sum_{k=0}^{N_b-1} b_k x(n-k) + \sum_{j=0}^{N_a-1} a_j y(n-j) \qquad (14.131)$$

where the b_k's are the coefficients that define the zeros of the filter and the a_k's define the poles. The LMS adaptive algorithm for IIR filters is derived in a similar manner as in the FIR case, although the recursive relation (Equation (14.131)) is used instead of the convolution sum (Equation (14.124)) to characterize the input–output relationship of the filter. If derivatives of the squared error function are calculated using the chain rule, with first-order dependencies taken into account and higher-order dependencies ignored, the result is

$$\nabla_{E[|e|^2]} = \left[-2e(n)\frac{\partial(y(n))}{\partial \mathbf{a}}, \; -2e(n)\frac{\partial(y(n))}{\partial \mathbf{b}} \right]$$

where

$$\frac{\partial y(n)}{\partial b_k} = x(n-k) + \sum_{j=1}^{N_a} a_j(n)\frac{\partial y(n-j)}{\partial b_k} \quad k = 0,\dots,N_{b-1} \tag{14.132a}$$

and

$$\frac{\partial y(n)}{\partial a_k} = y(n-k) + \sum_{j=1}^{N_a} a_j(n)\frac{\partial y(n-j)}{\partial a_k} \quad k = 0,\dots,N_{a-1} \tag{14.132b}$$

This procedure does not result in a closed form expression for the gradient but it does produce a recursive relation by which the gradients can be generated using Equation (14.132).

The output error formulation prevents bias in the solution due to noise in the desired signal, but the effect of the feedback terms is to make the problem nonlinear in the feedback coefficient parameters. Also, the current filter parameters now depend directly upon previous filter coefficients, which are time-varying. This often leads to MSE surfaces that are not quadratic in nature. There are many examples in the literature for which IIR MSE surfaces contain local minima in addition to the global minimum. In these cases, gradient search techniques may become entrapped in local minima, resulting in poor performance due to improper convergence of the adaptive filter.

Nonlinear Adaptive Systems

There are many applications where voice signals, audio signals, images, or video signals are subjected to nonlinear processing, and which require nonlinear adaptive compensation to achieve the proper system identification and parameter extraction. For example, a generic nonlinear communication system is shown in Figure 14.30. The overall communication channel between the transmitter and the receiver is often nonlinear, since the amplifiers located in the (satellite) repeaters usually operate at or near saturation in order to conserve power. The saturation nonlinearities of the amplifiers introduce nonlinear distortions in the signal they process. The path from the transmitter to the repeater as well as from the repeater to the receiver may each be modeled as a linear system. The amplifier characteristics are usually modeled using memoryless nonlinearities. In general, the static nonlinearity is comprised of a linear term and higher-order polynomial terms; hence, the output of such a system can be represented as the sum of a linear part and a nonlinear part. Such systems can be modeled by connecting nonlinear and linear filter modules into a series cascade configuration. In particular, many nonlinear systems can be represented by the LNL model shown in Figure 14.31 (Yao, 1994).

FIGURE 14.30 Model of a typical nonlinear communication channel.

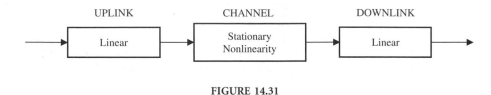

FIGURE 14.31

A similar nonlinear model can be used for magnetic recording channels, where the interaction between the electronic bit stream and the magnetic recording medium via the read/write heads exhibits nonlinear behavior. As in the case of IIR adaptive structures, nonlinear adaptive filters can also produce multimodal error surfaces on which stochastic gradient optimization strategies may fail to reach the global minimum due to premature entrapment. Neural networks and Volterra nonlinear adaptive filters are known for their tendency to generate multimodal error surfaces (Mathews, 2000).

Structured Stochastic Optimization Algorithms

Stochastic optimization algorithms aim at increasing the probability of encountering the global minimum without performing an exhaustive search of the entire parameter space. Unlike gradient-based techniques, the performance of stochastic optimization techniques in general is not dependent upon the filter structure. Therefore, these types of algorithms are capable of globally optimizing any class of adaptive filter structures or objective functions by assigning the parameter estimates to represent filter tap weights, neural network weights, or any other possible parameter of the unknown system model (even the exponents of polynomial terms).

The foundation of a *structured* stochastic search is to *intelligently* generate and modify the randomized estimates in a manner that efficiently searches the error space, based on some previous or collective information generated by the search (Engelbrecht, 2002). Several different structured stochastic optimization techniques can be found in adaptive filtering literature, most notably simulated annealing (Aarts, 1989), evolutionary algorithms such as the genetic algorithm (Goldberg, 1989) and swarm intelligence algorithms such as particle swarm optimization (Eberhardt, 1995). One interesting item to note is that all of the prominent structured stochastic techniques are inspired by a natural or biological process. This can be attributed to the observance that such natural processes exhibit a sense of structure and stability achieved through some sort of randomness or chaos. This section reviews the two prominent population-based structured stochastic optimization strategies, evolutionary algorithms, and particle swarm optimization (PSO), due to their superior convergence properties for adaptive filtering applications. Special emphasis is placed on PSO due to its relative novelty.

Evolutionary Algorithms

Evolutionary algorithms (EA) begin with a random set of possible solutions (the unknown parameters to be optimized), termed the population. Each possible solution in the population is termed an individual. Each individual's set of parameters is termed a chromosome or genome, and each parameter is termed a gene. Depending on the nature of the problem, the chromosomes may represent real numbers or can be encoded as binary strings.

At every generation, the fitness of each individual is evaluated by a predetermined fitness function that is assumed to have an extremum at the desired optimal solution. An individual with a fitness value closer to that of the optimal solution is considered better fit than an individual with a fitness value farther from that of the optimal solution. The population is then evolved based on a set of principles rooted in evolutionary theory such as natural selection, survival of the fittest, and mutation. Natural selection is the mating of the fittest individuals (*parents*) within the population to produce a new individual (*offspring*). This equates to randomly swapping corresponding parameters (*crossover*) between the parents to produce a new, potentially fitter individual. The new offspring then replace the least-fit individuals in the population, which is the survival of the fittest facet of the evolution. A portion of the population is then randomly mutated in order to add new parameters to the search. The expectation is that only the offspring that inherit the best parameters from the parents will survive and the population will continually converge to the best possible fitness that represents the optimal or suitable solution. Several EA paradigms exist, such as the genetic algorithm, evolutionary programming, and evolutionary strategies, each emphasizing only specific evolutionary constructs, encoding, and operators.

Particle Swarm Optimization (PSO)

Particle swarm optimization was first developed in 1995 (Eberhart, 1995) rooted on the notion of swarm intelligence of insects, birds, etc. The algorithm attempts to mimic the natural process of group communication of individual knowledge that occurs when such swarms flock, migrate, forage, etc., in order to achieve some optimum property such as configuration or location. The premise is to efficiently search the solution space by swarming the particles toward the best-fit solution encountered in previous epochs with the intent of encountering better solutions through the course of the process and eventually converging on a single minimum error solution.

Similar to EAs, conventional PSO begins with a random population of individuals, here termed a swarm of particles. As with EAs, each particle in the swarm is a different possible set of the unknown parameters to be optimized; therefore the particles' parameters can be real-valued or encoded, depending on the particular circumstances.

The conventional PSO algorithm begins by initializing a random swarm of M particles, each having R unknown parameters to be optimized. At each epoch, the fitness of each particle is evaluated according to the selected fitness function. The algorithm stores and progressively replaces the most-fit parameters of each particle ($pbest_i$, $i = 1, 2, \ldots, M$) as well as a single most-fit particle ($gbest$) as better fit parameters are encountered. The parameters of each particle (p_i) in the swarm are updated at each epoch (n) according to the following equations.

$$\overline{vel}_i(n) = w * \overline{vel}_i(n-1) + acc_1 * \mathrm{diag}[e_1, e_2, \ldots, e_R]_{i1} * (gbest - p_i(n-1))$$
$$+ acc_2 * \mathrm{diag}[e_1, e_2, \ldots, e_R]_{i2} * (pbest_i - p_i(n-1)) \tag{14.133}$$

$$p_i(n) = p_i(n-1) + \overline{vel}_i(n) \tag{14.134}$$

where $\overline{vel}_i(n)$ is the velocity vector of particle i, e_r is a vector of random values ϵ (0, 1), acc_1 and acc_2 are the acceleration coefficients toward $gbest$, $pbest_i$ respectively, and w is the inertia weight.

It can be gathered from the update equations that the trajectory of each particle is influenced in a direction determined by the previous velocity and the location of $gbest$ and $pbest_i$. Each particle's previous position ($pbest_i$) and the swarm's overall best position ($gbest$) are meant to represent the notion of individual experience memory and group knowledge of a "leader or queen," respectively, that emerges during the natural swarming process. The acceleration constants are typically chosen in the range ϵ (0, 2) and serve dual purposes in the algorithm. For one, they control the relative influence toward $gbest$ and $pbest_i$ respectively, by scaling each resulting distance vector, as illustrated for a two-dimensional case in Figure 14.32. Second, the two acceleration coefficients combined form what is analogous to the step size of an adaptive algorithm. Acceleration coefficients closer to 0 will produce fine searches of a region, while coefficients closer to 1 will result in lesser exploration and faster convergence. Setting the acceleration greater than 1 allows the particle to possibly overstep $gbest$ or $pbest$, resulting in a broader search. The random e_i vectors have R different components, which are randomly chosen from a uniform distribution in the interval (0, 1). This allows the particle to take constrained randomly directed steps in a bounded region between $gbest$ and $pbest_i$, as shown in Figure 14.32.

A single particle update is graphically illustrated in two dimensions in Figure 14.33. The new particle coordinates can lie anywhere within the bounded region, depending upon the weights and random components associated with each vector. The particle update bounds in Figure 14.33 are basically composed of all of the bounded regions for each vector, as shown in Figure 14.32, with the addition of the nonrandom

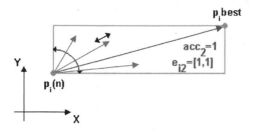

FIGURE 14.32   Example of scaling and random direction bounds for the vectors.

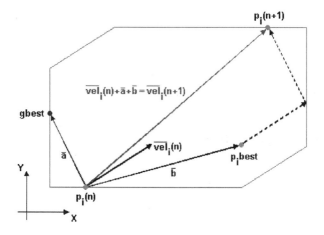

FIGURE 14.33   Example of the possible search region for a single particle.

velocity component. When a new *gbest* is encountered during the update process, all other particles begin to swarm toward the new *gbest*, continuing the directed global search along the way.

The search regions continue to decrease as new *pbest$_i$*'s are found within the search regions. When all of the particles in the swarm have converged to *gbest*, the *gbest* parameters characterize the minimum error solution obtained by the algorithm.

One of the key advantages of PSO is the ease of implementation in both the context of coding and parameter selection. This is much simpler and intuitive to implement than complex, probability-based selection and mutation operators required for evolutionary algorithms.

Modified Particle Swarm Optimization

Certain enhancements of the standard PSO algorithm are known to improve the overall efficiency and performance of the stochastic search procedure. The inclusion of three particular enhancements provides substantial performance improvements in virtually all variations of PSO. These three enhancements are (1) mutation, (2) re-randomization about *gbest*, and (3) adaptive inertia operations. When these are added to the basic PSO algorithm, the result is referred to as the modified PSO (MPSO). The MPSO algorithm is designed to balance convergence speed and search quality trade-offs, and by so doing provides significantly improved performance compared to conventional PSO.

Illustrative Examples

In these examples the properties of the aforementioned algorithms are demonstrated for both IIR and nonlinear adaptive filters in a system identification configuration, using a windowed mean squared error between the desired training signal and the adaptive filter output as the fitness function. The algorithms were initialized with the same population of real-valued parameters and allowed to evolve. The population sizes were selected to provide reasonable performance, while revealing the performance characteristics of each algorithm. The window length was set to 100 in each case. Unless specified otherwise, the input signals are zero-mean Gaussian white noise with unit variance. For each simulation, the MSE is averaged over 50 independent Monte-Carlo trials.

Matched High-Order IIR Filter, White Noise Input

In this example, the following five algorithms were compared:

Algorithm	Description
PSO	The conventional PSO algorithm
MPSO	The modified PSO algorithm, with matching PSO parameters
PSO–LMS	The PSO–LMS hybrid algorithm, with parameters matching the conventional PSO
CON–LMS	The congregational LMS algorithm is implemented where a population of independent LMS estimates is updated at each epoch
GA	The genetic algorithm using a ranked elitist strategy

The PSO algorithm is the basic PSO algorithm described by Equations (11.133) and (11.134), while the MPSO is a modified PSO algorithm that includes the three enhancements described earlier (rerandomization, mutation, and an adaptive inertial parameter). The PSO–LMS is a hybrid algorithm that combines the standard PSO with an LMS update term in an attempt to pull the PSO solution toward the global minimum. The CON–LMS constitutes the simultaneous operation of an entire population of LMS filters, each started with different random initial conditions. This population-based LMS algorithm was used to place the computational complexity of the LMS and the other algorithms on the same order, although no swarm intelligence is included in this algorithm. Finally, the GA algorithm is the standard genetic algorithm that forms a baseline against which to compare the performance of the others.

The unknown system to be identified in this example is a fifth-order, low-pass Butterworth filter (Mars, 1996).

$$H_{\text{PLANT}}(z^{-1}) = \frac{0.1084 + 0.5419z^{-1} + 1.0837z^{-2} + 1.0837z^{-3} + 0.5419z^{-4} + 0.1084z^{-5}}{1 + 0.9853z^{-1} + 0.9738z^{-2} + 0.3864z^{-3} + 0.1112z^{-4} + 0.0113z^{-5}}$$

(14.135)

$$H_{\text{AF}}(z^{-1}) = \frac{p_i^1 + p_i^2 z^{-1} + p_i^3 z^{-2} + p_i^4 z^{-3} + p_i^5 z^{-4} + p_i^6 z^{-5}}{1 + p_i^7 z^{-1} + p_i^8 z^{-2} + p_i^9 z^{-3} + p_i^{10} z^{-4} + p_i^{11} z^{-5}}$$

(14.136)

The learning curves for this example are given in Figure 14.34 and Figure 14.35. The simulations having the larger population illustrate the case where the population size is sufficient and all of the algorithms rarely become trapped in a local minimum.

Nonlinear LNL System Identification

In this example, the identification of an LNL nonlinear system taken from Mathews (2000) is performed using an unmatched LNL adaptive filter. The LNL plant consists of a fourth-order Butterworth low-pass filter (Equation (14.137)), followed by a fourth-power memoryless nonlinear operator, followed by a fourth-order Chebyshev lowpass filter (Equation (14.138)). This system is a common model for satellite communication systems in which the linear filters model the dispersive transmission paths to and from the satellite, and the

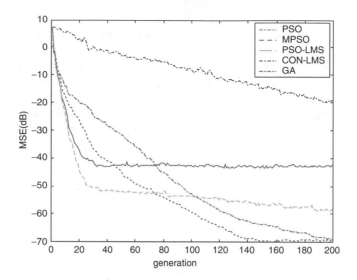

FIGURE 14.34   Population = 50.

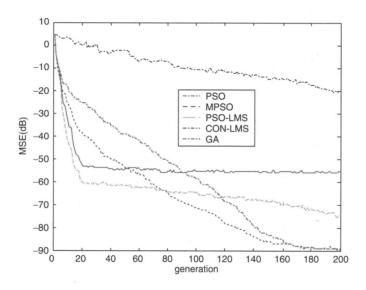

FIGURE 14.35   Population = 100.

nonlinearity models the traveling wave tube (TWT) transmission amplifiers operating near the saturation region:

$$H_B(z^{-1}) = \frac{(0.2851 + 0.5704z^{-1} + 0.2851z^{-2})(0.2851 + 0.5701z^{-1} + 0.2851z^{-2})}{(1 - 0.1024z^{-1} + 0.4475z^{-2})(1 - 0.0736z^{-1} + 0.0408z^{-2})} \qquad (14.137)$$

$$H_C(z^{-1}) = \frac{(0.2025 + 0.288z^{-1} + 0.2025z^{-2})(0.2025 + 0.0034z^{-1} + 0.2025z^{-2})}{(1 - 1.01z^{-1} + 0.5861z^{-2})(1 - 0.6591z^{-1} + 0.1498z^{-2})} \qquad (14.138)$$

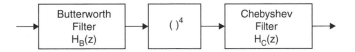

FIGURE 14.36 Nonlinear LNL system to be identified.

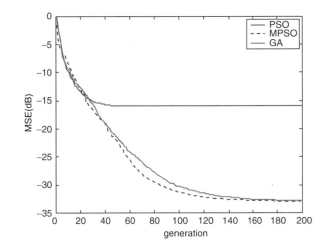

FIGURE 14.37 Population $= 50$.

The LNL adaptive filter structure is given as follows:

$$H_B(z^{-1}) = \frac{p_i^1 + p_i^2 z^{-1} + p_i^3 z^{-2} + p_i^4 z^{-3}}{1 + p_i^5 z^{-1} + p_i^6 z^{-2} + p_i^7 z^{-3}} \tag{14.139}$$

$$\text{nonlinearity} = p_i^8 [H_B(z^{-1})]^4 \tag{14.140}$$

$$H_C(z^{-1}) = \frac{p_i^9 + p_i^{10} z^{-1} + p_i^{11} z^{-2} + p_i^{12} z^{-3}}{1 + p_i^{13} z^{-1} + p_i^{14} z^{-2} + p_i^{15} z^{-3}} \tag{14.141}$$

The learning characteristics for this example are shown in Figure 14.37 for the PSO, MPSO, and GA algorithms. Note that both the GA and MPSO algorithms are effective in quickly finding the minimum mean squared error solution, while the conventional PSO algorithm stagnates at an error condition that is considerably above the minimum mean squared error solution. Note that in this example the minimum mean squared error is approximately -33 dB, because the adaptive filter is of insufficient order to accurately match the unknown system.

The results of these two examples demonstrate that the structured stochastic algorithms are capable of quickly and effectively adapting the coefficients of both IIR and nonlinear structures. From the simulation results, it is observed that, with a sufficient population size, all of the structured stochastic algorithms are capable of converging rapidly to below -20 dB in most instances, which is an order of magnitude faster than most existing gradient-based techniques. In all cases, the congregational LMS algorithm exhibits the slowest convergence rate due to the fact that there is no information transfer between the estimates. The LMS algorithm is capable of eventually attaining the noise floor when the number of generations is increased, assuming that they are not trapped in a local minimum.

Since the GA does not have an explicit step size, the convergence rate can only be controlled to a limited extent through the crossover and mutation operations, and the algorithm must evolve at its own intrinsic rate. Because of the nature of the algorithm, these GA operators become increasingly taxed as the population decreases, resulting in depreciating performance. However, as the population size increases, the performance gap between the GA and MPSO begins to diminish for large parameter spaces.

Though conventional PSO exhibits a fast convergence initially, it fails to improve further because the swarm quickly becomes stagnant, converging to a suboptimal solution in all instances. However, with the same set of algorithm parameters, the MPSO particles do not stagnate, allowing it to reach the noise floor. Smaller acceleration coefficients can be used with conventional PSO to allow it to approach the noise floor, forsaking the rapid convergence rate. With the introduction of a simple mutation-type operator, adaptive inertia weights, and rerandomization, MPSO can retain the favorable convergence rate with a smaller population while still achieving the noise floor and avoiding local minima. This can offer considerable savings in cases where computational complexity is an issue.

Defining Terms

Adaptive linear prediction: Process by which the value of a signal $s(n)$ at time n is predicted by a linear combination of N past values $s(n-1)$, $s(n-2)$, ..., $s(n-N)$ while minimizing the mean squared error between the predicted $\hat{s}(n)$ and the true value of $s(n)$.

Adaptive noise canceler: An adaptive filter configuration that uses the adaptive filter to estimate an additive noise that corrupts an information signal and removes the optimal mean squared estimate of the noise component from the contaminated information bearing signal.

Adaptive echo canceler: An adaptive filter configuration that uses the adaptive filter to estimate an echo leakage path so that an optimal mean squared estimate of the echo signal can be removed from the contaminated information bearing signal.

Evolutionary algorithm: A structured stochastic optimization algorithm based on a population of solutions that evolve from one generation to the next based on a set of principles rooted in evolution theory such as natural selection, survival of the fittest, and mutation.

Finite impulse response (FIR) adaptive filter: A class of adaptive digital filters whose unit pulse response consists of a finite number of nonzero values.

Genetic algorithm: A type of evolutionary algorithm that attempts to mimic the laws of genetics while trying to achieve some optimal property through the principles of evolution.

Infinite impulse response (IIR) adaptive filter: A class of adaptive digital filters whose unit pulse response consists of an infinite number of nonzero values.

Least mean squares (LMS) algorithm: A popular algorithm used to adjust the tap weights of an adaptive filter based on an approximation to a steepest decent strategy whereby the mean squared error is approximated by the instantaneous squared error.

LNL nonlinear system model: A representation of a nonlinear system that consists of a series cascade connection of a linear module, a memoryless nonlinear module, and a second linear module.

Mean square error (MSE) surface: A surface in an $N+1$-dimensional hyperspace that represents the expected value of the squared error as a function of N parameters that characterize the related adaptive system.

Particle swarm optimization (PSO) algorithm: A structured stochastic optimization algorithm that attempts to mimic the behavior of swarming insects while trying to achieve some optimal property through the principles of swarm intelligence.

Structured stochastic search algorithm: A stochastic search algorithm that is structured so as to intelligently generate and modify the randomized estimates in a manner that efficiently searches the error space based on some previous or collective information generated by the search.

System identification: A method of adaptively modeling an unknown system by driving the unknown system and the adaptive system with the same excitation signal, comparing their outputs to form an error signal, and then iteratively adjusting the parameters of the adaptive system until the error signal is minimized according to some prespecified criterion.

Training signal: A signal against which the output of an adaptive system is compared in order to form an error signal to be minimized by the learning process.

Weiner solution: The set of parameters that characterize an adaptive system when the output error has been minimized according to the mean squared criterion.

References

E. Aarts and J. Korst, *Simulated Annealing and Boltzmann Machines*, New York: Wiley, 1989.

R.C. Eberhart and J. Kennedy, "A new optimizer using particle swarm theory," *Proc. Sixth Int. Symp. Micromachine Human Sci.*, Nagoya, Japan, pp. 39–43, 1995.

S.R. Paulo, *Adaptive Filtering*, Dordecht, Germany: Kluwer, 2002.

A.P. Engelbrecht, *Computational Intelligence: An Introduction*, New York: Wiley, 2002.

B. Farhang-Boroujeny, *Adaptive Filters Theory and Applications*, New York: Wiley, 1999.

D.E. Goldberg, *Genetic Algorithms in Search, Optimization, and Machine Learning*, Reading, MA: Addison-Wesley, 1989.

S. Haykin, *Adaptive Filter Theory*, 4th ed., Englewood Cliffs, NJ: Prentice-Hall, 2002.

W.K. Jenkins, A.W. Hull, J.C. Strait, B.A. Schnaufer, and X. Li, *Advanced Concepts in Adaptive Signal Processing*, Dordecht, Germany: Kluwer, 1996.

J. Kennedy, R.C. Eberhart, and Y. Shi, *Swarm Intelligence*, San Francisco, CA: Morgan Kaufmann Publishers, 2001.

P. Mars, J.R. Chen, and R. Nambiar, *Learning Algorithms: Theory and Applications in Signal Processing, Control, and Communications*, Boca Raton, FL: CRC Press, 1996.

V.J. Mathews and G.L. Sicuranze, *Polynomial Signal Processing*, New York: Wiley, 2000.

A.E. Nordsjo and L.H. Zetterberg, "Identification of certain time-varying nonlinear Wiener and Hammerstein systems," *IEEE Trans. Signal Proc.*, vol. 49, no. 3, pp. 577–592, 2001.

K.S. Tang, K.F. Man, and Q. He, "Genetic algorithms and their applications," *IEEE Signal Process. Mag.*, pp. 22–37, 1996.

B. Widrow and S.D. Stearns, *Adaptive Signal Processing*. Englewood Cliffs, NJ: Prentice-Hall, 1985.

L. Yao and W.A. Sethares, "Nonlinear Parameter Estimation via the Genetic Algorithm," *IEEE Trans. Signal Process.*, 42, April, 1994.

15

Speech Signal Processing

Jerry D. Gibson
University of California

Bo Wei
Southern Methodist University

Hui Dong
University of California

Yariv Ephraim
George Mason University

Israel Cohen
Israel Institute of Technology

Jesse W. Fussell
U.S. Department of Defense

Lynn D. Wilcox
Rice University

Marcia A. Bush
Xerox Palo Alto Research Center

15.1 Coding, Transmission, and Storage

Jerry D. Gibson, Bo Wei, and Hui Dong

The goal of speech coding, or speech compression, is to represent speech in digital form with as few bits as possible while maintaining the intelligibility and quality required for the particular application. Interest in speech coding is motivated by the desire or requirement to conserve bit rate or bandwidth. There is always a tradeoff between lowering the bit rate while maintaining quality and intelligibility; however, depending on the application, many other constraints also must be considered, such as complexity, delay, and performance with bit errors or packet losses [1]. Two networks that have been developed primarily with voice communications in mind are the public switched telephone network (PSTN) and digital cellular networks. Additionally, with the pervasiveness of the Internet, voice over Internet protocol (VoIP) is growing rapidly and is expected to continue to do so over the near future. A new and powerful development for data communications is the emergence of wireless local area networks (WLANs) in the embodiment of the 802.11a, b, g standards, collectively referred to as Wi-Fi. Because of the proliferation and expected expansion of Wi-Fi networks, considerable attention is now being turned to voice over Wi-Fi, with some companies already offering proprietary networks, handsets, and solutions. Each of these networks has its own set of requirements, and these are discussed in the section "Networks for Voice Communications."

Speech and audio coding can be classified according to the bandwidth occupied by the input and the reproduced source. Narrowband or telephone bandwidth speech occupies the band from 200 to 3400 Hz, while wideband speech is contained in the range of 50 Hz to 7 kHz. High-quality audio is generally taken to cover the range of 20 Hz to 20 kHz. The discussion in this chapter addresses narrowband and wideband speech, with an emphasis on narrowband speech. The most common approaches to narrowband speech coding today primarily center around two paradigms, namely, waveform-following coders and analysis-by-synthesis methods. Waveform-following coders attempt to reproduce the time-domain speech waveform as accurately as possible, while analysis-by-synthesis methods utilize the linear prediction model and a perceptual distortion measure to reproduce only those characteristics of the input speech that are determined to be most important. Familiar waveform-following methods are logarithmic pulse code modulation (log-PCM) and adaptive differential pulse code modulation (ADPCM). The most common analysis-by-synthesis method is called code-excited linear prediction (CELP) [1–3]. General structures for speech coding are developed in the section "Narrowband Speech Coding Methods."

In many important applications, it is usual for the designated standards body to specify standards for speech coding in that particular application. There was almost an exponential growth of speech coding standards in the 1990s. For each of the networks examined in the section "Networks for Voice Communications," we discuss the most prominent speech coding standards and their characteristics.

In order to compare the various speech coding methods and standards, it is necessary to have methods for establishing the quality and intelligibility produced by a speech coder. It is a difficult task to find objective measures of speech quality, and often the only acceptable approach is to perform subjective listening tests. However, there have been some recent successes in developing objective quantities, experimental procedures, and mathematical expressions that have a good correlation with speech quality and intelligibility. We provide an overview of these methods in the section "Speech Quality and Intelligibility."

The chapter is completed by discussing current directions in speech coding, including variable rate coding in the section "Variable Rate Coding" and wideband speech coding in the section "Wideband Speech Coding."

Speech Quality and Intelligibility

To compare the performance of two speech coders, it is necessary to have some indicator of the intelligibility and quality of the speech produced by each coder. The term *intelligibility* usually refers to whether the output speech is easily understandable, while the term *quality* is an indicator of how natural the speech sounds. It is possible for a coder to produce highly intelligible speech that is low quality in that the speech may sound very machine-like and the speaker unidentifiable. However, it is unlikely that unintelligible speech would be called high quality, but there are situations in which perceptually pleasing speech does not have high intelligibility. We briefly discuss here the most common measures of intelligibility and quality used in formal tests of speech coders. We also highlight some newer performance indicators that attempt to incorporate the effects of the network on speech coder performance in particular applications [4,5,12].

MOS

The mean opinion score (MOS) is an often-used performance measure [4]. To establish a MOS for a coder, listeners are asked to classify the quality of the encoded speech in one of five categories: excellent (5), good (4), fair (3), poor (2), or bad (1). Alternatively, the listeners may be asked to classify the coded speech according to the amount of perceptible distortion present, i.e., imperceptible (5), barely perceptible but not annoying (4), perceptible and annoying (3), annoying but not objectionable (2), or very annoying and objectionable (1). The numbers in parentheses are used to assign a numerical value to the subjective evaluations, and the numerical ratings of all listeners are averaged to produce a MOS for the coder. A MOS between 4.0 and 4.5 usually indicates high quality.

It is important to compute the variance of MOS values. A large variance, which indicates an unreliable test, can occur because participants do not know what categories such as "good" and "bad" mean. It is sometimes useful to present examples of good and bad speech to the listeners before the test to calibrate the five-point scale. MOS values can and will vary from test to test and so it is important not to put too much emphasis on particular numbers when comparing MOS values across different tests. We will quote some MOS values for common speech coders in later sections.

EMBSD

A relatively new objective measure that has a high correlation with MOS is the enhanced modified bark spectral distance (EMBSD) measure [5]. The EMBSD is based on the bark spectral distance measure that relates to perceptually significant auditory attributes. A value of zero for the EMBSD indicates no distortion and a higher value indicates increasing distortion. The G.729 codec has been tested to have an EMBSD of 0.9, indicating low distortion in the reconstructed speech. The EMBSD values are often mapped into MOS values, since acceptable MOS values are more readily known.

DRT

The diagnostic rhyme test (DRT) was devised to test the intelligibility of coders known to produce speech of lower quality. Rhyme tests are so named because the listener must determine which consonant was spoken when presented with a pair of rhyming words; that is, the listener is asked to distinguish between word pairs such as meat–beat, pool–tool, saw–thaw, and caught–taught. Each pair of words differs on only one of six phonemic attributes: voicing, nasality, sustention, sibilation, graveness, and compactness. Specifically, the listener is presented with one spoken word from a pair and asked to decide which word was spoken. The final DRT score is the percent responses computed according to $P = (R-W) \times 100/T$, where R is the number correctly chosen, W is the number of incorrect choices, and T is the total of word pairs tested. Usually, $75 \leq DRT \leq 95$, with a "good" system being about 90.

DAM

The diagnostic acceptability measure (DAM) developed by Dynastat is an attempt to make the measurement of speech quality more systematic. For the DAM, it is critical that the listener crews be highly trained and repeatedly calibrated in order to get meaningful results. The listeners are each presented with encoded sentences taken from the Harvard 1965 list of phonetically balanced sentences, such as "Cats and dogs each hate the other" and "The pipe began to rust while new." The listener is asked to assign a number between 0 and 100 to characteristics in three classifications: signal qualities, background qualities, and total effect. The ratings of each characteristic are weighted and used in a multiple nonlinear regression. Finally, adjustments are made to compensate for listener performance. A typical DAM score is 45 to 55%, with 50% corresponding to a "good" system.

The perception of "good-quality" speech is a highly individual and subjective area. As such, no single performance measure has gained wide acceptance as an indicator of the quality and intelligibility of speech produced by a coder. Further, there is no substitute for subjective listening tests under the actual environmental conditions expected in a particular application.

PESQ

A new and important objective measure is the perceptual evaluation of speech quality (PESQ) method in ITU Recommendation P.862, which attempts to incorporate more than just speech codecs but also end-to-end network measurements [12]. The PESQ has been shown to have good accuracy for the factors listed in Table 15.1. It is clear that this is a very ambitious and promising testing method. There are parameters for which the PESQ is known to provide inaccurate predictions or is not intended to be used with, such as listening levels, loudness loss, effect of delay in conversational tests, talker echo, and two-way communications. The PESQ also has not been validated to test for packet loss and packet-loss concealment with PCM codecs, temporal and amplitude clipping of speech, talker dependencies, music as input to a codec, CELP and hybrid codecs < 4 kbits/sec, and MPEG-4 HVXC, among others.

The PESQ is being used more often but it is not freely available, so cost has been a hindrance to its widespread use so far.

TABLE 15.1 Factors for which the PESQ has Demonstrated Acceptable Accuracy

Test Factors
Speech input levels to a codec
Transmission channel errors
Packet loss and packet loss concealment with CELP codecs
Bit rates for multiple bit rate codecs
Transcodings
Environmental noise at the sending side
Effect of varying delay in listening only tests
Short-term time warping of the audio signal
Long-term time warping of the audio signal

Coding Technologies
Waveform codecs such as G.711, G.726, G.727
CELP and hybrid codecs at rates $\geq$ 4 kbits/sec, such As G.728, G.729, G.723.1
Other codecs: GSM-FR, GSM-HR, GSM-EFR, GSM-AMR, CDMA-EVRC, TDMA-ACELP, TDMA-VSELP, TETRA

Applications
Codec evaluation
Codec selection
Live network testing using a digital or analog connection to the network
Testing of emulated and prototype networks

The E-Model

Another relatively recent objective method for speech-quality evaluation is the E-Model in ITU Recommendation G.107 and G.108. The E-Model attempts to assess the "mouth-to-ear" quality of a telephone connection and is intended to be used in network planning. The E-Model has components for representing the effects of "equipment" and different types of impairments. The equipment effects can be mapped into packet losses so that the E-Model can be helpful in voice-over-IP evaluations.

Narrowband Speech Coding Methods

Logarithmic PCM (log PCM) and ADPCM are waveform-following speech coders and have found widespread applications. Log PCM at 64 kbits/sec is the speech codec (coder/decoder) used in the long-distance public switched telephone network at a rate of 64 kbits/sec. It is a simple coder and it achieves what is called toll quality, which is the standard level of performance against which all other narrowband speech coders are judged. Log PCM uses a nonlinear quantizer to reproduce low-amplitude signals, which are important to speech perception. There are two closely related types of log-PCM quantizer used in the world: μ-law, which is used in North America and Japan, and A-law, which is used in Europe, Africa, Australia, and South America. Both achieve toll-quality speech, and in terms of the MOS value it is usually between 4.0 and 4.5 for log-PCM [2].

ADPCM operates at 32 kbits/sec or lower, and it achieves performance comparable to log-PCM by using a linear predictor to remove short-term redundancy in the speech signal before quantization. The most common form of ADPCM uses what is called backward adaptation of the predictors and quantizers to follow the waveform closely. Backward adaptation means that the predictor and quantizer are adapted based upon past reproduced values of the signal that are available at the encoder and decoder. No predictor or quantizer parameters are sent along with the quantized waveform values (called forward adaptation). By subtracting a predicted value from each input sample, the dynamic range of the signal to be quantized is reduced, and hence, good reproduction of the signal is possible with fewer bits [1].

Analysis-by-synthesis (AbS) methods are a considerable departure from waveform-following techniques. The most common and most successful analysis-by-synthesis method is code-excited linear prediction (CELP). In CELP speech coders, a segment of speech is synthesized using the linear prediction model along with a long-term redundancy predictor for all possible excitations in what is called a codebook. For each excitation, an error signal is calculated and passes through a perceptual weighting filter. The excitation that produces the best perceptually weighted coded speech is selected for use at the decoder. Hence the name analysis-by-synthesis. The predictor parameters and the excitation codeword are sent to the receiver to decode the speech.

The analysis-by-synthesis procedure is very complicated, and it is fortunate that algebraic codebooks, which have mostly zero values and only a few nonzero pulses, have been discovered and work well [1,2].

Networks for Voice Communications

In this section, we describe the relevant details of current voice codecs in the context of the network for which they were designed. In particular, we develop the voice codecs for the PSTN, digital cellular networks, VoIP, and voice over wireless local area networks (voice over Wi-Fi).

The Public Switched Telephone Network (PSTN)

The most familiar form of network for voice communications is the PSTN, which consists of a wired, time division multiplexed (TDM), circuit-switched backbone network with (often) copper wire pair local loops [6]. The PSTN was designed and evolved with voice communications as a primary application. The voice codec most often used in the PSTN is 64 kbits/sec logarithmic pulse code modulation (log-PCM) designated by the ITU-T as G.711, and which is taken as the standard for toll-quality voice transmission. The TDM links in the PSTN are very reliable with bit error rates (BERs) of 10^{-6} to 10^{-9}. As a result, bit error concealment is not an issue for G.711 transmission over TDM PSTN links, even though bit errors in G.711 encoded voice generate very irritating "pops" in reconstructed speech. Furthermore, G.711 is designed with several asynchronous tandems in mind, since it was possible to encounter several analog switches during a long-distance telephone call prior to the mid-1980s. Even eight asynchronous tandems of G.711 with itself has been shown to still maintain a MOS greater than 4.0 when a single encoding is 4.4 to 4.5.

Other voice codecs have been standardized for the PSTN over the years. These include G.721 (now G.726), G.727, G.728, G.729, and G.729A for narrowband (telephone bandwidth) speech (200 to 3400 Hz) and G.722, G.722.1 [13], and G.722.2 [14] for wideband speech (50 Hz to 7 kHz). Table 15.2 summarizes the rate, performance, complexity, and algorithmic delay of each of the narrowband speech codec standards. It is emphasized that the MOS values shown are approximate since they are taken from several different sources. It is recognized that MOS values for a given codec can (and will) change from test to test, but the values given here provide an approximate ordering and indication of codec performance.

Note that for the waveform-based and low-delay codecs, the principal mode of error dissipation was not error concealment *per se*, but error dissipation due to fading the "memories" of the adaptive algorithms involved. Table 15.3 presents some results concerning the tandem performance of the narrowband speech codecs. Note that asynchronous tandem connections of these codecs with themselves do not cause an unacceptable loss in performance compared to a single encoding, although the MOS for four tandems of G.726 and three tandems of G.729 drops considerably.

Digital Cellular Networks

Digital cellular networks provide wireless voice connectivity by combining high-quality voice codecs, unequal forward error correction of sensitive bits, error detection and concealment of uncorrected errors, and interleaving [7,8]. Table 15.4 contains the rate, performance, complexity, and algorithmic delay for selected digital cellular speech codecs. As in previous tables, the MOS values and complexity in MIPS are representative

TABLE 15.2　Comparison of Voice Codecs for the PSTN

Codec	Rate (kbits/sec)	MOS	Complexity (MIPS)	Frame Size/Look-Ahead (msec)
G.711	64	4.0+	<<1	0.125
G.721/726	32	~4.0	1.25	0.125
G.728	16	3.9	30	0.625
G.729	8	4.0	20	10/5
G.729A	8	4.0	12	10/5
G.723.1	5.3/6.3	3.7/3.9	11	30/7.5

TABLE 15.3   Representative Asynchronous
Tandem Performance of Selected PSTN Codecs

Voice Codec	Mean Opinion Score (MOS)
G.711×4	>4.0
G.726×4	2.91
G.729×2	3.27
G.729×3	2.68
G.726+G.729	3.56
G.729+G.726	3.48

TABLE 15.4   Comparison of Selected Digital Cellular Voice Codecs

Codec	Rate (kbits/sec)	MOS	Complexity (MIPS)	Frame Size/ Look-Ahead (msec)
IS-641	7.4	4.09	14	20/5
IS-127-2 EVRC	8.55, 4.0, 0.8	3.82	20	20
GSM-EFR	12.2	4.26	14	20
NB-AMR	4.75–12.2	3.4–4.2	14	20/5
IS-893, cdma2000	8.5, 4.0, 2.0, 0.8	3.93 at 3.6 kbits/sec ADR	18	20

numbers taken from several sources. It is evident from Table 15.3 that the voice codecs have good performance in ideal conditions without large algorithmic delay or excessive complexity.

As mentioned before, the bits to be transmitted are typically classified into categories for unequal error protection, as well as for enabling different modes of error concealment for errors detected after forward error correction. In Table 15.4, the rate is for the voice codec only without error correction/detection, and the frame size/look-ahead numbers in the table do not include delay due to interleaving.

Table 15.5 shows available results for multiple tandem encodings of the codecs in Table 15.4, including results from tandem connections of these codecs with the PSTN backbone codecs in Table 15.2. It is clear that tandem encodings result in a drop in performance as seen in the lower MOS values. Furthermore, tandem encodings add to the end-to-end delay because of the algorithmic delays in decoding and reencoding. Tandem encodings are not discussed often within digital cellular applications since the codec for the backbone wireline network is often assumed to be G.711. However, it is recognized that tandem encodings with codecs other than G.711 can lead to a loss in performance and that tandem encodings constitute a significant problem for end-to-end voice quality. In particular, transcoding at network interfaces and source coding distortion accumulation due to repeated coding have been investigated with the goal of obtaining a transparent transition between certain speech codecs. Some system-wide approaches also have been developed for specific networks. The general tandeming/transcoding problem remains open.

TABLE 15.5   Representative Asynchronous Tandem Performance
of Selected Digital Cellular Codecs

Voice Codec	Mean Opinion Score (MOS)
IS-641×2	3.62
GSM-EFR×2	4.13
IS-641 + G.729	3.48
GSM-FR + G.729	3.05
GSM-EFR + G.729	3.53
GSM-EFR + G.729+G.729	3.21
IS-641+G.729 + G.729	3.10

Digital cellular networks perform exceedingly well given the difficulty of the task. The melding of voice coder design, forward error correction and detection, unequal error protection, and error concealment in digital cellular has important lessons for designing voice communications systems for VoIP and voice over Wi-Fi.

Voice Over Internet Protocol (VoIP)

VoIP has been evolving for more than ten years, but it is now projected finally to be taking off as a voice communications service [9]. Among the issues in developing good VoIP systems are voice quality, latency, jitter, packet loss performance, packetization, and the design of the network. Broadly speaking, the voice codec in VoIP systems should achieve toll or near toll quality, have as low a delay as possible, and have good resilience to packet loss. ITU-T Recommendation G.114 provides specifications for delay when echo is properly controlled. In particular, one-way transmission time (including processing and propagation delay) is categorized as: (a) 0 to 150 msec; acceptable for most user applications; (b) 150 to 400 msec; acceptable depending upon the transmission time impact; (c) above 400 msec; unacceptable for general network planning purposes.

Furthermore, guidelines are given in G.114 for calculating the delay based upon the codec frame size and look ahead and the application. For example, it is assumed that there is a delay of one frame before processing can begin and then that the time required to process a frame of speech is the same as the frame length. As a result, if there is no additional delay at the interface of the codec and the network, the mean one-way delay of a codec is $2 \times$ frame size $+$ look-ahead. Thus, for G.729 the codec delay is 25 msec, and for G.723.1, the codec delay is 67.5 msec. Another one frame delay may be incurred due to clocking out bits to match the network in a wireline environment or due to placing more than one frame per packet. From this calculation, it is evident that frame size is a critical parameter, and the delay totals that appear in publications with respect to the standard codecs may be one, two, or three frames plus look-ahead. It is noted that none of these delays includes a delay due to a jitter buffer.

Interestingly, voice codecs used in prominent VoIP products are all ported from previous standards and other applications. Some of the earliest VoIP applications used G.723.1, which was originally intended for video telephony, but the relatively long frame size and look-ahead and the somewhat lower quality led developers to consider other alternatives. Today's VoIP product offerings typically include G.711, G.729, and G.722, in addition to G.723.1. See Table 15.6 for a summary of the relevant properties offered by each coder. G.723.1 is often favored for videophone applications since the delay of the video coding, rather than the voice codec, sets the delay of the videophone operation.

The coders in Table 15.6, as a set, offer a full range of alternatives in terms of rate, voice quality, complexity, and delay. What is not evident from this table is how effectively one can conceal packet losses with each of these coders. Packet-loss concealment is particularly important since in order to reduce latency, retransmissions are not allowed in VoIP.

Rather recently, a packet-loss concealment algorithm has been developed for G.711. Based upon 10-msec packets and assuming the previous frame was received correctly, the method generates a synthesized or concealment waveform from the last pitch period with no attenuation. If the next packet is lost as well, the method uses multiple pitch periods with a linear atttentuation at a rate of 20% per 10 msec. After 60 msec, the synthesized signal is zero.

G.729 and G.723.1 suffer from the problem that the predictor parameters (line spectral frequencies) are predictively encoded from frame to frame. For G.729, the basic approach to packet-loss concealment if a

TABLE 15.6 Properties of Common VoIP Codecs

Codec	Relevant Properties
G.711	Low delay, toll quality, low complexity, higher rate
G.729	Toll quality, acceptable delay, low rate, acceptable complexity
G.723.1	Low rate, acceptable quality, relatively high delay
G.722	Wideband speech, low delay, low complexity, higher rate

single 10-msec frame is erased is: (i) generate a replacement excitation based upon the classification of the previous frame as voiced or unvoiced, (ii) repeat the synthesis filter parameters from the previous frame, and (iii) attenuate the memory of the gain predictor. It seems intuitive that a speech codec that allowed interpolation of erased frames would perform better than using only past information, and this is indeed true. However, the frame-to-frame predictive coding of the short-term predictor parameters in precludes using interpolation. Note, however, that interpolation implies additional delay in reconstructing the speech, and so the performance improvement provided by interpolation schemes must include the effects of any additional delay.

Voice over Wi-Fi

Wireless local area networks (WLANs), such as 802.11b, 802.11a, and 802.11g, are becoming extremely popular for use in businesses, homes, and public places. As a result, there is considerable interest in developing VoIP for Wi-Fi, which we designate here as voice over Wi-Fi. The issues involved for voice over Wi-Fi would seem to be very much the same as for VoIP over the Internet; and it is certainly true that voice quality, latency, jitter, packet-loss performance, and packetization all play a critical role. However, the physical link in Wi-Fi is wireless, and as a result, bit errors will commonly occur and this, in turn, affects link protocol design and packet loss concealment.

One issue in particular is how to handle packets with bit errors. One approach would be to detect bit errors in a packet using a cyclic redundancy check (CRC) error detection code, and request a retransmission if a bit error is detected in the packet. In fact, the IEEE 802.11 MAC layer defines two different access methods. One method, called the distributed coordination function (DCF), is basically carrier sense multiple access with collision avoidance (CSMA/CA), and the basic access scheme is, if the channel is sensed to be idle, the node starts its transmissions. A CRC is computed over the entire received packet and an acknowledgment is sent if the packet is error-free. If an error is detected in the packet, a retransmission is requested. Up to seven retransmissions for short packets and up to four retransmissions for large packets are allowed. This method clearly adds to latency and is in contrast to avoiding retransmissions altogether in Internet VoIP, but how does one deal with bit errors? The answer lies in a combination of those techniques used in digital cellular in conjunction with different packetization and packet-loss concealment methods. This is an area for current research and development.

While voice over Wi-Fi is projected to be an exponentially growing market in the next five years, it is just now in its formative stages; however, there are some proprietary products and systems being offered. We mention only two here. First, Cisco announced their Wireless IP Phone 7920 for 802.11b. The voice codecs available in this phone are G.711 and G.729A. Spectralink offers voice over Wi-Fi voice systems for businesses. This system also operates over 802.11b and uses G.711 and G.729A as the candidate voice codecs. Additionally, Spectralink implements a proprietary protocol that gives voice priority over data within the confines of the 802.11 standard. In particular, their protocol specifies zero back-off after each packet transmission if the next packet is voice (this is in contrast to the requirement of random back-off after each transmission, which would result in variable delays of packets). Further, a priority queuing method (not specified in 802.11) is used to push voice packets to the head of the transmission queue.

Numerous research efforts have been conducted to analyze and improve the performance and capacity of the IEEE 802.11 MAC protocol for WLANs and real-time traffic. An efficient way to transmit voice over WLANs is to employ a reservation scheme that guarantees delay and bandwidth. Work is underway on a new standard, 802.11e, which is designed to support delay-sensitive applications with multiple managed levels of QoS (quality of service) for data, voice, and video.

Variable Rate Coding

For more than 30 years, researchers have been interested in removing silent periods in speech to reduce the average bit rate [10]. This was successfully accomplished for some digital cellular coders where silence was removed and coded with a short length code and then replaced at the decoder with "comfort noise." Comfort noise is needed because the background sounds for speech coders are seldom pure silence and inserting pure silence generates unwelcome artifacts at the decoder. The result, of course, is a variable rate speech coder. One of the first variable rate speech coders with more than a silence removal mode was the Qualcomm IS-96 QCELP coder, which operated at the rates of 0.8, 2, 4, and 8 kbits/sec, depending upon the classification of the input speech. This coder was part of the CDMA standard in North America, but it did not have good performance even at the highest supported rate, achieving an MOS of about 3.3. A replacement coder for IS-96 QCELP is the IS-127 Enhanced Variable Rate Coder (EVRC) that has three possible bit rates of 0.8, 4, and 8 kbits/sec, depending upon voice activity or a command from the network. The IS-127 EVRC coder achieves an MOS of about 3.8 at the highest rate of 8 kbits/sec, but is not operated at lower average data rates because of low voice quality.

A more recent variable rate speech codec is the IS-893 Selectable Mode Vocoder (SMV), which has six different modes that produce different average data rates and voice quality. The highest quality mode, Mode 0, can achieve a higher MOS than the IS-127 EVRC at an average data rate of 3.744 kbit/sec, and Mode 1 also typically outperforms IS-127. The SMV coder is part of the IS-893 cdma2000 standard.

Wideband Speech Coding

Even though we are quite comfortable communicating using telephone bandwidth speech (200 to 3400 Hz), there has been considerable recent interest in compression methods for wideband speech covering the range of 50 Hz to 7 kHz [1,11]. The primary reasons for this interest are that wideband speech improves intelligibility, naturalness, and speaker identifiability. Originally, the primary application of wideband speech coding was to videoconferencing, and the first standard, G.722, separated the speech into two subbands and used ADPCM to code each band. The G.722 codec is relatively simple and produces good quality speech at 64 kbits/sec, and lower quality speech at the two other possible codec rates of 56 and 48 kbits/sec. The G.722 speech codec is still widely available in the H.323 videoconferencing standard, and it is often provided as an option in VoIP systems.

Two recently developed wideband speech coding standards, designated as G.722.1 and G.722.2, utilize coding methods that are quite different from G.722, as well as completely different from each other. The G.722.1 standard employs a filter bank/transform decomposition called the modulated lapped transform (MLT) and operates at the rates of 24 and 32 kbits/sec. A categorization procedure is used to determine the quantization step sizes and coding parameters for each region. The coder has an algorithmic delay of 40 msec, which does not include any computational delay. Since G.722.1 employs filter bank methods, it performs well for music and less well for speech. In one MOS test using British English, the G.722.1 coder at 24 kbits/sec achieved an MOS value of 4.1.

G.722.2 is actually an ITU-T designation for the adaptive multirate wideband (AMR-WB) speech coder standardized by the 3GPP. This coder operates at rates of 6.6, 8.85, 12.65, 14.25, 15,85, 18.25, 19.85, 23.05, and 23.85 kbits/sec and is based upon an algebraic CELP (ACELP) analysis-by-synthesis codec. Since ACELP utilizes the linear prediction model, the coder works well for speech but less well for music, which does not fit the linear prediction model. As noted in Table 15.7, G.722.2 achieves good speech quality at rates greater than 12.65 kbits/sec and performance equivalent to G.722 at 64 kbits/sec with a rate of 23.05 kbits/sec and higher. For speech, one MOS test for the French language showed the G.722.2 codec achieving an MOS value of 4.5 at the 23.85 kbits/sec rate and an MOS value of around 4.2 for the 12.65 kbits/sec rate.

TABLE 15.7 Characteristics of Some Wideband Speech Coding Standards

Standard	Bit Rate (kbit/sec)	Coding Method	Quality	Frame Size/ Look-Ahead	Complexity
G.722	48, 56, 64	Subbband ADPCM	Commentary grade		10 MIPS
G.722.1	24 and 32	Transform	Good music/poorer for speech		<15 MIPS
G.722.2	23.85	ACELP	Good speech/poor music		<40 MIPS

The MPEG-4 Natural Audio Coding Tool

The MPEG-4 audio coding standard specifies a complete toolbox of compression methods for everything including low bit rate narrowband speech, wideband speech, and high-quality audio. It offers several functionalities not available in other standards as well, such as bit rate scalability (also called SNR scalability) and bandwidth scalability. We provide an overview here with an emphasis on narrowband and wideband speech coding. Tables 15.8 and 15.9 summarize the many options available in the MPEG-4 toolbox [15].

The harmonic vector excitation coder (HXVC) tool performs a linear prediction analysis and calculates the prediction error signal. The prediction error is then transformed into the frequency domain where the pitch and envelope are analyzed. The envelope is quantized using vector quantization for voiced segments and a search for an excitation is performed for unvoiced speech.

The CELP coder in the MPEG-4 toolbox utilizes either a multipulse excitation (MPE) or a regular pulse excitation (RPE), which were both predecessors of code-excited systems. The MPE provides for better quality but it is more complicated than RPE. The coder also uses a long-term pitch predictor rather than an adaptive codebook.

Note from Table 15.9 that there are 28-bit rates from 3850 bits/sec to 12.2 kbits/sec for narrowband speech, and 30-bit rates from 10.9 kbits/sec to 23.8 kbits/sec for wideband speech. The larger changes in rate come about by changes in the frame size, which, of course, leads to greater delay.

The functionalities built into the HVXC and CELP speech coding tools are impressive. The HVXC speech coder has a multibit rate capability and a bit rate scalability option. For multibit rate coding, the bit rate is chosen from a set of possible rates upon call setup. For HVXC, there are only two rates, 2 kbits/sec and 4 kbits/sec. For CELP, bit rates are selectable in increments as small as 200 bits/sec. Bit rate scalability can be useful in multicasting as well as many other applications. The bit rate scalability options in MPEG-4 are many indeed. The HVXC and CELP coders can be used to generate core bit streams that are enhanced by their own coding method or by one of the other coding methods available in the MPEG-4 natural audio coding tool. In bit rate scalability, the enhancement layers can be added in rate increments of 2 kbits/sec for narrowband speech, and in increments of 4 kbits/sec for wideband speech. For bandwidth scalable coding, the enhancement bit stream increments depend upon the total coding rate. Table 15.10 summarizes the enhancement bit streams for bandwidth scalability in relation to the core bit stream rates.

The speech quality produced by the MPEG-4 natural audio coding tool is very good in comparison to other popular standards, especially considering the range of bit rates covered. For example, at 6 kbits/sec the MPEG-4 tool produces an MOS comparable to G.723.1, and at 8.3 kbits/sec, the MOS value achieved is

TABLE 15.8   Specifications of the HVXC Speech Coding Tool

Sampling Frequency	8 kHz
Bandwidth	300–3400 Hz
Bit Rate (bits/sec)	2000 and 4000
Frame Size	20 msec
Delay	33.5–56 msec
Features	Multibit rate coding/bit rate scalability

TABLE 15.9   Specifications of the CELP Speech Coding Tool

Sampling Frequency	8 kHz	16 kHz
Bandwidth	300–3400 Hz	50–7000 Hz
Bit Rate (bits/sec)	3850–12,200 (28-bit rates)	10,900–23,800 (30-bit rates)
Frame Size	10–40 msec	10–20 msec
Delay	15–45 msec	15–26.75 msec
Features	Multibit rate coding/bit rate scalability/bandwidth scalability	

TABLE 15.10   Bandwidth Scalable Bit Rate Options

Core Bit Rate (bits/sec)	Enhancement Layer Bit Rates (bits/sec)
3850–4650	9200, 10,400, 11,600, 12,400
4900–5500	9467, 10,667, 11,867, 12,667
5700–10,700	10,000, 11,200, 12,400, 13,200
11,000–12,200	11,600, 12,800, 14,000, 14,800

comparable to G.729, while at 12 kbits/sec, it performs as well as the GSM EFR at 12.2 kbits/sec. The bit rate scalable modes perform slightly poorer than G.729 at 8 bits/sec and the GSM EFR at 12 kbits/sec. Thus, as should be expected, bit rate scalability functionality extracts a penalty in coder performance.

Summary and Conclusions

Speech coding has become an integral part of our communications backbone services. New speech coders continue to be designed and standardized. Current efforts emphasize functionalities such as bit-rate scalability, bandwidth scalability, and selectable multiple bit rates. As VoIP becomes more important and voice over Wi-Fi is introduced, new challenges in terms of asynchronous tandem connections of speech coders move to the forefront. Also, reducing latency and jitter and improving packet-loss concealment methods are paramount to maintaining voice services at the level we have become to expect.

Defining Terms

Analysis-by-synthesis: Constructing several versions of a waveform and choosing the best match.

Asynchronous tandeming: A series connection of speech coders that requires digital-to-analog conversion followed by resampling and reencoding.

Bandwidth scalable coding: A core layer bit stream that represents narrowband speech that can be enhanced to a wideband signal using an incremental bit stream plus the core.

Bit-rate scalable coding: A core layer bit stream that represents speech that can be represented with greater accuracy by adding an incremental bit stream to the core layer.

Mean Opinion Score (MOS):   A popular method for classifying the quality of encoded speech based on a five-point scale.

Narrowband speech: A speech signal that occupies the band from 200 to 3400 Hz.

Predictive coding: Coding of time-domain waveforms based on a (usually) linear prediction model.

Standard: An encoding technique adopted by an industry to be used in a particular application.

Variable-rate coders: Coders that output different amounts of bits based on the time-varying characteristics of the source.

Wideband speech: A speech signal that occupies the band from 50 to 7000 Hz.

References

1. J.D. Gibson, T. Berger, T. Lookabaugh, D. Lindbergh, and R.L. Baker, *Digital Compression for Multimedia: Principles & Standards*, Los Altos, CA: Morgan-Kaufmann, 1998.

2. A.S. Spanias, "Speech coding standards," in *Multimedia Communications: Directions and Innovations*, J.D. Gibson, Ed., New York: Academic Press, 2001, pp. 25–44, chap. 3.

3. W.B. Kleijn and K.K. Paliwal, Eds., *Speech Coding and Synthesis*, Amsterdam, The Netherlands: Elsevier, 1995.

4. P. Kroon, "Evaluation of speech coders," chap. 13 in Ref. [3], pp. 467–494.

5. W. Yang, M. Benbouchta, and R. Yantorno, "Performance of the modified bark spectral distortion as an objective speech quality measure," *Proc. ICASSP*, Seattle, 1998, pp. 541–544.

6.   J.C. Bellamy, *Digital Telephony*, New York: Wiley, 2000.

7.   T.S. Rappaport, *Wireless Communications: Principles and Practice*, 2nd ed., Englewood Cliffs, NJ: Prentice-Hall, 2002.

8.   R. Steele and L. Hanzo, Eds., *Mobile Radio Communications*, 2nd ed., New York: Wiley, 1999.

9.   B. Goode, "Voice over Internet protocol (VoIP)," *Proc. IEEE*, 90, September, 1495–1517, 2002.

10.   A. Das, E. Paksoy, and A. Gersho, "Multimode and variable-rate coding of speech," chap. 7 in Ref. [3], pp. 257–288.

11.   J.-P. Adoul and R. Lefebvre, "Wideband speech coding," chap. 8 in Ref. [3], pp. 289–310.

12.   ITU-T Recommendation P. 862, "Perceptual evaluation of speech quality (PESQ), an objective method for end-to-end speech quality assessment of narrowband telephone networks and speech codecs," Feburary 2001.

13.   ITU, "Coding at 24 and 32 Kbit/s for hands-free operation in systems with low frame loss," September 1999.

14.   ITU-T Recommendation G.722.2, Wideband coding of speech at around 16 kbit/s using adaptive multi-rate wideband (AMR-WB), 2002.

15.   K. Brandenburg, O. Kunz, and A. Sugiyama, "MPEG-4 natural audio coding," *Signal Process. Image Commun.*, 15, 423–444, 2000.

15.2   Recent Advancements in Speech Enhancement

Yariv Ephraim and Israel Cohen

Speech enhancement is a long-standing problem with numerous applications ranging from hearing aids to coding and automatic recognition of speech signals. In this section we focus on enhancement from a single microphone, and assume that the noise is additive and statistically independent of the signal. We present the principles that guide researchers working in this area, and provide a detailed design example. The example focuses on minimum mean square error estimation of the clean signal's log-spectral magnitude. This approach has attracted significant attention in the past 20 years. We also describe the principles of a Monte-Carlo simulation approach for speech enhancement.

Introduction

Enhancement of speech signals is required in many situations in which the signal is to be communicated or stored. Speech enhancement is required when either the signal or its receiver is degraded. For example, hearing impaired individuals require enhancement of perfectly normal speech to fit their individual hearing capabilities. Speech signals produced in a room generate reverberations, which may be quite noticeable when a hands-free single channel telephone system is used and binaural listening is not possible. A speech coder may be designed for clean speech signals while its input signal may be noisy. Similarly, a speech recognition system may be operated in an environment different from that it was designed to work in. This short list of examples illustrates the extent and complexity of the speech enhancement problem.

Here, we focus on enhancement of noisy speech signals for improving their perception by human. We assume that the noise is additive and statistically independent of the signal. In addition, we assume that the noisy signal is the only signal available for enhancement. Thus, no reference noise source is assumed available. This problem is of great interest, and has attracted significant research effort for over 50 years. A successful algorithm may be useful as a preprocessor for speech coding and speech recognition of noisy signals.

The perception of a speech signal is usually measured in terms of its quality and intelligibility. The *quality* is a subjective measure that reflects on individual preferences of listeners. *Intelligibility* is an objective measure that predicts the percentage of words that can be correctly identified by listeners. The two measures are not correlated. In fact, it is well known that intelligibility can be improved if one is willing to sacrifice quality. This can be achieved, for example, by emphasizing high frequencies of the noisy signal [35]. It is also well known that improving the quality of the noisy signal does not necessarily elevate its intelligibility. On the contrary, quality improvement is usually associated with loss of intelligibility relative to that of the noisy signal. This is due to the distortion that the clean signal undergoes in the process of suppressing the input noise. From a pure information theoretic point of view, such loss in "information" is predicted by the *data processing theorem* [10]. Loosely speaking, this theorem states that one can never learn from the enhanced signal more than can be learnt from the noisy signal about the clean signal.

A speech-enhancement system must perform well for all speech signals. Thus, from the speech-enhancement system point of view, its input is a random process whose sample functions are randomly selected by the user. The noise is naturally a random process. Hence, the speech-enhancement problem is a statistical estimation problem of one random process from the sum of that process and the noise. Estimation theory requires statistical models for the signal and noise, and a distortion measure which quantifies the similarity of the clean signal and its estimated version. These two essential ingredients of estimation theory are not explicitly available for speech signals. The difficulties are with the lack of a precise model for the speech signal and a perceptually meaningful distortion measure. In addition, speech signals are not strictly stationary. Hence, adaptive estimation techniques, which do not require explicit statistical model for the signal, often fail to track the changes in the underlying statistics of the signal.

In this section, we survey some of the main ideas in the area of speech enhancement from a single microphone. We begin in the section "Statistical Models and Estimation" by describing some of the most promising statistical models and distortion measures which have been used in designing speech-enhancement systems. In the section "MMSE Spectral Magnitude Estimation," we present a detailed design example for a speech-enhancement system which is based on minimum mean squared error (MSE) estimation of the speech spectral magnitude. This approach integrates several key ideas from the section "Statistical Models and Estimation", and has attracted much attention in the past 20 years. In the section "Monte-Carlo Simulation", we present the principles of a Monte-Carlo simulation approach to speech enhancement. Some concluding comments are given in "Comments."

Statistical Models and Estimation

Enhancement of noisy speech signals is essentially an estimation problem in which the clean signal is estimated from a given sample function of the noisy signal. The goal is to minimize the expected value of some distortion measure between the clean and estimated signals. For this approach to be successful, a perceptually meaningful distortion measure must be used, and a reliable statistical model for the signal and noise must be specified. At present, the best statistical model for the signal and noise and the most perceptually meaningful distortion measure are not known. Hence, a variety of speech-enhancement approaches have been proposed. They differ in the statistical model, distortion measure, and in the manner in which the signal estimators are being implemented. In this section, we briefly survey the most commonly used statistical models, distortion measures, and the related estimation schemes.

Linear Estimation

Perhaps the simplest scenario is obtained when the signal and noise are assumed statistically independent Gaussian processes, and the MSE distortion measure is used. For this case, the optimal estimator of the clean signal is obtained by the Wiener filter. Since speech signals are not strictly stationary, a sequence of Wiener filters is designed and applied to vectors of the noisy signal. Suppose that Y_t and W_t represent, respectively, l-dimensional vectors from the clean signal and the noise process where $t = 0, 1, 2, \ldots$. Let $Z_t = Y_t + W_t$ denote the corresponding noisy vector. Let R_{Y_t} and R_{W_t} denote the covariance matrices of Y_t and W_t, respectively. Then, the minimum mean squared error (MMSE) estimate of the signal Y_t is obtained by applying the Wiener filter to the noisy signal Z_t as follows:

$$\hat{Y}_t = \left[R_{Y_t}(R_{Y_t} + R_{W_t})^{-1} \right] Z_t \tag{15.1}$$

Remarkably, this simple approach is one of the most effective speech-enhancement approaches known today. The key to its success is reliable estimation of the covariance matrices of the clean signal and of the noise process. Many variations on this approach have been developed and were nicely summarized by

Lim and Oppenheim [26]. When R_{Y_t} is estimated by subtracting an estimate of the covariance matrix of the noise vector, say $\hat{R}_{W_t}$, from an estimate of the covariance matrix of the noisy vector, say $\hat{R}_{Z_t}$, then the Wiener filter at time t becomes $(\hat{R}_{Z_t} - \hat{R}_{W_t})\hat{R}_{Z_t}^{-1}$. The subtraction is commonly performed in the frequency domain where it is simpler to control the positive definiteness of the estimate of R_Y. This approach results in the simplest form of the family of "spectral subtraction" speech-enhancement approaches [26].

MMSE estimation under Gaussian assumptions leads to linear estimation in the form of Wiener filtering given in Equation (15.1). The same filter could be obtained if the Gaussian assumptions are relaxed, and the best *linear* estimator in the MMSE sense is sought. If we denote the linear filter for Y_t by the $l \times l$ matrix H_t, then the optimal H_t is obtained by minimizing the MSE given by $E\{\|Y_t - H_t Z_t\|^2\}$. Here, $E\{\cdot\}$ denotes expected value, and $\|\cdot\|$ denotes the usual Euclidean norm. Note that when the filter H_t is applied to the noisy signal Z_t, it provides a residual signal given by

$$Y_t - \hat{Y}_t = Y_t - H_t Z_t = (I - H_t)Y_t + H_t W_t \qquad (15.2)$$

The term $(I - H_t)Y_t$ represents the distortion caused by the filter, and the term $H_t W_t$ represents the residual noise at the output of the filter. Since the signal and noise are statistically independent, the MSE error is the sum of two terms, the distortion energy $\overline{\epsilon_d^2} = E\{\|(I - H_t)Y_t\|^2\}$ and the residual noise energy $\overline{\epsilon_n^2} = E\{\|H_t W_t\|^2\}$. The Wiener filter minimizes $\overline{\epsilon_d^2} + \overline{\epsilon_n^2}$ over all possible filters H_t. An alternative approach proposed by Ephraim and Van-Trees [18] was to design the filter H_t by minimizing the distortion energy $\overline{\epsilon_d^2}$ for a given level of acceptable residual noise energy $\overline{\epsilon_n^2}$. This approach allows the design of a filter that controls the contributions of the two competing components $\overline{\epsilon_d^2}$ and $\overline{\epsilon_n^2}$ to the MSE. The resulting filter is similar to that in Equation (15.1) except that R_{W_t} is replaced by $\mu_t R_{W_t}$ where μ_t is the Lagrange multiplier of the constrained optimization problem. The idea was extended to filter design which minimizes the distortion energy for a given desired spectrum of the residual noise. This interesting optimization problem was solved by Lev-Ari and Ephraim [25]. The estimation criterion was motivated by the desire to adjust the spectrum of the residual noise so that it is least audible.

In [18], the two estimation criteria outlined above were applied to enhancement of noisy speech signals. It was noted that there is strong empirical evidence that supports the notion that covariance matrices of many speech vectors are not full rank matrices. This notion is also supported by the popular sinusoidal model for speech signals, in which a speech vector with $l = 200$ to 400 samples at an 8 kHz sampling rate is spanned by fewer than l sinusoidal components. As such, some of the eigenvalues of R_{Y_t} are practically zero, and the vector Y_t occupies a subspace of the Euclidean space $\mathcal{R}^l$. A white noise, however, occupies the entire space $\mathcal{R}^l$. Thus, the Euclidean space $\mathcal{R}^l$ may be decomposed into a "signal subspace" containing signal plus noise, and a complementary "noise subspace" containing noise only. Thus, in enhancing a noisy vector Z_t, one can first null out the component of Z_t in the noise subspace and filter the noisy signal in the signal subspace. The decomposition of Z_t into its signal subspace component and noise subspace component can be performed by applying the Karhunen–Loève transform to Z_t.

Spectral Magnitude Estimation

In the section "Linear Estimation" we focused on MMSE estimation of the waveform of the speech signal. This estimation may be cast in the frequency domain as follows. We use $(\cdot)'$ to denote conjugate transpose. Let D' denote the discrete Fourier transform (DFT) matrix. Let $\mathbf{Z}_t = \frac{1}{\sqrt{l}}D'Z_t$ denote the vector of spectral components of the noisy vector Z_t. For convenience, we have chosen to use normalized DFT. We denote the kth spectral component of the noisy vector Z_t by $\mathbf{Z}_{tk}$. Let Λ_{Z_t} be a diagonal matrix with the variances of the spectral components $\{\mathbf{Z}_{tk}, k = 0, 1, \dots, l-1\}$ on its main diagonal. Assume, for simplicity, that R_{Y_t} and R_{W_t} are circulant matrices [24]. This means that $R_{Y_t} = \frac{1}{l}D\Lambda_{Y_t}D'$ and $R_{W_t} = \frac{1}{l}D\Lambda_{W_t}D'$. Let $\hat{\mathbf{Y}}_t = \frac{1}{\sqrt{l}}D'\hat{Y}_t$ be the normalized DFT of the MMSE estimate $\hat{Y}_t$. Under these assumptions, Equation (15.1) becomes

$$\hat{\mathbf{Y}}_t = \left[\Lambda_{Y_t}(\Lambda_{Y_t} + \Lambda_{W_t})^{-1}\right]\mathbf{Z}_t \qquad (15.3)$$

This filter performs MMSE estimation of the spectral components $\{\mathbf{Y}_{tk}\}$ of the clean vector Y_t. It is commonly believed, however, that the human auditory system is more sensitive to the short-term spectral magnitude $\{|\mathbf{Y}_{tk}|, k = 0, 1, \ldots, l - 1\}$ of the speech signal than to its short-term phase $\{\arg(\mathbf{Y}_{tk}), k = 0, 1, \ldots, l - 1\}$. This has been demonstrated by Wang and Lim [37] in a sequence of experiments. They have synthesized speech signals using short-term spectral magnitude and phase derived from two noisy versions of the same speech signal at different signal-to-noise ratios (SNRs). Thus, they could control the amount of noise in the spectral magnitude and in the phase. Hence, it was suggested that better enhancement results could be obtained if the spectral magnitude of a speech signal rather than its waveform is directly estimated. In this situation, the phase of the noisy signal is combined with the spectral magnitude estimator in constructing the enhanced signal. Maximum likelihood estimates of the short-term spectral magnitude of the clean signal were developed by McAulay and Malpass [32] for additive Gaussian noise. An MMSE estimator of the short-term spectral magnitude of speech signal was developed by Ephraim and Malah [14]. The spectral components of the clean signal and of the noise process were assumed statistically independent Gaussian random variables. Under the same assumptions, the MMSE estimator of the short-term complex exponential of the clean signal, $\exp(j\arg(\mathbf{Y}_{tk}))$, which does not affect the spectral magnitude estimator (i.e., has a unity modulus), was shown in Ref. [14] to be equal to the complex exponential of the noisy signal. This confirmed the intuitive use of the noisy phase in systems which capitalize on spectral magnitude estimation.

It is further believed that the human auditory system compresses the signal's short-term spectral magnitude in the process of its decoding. It was suggested that a form of logarithmic compression is actually taking place. Hence, better enhancement of the noisy signal should be expected if the logarithm of the short-term spectral magnitude is directly estimated. An MMSE estimator of the log-spectral magnitude of speech signal was developed by Ephraim and Malah [15] under the same Gaussian assumptions described above. This approach has attracted much interest in recent years and will be presented in more detail in the section "MMSE Spectral Magnitude Estimation."

The Gaussian Model

The assumption that spectral components of the speech signal at any given frame are statistically independent Gaussian random variables underlies the design of many speech-enhancement systems. In this model, the real and imaginary parts of each spectral component are statistically independent identically distributed Gaussian random variables. We have mentioned here the Wiener filter for MMSE estimation of the spectral components of the speech signal, and the MMSE estimators for the spectral magnitude and for the logarithm of the spectral magnitude of the clean signal. The Gaussian assumption is mathematically tractable, and it is often justified by a version of the central limit theorem for correlated signals ([4], Theorem 4.4.2). The Gaussian assumption for the real and imaginary parts of a speech spectral component has been challenged by some authors [30,33]. In [33], for example, the spectral magnitude was claimed to have a Gamma distribution. In [30], the real and imaginary parts of a spectral component were assumed statistically independent Laplace random variables. We now show that the Gaussian and other models are not necessarily contradictory.

The assumption that a spectral component is Gaussian is always conditioned on knowledge of the variance of that component. Thus, the Gaussian assumption is attributed to the conditional probability density function (pdf) of a spectral component given its variance. A conditionally Gaussian spectral component may have many different marginal pdfs. To demonstrate this point, consider the spectral component $\mathbf{Y}_{tk}$ and its variance $\sigma_{\mathbf{Y}_{tk}}^2$. Let the real part of $\mathbf{Y}_{tk}$ be denoted by Y. Let the variance $\sigma_{\mathbf{Y}_{tk}}^2/2$ of the real part of $\mathbf{Y}_{tk}$ be denoted by V. Assume that the conditional pdf of Y given V is Gaussian. Denote this pdf by $p(y|v)$. Assume that the variance V has a pdf $p(v)$. Then the marginal pdf of Y is given by

$$p(y) = \int p(y|v)p(v)\mathrm{d}v \tag{15.4}$$

The pdf of Y is thus a continuous mixture of Gaussian densities. This pdf may take many different forms, which are determined by the specific prior pdf assumed for V. For example, suppose that V is exponentially distributed with expected value $2\lambda^2$, i.e., assume that:

$$p(y|v) = \frac{e^{-\frac{y^2}{2v}}}{\sqrt{2\pi v}} \quad \text{and} \quad p(v) = \frac{e^{-\frac{v}{2\lambda^2}}}{2\lambda^2} u(v) \tag{15.5}$$

where $u(\sigma)$ is a unit step-function. Substituting Equation (15.5) into Equation (15.4) and using ([23], Equation (3.325)) shows that

$$p(y) = \frac{1}{2\lambda} e^{-\frac{|y|}{\lambda}} \tag{15.6}$$

or that Y has a Laplace pdf just as was assumed [30]. This argument shows that estimators for a spectral component of speech signal obtained under non-Gaussian models may be derived using the conditional Gaussian pdf and an appropriately chosen pdf for the variance of the spectral component. In our opinion, using the conditional Gaussian model is preferable, since it is much better understood and it is significantly easier to work with.

The variance of a spectral component must be assumed a random variable, since speech signals are not strictly stationary. Thus, the variance sequence $\{\sigma^2_{\mathbf{Y}_{tk}}, t = 1, 2, \ldots\}$ corresponding to the sequence of spectral components $\{\mathbf{Y}_{tk}, t = 1, 2, \ldots\}$ at a given frequency k, is not known in advance and is best described as a random sequence. In [14] and [15], the variance of each spectral component of the clean signal was estimated and updated from the noisy signal using the decision-directed estimator. In [13], the variance sequence was assumed a Markov chain and it was estimated online from the noisy signal. In [8], a recursive formulation of the variance estimator is developed following the rational of Kalman filtering.

A closely related statistical model for speech enhancement is obtained by modeling the clean speech signal as a hidden Markov process (HMP). An overview of HMPs may be found in [19]. Speech enhancement systems using this model were first introduced by Ephraim, Malah, and Juang [16]. An HMP is a bivariate process of state and observation sequences. The state sequence is a homogeneous Markov chain with a given number of states, say M. The observation sequence is conditionally independent given the sequence of states. This means that the distribution of each observation depends only on the state at the same time and not on any other state or observation. Let $S^n = \{S_1, \ldots, S_n\}$ denote the state sequence where we may assume without loss of generality that $S_t \in \{1, \ldots, M\}$. Let $Y^n = \{Y_1, \ldots, Y_n\}$ denote the observation sequence where each Y_t is a vector in a Euclidean space $\mathcal{R}^l$. The joint density of (S^n, Y^n) is given by

$$p(s^n, y^n) = \prod_{t=1}^{n} p(s_t|s_{t-1}) p(y_t|s_t) \tag{15.7}$$

where $p(s_1|s_0) = p(s_1)$. When $S_t = j$, we replace $p(y_t|s_t)$ by $p(y_t|j)$. In [16] and [17], $p(y_t|j)$ was assumed to be the pdf of a vector from a zero mean Gaussian autoregressive process. The parameter of the process, i.e., the autoregressive coefficients and gain, depends on the state j. This parameter characterizes the power spectral density of the signal in the given vector. Thus, $p(y_t|j)$ was assumed in [16,17] to be conditionally Gaussian, given the power spectral density of the signal. There are M power spectral density prototypes for all vectors of the speech signal. The HMP assumes that each vector of the speech signal is drawn with some probability from one of the M autoregressive processes. The identity of the autoregressive process producing a particular vector is not known, and hence the pdf of each vector is a finite mixture of Gaussian autoregressive pdfs. In contrast, Equation (15.4) represents a mixture of countably infinite Gaussian pdfs. In the HMP model, spectral components of each vector of the speech signal are assumed correlated since each vector is assumed autoregressive, and consecutive speech vectors are weakly dependent since they inherit the memory of the Markov chain.

Signal Presence Uncertainty

In all models presented thus far in this section, the clean signal was assumed to be present in the noisy signal. Thus, we have always viewed the noisy signal vector at time t as $Z_t = Y_t + W_t$. In reality, however, speech contains many pauses while the noise may be continuously present. Thus, the noisy signal vector at time t may be more realistically described as resulting from two possible hypotheses: H_1 indicating signal presence and H_0 indicating signal absence. We have:

$$Z_t = \begin{cases} Y_t + W_t & \text{under } H_1 \\ W_t & \text{under } H_0 \end{cases} \tag{15.8}$$

This insightful observation was first made by McAulay and Malpass [32], who have modified their speech signal estimators accordingly. For MMSE estimation, let $E\{Y_t|Z_t, H_1\}$ denote the conditional mean estimate of Y_t when the signal is assumed present in Z_t. Let $P(H_1|Z_t)$ denote the probability of signal presence given the noisy vector. The MMSE of Y_t given Z_t is given by

$$E\{Y_t|Z_t\} = P(H_1|Z_t)E\{Y_t|Z_t, H_1\} \tag{15.9}$$

The model of speech presence uncertainty may be refined and attributed to spectral components of the vector Z_t [14]. This aspect will be dealt with in more detail in the section "MMSE Spectral Magnitude Estimation."

Multi-State Speech Model

The signal presence uncertainty model may be seen as a two-state model for the noisy signal. A five-state model for the clean signal was proposed earlier by Drucker [12]. The states in his model represent fricative, stop, vowel, glide, and nasal speech sounds. For enhancing a noisy signal, he proposed to first classify each vector of the noisy signal as originating from one of the five possible class sounds, and then to apply a class-specific filter to the noisy vector.

The HMP model for the clean signal described in the section "The Gaussian Model" is a multi-class model. When HMPs are used, the classes are not *a priori* defined, but they are rather created in a learning process from some training data of clean speech signals. The learning process is essentially a clustering process that may be performed using vector quantization techniques [22]. For example, each class may contain spectrally similar vectors of the signal. Thus, each class may be characterized by a prototype power spectral density which may be parameterized as an autoregressive process. Transitions from one spectral prototype to another are probabilistic and are performed in a Markovian manner. The noise process may be similarly represented. If there are M speech classes and N noise classes, then $M \times N$ estimators must be designed for enhancing noisy speech signals. Suppose that we are interested in estimating the speech vector Y_t given a sequence of noisy speech vectors $z^t = \{z_1, \ldots, z_t\}$. Let $p((i,j)|z^t)$ denote the probability of the signal being in state i and the noise being in state j given z^t. Then, the MMSE estimator of Y_t from z^t is given [17]:

$$E\{Y_t|z^t\} = \sum_{i=1}^{M} \sum_{j=1}^{N} p((i,j)|z^t)E\{Y_t|z^t, (i,j)\} \tag{15.10}$$

MMSE Spectral Magnitude Estimation

In this section we focus on MMSE estimation of the logarithm of the short-term spectral magnitude of the clean signal. We provide a design example of a speech enhancement system that relies on conditional Gaussian modeling of spectral components and on speech presence uncertainty. Recall that the kth spectral component of the clean speech vector Y_t is denoted by $\mathbf{Y}_{tk}$. The variance of Y_{tk} is denoted by $\sigma^2_{Y_{tk}}$. It is assumed that spectral components $\{\mathbf{Y}_{tk}\}$ with given variances $\{\sigma^2_{Y_{tk}} > 0\}$ are statistically independent Gaussian random variables. Similar assumptions are made for the spectral components of the noise process $\{\mathbf{W}_{tk}\}$.

The spectral component $\mathbf{Z}_{tk}$ of the noisy signal is given by

$$\mathbf{Z}_{tk} = \mathbf{Y}_{tk} + \mathbf{W}_{tk} \qquad (15.11)$$

Let H_1^{tk} and H_0^{tk} denote the hypotheses of speech presence and speech absence in the noisy spectral component $\mathbf{Z}_{tk}$, respectively. Let q_{tk} denote the probability of H_1^{tk}. The spectral components of the noisy signal $\{Z_{tk}\}$ are statistically independent Gaussian random variables given their variances $\{\sigma_{Z_{tk}}^2\}$.

We are interested in estimating the logarithm of the spectral magnitude of each component of the clean signal from all available spectral components of the noisy signal. Under the statistical model assumed here, given the variances of the spectral components and the probabilities of speech presence, estimation of $\log|Y_{tk}|$ is performed from $\mathbf{Z}_{tk}$ only. Since the variances of the spectral components and the probabilities of speech presence are not available, however, these quantities are estimated for each frequency k from the noisy spectral components observed up to time t, and the estimates are plugged in the signal estimate. We use $\widehat{\sigma_{Y_{tk}}^2}$ and $\widehat{\sigma_{W_{tk}}^2}$ to denote estimates of the variances of $\mathbf{Y}_{tk}$ and W_{tk}, respectively, and $\hat{q}_{tk}$ to denote an estimate of q_{tk}. We next present estimation of the signal and its assumed known parameter.

Signal Estimation

The signal estimator is conveniently expressed in terms of the *a priori* and *a posteriori* SNRs. These quantities are defined as

$$\xi_{tk} = \frac{\sigma_{Y_{tk}}^2}{\sigma_{W_{tk}}^2} \quad \text{and} \quad \gamma_{tk} = \frac{|Z_{tk}|^2}{\sigma_{W_{tk}}^2} \qquad (15.12)$$

respectively. We also define

$$\vartheta_{tk} = \frac{\xi_{tk}}{\xi_{tk}+1}\gamma_{tk} \qquad (15.13)$$

The estimates of ξ_{tk} and γ_{tk} used here are $\hat{\xi}_{tk} = \widehat{\sigma_{Y_{tk}}^2}/\widehat{\sigma_{W_{tk}}^2}$ and $\hat{\gamma}_{tk} = |Z_{tk}|^2/\widehat{\sigma_{W_{tk}}^2}$. To prevent estimation of the logarithm of negligibly small spectral magnitudes under the hypothesis that speech is absent in Z_{tk}, Cohen and Berdugo [6] proposed to estimate the conditional mean of the following function of $\mathbf{Y}_{tk}$:

$$f(\mathbf{Y}_{tk}) = \begin{cases} \log|\mathbf{Y}_{tk}|, & \text{under } H_1^{tk} \\ \log v_{tk}, & \text{under } H_0^{tk} \end{cases} \qquad (15.14)$$

where v_{tk} is a spectral threshold. They showed that

$$\widehat{|Y_{tk}|} = \exp\left\{ E\left\{ f(\mathbf{Y}_{tk})|Z_{tk}; \widehat{\sigma_{Y_{tk}}^2}, \hat{\xi}_{tk}, \hat{q}_{tk} \right\} \right\}$$
$$= \left[G(\hat{\xi}_{tk}, \hat{\gamma}_{tk})|Z_{tk}| \right]^{\hat{q}_{tk}} v_{tk}^{1-\hat{q}_{tk}} \qquad (15.15)$$

where

$$G(\xi, \gamma) = \frac{\xi}{\xi+1} \exp\left(\frac{1}{2} \int_{\vartheta}^{\infty} \frac{e^{-x}}{x}\,\mathrm{d}x \right) \qquad (15.16)$$

represents the spectral gain function derived by Ephraim and Malah [15] under H_1^{tk}. Note that this gain function depends on Z_{tk} and hence the estimator in Equation (15.5) is nonlinear even when the parameter of the statistical model is known. It was further proposed to replace v_{tk} in Equation (15.5) by $G_{\min}|Z_{tk}|$ where $G_{\min} \ll 1$ [6]. This substitution provides a constant attenuation of $|Z_{tk}|$ under H_0^{tk} rather than using a constant term that is independent of $|Z_{tk}|$. This practice is closely related to the "spectral floor" modification of the spectral subtraction method proposed by Berouti, Schwartz, and Makhoul [3]. The constant attenuation retains the naturalness of the residual noise when the signal is absent. Substituting this constant attenuation in Equation (15.5) gives

$$\widehat{|\mathbf{Y}_{tk}|} = [G(\hat{\xi}_{tk}, \hat{\gamma}_{tk})]^{\hat{q}_{tk}} G_{\min}^{1-\hat{q}_{tk}} |\mathbf{Z}_{tk}| \tag{15.17}$$

To form an estimator $\hat{\mathbf{Y}}_{tk}$ for the clean spectral component $\mathbf{Y}_{tk}$, the spectral magnitude estimator $\widehat{|Y_{tk}|}$ is combined with an estimator of the phase of Y_{tk}. Ephraim and Malah [14] proposed to use the MMSE estimator of the complex exponential of that phase. The modulus of the estimator was constrained to a unity so that it does not affect the optimality of the spectral magnitude estimator $\widehat{|\mathbf{Y}_{tk}|}$. They showed that the constrained MMSE estimator is given by the complex exponential of the noisy phase.

The integral in Equation (15.6) is the well-known exponential integral of ϑ, and it can be numerically evaluated, e.g., using the *expint* function in MATLAB. Alternatively, it may be evaluated by using the following computationally efficient approximation, which was developed by Martin et al. [31]:

$$\exp \mathrm{int}(\vartheta) = \int_{\vartheta}^{\infty} \frac{e^{-x}}{x} \, \mathrm{d}x \approx \begin{cases} -2.31 \log_{10}(\vartheta) - 0.6, & \text{for } \vartheta < 0.1 \\ -1.544 \log_{10}(\vartheta) + 0.166, & \text{for } 0.1 \le \vartheta \le 1 \\ 10^{-0.52\vartheta - 0.26}, & \text{for } \vartheta > 1 \end{cases} \tag{15.18}$$

Signal Presence Probability Estimation

In this section we address the problem of estimating the speech presence probability q_{tk}. Define a binary random variable V_{tk}, which indicates whether or not speech is present in the spectral component Z_{tk}:

$$V_{tk} = \begin{cases} 1 & \text{under } H_1^{tk} \\ 0 & \text{under } H_0^{tk} \end{cases} \tag{15.19}$$

Cohen and Berdugo [6] proposed to estimate q_{tk} as the conditional mean of V_{tk} given Z_{tk} and an estimate of the parameter of the statistical model. Specifically:

$$\hat{q}_{tk} = E\{V_{tk}|z_{tk}; \widehat{\sigma_{W_{tk}}^2}, \hat{\xi}_t\} = P(H_1^{tk}|z_{tk}; \widehat{\sigma_{W_{tk}}^2}, \hat{\xi}_t) \tag{15.20}$$

Using Bayes' rule, they expressed the conditional probability of H_1^{kt} in Equation (15.20) in terms of the Gaussian densities of Z_{tk} under the two hypotheses and some estimate of the prior probability of H_1^{tk}. They provided a scheme for estimating the prior probability from spectral components observed up to time $t - 1$. Let the prior probability estimate be denoted by $\hat{q}_{tk|t-1}$. Following this approach they showed that

$$\hat{q}_{tk} = \left[1 + \frac{1 - \hat{q}_{tk|t-1}}{\hat{q}_{tk|t-1}} (1 + \hat{\xi}_{tk}) \exp(-\hat{\vartheta}_{tk}) \right]^{-1} \tag{15.21}$$

where $\hat{\vartheta}_{tk}$ is the estimate of ϑ_{tk} defined in Equation (15.13) [6].

The estimator $\hat{q}_{tk|t-1}$ is based on the distribution of the *a priori* SNR, and the relation between the likelihood of speech absence in the time–frequency domain and the local and global averages of the *a priori* SNR. The speech absence probability is estimated for each frequency bin and each frame by a soft-decision approach, which exploits the strong correlation of speech presence in neighboring frequency bins of consecutive frames.

A *Priori* SNR Estimation

Reliable estimation of the speech spectral component variances is crucial for successful implementation of the signal estimator (Equation (15.7)). Ephraim and Malah [14] proposed a decision-directed variance estimator for their MMSE spectral magnitude estimator. The variance estimator at a given frame uses the signal spectral magnitude estimate from the previous frame along with the current noisy spectral component. Let $\hat{A}_{tk} = |\hat{Y}_{tk}|$ denote the MMSE signal spectral magnitude estimate from $\mathbf{Z}_{tk}$. The decision-directed estimate of the variance of Y_{tk} is given by

$$\widehat{\sigma^2_{Y_{tk}}} = \frac{1}{\hat{q}_{tk}}\left[\alpha\hat{A}^2_{t-1,k} + (1-\alpha)\max\left\{|\mathbf{Z}_{tk}|^2 - \widehat{\sigma^2_{W_{tk}}}, 0\right\}\right] \tag{15.22}$$

where $0 \leq \alpha \leq 1$ is an experimental constant. The estimator was also found useful when $\hat{A}_{tk}$ is the MMSE log-spectral magnitude estimator [15]. In the latter case, the estimator was used with $\hat{q}_{tk} = 1$, since the signal was assumed zero under the null hypothesis. While this estimator was found useful in practice, the division by $\hat{q}_{tk}$ may deteriorate the performance of the speech-enhancement system [34]. In some cases, it introduces interaction between the estimated $\hat{q}_{tk}$ and the *a priori* SNR, resulting in unnaturally structured residual noise [28].

Cohen and Berdugo [6] showed that a preferable variance estimator is obtained if $\hat{A}_{t-1,k}$ in Equation (15.22) is replaced by the estimator $\hat{A}_{t-1,k|H_1^{tk}}$ for the magnitude of $\mathbf{Y}_{t-1,k}$ obtained under the signal presence hypothesis, and the division by $\hat{q}_{tk}$ is not performed. The resulting estimator is given by

$$\widehat{\sigma^2_{Y_{tk}}} = \alpha\hat{A}^2_{t-1,k|H_1^{tk}} + (1-\alpha)\max\left\{|\mathbf{Z}_{tk}|^2 - \widehat{\sigma^2_{W_{tk}}}, 0\right\} \tag{15.23}$$

Expressing $\hat{A}_{t-1,k|H_1^{tk}}$ in terms of the gain function form (15.17), dividing by $\widehat{\sigma^2_{W_{tk}}}$, and imposing a lower bound $\xi_{\min} > 0$ on the *a priori* SNR estimate as proposed by Cappé [5], they obtained the following recursive estimator for ξ_{tk}:

$$\hat{\xi}_{tk} = \max\left\{\alpha\, G^2\left(\hat{\xi}_{t-1,k}, \hat{\gamma}_{t-1,k}\right)\hat{\gamma}_{t-1,k} + (1-\alpha)(\hat{\gamma}_{tk} - 1), \xi_{\min}\right\} \tag{15.24}$$

The parameters α and $\xi_{\min}$ control the trade-off between the noise reduction and the transient distortion introduced into the signal [5,14]. Greater reduction of the musical noise phenomenon is obtained by using a larger α and a smaller $\xi_{\min}$ at the expense of attenuated speech onsets and audible modifications of transient speech components. Typical values for α range between 0.9 and 0.99, and typical values for $\xi_{\min}$ range between -10 and -25 dB.

Noise Spectrum Estimation

In stationary noise environments, the noise variance of each spectral component is time invariant, i.e., $\sigma^2_{W_{tk}} = \sigma^2_{W_k}$ for all t. An estimator for $\sigma^2_{W_k}$ may be obtained from recursive averaging of $\{|\mathbf{Z}_{tk}|^2\}$ for all spectral components classified as containing noise only.

In nonstationary noise environments, an alternative approach known as the *minimum statistics* was proposed by Martin [27,29]. In this approach, minima values of a smoothed power spectral density estimate of the noisy signal are tracked, and multiplied by a constant that compensates the estimate for possible bias. We present here a recent algorithm, developed by Cohen and Berdugo [7,9], which is based on *minima controlled recursive averaging*. This noise variance estimator is capable of fast adaptation to abrupt changes in the noise spectrum.

Recall that H_0^{tk} and H_1^{tk} denote, respectively, speech absence and presence hypotheses in the noisy spectral component $\mathbf{Z}_{tk}$. A recursive estimate for the noise spectral variance can be obtained as follows:

$$\widehat{\sigma_{\mathrm{W}_{t+1,k}}^2} = \begin{cases} \mu\widehat{\sigma_{\mathrm{W}_{tk}}^2} + (1-\mu)\beta|\mathbf{Z}_{tk}|^2 & \text{under } H_0^{tk} \\ \widehat{\sigma_{\mathrm{W}_{tk}}^2} & \text{under } H_1^{tk} \end{cases} \qquad (15.25)$$

where $0 < \mu < 1$ is a smoothing parameter and $\beta \geq 1$ is a bias compensation factor [9]. The probability of H_1^{tk} is estimated here independently of $\hat{q}_{tk}$ in section "Signal Presence Probability Estimation," since the penalty in misclassification of the two hypotheses has different consequences when estimating the signal than when estimating the noise spectral variance. Generally, here we tend to decide H_0^{tk} with higher confidence than in section "Signal Presence Probability Estimation". Let $\tilde{q}_{tk}$ denote the estimate of the probability of H_1^{tk} in this section. A soft-decision recursive estimator can be obtained from Equation (15.25) by

$$\widehat{\sigma_{\mathrm{w}_{t+1,k}}^2} = \tilde{q}_{tk}\widehat{\sigma_{\mathrm{w}_{tk}}^2} + (1-\tilde{q}_{tk})\left[\mu\widehat{\sigma_{\mathrm{w}_{tk}}^2} + (1-\mu)\beta|\mathbf{Z}_{tk}|^2\right] = \tilde{\mu}_{tk}\widehat{\sigma_{\mathrm{w}_{tk}}^2} + (1-\tilde{\mu}_{tk})\beta|\mathbf{Z}_{tk}|^2 \quad (15.26)$$

where $\tilde{\mu}_{tk} = \mu + (1-\mu)\tilde{q}_{tk}$ is a time-varying smoothing parameter.

The probability $\tilde{q}_{tk}$ is estimated using Equation (15.21) when $\hat{q}_{tk|t-1}$ is substituted by a properly designed estimate $\tilde{q}_{tk|t-1}$. Cohen [9] proposed an estimator $\tilde{q}_{tk|t-1}$, which is controlled by the minima values of a smoothed power spectrum of the noisy signal. The estimation procedure comprises two iterations of smoothing and minimum tracking. The first iteration provides a rough voice activity detection in each frequency. Smoothing during the second iteration excludes relatively strong speech components, which makes the minimum tracking during speech activity more robust.

Summary of Algorithm

(i) For $t = 0$ and all k's, set $\widehat{\sigma_{\mathrm{W}_{tk}}^2} = |\mathbf{Z}_{0k}|^2$, $\hat{\gamma}_{-1,k} = 1$, $\hat{\xi}_{-1,k} = \xi_{\min}$. Set $t = 1$.

(ii) For each k:
- Calculate $\hat{\gamma}_{tk}$ from Equation (15.12), and $\hat{\xi}_{tk}$ from Equation (15.24).
- Calculate $\hat{q}_{tk|t-1}$ from [6], Equation 29, and $\hat{q}_{tk}$ from Equation (15.21).
- Calculate $G(\hat{\xi}_{tk}, \hat{\gamma}_{tk})$ from Equation (15.16), and $\widehat{|\mathbf{Y}_{tk}|}$ by using Equation (15.17).
- Calculate $\tilde{q}_{tk|t-1}$ from [9], Equation 28, and $\tilde{q}_{tk}$ from the analog of Equation (15.21).
- Update $\widehat{\sigma_{\mathrm{W}_{tk}}^2}$ by using Equation (15.26).

(iii) Set $t \rightarrow t + 1$ and go to step (ii) for enhancement of the next frame.

Monte-Carlo Simulation

The Monte-Carlo simulation approach for audio signal enhancement has been promoted by Vermaak et al. and Fong et al. [20,36]. In this section, we present the principles of this approach. The clean and noisy speech signals are represented by the sequences of scalar random variables $\{Y_t, t = 0, 1, \ldots\}$ and $\{Z_t, t = 1, 2, \ldots\}$, respectively. These signals are assumed to satisfy some time-varying state-space equations. The time-varying parameter of the system is denoted by $\{\theta_t, t = 1, 2, \ldots\}$. The system is characterized by three deterministically known nonlinear transition functions which we denote here by f, g and h. The explicit dependence of f on t, and of g and h on θ_t, is expressed by writing these functions as f_t, g_{θ_t} and h_{θ_t}, respectively. The innovation processes of the dynamical system are denoted by $\{U_t, t = 1, 2, \ldots\}$, $\{V_t, t = 1, 2, \ldots\}$ and $\{W_t, t = 1, 2, \ldots\}$. These three processes are assumed statistically independent iid processes. The state-space equations are given by

$$\theta_t = f_t(\theta_{t-1}, U_t)$$
$$Y_t = g_{\theta_t}(Y_{t-1}, V_t)$$
$$Z_t = h_{\theta_t}(Y_t, W_t) \tag{15.27}$$

for $t = 1, 2, \ldots$.

Assume first that the sample path of $\{\theta_t\}$ is known. For this case, the signal $\{Y_t\}$ can be recursively estimated from $\{Z_t\}$. To simplify notation, we present these recursions without explicitly showing the dependence of the various pdfs on the assumed known parameter path. We use lower case letters to denote realizations of the random variables in Equation (15.27). We also denote $z^t = \{z_1, \ldots, z_t\}$. The filtering and prediction recursions result from Markov properties of the signals in Equation (15.27) and from Bayes' rule. These recursions are, respectively, given by

$$p(y_t|z^t) = \frac{p(y_t|z^{t-1})p(z_t|y_t)}{\int p(y_t|z^{t-1})p(z_t|y_t)dy_t}, \; t = 1, \ldots, n \tag{15.28}$$

where $p(y_1|z_1^0) = p(y_1)$, and by

$$p(y_t|z^{t-1}) = \int p(y_t|y_{t-1})p(y_{t-1}|z^{t-1})dy_{t-1}, \; t = 2, \ldots, n \tag{15.29}$$

The smoothing recursion was derived by Askar and Derin in [2, Theorem 1], and it is given by

$$p(y_t|z^n) = p(y_t|z^t) \int \frac{p(y_{t+1}|z_t)p(y_{t+1}|z^n)}{p(y_{t+1}|z^t)} dy_{t+1} \tag{15.30}$$

for $t = n - 1, n - 2, \ldots, 1$, where $p(y_n|z^n)$ is given by Equation (15.28).

When the sample path of $\{\theta_t\}$ is given, or when the parameter is time-invariant and known ($\theta_t = \theta_0$ for all t), these recursions can be implemented with reasonable complexity for two well-known cases. First, when g and h are linear functions, $\{V_t\}$ and $\{W_t\}$ are Gaussian processes, and the initial distribution of Y_0 is Gaussian. In that case, $\{Y_t\}$ can be estimated using the Kalman filter or smoother. Second, when $\{Y_t\}$ takes finitely many values, then the integrals become summations and the recursions coincide with a version of the forward-backward recursions for hidden Markov processes (see [19], Equation 5.14 to Equation 5.16). For all other systems, the estimation problem is highly nonlinear and requires multidimensional integrations. No simple solution exists for these situations. Approximate solutions are often obtained using the extended Kalman filter. The latter applies Kalman filtering to locally linearized versions of the state-space equations.

When the sample path of $\{\theta_t\}$ is not known, but the three transition functions are linear and the innovation processes are Gaussian, maximum *a posteriori* estimation of $\{\theta_t\}$ is possible using the expectation-maximization (EM) algorithm. This was shown by Dembo and Zeitouni [11], who developed an EM algorithm for estimating $\{\theta_t\}$ when the signal $\{Y_t\}$ is a time-varying autoregressive process. The parameter estimator relies on Kalman smoothers for the clean signal $\{Y_t\}$ and its covariance at each EM iteration. Thus, an estimate of the clean signal is obtained as a by-product in this algorithm. A similar approach for maximum likelihood estimation of a deterministically unknown parameter was implemented and tested for speech enhancement by Gannot et al. [21].

The computational difficulties in estimating the parameter or the clean signal in Equation (15.27) have stimulated the use of Monte-Carlo simulations. A good tutorial on the subject was written by Arulampalam et al. [1]. In this approach, probability distributions are sampled and replaced by empirical distributions. Thus integrals involving the sampled pdfs can be straightforwardly evaluated using sums. Recursive sampling is often desirable to facilitate the approach. The filters or smoothers designed in this way are often referred to as *particle filters*. The "particles" refer to the point masses obtained from sampling the distribution which is of interest in the given problem. There is more than one way to simulate the filtering or smoothing recursions presented earlier. We focus here on the work in [20] and [36], where the approach has been applied to speech and audio signals and compared with the extended Kalman filter. In [20], Monte-Carlo approaches for filtering as well as smoothing were developed. We shall demonstrate here only the principles of the filtering approach.

Similar to the work of Dembo and Zeitouni [11], the signal in [20] was assumed a Gaussian time-varying autoregressive process, and the additive noise was assumed Gaussian. In fact, the reflection coefficients of the time-varying autoregressive process were assumed a Gaussian random walk process, which was constrained to the interval of $(-1, 1)$, but the nonlinear transformation from the reflection coefficients to the autoregressive coefficients was ignored. The logarithm of the gain of the autoregressive process was also modeled as a Gaussian random walk. The pdf $p(\theta_t|z^t)$ of θ_t given z^t was shown to be proportional to

$$p(\theta_t|z^t) \propto \int p(z_t|\theta^t, z^{t-1})p(\theta_t|\theta_{t-1})p(\theta^{t-1}|z^{t-1})\mathrm{d}\theta^{t-1} \tag{15.31}$$

This equation can be derived similarly to Equation (15.28). The goal now is to recursively sample $p(\theta_t|z^t)$ and estimate the signal using an efficient algorithm such as the Kalman filter. Suppose that at time t we have an estimate of $p(\theta^{t-1}|z^{t-1})$. This pdf can be sampled N times to produce N sample paths of θ^{t-1}. Let these sample paths be denoted by $\{\theta^{t-1}(i), i = 1, \ldots, N\}$. Next, for each $i = 1, \ldots, N$, the pdf $p(\theta_t|\theta_{t-1}(i))$ can be sampled N times to provide $\{\theta_t(1), \ldots, \theta_t(N)\}$. Augmenting the former and latter samples, we obtain N sample paths of θ^t given z^{t-1}. We denote these sample paths by $\{\theta^t(i), i = 1 \ldots, N\}$. The empirical distribution of θ^t given z^{t-1} is given by

$$q(\theta^t|z^{t-1}) = \frac{1}{N}\sum_{i=1}^{N} \delta(\theta^t - \theta^t(i)) \tag{15.32}$$

where $\delta(\cdot)$ denotes the Dirac function. Substituting Equation (15.32) for $p(\theta_t|\theta_{t-1})p(\theta^{t-1}|z^{t-1})$ in Equation (15.31) gives

$$p(\theta_t|z^t) \propto \sum_{i=1}^{N} p(z_t|\theta^t(i), z^{t-1})\delta(\theta_t - \theta_t(i)) \tag{15.33}$$

Next, it was observed that Z_t given $\theta^t(i)$ and z^{t-1} is Gaussian with conditional mean and covariance that can be calculated using the Kalman filter for estimating Y_t given $\theta^t(i)$ and z^{t-1}. Following this procedure, we now have an estimate of $p(\theta_t|z^t)$ which can be resampled to obtain a new estimate of θ^{t+1} and of Y^{t+1} at time $t+1$, and so on. Note that the estimate of the signal is obtained as a by-product in this procedure.

Comments

We have reviewed traditional as well as more recent research approaches to enhancement of noisy speech signals. The section was not intended to be comprehensive but rather to provide a general overview of the area. We have emphasized the methodology and principles of the various approaches, and presented in some more detail one design example of a speech-enhancement system.

Defining Terms

Speech enhancement: A subject dealing with processing of speech signals, in particular of noisy speech signals, aimed at improving their perception by human or their correct decoding by machines.

Quality: A subjective measure of speech perception reflecting individual preferences of listeners.

Intelligibility: An objective measure that predicts the percentage of spoken words (often meaningless) that can be correctly transcribed.

Statistical model: A set of assumptions, formulated in mathematical terms, on the behavior of many examples of signal and noise samples.

Distortion measure: A mathematical function that quantifies the dissimilarity of two speech signals such as the clean and processed signals.

Signal estimator: A function of the observed noisy signal that approximates the clean signal by minimizing a distortion measure based on a given statistical model.

Wiener filter: An optimal linear signal estimator in the minimum mean squared error sense.

Monte Carlo simulation: A statistical approach to develop signal estimators by sampling their statistical model.

Hidden Markov process: A Markov chain observed through a noisy communication channel.

References

1.   M.S. Arulampalam, S. Maskell, N. Gordon, and T. Clapp, "A tutorial on particle filters for online nonlinear/non-Gaussian Bayesian tracking signal processing," *IEEE Trans. Signal Proc.*, 50, February, 174–188, 2002.

2.   M. Askar and H. Derin, "A recursive algorithm for the Bayes solution of the smoothing problem," *IEEE Trans. Automatic Control*, 26, 558–561, 1981.

3.   M. Berouti, R. Schwartz, and J. Makhoul, "Enhancement of speech corrupted by acoustic noise," in *Proc. IEEE Int. Conf. Acoust. Speech Signal Proc.*, 1979, pp. 208–211.

4.   D.R. Brillinger, *Time Series: Data Analysis and Theory*, Philadelphia, PA: SIAM, 2001.

5.   O. Cappé, "Elimination of the musical noise phenomenon with the Ephraim and Malah noise suppressor," *IEEE Trans. Speech and Audio Proc.*, 2, April, 345–349, 1994.

6.   I. Cohen and B. Berdugo, "Speech enhancement for non-stationary noise environments," *Signal Process.*, 81, 2403–2418, 2001.

7.   I. Cohen and B. Berdugo, "Noise estimation by minima controlled recursive averaging for robust speech enhancement," *IEEE Sig. Proc. Lett.*, 9, January, 12–15, 2002.

8.   I. Cohen, "Relaxed statistical model for speech enhancement and *a priori* SNR estimation," *IEEE Trans. Speech Audio Process.*, Vol. 13, September, 2005.

9.   I. Cohen, "Noise spectrum estimation in adverse environments: improved minima controlled recursive averaging," *IEEE Trans. Speech Audio Process.*, 11, September, 466–475, 2003.

10.   T.M. Cover and J.A. Thomas, *Elements of Information Theory*, New York: Wiley, 1991.

11.   A. Dembo and O. Zeitouni, "Maximum a posteriori estimation of time-varying ARMA processes from noisy observations," *IEEE Trans. Acoust. Speech Signal Process.*, 36, April, 471–476, 1988.

12.   H. Drucker, "Speech processing in a high ambient noise environment," *IEEE Trans. Audio Electroacoust.*, AU-16, June, 165–168, 1968.

13.   Y. Ephraim and D. Malah, "Signal to noise ratio estimation for enhancing speech using the Viterbi algorithm," Technion, EE Pub. No. 489, March 1984.

14.   Y. Ephraim and D. Malah, "Speech enhancement using a minimum mean square error short time spectral amplitude estimator," *IEEE Trans. Acoust. Speech Signal Process.*, ASSP-32, December, 1109–1121, 1984.

15.   Y. Ephraim and D. Malah, "Speech enhancement using a minimum mean square error log-spectral amplitude estimator," *IEEE Trans. Acoust. Speech Signal Process.*, ASSP-33, April, 443–445, 1985.

16.   Y. Ephraim, D. Malah, and B.-H. Juang, "On the application of hidden Markov models for enhancing noisy speech," *IEEE Trans. Acoust. Speech Signal Process.*, ASSP-37, December, 1846–1856, 1989.

17. Y. Ephraim, "A Bayesian estimation approach for speech enhancement using hidden Markov models," *IEEE Trans. Signal Process.*, 40, April, 725–735, 1992.

18. Y. Ephraim and H.L. Van Trees, "A signal subspace approach for speech enhancement," *IEEE Trans. Speech Audio Proc.*, 3, July, 251–266, 1995.

19. Y. Ephraim and N. Merhav, "Hidden Markov processes," *IEEE Trans. Inform. Theory*, 48, June, 1518–1569, 2002.

20. W. Fong, S.J. Godsill, A. Doucet, and M. West, "Monte Carlo smoothing with application to audio signal enhancement," *IEEE Trans. Signal Process.*, 50, February, 438–449, 2002.

21. S. Gannot, D. Burshtein, and E. Weinstein, "Iterative and sequential Kalman filter-based speech enhancement algorithms," *IEEE Trans. Speech Audio Process.*, 6, July, 373–385, 1998.

22. A. Gersho and R.M. Gray, *Vector Quantization and Signal Compression*, Boston, MA: Kluwer Academic Publishers, 1991.

23. I.S. Gradshteyn and I.M. Ryzhik, *Table of Integrals, Series, and Products*, New York: Academic Press, 2000.

24. R.M. Gray, "Toeplitz and circulant matrices: II," *Stanford Electron. Lab. Tech. Rep. 6504-1*, April 1977.

25. H. Lev-Ari and Y. Ephraim, "Extension of the signal subspace speech enhancement approach to colored noise," *IEEE Signal Proc. Let.*, 10, April, 104–106, 2003.

26. J.S. Lim and A.V. Oppenheim, "Enhancement and bandwidth compression of noisy speech," *Proc. IEEE*, 67, December, 1586–1604, 1979.

27. R. Martin, "Spectral subtraction based on minimum statistics," in *Proc. 7th Eur. Signal Process. Conf., EUSIPCO-94*, September, 1994, pp. 1182–1185.

28. R. Martin, I. Wittke, and P. Jax, "Optimized estimation of spectral parameters for the coding of noisy speech," *Proc. IEEE Int. Conf. Acoust. Speech Signal Process.*, 9, July, 1479–1482, 2001.

29. R. Martin, "Noise power spectral density estimation based on optimal smoothing and minimum statistics," *IEEE Trans. Speech and Audio Processing*, 9, July, 504–512, 2001.

30. R. Martin and C. Breithaupt, "Speech enhancement in the DFT domain using Laplacian speech priors," *Proc. 8th Int. Workshop Acoustic Echo Noise Control (IWAENC)*, Kyoto, Japan, 8.11, pp. 87.90, September 2003.

31. R. Martin, D. Malah, R.V. Cox, and A.J. Accardi, "A noise reduction preprocessor for mobile voice communication," *EURASIP J. Appl. Signal Process.*, 8, July, 1046–1058, 2004.

32. R.J. McAulay and M.L. Malpass, "Speech enhancement using a soft-decision noise suppression filter," *IEEE Trans. Acoust. Speech Signal Process.*, ASSP-28, April, 137–145, 1980.

33. J. Porter and S. Boll, "Optimal estimators for spectral restoration of noisy speech," *IEEE Int. Conf. Acoust. Speech Signal Proc.*, 9, March, 53–56, 1984.

34. I.Y. Soon, S.N. Koh, and C.K. Yeo, "Improved noise suppression filter using self-adaptive estimator of probability of speech absence," *Signal Process.*, vol. 75, no. 2, pp. 151–159, 1999.

35. I.B. Thomas and A. Ravindran, "Intelligibility enhancement of already noisy speech signals," *J. Audio Eng. Soc.*, 22, May, 234–236, 1974.

36. J. Vermaak, C. Andrieu, A. Doucet, and S.J. Godsill, "Particle methods for Bayesian modeling and enhancement of speech signals," *IEEE Trans. Speech Audio Process.*, 10, March, 173–185, 2002.

37. D.L. Wang and J.S. Lim, "The unimportance of phase in speech enhancement," *IEEE Trans. Acoust. Speech Signal Process.*, ASSP-30, August, 679–681, 1982.

Further Information

The following is a noncomprehensive list of references for further reading on the subject. The edited book by Lim [R1] provides a collection of key papers in the area of speech enhancement. The book by Quatieri [R4] provides extensive background for speech processing including speech enhancement. The National Academy Press Report [R2] details the state of the art of speech enhancement at the time of publication. It also addresses evaluation of speech enhancement systems.

[R1] J.S. Lim, Ed., *Speech Enhancement*. Prentice-Hall, Inc, Englewood Cliffs, NJ, 1983.

[R2] J. Makhoul, T.H. Crystal, D.M. Green, D. Hogan, R.J. McAulay, D.B. Pisoni, R.D. Sorkin, and T.G. Stockham, *Removal of Noise From Noise-Degraded Speech Signals*. Panel on removal of noise from a speech/noise, National Research Council, National Academy Press, Washington, D.C., 1989.

[R3] Y. Ephraim, "Statistical model based speech enhancement systems," *Proc. IEEE*, vol. 80, pp. 1526–1555, Oct. 1992.

[R4] T.F. Quatieri, *Discrete-Time Speech Signal Processing: Principles and Practice*. Prentice-Hall, Englewood Cliffs, NJ, 2001.

[R5] Y. Ephraim, H. Lev-Ari, W.J.J. Roberts, "A Brief Survey of Speech Enhancement," *The Electrical Engineering Handbook*, 3rd ed., CRC Press, Boca Raton, FL, 2005.

[R6] J. Benesty, S. Makino, and J. Chen, Eds., *Speech Enhancement*, Springer, Berlin, 2005.

15.3   Analysis and Synthesis

Jesse W. Fussell

After an acoustic speech signal is converted to an electrical signal by a microphone, it may be desirable to analyze the electrical signal to estimate some time-varying parameters which provide information about a model of the speech production mechanism. **Speech analysis** is the process of estimating such parameters. Similarly, given some parametric model of speech production and a sequence of parameters for that model, **speech synthesis** is the process of creating an electrical signal which approximates speech. While analysis and synthesis techniques may be done either on the continuous signal or on a sampled version of the signal, most modern analysis and synthesis methods are based on digital signal processing.

A typical speech production model is shown in Figure 15.1. In this model the output of the excitation function is scaled by the gain parameter and then filtered to produce speech. All of these functions are time-varying.

For many models, the parameters are varied at a periodic rate, typically 50 to 100 times per second. Most speech information is contained in the portion of the signal below about 4 kHz.

The excitation is usually modeled as either a mixture or a choice of random noise and periodic waveform. For human speech, voiced excitation occurs when the vocal folds in the larynx vibrate; unvoiced excitation occurs at constrictions in the vocal tract which create turbulent air flow (Flanagan, 1965). The relative mix of these two types of excitation is termed **voicing.** In addition, the periodic excitation is characterized by a fundamental frequency, termed **pitch** or F0. The excitation is scaled by a factor designed to produce the proper amplitude or level of the speech signal. The scaled excitation function is then filtered to produce the proper spectral characteristics. While the filter may be nonlinear, it is usually modeled as a linear function.

Analysis of Excitation

In a simplified form, the excitation function may be considered to be purely periodic for voiced speech, or purely random for unvoiced. These two states correspond to voiced phonetic classes such as vowels and nasals and unvoiced sounds such as unvoiced fricatives. This binary voicing model is an oversimplification for sounds such as voiced fricatives, which consist of a mixture of periodic and random components. Figure 15.2 is an example of a time waveform of a spoken /i/ phoneme, which is well modeled by only periodic excitation.

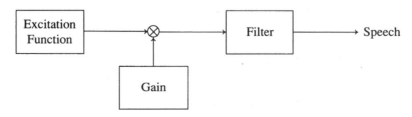

FIGURE 15.1   A general speech production model.

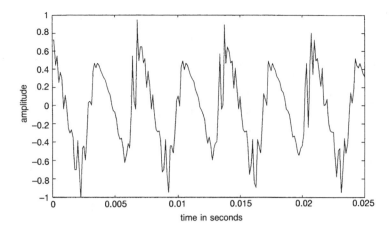

FIGURE 15.2   Waveform of a spoken phoneme /i/ as in "beet."

Both time-domain and frequency-domain analysis techniques have been used to estimate the degree of voicing for a short segment or frame of speech. One time domain feature, termed the zero crossing rate, is the number of times the signal changes sign in a short interval. As shown in Figure 15.2, the zero crossing rate for voiced sounds is relatively low. Since unvoiced speech typically has a larger proportion of high-frequency energy than voiced speech, the ratio of high-frequency to low-frequency energy is a frequency domain technique that provides information on voicing.

Another measure used to estimate the degree of voicing is the autocorrelation function, which is defined for a sampled speech segment, S, as

$$\text{ACF}(\tau) = \frac{1}{N} \sum_{n=0}^{N-1} s(n)s(n - \tau) \qquad (15.34)$$

where $s(n)$ is the value of the nth sample within the segment of length N. Since the autocorrelation function of a periodic function is itself periodic, voicing can be estimated from the degree of periodicity of the autocorrelation function. Figure 15.3 is a graph of the nonnegative terms of the autocorrelation function for a

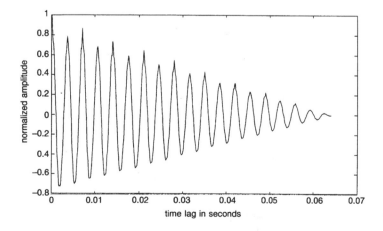

FIGURE 15.3   Autocorrelation function of one frame of /i/.

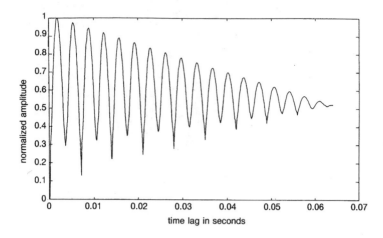

FIGURE 15.4   Absolute magnitude difference function of one frame of /i/.

64 msec frame of the waveform of Figure 15.2. Except for the decrease in amplitude with increasing lag that results from the rectangular window function which delimits the segment, the autocorrelation function is seen to be quite periodic for this voiced utterance.

If an analysis of the voicing of the speech signal indicates a voiced or periodic component is present, another step in the analysis process may be to estimate the frequency (or period) of the voiced component. There are a number of ways in which this may be done. One is to measure the time lapse between peaks in the time domain signal. For example in Figure 15.2 the major peaks are separated by about 0.0071 sec, for a fundamental frequency of about 141 Hz. Note, it would be quite possible to err in the estimate of fundamental frequency by mistaking the smaller peaks that occur between the major peaks for the major peaks. These smaller peaks are produced by resonance in the vocal tract which, in this example, happen to be at about twice the excitation frequency. This type of error would result in an estimate of pitch approximately twice the correct frequency.

The distance between major peaks of the autocorrelation function is a closely related feature that is frequently used to estimate the pitch period. In Figure 15.3, the distance between the major peaks in the autocorrelation function is about 0.0071 sec. Estimates of pitch from the autocorrelation function are also susceptible to mistaking the first vocal track resonance for the glottal excitation frequency.

The absolute magnitude difference function (AMDF), defined as

$$\text{AMDF}(\tau) = \frac{1}{N} \sum_{n=0}^{N-1} |s(n) - s(n - \tau)| \tag{15.35}$$

is another function often used in estimating the pitch of voiced speech. An example of the AMDF is shown in Figure 15.4 for the same 64-msec frame of the /i/ phoneme. However, the minima of the AMDF are used as an indicator of the pitch period. The AMDF has been shown to be a good pitch period indicator (Ross et al., 1974) and does not require multiplication.

Fourier Analysis

One of the more common processes for estimating the spectrum of a segment of speech is the Fourier transform (Oppenheim and Schafer, 1975). The Fourier transform of a sequence is mathematically defined as

$$S(e^{j\omega}) = \sum_{n=-\infty}^{\infty} s(n)e^{-j\omega n} \tag{15.36}$$

where $s(n)$ represents the terms of the sequence. The short-time Fourier transform of a sequence is a time-dependent function, defined as

$$S_m(e^{j\omega}) = \sum_{n=-\infty}^{\infty} w(m-n)\, s(n)e^{-j\omega n} \tag{15.37}$$

where the window function $w(n)$ is usually zero except for some finite range, and the variable m is used to select the section of the sequence for analysis. The discrete Fourier transform (DFT) is obtained by uniformly sampling the short-time Fourier transform in the frequency dimension. Thus an N-point DFT is computed using Equation (15.38),

$$S(k) = \sum_{n=0}^{N-1} s(n)e^{-j2\pi nk/N} \tag{15.38}$$

where the set of N samples, $s(n)$, may have first been multiplied by a window function. An example of the magnitude of a 512-point DFT of the waveform of the /i/ is shown in Figure 15.5. Note for this figure, the 512 points in the sequence have been multiplied by a Hamming window defined by

$$\begin{aligned} w(n) &= 0.54 - 0.46 \ \cos(2\pi n/(N-1)) \quad 0 \leqslant n \leqslant N-1 \\ &= 0 \qquad\qquad\qquad\qquad\qquad\qquad \text{otherwise} \end{aligned} \tag{15.39}$$

Since the spectral characteristics of speech may change dramatically in a few milliseconds, the length, type, and location of the window function are important considerations. If the window is too long, changing spectral characteristics may cause a blurred result; if the window is too short, spectral inaccuracies result. A Hamming window of 16 to 32 msec duration is commonly used for speech analysis.

Several characteristics of a speech utterance may be determined by examination of the DFT magnitude. In Figure 15.5, the DFT of a voiced utterance contains a series of sharp peaks in the frequency domain. These peaks, caused by the periodic sampling action of the glottal excitation, are separated by the fundamental

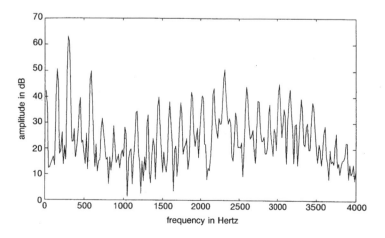

FIGURE 15.5 Magnitude of 512-point FFT of Hamming windowed /i/.

frequency which is about 141 Hz, in this example. In addition, broader peaks can be seen, for example at about 300 Hz and at about 2300 Hz. These broad peaks, called formants, result from resonances in the vocal tract.

Linear Predictive Analysis

Given a sampled (discrete-time) signal $s(n)$, a powerful and general parametric model for time-series analysis is

$$s(n) = -\sum_{k=1}^{p} a(k)s(n-k) + G\sum_{l=0}^{q} b(l)\,u(n-l) \tag{15.40}$$

where $s(n)$ is the output and $u(n)$ is the input (perhaps unknown). The model parameters are $a(k)$ for $k = 1$, p, $b(l)$ for $l = 1$, q, and G. $b(0)$ is assumed to be unity. This model, described as an autoregressive moving average (ARMA) or pole-zero model, forms the foundation for the analysis method termed linear prediction. An autoregressive (AR) or all-pole model, for which all of the "b" coefficients except $b(0)$ are zero, is frequently used for speech analysis (Markel and Gray, 1976).

In the standard AR formulation of linear prediction, the model parameters are selected to minimize the mean-squared error between the model and the speech data. In one of the variants of linear prediction, the autocorrelation method, the minimization is carried out for a windowed segment of data. In the autocorrelation method, minimizing the mean-square error of the time-domain samples is equivalent to minimizing the integrated ratio of the signal spectrum to the spectrum of the all-pole model. Thus, linear predictive analysis is a good method for spectral analysis whenever the signal is produced by an all-pole system. Most speech sounds fit this model well.

One key consideration for linear predictive analysis is the order of the model, p. For speech, if the order is too small, the formant structure is not well represented. If the order is too large, pitch pulses as well as formants begin to be represented. Tenth- or twelfth-order analysis is typical for speech. Figure 15.6 and Figure 15.7 provide examples of the spectrum produced by eighth-order and sixteenth-order linear predictive analysis of the /i/ waveform of Figure 15.2. Figure 15.6 shows there to be three formants at frequencies of about 300, 2300, and 3200 Hz, which are typical for an /i/.

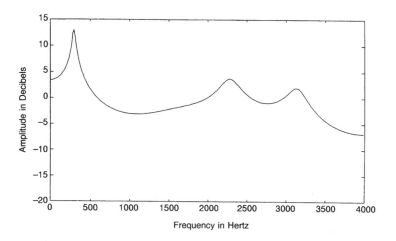

FIGURE 15.6   Eighth-order linear predictive analysis of an "i."

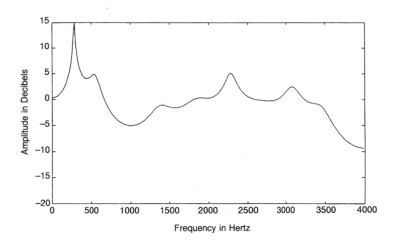

FIGURE 15.7 Sixteenth-order linear predictive analysis of an "i."

Homomorphic (Cepstral) Analysis

For the speech model of Figure 15.1, the excitation and filter impulse response are convolved to produce the speech. One of the problems of speech analysis is to separate or deconvolve the speech into these two components. One such technique is called homomorphic filtering (Oppenheim and Schafer, 1968). The char-acteristic system for a system for homomorphic deconvolution converts a convolution operation to an addition operation. The output of such a characteristic system is called the complex **cepstrum**. The complex cepstrum is defined as the inverse Fourier transform of the complex logarithm of the Fourier transform of the input. If the input sequence is minimum phase (i.e., the z-transform of the input sequence has no poles or zeros outside the unit circle), the sequence can be represented by the real portion of the transforms. Thus, the real cepstrum can be computed by calculating the inverse Fourier transform of the log-spectrum of the input.

Figure 15.8 shows an example of the cepstrum for the voiced /i/ utterance from Figure 15.2. The cepstrum of such a voiced utterance is characterized by relatively large values in the first one or two milliseconds as well as by pulses of decaying amplitudes at multiples of the pitch period. The first two of these pulses can clearly be

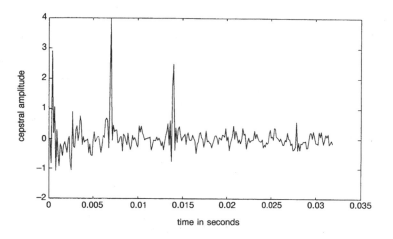

FIGURE 15.8 Real cepstrum of /i/.

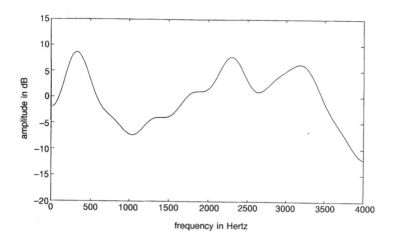

FIGURE 15.9   Smoothed spectrum of /i/ from 16 points of cepstrum.

seen in Figure 15.8 at time lags of 7.1 and 14.2 msec. The location and amplitudes of these pulses may be used to estimate pitch and voicing (Rabiner and Schafer, 1978).

In addition to pitch and voicing estimation, a smooth log magnitude function may be obtained by windowing or "liftering" the cepstrum to eliminate the terms that contain the pitch information. Figure 15.9 is one such smoothed spectrum. It was obtained from the DFT of the cepstrum of Figure 15.7 after first setting all terms of the cepstrum to zero except for the first 16.

Speech Synthesis

Speech synthesis is the creation of speech-like waveforms from textual words or symbols. In general, the speech synthesis process may be divided into three levels of processing (Klatt, 1982). The first level transforms the text into a series of acoustic phonetic symbols, the second transforms those symbols to smoothed synthesis parameters, and the third level generates the speech waveform from the parameters. While speech synthesizers have been designed for a variety of languages and the processes described here apply to several languages, the examples given are for English text-to-speech.

In the first level of processing, abbreviations such as "Dr." (which could mean "doctor" or "drive"), numbers ("1492" could be a year or a quantity), special symbols such as "$," upper-case acronyms (e.g., NASA), and nonspoken symbols such as (apostrophe) are converted to a standard form. Next, prefixes and perhaps suffixes are removed from the body of words prior to searching for the root word in a lexicon, which defines the phonetic representation for the word. The lexicon includes words that do not obey the normal rules of pronunciation, such as "of." If the word is not contained in the lexicon, it is processed by an algorithm that contains a large set of rules of pronunciation.

In the second level, the sequences of words consisting of phrases or sentences are analyzed for grammar and syntax. This analysis provides information to another set of rules that determines the stress, duration, and pitch to be added to the phonemic representation. This level of processing may also alter the phonemic representation of individual words to account for coarticulation effects. Finally, the sequences of parameters that specify the pronunciation are smoothed in an attempt to mimic the smooth movements of the human articulators (lips, jaw, velum, and tongue).

The last processing level converts the smoothed parameters into a time waveform. Many varieties of waveform synthesizers have been used, including formant, linear predictive, and filter-bank versions. These waveform synthesizers generally correspond to the synthesizers used in speech-coding systems, as previously described.

Defining Terms

Cepstrum: Inverse Fourier transform of the logarithm of the Fourier power spectrum of a signal. The complex cepstrum is the inverse Fourier transform of the complex logarithm of the Fourier tranform of the signal.

Pitch: Frequency of glottal vibration of a voiced utterance.

Spectrum or power density spectrum: Amplitude of a signal as a function of frequency, frequently defined as the Fourier transform of the autocovariance of the signal.

Speech analysis: Process of extracting time-varying parameters from the speech signal which represent a model for speech production.

Speech synthesis: Production of a speech signal from a model for speech production and a set of time-varying parameters of that model.

Voicing: Classification of a speech segment as being voiced (i.e., produced by glottal excitation), unvoiced (i.e., produced by turbulent air flow at a constriction), or some mix of those two.

References

1. J. Allen, "Synthesis of speech from unrestricted text," *Proc. IEEE,* vol. 64, no. 4, pp. 433–442, 1976.
2. J.L. Flanagan, *Speech Analysis, Synthesis and Perception,* Berlin: Springer-Verlag, 1965.
3. D.H. Klatt, "The Klattalk Text-to-Speech System," IEEE Int. Conf. on Acoustics, Speech and Signal Proc., pp. 1589–1592, Paris, 1982.
4. J.D. Markel and A.H. Gray, Jr., *Linear Prediction of Speech,* Berlin: Springer-Verlag, 1976.
5. A.V. Oppenheim and R.W. Schafer, "Homomorphic analysis of speech," *IEEE Trans. Audio Electroacoust.,* pp. 221–226, 1968.
6. A.V. Oppenheim and R.W. Schafer, *Digital Signal Processing,* Englewood Cliffs, NJ: Prentice-Hall, 1975.
7. D. O'Shaughnessy, *Speech Communication,* Reading, MA: Addison-Wesley, 1987.
8. L.R. Rabiner and R.W. Schafer, *Digital Processing of Speech Signals,* Englewood Cliffs, NJ: Prentice-Hall, 1978.
9. M.J. Ross, H.L. Shaffer, A. Cohen, R. Freudberg, and H.J. Manley, "Average magnitude difference function pitch extractor," *IEEE Trans. Acoustics, Speech and Signal Proc.,* vol. ASSP-22, pp. 353–362, 1974.
10. R.W. Schafer and J.D. Markel, *Speech Analysis,* New York: IEEE Press, 1979.

Further Information

The monthly journal *IEEE Transactions on Signal Processing,* formerly *IEEE Transactions on Acoustics, Speech and Signal Processing,* frequently contains articles on speech analysis and synthesis. In addition, the annual conference of the IEEE Signal Processing Society, the International Conference on Acoustics, Speech, and Signal Processing, is a rich source of papers on the subject.

15.4 Speech Recognition

Lynn D.Wilcox and Marcia A. Bush

Speech recognition is the process of translating an acoustic signal into a linguistic message. In certain applications, the desired form of the message is a verbatim transcription of a sequence of spoken words. For example, in using speech-recognition technology to automate dictation or data entry to a computer, transcription accuracy is of prime importance. In other cases, such as when speech recognition is used as an interface to a database query system or to index by keyword into audio recordings, word-for-word transcription is less critical. Rather, the message must contain only enough information to reliably communicate the speaker's goal. The use of speech-recognition technology to facilitate a dialog between person and computer is often referred to as "spoken language processing."

Speech recognition by machine has proven an extremely difficult task. One complicating factor is that, unlike written text, no clear spacing exists between spoken words; speakers typically utter full phrases or sentences without pause. Furthermore, acoustic variability in the speech signal typically precludes an unambiguous mapping to a sequence of words or subword units, such as phones.[1] One major source of variability in speech is coarticulation, or the tendency for the acoustic characteristics of a given speech sound or phone to differ depending upon the phonetic context in which it is produced. Other sources of acoustic variability include differences in vocal-tract size, dialect, speaking rate, speaking style, and communication channel.

Speech-recognition systems can be constrained along a number of dimensions in order to make the recognition problem more tractable. Training the parameters of a recognizer to the speech of the user is one way of reducing variability and, thus, increasing recognition accuracy. Recognizers are categorized as speaker-dependent or speaker-independent, depending upon whether or not full training is required by each new user. Speaker-adaptive systems adjust automatically to the voice of a new talker, either on the basis of a relatively small amount of training data or on a continuing basis while the system is in use.

Recognizers can also be categorized by the speaking styles, vocabularies, and language models they accommodate. **Isolated word recognizers** require speakers to insert brief pauses between individual words. **Continuous speech recognizers** operate on fluent speech, but typically employ strict language models, or grammars, to limit the number of allowable word sequences. **Wordspotters** also accept fluent speech as input. However, rather than providing full transcription, wordspotters selectively locate relevant words or phrases in an utterance. Wordspotting is useful both in information-retrieval tasks based on keyword indexing and as an alternative to isolated word recogniton in voice command applications.

Speech-Recognition System Architecture

Figure 15.10 shows a block diagram of a speech-recognition system. Speech is typically input to the system using an analog transducer, such as a microphone, and converted to digital form. **Signal pre-processing** consists of computing a sequence of acoustic feature vectors by processing the speech samples in successive time intervals. In some systems, a clustering technique known as vector quantization is used to convert these continuous-valued features to a sequence of discrete codewords drawn from a codebook of acoustic prototypes. Recognition of an unknown utterance involves transforming the sequence of feature vectors, or codewords, into an appropriate message. The recognition process is typically constrained by a set of acoustic models which correspond to the basic units of speech employed in the recognizer, a lexicon which defines the vocabulary of the recognizer in terms of these basic units, and a language model which specifies allowable sequences of vocabulary items. The acoustic models, and in some cases the language model and lexicon, are learned from a set of representative training data. These components are discussed in greater detail in the remainder of this section, as are the two recognition paradigms most frequently employed in speech recognition: **dynamic time warping** and **hidden Markov models.**

Signal Pre-Processing

An amplitude waveform and speech spectrogram of the sentence "Two plus seven is less than ten" is shown in Figure 15.11. The spectrogram represents the time evolution (horizontal axis) of the frequency spectrum (vertical axis) of the speech signal, with darkness corresponding to high energy. In this example, the speech has been digitized at a sampling rate of 16 kHz, or roughly twice the highest frequency of relevant energy in a high-quality speech signal. In general, the appropriate sampling rate is a function of the communication channel. In telecommunications, for example, a bandwidth of 4 kHz and, thus, a Nyquist sampling rate of 8 kHz, is standard.

---

[1]Phones correspond roughly to pronunciations of consonants and vowels.

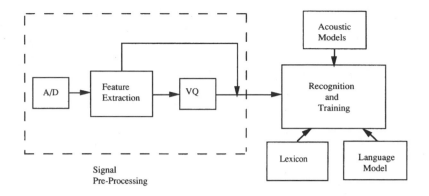

FIGURE 15.10   Architecture for a speech-recognition system.

The speech spectrum can be viewed as the product of a source spectrum and the transfer function of a linear, time-varying filter which represents the changing configuration of the vocal tract. The transfer function determines the shape, or envelope, of the spectrum, which carries phonetic information in speech. When excited by a voicing source, the formants, or natural resonant frequencies of the vocal tract, appear as black bands running horizontally through regions of the speech spectrogram. These regions represent voiced segments of speech and correspond primarily to vowels. Regions characterized by broadband high-frequency energy, and by extremely low energy, result from noise excitation and vocal-tract closures, respectively, and are associated with the articulation of consonantal sounds.

Feature extraction for speech recognition involves computing sequences of numeric measurements, or feature vectors, which typically approximate the envelope of the speech spectrum. Spectral features can be extracted directly from the discrete Fourier transform (DFT) or computed using linear predictive coding (LPC) techniques. Cepstral analysis can also be used to deconvolve the spectral envelope and the periodic voicing source. Each feature vector is computed from a frame of speech data defined by windowing N samples of the signal. While a better spectral estimate can be obtained using more samples, the interval must be short enough so that the windowed signal is roughly stationary. For speech data, N is chosen such

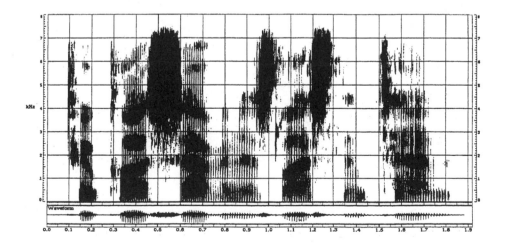

FIGURE 15.11   Speech spectrogram of the utterance "Two plus seven is less than ten." (*Source:* V.W. Zue, "The use of speech knowledge in automatic speech recognition," *Proc. IEEE*, vol. 73, no. 11, pp. 1602–1615, © 1985 IEEE. With permission.)

that the length of the interval covered by the window is approximately 25 to 30 msec. The feature vectors are typically computed at a frame rate of 10 to 20 msec by shifting the window forward in time. Tapered windowing functions, such as the Hamming window, are used to reduce dependence of the spectral estimate on the exact temporal position of the window. Spectral features are often augmented with a measure of the short time energy of the signal, as well as with measures of energy and spectral change over time (Lee, 1988).

For recognition systems that use discrete features, vector quantization can be used to quantize continuous-valued feature vectors into a set or codebook of K discrete symbols, or codewords (Gray, 1984). The K codewords are characterized by prototypes $y^1 \ldots y^K$. A feature vector x is quantized to the kth codeword if the distance from x to y^k, or $d(x, y^k)$, is less than the distance from x to any other codeword. The distance $d(x, y)$ depends on the type of features being quantized. For features derived from the short-time spectrum and cepstrum, this distance is typically Euclidean or weighted Euclidean. For LPC-based features, the Itakura metric, which is based on spectral distortion, is typically used (Furui, 1989).

Dynamic Time Warping

Dynamic time warping (DTW) is a technique for nonlinear time alignment of pairs of spoken utterances. DTW-based speech recognition, often referred to as "template matching," involves aligning feature vectors extracted from an unknown utterance with those from a set of exemplars or templates obtained from training data. Nonlinear feature alignment is necessitated by nonlinear time-scale warping associated with variations in speaking rate.

Figure 15.12 illustrates the time correspondence between two utterances, A and B, represented as feature-vector sequences of unequal length. The time-warping function consists of a sequence of points $F = c_1, \ldots, c_K$ in the plane spanned by A and B, where $c_k = (i_k, j_k)$. The local distance between the feature vectors a_i and b_j on the warping path at point $c = (i, j)$ is given as

$$d(c) = d(a_i, b_j) \tag{15.41}$$

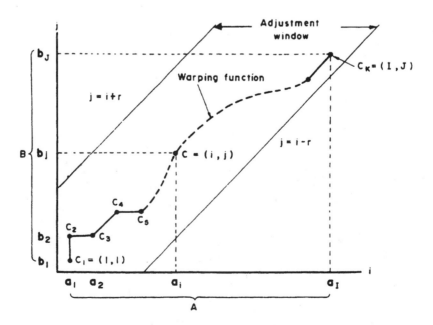

FIGURE 15.12   Dynamic time warping of utterances A and B. (*Source:* S. Furui, *Digital Speech Processing, Synthesis and Recognition*, New York: Marcel Dekker, 1989. With permission.)

The distance between A and B aligned with warping function F is a weighted sum of the local distances along the path:

$$D(F) = \frac{1}{N} \sum_{k=1}^{k} d(c_k) w_k \qquad (15.42)$$

where w_k is a nonnegative weighting function and N is the sum of the weights. Path constraints and weighting functions are chosen to control whether or not the distance $D(F)$ is symmetric and the allowable degree of warping in each direction. Dynamic programming is used to efficiently determine the optimal time alignment between two feature-vector sequences (Sakoe and Chiba, 1978).

In DTW-based recognition, one or more templates are generated for each word in the recognition vocabulary. For speaker-dependent recognition tasks, templates are typically created by aligning and averaging the feature vectors corresponding to several repetitions of a word. For speaker-independent tasks, clustering techniques can be used to generate templates which better model pronunciation variability across talkers. In isolated word recognition, the distance $D(F)$ is computed between the feature-vector sequence for the unknown word and the templates corresponding to each vocabulary item. The unknown is recognized as that word for which $D(F)$ is a minimum. DTW can be extended to connected word recognition by aligning the input utterance to all possible concatenations of reference templates. Efficient algorithms for computing such alignments have been developed (Furui, 1989); however, in general, DTW has proved most applicable to isolated word-recognition tasks.

Hidden Markov Models[1]

Hidden Markov modeling is a probabilistic pattern matching technique which is more robust than DTW at modeling acoustic variability in speech and more readily extensible to continuous speech recognition. As shown in Figure 15.13, hidden Markov models (HMMs) represent speech as a sequence of states, which are assumed to model intervals of speech with roughly stationary acoustic features. Each state is characterized by an output probability distribution which models variability in the spectral features or observations associated with that state. Transition probabilities between states model durational variability in the speech signal. The probabilities, or parameters, of an HMM are trained using observations (VQ codewords) extracted from a representative sample of speech data. Recognition of an unknown utterance is based on the probability that the speech was generated by the HMM.

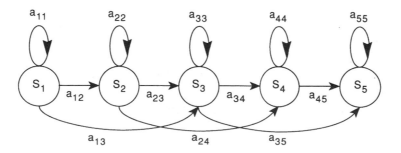

FIGURE 15.13 A typical HMM topology.

---

[1]Although the discussion here is limited to HMMs with discrete observations, output distributions such as Gaussians can be defined for continuous-valued features.

More precisely, an HMM is defined by:

1. A set of N states $\{S_1 \ldots S_N\}$, where q_t is the state at time t
2. A set of K observation symbols $\{v_1 \ldots v_K\}$, where O_t is the observation at time t
3. A state transition probability matrix $A = \{a_{ij}\}$, where the probability of transitioning from state S_i at time t to state S_j at time $t + 1$ is $a_{ij} = P(q_{t+1} = S_j | q_t = S_i)$
4. A set of output probability distributions B, where for each state j, $b_j(k) = P(O_t = v_k | q_t = S_j)$
5. An initial state distribution $\pi = \{\pi_i\}$, where $\pi_i = P(q_1 = S_i)$

At each time t, a transition to a new state is made, and an observation is generated. State transitions have the Markov property, in that the probability of transitioning to a state at time t depends only on the state at time $t - 1$. The observations are conditionally independent given the state, and the transition probabilites are not dependent on time. The model is called hidden because the identity of the state at time t is unknown; only the output of the state is observed. It is common to specify an HMM by its parameters $\lambda = (A, B, \pi)$.

The basic acoustic unit modeled by the HMM can be either a word or a subword unit. For small recognition vocabularies, the lexicon typically consists of whole-word models similar to the model shown in Figure 15.13. The number of states in such a model can either be fixed or be made to depend on word duration. For larger vocabularies, words are more often defined in the lexicon as concatenations of phone or triphone models. Triphones are phone models with left and right context specified (Lee, 1988); they are used to model acoustic variability which results from the coarticulation of adjacent speech sounds.

In isolated word recognition tasks, an HMM is created for each word in the recognition vocabulary. In continuous speech recognition, however, a single HMM network is generated by expressing allowable word strings or sentences as concatenations of word models, as shown in Figure 15.14. In wordspotting, the HMM network consists of a parallel connection of keyword models and a background model which represents the speech within which the keywords are embedded. Background models, in turn, typically consist of parallel connections of subword acoustic units such as phones (Wilcox and Bush, 1992).

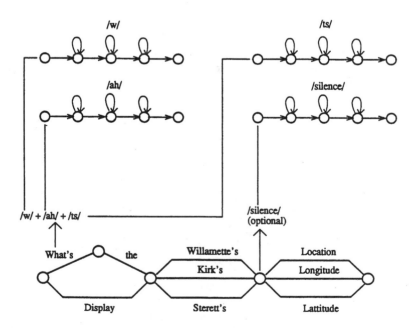

FIGURE 15.14 Language model, lexicon, and HMM phone models for a continuous speech-recognition system. (*Source:* K.F. Lee, "Large-Vocabulary Speaker-Independent Continuous Speech Recognition: The SPHINX System," Ph.D. dissertation, Computer Science Dept., Carnegie Mellon, April 1988. With permission.)

The language model or grammar of a recognition system defines the sequences of vocabulary items that are allowed. For simple tasks, deterministic finite-state grammars can be used to define all allowable word sequences. Typically, however, recognizers make use of stochastic grammars based on n-gram statistics (Jelinek, 1985). A bigram language model, for example, specifies the probability of a vocabulary item given the item which precedes it.

Isolated word recognition using HMMs involves computing, for each word in the recognition vocabulary, the probability $P(O|\lambda)$ of the observation sequence $O = O_1 \dots O_T$. The unknown utterance is recognized as the word which maximizes this probability. The probability $P(O|\lambda)$ is the sum over all possible state sequences $Q = q_1 \dots q_T$ of the probability of O and Q given λ, or:

$$p(O|\lambda) = \sum_Q P(O, Q|\lambda) = \sum_Q P(O|Q, \lambda)P(Q|\lambda) = \sum_{q_1 \dots q_T} \pi_{q_1} b_{q_1}(O_1) a_{q_1 q_2} b_{q_2}(O_2) \dots \quad (15.43)$$

Direct computation of this sum is computationally infeasible for even a moderate number of states and observations. However, an iterative algorithm known as the **forward–backward** procedure (Rabiner, 1989) makes this computation possible. Defining the forward variable α as

$$\alpha_t(i) = P(O_1 \dots O_t, q_t = S_i|\lambda) \quad (15.44)$$

and initializing $\alpha_1(i) = \pi_i b_i(O_1)$, subsequent $\alpha_t(i)$ are computed inductively as

$$\alpha_{t+1}(j) = \sum_{i=1}^{N} \alpha_t(i) a_{ij} b_j(O_{t+1}) \quad (15.45)$$

By definition, the desired probability of the observation sequence given the model λ is

$$P(O/\lambda) = \sum_{i=1}^{N} \alpha_T(i) \quad (15.46)$$

Similarly, the backward variable β can be defined:

$$\beta_t(i) = P(O_{t+1} \dots O_T|q_t = S_i, \lambda) \quad (15.47)$$

The βs are computed inductively backward in time by first initializing $\beta_T(j) = 1$ and computing

$$\beta_t(i) = \sum_{j=1}^{N} a_{ij} b_j(O_{t+1}) \beta_{t+1}(j) \quad (15.48)$$

HMM-based continuous speech recognition involves determining an optimal word sequence using the **Viterbi** algorithm. This algorithm uses dynamic programming to find the optimal state sequence through an HMM network representing the recognizer vocabulary and grammar. The optimal state sequence $Q^* = (q_1^* \dots q_T^*)$ is defined as the sequence which maximizes $P(Q|O,\lambda)$, or equivalently $P(Q,O|\lambda)$. Let $\delta_t(i)$ be the joint probability of the optimal state sequence and the observations up to time t, ending in state S_i at time t. Then:

$$\delta_t(i) = \max P(q_1 \dots q_{t-1}, q_t = S_i, O_1 \dots O_t|\lambda) \quad (15.49)$$

where the maximum is over all state sequences $q_1 \dots q_{t-1}$. This probability can be updated recursively by extending each partial optimal path using

$$\delta_{t+1}(j) = \max_{i} \delta_t(i) a_{ij} b_j(O_{t+1}) \tag{15.50}$$

At each time t, it is necessary to keep track of the optimal precursor to state j, that is, the state which maximized the above probability. Then, at the end of the utterance, the optimal state sequence can be retrieved by backtracking through the precursor list.

   Training HMM-based recognizers involves estimating the parameters for the word or phone models used in the system. As with DTW, several repetitions of each word in the recognition vocabulary are used to train HMM-based isolated word recognizers. For continuous speech recognition, word or phone exemplars are typically extracted from word strings or sentences (Lee, 1988). Parameters for the models are chosen based on a maximum likelihood criterion; that is, the parameters λ maximize the likelihood of the training data O, $P(O|\lambda)$. This maximization is performed using the **Baum-Welch** algorithm (Rabiner, 1989), a re-estimation technique based on first aligning the training data O with the current models, and then updating the parameters of the models based on this alignment.

   Let $\xi_t(i,j)$ be the probability of being in state S_i at time t and state S_j at time $t + 1$ and observing the observation sequence O: Using the forward and backward variables $\alpha_t(i)$ and $\beta_t(j)$, $\xi_t(i,j)$ can be written as

$$\xi_t(i,j) = P(q_t = S_i, q_{t+1} = S_j | O, \lambda) = \frac{\alpha_t(i) a_{ij} \beta_{t+1}(j) b_j(O_{t+1})}{\displaystyle\sum_{ij=1}^{N} \alpha_t(i) a_{ij} \beta_{t+1}(j) b_j(O_{t+1})} \tag{15.51}$$

An estimate of a_{ij} is given by the expected number of transitions from state S_i to state S_j divided by the expected number of transitions from state S_i. Define $\gamma_t(i)$ as the probability of being in state S_i at time t, given the observation sequence O:

$$\gamma_t(i) = P(q_t = S_i | O, \lambda) = \sum_{j=1}^{N} \xi_t(i,j) \tag{15.52}$$

Summing $\gamma_t(i)$ over t yields a quantity which can be interpreted as the expected number of transitions from state S_i. Summing $\xi_t(i,j)$ over t gives the expected number of transitions from state i to state j. An estimate of a_{ij} can then be computed as the ratio of these two sums. Similarly, an estimate of $b_j(k)$ is obtained as the expected number of times being in state j and observing symbol v_k divided by the expected number of times in state j:

$$\hat{a}_{ij} = \frac{\displaystyle\sum_{t=1}^{T-1} \xi_t(i,j)}{\displaystyle\sum_{t=1}^{T} \gamma_t(i)} \qquad \hat{b}_j(k) = \frac{\displaystyle\sum_{t:O_t = y_k} \gamma_t(j)}{\displaystyle\sum_{t=1}^{T} \gamma_t(j)} \tag{15.53}$$

State-of-the-Art Recognition Systems

Dictation-oriented recognizers that accommodate isolated word vocabularies of many thousands of words in speaker-adaptive manner are currently available commercially. So too are speaker-independent, continuous speech recognizers for small vocabularies, such as the digits; similar products for larger (1000-word) vocabularies with constrained grammars are imminent. Speech recognition research is aimed, in part, at the development of more robust pattern classification techniques, including some based on neural networks (Lippmann, 1989) and on the development of systems that accommodate more natural spoken language dialogs between human and machine.

Defining Terms

Baum–Welch: A re-estimation technique for computing optimal values for HMM state transition and output probabilities.

Continuous speech recognition: Recognition of fluently spoken utterances.

Dynamic time warping (DTW): A recognition technique based on nonlinear time alignment of unknown utterances with reference templates.

Forward–backward: An efficient algorithm for computing the probability of an observation sequence from an HMM.

Hidden Markov model (HMM): A stochastic model which uses state transition and output probabilities to generate observation sequences.

Isolated word recognition: Recognition of words or short phrases preceded and followed by silence.

Signal pre-processing: Conversion of an analog speech signal into a sequence of numeric feature vectors or observations.

Viterbi: An algorithm for finding the optimal state sequence through an HMM given a particular observation sequence.

Wordspotting: Detection or location of keywords in the context of fluent speech.

References

S. Furui, *Digital Speech Processing, Synthesis, and Recognition,* New York: Marcel Dekker, 1989.

R.M. Gray, "Vector quantization," *IEEE ASSP Magazine,* vol. 1, no. 2, pp. 4–29, April 1984.

F. Jelinek, "The development of an experimental discrete dictation recognizer," *Proc. IEEE,* vol. 73, no. 11, pp. 1616–1624, Nov. 1985.

K.F. Lee, "Large-Vocabulary Speaker-Independent Continuous Speech Recognition: The SPHINX System," Ph.D. dissertation, Computer Science Department, Carnegie Mellon University, April 1988.

R.P. Lippmann, "Review of neural networks for speech recognition," *Neural Computation,* vol. 1, pp. 1–38, 1989.

L.R. Rabiner, "A tutorial on hidden Markov models and selected applications in speech recognition," *Proc. IEEE,* vol. 77, no. 2, pp. 257 285, Feb. 1989.

H. Sakoe and S. Chiba, "Dynamic programming algorithm optimization for spoken word recognition," *IEEE Transactions on Acoustics, Speech and Signal Processing,* vol. 26, no. 1, pp. 43–49, Feb. 1978.

L.D. Wilcox and M.A. Bush, "Training and search algorithms for an interactive wordspotting system," in *Proceedings, International Conference on Acoustics, Speech and Signal Processing,* San Francisco, March 1992, pp. II-97–II-100.

V.W. Zue, "The use of speech knowledge in automatic speech recognition," *Proc. IEEE,* vol. 73, no. 11, pp. 1602–1615, Nov. 1985.

Further Information

Papers on speech recognition are regularly published in *IEEE Speech and Audio Transactions* (formerly part of the *IEEE Transactions on Acoustics, Speech and Signal Processing*) and in the journal *Computer Speech and Language.* Speech recognition research and technical exhibits are presented at the annual IEEE International Conference on Acoustics, Speech and Signal Processing (ICASSP), the biannual European Conference on Speech Communication and Technology (Eurospeech), and the biannual International Conference on Spoken Language Processing (ICSLP), all of which publish proceedings. Commercial applications of speech recognition technology are featured at annual American Voice Input-Output Society (AVIOS) and Speech Systems Worldwide meetings. A variety of standardized databases for speech recognition system development are available from the National Institute of Standards and Technology in Gaithersburg, MD.

16

Text-to-Speech (TTS) Synthesis

Juergen Schroeter

AT&T Labs

16.1   Introduction

The goal of text-to-speech (TTS) synthesis is to convert arbitrary input text to intelligible and natural sounding speech so as to transmit information from a machine to a person. Therefore, TTS goes beyond simple "cut-and-paste" systems used, for example, in some telecom applications to read back a phone number. Such systems string together words spoken in isolation and the artifacts of such a scheme are often perceptible. The methodology used in TTS is to exploit acoustic representations of speech for synthesis, together with linguistic analyses of text to extract correct pronunciations ("content"; *what* is being said) and prosody in context ("melody" of a sentence; *how* it is being said). Synthesis systems are commonly evaluated in terms of three characteristics: *accuracy* of rendering the input text (does the TTS system pronounce, e.g., acronyms, names, URLs, email addresses as a knowledgeable human would?), *intelligibility* of the resulting voice message (measured as a percentage of a test set that is understood), and perceived *naturalness* of the resulting speech (does the TTS sound like a recording of a live human?). Today, applications of TTS are in automated telecom services (e.g., name and address rendering), as a part of a network voice server for e-mail (e-mail by phone), in directory assistance, as an aid in providing up-to-the-minute information to a telephone user (e.g., business locator services, banking services, helplines), in computer games, and last but not least, in aids to the handicapped (e.g., cosmologist Steven Hawking). For a much more detailed overview of TTS and its applications, see Reference 1 and 2.

16.2   Overview of TTS

A block diagram of a general TTS engine is depicted in Figure 16.1. We distinguish a TTS "front-end" (i.e., the part of the system closer to the text input) from a TTS "back-end" (i.e., the part of the system that is closer to the speech output). Input text, optionally enriched by tags that control prosody or other characteristics, enters the front-end where a text analysis module detects the document structure (in terms of, e.g., lists vs. running text, paragraph breaks, sentence breaks, etc.), followed by text normalization (expansion to literal word tokens,

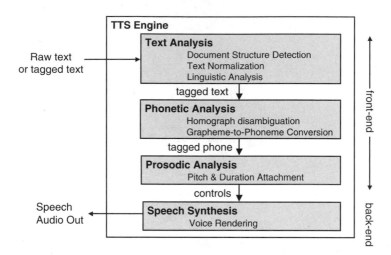

FIGURE 16.1 Block diagram of a general text-to-speech system.

encompassing transcription of acronyms, abbreviations, currency, dates, times, URLs, etc.), and further linguistic analysis that enables other tasks down the line. The tagged text then enters a phonetic analysis module that performs *homograph* disambiguation, and *grapheme*-to-*phoneme* conversion. The latter process is also called "letter-to-sound" conversion. The string of tagged *phones* enters a prosodic analysis module that determines *pitch*, duration (and amplitude) targets for each phone. Finally, the string of symbols that was derived from a given input sentence is passed on to the speech synthesis module where it controls the voice rendering that corresponds to the input text.

16.3 Front-End Issues

The text analysis and normalization module in the front-end determines to a large extend the "what" and "how" of the resulting synthetic speech. Note that punctuation (e.g., "." and "?") is not infallible. For example, the TTS system should not misinterpret the dot after "in" in the example "The table is 36.5 in. long" as the end of the sentence. In addition, punctuation and other special characters can be part of a time (e.g., 7:30 pm), or date (e.g., 5/25/2004), or currency expression ($10 = "ten dollars"). Text normalization is difficult because it is context sensitive (e.g., $1.5 million = "one point five million dollars").

Abbreviations and acronyms fall in either of two categories. The first category contains a finite set of known "mappings" such as "Dr." in the sentence "Dr. Smith lives on Smith Dr." Note that a mapping may be ambiguous (Dr. can be "doctor" or "drive"). More difficult to handle, however, is the open category of abbreviations that people invent on the fly. COMM could mean "communications," "committee," etc. Other possible short forms are CMNCTNS or COMMS. Also, a specific abbreviation can have different expansions depending on the task or topic domain. For example, DC could mean "direct current" in this book, but could mean "District of Columbia" as part of an address. Therefore, it may be necessary to handle certain text normalization tasks in the form of a domain-specific text "filter" that would alter the raw text before it is passed on to the TTS system depicted in Figure 16.1. Applications like e-mail reading or web page reading, for example, also require text filters to strip out mundane header or formatting information. Even the "simple" reading of numbers can be difficult, such as "370," where the 370 can be part of a phone number (370-1111, read as "three-seven-zero...") or part of a name (e.g., IBM370, read as "i-b-m-three-seventy"). Note that the performance of the text analysis and normalization module affects the *accuracy* rating of a TTS system.

Linguistic analysis in the front-end encompasses the determination of parts-of-speech (POS), word sense, emphasis, appropriate speaking style, and speech acts (such as, e.g., greetings, apologies, etc.). A linguistic parser could be used, but typically only a shallow analysis is done for reasons of computational speed.

Grapheme-to-phoneme conversion involves word pronunciation. The mapping from orthography to phonemes can be difficult because of context dependence. This is usually handled by trained classification and regression (decision) trees (CARTs) that capture the probabilities of specific conversions given the context or the fact that a specific word is a homomorph. Grammatical (POS) analysis helps ("*Wind* down and let the *wind* blow," "*Lead* the way down by following the way *lead* drops.") At the word level, pronunciation dictionaries combined with *morphological* decomposition are used. Finally, letter-to-sound rules are employed as a fall-back in most TTS systems. However, names are a difficult problem because even people with the same name may pronounce it differently and the identification of a name in the text is also difficult (e.g., "Begin met the president" vs. "Begin the work now!"). Finally, note that pronunciation is very much speaker- and accent-dependent, even within the same language. Because of these difficulties, automation of creating speaker-dependent pronunciation dictionaries currently is an active area of research.

Prosody determines *how* a sentence is spoken in terms of melody, phrasing, rhythm, accent locations, and emotions. Prosody may even carry meaning, even in nontonal languages (e.g., "I like the Eggs Benedict" vs. "I like the eggs, Benedict"). Therefore, prosody affects naturalness *and* intelligibility ratings of a TTS system. Prosody is difficult to annotate automatically. Mainly for use in manual annotation efforts, so-called tone and break indices (ToBI) have emerged as the standard [3]. Speech-signal correlates of prosody are the presence and duration of pauses, the pitch (fundamental frequency value, in particular, its dynamic behavior), as well as phone durations and amplitudes. As in the case of pronunciation, prosody can be dependent on the speaker gender, on the specific speech act or application task, and even on the individual speaker. This may be one reason why researchers now seem to move to using data-driven prosody approaches that employ large speech databases of a single speaker over rule-based systems [4].

16.4 Back-End Issues

An important decision a designer of a TTS system needs to make is which synthesis method to use. In the literature [5] we find two basic categories of methods: *synthesis by rule* and *concatenative synthesis*. Synthesis by rule exploits the expert knowledge of speech scientists in speech production and perception by putting the human expert in the design (and quality) loop. Concatenative synthesis employs recordings of a human speaker, inherently putting more emphasis on data. Experts generally disagree on which is the more promising approach although, today, the concatenative approach clearly produces higher quality synthetic speech.

Articulatory synthesis uses computational models of the articulators (e.g., tongue, lips, etc.) and the glottis (voice box) to synthesize speech. Instead of describing the speech signal itself, it employs control parameters such as tongue position and movement, lip and glottal opening, etc., that are meaningful in speech production. Therefore, articulatory synthesis has appeal to speech scientists that explore speech-production theories and related topics. Mathematically, articulatory synthesizers can be as simple as describing the vocal tract as a straight tube of variable cross-section [6], or as complicated as solving the full blown Navier–Stokes equations [7]. Early articulatory synthesizers clearly followed the synthesis by rule approach [8,9]; recently, researchers have adopted some data-driven concepts [10]. Although the method of choice for a few specific uses, articulatory synthesis has so far not delivered synthesis that can be mistaken as a recording of a specific speaker. In addition, deriving rules or dynamic system control parameters for it is computationally expensive and/or requires speech production measurements (e.g., dynamic MRI [11], tracing of x-ray pellets [12]) that may be considered as somewhat "invasive." However, a highly accurate articulatory synthesizer still carries the promise of producing completely "tunable" (e.g., made to sound like different speakers, speaking styles, etc.) high-quality speech that then could be used to "train" other, more practical synthesizers.

Formant Synthesis

Within the rule-based synthesis category, *formant synthesis* is the most prominent example. Well-known examples include efforts by Dennis Klatt [13,14], whose TTS system later was productized as "MITalk" and "DECTalk," and is now maintained and sold commercially by Scansoft. In its simplest form, formant synthesis employs second-order filter sections either in cascade/series, or in a parallel structure. Figure 16.2 depicts these

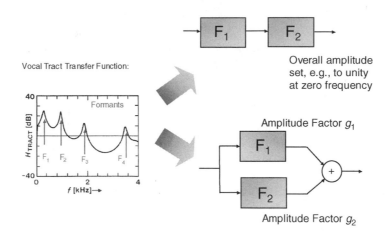

FIGURE 16.2 The vocal-tract transfer function (left) shows four formants in the frequency range from 0 to 4 kHz. It can be approximated in two different ways using second-order filter sections: serial (top right) or parallel (bottom right).

two options. Starting from the vocal-tract transfer function on the left that relates volume flow at the lips (output) to volume flow at the glottis (input), the task is to approximate all vocal-tract resonances (peaks in the transfer function, "formants") by a network of second-order filters, depicted on the right. It can be shown that the series representation of filters (top right) approximates a nonnasal (no nasal side branch) vocal tract reasonably well. For example, given about three to five filter sections (each matching one formant), we are able to approximate vocal-tract transfer functions for vowels even between the formant frequencies (peaks). In this approach, we only need to specify formant frequencies and bandwidths, plus an overall gain factor. However, for nasal sounds, as well as for any fricative or mixed (voiced/unvoiced) sounds, a series representation of second-order filters may not be good enough. Parallel filters (shown in the bottom right of Figure 16.2) create the flexibility to approximate *any* speech spectrum (see, e.g., Reference 15), but require individual gain factors, in addition to formant frequencies and bandwidths. As a side effect, parallel filters also introduce spectral zeros between the formant frequencies that may be cancelled by special correction filters.

The source excitation can be either voice pulses (approximating glottal pulses) or noise (approximating fricative or aspirative noise sources in the vocal tract), or mixtures of the two. For example, for the word "two" the /t/ phoneme pattern might specify a sudden (burst) onset of noise excitation of a filter with a flat characteristic followed by the /uw/ pattern of voice excitation of three formant filters with peak frequencies at 300, 850, and 2250 Hz, representing formants F1, F2, and F3, respectively. Note that in this case, one usually also sees a transition region between the burst and the vowel that is filled with aspiration noise but already shows the formant structure of the following vowel.

Deriving rules for synthesizing running speech is the main problem for formant synthesis. These rules specify the timings of source (voiced/unvoiced) and the dynamic values of all filter parameters. This is difficult to do by hand, even for simple words, let alone full sentences at high intelligibility. Automatic derivation of these rules can be achieved by analysis-by-synthesis approaches that try to mimic input speech [16]. A commercial system that makes use of articulatory and acoustic-phonetic knowledge and drives the Klatt synthesizer is HLSyn [17].

Formant synthesizers have moderate computational requirements. Some implementations allow a full TTS system to run within 2 MB of memory or less. Therefore, embedded applications, such as in handheld devices, for example, talking dictionaries, calendars, etc., or even in cellphones (e.g., for reading back names and key presses for car drivers who should not take their eyes off the road) are enabled by formant synthesis. Voice quality can be controlled to a high degree, but it is usually impossible to match the voice quality of a given target speaker. Intelligibility, however, is usually high. Finally, formant synthesis is highly appropriate for creating speech stimuli for research in speech perception, given the high level of control such experiments require.

Concatenative Synthesis

Concatenative synthesis uses actual short segments of recorded speech that were cut from recordings and stored in an inventory ("voice database"), either as "waveforms" (uncoded) or encoded by a suitable speech coding method (see the "Speech Signal Representation for Concatenative Synthesis"). A block diagram of a typical concatenative TTS system is shown in Figure 16.3. As described in more detail above, and previously depicted in Figure 16.1, the front-end on the left converts a given input text string into a string of phonetic symbols and prosody (fundamental frequency, duration, and amplitude) targets. The front-end employs a set of rules and/or a pronunciation dictionary. Together with a string of phonetic symbols, it produces target values for fundamental frequency (pitch), phoneme durations, and amplitudes. The center block in Figure 16.3 assembles the units according to the list of targets set by the front-end. These units are selected from a store (top center in Figure 16.3) that holds the inventory of available sound units.

Different types of speech units may be stored in the inventory of a concatenative TTS system. Storing whole word units is impractical for general TTS because of the tremendous demands on a voice talent that would have to read a few hundreds of thousands of words in a consistent voice and manner. Even if recorded successfully in multiple sessions spread over several weeks, a lack of coarticulation and phonetic recoding at word boundaries may result in unnatural sounding speech. On the other extreme, using phones (e.g., about 50 for English) is also unsatisfactory because of the large coarticulatory effects that exist between adjacent phones. Therefore, transitions from one unit to the next may be audible as "glitches" that introduce perceptually disruptive discontinuities. Intuitively, longer units are more likely to result in higher quality synthesis, given that the rate of concatenations (how many unit-to-unit transitions occur per second of speech) is lower than in the case of shorter units. However, we need a larger set of longer units to "cover" any application domain, for example, travel (with TTS-generated prompts such as "From which airport do you want to leave?"), because of the tremendous multiplicity of possible unit variants [18]. Given these contradictory requirements, most practical TTS implementations until the mid-1990s compromised by using one of two types of inventory units: the diphone and the demisyllable.

A diphone is the snippet of speech from the middle of one phone to the middle of the next phone. (Note that the average length of a diphone is identical to that of a phone!) The middle of a phone tends to be its acoustically most stable region. Therefore, diphones represent acoustic transitions from the stable midsection of one phone to the next. A minimum inventory of about 1000 diphones is required to synthesize unrestricted English text [19]. Because concatenative synthesis preserves the acoustic detail of natural speech, diphone synthesis is generally highly intelligible. A disadvantage of strict diphone synthesis is that coarticulation is only

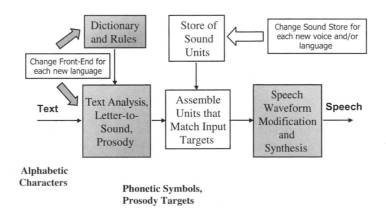

FIGURE 16.3 Block diagram of a concatenative text-to-speech system showing the front-end (left), and back-end (right), plus the sound store in the center that holds the voice inventory.

provided with the immediately preceding and following phonemes, whereas some phonemes can affect the articulation over several phonemes. Demisyllables are alternative units for concatenative synthesis [20]. A demisyllable encompasses half a syllable; that is, either the syllable-initial portion up to the first half of the syllable nucleus, or the syllable-final portion starting from the second half of the syllable nucleus. The number of demisyllables in English is roughly the same as the number of diphones. Because demisyllable units are usually longer than diphones, and allow for better capture of coarticulation effects compared to diphones, they should pose fewer concatenation problems.

Note that a typical database (a database that covers all possible diphone units in a minimum amount of sentences) might contain as little as 30 minutes of speech of a single speaker (voice), given that the units must be modified by signal processing to match the front-end predictions and to smooth over the concatenation points. In the following, we will look into some of the signal representations used in TTS. Note, however, that the latest high-quality TTS systems all employ voice databases containing many hours of recorded speech, because *not* having to modify a segment of speech (because you have the right one in the inventory) will always produce higher quality speech synthesis.

Speech Signal Representation for Concatenative Synthesis

A good speech signal representation for concatenative synthesis approximates the following set of requirements:

1. The speech signal can be stored in a highly compressed (i.e., coded) form so that a large voice database can be used even under tight memory limitations. Coder and decoder are of low computational complexity.
2. Coding/decoding is perceptually transparent. Since we would like to mimic all the voice characteristics of a real person, subjecting the speech signal to "vocoder"-like degradations will not lead to speech synthesis of high naturalness.
3. Coding algorithms (see Chapter 15) have to allow for "random access." Since most speech coders contain some sort of autoregressive memory, all state variables of the coder have to be made available at concatenation points since the decoder will have to switch between units of speech that are very unlikely to have been recorded consecutively in time.
4. An ideal speech representation must allow for natural-sounding modifications of pitch, duration, and amplitude. This is particularly important for small inventories with one, or just a few, "typical" examples for each unit. Unfortunately, experience shows that, for most signal processing algorithms, modifying pitch more than a few percent may destroy the perceived naturalness; that is, a pitch-modified speech signal is likely to be perceptually very different from a speech signal that has been recorded without modifications from the speaker producing the desired pitch value directly. (This is the reason why "singing TTS" does not sound like an opera star.)
5. For some advanced applications, it even might be desirable to allow for fine-tuning of the voice, for example, to add more aspiration, mellowness, or let the voice "scream" when needed. Instead of recording different voice inventories for different speaking "styles," advanced "voice conversion" might be used to approximate an "angry" voice using a "happy" (or "neutral") voice as a starting point. Today, algorithms for voice conversion (usually concerned with converting the speech of one speaker to sound like speech from another speaker) still do not produce consistently good enough results for sounding like the "real thing," but might be sufficient for applications such as computer games where even the original voice does not sound "human."

Given this set of requirements, in the following we will briefly touch on three classes of speech signal processing algorithms with their (native) speech representations: low-complexity time-domain algorithms such as TD-PSOLA and its variants, LPC-based algorithms, and frequency domain-based speech representations.

Time-domain pitch-synchronous overlap add (TD-PSOLA [21]) consists of cutting exactly two pitch periods from a voiced speech signal (a vowel), windowing each segment with a Hanning window centered on

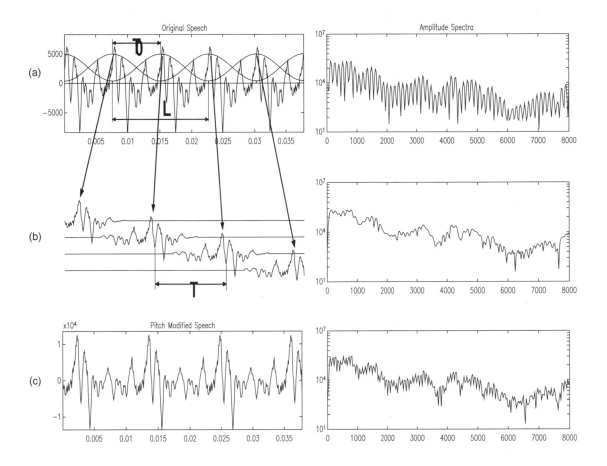

FIGURE 16.4 Pitch modification using TD-PSOLA. The lengths of individual pitch epochs are modified by adding up neighboring segments to form the pitch-modified output speech. (*Source:* T. Dutoit, (1997). *An Introduction to Text-to-Speech Synthesis.* Kluwer Academic Publishers, Dordrecht; Figure 10.1. With permission.)

one glottal closure (maximum excitation) point and weighing down the signal in the vicinity of the previous and next glottal closure points. This process is depicted in Figure 16.4(a) and (b). For all panels (a) through (c) on the left, we see time-domain signals, and on the right, spectra. The extracted and tapered signal traces (b) are recombined (via "overlap-add") in (c), after shortening (for increased pitch) or padding with zero amplitude signal samples (for lowering pitch). The resulting reconstructed signal of a different pitch value is shown in panel (c). TD-PSOLA, although extremely efficient and widely used, can introduce audible glitches at concatenation points because it has no inherent way of smoothing the transition, which happens abruptly. A variant of TD-PSOLA may use the LPC filter excitation ("PSRELP" [22]) instead of the speech waveform directly. This allows for smoothing of the spectral envelope at concatenation points, but still leaves the task of smoothing the LPC excitation itself.

Some TTS systems employ modified LPC-based (see Chapter 15) coder/decoders such as CELP or multi-pulse [23], and/or coders that employ glottal model excitation pulses. The latter approximate true glottal flow waveforms, but can be synthesized given a small set of parameters. Since such parameters can be smoothed over a few pitch periods, the concatenation problem can be addressed. However, it is very difficult to achieve a "transparent" coder/decoder system using this approach. Results usually sound a bit "buzzy."

Finally, so-called "hybrid" speech representations such as the harmonic-plus-noise model (HNM) [24] make use of the fact that the speech spectrum of a voiced sound tends to be composed of two distinct

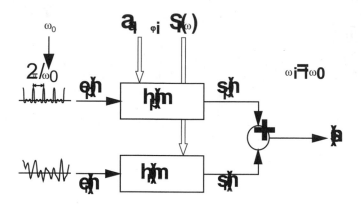

FIGURE 16.5 A hybrid harmonic/stochastic synthesizer. (*Source*: T. Dutoit, (1997). *An Introduction to Text-to-Speech Synthesis*, Dordrecht: Kluwer Academic Publishers, Figure 9.2. With permission.)

components: a harmonic (i.e., periodic) part, mostly at lower frequencies and highly relevant for representing a specific speaker, and a stochastic (noise-like) part, mostly at higher frequencies. Consequently, *two* separate synthesizers might be used as depicted in Figure 16.5: a harmonic (or sinusoidal [25]) synthesizer, shown at the top, and an LPC-based synthesizer using a high-pass filtered stochastic (noise) excitation at the bottom of the figure. Note that the harmonic synthesizer is controlled by parameters like the fundamental frequency ω_0, the amplitudes α_i and phases ϕ_i for the ith harmonic, and the parameters of an optional time-varying filter with the impulse response $h_p(n, m)$. It synthesizes the harmonic speech waveform $s_p(n)$. The stochastic synthesizer (bottom) consists of a time-varying filter with the impulse response $h_r(n, m)$, excitation signal $e_r(n)$, and creates the output speech waveform $s_r(n)$. Both speech components are then added to form the full-band speech signal $s(n)$. Note that the HNM and similar approaches allow for sophisticated spectral and excitation smoothing at concatenation points. Another advantage is that hybrid synthesizers, such as HNM, can exploit any relevant knowledge of speech perception. However, one drawback of the hybrid approach is their relatively high computational complexity.

Voice Creation for Concatenative Synthesis

As stated above, concatenative speech synthesis exploits recorded speech that forms the content of the speech inventory. An example fragment from such a database is shown in Figure 16.6. The top panel shows the time waveform of the recorded speech signal for the words "pink silk dress," the middle panel shows the spectrogram ("voice print"), and the bottom panel shows the annotations that are needed to make the recorded speech useful for concatenative synthesis. For the word "dress," we have highlighted the phone /s/ and the diphone /eh-s/ that encompasses the latter half of the /eh/ and the first half of the /s/ of the word "dress." Until recently, expert labelers were employed to examine waveform and spectrogram, as well as using their sophisticated listening skills, in order to produce annotations ("labels") such as those shown in the bottom panel of the figure. The set consists of word labels (time markings for the end of words), tone labels (symbolic representations of the "melody" of the utterance and syllable and stress labels, all labeled in the ToBI standard [3]), phone labels, and break indices (that distinguish between breaks between words, subphrases, and sentences, for example). Experience shows that expert labelers need approximately 100 to 250 sec of work time to label one second of speech with the set depicted in Figure 16.6. For a diphone-based synthesizer, this might be a reasonable investment, given that a "diphone-rich" database (a database that covers all possible diphones in a minimum amount of sentences) might be as short as 30 min. Clearly, manual labeling would be impractical for much larger databases (dozens of hours), and/or if we were interested in creating

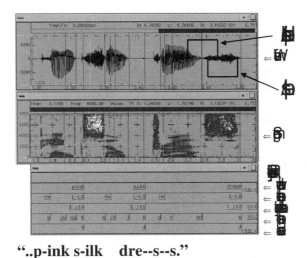

"..p-ink s-ilk   dre--s--s."

FIGURE 16.6 Screenshot of a speech labeler's screen showing waveform (top), spectrogram (middle), and edited labels (bottom) for the speech segment corresponding to the text "pink silk dress."

many voices, factory-style. In such cases, we would require fully automatic labeling, using speech-recognition tools. Fortunately, these tools have become so good that speech synthesized from an automatically labeled speech database may be of higher quality than speech synthesized from the same database that has been labeled manually [26].

Automatic labeling tools fall into two categories: automatic phonetic labeling tools to create the necessary phone labels and automatic prosodic labeling tools to create the necessary tone and stress labels, as well as break indices. Automatic phonetic labeling is adequate, provided it is done with a speech recognizer in "forced alignment mode" (i.e., with the help of the known text message so that the recognizer is only allowed to choose the proper phone boundaries but not the phoneme identities). The speech recognizer also needs be speaker-dependent (i.e., be trained on the given voice), and has to be properly bootstrapped from a small manually labeled corpus for best results. Finally, it should be obvious that both the labeling tool and the TTS system that is the target for the voice database to be labeled should use identical phone labels and unit definition conventions. Automatic prosodic labeling tools work from a set of linguistically motivated acoustic features (e.g., normalized durations, maximum/average pitch ratios) plus some binary features looked up in the lexicon (e.g., word-final vs. word-initial stress) [27], plus the output from the phonetic labeling.

Recording large voice databases can bring with it daunting organizational tasks that should not be underestimated. Selecting the right voice talent, choosing the optimal material to record, providing a consistent recording environment (low background noise, negligible acoustic reflections from walls, tables, manuscript, constant microphone position relative to mouth, etc.), and ensuring a correct and consistent speaking style are all part of the effort.

Unit Selection Synthesis

With the availability of good automatic speech labeling tools, concatenative speech synthesis has now embraced the use of multi-hour voice databases. With the availability of several (potentially hundreds of thousand) instances of a specific type unit (only different in pitch, duration, linguistic context at recording time), so-called *Unit-Selection Synthesis* has become viable for obtaining high-quality TTS. Based on early research done at ATR in Japan [28–30], this method enables the use of large speech databases recorded using specific, carefully crafted and controlled speaking styles (e.g., angry, happy, apologetic, etc.). In addition, a given database may be focused on narrow-domain applications (such as "travel reservations" or "telephone

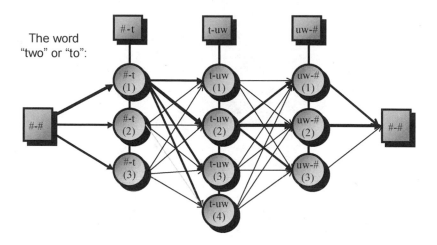

FIGURE 16.7 Viterbi search of diphone inventory to select units for synthesizing the word "two" or "to" in isolation (i.e., between silence /#/ segments). Arrow line width symbolizes appropriateness of a given transition (thicker is better) evaluated, e.g., in terms of spectral match across unit boundaries. Candidates in each of the three columns are assumed to be ordered according to their suitability in terms of pitch, context match between recording and synthesis, etc.

number synthesis") that commonly allow for the use of smaller databases for a preset level of quality, or it may be used for general applications like e-mail or news reading (requiring a larger voice database). In the latter case, unit-selection synthesis can require on the order of ten hours of recording of spoken general material to achieve a desired level of quality, and several dozen hours for "natural quality" (that can be mistaken for direct recordings). In contrast with earlier concatenative synthesizers, unit-selection synthesis automatically picks the optimal synthesis units (on the fly) from an inventory that may contain thousands of tokens of a specific unit, and concatenates the selected units to produce the synthetic speech. This is in stark contrast with the fact that as late as the mid-1990s voice inventories for concatenative TTS were always carefully crafted by hand, trying to find the one or few units of each type that lead to optimum results in all possible synthesis scenarios (contexts). In more than one sense, the unit-selection approach has succeeded in putting "the expert into the box," that is, has automated the process of finding the optimal sequence of inventory units given a "search query" of tagged phoneme strings. One important difference between "old" and "new" TTS is that unit-selection synthesis "knows" the text to be synthesized at selection time, while previous methods tried to satisfy more global selection criteria *without* explicit knowledge of the text to be synthesized.

The unit selection "database query" process is outlined in Figure 16.7, which shows how the method must dynamically find the best path through the unit-selection network corresponding to the diphones for the word "two." (Note that, in practice, a better choice would be to use half-phones instead of diphones because half-phones allow the search algorithm to create diphones on the fly from two half-phones that were not recorded in sequence.) For any query, the optimal choice of units selected from the database depends on factors such as spectral similarity at unit boundaries (components of the "join cost" between two units) and on matching prosodic targets set by the front-end (components of the "target cost" of each unit).

16.5   TTS Evaluation

Evaluation of TTS systems is currently a much discussed research topic. Here, we can only give broad guidelines. The three quality criteria mentioned in the beginning of this section—accuracy, intelligibility, and naturalness—overlap with respect to which part of a TTS system impacts which criteria. Accuracy, that is, the ability to read a given input text the way a knowledgeable human reader would, is the one criterion solely homed in the TTS front-end. Contrary to this, unsatisfactory intelligibility or naturalness is much harder to pinpoint. Between the latter two, designers of formant synthesizers should aim at maximizing intelligibility, accepting the fact that they are unlikely to deliver high naturalness. However, some of the newer concatenative synthesis systems may overemphasize naturalness with the undesired result of less-than-stellar intelligibility.

Accuracy can be evaluated by running a test corpus of task-relevant acronyms, abbreviations, embedded in context-defining input text through the TTS front-end and judging the generated output *text* (or transcribing the output speech back to text). Evaluating intelligibility and/or naturalness requires conducting elaborate listening tests. Intelligibility tests, known from testing speech coders, present word or sentence lists and let subjects transcribe words they heard [31] on TTS-related testing. Unfortunately, TTS intelligibility has many more aspects to it than are covered by standard word-list driven intelligibility tests. On a sentence level, for example, a wrong prosody can destroy intelligibility/comprehension. So far, a generally accepted standard intelligibility test for TTS systems is lacking. For overall quality evaluation, the International Telecommunication Union (ITU) recommends a specific method [32] that, in this author's opinion, is suitable for also testing "naturalness." Such tests involve five (or more) point rating scales for characteristics such as "overall impression," "listening effort," "comprehension," etc. Alternatively, listeners might be asked to state their preference (A/B tests) of which one of two systems sounds "better." Selection of the test material is also very relevant. For example, some systems sound very good for short sentences spoken in isolation, but show weaknesses when longer paragraphs of text are being rendered. Generally, a rule-of-thumb is to use material from the intended application. For more details on quality testing, see [33].

16.6 Conclusions

This chapter has highlighted some aspects of TTS synthesis with a slant toward catering to electrical engineers. Many aspects, such as, for example, prosody generation, natural language processing, and others, have been skimmed only for space reasons. It is clear that TTS systems have come a long way toward delivering high-quality output to listeners that sometimes fools them to believe that they are listening to recordings. This said, it is also clear that we are still far from delivering the perfect synthesis for all possible applications. Shorter term, the best way toward high quality synthesis seems to be to tailor TTS specifically for a given application. Both the front-end and the back-end of a TTS system can be optimized for this purpose. For example, including and maintaining a pronunciation dictionary of names of all prescription drugs on the market could be essential for using TTS in a healthcare application. Eliciting the desired voice characteristics from a voice talent that is being recorded for a unit-selection synthesis voice database could be essential for customers accepting an automated dialog system that speaks with a TTS voice. Last but not least, typical engineering choices, such as trading off memory vs. speed, quality vs. complexity, and time in development vs. market pressures, are also very relevant for TTS systems.

Defining Terms

Homographs: Words that are spelled the same but pronounced differently such as, e.g., the present and past tense form of the verb "to read."

Graphemes: "Functional spelling units" encompassing one or more letters of the text input; a grapheme in the text input corresponds to a single phoneme.

Morphology: Deals with the units of meaning.

Pitch: The fundamental frequency of vibration of the vocal cords that produce voiced sounds such as vowels.

Phones: Characterize any sound that can be produced by a human vocal tract; if a phone is part of a specific language, it becomes a phoneme of that language.

Phonemes: The elementary sounds of a language, such as /ow/ in the word "boat."

References

1. C. Bickley, A. Syrdal, and J. Schroeter, "Speech Synthesis," in *The Acoustics of Speech Communication*, J.M. Picket, Ed., Boston, MA: Allyn and Bacon, 1998.

2. A. Syrdal, R. Bennett, and S. Greenspan, Eds., *Applied Speech Technology*, Boca Raton, FL: CRC Press, 1995.

3. K. Silverman, M. Beckman, J. Pierrehumbert, M. Ostendorf, C. Wightman, P. Price, and J. Hirschberg, *ToBI: A Standard Scheme for Labeling Prosody*, ICSLP, 1992, pp. 867–879, Banff.

4. Y. Sagisaka, N. Campbell, and N. Higuchi, Eds., *Computing Prosody – Computational Models for Processing Spontaneous Speech*, Berlin: Springer, 1997.

5. T. Dutoit, *An Introduction to Text-to-Speech Synthesis*, Dordrecht: Kluwer Academic Publishers, 1997.

6. J. Schroeter, and M.M. Sondhi, "Speech coding based on physiological models of speech production," in *Advances in Speech Signal Processing*, S. Furui and M. Mohan Sondhi, Eds., New York: Marcel Dekker, 1991, pp. 231–268.

7. G. Richard, M. Liu, D. Sinder, H. Duncan, Q. Lin, J. Flanagan, S. Levinson, D. Davis, and S. Slimon, "Numerical simulations of fluid flow in the vocal tract," in *Proc. of Eurospeech*, Madrid, Spain, September 18–21, 1995.

8. K. Ishizaka and J.L. Flanagan, "Synthesis of voiced sounds from a two-mass model of the vocal cords," *Bell Syst. Tech. J.*, vol. 51, no. 6, pp. 133–1268, 1972.

9. C.H. Coker, "A model of articulatory dynamics and control," *Proc. IEEE*, vol. 64, no. 4, pp. 452–460, 1976.

10. M.M. Sondhi and D.J. Sinder, "Articulatory modeling: a role in concatenative text-to-speech synthesis," in *Text to Speech Synthesis: New Paradigms and Advances*, A. Alwan and S. Narayanan, Eds., Englewood Cliffs, NJ: Prentice-Hall, 2003.

11. S. Narayanan, K. Nayak, S. Lee, A. Sethy, and D. Byrd, "An approach to real-time magnetic resonance imaging for speech production," *J. Acoust. Soc. Am.*, in press, 2004.

12. S. Kiritani, S.K. Itoh, and O. Fujimura, "Tongue-pellet tracking by a computer controlled X-ray microbeam system," *J. Acoust. Soc. Am.*, vol. 57, 1516–1520, 1975.

13. D.H. Klatt, "Software for a cascade/parallel formant synthesizer," *J. Acoust. Soc. Am.*, vol. 67, no. 3, 971–995, 1980.

14. J. Allen, M.S. Hunnicutt, and D. Klatt, *From Text to Speech, The MITalk System*, Cambridge: Cambridge University Press, 1987.

15. J.N. Holmes, "Formant Synthesizers: Cascade or Parallelm" *Speech Commun.*, 2, 251–273, 1983.

16. S. Parthasarathy and C. Coker, "On automatic estimation of articulatory parameters in a text-to-speech system," *Comput. Speech Language*, 6, 37–75, 1992.

17. H.M. Hanson, and K.N. Stevens, "A quasiarticulatory approach to controlling acoustic source parameters in a Klatt-type formant synthesizer using HLsyn," *J. Acoust. Soc. Am.*, 112, 1158–1182, 2002.

18. J.P.H. van Santen, "Combinatorial issues in text-to-speech synthesis," in: *EuroSpeech'97: 5th Eur. Conf. Speech Commun. Technol.*, 5, 2511–2514, 1997.

19. J.P. Olive, "Rule synthesis of speech from diadic units," in *Proc. ICASSP-77*, 1977, pp. 568–570.

20. O. Fujimura and J. Lovins, "Syllables as concatenative phonetic elements," in *Syllables and Segments*, A. Bell and J.B. Hooper, Eds., New York: North-Holland, 107–120, 1978.

21. E. Moulines and F. Charpentier, "Pitch-synchronous waveform processing techniques for text-to-speech synthesis using diphones," *Speech Commun.*, vol. 9, no. 5–6, pp. 453–467, 1990.

22. M. Macchi, M.J. Altom, D. Kahn, S. Singhal, and M. Spiegel, "Intelligibility as a function of speech coding method for template-based speech synthesis," in *Proc. Eurospeech'93*, Berlin, Germany, 1993, pp. 893–896.

23. W. Kleijn and K. Paliwal, Eds., *Speech Coding and Synthesis*, Amsterdam: Elsevier, 1995.

24. Y. Stylianou, "Applying the harmonic plus noise model in concatenative speech synthesis," *IEEE Trans. Speech Audio Process.*, vol. 9, no. 1, pp. 21–29, 2001.

25. T.F. Quartieri and R.J. McAulay, "Shape Invariant time-scale and pitch modification of speech," *IEEE Trans. Signal Process.*, vol. 40, no. 3, pp. 497–510, 1992.

26. M.J. Makashay, C.W. Wightman, A.K. Syrdal, and A.D. Conkie, "Perceptual evaluation of automatic segmentation in text-to-speech synthesis," in *Proc. ICSLP 2000*, Beijing, China, October, 2000.

27. C.W. Wightman, A.K. Syrdal, G. Stemmer, A.D. Conkie, and M.C. Beutnagel, "Perceptually based automatic prosody labeling and prosodically enriched unit selection improve concatenative text-to-speech synthesis," in *Proc. ICSLP 2000*, Beijing, China, October, 2000.

28. Y. Sagisaka, N. Kaiki, N. Iwahashi, and K. Mimura, K., "ATR – *v*-TALK speech synthesis system," in *Proc. Int. Conf. Speech Language Process. 92*, Banff, Canada, vol. 1, 1992, pp. 483–486.

29. A. Hunt and A.W. Black, "Unit selection in a concatenative speech synthesis system using a large speech database," in *Proc. ICASSP-96*, 1996, pp. 373–376.

30. M. Beutnagel, A. Conkie, J. Schroeter, Y. Stylianou, and A. Syrdal, "The AT&T Next-Gen TTS System," in *Proc. Joint Meet. ASA EAA DEGA*, Berlin, Germany, March 1999, available on-line at http://www.research.att.com/projects/tts/pubs.html.

31. M.F. Spiegel, M.J. Altom, and M.J. Macchi, "Comprehensive assessment of the telephone intelligibility of synthesized and natural speech," *Speech Commun.*, 9, 279–291, 1990.

32. ITU-T Recommendation P.85, "A Method for Subjective Performance Assessment of the Quality of Speech Output Devices," International Telecommunications Union publication, 1994.

33. Y.V. Alvarez and M. Huckvale, "The reliability of the ITU-T P.85 standard for the evaluation of text-to-speech systems," in *Proc. ICSLP2002*, Denver, Colorado, Session TuB9p.8, 329-332, 16–20 September 2002.

Further Information

There is a treasure of information on speech synthesis available on the Web.

For information on how TTS fits in and what it contributes to Language Technology, see, for example: http://cslu.cse.ogi.edu/HLTsurvey/HLTsurvey.html (Chapter 5, spoken output technologies).

To get started on speech synthesis, readers may want to explore the Festival project: http://www.cstr.ed.ac.uk/projects/festival/, and the follow-up project at http://www.festvox.org.

People interested in the earlier efforts in speech synthesis find historic examples at http://www.cs.indiana.edu/rhythmsp/ASA/Contents.html and at http://www.mindspring.com/~ssshp/ssshp_cd/ss_home.htm.

Finally, readers new to the topic might enjoy the interactive demos available at http://www.research.att.com/projects/tts.

17

Spectral Estimation and Modeling

S. Unnikrishna Pillai
Polytechnic University

Theodore I. Shim
Polytechnic University

Stella N. Batalama
State University of New York at Buffalo

Dimitri Kazakos
University of Idaho

Ping Xiong
State University of New York at Buffalo

David D. Sworder
University of California

John E. Boyd
Cubic Defense Systems

17.1 Spectral Analysis

S. Unnikrishna Pillai and Theodore I. Shim

Historical Perspective

Modern spectral analysis dates back at least to Sir Isaac Newton (1671), whose prism experiments with sunlight led him to discover that each color represented a particular wavelength of light and that the sunlight contained all wavelengths. Newton used the word *spectrum*, a variant of the Latin word *specter*, to describe the band of visible light colors.

In the early eighteenth century, Bernoulli discovered that the solution to the wave equation describing a vibrating string can be expressed as an infinite sum containing weighted sine and cosine terms. Later, the French engineer Joseph Fourier in his *Analytical Theory of Heat* (Fourier, 1822) extended Bernoulli's wave equation results to arbitrary periodic functions that might contain a finite number of jump discontinuities. Thus, for some $T_0 > 0$, if $f(t) = f(t + T_0)$ for all t, then $f(t)$ represents a periodic signal with period T_0 and in the case of real signals, it has the Fourier series representation:

$$f(t) = A_0 + 2 \sum_{k=1}^{\infty} (A_k \cos k\omega_0 t + B_k \sin k\omega_0 t)$$

where $\omega_0 = 2\pi/T_0$, and

$$A_k = \frac{1}{T_0} \int_0^{T_0} f(t) \cos k\omega_0 t dt, \qquad B_k = \frac{1}{T_0} \int_0^{T_0} f(t) \sin k\omega_0 t dt$$

with A_0 representing the dc term ($k = 0$). Moreover, the infinite sum on the right-hand side of the above expression converges to $[f(t_{-0}) + f(t_{+0})]/2$. The total power P of the periodic function satisfies the relation:

$$P = \frac{1}{T_0} \int_0^{T_0} |f(t)|^2 dt = A_0^2 + 2 \sum_{k=1}^{\infty} (A_k^2 + B_k^2)$$

implying that the total power is distributed only among the dc term, the fundamental frequency $\omega_0 = 2\pi/T_0$ and its harmonics $k\omega_0$, $k \geqslant 1$, with $2(A_k^2 + B_k^2)$ representing the power contained at the harmonic $k\omega_0$. For every periodic signal with finite power, since $A_k \to 0$, $B_k \to 0$, eventually the overharmonics become of decreasing importance.

The British physicist Schuster (1898) used this observation to suggest that the partial power $P_k = 2(A_k^2 + B_k^2)$ at frequency $k\omega_0$, $k = 0 \to \infty$, be displayed as the spectrum. Schuster termed this method the *periodogram*, and information over a multiple of periods was used to compute the Fourier coefficients and/or to smooth the periodogram, since depending on the starting time, the periodogram may contain irregular and spurious peaks. A notable exception to periodogram was the linear regression analysis introduced by the British statistician Yule (1927) to obtain a more accurate description of the periodicities in noisy data. Because the sampled periodic process $x(k) = \cos k\omega_0 T$ containing a single harmonic component satisfies the recursive relation:

$$x(k) = ax(k-1) - x(k-2)$$

where $a = 2 \cos \omega_0 T$ represents the harmonic component, its noisy version $x(k) + n(k)$ satisfies the recursion:

$$x(k) = ax(k-1) - x(k-2) + n(k)$$

Yule interpreted this time series model as a recursive harmonic process driven by a noise process and used this form to determine the periodicity in the sequence of sunspot numbers. Yule further generalized the above recursion to

$$x(k) = ax(k-1) + bx(k-2) + n(k)$$

where a and b are arbitrary, to describe a truly autoregressive process and since for the right choice of a, b the least-square solution to the above autoregressive equation is a damped sinusoid, this generalization forms the basis for the modern-day parametric methods.

Modern Spectral Analysis

Norbert Wiener's classic work on Generalized Harmonic Analysis (Wiener, 1930) gave random processes a firm statistical foundation, and with the notion of ensemble average several key concepts were then introduced.

The formalization of modern-day probability theory by Kolmogorov and his school also played an indispensable part in this development. Thus, if $x(t)$ represents a continuous-time stochastic (random) process, then for every fixed t, it behaves like a **random variable** with some **probability density function** $f_x(x, t)$. The ensemble average or **expected value** of the process is given by

$$\mu_x(t) = E[x(t)] = \int_{-\infty}^{\infty} x f_x(x, t) dx$$

and the statistical correlation between two time instants t_1 and t_2 of the random process is described through its **autocorrelation function**:

$$R_{xx}(t_1, t_2) = E[x(t_1)x^*(t_2)] = \int_{-\infty}^{\infty} \int_{-\infty}^{\infty} x_1 x_2^* f_{x_1 x_2}(x_1, x_2, t_1, t_2) dx_1 dx_2 = R_{xx}^*(t_2, t_1)$$

where $f_{x1x2}(x_1, x_2, t_1, t_2)$ represents the joint probability density function of the random variable $x_1 = x(t_1)$ and $x_2 = x(t_2)$ and * denotes the complex conjugate transpose in general. Processes with autocorrelation functions that depend only upon the difference of the time intervals t_1 and t_2 are known as wide sense stationary processes. Thus, if $x(t)$ is wide sense stationary, then

$$E[x(t + \tau)x^*(t)] = R_{xx}(\tau) = R_{xx}^*(-\tau)$$

To obtain the distribution of power vs. frequency in the case of a **stochastic process,** one can make use of the Fourier transform based on a finite segment of the data. Letting

$$P_T(\omega) = \frac{1}{2T} \left| \int_{-T}^{T} x(t) e^{-j\omega t} dt \right|^2$$

represent the power contained in a typical realization over the interval $(-T, T)$, its ensemble average value as $T \to \infty$ represents the true power contained at frequency ω. Thus, for wide sense stationary processes:

$$S(\omega) = \lim_{T \to \infty} E[P_T(\omega)] = \lim_{T \to \infty} \int_{-T}^{T} \int_{-T}^{T} R_{xx}(t_1 - t_2) e^{-j\omega(t_1 - t_2)} dt_1 dt_2$$

$$= \lim_{T \to \infty} \int_{-2T}^{2T} R_{xx}(\tau) \left(1 - \frac{|\tau|}{2T} \right) e^{-j\omega\tau} d\tau = \int_{-\infty}^{\infty} R_{xx}(\tau) e^{-j\omega\tau} d\tau \geqslant 0 \qquad (17.1)$$

Moreover, the inverse relation gives

$$R_{xx}(\tau) = \frac{1}{2\pi} \int_{-\infty}^{\infty} S(\omega) e^{j\omega\tau} d\omega \qquad (17.2)$$

and hence

$$R_{xx}(0) = E[|x(t)|^2] = P = \frac{1}{2\pi} \int_{-\infty}^{\infty} S(\omega) d\omega$$

Thus, $S(\omega)$ represents the power spectral density and from Equation (17.1) and Equation (17.2), the power spectral density and the autocorrelation function form a Fourier transform pair, the well-known Wiener–Khinchin theorem.

If $x(kT)$ represents a discrete-time wide sense stationary stochastic process, then

$$r_k = E\{x((n+k)T)x^*(nT)\} = r_{-k}^*$$

and the power spectral density is given by

$$S(\omega) = \sum_{k=-\infty}^{\infty} r_k e^{-jk\omega T}$$

or in terms of the normalized variable $\theta = \omega T$:

$$S(\theta) = \sum_{k=-\infty}^{\infty} r_k e^{-jk\theta} = S(\theta + 2\pi k) \geqslant 0 \qquad (17.3)$$

and the inverse relation gives the autocorrelations to be

$$r_k = \frac{1}{2\pi} \int_{-\pi}^{\pi} S(\theta) e^{jk\theta} d\theta = r_{-k}^*$$

Thus, the power spectral density of a discrete-time process is periodic. Such a process can be obtained by sampling a continuous-time process at $t = kT$, $|k| = 0 \rightarrow \infty$, and if the original continuous-time process is band-limited with a two-sided bandwidth equal to $2\omega_b = 2\pi/T$, then the set of discrete samples so obtained is equivalent to the original process in a mean-square sense.

As Schur (1917) has observed, for discrete-time stationary processes, that the nonnegativity of the **power spectrum** is equivalent to the nonnegative definiteness of the Hermitian Toeplitz matrices, i.e.:

$$S(\theta) \geqslant 0 \Leftrightarrow \mathbf{T}_k = \begin{bmatrix} r_0 & r_1 & \cdots & r_k \\ r_1^* & r_0 & \cdots & r_{k-1} \\ \vdots & \vdots & \ddots & \vdots \\ r_k^* & r_{k-1} & \cdots & r_0 \end{bmatrix} = \mathbf{T}_k^* \geqslant 0, \quad k = 0 \rightarrow \infty \qquad (17.4)$$

If $x(nT)$ is the output of a discrete-time linear time-invariant causal system driven by $w(nT)$, then we have the following representation:

$$w(nT) \rightarrow H(z) = \sum_{k=0}^{\infty} h(kT)z^k \rightarrow x(nT) = \sum_{k=0}^{\infty} h(kT)w((n-k)T) \qquad (17.5)$$

In the case of a stationary input, the output is also stationary, and its power spectral density is given by

$$S_x(\theta) = |H(e^{j\theta})|^2 S_w(\theta) \qquad (17.6)$$

where $S_w(\theta)$ represents the power spectral density of the input process. If the input is a white noise process, then $S_w(\theta) = \sigma^2$ and

$$S_x(\theta) = \sigma^2 |H(e^{j\theta})|^2$$

Clearly if $H(z)$ is rational, so is $S_x(\theta)$. Conversely, given a power spectral density:

$$S_x(\theta) = \sum_{k=-\infty}^{\infty} r_k e^{jk\theta} \geqslant 0 \qquad (17.7)$$

that satisfies the integrability condition

$$\int_{-\pi}^{\pi} S_x(\theta)\mathrm{d}\theta < \infty \qquad (17.8)$$

and the physical realizability (Paley–Wiener) criterion

$$\int_{-\pi}^{\pi} \ln S_x(\theta)\mathrm{d}\theta > -\infty \qquad (17.9)$$

there exists a unique function $H(z)$ that is analytic together with its inverse in $|z| < 1$ (minimum phase factor) such that

$$H(z) = \sum_{k=0}^{\infty} b_k z^k, \qquad |z| < 1 \qquad (17.10)$$

and

$$S_x(\theta) = \lim_{r \to 1-0} |H(re^{j\theta})|^2 = |H(e^{j\theta})|^2, \text{a.e.}$$

$H(z)$ is known as the Wiener factor associated with $S_x(\theta)$ and as Equation (17.5) shows, when driven by white noise, it generates a stochastic process $x(nT)$ from past samples and its power spectral density matches with the given $S_x(\theta)$.

In this context, given a finite set of autocorrelations $r_0, r_1, \ldots, r_n$, the spectral extension problem is to obtain the class of all extensions that match the given data, i.e., such an extension $K(\theta)$ must automatically satisfy

$$K(\theta) \geqslant 0$$

and

$$\frac{1}{2\pi} \int_{-\pi}^{\pi} K(\theta)e^{jk\theta}\mathrm{d}\theta = r_k, \qquad k = 0 \to n$$

in addition to satisfying Equation (17.8) and Equation (17.9).

The solution to this problem is closely related to the trigonometric moment problem, and it has a long and continuing history through the works of Schur (1917), Nevanlinna, Akheizer and Krein (1962), Geronimus (1954), and Shohat and Tamarkin (1970), to name a few. If the given autocorrelations are such that the matrix $\mathbf{T}_n$ in Equation (17.4) is singular, then there exists an $m \leqslant n$ such that $\mathbf{T}_{m-1}$ is positive definite ($\mathbf{T}_{m-1} > 0$) and $\mathbf{T}_m$ is singular (det $\mathbf{T}_m = 0$, det (.) representing the determinant of (.)). In that case, there exists a unique vector $\mathbf{X} = (x_0, x_1, \ldots, x_m)^{\mathrm{T}}$ such that $\mathbf{T}_m \mathbf{X} = 0$ and further, the autocorrelations have a unique extension given by

$$c_k = \sum_{i=1}^{m} P_i e^{jk\theta_i}, \qquad |k| = 0 \to \infty \qquad (17.11)$$

where $e^{j\theta_i}$, $i = 1 \rightarrow m$ are the m zeros of the polynomial $x_0 + x_1 z + \ldots + x_m z^m$ and $P_i > 0$. This gives

$$
\mathbf{T}_{m-1} = \mathbf{A}
\begin{bmatrix}
P_1 & 0 & \cdots & 0 \\
0 & P_2 & \cdots & 0 \\
\vdots & \vdots & \ddots & \vdots \\
0 & 0 & \cdots & P_m
\end{bmatrix}
\mathbf{A}^* \tag{17.12}
$$

where $\mathbf{A}$ is an $m \times m$ Vandermonde matrix given by

$$
\mathbf{A} =
\begin{bmatrix}
1 & 1 & \cdots & 1 \\
\lambda_1 & \lambda_2 & \cdots & \lambda_m \\
\lambda_1^2 & \lambda_2^2 & \cdots & \lambda_m^2 \\
\vdots & \vdots & \cdots & \vdots \\
\lambda_1^{m-1} & \lambda_2^{m-1} & \cdots & \lambda_m^{m-1}
\end{bmatrix}, \quad \lambda_i = e^{j\theta_i}, \quad i = 1 \rightarrow m
$$

and Equation (17.12) can be used to determine $P_k > 0$, $k = 1 \rightarrow m$. The power spectrum associated with Equation (17.11) is given by

$$
S(\theta) = \sum_{k=1}^{m} P_k \delta(\theta - \theta_k)
$$

and it represents a discrete spectrum that corresponds to pure uncorrelated sinusoids with signal powers P_1, $P_2, \cdots, P_m$.

If the given autocorrelations satisfy $\mathbf{T}_n > 0$, from Equation (17.4), every unknown r_k, $k \geqslant n + 1$, must be selected so as to satisfy $\mathbf{T}_k > 0$, and this gives

$$
|r_{k+1} - \zeta_k|^2 \leqslant R_k^2 \tag{17.13}
$$

where $\zeta_k = \mathbf{f}_k^\mathrm{T} \mathbf{T}_k^{-1} \mathbf{b}_k$, $\mathbf{f}_k = (r_1, r_2, \ldots, r_k)^\mathrm{T}$, $\mathbf{b}_k = (r_k, r_{k-1}, \ldots, r_1)$ and $R_k = \det \mathbf{T}_k / \det \mathbf{T}_{k-1}$.

From Equation (17.13), the unknowns could be anywhere inside a sequence of circles with center ζ_k and radius R_k, and as a result, there are an infinite number of solutions to this problem. Schur and Nevanlinna have given an analytic characterization to these solutions in terms of bounded function extensions. A bounded function $\rho(z)$ is analytic in $|z| < 1$ and satisfies the inequality $|\rho(z)| \leqslant 1$ everywhere in $|z| < 1$.

In a network theory context, Youla (1980) has also given a closed form parametrization to this class of solutions. In that case, given $r_0, r_1, \ldots, r_n$, the minimum phase transfer functions satisfying Equation (17.8) and Equation (17.9) are given by

$$
H_\rho(z) = \frac{\Gamma(z)}{P_n(z) - z\rho(z) \tilde{P}_n(z)} \tag{17.14}
$$

where $\rho(z)$ is an *arbitrary* bounded function that satisfies the inequality (Paley–Wiener criterion):

$$
\int_{-\pi}^{\pi} \ln\left[1 - |\rho(e^{j\theta})|^2\right] d\theta > -\infty
$$

and $\Gamma(z)$ is the minimum phase factor obtained from the factorization:

$$
1 - |\rho(e^{j\theta})|^2 = |\Gamma(e^{j\theta})|^2
$$

Further, $P_n(z)$ represents the Levinson polynomial generated from $r_0 \rightarrow r_n$ through the recursion:

$$\sqrt{1 - |s_n|^2} P_n(z) = P_{n-1}(z) - z s_n \tilde{P}_{n-1}(z)$$

that starts with $P_0(z) = 1/\sqrt{r_0}$, where

$$s_n = \left\{ P_{n-1}(z) \sum_{k=1}^{n} r_k z^k \right\}_n P_{n-1}(0) \qquad (17.15)$$

represents the reflection coefficient at stage n. Here, $\{ \}_n$ denotes the coefficient of z^n in the expansion $\{ \}$, and $\tilde{P}_n(z) \overset{\Delta}{=} z^n P_n^*(1/z^*)$ represents the polynomial reciprocal to $P_n(z)$. Notice that the given information $r_0 \rightarrow r_n$ enters $P_n(z)$ through Equation (17.5). The power spectral density:

$$K(\theta) = |H_\rho(e^{j\theta})|^2$$

associated with Equation (17.14) satisfies all the interpolation properties described before. In Equation (17.14), the solution $\rho(z) \equiv 0$ gives $H(z) = 1/P_n(z)$, a pure AR(n) system that coincides with Burg's maximum entropy extension. Clearly, if $H_\rho(z)$ is rational, then $\rho(z)$ must be rational and, more interestingly, every rational system must follow from Equation (17.14) for a specific rational bounded function $\rho(z)$. Of course, the choice of $\rho(z)$ brings in extra freedom, and this can be profitably used for system identification as well as rational and stable approximation of nonrational systems (Pillai and Shim, 1993).

Defining Terms

Autocorrelation function: The expected value of the product of two random variables generated from a random process for two time instants; it represents their interdependence.

Expected value (or mean) of a random variable: Ensemble average value of a random variable that is given by integrating the random variable after scaling by its probability density function (weighted average) over the entire range.

Power spectrum: A nonnegative function that describes the distribution of power vs. frequency. For wide sense stationary processes, the power spectrum and the autocorrelation function form a Fourier transform pair.

Probability density function: The probability of the random variable taking values between two real numbers x_1 and x_2 is given by the area under the nonnegative probability density function between those two points.

Random variable: A continuous or discrete valued variable that maps the set of all outcomes of an experiment into the real line (or complex plane). Because the outcomes of an experiment are inherently random, the final value of the variable cannot be predetermined.

Stochastic process: A real valued function of time t, which for every fixed t behaves like a random variable.

References

N.I. Akheizer and M. Krein, *Some Questions in the Theory of Moments*, American Math. Soc. Monogr., 2, 1962.

J.B.J. Fourier, *Theorie Analytique de la Chaleur (Analytical Theory of Heat)*, Paris, 1822.

L.Y. Geronimus, *Polynomials Orthogonal on a Circle and Their Applications*, American Math. Soc., Translation, 104, 1954.

I. Newton, *Philos. Trans.*, vol. IV, p. 3076, 1671.

S.U. Pillai and T.I. Shim, *Spectrum Estimation and System Identification*, New York: Springer-Verlag, 1993.

I. Schur, "Uber Potenzreihen, die im Innern des Einheitzkreises Beschrankt Sind," *Journal fur Reine und Angewandte Mathematik,* vol. 147, pp. 205–232, 1917.

J.A. Shohat and J.D. Tamarkin, *The Problem of Moments,* American Math. Soc., Math. Surveys, 1, 1970.

N. Wiener "Generalized harmonic analysis," *Acta Math.,* vol. 55, pp. 117–258, 1930.

D.C. Youla, "The FEE: A New Tunable High-Resolution Spectral Estimator: Part I," Technical note, no. 3, Dept. of Electrical Engineering, Polytechnic Institute of New York, Brooklyn, New York; also RADC Report, RADC-TR-81–397, AD A114996, 1982, 1980.

G.U. Yule, "On a method of investigating periodicities in disturbed series, with special reference to Wolfer's sunspot numbers," *Philos. Trans. R. Soc. London, Ser. A*, vol. 226, pp. 267–298, 1927.

17.2   Parameter Estimation

Ping Xiong, Stella N. Batalama, and Dimitri Kazakos

Parameter estimation is the operation of assigning a value in a continuum of alternatives to an unknown parameter based on a set of observations that involve some function of the parameter. **Estimate** is the value assigned to the parameter and **estimator** is the function of the observations that yields the estimate.

Applications of parameter estimation are broad and numerous, arising from problems in radar, sonar, communications, signal processing, and many more. Examples include target location/range estimation in radar, channel estimation in wireless communications, medical image processing for early diagnosis, global positioning systems (GPS), seismic engineering, and environment identification in outer space explorations, just to name a few.

The basic elements in parameter estimation are a vector parameter θ^m and a vector parameter space $\mathcal{E}^m$, a stochastic process $X(t)$ parameterized by θ^m, an optimization criterion, and a cost or penalty function. The estimate $\hat{\theta}^m(x^n)$ based on the observation vector $x^n = [x_1, x_2, \ldots, x_n]$ is the solution of an optimization problem that is formulated according to the cost function. Usually the optimization criterion coincides with the estimator's performance measure of interest. However, when the latter is mathematically intractable or of high complexity, we may prefer to proceed with a rather simple optimization criterion and then evaluate the performance of the estimator under the measure of interest. In the following, the function $f(x^n|\theta^m)$ will denote the joint conditional probability density function (pdf) of the random variables $x_1, \ldots, x_n$.

There are several parameter estimation schemes. If the stochastic process $X(t)$ is parametrically known, i.e., if the conditional probability density functions are known for each fixed value of the vector parameter θ^m, then the corresponding parameter estimation scheme is called **parametric**. If $X(t)$ is nonparametrically described, i.e., given $\theta^m \in \mathcal{E}^m$ any joint probability density function of $X(t)$ is a member of some nonparametric class of probability density functions, then **nonparametric** estimation schemes arise. When the parameter is modeled as deterministic, then we deal with *classical* or *deterministic* or *nonrandom* parameter estimation. If the parameter is modeled as random, then the estimation process is called *Bayesian* or *random*.

Let Γ^n denote the *n*-dimensional observation space. Then an estimator $\hat{\theta}^m(x^n)$ of a vector parameter θ^m is a function from the observation space, Γ^n, to the parameter space, $\mathcal{E}^m$. Since this is a function of random variables, it is itself a random variable (scalar or vector). Throughout this section, we assume real parameters and observations (generalizations to the complex case are often straightforward).

There are certain stochastic properties of estimators that quantify their quality. The most important measures of quality are the **bias** and **variance**. An estimator is **unbiased** if its expected value is equal to the true parameter value, i.e., if

$$E\{\hat{\theta}^m(x^n)\} = \theta^m \tag{17.16}$$

where the expectation is taken with respect to the probability density function $f_{\theta^m}(x^n)$ (the subscript θ^m signifies that the pdf is parameterized by θ^m). In the case where the observation space is $\Re^n$ and the parameter is a scalar, it is

$$E\{\hat{\theta}(x^n)\} = \int_{\mathcal{R}^n} \hat{\theta}(x^n) f_\theta(x^n) \mathrm{d}x^n \tag{17.17}$$

The bias of the estimator is defined as $B(\theta^m) = E\{\hat{\theta}^m(x^n)\} - \theta^m$. Thus, the bias measures the distance between the expected value of the estimator and the true value of the parameter. Usually, it is of interest to know the conditional variance of an unbiased estimator given by

$$E\left\{\left\|\hat{\theta}^m(x^n) - E\{\hat{\theta}^m(x^n)\}\right\|^2\right\} \tag{17.18}$$

The bias of the estimator $\hat{\theta}^m(x^n)$ and the conditional variance generally represent a trade-off. Indeed, an unbiased estimator may induce relatively large variance. However, the introduction of some low-level bias may then result in significant reduction of the induced variance. In general, the bias vs. variance trade-off should be studied carefully for the correct evaluation of any given parameter estimator. A parameter estimator is called **efficient** if the conditional variance equals a lower bound known as the Cramér–Rao bound. It will be useful to present briefly this bound.

The Cramér–Rao bound gives a theoretical minimum for the variance of any estimator. More specifically, let $\hat{\theta}(x^n)$ be the estimate of a scalar parameter θ given the observation vector x^n. Let $f_\theta(x^n)$ be given, twice continuously differentiable with respect to θ, and satisfy also some other mild regularity conditions. Then:

$$E\left\{[\hat{\theta}(x^n) - \theta]^2\right\} \geqslant E\left\{\left[\frac{\partial}{\partial\theta}\log f_\theta(x^n)\right]^2\right\}^{-1} \tag{17.19}$$

If an estimate $\hat{\theta}(x^n)$ is *biased*, then $E\{\hat{\theta}(x^n)\} = \theta + B(\theta)$, and the Cramér–Rao lower bound takes the form:

$$E\left\{[\hat{\theta}(x^n) - \theta]^2\right\} > \frac{\left(1 + \frac{\mathrm{d}B(\theta)}{\mathrm{d}\theta}\right)^2}{E\left\{\left[\frac{\partial}{\partial\theta}\log f_\theta(x^n)\right]^2\right\}} \tag{17.20}$$

Many times it is of interest to examine the properties of an estimator as the number of independent samples n approaches infinity. In this context, an estimator is said to be **consistent** if

$$\hat{\theta}(x^n) \to \theta \text{ as } n \to \infty \tag{17.21}$$

Since the estimate $\hat{\theta}(x^n)$ is a random variable, we have to specify in what sense Equation (17.21) holds. Thus, if the above limit holds w.p. 1, we say that the estimator is strongly consistent or consistent w.p. 1. In a similar way we can define a *weakly consistent* estimator. As far as the asymptotic distribution of $\theta(x^n)$ as $n \to \infty$ is concerned, it turns out that the central limit theorem can often be applied to $\hat{\theta}(x^n)$ to infer that $\sqrt{n}[\hat{\theta}(x^n) - \theta]$ is asymptotically normal with zero mean as $n \to \infty$.

Finally, **penalty or cost function** $c[\hat{\theta}^m(x^n), \theta^m]$ is a scalar, nonnegative function whose values vary as θ^m varies in the parameter space $\mathcal{E}^m$ and as the sequence x^n takes different values in the observation space Γ^n. The conditional expected penalty $c(\hat{\theta}^m, \theta^m)$ induced by the parameter estimate $\hat{\theta}^m(x^n)$ and the penalty function $c[\hat{\theta}^m(x^n), \theta^m]$ is defined as

$$c(\hat{\theta}^m, \theta^m) = E\left\{c[\hat{\theta}^m(x^n), \theta^m] | \theta^m\right\} = \int_{\Gamma^n} c[\hat{\theta}^m(x^n), \theta^m] f(x^n | \theta^m) \mathrm{d}x^n \tag{17.22}$$

where the expectation is taken with respect to the joint conditional probability $f(x^n|\theta^m)$. If the *a priori* density function $p(\theta^m)$ is available, then the expected penalty $c(\hat{\theta}^m, p)$ can be evaluated as

$$c(\hat{\theta}^m, p) = \int_{\mathcal{E}^m} c(\hat{\theta}^m, \theta^m) p(\theta^m) d\theta^m = \int_{\mathcal{E}^m} \int_{\Gamma^n} c[\hat{\theta}^m(x^n), \theta^m] f(x^n|\theta^m) dx^n p(\theta^m) d\theta^m \quad (17.23)$$

Bayesian Estimation Schemes

Bayesian estimation schemes utilize:

- A parametrically known stochastic process parameterized by θ^m, or, in other words, a given joint conditional density function $f(x^n|\theta^m)$ defined on the observation space Γ^n, where θ^m is a well-defined parameter vector
- A realization x^n from the underlying active process (it is assumed that the process remains unchanged throughout the whole observation period)
- A density function $p(\theta^m)$ defined on the parameter space $\mathcal{E}^m$
- A penalty scalar function $c[\theta^m, \hat{\theta}^m(x^n)]$ defined for each data sequence x^n, parameter vector θ^m, and parameter estimate $\hat{\theta}^m(x^n)$

The Bayesian estimate, $\hat{\theta}_o^m = \operatorname{argmin}_{\hat{\theta}^m} c(\hat{\theta}^m, p)$, minimizes the expected penalty $c(\hat{\theta}^m, p)$, and is called the *optimal Bayesian estimate at p.*

If the penalty/cost function has the form $c[\theta^m, \hat{\theta}^m] = 1 - \delta(\|\theta^m - \hat{\theta}^m\|)$, where $\delta(\cdot)$ is the Dirac delta function, then the optimal Bayesian estimate is called **maximum *a posteriori* (MAP) estimate** since it maximizes the *a posteriori* probability, i.e., $\hat{\theta}_{MAP}^m = \operatorname{argmin}_{\hat{\theta}^m} f(\theta^m|x^n)$. If the penalty/cost function is given by $\|\theta^m - \hat{\theta}^m\|^2$, then the Bayesian estimate is called the **minimum mean-square error (MMSE) estimate** and equals the conditional mean $E\{\theta^m|x^n\}$. The following section presents further details on MMSE estimation.

Minimum Mean-Square Error (MMSE) Estimator

For simplicity in presentation, we consider the case of estimating a single continuous-type random variable θ with density $p(\theta)$. We also assume that the dimension of the observation space is 1. The penalty function is the square of the estimation error $(\theta - \hat{\theta})^2$ and the optimality criterion is the minimization of the mean (expected) square value of the estimation error.

Example 1: Let us consider the case of estimating a random variable θ by a constant $\hat{\theta}$. The MMSE estimate $\hat{\theta}$ minimizes the mean-square error (MSE):

$$e = E\{(\theta - \hat{\theta})^2\} = \int_{-\infty}^{\infty} (\theta - \hat{\theta})^2 p(\theta) d\theta \quad (17.24)$$

In other words:

$$\left.\frac{de}{d\theta}\right|_{\theta=\hat{\theta}} = 0 \Rightarrow \left.\int_{-\infty}^{\infty} 2(\theta - \hat{\theta}) p(\theta) d\theta\right|_{\theta=\hat{\theta}} = 0 \quad (17.25)$$

that is,

$$\hat{\theta} = E\{\theta\} = \int_{-\infty}^{\infty} \theta p(\theta) d\theta \quad (17.26)$$

Example 2: Let us now consider the case where θ is to be estimated by a function $\hat{\theta}(x)$ of the random variable (observation) x. In this case, the MSE takes the form:

$$e = E\{(\theta - \hat{\theta})^2\} = \int_{-\infty}^{\infty} \int_{-\infty}^{\infty} [\theta - \hat{\theta}(x)]^2 p(\theta, x) \mathrm{d}\theta \mathrm{d}x \tag{17.27}$$

where $p(\theta, x)$ is the joint density of the random variables θ and x. It can be shown that the function that minimizes the MSE is

$$\hat{\theta}(x) = E\{\theta|x\} = \int_{-\infty}^{\infty} \theta p(\theta|x) \mathrm{d}\theta \tag{17.28}$$

That is, the MMSE estimate is the conditional mean $E\{\theta|x\}$.

If x and θ are jointly Gaussian, then the above conditional mean is a linear function of x, and it is easy to obtain the MMSE estimate. However, in general, the conditional mean is a nonlinear function of x; thus $\hat{\theta}(x)$ is, in general, nonlinear. A popular, although suboptimum, Bayesian scheme is the linear MMSE estimator that provides the estimate that minimizes the expected quadratic penalty in Equation (17.27) within the class of linear parameter estimates.

Example 3: The linear estimation problem involves estimation of a random variable θ in terms of a linear function of x, i.e., $\hat{\theta}(x) = Ax + B$. In this case we need to find the constants A and B that minimize the MSE:

$$e = E\{[\theta - (Ax + B)]^2\} \tag{17.29}$$

A fundamental principle in MMSE estimation is the **orthogonality principle**. This principle states that the optimum linear MMSE estimate $Ax + B$ of θ is such that the estimation error $\theta - (Ax + B)$ is orthogonal to the data x, i.e.:

$$E\{[\theta - (Ax + B)]x\} = 0 \tag{17.30}$$

Using the above principle, we can prove that e is minimum if

$$A = \frac{\rho \sigma_\theta}{\sigma_x} \quad \text{and} \quad B = \mu_\theta - A\mu_x \tag{17.31}$$

where

$$\mu_x = E\{x\}, \quad \mu_\theta = E\{\theta\}$$
$$\sigma_x^2 = E\{(x - \mu_x)^2\}, \quad \sigma_\theta^2 = E\{(\theta - \mu_\theta)^2\}$$
$$\rho = \frac{E\{(x - \mu_x)(\theta - \mu_\theta)\}}{\sigma_x \sigma_\theta}$$

that is, μ_x, μ_θ, σ_x^2 and σ_θ^2 are the mean and variance of x and θ, respectively, while ρ is the correlation coefficient of x and θ. Thus, the MSE takes the form $e = \sigma_\theta^2 (1 - \rho^2)$.

The estimate

$$\hat{\theta}(x) = Ax + B \tag{17.32}$$

is called the **nonhomogeneous linear estimate** of θ. If θ is estimated by a function $\hat{\theta}(x) = Ax$, then the estimate is called **homogeneous**. Parameter A in Equation (17.32) can be evaluated using the orthogonality principle:

$$A = \frac{E\{x\theta\}}{E\{x^2\}} \tag{17.33}$$

We note that if the random variables θ and x are Gaussian zero mean, then the optimum nonlinear estimate of θ equals the linear estimate. In other words, if $\hat{\theta}(x) = E\{\theta|x\}$ is the optimum nonlinear estimate of θ and $\hat{\theta}(x) = Ax$ is the optimum linear estimate, then $E\{\theta|x\} = Ax$. This is true because the random variables θ and x have zero mean, $E\{\theta\} = 0$, $E\{x\} = 0$, and thus the linear estimate $\hat{\theta}$ has zero mean too, $E\{\hat{\theta}\} = 0$. The latter implies that the linear estimation error $e = \theta - \hat{\theta}$ has also zero mean, $E\{e\} = E\{\theta - \hat{\theta}\} = 0$. However, the orthogonality principle implies that the linear estimation error e is orthogonal to the data, $E\{ex\} = 0$. Since e is Gaussian, it is independent of x and thus $E\{e|x\} = E\{e\} = 0$, which is equivalent to the following:

$$E\{\theta - \hat{\theta}|x\} = 0 \Rightarrow E\{\theta|x\} - E\{\hat{\theta}|x\} = 0 \tag{17.34}$$

$$\Rightarrow E\{\theta|x\} = E\{\hat{\theta}|x\} \Rightarrow \hat{\theta}(x) = ax \Rightarrow \hat{\theta}(x) = \hat{\theta} \tag{17.35}$$

That is, the nonlinear and the linear estimates coincide. In addition, since the linear estimation error $e = \theta - \hat{\theta}$ is independent of the data x, so is the mean-square error; that is,

$$E\{(\theta - \hat{\theta})^2|x\} = E\{(\theta - \hat{\theta})^2\} = V \tag{17.36}$$

Thus, the conditional mean of θ given the data x is equal to the MMSE estimate of θ while the conditional variance is equal to the MSE. We emphasize that the optimum linear MMSE estimate Ax and the corresponding MSE in Equation (17.36) are the conditional mean and conditional variance of the Gaussian random variable θ. The latter makes the evaluation of the pdf $f(\theta|x)$ an easy task, i.e., $f(\theta|x) = \frac{1}{\sqrt{2\pi V}} \exp\left\{\frac{-(\theta - Ax)^2}{2V}\right\}$.

Minimax Estimator

Minimax estimation schemes utilize:

- A parametrically known stochastic process parameterized by θ^m
- A realization x^n from the underlying active process
- A scalar penalty function $c[\theta^m, \hat{\theta}^m(x^n)]$ for each data sequence x^n, parameter vector θ^m, and parameter estimate $\hat{\theta}^m(x^n)$

In minimax estimation we know a family $\{p_i(\theta^m)\}$ of *a priori* densities and we seek an estimate that minimizes the maximum penalty within this family. The minimax estimates are solutions of saddle-point game formalizations, with payoff function the expected penalty $c(\hat{\theta}^m, p)$ that is parameterized by the parameter estimate $\hat{\theta}^m$ and the *a priori* parameter density function p. Specifically, we want to find a pair $(\hat{\theta}^m_*, p_*)$ such that:

$$\forall p(\theta^m), \theta^m \in \mathcal{E}^m \quad c(\hat{\theta}^m_*, p) \leqslant c(\hat{\theta}^m_*, p_*) \leqslant c(\hat{\theta}^m, p_*) \quad \forall \hat{\theta}^m(x^n) \in \mathcal{E}^m \quad \forall x^n \in \Gamma^n \tag{17.37}$$

In other words, if a minimax estimate $\hat{\theta}^m_*$ exists, it is an optimal Bayesian estimate at some least favorable *a priori* distribution p_*.

Minimum Variance Unbiased Estimator

The unbiased estimator that has minimum variance is called the minimum variance unbiased estimator (MVUE). An MVUE does not always exist. Under certain conditions, an MVUE can be found either as the solution that maximizes the likelihood function or through the use of sufficient statistics. In particular, if an efficient estimator exists, then it is also MVUE and can be found as the unique solution that maximizes the likelihood function (the maximum likelihood estimator will be discussed in detail in the next section). However, if an efficient estimate does not exist, the ML estimator is not MVUE. An alternative way to find an MVUE, provided it exists, is through the use of sufficient statistics.

A **sufficient statistic** is a set of the observations (or functions of the data) that carries all the information about the unknown parameter. Thus, given a sufficient statistic, the distribution of the data no longer depends on the unknown parameter. A sufficient statistic is said to be *minimal* if of all sufficient statistics it provides the greatest possible reduction of the data. Furthermore, a sufficient statistic is *complete* if there is only one function of the statistic that is unbiased.

Identifying sufficient statistics can be difficult. A useful tool for this task is the Neyman–Fisher factorization theorem. The Neyman–Fisher factorization theorem states that if we can factor the density function $f_{\theta^m}(x^n)$ as $f_{\theta^m}(x^n) = g(\theta^m, T(x^n))h(x^n)$, where g depends on x^n only through $T(x^n)$ and h depends only on x^n, then $T(x^n)$ is a sufficient statistic for θ^m. Conversely, if $T(x^n)$ is a sufficient statistic for θ^m, then the density can be factored as above.

Once we find a sufficient statistic $T(x^n)$, we can apply the Rao–Blackwell–Lehmann–Scheffe theorem to find the MVUE estimator. The Rao–Blackwell–Lehmann–Scheffe theorem states that if $\tilde{\theta}^m$ is an unbiased estimator of θ^m and $T(x^n)$ is a sufficient statistic for θ^m, and if $T(x^n)$ is complete, then $\hat{\theta}^m = E\{\tilde{\theta}^m \,|\, T(x^n)\}$ is the MVUE estimator. Alternatively, if $T(x^n)$ is complete and we can find some function g such that $\hat{\theta}^m = g(T(x^n))$ is an unbiased estimator of θ^m, then $\hat{\theta}^m$ is the MVUE estimator. The latter is usually easier to evaluate and thus, it is favored in practice.

Example 4: In wireless communications, empirical measurements suggest that Rayleigh distribution is a suitable model of the wireless channel. Let us suppose that we receive n independent identically distributed (iid) data x_i, $i = 1, \ldots, n$. Let their common distribution be Rayleigh, given by

$$p_\theta(x_i) = \frac{x_i}{\theta} \exp\left\{ -\frac{x_i^2}{2\theta} \right\}, \quad x_i \geq 0 \tag{17.38}$$

and let θ be the parameter that needs to be estimated.

The joint probability density of the data parameterized by θ is then

$$p_\theta(x^n) = \prod_{i=1}^{n} \frac{x_i}{\theta} \exp\left\{ -\frac{x_i^2}{2\theta} \right\} = \exp\left\{ -\frac{\sum_{i=1}^{n} x_i^2}{2\theta} - n\log\theta \right\} \prod_{i=1}^{n} x_i \tag{17.39}$$

The Neyman–Fisher factorization theorem implies that $T(x^n) = \sum_{i=1}^{n} x_i^2$ is a sufficient statistic. Furthermore, $T(x^n)$ is complete because the Rayleigh density function is a member of the exponential family. Since $E\{T(x^n)\} = 2n\theta$, it is implied that $\hat{\theta} = \frac{1}{2n} \sum_{i=1}^{n} x_i^2$ is unbiased. Thus, $\hat{\theta}$ is the MVUE estimator.

When the MVUE cannot be obtained by either one of the approaches presented above, then we utilize suboptimum estimators. A popular such estimator that is unbiased, linear with respect to the data, and has minimum variance is the *best linear unbiased estimator (BLUE)*. The major advantage of BLUE is that it can be obtained from the first- and second-order moments of the data (rather than the distribution).

Example 5: In this example, we would like to estimate a signal θ that is embedded in Laplacian noise. Again, n iid measurements x_i, $i = 1, \ldots, n$, are taken. The signal model and the joint density of the data are as follows:

$$p_\theta(x_i) = \frac{1}{2} \exp(-|x_i - \theta|) \tag{17.40}$$

$$p_\theta(x^n) = \frac{1}{2^n} \exp\left(-\sum_{i=1}^{n} |x_i - \theta| \right) \tag{17.41}$$

The Cramér–Rao lower bound is equal to $\frac{1}{n}$. The BLUE estimator is such that:

$$\hat{\theta}_{\mathrm{BL}} = \arg \min_{\hat{\theta}} E\left\{(\hat{\theta} - E\{\hat{\theta}\})^2\right\} \tag{17.42}$$

$$\text{subject to } \hat{\theta} = a^T x^n \quad \text{and} \quad E\{\hat{\theta}\} = \theta \tag{17.43}$$

The unbiased constraint leads to $E\{a^T x^n\} = a^T E\{x^n\} = a^T s\theta = \theta$, that is, $a^T s = 1$, where $s = [1, 1, \ldots, 1]^T$ is a vector with all elements equal to 1. The variance of the estimator is $\mathrm{Var}(\hat{\theta}_{\mathrm{BL}}) = a^T C_{x^n} a$, where the data covariance matrix is $C_{x^n} = E\left\{(x^n - \theta s)(x^n - \theta s)^T\right\} = 2I$. Solving the convex quadratic optimization problem we obtain $a = \dfrac{1}{n}s$, thus $\hat{\theta}_{\mathrm{BL}} = \dfrac{1}{n}s^T x^n = \dfrac{1}{n}\sum_{i=1}^{n} x_i$, which is just the sample mean of the observations. We note that $\mathrm{Var}(\hat{\theta}_{\mathrm{BL}}) = \dfrac{2}{n}$, and this value is always larger than the Cramér–Rao bound.

Maximum Likelihood Estimator

Let $X(t)$ be a random process parameterized by θ^m, where θ^m is an unknown but fixed parameter vector of finite dimension m (e.g., $\theta^m \in \mathcal{R}^m$). We assume that the joint probability density function $f_{\theta^m}(x_1, \ldots, x_n)$ is known, where $x^n = [x_1, \ldots, x_n]$ is a realization (or observation vector or sample vector) of the process $X(t)$. The problem is to find an estimate of the parameter vector θ^m based on the realization of $X(t)$. (We note that the dimension of the parameter vector θ^m remains fixed throughout the observation period.)

The intuition behind the maximum likelihood method is that we choose those parameters $[\theta_1, \ldots, \theta_m]$ from which the observed sample vector is most likely to have come. That is, the estimator of θ^m is selected so that the observed sample vector becomes "as likely as possible." In this context, we define the likelihood function $l(\theta^m)$ as the deterministic function that is obtained by the joint probability density function $f_{\theta^m}(x^n)$ after setting the variables $x_1, \ldots, x_n$ equal to their observed value. In other words, in the likelihood function, the parameter θ^m is variable and the sample vector x^n is fixed, while in the joint probability density function, x^n is variable and θ^m is fixed. The maximum likelihood (ML) estimate, $\hat{\theta}_{ML}^m$, of θ^m is the value of the parameter vector for which the likelihood function is maximized. In many cases it is more convenient to work with the natural logarithm of $l(\theta^m)$, $L(\theta^m) = \log l(\theta^m)$, rather than the likelihood function. This function is called the log-likelihood function. Since the logarithm is a monotonic function and the likelihood function is nonnegative, it follows that both $l(\theta^m)$ and $L(\theta^m)$ achieve maximum at the same value of parameter vector θ^m. Thus, the log-likelihood function is maximized for the value of the vector parameter θ^m for which the gradient with respect to θ^m is equal to zero; that is,

$$\hat{\theta}_{ML}^m : \quad \frac{\partial L(\theta^m)}{\partial \theta_i} = 0, \quad i = 1, \ldots, m \tag{17.44}$$

There is a close connection between the ML estimate $\hat{\theta}_{ML}^m = \arg \max_{\theta^m} f_{\theta^m}(x^n)$ and the MAP estimate $\hat{\theta}_{MAP}^m = \arg \max_{\theta^m} f(\theta^m | x^n)$. Since $f(\theta^m | x^n) = f(x^n | \theta^m)p(\theta^m)/p(x^n)$, $\hat{\theta}_{MAP}^m = \arg \max_{\theta^m}(\log f(x^n | \theta^m) + \log p(\theta^m))$. The cost function of MAP estimation has a similar form to the cost function of ML estimation, except for a term $\log p(\theta^m)$ that represents our knowledge of the *a priori* density of the parameter θ^m. We recall that in MAP estimation, the parameter is modeled as random while in MLE, the parameter is treated as deterministic unknown.

It can be shown that when the process $X(t)$ is memoryless and stationary (i.e., when $x_1, \ldots, x_n$ are independent identically distributed), then ML estimators are *consistent, asymptotically efficient, and asymptotically Gaussian.* That is, even when an efficient estimator does not exist, the ML estimator is still nearly optimal. These asymptotic properties, coupled with the fact that whenever an efficient estimate exists it is the MLE, justify the popularity of ML estimation.

Example 6: We return to Example 5. However, instead of BLUE, we are interested in obtaining the MLE $\hat{\theta}_{\text{ML}}$:

$$\hat{\theta}_{\text{ML}} = \arg\max_{\theta} \frac{1}{2^n} \exp\left(-\sum_{i=1}^{n} |x_i - \theta|\right) = \arg\min_{\theta} \sum_{i=1}^{n} |x_i - \theta| \tag{17.45}$$

Thus $\hat{\theta}_{\text{ML}}$ is the median of x_i, $i = 1, \ldots, n$.

Example 7: Let x_i, $i = 1, \ldots, n$, be Gaussian independent random variables with mean θ and variance σ_i^2: $x_i \sim N(\theta, \sigma_i^2)$. We would like to estimate the mean θ and evaluate the Cramér–Rao bound. Since θ is unknown but fixed, we will use the maximum likelihood estimation scheme. The probability density function of the random variable x_i is given by

$$f_\theta(x_i) = \frac{1}{\sqrt{2\pi}\sigma_i} \exp\left\{\frac{-(x_i - \theta)^2}{2\sigma_i^2}\right\} \tag{17.46}$$

Since x_i, $i = 1, \ldots, n$, are independent, the joint density function is given by

$$f_\theta(x_i, \ldots, x_n) = \prod_{i=1}^{n} \frac{1}{\sqrt{2\pi}\sigma_i} \exp\left\{\frac{-(x_i - \theta)^2}{2\sigma_i^2}\right\} \tag{17.47}$$

which is exactly the likelihood function for this estimation problem. The log-likelihood function is

$$\log f_\theta(x_i, \ldots, x_n) = -\frac{n}{2}\log(2\pi) - \sum_{i=1}^{n} \log(\sigma_i) - \frac{1}{2}\sum_{i=1}^{n} \frac{(x_i - \theta)^2}{\sigma_i^2} \tag{17.48}$$

Then the value of θ that maximizes the log-likelihood function is given by

$$\hat{\theta}_{\text{ML}}(x^n) = \frac{1}{\sum_{i=1}^{n} \frac{1}{\sigma_i^2}} \sum_{i=1}^{n} \frac{x_i}{\sigma_i^2}. \tag{17.49}$$

We note that when the variances are equal, the maximum likelihood estimate coincides with the sample mean. The Cramér–Rao bound can be found as follows:

$$E\left\{\left[\frac{\mathrm{d}}{\mathrm{d}\theta}\log f_\theta(x^n)\right]^2\right\}^{-1} = -E\left\{\frac{\mathrm{d}^2}{\mathrm{d}\theta^2}\log f_\theta(x^n)\right\} = \sum_{i=1}^{n} \frac{1}{\sigma_i^2} \tag{17.50}$$

In conclusion, we see that for Gaussian data the sample mean estimate is efficient, because it coincides with the maximum likelihood estimate and achieves the Cramér–Rao bound. However, in the presence of *outliers*, the sample mean performs poorly even when the fraction of outliers is small. This observation gave birth to the branch of statistics called robust statistics.

Other Parameter Estimation Schemes

The Bayesian, minimax, minimum variance unbiased, and maximum likelihood estimation schemes described above are members of the class of parametric parameter estimation procedures. All of these schemes require knowledge of the underlying probability density function of the observations (as well as the probability density function of the parameter in the case of Bayesian schemes). This requirement may not be satisfied in many applications and as such, suboptimum estimators that are easy to derive and simple to realize are also of practical interest.

The common characteristic of all estimation procedures described in this section is the availability of some parametrically known stochastic process that generates the observation sequence x^n. When the stochastic process that generates x^n is nonparametrically described for every given parameter value θ^m, nonparametric estimation schemes arise. Nonparametric estimation schemes may result as solutions of certain saddle-point games, whose payoff function originates from parametric maximum likelihood formalizations.

Defining Terms

Bayesian estimator: An estimation scheme in which the parameter to be estimated is modeled as a random variable with known probability density function.

Bias: The difference between the mean value of the estimate and its true value.

Consistent estimator: An estimator whose value converges to the true parameter value as the sample size tends to infinity. If the convergence holds w.p. 1; then the estimator is called strongly consistent or consistent w.p. 1.

Efficient estimator: An estimator whose variance achieves the Cramér–Rao bound.

Estimate: Our best guess of the parameter of interest based on a set of observations.

Estimator: A mapping from the data space to the parameter space that yields the estimate.

Homogeneous linear estimator: An estimator that is a homogeneous linear function of the data.

Linear MMSE estimator: The estimator that minimizes the mean-square error under the constraint that it is a linear function of the data.

Maximum *a posteriori* estimator: The estimator that maximizes the *a posteriori* density of the parameter conditioned on the observed data.

Maximum likelihood estimator: The estimator that maximizes the probability density function of the data conditioned on the parameter.

Minimum mean-square error estimator: The estimation scheme in which the cost function is the mean-square error.

Minimax estimator: The optimum Bayesian estimator for the least favorable prior distribution.

Nonhomogeneous linear estimator: An estimator that is a nonhomogeneous linear function of the data.

Nonparametric estimator: An estimation scheme in which no parametric description of the statistical model is available.

Orthogonality principle: The fundamental principle for MMSE estimates. It states that the estimation error is orthogonal to the data.

Parameter estimation: The procedure by which we combine all available data to obtain our best guess about a parameter of interest.

Parametric estimator: An estimation scheme in which the statistical description of the data is given according to a parametric family of statistical models.

Penalty or cost function: A nonnegative scalar function that represents the cost incurred by an inaccurate value of the estimate.

Robust estimator: An estimation scheme in which we optimize performance for the least favorable statistical environment among a specified statistical class.

Sufficient statistic: A set of the observations (or functions of the data) that carries all the information about the unknown parameter.

Unbiased estimator: An estimator whose mean value is equal to the true parameter value.

References

D. Kazakos and P. Papantoni-Kazakos, *Detection and Estimation*. New York: Computer Science Press, 1990.

H.L. Van Trees, *Detection, Estimation, and Modulation Theory*, Part I. New York: Wiley, 1968.

M. Kay, *Fundamentals of Statistical Signal Processing: Estimation Theory*. Englewood Cliffs, NJ: Prentice-Hall, 1993.

S. Haykin, *Adaptive Filter Theory*. Englewood Cliffs, NJ: Prentice-Hall, 1991.

A. Papoulis, *Probability, Random Variables, and Stochastic Processes*. New York: McGraw-Hill, 1984.

L. Ljung and T. Soderstrom, *Theory and Practice of Recursive Identification*. Cambridge, MA: The MIT Press, 1983.

Further Information

IEEE Transactions on Information Theory is a bimonthly journal that publishes papers on theoretical aspects of estimation theory and in particular on transmission, processing, and utilization of information.

IEEE Transactions on Signal Processing is a monthly journal which presents applications of estimation theory to speech recognition and processing, acoustical signal processing, and communication.

IEEE Transactions on Communications is a monthly journal presenting applications of estimation theory to data communication problems, synchronization of communication systems, and channel equalization.

17.3 Multiple-Model Estimation and Tracking

David D. Sworder and John E. Boyd

Notation

1. S designates an integer index set $\{1, ..., S\}$.
2. $\mathbf{e}_i$ is the ith canonical unit vector in a space whose dimension is apparent from the context.
3. A subscript may identify time, the component of a vector, or the element of an indexed family with the meaning determined by context; e.g., m_i is the conditional mean of the ith distribution of an indexed family.
4. A superscript may denote an estimate before $(-)$ or after $(+)$ an update; e.g., m_i^+ is the conditional value after an observation update.
5. When a process, $\{x_t\}$, is sampled every T sec, the discrete sequence so generated is written $\{x[k]\}$.
6. Conditional expectation is denoted with a circumflex with the relevant observation apparent from context; e.g., $\{\hat{x}[k]\}$.
7. A Gaussian random variable with mean $\hat{x}_t$ and covariance P_{xx} is indicated by $x \sim \mathbf{N}(\hat{x}_t, P_{xx})$ with the same symbol used for the density function itself.
8. If A is a positive symmetric matrix and x a compatible vector, $x'Ax$ is denoted $\|x\|_A^2$.
9. If P is a positive covariance matrix, D denotes its positive symmetric inverse with F the positive symmetric square root of D. If P is a *covariance* matrix, D is called the *information* matrix.

Introduction

A dynamic model of a system describes how the system state changes over time when acted upon by an external forcing function. In a tracking problem, the conventional state vector (called the base state), x_t, consists of positions, velocities, rotation angles and the like. The dynamic model generates broad motion templates that are recognizable by the estimator in the observation. An estimation algorithm uses the observations and the model to improve its assessment of the base state, to compute uncertainty regions for the state, and to predict the future evolution of the system. The exogenous processes in the model may have both unstructured and structured components. The former is commonly expressed by Gaussian white noise.

A comprehensive description of current modeling practice is presented in Li and Jilkov [1]. The most familiar dynamic models use a mixed-time representation in which the time-continuous base state is represented by a stochastic differential equation and a state measurement is taken every T sec. When the base state is also sampled, a time-discrete model results. Often the equations of plant dynamics and the observation are linearized about the estimated base state to yield a linear-Gauss–Markov (LGM) model:

$$x[k + 1] = A\dot{x}[k] + \sqrt{P^w}w[k + 1] \tag{17.51}$$

$$y[k] = Hx[k] + \sqrt{P^n}n[k] \tag{17.52}$$

where $\{w[k]\}$ and $\{n[k]\}$ are independent unit Gaussian white sequences. The positive matrices P^w and P^n are noise scaling factors. The measurements, $\{y[k]\}$, generate an aggregate observation process, $y[k]$, and state estimates are based upon $y[k]$.

Equation (17.51) and Equation (17.52) underlie the Kalman filter—and its nonlinear version, the extended Kalman filter (EKF).The EKF is a predictor-corrector that extrapolates along the drift direction of the unforced system and then corrects the forward estimate by integrating the new measurement. The EKF mapping is written:

Extrapolation:

$$\hat{x}^-[k + 1] = A\hat{x}[k] \tag{17.53}$$

$$P_{xx}^-[k + 1] = AP_{xx}[k]A' + P^w \tag{17.54}$$

Correction:

$$\Delta\hat{x}[k + 1] = K[k + 1]v[k + 1] \tag{17.55}$$

$$\Delta P_{xx}[k + 1] = -K[k + 1]P_{yy}[k + 1]K[k + 1]' \tag{17.56}$$

The innovations process, $\{v[k] = y[k] - H\hat{x}[k]\}$, is a white process. The covariance of innovations is given by $P_{yy}[k] = HP_{xx}^-[k]H' + P^n$ with inverse $D_{yy}[k]$. The Kalman gain is given by $K[k] = P_{xx}^-[k + 1]H'D_{yy}[k + 1]$.

The $y[k]$-conditional distribution of $x[k]$ is $\mathbf{N}(\hat{x}[k], P_{xx}[k])$ [2]. The EKF locates the base state at $\hat{x}[k]$, and assesses its error with $P_{xx}[k]$.

To illustrate the use (and misuse) of the EKF, consider tracking a maneuvering aircraft flying the path shown in Figure 17.1 at constant altitude (the solid curve beginning at [35, 10] km). The base-state consists of $\{X, Y\}$, the position coordinates, and $\{V_x, V_Y\}$, the associated velocities. The target is subject to two accelerations: a wide-band, omnidirectional acceleration described by the Brownian motion $\{w_x, w_y\}$ with intensity P^w; and a lateral acceleration represented by the turn rate process $\{\Phi_t\}$. The speed is slowly varying when the turn rate is constant, so the omnidirectional acceleration is small, the intensity is about 0.1 g. The jinking behavior can be captured by positing a turn rate process of value $\pm 0.2 r/s$. Further, suppose that when the aircraft turns, it slows by 40% with a return to nominal speed after a turn.

At the origin of the coordinate system there is a radar measuring range and bearing every second with (1σ)-errors of 40 m in range and 1.75 mr in bearing. The measurement can be linearized about $\hat{x}_t$. This ancillary linearization is acceptable when the sensor nonlinearities are smooth and the estimation errors reasonably small.

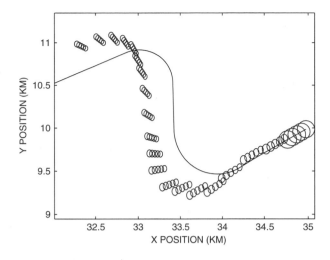

FIGURE 17.1 The path of a target with error ellipses generated by an EKF.

The most rudimentary approach to the tracking problem would be to ignore the turn process and design an EKF based upon the specification of radar quality given above. Figure 17.1 displays the target path along with the computed (1σ)-error ellipses (shown every 0.2 sec for clarity) centered at the location estimates. At first, when the dynamic hypotheses of the EKF match the motion, tracking uncertainty is reduced (the ellipses shrink) with each radar measurement and the path lies within or next to the envelope of the (1σ)-error circles. When the target turns to the right and slows, EKF tacks away from the true path because the gain is too small to bring $\hat{x}_t$ back to x_t. The error ellipses evidence no sensitivity to the growth in the size of the measurement residuals. The residuals may exceed 10σ, a near impossibility if the errors were truly Gaussian. The performance of the EKF could be improved by injecting pseudo-noise although this approach has its own shortcomings.

The EKF fails in this application because the model does not exhibit the full extent of dynamic variability.

Specifically, the target motion has distinct regimes which require individuated models. A broader representation of the motion would use a family of local models, one for each motion mode. Let $\phi_i[k]$ be a pointer vector to the event *the aircraft is in motion mode* i : $\phi_i[k] = e_i$ implies the current mode is the *i*th. Then:

$$x[k + 1] = \sum_{i \in S} \left(A_i x[k] + \sqrt{P_i^w} w[k + 1] \right) \phi_i[k] \qquad (17.57)$$

$$y[k] = \sum_{i \in S} \left(H_i x[k] + \sqrt{P_i^n} n[k] \right) \phi_i[k] \qquad (17.58)$$

where the Gaussian white sequences are as before. A single mode plant ($S = 1$) is called unimodal (orunimorphic) to distinguish it from polymodal (polymorphic) case.

The modal process, $\{\phi_t\}$, is commonly represented by a Markov process on the unit vectors with generator Q' selected to match the mean sojourn times and the transition likelihoods. The time-sampled modal process, $\{\phi[k]\}$, is characterized by a transition rate matrix, $\Pi_{ij} = \mathcal{P}\left(\phi_j \to \phi_i\right)$, derived from Q in the usual way. The discernibility matrix, $\mathbf{D}_{ij} = \mathcal{P}(\phi[k] = e_j \Rightarrow z[k] = e_i)$, quantifies the fidelity of the modal measurement; e.g., $\mathbf{D}_{ii} = \mathcal{P}(\phi[k] = e_i)$ yields the measurement $z[k] = e_i$):

$$\phi[k + 1] = \Pi \phi[k] + \omega[k + 1] \qquad (17.59)$$

$$[k] = \mathbf{D}\phi[k] + \eta[k] \qquad\qquad (17.60)$$

where $\{\eta[k]\}$ and $\{\omega[k]\}$ are white (discrete martingale increment sequences in this case). The measurements $\{z[k]\}$ generate an aggregate observation process, $\mathcal{Z}[k]$, which can be combined with $\mathcal{Y}[k]$ to yield a composite observation process, $\mathcal{G}[k]$. This observation model (often absent the modal measurement) has been used in the development of many polymorphic estimators and tracking algorithms. The integrated base- and modal-state dynamics and measurement equations in Equation (17.57) to Equation (17.60) are called a time-discrete hybrid system and are described in detail in the recent book by Sworder and Boyd [3].

The modal process points toward the global status of the system. During operation, the system will operate in one regime for a time ($\phi[k] = e_i$), and then suddenly shift ($\phi[k+1] = e_p$) to another. In most applications, the discontinuous sample paths of $\{\phi_t\}$ are an approximation to the continuous, albeit abrupt, modal transitions that actually occur. Figure 17.2 illustrates a situation that requires the flexibility of a polymorphic model. Let us follow the north-bound car entering the intersection from the lower left corner of the picture. The vehicle has begun a left turn and will head west. Suppose the tracker has a road map showing the direction of the street segments and the location of intersections. The map is, however, too coarse to show the lane structure. As the vehicle travels north (e.g., $\phi[k] = e_1$), there is considerably more uncertainty in position and velocity in the N–S direction (along the road) than there is in E–W (across the road). This is represented in Equation (17.57) by selecting P_1^w so that the axis of the uncertainty ellipse is far longer in N–S than it is in E–W. If the modal state distinguishes direction of motion on a (N–S, E–W) grid, then four local kinematic models delineate motion in all of the permitted directions, $S = 4$.

However, the problem of tracking on a road grid is more complicated than simply specifying a local model for each directional motion. As the car enters the intersection, it turns west. There is a modal transition at the intersection entry time (e.g., $\phi[k] = e_1 \rightarrow \phi[k+1] = e_4$), and there will be a concomitant reduction in speed if the E–W roads are known to be impaired. These transition events require subtle adjustments in the tracker. The error covariance matrix, P_{xx}, quantifies the tracker's internal uncertainty regarding its kinematic estimates. As pointed out earlier, this will initially be more diffuse in N–S than in E–W. But when the vehicle turns west, the N–S uncertainty is reduced to make the estimate map-compliant. The E–W uncertainty begins to grow from its small value at the intersection. Such adjustments are beyond the capacity of a unimodal model, which has neither the needed flexibility in P^w nor the recognition of mode-induced, base-state discontinuities.

Integrating the modal transitions into a polymorphic model requires engineering judgment. When the car is traveling in a region of restricted access, e.g., between intersections, modal change is unlikely, but at a junction, the car could turn either east or west, or continue north—if a U-turn is not permitted. The modal transition rates must be adjusted accordingly. Thus, Π in Equation (17.59) is actually strongly dependent upon $x[k]$. However,

FIGURE 17.2   A car entering an intersection.

$x[k]$ is not precisely known to the tracker, and the filter must improvise. Here again, the multiple model representation has a significant advantage over the unimodal. An EKF establishes the base state with the single distribution, $\mathbf{N}(\hat{x}[k], P_{xx}[k])$. A polymorphic algorithm uses a family of parallel models and estimators that can be adjusted independently.

Multiple-Model Estimation

A time-discrete hybrid model supposes that modal changes are coincident with a sample times. A polymorphic estimator creates a family of local estimates, each one associated with a different modal sequence of length L ending at the measurement time, $t = kT$. Denote the set of all length-L sequences by κ. Then, each local estimator gauges the base state with a Gaussian distribution with (mean, covariance)-pair, $(m_i^+[k], P_i^+[k])$; $i \in \kappa$, where i runs over the set κ. The index variable is a partial history of a hypothetical modal path extending back to $t = 0$. Most of the current fixed structure multiple-model (FSMM) estimators are based upon a family of local linear representations as displayed in Equation (17.57) and Equation (17.58) [4]. Such estimators include an affiliated group of EKFs as submodules, each tuned to one of the modal hypotheses encapsulated in κ. To keep track of a specific local filter it is advantageous to have a simple notation for modal L-segments. Let us focus on kth-time interval, $t \in [kT, (k+1)T)$. Denote the current modal state by e_i; i.e., $\phi[k] = e_i$. Denote the predecessor mode by e_j; i.e., $\phi[k-1] = e_j$. A modal fragment of length L counting back from $t = (k+1)T^-$ is written $i = \left[e_i, e_j, ..., e_l \right]$ where i is an $S \times L$ array of unit vectors. The collection of all such arrays identifies with κ. An even more compact notation for the path L-fragments is created by simply listing the modal values in sequence as a single number; e.g., $i = ij...l$ is an L-digit, S-radix number with κ again used to represent the aggregate.

From i, other modal fragments can be created. The $(L-1)$-fragment preceding the current time is denoted i^-; e.g., $i^- = \left[e_j, ..., e_l \right]$. Alternatively, an element of κ can be extended forward by prefixing the L-fragment with any of the possible modes on $t \in [k+1]T, (k+2)T)$. This set of $(L+1)$-fragments, κ^+, carries the index variable $i^+ = pij...l$ where e_p is the mode at $t = (k+1)T$.

This extended notation might seem extravagant. The local model in Equation (17.57) is an explicit function of i alone, and the other digits in κ appear superfluous. Although plausible, this is not correct. The initialization in Equation (17.57) at $t = kT$ actually depends on the modal chronology. Thus, even though there are only S forms of extrapolation, there are far more modal path hypotheses, and account must be taken of this variety. This is evident in the interpretation of the measurement equation, Equation (17.58). Sensor linearization takes place about a base-state estimate. With the dynamic memory implicit in Equation (17.57), there are more such estimates than are symbolized by the index i. Actually, Equation (17.58) should be more correctly written with an expanded notation: $i \rightarrow i$.

The common FSMM estimators differ in the depth of the modal sequence retained. Some do not look back at all: $L = 1$ and $i = i$. Some only look back to an immediate predecessor: $L = 2$ and $i = ij$. Others look more deeply, and it is such a one, the Gaussian wavelet estimator (GWE) that we will consider here. The GWE has the flexibility to accommodate arbitrary values of L, and it easily fuses the modal measurements (if any) as well. Unfortunately, as L increases, the number of EKF submodules grows as S^L. In the example that follows, we will let $L = 3 : i = ijk$, and κ is the set of all three-digit S-radix numbers.

The GWE approximates the conditional distribution of the system state with a highly partitioned Gaussian sum [5]. At the beginning of the kth time interval, the joint base-state, modal-state conditional density is given by

$$p[k](\zeta) = \sum_{i \in k} \alpha_i[k] \mathbf{N}_\zeta(m_i - [k], P_i - [k]) \tag{17.61}$$

All of the coefficients in Equation (17.61) are functions of the observation $\mathcal{G}[k]$.

The algorithm is a predictor-corrector that maps a family of estimates forward from $\kappa[k] \rightarrow \kappa[k]^+$; i.e., it maps the L-deep past–present to the $(L+1)$-deep past–present–future. Gaussian merging is then used to reduce the size of this extended index set to that of the nominal index set: $\kappa[k]^+ \rightarrow \kappa[k+1]$. Extrapolation begins by first initializing $(m_\iota[k], P_\iota[k])$ at time $t = kT$. In more elementary FSMM algorithms, the statistics from the preceding interval are simply carried forward: $(m_\iota[k], P_\iota[k]) = (m_\iota - [k], P_\iota - [k]); \iota \in k$. However, if there is supplementary information at $t = kT$, $(m_\iota[k], P_\iota[k])$ should integrate these data. In the example that follows, initialization depends upon only (i, j), and the initialization transformation will be labeled $\mathfrak{I}_{ij}$:

$$(m_\iota - [k], P_\iota - [k]) \xrightarrow{\mathfrak{I}_{ij}} (m_\iota[k], P_\iota[k]; \iota \in \kappa) \qquad (17.62)$$

The GWE employs a family of tuned EKFs for extrapolation and correction:

$$(m_\iota[k], P_\iota[k]) \xrightarrow{\text{EKF}_\iota} (m_\iota^+[k+1], P_\iota^+[k+1]) \qquad (17.63)$$

If there is supplementary terminal information, $(m_\iota^+[k+1], P_\iota^+[k+1])$ should integrate these data at $t = (k+1)T$. In the example that follows, this depends upon $i \in \mathbf{S}$ alone, and the closure transformation will be labeled $\mathfrak{I}_i$:

$$\left(m_\iota^+[k+1], P_\iota^+[k+1]\right) \xrightarrow{\mathfrak{I}_i} \left(m_\iota^+[k+1], P_\iota^+[k+1]\right) \qquad (17.64)$$

Again, $\mathfrak{I}_i$ is the identity transformation for the conventional multiple-model estimators.

Usually the modal transition rate matrix, Π, is assumed to be known and constant. Indeed, the development of conventional FSMM trackers is premised on the assumption that $\{\phi[k]\}$ is a Markov process; i.e., constant Π. The FSMM estimators distinguish themselves by the way they update the modal probabilities, $\{\alpha_\iota\}$, and the way they control the growth in number of modal hypotheses. The simplest way to compute the likelihood of the various modal states is to ignore modal switching; i.e., $\Pi = I$. The more sophisticated estimators require modal mixing. Since the computation of the conditional state distribution is only approximate, various forms of merging have been proposed. The GWE uses a linear form to update the conditional probabilities of the modal $(L + 1)$-segments. First $y[k+1]$ is assimilated, and then $z[k+1]$ fusing takes place. The result is smoothed:

$$\alpha_\iota^-[k+1] = \alpha_\iota[k]|F_\iota^y|\exp -\frac{1}{2}\left(\|y[k+1]\|_{D_\iota^n}^2 - \Delta\|d_\iota\|_{P_\iota^y}^2\right) \qquad (17.65)$$

$$\alpha_\iota + [k+1] = \alpha_\iota^-[k+1]\Pi_{pi}z[k+1]'\mathbf{D}_p \qquad (17.66)$$

where the observation covariances are $\{P_\iota^y; \iota \in \kappa\}$. The updated probability of the modal segment $\iota \in \kappa[k]$ is $\{\alpha_\iota^+\}$ where

$$\alpha_\iota^+[k+1] = \sum_{\iota^+ \in \kappa_{\iota[\kappa]}^+} \alpha_\iota + [k+1] \qquad (17.67)$$

The various $\{\alpha_\iota\}$ are then renormalized.

The expanded conditional distribution of the hybrid state is by an S^L-fold Gaussian sum:

$$p[k+1]^- = \sum_{\iota \in \kappa[k]} \alpha_\iota^+ \mathbf{N}_\zeta(m_\iota^+, P_\iota^+) \tag{17.68}$$

Various moments of interest can be computed from Equation (17.68); e.g., the value of $\mathcal{G}[k+1]$-mean of the target state:

$$\hat{\phi}_i[k] = \sum_{\kappa_i} \alpha_\iota^+[k+1] \tag{17.69}$$

and

$$\hat{x}[k+1] = \sum_{\kappa} \alpha_\iota^+[k+1] m_\iota^+[k+1] \tag{17.70}$$

To reduce Equation (17.68) to the form given in Equation (17.61), the number of terms in the density must be reduced. This may be done by pruning or merging, with the latter used here (see Sworder et al. [6]). The set κ can be partitioned into disjoint subsets, $\{\kappa_\iota -\}$ such that $\iota \in \kappa_\iota -$ has the form $[\iota^-[k+1], \mathbf{e}_l]$ for some $l \in S$: the elements of $\kappa_\iota -$ are identified by their first $L-1$ digits. Mixture is achieved by averaging over the most distant modal hypothesis using the conventional Gaussian sum merging formula:

$$\alpha_\iota[k+1] = \sum_{\iota+ \in \kappa_{\iota[k+1]}^+} \alpha_\iota+[k+1]; \quad m_\iota -[k+1] = \sum_{\iota \in \kappa_{\iota -[k+1]}} m_\iota^+[k+1] \alpha_\iota^+[k+1] \tag{17.71}$$

$$P_\iota -[k+1] = \sum_{\iota \in \kappa_{\iota -[k+1]}} \left(P_\iota^+[k+1] + (m_\iota^+[k+1] - m_\iota -[k+1])(\cdot)' \right) \alpha_\iota^!{[k+1]} \tag{17.72}$$

The GWE recurrence is now complete.

Map-Enhanced Tracking

To illustrate the flexibility of a polymorphic model, let us study in detail the problem of tracking a vehicle on a known road network. Figure 17.2 illustrates some of the broad uncertainties that arise. The subject automobile is limited in its lateral motion by roadbed constraints. The local models express this by making P_i^w a function of direction. At a junction there may be a change in direction and velocity. The forward motion depends on the both the current direction and the predecessor direction from which the vehicle accelerates after a turn.

Figure 17.3 illustrates an extended route. The left panel shows the actual path (the solid curve) of a vehicle initially at $(-500, -400)$ m (labeled "start"). The target moves north until it reaches a crossroad at $(-500, -200)$ m at which time the target turns east. It progresses east until it comes to an intersection and turns south. The vehicle follows the indicated path, ending northbound at $(-700, 450)$ m. The turns shown occur at junctions, but there may be junctions that are not manifest in the figure because the target passed through them.

This irregular motion can be described using a four-mode polymorphic model. List the nominal speeds in the N, S, E, W sequence: $\mathcal{V} = \{20, 20, 10, 10\}$ m/sec. In each model the kinematic state consists of the velocity vector and its integral, the position. For example, if $\phi_t = \mathbf{e}_1$, the car is moving north on a road with nominal speed 20 m/sec. If the target continues in a specific direction, $\phi_t \equiv \mathbf{e}_i$, the primary acceleration is white and along the road direction: $dV_t = \sqrt{W_i} dw_t$ with the major axis of W_i along the ith road direction. Specifically, W_i is

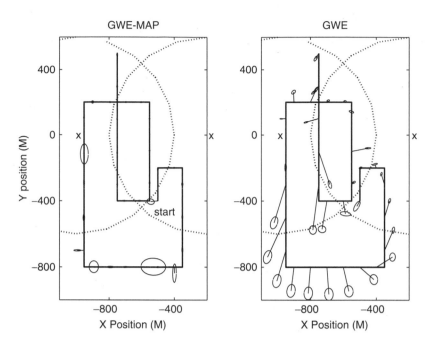

FIGURE 17.3 A motion of a ground target on a grid with several intersections. The left panel shows the response of a map-based tracker. The right panel uses a tracker without a map. The design range of each sensor (shown by "x") is 600 m. A line from the true target location is drawn to a (1σ)-error ellipse centered on the mean target location.

such that the N $-$ S(1σ)-speed uncertainty grows at the rate 0.6 m/sec^2 along the road and 0.2 m/sec^2 laterally. Similarly, in the E–W direction the (1σ)-speed uncertainty grows at the rate 0.3 m/sec^2 along the road and 0.1 m/sec^2 in the transverse direction.

While a Brownian excitation oriented in the road direction suffices if there is no change in direction, when the direction does change, there is a velocity discontinuity. When $\phi_t = \mathbf{e}_i \neq \phi_{t-}$, the vehicle position is continuous but the velocity jumps to that associated with the new forward direction.

The sensor suite contains a pair of range-bearing, (r_t, θ_t), sensors of variable quality situated as shown in the figure, and an acoustic target-speed indicator. The nominal (1σ)-sensor quality is: 54 m in range and 17 mr in bearing. The time interval between observations is 5 sec. Outside a fixed design range, the sensors degrade to 160 m in range and 50 mr in bearing. The acoustic measurement, $z[k]$, is simultaneous with the location measurements. Given the speed variability within a mode, the quality of acoustic sensor is only fair: speed is classified correctly 75% of the time. The velocity classification errors parallel the nominal speed differences; e.g., a north velocity is more likely to be classified as south than as east. Of course, a speed sensor cannot distinguish direction: a north motion is equally likely to be classified as south.

In the time-discrete model presented in Equation (17.57), the $\{A_i\}$ matrices are the same for all $i \in S$. The wideband disturbance individuates the modes. For the purpose of this example, we will use a three-deep GWE: $L = 3$. Extrapolation requires 64 EKFs tuned to the alternative modal sequences. The measurements from the (r_t, ψ_t)-sensor grid are converted to the common $(X - Y)$-coordinate system and grouped as the vector $y[k]$. The quality of the range-bearing measurement depends on the range to the vehicle with severe degradation outside the dotted circles. Equation (17.58) involves 64 separate linearizations with the decision on the additive noise based on m_t^-.

The polymorphic model delineates the formal kinematic structure of this encounter. More subtle is the selection of the modal transition rates. The GWE utilizes Π to alert it to the possibility of a turn. An adaptive GWE uses the road map to adjust the transition rates. If the vehicle is close to a junction, Π should gauge the likelihood of the various continuation directions. Alternatively, in a region of limited access, there is little likelihood of a turn. To quantify *close to a junction*, list the locations of the junctions: $\chi_r; r \in N_{cr}$. Consider the ιth local filter with $\mathcal{G}[k+1]^-$-position statistic $(m_\iota^+[k+1], P_\iota^+[k+1])_p$; the (mean, variance) of target location. This planar statistic can be used to infer the distance of the ι-local target state to intersection r:

Denote by $\varrho_\iota[r, k+1] = \sqrt{\left(\|\chi_r - (m_\iota^+[k+1])_p\|^2_{D_\iota^+[k+1]} \right)}$. Denote the minimum distance to a junction

measured in standard units by $\bar{\varrho}_\iota[k+1]$. We will suppose here that if $\bar{\varrho}_\iota[k+1] > 3$ (the local filter places the target more than $[3\sigma]$ from any crossroad), $\Pi \approx I$ (the target is unlikely to turn); if $\bar{\varrho}_\iota[K+1] \leq 3$ (the local filter places the target within a (3σ) neighborhood of a crossroad), the likelihood of selecting each of the three forward directions is equal.

Each of the local filters in the GWE must be initialized according to the map. Since the conditioning at the end of every interval, $\mathcal{J}$, translates the estimate to the map $(m_\iota - [k], P_\iota - [k])$ will be map-compliant. If ι is such that $\iota = iil$, then there is no change in direction and the local statistics are unchanged. When $\iota = ijl; i \neq j$, the velocity is reinitialized. The local covariance, $P_\iota[k]$, is adjusted in accord with the remaining kinematic uncertainties.

At the end of a sample interval, there are several ways that the map can be used to correct the estimates generated by the local EKFs [7]. The simplest one will be used here: $(m_\iota^+[k+1], P_\iota^+[K+1])$ is projected back to the road in the map coordinates. For example, suppose $\iota = 1jl$; i.e., the target is moving north. Then $\mathcal{J}_i$ moves the X-coordinate back to the E–W location of the road and sets $V_x = 0$. $P_\iota^+[k+1]$ is adjusted to eliminate the E–W uncertainty in location and velocity while the retaining uncertainty in N–S.

The utility of map-integration can be seen by contrasting the performance of the adaptive GWE (labeled GWE-M using the map) and the basic GWE (ignoring the map). Figure 17.3 shows an example when the vehicle is within the high quality region of the sensor half of the time (the design range of the sensors is 600 m). Look first at the left panel. When the path is within the design range of either sensor, the tracking error of GWE-M is minimal. In only one instance is this tracker uncertain about the target motion. When the range-to-target extends to 1 km, the GWE-M becomes more tentative; there are a few bigger error ellipses and even some confusion about a turn. Part of the reason for the modal uncertainty exhibited by the GWE-M after the first west turn is due to a peculiarity of the modal modeling. Immediately after the turn, the local error covariances are relatively large. If $\bar{\varrho}_\iota[k+1] \leq 3$, GWE-M will prepare to turn even though the crossroad is behind the target.

The basic GWE experiences difficulty when the range-to-target exceeds a couple of hundred meters. Only when the target is near sensor-W is tracking uniformly well. Even when the target is within the high-quality region of both locations sensors, large errors are seen. The computed error ellipses are far too small to adequately describe the uncertainty; the tracking error consistently exceeds a computed 2σ value, and the model-based statistics should be considered unsatisfactory. The tight lattice structure of the path is not apparent to this tracker. The GWE smooths the southernmost segment of the path, swinging 200 m south of the actual road.

Increasing the design range for the location sensors will reduce tracking error. Figure 17.4 shows a sample of the response of the trackers when the entirety of the path is included in the high-quality measurement region of both sensors. The GWE-M has no significant error, and the error ellipses are so small as to be invisible on the scale of the plot. The basic GWE has difficulty recognizing a turn even at close range. Its error ellipses are so small that several measurements are required to identify directional changes. Again, the lack of a map leads to a curved path estimate for the GWE.

These general behaviors are compounded in Figure 17.5. The design range of the location sensors is now only 300 m. The regions of good measurements are disjoint and the target spends most of its time outside either region. Even so, the map-compliant GWE does well. It again imagines turns to be plausible immediately post-crossroad. It exhibits one misplaced change-in-direction estimate in the midst of the southernmost segment of the path. This is due to the imprecision of the Π-matrix and the poor quality of the measurements. Even here, the large error ellipse indicates that the GWE-M identifies this estimate as an anomaly—truth is within a 2σ ellipse. The quality of the basic GWE is simply unacceptable.

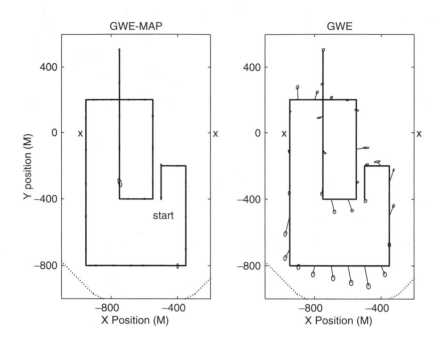

FIGURE 17.4   A motion of a ground target on a grid with several crossroads. The design range of the sensors is 1200 m. The full encounter is within the high-quality region of both sensors.

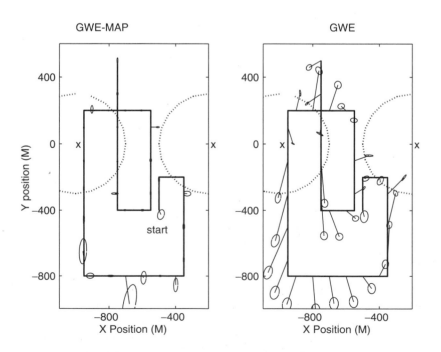

FIGURE 17.5   A motion of a ground target on a grid with several crossroads. The design range of the sensors is 300 m. The target is in a low-quality measurement region for most of the encounter.

The improvement has been shown to be fairly constant across a large range of sensor qualities. As the design range is reduced, the measurement error increases. The GWE-M maintains a 10 dB advantage over the GWE in mean-radial-error. Of course, this GWE could be improved by adding more pseudo-noise to the model—making P^n larger. However, that would increase the size of the error ellipses and dilute its predictions.

Conclusions

The intrinsic value of a model-based estimator is often difficult to quantify. Many such trackers are grounded on an LGM-model—or models in the case of the FSMM estimators. The most used of these algorithms is the EKF. The ubiquity of the EKF leads to its use in situations in which its modeling paradigm is crude; e.g., hard nonlinearities are smoothed, etc. The loose description of the tracking environment is sometimes concealed by arbitrarily adding pseudo-noise. Pseudo-noise desensitizes the estimates to the modeling errors but sacrifices the specificity that careful modeling could provide. Additionally, the computed error covariances are broader than they should be. This lack of refinement leads many designers to dismiss LGM-based methods.

The multiple-model approaches avoid some of these criticisms by representing complex kinematic environments with a family of LGM models. Each model provides a local description of the target/sensor at a different point in the neighborhood of the true base-state. In this way, several parallel system representations are maintained, each with its restricted domain of applicability. The FSMM estimator acts as a "self-adjusting variable bandwidth filter," weighting current measurements more when the uncertainty is greater [8].

Further Reading

There are several sources of further reading on FSMM estimation. The survey papers by Li and Jilkov [4] cover LGM-modeling and FSMM estimation with a large number of well-chosen references. The basic issues of hybrid estimation are examined in a recent book [3]. Applications of multiple model estimation are found in many journals and conference proceedings [9–11].

References

1. X. Li and V. Jilkov, "Survey of maneuvering target tracking. Part 1: Dynamic models," *IEEE Trans. On Aerospace and Electronic Systems* 39, pp. 1333–1364, Oct 2003.
2. M. Grewal and A. Andrews, *Kalman Filtering: Theory and Practice*. Prentice-Hall, Englewood Cliffs, NJ, 1993.
3. D. Sworder and J. Boyd, *Estimation Problems in Hybrid Systems*. Cambridge, UK: Cambridge University Press, 1999.
4. X. Li and V. Jilkov, "A survey of maneuvering target tracking– part V: Multiple- model methods," *IEEE Trans. on Aerospace and Electronic Systems*, (to appear), 2005.
5. D. Sworder and J. Boyd, "Sensor fusion in estimation algorithms," *J. Franklin Inst.* 339, pp. 375–386, Aug 2002.
6. D. Sworder, J. Boyd, and C. Leondes, "Multiple model methods in path following-II," *J. Math. Anal. Appl.* 281, pp. 86–98, Jan 2003.
7. B. Pannetier and K. Benameur, "Ground moving target tracking with road constraint," *Proc. SPIE: Signal and Data Processing of Small Targets* 5428, May 2004.
8. X. Li and Y. Bar-Shalom, "Design of an interacting multiple model algorithm for air traffic control tracking," *IEEE Trans. Control Syst. Technol.* 1, pp. 186–194, Sept 1993.
9. D. Sworder and J. Boyd, "Enhanced recognition in hybrid systems," *IEEE Trans. Signal Proc.* 50, pp. 981–984, April 2002.
10. D. Sworder and J. Boyd, "Tracking ground vehicles on a grid," *Proc. SPIE: Signal and Data Processing of Small Targets* 5428, May 2004.
11. D. Sworder and J. Boyd, "Measurement rate reduction in hybrid systems," *J. Guid. Control Dynam.* 24, pp. 411–414, April 2001.

18
Multidimensional Signal Processing

Yun Q. Shi
New Jersey Institute of Technology

Wei Su
U.S. Army RDECOM CERDEC

Chih-Ming Chen
*National Taiwan University of
Science and Technology*

Sarah A. Rajala
North Carolina State University

N.K. Bose
Pennsylvania State University

L.H. Sibul
Pennsylvania State University

18.1 Digital Image Processing

Yun Q. Shi, Wei Su, and Chih-Ming Chen

In the early 1960s, the joint advent of the semiconductor computer and the space program formally brought the field of digital image processing into public focus. Swiftly, digital image processing found many applications other than the space program and became an active research area and a graduate course offered in many universities in the 1970s and early 1980s. With the tremendous advancements continuously made in the very large scale integration (VLSI) computer and information processing, images, image sequences, and video have become indispensable elements of our modern daily life. The trend is powerfully continuing nowadays. Examples include the high availability of laptop personal computers (PCs), digital cameras, scanners, mobile phones, and Internet in our daily life. Examples also include popularly utilized international image and video compression standards such as Joint Photographic Experts Group (JPEG), JPEG2000, Moving Picture Experts Group (MPEG)-1, MPEG-2, MPEG-4, and International Telecommunication Union (ITU) video coding standards H.263 and H.26L. Therefore, it is not surprising to know that imaging technologies are considered one of the 20 Greatest Engineering Achievements made by mankind in the 20th century (www.greatachievements.org).

Digital Image Generation

When three-dimensional (3D) world scenery is perceived by our human visual system (HVS), what happens in our HVS is that the light of the scenery has been projected through lens in our eyes onto our retina. When 3D scenery is projected via an optical system onto film in an analogue camera, an analogue image is captured. Digital images are different from analogue images in that both the coordinates of picture elements (referred to as *pixels*) and the image brightness or intensity (referred to as grayscale levels for multitone or gray level images) are digitized. That is, these quantities are all represented with finitely many integers. Two typical scenarios are described here. One is the digital camera, where the incoming light projected to the charge-coupled device (CCD) sensors. The two-dimensional (2D) (often rectangular-shaped) array of CCD sensors converts the incoming light into electronic signals, and the read-out mechanism and analog/digital (A/D) converter then formulate a digital image. Another is the digital scanner (often flatbed-shaped). When the hardcopy of an image is being scanned, a line of light source is moving parallel to the plane in which the hardcopy is placed, and the sensors are converting the light into electronic signals, thus resulting in a digital image. In both scenarios, the processes of sampling (discretization of the input signal in image plane) and quantization (discretization of the input signal in magnitude) are involved in formulating 2D digital images. Both sampling and quantization have been well addressed in many texts on digital signal processing. Images as 2D signals follow the general principle of the sampling and quantization theory. Hence, sampling and quantization will not be discussed in detail in this chapter. The readers are referred to Gonzalez and Woods (2002), Shi and Sun (1999), and Section 18.2 in this handbook.

Note that not only visible light but also some other energy formats can be used to generate digital images via some proper transducers. Examples in this regard include ultrasound, infrared radiation, and electromagnetic waves.

Image, Image Sequence, and Imaging Space

Without loss of generality, we discuss visible light imaging in this section. Consider a sensor located in a specific position in 3D space. It generates images about the scene one after another. As time goes by the images form a sequence. The set of these images can be represented with a brightness function denoted by $g(x, y, t)$, where x and y are coordinates on image plane, t represents time moments, and g represents grayscale levels of the pixel at (x, y). Now a generalization of the above basic outline is considered. A sensor as a solid article (which has three free dimensions) can be translated and rotated (two free dimensions). It is noted that here the rotation of a sensor about its optical axis is not counted since the images generated will remain unchanged when this type of rotation takes place. So we can obtain a variety of images when a sensor is translated to different coordinates and rotated to different angles in the 3D space. Equivalently, we can imagine that there are infinitely many sensors in the 3D space which occupy all of possible spatial coordinates and assume all of possible orientations at each coordinate, i.e., they are located in all possible positions. At one specific moment all of these images form a set of images. When time varies these sets of images form a much bigger set of images. Clearly, it is impossible to describe such a set of images by using the above-mentioned brightness function $g(x, y, t)$. Instead, it should be described by a more general brightness function $g(x, y, t, \vec{s})$ where $\vec{s}$ indicates the sensor's position in the 3D space, i.e., the coordinates of the sensor center and the orientation of the optical axis of the sensor. As mentioned previously, $\vec{s}$ is a 5D vector. That is, $\vec{s} = (\tilde{x}, \tilde{y}, \tilde{z}, \beta, \gamma)$, where $\tilde{x}, \tilde{y}$, and $\tilde{z}$ represent the coordinates of the optical center of the sensor in the 3D space; and β and γ represent the orientation of the optical axis of the sensor. More specifically, each sensor may be considered associated with a 3D Cartesian coordinate system such that its optical center is located on the origin and its optical axis is aligned with the OZ axis. We choose in the 3D space a 3D Cartesian coordinate system as the reference coordinate system. Hence, a sensor with its associated Cartesian coordinate system coincident with the reference coordinate system has its position in the 3D space denoted by $\vec{s} = (0, 0, 0, 0, 0)$. An arbitrary sensor position denoted by $\vec{s} = (\tilde{x}, \tilde{y}, \tilde{z}, \beta, \gamma)$ can be described as follows. The sensor's associated Cartesian coordinate system has been first shifted from the reference coordinate system in 3D space with its origin settled at $(\tilde{x}, \tilde{y}, \tilde{z})$ in the reference coordinate system and has then been rotated, with the rotation angles β and γ being the same as Euler angles. Figure 10.1 in Shi and Sun (1999) shows the reference coordinate system and an arbitrary

Cartesian coordinate system (indicating an arbitrary sensor position). There oxy and $o'x'y'$ represent, respectively, the related image planes.

Hence, all possible forms assumed by the general brightness function $g(x, y, t, \vec{s})$ construct an imaging space. If we consider a sensor at a fixed position, we then consider a temporal image sequence, i.e., $g(x, y, t)$. However, if we consider a fixed moment, all images represented by $g(x, y, \vec{s})$ are a spatial image sequence. Each picture, which is taken by a sensor located in a particular position at a specific moment, is merely a special cross-section of this imaging space, and can be represented by $g(x, y)$ (Shi and Sun, 1999).

Grayscale, Binary, Halftone, and Color Images

So far, what we considered are grayscale images, i.e., multitone images. That is, each pixel will assume one of finitely many grayscale levels. Often, each pixel is represented by 8 bits. That is, the allowed range of grayscale levels is from 0 to 255. Conventionally, 0 represents the darkest grayscale, while 255 the brightest grayscale. In binary images, say, text document images, each pixel is only allowed to assume two different levels, either 0 or 1, representing black and white, respectively. In what is known as halftone images, each pixel's grayscale is represented by a group of white and black elements. Imagine a pixel is represented by a group of white elements, hence representing the brightest grayscale level, while another pixel is represented by a group of black elements, hence representing the darkest grayscale level. In between, many different grayscale levels can be denoted. Halftone images are often used in newspaper and text production.

Color images have been used more and more frequently in our daily life, say, from color television (TV) frames to those generated by using digital cameras. For display purposes color images often consist of three prime color components, i.e., red (R), green (G), and blue (B), each represented by 8 bits. A specific color is then associated with a specific combination of different amounts of these three primary colors. In color image compression and manipulation, however, some other color models are preferred and are often used. That is, instead of the RGB model mentioned above, the color is characterized by three features: intensity, hue, and saturation. Here, intensity is the counterpart of grayscale used to describe brightness of image pixels, while hue and saturation are two features used to describe the so-called chromaticity. Roughly speaking, hue has something to do with the dominant color and saturation has something to do with relative purity of the dominant color. The separation of intensity from chromaticity is shown to facilitate color image processing including color image compression. Several color models along this line have been proposed and used, say, HSI, YUV, YIQ, YDbDr, and YCbCr. The YUV model is used in most European countries for phase alternating line (PAL) TV systems, the YIQ model is used in North America and Japan for the National Television Systems Committee (NTSC) TV systems, and YDbDr is used in France, Russia, and some other European countries for the Sequential Couleur a Memoire (SECAM) TV system. The YCbCr model is used in international compression standards for image (JPEG) and video (MPEG). The YUV, YIQ, YDbDr, and YCbCr models can be derived from the RGB model. For more detail, readers are referred to Shi and Sun (1999).

Image Quality Measurement

To evaluate the effectiveness of some digital image processing techniques, we often need to compare the image quality before and after a digital image processing procedure. For instance, in image and video compression, which is going to be discussed later in this chapter, image and video quality is an important factor in dealing with image and video compression. In evaluating two different compression methods, for example, we have to base the evaluation on some definite image and video quality. When both methods achieve the same quality of reconstructed image and video, the one that requires less data is considered to be superior. Alternatively, with the same amount of data to represent an image, the method providing a higher quality reconstructed image or video is considered the better method. Note that here we have not considered other performance criteria, such as computational complexity.

Surprisingly, however, it turns out that the measurement of image and video quality is not straightforward. There are two types of visual quality assessment. One is objective assessment (using electrical measurements), and the other is subjective assessment (using human observers). Each has its merits and drawbacks. A combination of these two methods is now widely utilized in practice.

Subjective Quality Measurement

It is natural that the visual quality of reconstructed video frames should be judged by human viewers if they are to be the ultimate receivers of the data. Therefore, subjective visual quality measure plays an important role in visual communications.

In subjective visual quality measurement, a set of video frames is generated with varying coding parameters. Observers are invited to subjectively evaluate the visual quality of these frames. Specifically, observers are asked to rate the pictures by giving some measure of picture quality. Alternatively, observers are requested to provide some measure of impairment to the pictures. A five-scale rating system of the degree of impairment, used by Bell Laboratories, is listed below (Sakrison, 1979). It has been adopted as one of the standard scales in CCIR Recommendation 500–3 (CCIR 1986). Note that CCIR is now International Telecommunications Union-Recommendations (ITU-R):

Impairment is not noticeable.
Impairment is just noticeable.
Impairment is definitely noticeable, but not objectionable.
Impairment is objectionable
Impairment is extremely objectionable.

In the subjective evaluation, there are a few things worth mentioning. In most applications there is a whole array of pictures simultaneously available for evaluation. These pictures are generated with different encoding parameters. By keeping some parameters fixed while making one parameter (or a subset of parameters) free to change, the resulting quality rating can be used to study the effect of the one parameter (or the subset of parameters) on encoding. An example using this method to study the effect of varying numbers of quantization levels on image quality can be found in (Gonzalez and Woods, 2002).

Another possible way to study this is to identify pictures with the same subjective quality measure from the whole array of pictures. From this subset of test pictures, we can produce, in the encoding parameter space, isopreference curves that can be used to study the effect of the parameter(s) under investigation. An example using this method to study the effect of varying both image resolution and numbers of quantization levels on image quality can be found in Huang (1965).

In the rating, a whole array of pictures is usually divided into columns with each column sharing some common conditions. The evaluation starts within each column with a pairwise comparison. This is because a pairwise comparison is relatively easy for the eyes. As a result, pictures in one column are arranged in an order according to visual quality, and quality or impairment measures are then assigned to the pictures in the one column. After each column has been rated, a unification between columns is necessary. That is, different columns need to have a unified quality measurement. As pointed out in Sakrison (1979), this task is not easy since it means we may need to equate impairment that results from different types of errors.

One thing is understood from the above discussion; subjective evaluation of visual quality is costly. It needs a large number of pictures and observers. The evaluation takes a long time because human eyes are easily fatigued and bored. Some special measures have to be taken in order to arrive at an accurate subjective quality measure. Examples in this regard include averaging subjective ratings and taking their deviation into consideration. For further details on subjective visual quality measurement, readers may refer to Sakrison (1979), Hidaka and Ozawa (1990), and Webster et al. (1993).

Objective Quality Measurement

Here, we first introduce the concept of signal-to-noise ratio (SNR), which is a popularly utilized objective quality assessment. Then we present a promising new objective visual quality assessment technique based on human visual perception.

Signal-to-Noise Ratio

Consider Figure 18.1, where $f(x, y)$ is the input image to a processing system. The system can be a low-pass filter, a subsampling system, or a compression system. It can even represent a process in which additive white Gaussian noise corrupts the input image. Then $g(x, y)$ is the output of the system. In evaluating the quality of

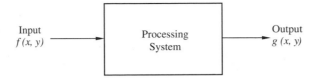

FIGURE 18.1 An image processing system.

$g(x, y)$, we define an error function $e(x, y)$ as the difference between the input and the output. That is:

$$e(x, y) = f(x, y) - g(x, y) \tag{18.1}$$

The mean square error is defined as E_{ms}:

$$E_{ms} = \frac{1}{MN} \sum_{x=0}^{M-1} \sum_{y=0}^{N-1} e(x, y)^2 \tag{18.2}$$

where M and N are the dimensions of the image in the horizontal and vertical directions. Note that it is sometimes denoted by *MSE*. The root mean square error is defined as E_{rms}:

$$E_{rms} = \sqrt{E_{ms}} \tag{18.3}$$

It is sometimes denoted by *RMSE*.

As noted earlier, SNR is widely used in objective quality measurement. Depending whether mean square error or root mean square error is used, the SNR may be called the mean square signal-to-noise ratio, SNR_{ms}, or the root mean square signal-to-noise ratio, SNR_{rms}. We have:

$$SNR_{ms} = 10 \log_{10} \left(\frac{\sum_{x=0}^{M-1} \sum_{y=0}^{N-1} g(x, y)^2}{MN \cdot E_{ms}} \right) \tag{18.4}$$

and

$$SNR_{rms} = \sqrt{SNR_{ms}} \tag{18.5}$$

In image and video data compression, another closely related term, peak signal-to-noise ratio (PSNR), which is essentially a modified version of SNR_{ms}, is widely used. It is defined as follows:

$$PSNR = 10 \log_{10} \left(\frac{255^2}{E_{ms}} \right) \tag{18.6}$$

The interpretation of the SNR is: The larger the SNR (SNR_{ms}, SNR_{rms}, or PSNR), the better the quality of the processed image, $g(x, y)$. That is, the closer the processed image $g(x, y)$ is to the original image $f(x, y)$. This seems correct, however, from our above discussion about the features of the HVS, we know that the HVS does not respond to visual stimuli in a straightforward way. Refer to the two-unit cascade model of our HVS shown in Figure 18.2. Its low-level processing unit is known to be nonlinear. Several masking phenomena exist, which will be discussed later in this section. Each confirms that the visual perception of the HVS is not simple. It is worth noting that our understanding of the high-level processing unit of the HVS is far from complete.

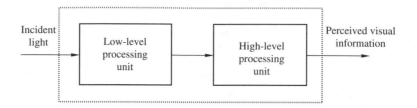

FIGURE 18.2   A two-unit cascade model of the human visual system (HVS).

Therefore, we can understand that the SNR does not always provide us with reliable assessments of image quality. One good example is presented in (Chapter 1, page 14, Shi and Sun, 1999), which uses the so-called IGS quantization technique to achieve high compression (using only 4 bits for quantization instead of the usual 8 bits) without introducing noticeable false contouring. In this case, the subjective quality is higher, and the SNR decreases due to low-frequency quantization noise and additive high-frequency random noise. Another example drawn from our discussion about the masking phenomena is that some additive noise in bright areas or in highly textured regions may be masked, while some minor artifacts in dark and uniform regions may turn out to be quite annoying. In this case, the SNR cannot truthfully reflect visual quality.

On the one hand, we see that an objective quality measure does not always provide reliable picture quality assessment. On the other hand, however, its implementation is much faster and easier than that of the subjective quality measure. Furthermore, objective assessment is repeatable. Owing to these merits, objective quality assessment is still widely used despite this drawback.

It is noted that combining subjective and objective assessments has been a common practice in international coding-standard activity. A new development along this new direction has been reported in Webster et al. (1993) and Shi and Sun (1999).

Image Transformation

A digital image as a 2D signal can be processed in either spatial domain (i.e., directly manipulated in the form of $g(x, y)$) or transform domain. That is, similar to signal processing in general, an image can also be processed in the transform domain such as in the discrete Fourier transform (DFT) domain for some consideration. One example in this regard is that we can easily observe in the DFT domain that image energy is highly concentrated in the dc and low-frequency ac components of the image. Interested readers are referred to an illustrative example contained in Figure 4.11 (Gonzalez and Woods, 2002). On contrary, this feature is not obvious in the spatial domain. Another example is image compression that can be effectively implemented in the discrete cosine transform (DCT) as in JPEG, or discrete wavelet transform (DWT) domain as in JPEG2000, more efficiently than in the spatial domain (Shi and Sun, 1999).

Some other useful image transformations include Hotelling transform (Shi and Sun, 1999), which finds applications in pattern recognition and image compression; Hough transform (Gonzalez and Woods, 2002), which is useful in pattern recognition; discrete Walsh transform, and discrete Hadamard transform (Shi and Sun, 1999), which find applications in image transmission.

Image Enhancement and Image Restoration

The purpose of both image enhancement and image restoration is to improve the quality of images for further image processing or for other applications. The difference between image enhancement and image restoration can be summarized as follows. Image enhancement is carried out in an empirical and intuitive manner. However, image restoration first establishes a mathematical model for the degradation the image has gone through, and then restores the original image based on the model. Three important processing techniques in enhancement and restoration are depicted below as specific examples.

Point Processing

Point processing means that each pixel has its grayscale level changed individually. Hence, it changes independently of its neighboring pixels. If we denote the original image grayscale level as "o," and the grayscale level after point processing as "p." Then we have $p = T(o)$, where T stands for the transformation. Therefore, once a mapping function T is specified, then a point processing is completely specified.

Below we use histogram modification as an example to illustrate the point processing. A *histogram* is a curve such that its horizontal axis represents the grayscale levels. Let us consider an 8-bit grayscale image. That is, each pixel has its grayscale level represented by 8 bits. Hence, each pixel's grayscale level is one of integers from 0 to 255. The vertical axis of the histogram is either the total number of pixels assuming the corresponding grayscale level or the percentage number of the total pixels assuming the corresponding grayscale level. Histograms are statistics often used to depict a grayscale image's brightness distribution. Given a specific histogram, one can have a general idea about the image's grayscale distribution. For instance, from a histogram one can tell if the image is bright or dark in general and what grayscale levels other pixels are assuming. Histogram modification is a type of image processing technique which changes the appearance of the image's histogram. One well-known histogram modification technique is called *histogram equalization*. That is, after the processing, the histogram looks more evenly distributed than before. In an ideal case, histogram equalization leads to a perfect horizontal histogram. Owing to digitization, however, histogram equalization actually leads to a more flat histogram than before.

Clearly, histogram modification including histogram equalization belongs to point processing. That can be depicted as $p = T(o)$.

Window Processing

In contrast to the point processing discussed above, window processing proceeds according to the neighborhood of a pixel under consideration. Frequently, by a window, it is meant a three by three, often written as 3×3, square neighborhood with the pixel under consideration in the center of the square. The window is not necessarily 3×3; it could be 5×5 or 7×7, and so on. That is, the grayscale level of the central pixel of the window after window processing will be replaced by a new grayscale level, which is a function of the grayscale levels assumed by all the pixels within the window. *Mean filter* and *median filter* are examples of window processing. For instance, one often runs a mean filter pixel-by-pixel from the top to bottom and from the left to the right. For each pixel under consideration, a 3×3 window is opened with the pixel at the center of the window. The central pixel's grayscale level is then replaced by the mean value of nine grayscale levels within the 3×3 window. In this way, the image is low-pass filtered. For a median filter, the central pixel's grayscale level is replaced by the median value of the nine grayscale levels occupied by the 3×3 window. Edge detection can be conducted by running a gradient operator. There are several possible gradient operators available. Examples include 3×3 Sobel and Prewitt gradient operators. Readers are referred to Pratt (1991) and Gonzalez and Woods (2002).

Motion Deblurring

When a relative motion between camera and scenery takes place, the picture taken will be blurred. To restore the image, we can establish a model for the experienced motion. Then based on the model, one can deblur the image. In this process, one can observe that a motion model has been established and used to deblur the image. This is an example of *image restoration* using a motion model. Readers are referred to other work (Pratt, 1991; Shi and Sun, 1999; Gonzalez and Woods, 2002) for more information on image restoration and motion compensation for image sequence processing.

Image Compression

Image and video (a sequence of video frames) data compression refers to a process in which the amount of data used to represent image and video is reduced to meet a bit rate requirement (below or at most equal to the maximum available bit rate), while the quality of the reconstructed image or video satisfies a requirement for a certain application and the complexity of computation involved is affordable for the application.

FIGURE 18.3   Image and video compression for visual transmission and storage.

The block diagram in Figure 18.3 shows the functionality of image and video data compression in visual transmission and storage. Image and video data compression has been found to be necessary in these important applications because the huge amount of data involved in these and other applications usually exceeds the capability of today's hardware despite rapid advancements in semiconductor, computer, and other industries.

The required quality of the reconstructed image and video is application-dependent. In medical diagnosis and some scientific measurements, we may need the reconstructed image and video to mirror the original image and video. In other words, only reversible, information-preserving schemes are allowed. This type of compression is referred to as *lossless* compression. In applications such as motion picture and TV, a certain amount of information loss is allowed. This type of compression is called *lossy* compression.

This section first addresses the necessity as well as the feasibility of image and video data compression. Thereafter, fundamental image compression techniques such as quantization, codeword assignment, differential coding, and transform coding are introduced. With these fundamentals, the international still image coding standards JPEG and JPEG2000, and video coding standards MPEG-1, MPEG-2, and MPEG-4 can be understood. However, it is not possible for this section to cover all of these standards. Readers are referred to books on image and video compression (Shi and Sun, 1999).

Practical Needs for Image and Video Compression

Needless to say, visual information is of vital importance for human beings to perceive, recognize, and understand the surrounding world. With the tremendous progress that has been made in advanced technologies, particularly in VLSI circuits, increasingly powerful computers and computations, it is becoming more possible than ever for video to be widely utilized in our daily life. Examples include videophony, videoconferencing, high-definition TV (HDTV), and digital video disk, also known as digital versatile disk (DVD), to name a few.

Video as a sequence of video frames, however, involves a huge amount of data. Let us take a look at an illustrative example. Assume the present switch telephone network (PSTN) modem can operate at a maximum bit rate of 56,600 bits per second. Assume each video frame has a resolution of 288 by 352 (288 lines and 352 pixels per line), which is comparable with that of a normal TV picture and is referred to as common intermediate format (CIF). Each of the three primary colors red, green, blue (RGB) is represented for one pixel with 8 bits, as usual, and the frame rate in transmission is 30 frames per second to provide a continuous motion video. The required bit rate, then, is $288 \times 352 \times 8 \times 3 \times 30 = 72,990,720$ bits per second. Therefore, the ratio between the required bit rate and the largest possible bit rate is about 1289. This implies that we have to compress the video data by at least 1289 times in order to accomplish the transmission described in this example. Note that an audio signal has not been accounted for yet in this illustration.

With increasingly demanding video services such as 3D movies and 3D games, and high video quality such as HDTV, advanced image and video data compression is necessary. It becomes an enabling technology to bridge the gap between the required huge amount of video data and the limited hardware capability.

Feasibility of Image and Video Compression

Here we shall see that image and video compression is not only a necessity for rapid growth of digital visual communications, but it is also feasible. Its feasibility rests with two types of redundancies, i.e., statistical redundancy and psychovisual redundancy. By eliminating these redundancies, we can achieve image and video compression.

Statistical Redundancy

Statistical redundancy can be classified into two types: interpixel redundancy and coding redundancy. By interpixel redundancy we mean that pixels of an image frame, and pixels of a group of successive image or video frames, are not statistically independent. On the contrary, they are correlated to various degrees. This type of interpixel correlation is referred to as interpixel redundancy. Interpixel redundancy can be divided into two categories: spatial redundancy and temporal redundancy. By coding redundancy we mean the statistical redundancy associated with coding techniques.

Spatial Redundancy

Spatial redundancy represents the statistical correlation between pixels within an image frame. Hence, it is also called intraframe redundancy. It is well known that for most properly sampled TV signals, the normalized autocorrelation coefficients along a row (or a column) with a one-pixel shift is very close to the maximum value 1. That is, the intensity values of pixels along a row (or a column) have a very high autocorrelation (close to the maximum autocorrelation) with these of pixels along the same row (or the same column) but shifted by a pixel. This does not come as a surprise because most of the intensity values change continuously from pixel to pixel within an image frame except for the edge regions. That is, often intensity values change gradually from one pixel to the other along a row and along a column.

Spatial redundancy implies that the intensity value of a pixel can be *guessed* from that of its neighboring pixels. In other words, it is not necessary to represent each pixel in an image frame independently. Instead, one can predict a pixel from its neighbors. Predictive coding, also known as differential coding, is based on this observation. The direct consequence of recognition of spatial redundancy is that by removing a large amount of the redundancy (or utilizing the high correlation) within an image frame, we may save a lot of data in representing the frame, thus achieving data compression.

Temporal Redundancy

Temporal redundancy is concerned with the statistical correlation between pixels from successive frames in a temporal image or video sequence. Therefore, it is also called interframe redundancy. Consider a temporal image sequence. That is, a camera is fixed in the 3D world and it takes pictures of a scene one by one as time progresses. As long as the time interval between two consecutive pictures is short enough, i.e., the pictures are taken densely enough, we can imagine that the similarity between two neighboring frames is strong. It is stated in Mounts (1969) that for a videophone-like signal with moderate motion in the scene, on average, less than 10% of pixels change their grayscale levels between two consecutive frames by an amount of 1% of the peak signal. The high interframe correlation was reported in Kretzmer (1952). There, the autocorrelation between two adjacent frames was measured for two typical motion-picture films. The measured autocorrelations were 0.80 and 0.86. In summary, pixels within successive frames usually bear a strong similarity or correlation. As a result, we may predict a frame from its neighboring frames along the temporal dimension. This is referred to as interframe predictive coding. A more precise, hence more efficient, interframe predictive coding scheme, which has been in development since the 1980s, uses motion analysis. That is, it considers that the changes from one frame to the next are mainly due to the motion of some objects in the frame. Taking this motion information into consideration, we refer to the method as motion compensated predictive coding. Removing a large amount of temporal redundancy leads to a great deal of data compression. At present, all the international video coding standards have adopted motion compensated predictive coding, which has been a vital factor in the increased use of digital video in digital media.

Coding Redundancy

As we discussed, interpixel redundancy is concerned with the correlation between pixels. That is, some information associated with pixels is redundant. The psychovisual redundancy, which is discussed next in this subsection, is related to the information that is psychovisually redundant, i.e., to which the HVS is not sensitive. It is clear that both the interpixel and psychovisual redundancies are somehow associated with some information contained in image and video. Eliminating these redundancies, or utilizing these correlations, by using fewer bits to represent the information results in image and video data compression. In this sense, the coding redundancy is different. It has nothing to do with information redundancy but rather with the representation of information, i.e., coding itself.

It is well known that, for example, compared with the fixed-length binary coding, also sometimes referred to as straightforward binary coding or natural binary coding (NBC), variable-length coding (VLC) is more efficient. This implies that for the same set of symbols, different codes may perform differently. Some may be more efficient than others. Huffman coding and arithmetic coding, two variable-length coding techniques, are popularly utilized in the image and video coding standards such as JPEG, JPEG2000, MPEG-1, MPEG-2, and MPEG-4.

Psychovisual Redundancy

While interpixel redundancy inherently rests in image and video data, psychovisual redundancy originates from the characteristics of the HVS. It is known that the HVS perceives the outside world in a rather complicated way. Its response to visual stimuli is not a linear function of the strength of some physical attributes of the stimuli, such as intensity and color. HVS perception is different from camera sensing. In the HVS, visual information is not perceived equally; some information may be more important than other information. This implies that if we apply less data to represent less important visual information, perception will not be affected. In this sense, we see that some visual information is psychovisually redundant. Eliminating this type of psychovisual redundancy leads to data compression.

In order to understand this type of redundancy, let us study some properties of the HVS. We can model the human vision system as a cascade of two units (Lim, 1990), as depicted in Figure 18.2. The first one is a low-level processing unit, which converts incident light into a neural signal. The second one is a high-level processing unit, which extracts information from the neural signal. While much research has been carried out to investigate the low-level processing, the high-level processing remains relatively unknown. The low-level processing unit is known as a nonlinear system (approximately logarithmic). While a great body of literature exists, we will limit our discussion only to image and video compression-related results. That is, several aspects of HVS which are closely related to image and video compression are discussed in this subsection. They are luminance masking, texture masking, frequency masking, temporal masking, and color masking. Their relevance in image and video compression is addressed. Finally, a summary is provided, in which it is pointed out that all of these features can be unified as one: differential sensitivity. This seems to be the most important feature of the human visual perception.

Luminance Masking

Luminance masking concerns the brightness perception of the HVS, which is the most fundamental aspect among the five to be discussed here. Luminance masking is also referred to as *luminance dependence* (Connor et al., 1972) and *contrast masking* (Legge and Foley, 1980; Watson, 1987).

Consider the monochrome image shown in Figure 18.4. There, a uniform disk-shaped object with a gray level (intensity value) I_1 is imposed on a uniform background with a gray level I_2. Now the question is: Under what circumstances can the disk-shaped object be discriminated from the background by the HVS? That is, we want to study the effect of one stimulus (the background in this example, the masker) on the detectability of another stimulus (in this example, the disk). Two extreme cases

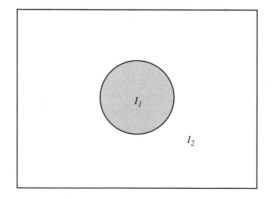

FIGURE 18.4 A uniform object with grayscale level I_1 imposed on a uniform background with grayscale level I_2.

are obvious. That is, if the difference between the two gray levels is quite large, the HVS has no problem with discrimination, or in other words the HVS notices the object from the background. If, however, the two gray levels are the same, the HVS cannot identify the existence of the object. What we are concerned with here is the critical threshold in the gray level difference for discrimination to take place.

If we define the threshold ΔI as such a gray level difference $\Delta I = I_1 - I_2$ that the object can be noticed by the HVS with a 50% chance, then we have the following relation, known as *contrast sensitivity function*, according to Weber's law:

$$\frac{\Delta I}{I} \approx \text{constant} \tag{18.7}$$

where the constant is about 0.02. Weber's law states that for a relatively very wide range of I, the threshold for discrimination, ΔI, is directly proportional to the intensity I. The implication of this result is that when the background is bright, a larger difference in grayscale levels is needed for the HVS to discriminate the object from the background. However, the intensity difference required could be smaller if the background is relatively dark. It is noted that Equation (18.7) implies a logarithmic response of the HVS, and Weber's law holds for all other human senses as well.

The direct impact that luminance masking has on image and video compression is related to quantization. Roughly speaking, quantization is a process that converts a continuously distributed quantity into a set of finitely many distinct quantities. The number of these distinct quantities (known as quantization levels) is one of the keys in quantizer design. It significantly influences the resulting bit rate and the quality of the reconstructed image and video. An effective quantizer should be able to minimize the visibility of quantization error. The contrast sensitivity function provides a guideline in analysis of the visibility of quantization error. Therefore, it can be applied to quantizer design. Luminance masking suggests a nonuniform quantization scheme that takes the contrast sensitivity function into consideration. One such example was presented in Watson (1987).

Texture Masking

Texture masking is sometimes also called *detail dependence* (Connor et al., 1972), *spatial masking* (Netravali, 1977; Lim, 1990), or *activity masking* (Mitchell et al., 1997). It states that the discrimination threshold increases with increasing picture detail. That is, the stronger the texture, the larger the discrimination threshold. It is known that when the number of quantization levels decreases from 256 (8 bits per pixel) to 16 (4 bits per pixel), for instance, the unnatural contours, caused by the coarse quantization (using only 4 bits to represent a pixel's grayscale levels), can be noticed in the relative uniform regions. (For a specific example, readers are referred to Figure 1.9 in Shi and Sun, 1999). This phenomenon was first noted by Goodall (1951) and is called *false contouring* (Gonzalez and Woods, 2002). Now we see that the false contouring can be explained by using texture masking since texture masking indicates that the human eye is more sensitive to the smooth region than to the textured region, where intensity exhibits a high variation. A direct impact on image and video compression is that the number of quantization levels, which affects bit rate significantly, should be adapted according to the intensity variation of image regions.

Frequency Masking

While the above two characteristics are picture-dependent in nature, frequency masking is picture-independent. It states that the discrimination threshold increases with frequency increase. It is also referred to as *frequency dependence*.

Owing to frequency masking, in the transform domain, say, the discrete cosine transform (DCT) domain, we can drop some high frequency coefficients with small magnitudes to achieve data compression without noticeably affecting the perception of the HVS. Note that the frequency masking has been used in JPEG standard. This leads to what is called transform coding, which is discussed later in this section.

Temporal Masking

Temporal masking is another picture-independent feature of the HVS. It states that it takes a while for the HVS to adapt itself to the scene when the scene changes abruptly. During this transition the HVS is not sensitive to details. The masking takes place both before and after the abrupt change. It is called forward temporal masking if it happens after the scene change. Otherwise, it is referred to backward temporal masking (Mitchell et al., 1997). This implies that one should take temporal masking into consideration when allocating data in image and video coding.

Color Masking

In physics, it is known that any visible light corresponds to an electromagnetic spectral distribution. Therefore, a color, as a sensation of visible light, is an energy with an intensity as well as a set of wavelengths associated with the electromagnetic spectrum. Obviously, intensity is an attribute of visible light. The composition of wavelengths is another attribute: chrominance. There are two elements in the chrominance attribute: *hue* and *saturation*. The hue of a color is characterized by the dominant wavelength in the composition. Saturation is a measure of the purity of a color. A pure color has a saturation of 100%, whereas white light has a saturation of 0%.

It is known that the HVS is much more sensitive to the luminance component than to the chrominance components. Following van Ness and Bouman (1967) and Mullen (1985), Mitchell et al. (1997) included a figure to quantitatively illustrate the above statement. A modified version is shown in Figure 1.10 in Shi and Sun (1999). The figure indicates that the HVS is much more sensitive to luminance than to chrominance. The direct impact of color masking on image and video coding is that by utilizing this psychovisual feature, we can allocate more bits to the luminance component than to the chrominance components. This leads to a common practice in color image and video coding: using full resolution for the intensity component, while using a 2×1 subsampling both horizontally and vertically for the two chrominance components. This has been adopted in related international video coding standards, such as MPEG-1 and MPEG-2.

Summary: Differential Sensitivity

Here, let us summarize what we have discussed so far. We see that luminance masking, also known as contrast masking, is of fundamental importance among several types of masking. It states that the sensitivity of the eyes to a stimulus depends on the intensity of another stimulus. Thus it is a differential sensitivity. Both texture (detail or activity) and frequency of another stimulus significantly influence this differential sensitivity. The same mechanism exists in color perception, where the HVS is much more sensitive to luminance than to chrominance. Therefore, we conclude that differential sensitivity is the key in studying human visual perception. These features can be utilized to eliminate psychovisual redundancy, and thus compress image and video data. It is noted that this differential sensitivity feature of the HVS is common to human perception. For instance, there is also forward and backward temporal masking in human audio perception.

Quantization and Codeword Assignment

Recall Figure 18.3, in which the functionality of image and video compression in the applications of visual communications and storage is depicted. In this subsection, we are mainly concerned with source encoding and source decoding. To this end, we take a step further. That is, we show block diagrams of a source encoder and decoder in Figure 18.5. As shown in Figure 18.5(a), there are three components in source encoding: transformation, quantization, and codeword assignment. After the transformation, some form of an input information source is presented to a quantizer. In other words, the transformation block decides which types of quantities from the input image and video are to be encoded. It is not necessary that the original image and video waveform be quantized and coded; we will show that some formats obtained from the input image and video are more suitable for encoding. An example is the difference signal. From the discussion of interpixel correlation in the previous subsection, it is known that a pixel is normally highly correlated with its immediately horizontal or vertical neighboring pixel. Therefore, a better strategy is to encode the difference of grayscale levels between a pixel and its neighbor. Since these data are highly correlated, the difference usually

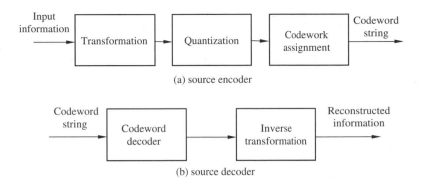

FIGURE 18.5 Block diagram of a source encoder and a source decoder.

has a smaller dynamic range. Consequently, the encoding is more efficient. This idea is discussed later in this section. Another example is what is called transform coding, which is also addressed later in this section. There, instead of encoding the original input image and video, we encode a transform of the input image and video. Since the redundancy in the transform domain is reduced greatly, the coding efficiency is much higher compared to directly encoding the original image and video.

Note that the term *transformation* in Figure 18.5(a) is sometimes referred to as *mapper* and *signal processing* in the literature (Li and Zhang, 1995; Gonzalez and Woods, 2002). Quantization refers to a process that converts input data into a set of finitely different values. Often, the input data to a quantizer are continuous in magnitude.

Hence, quantization is essentially discretization in magnitude, which is an important step in the lossy compression of digital image and video. (The reason that the term *lossy compression* is used here will be shown shortly.) The input and output of quatization can be either scalars or vectors. The quantization with scalar input and output is called *scalar quantization*, whereas that with vector input and output is referred to as *vector quantization*. Here we will only discuss scalar quantization.

After quantization, codewords are assigned to the finitely many different values, the output of the quantizer. Natural binary code (NBC) and variable-length code (VLC), introduced previously in this section, are two examples of codeword assignment. Other examples are the widely utilized entropy code (including Huffman code and arithmetic code), dictionary code, and run-length code (RLC) (frequently used in facsimile transmission).

The source decoder, as shown in Figure 18.5(b), consists of two blocks: codeword decoder and inverse transformation. They are counterparts of the codeword assignment and transformation in the source encoder. Note that there is no block that corresponds to quantization in the source decoder. The implication of this observation is the following. First, quantization is an irreversible process. That is, in general there is no way to find the original value from the quantized value. Second, quantization is, therefore, a source of information loss. In fact, quantization is a critical stage in image and video compression. It has significant impact on the distortion of reconstructed image and video as well as the bit rate of the encoder. Obviously, coarse quantization results in more distortion and lower bit rate than fine quantization.

It is noted that the uniform quantization is the simplest yet the most important case that has been used in various image and video coding standards. Pulse code modulation (PCM) is the best established and most frequently implemented digital coding method that involves quantization.

Differential Coding

Instead of encoding a signal directly, the *differential coding* technique codes the difference between the signal itself and its prediction. Therefore it is also known as *predictive coding*. By utilizing spatial and/or temporal interpixel correlation, differential coding is an efficient and yet computationally simple coding technique.

There are two components of differential coding, prediction and quantization. When the difference signal (also known as prediction error) is quantized, the differential coding is called differential pulse code modulation (DPCM). Delta modulation (DM) is a special case of DPCM. If quantization is not included, the differential coding is referred to as information-preserving differential coding. For more information, the readers are referred to Shi and Sun (1999).

Transform Coding

As introduced above, differential coding achieves high coding efficiency by utilizing the correlation between pixels existing in image frames. Transform coding (TC), the focus of this subsection, is another efficient coding scheme based on utilization of interpixel correlation. As we see, TC has become a fundamental technique recommended by the international still image coding standards JPEG and JPEG2000. Moreover, TC has been found to be efficient in coding prediction error in motion compensated predictive coding. As a result, TC was also adopted by the international video coding standards such as H.261, H.263, and MPEG-1, MPEG-2, and MPEG-4.

Recall the block diagram of source encoders shown in Figure 18.5. There are three components in a source encoder: transformation, quantization, and codeword assignment. It is the transformation component that decides which format of input source is quantized and encoded. In DPCM, for instance, the difference between an original signal and a predicted version of the original signal is quantized and encoded. As long as the prediction error is small enough, i.e., the prediction resembles the original signal well (by using correlation between pixels), differential coding is efficient.

In transform coding, the main idea is that if the transformed version of a signal is less correlated compared with the original signal, then quantizing and encoding the transformed signal may lead to data compression. At the receiver, the encoded data are decoded and transformed back to reconstruct the signal. Therefore, in transform coding, the transformation component illustrated in Figure 18.5 is a transform. Quantization and codeword assignment are carried out with respect to the transformed signal or, in other words, carried out in the transformed domain.

Image Segmentation

In image segmentation, we consider how to segment an image into different segments. For instance, a videophone picture may be segmented into two segments: one is the speaker's head, face, and shoulder, and another is the background. Apparently, image segmentation is useful for image representation and image analysis.

Image Representation

Image representation deals with how to represent an image. An image or an object in an image can be represented by its boundary or the region occupied by the image or the object.

Image Analysis

Image analysis deals with how to analyze an image or objects in an image so that human being can have a high-level understanding about the image and scenery.

In this section, we have discussed some subjects in digital image processing such as image enhancement, image restoration, and image compression. In some sense, all of these tasks belong to low-level image processing. Image segmentation can thus be considered as the middle-level image processing task, which is a necessary step for image representation and image analysis, while image representation and image analysis can be viewed as high-level image processing tasks.

Note that here by low-level, middle-level, and high-level processing we only mean the division of image processing tasks from a human point of view. That is, a comprehensive image processing procedure can often start with low-level processing, say, image enhancement and/or image restoration in order to make image more suitable for the later processing. Or, an image may need to be compressed prior to other processing.

These low-level image processings may be viewed as preprocessing of a given image. The preprocessed image then needs to be segmented. The segmented image needs to be further represented and analyzed for some applications. Also, it is noted that for some applications, there is no need for image segmentation, image representation, or image analysis. For instance, in an application, we may only need to compress a given image and then store it in a computer. The last point that we would like to bring to the reader's attention is that by no means do we imply that the image processing techniques in the low and middle levels are trivial. In fact, image compression remains challenging. There are still many theoretical issues remaining open. Any significant progress in image compression will bring great advancement in information technologies. Also, image segmentation has been seen generally as a difficult task for several decades. In summary, the division of low-, middle-, and high-level processing tasks is only from the human's point of view based on what level of the role image processing tasks under consideration are playing.

References

T. Berger, *Rate Distortion Theory*, Englewood Cliffs, NJ: Prentice-Hall, 1971.

CCIR Recommendation 500-3, "Method for the subjective assessment of the quality of television pictures," in *Recommendations and Reports of the CCIR*, XVIth Plenary Assembly, vol. XI, Part 1, 1986.

D.J. Connor, R.C. Brainard, and J.O. Limb, "Interframe coding for picture transmission," *Proc. IEEE*, vol. 60, no. 7, pp. 779–790, 1972.

R.C. Gonzalez and R.E. Woods, *Digital Image Processing*, 2nd ed., Upper Saddle River, NJ: Prentice-Hall, 2002.

W.M. Goodall, "Television by pulse code modulation," *Bell Syst. Tech. J.*, 33–49, 1951, January.

T. Hidaka and K. Ozawa, "Subjective assessment of redundancy-reduced moving images for interactive application: test methodology and report," *Signal Process.: Image Commun.*, 2, 201–219, 1990.

T.S. Huang, "PCM picture transmission," *IEEE Spectrum*, vol. 2, no. 12, pp. 57–63, 1965.

G.E. Legge and J.M. Foley, "Contrast masking in human vision," *J. Opt. Soc. Am.*, vol. 70, no. 12, pp. 1458–1471, 1980 (December).

W. Li and Y.-Q. Zhang, "Vector-based signal processing and quantization for iamge and video compression," *Proc. IEEE*, vol. 83, no. 2, pp. 317–335, 1995.

J.S. Lim, *Two-Dimensional Signal and Image Processing*, Englewood Cliffs, NJ: Prentice-Hall, 1990.

E.R. Kretzmer, "Statistics of television signal," *Bell Syst. Tech. J.*, vol. 31, no. 4, pp. 751–763, 1952.

J.L. Mitchell, W.B. Pennebaker, C.E. Fogg, and D.J. LeGall, *MPEG Video Compression Standard*, New York: Chapman & Hall, 1997.

F.W. Mounts, "A video encoding system with conditional picture-element replenishment," *Bell Syst. Tech. J.*, vol. 48, no. 7, pp. 2545–2554, 1969.

K.T. Mullen, "The contrast sensitivity of human color vision to red-green and blue-yellow chromatic gratings," *J. Physiol.*, 359, 381–400, 1985.

A.N. Netravali and B. Prasada, "Adaptive quantization of picture signals using spatial masking," *Proc. IEEE*, vol. 65, pp. 536–548, 1977.

W.K. Pratt, *Digital Image Processing*, 2nd ed., New York: Wiley, 1991.

D.J. Sakrison, "Image coding applications of vision model," in *Image Transmission Techniques*, W.K. Pratt Ed., London: Academic Press, 1979, pp. 21–71.

Y.Q. Shi and H. Sun, *Image and Video Compression for Multimedia Engineering: Fundamentals, Algorithms, and Standards*, Boca Raton, FL: CRC Press, 1999.

F.I. Van Ness and M.A. Bouman, "Spatial modulation transfer in the human eye," *J. Opt. Soc. Am.* vol. 57 no. 3, pp. 401–406, 1967.

A.B. Watson, "Efficiency of a model human image code," *J. Opt. Soc. Am. A.* vol. 4, no. 12, pp. 2401–2417, 1987.

A.A. Webster, C.T. Jones, and M.H. Pinson, "An objective video quality assessment system based on human perception," in *Proceedings of Human Vision, Visual Processing and Digital Display IV*, J.P. Allebach and B.E. Rogowitz, Eds., SPIE, vol. 1913, pp. 15–26, 1993.

18.2 Video Signal Processing

Sarah A. Rajala

Video signal processing is the area of specialization concerned with the processing of time sequences of image data, i.e., digital video. Because of significant advances in computing power and increases in available transmission bandwidth, there has been a proliferation of potential applications in the area of video signal processing, including full-motion digital video to the desktop. Thus a desktop workstation can serve as a personal computer, videophone, high-definition television, and/or fax machine. Other applications of digital video include aviation, law enforcement, medicine, military, multimedia, and wireless communications. As diverse as the applications may seem, it is possible to specify a set of fundamental principles and methods that can be used to develop the applications.

Video signal processing is concerned with the manipulation of digital video data. Processing can include sampling, filtering, format conversion, motion estimation, compression, and coding. Considerable understanding of a video signal processing system can be gained by representing the system with the block diagram given in Figure 18.6. Light from a real-world scene is captured by a *scanning system* and causes an image frame $f(x, y, t_0)$ to be formed on a focal plane. A video signal is a sequence of image frames that are created when a scanning system captures a new image frame at periodic intervals in time. In general, each frame of the video sequence is a function of two spatial variables x and y and one temporal variable t. An integral part of the scanning system is the process of converting the original analog signal into an appropriate digital representation. The conversion process includes the operations of sampling and quantization. *Sampling* is the process of converting a continuous-time/space signal into a discrete-time/space signal. *Quantization* is the process of converting a continuous-valued signal into a discrete-valued signal.

Once the video signal has been sampled and quantized, it can be processed digitally. Processing can be performed on special-purpose hardware or general-purpose computers. The type of processing performed depends on the particular application. For example, if the objective is to generate high-definition television, the processing would typically include compression and motion estimation. In fact, in most of the applications listed above, these are the fundamental operations. *Compression* is the process of compactly representing the information contained in an image or video signal. *Motion estimation* is the process of estimating the displacement of the moving objects in a video sequence. The displacement information can then be used to interpolate missing frame data or to improve the performance of compression algorithms.

After the processing is complete, a video signal is ready for transmission over some channel or storage on some medium. If the signal is transmitted, the type of channel will vary depending on the application. For example, today's analog television signals are transmitted one of three ways: via satellite, terrestrially, or by cable. All three channels have limited transmission bandwidths and the signals can be adversely affected because of the imperfect frequency responses of the channels. Alternatively, with a digital channel, the primary limitation will be the bandwidth.

The final stage of the block diagram shown in Figure 18.6 is the display. Of critical importance at this stage is the human observer. Understanding how humans respond to visual stimuli, i.e., the psychophysics of vision, will not only allow for better evaluation of the processed video signals but will also permit the design of better systems.

FIGURE 18.6 Video signal processing system block diagram.

Sampling

If a continuous-time video signal satisfies certain conditions, it can be exactly represented by and be reconstructed from its sample values. The conditions that must be satisfied are specified in the sampling theorem. The sampling theorem can be stated as follows.

Sampling Theorem

Let $f(x, y, t)$ be a band-limited signal with $F(\omega_x, \omega_y, \omega_t) = 0$ for $|\omega_x| > \omega_{xM}$, $|\omega_y| > \omega_{yM}$, and $|\omega_t| > \omega_{tM}$. Then $f(x, y, t)$ is uniquely determined by its samples $f(jX_S, kY_S, lT_S) = f(j, k, l)$, where $j, k, l = 0, \pm 1, \pm 2, \ldots$ if

$$\omega_{sx} > 2\omega_{xM}, \omega_{sy_1 t} > 2\omega_{yM}, \text{ and } \omega_{sy_1 t} > 2\omega_{tM}$$

and

$$\omega_{sx} = 2\pi/X_S, \omega_{sy} = 2\pi/Y_S, \text{ and } \omega_{st} = 2\pi/T_S$$

X_S is the sampling period along the x direction, $\omega_{sx} = 2\pi/X_S$ is the spatial sampling frequency along the x direction, Y_S is the sampling period along the y direction, $\omega_{sy} = 2\pi/Y_S$ is the spatial sampling frequency along the y direction, T_S is the sampling period along the temporal direction, and $\omega_{st} = 2\pi/T_S$ is the temporal sampling frequency.

Given these samples, $f(j, k, l)$ can be reconstructed by generating a periodic impulse train in which successive impulses have amplitudes that are successive sample values. This impulse train is then processed through an ideal low-pass filter with appropriate gain and cut-off frequencies. The resulting output signal will be exactly equal to $f(j, k, l)$ (Oppenheim et al., 1997).

If the sampling theorem is not satisfied, *aliasing* will occur. Aliasing occurs when the signal is undersampled and therefore no longer recoverable by low-pass filtering. Figure 18.7(a) shows the frequency spectrum of a sampled band-limited signal with no aliasing. Figure 18.7(b) shows the frequency response of the same signal with aliasing. The aliasing occurs at the points where there is overlap in the diamond-shaped regions. For video signals, aliasing in the temporal direction will give rise to flicker on the display. For analog television systems, the standard temporal sampling rate is 30 Hz in the United States and Japan and 25 Hz in Europe with effective temporal refresh rates of 60 Hz and 50 Hz, respectively, through the use of interlace scanning. However, 72 Hz has become a *de facto* standard in the computer industry.

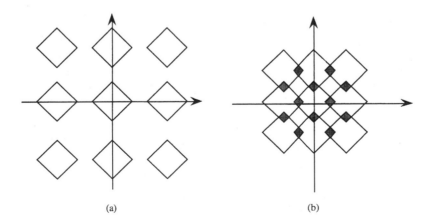

(a)   (b)

FIGURE 18.7   (a) Frequency spectrum of a sampled signal with no aliasing; (b) frequency spectrum of a sampled signal with aliasing.

If the sampling rate (spatial and/or temporal) of a system is fixed, a standard approach for minimizing the effects of aliasing for signals that do not satisfy the sampling theorem is to use a presampling filter. Presampling filters are low-pass filters whose cutoff frequencies are chosen to be less than ω_{xM}, ω_{yM}, and ω_{tM}. Although the signal still will not be able to be reconstructed exactly, the degradations are less annoying. Another problem in a real system is the need for an ideal low-pass filter to reconstruct an analog signal. An ideal filter is not physically realizable, so in practice an approximation must be made. Several very simple filter structures are common in video systems: sample and hold, bilinear, and raised cosine.

Quantization

Quantization is the process of converting the continuous-valued amplitude of the video signal into a discrete-valued representation, i.e., a finite set of numbers. The output of the quantizer is characterized by quantities that are limited to a finite number of values. The process is a many-to-one mapping, and thus there is a loss of information. The quantized signal can be modeled as

$$f_q(j, k, l) = f(j, k, l) - e(j, k, l)$$

where $f_q(j, k, l)$ is the quantized video signal and $e(j, k, l)$ is the quantization noise. If too few bits per sample are used, the quantization noise will produce visible false contours in the image data.

The quantizer is a mapping operation that generally takes the form of a staircase function (see Figure 18.8). A rule for quantization can be defined as follows: Let $\{d_k, k = 1, 2, \ldots, N + 1\}$ be the set of decision levels with d_1 the minimum amplitude value and d_N the maximum amplitude value of $f(j, k, l)$. If $f(j, k, l)$ is contained in the interval (d_k, d_{k+1}), then it is mapped to the kth reconstruction level r. Methods for designing quantizers can be broken into two categories: uniform and nonuniform. The input–output function for a typical uniform quantizer is shown in Figure 18.8. The mean square value of the quantizing noise can be easily calculated if it is assumed that the amplitude probability distribution is constant within each quantization step. The quantization step size for a uniform quantizer is

$$q = \frac{d_{N+1} - d_1}{N}$$

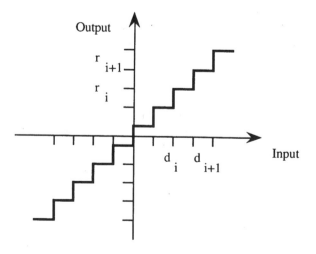

FIGURE 18.8   Characteristics of a uniform quantizer.

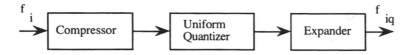

FIGURE 18.9 Nonuniform quantization using a compandor.

and all errors between $q/2$ and $-q/2$ are equally likely. The mean square quantization error D is given by

$$\langle e^2(j, k, l) \rangle = \int_{-q/2}^{q/2} \frac{f^2}{q} \, df = \frac{q^2}{12}$$

If one takes into account the exact amplitude probability distribution, an optimal quantizer can be designed. Here, the objective is to choose a set of decision levels and reconstruction levels that will yield the minimum quantization error. If f has a probability density function $p_f(f)$, the mean square quantization error is

$$\langle e^2(j, k, l) \rangle = \sum_{i=1}^{N} \int_{d_i}^{d_{i+1}} (f - r_i)^2 p_f(f) df$$

where N is the number of quantization levels. To minimize, the mean square quantization error is differentiated with respect to d_i and r_i. This results in the Max quantizer:

$$d_i = \frac{r_i + r_{i-1}}{2}$$

and

$$r_i = \frac{\int_{d_i}^{d_{i+1}} f p_f(f) df}{\int_{d_i}^{d_{i+1}} p_f(f) df}$$

Thus, the quantization levels need to be midway between the reconstruction levels, and the reconstruction levels are at the centroid of that portion of $p_f(f)$ between d_i and d_{i+1}. Unfortunately, these requirements do not lead to an easy solution. Max used an iterative numerical technique to obtain solutions for various quantization levels assuming a zero-mean Gaussian input signal. These results and the quantization levels for other standard amplitude distributions can be found in Jain (1989).

A more common and less computationally intensive approach to nonuniform quantization is to use a compandor (compressor–expander). The input signal is passed through a nonlinear compressor before being quantized uniformly. The output of the quantizer must then be expanded to the original dynamic range (see Figure 18.9). The compression and expansion functions can be determined so that the compandor approximates a Max quantizer.

For more information on vector quantization, see Tekalp (1995) or Wang et al. (2002).

Vector Quantization

Quantization does not have to be done on a single pixel at a time. In fact, better results can be achieved if the video data are quantized on a vector (block) basis. In vector quantization, the image data are first processed into a set of vectors. A codebook (set of code words or templates) that best matches the data to be quantized is then generated. Each input vector is then quantized to the closest code word. Compression is achieved by transmitting only the indices for the code words. At the receiver, the images are reconstructed using a table

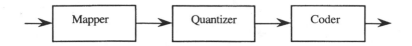

FIGURE 18.10   Three-stage model of an encoder.

look-up procedure. Two areas of ongoing research are finding better methods for designing the codebooks and developing better search and update techniques for matching the input vectors to the code words.

For more information on vector quantization, see Tekalp (1995) or Wang et al. (2002).

Video Compression

Digital representations of video signals typically require a very large number of bits. If the video signal is to be transmitted and/or stored, compression is often required. Applications include conventional and high-definition television, videophone, video conferencing, multi-media, remote-sensed imaging, and magnetic resonance imaging. The objective of compression (source encoding) is to find a representation that maximizes picture quality while minimizing the data per picture element (pixel). A wealth of compression algorithms have been developed during the past 30 years for both image and video compression. However, the ultimate choice of an appropriate algorithm is application-dependent. The following summary will provide some guidance in that selection process.

Compression algorithms can be divided into two major categories: information-preserving, or lossless, and lossy techniques. Information-preserving techniques introduce no errors in the encoding/decoding process; thus, the original signal can be reconstructed exactly. Unfortunately, the achievable compression rate, i.e., the reduction in bit rate, is quite small, typically on the order of 3:1. However, lossy techniques introduce errors in the coding/decoding process; thus, the received signal cannot be reconstructed exactly. The advantage of the lossy techniques is the ability to achieve much higher compression ratios. The limiting factor on the compression ratio is the required quality of the video signal in a specific application.

One approach to compression is to reduce the spatial and/or temporal sampling rate and the number of quantization levels. Unfortunately, if the sampling is too low and the quantization too coarse, aliasing, contouring, and flickering will occur. These distortions are often much greater than the distortions introduced by more sophisticated techniques at the same compression rate. Compression systems can generally be modeled by the block diagram shown in Figure 18.10. The first stage of the compression system is the mapper. This is an operation in which the input pixels are mapped into a representation that can be more effectively encoded. This stage is generally reversible. The second stage is the quantizer and performs the same type of operation as described earlier. This stage is not reversible. The final stage attempts to remove any remaining statistical redundancy. This stage is reversible and is typically achieved with one of the information-preserving coders.

Information-Preserving Coders

The data rate required for an original digital video signal may not represent its average information rate. If the original signal is represented by M possible independent symbols with probabilities p_i, $i = 0, 1, \ldots, M-1$, then the information rate as given by the first-order entropy of the signal H is

$$H = -\sum_{i=1}^{M-1} p_i \log_2 p_i \text{ bits per sample}$$

According to Shannon's coding theorem (see Jain, 1989), it is possible to perform lossless coding of a source with entropy H bits per symbol using $H + \varepsilon$ bits per symbol. ε is a small positive quantity. The maximum obtainable compression rate C is then given by

$$C = \frac{\text{average bit rate of the original data}}{\text{average bit rate of the encoded data}}$$

Huffman Coding

One of the most efficient information-preserving (entropy) coding methods is Huffman coding. Construction of a Huffman code involves arranging the symbol probabilities in decreasing order and considering them as leaf nodes of a tree. The tree is constructed by merging the two nodes with the smallest probability to form a new node. The probability of the new node is the sum of the two merged nodes. This process is continued until only two nodes remain. At this point, 1 and 0 are arbitrarily assigned to the two remaining nodes. The process now moves down the tree, decomposing probabilities and assigning 1s and 0s to each new pair. The process continues until all symbols have been assigned a code word (string of 1s and 0s). An example is given in Figure 18.11. Many other types of information-preserving compression schemes exist (see, for example, Tekalp, 1995; Gonzalez and Woods, 2002; Wang et al., 2002), including arithmetic coding, Lempel–Ziv algorithm, shift coding, and run-length coding.

Predictive Coding

Traditionally one of the most popular methods for reducing the bit rate has been predictive coding. In this class, differential pulse-code modulation (DPCM) has been used extensively. A block diagram for a basic DPCM system is shown in Figure 18.12. In such a system, the difference between the current pixel and a predicted version of that pixel gets quantized, coded, and transmitted to the receiver. This difference is referred to as the prediction error and is given by

$$e_i = f_i - \hat{f}_i$$

The prediction is based on previously transmitted and decoded spatial and/or temporal information and can be linear or nonlinear, fixed or adaptive. The difference signal e_i is then passed through a quantizer. The signal at the output of the quantizer is the quantized prediction error e_{iq}, which is entropy encoded transmission. The first step at the receiver is to decode the quantized prediction error. After decoding, d_{iq} is added to the predicted value of the current pixel $\hat{f}_i$ to yield the reconstructed pixel value. Note that as long as a quantizer is included in the system, the output signal will not exactly equal the input signal.

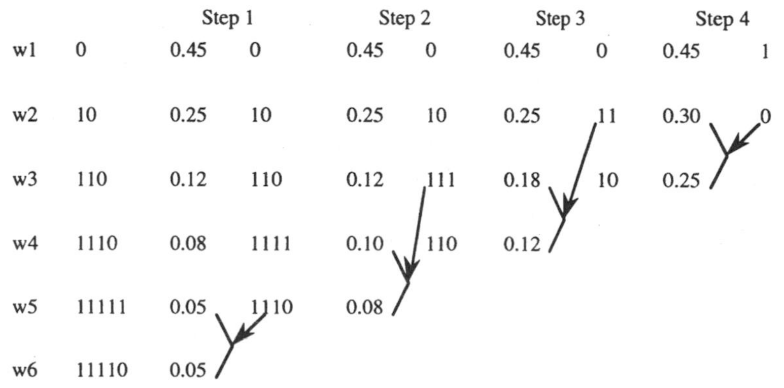

FIGURE 18.11 An example of constructing a Huffman code.

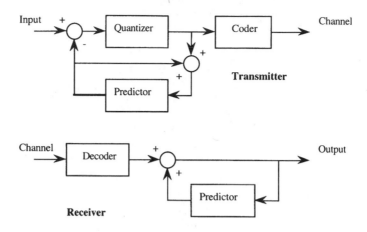

FIGURE 18.12   Block diagram of a basic DPCM system.

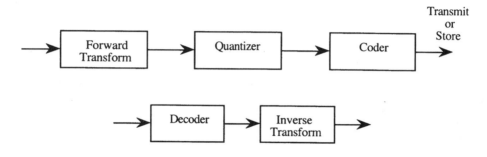

FIGURE 18.13   Transform coding system.

The predictors can include pixels from the present frame as well as those from previous frames (see Figure 18.13). If the motion and the spatial detail are not too high, frame (or field) prediction works well. If the motion is high and/or the spatial detail is high, intrafield prediction generally works better. A primary reason is that there is less correlation between frames and fields when the motion is high.

For more information on predictive coding, see Tekalp (1995) or Jain (1989).

Motion-Compensated Predictive Coding

Significant improvements in image quality, at a fixed compression rate, can be obtained when adaptive prediction algorithms take into account the frame-to-frame displacement of moving objects in the sequence. Alternatively, one could increase the compression rate for a fixed level of image quality. The amount of increase in performance will depend on one's ability to estimate the motion in the scene. Techniques for estimating the motion are described in a later subsection.

Today, motion-compensated prediction algorithms are integral components of modern video-compression coders such as those standardized in the ISO MPEG standards or ITU H.261 or H.263. Motion prediction can be divided into two basic categories. One category estimates the motion on a block-by-block basis and the other estimates the motion one pixel at a time. For the block-based methods, an estimate of the displacement is obtained for each block in the image. The block matching is achieved by finding the maximum correlation between a block in the current frame and a somewhat larger search area in the previous frame. A number of researchers have proposed ways to reduce the computational complexity, including using a simple matching criterion and using logarithmic searches for finding the peak value of the correlation.

The second category obtains a displacement estimate at each pixel in a frame. These techniques are referred to as pel recursive methods. They tend to provide more accurate estimates of the displacement but at the expense of higher complexity. Both categories of techniques have been applied to video data; however, block matching is used more often in real systems. The primary reason is that more efficient implementations have been feasible. It should be noted, however, that every pixel in a block will be assigned the same displacement estimate. Thus, the larger the block size the greater the potential for errors in the displacement estimate for a given pixel.

More details can be found in Tekalp (1995) or Wang et al. (2002).

Transform Coding

In transform coding, the video signal $f(x, y, t)$ is subjected to an invertible transform, then quantized and encoded (see Figure 18.13). The purpose of the transformation is to convert statistically dependent picture elements into a set of statistically independent coefficients. In practice, one of the separable fast transforms in the class of unitary transforms is used, e.g., cosine, Fourier, or Hadamard. In general, the transform coding algorithms can be implemented in 2D or 3D. However, because of the real-time constraints of many video signal processing applications, it is typically more efficient to combine a 2D transform with a predictive algorithm in the temporal direction, e.g., motion compensation.

For 2D transform coding, the image data are first subdivided into blocks. Typical block sizes are 8×8 or 16×16. The transform independently maps each image block into a block of transform coefficients; thus, the processing of each block can be done in parallel. At this stage the data have been mapped into a new representation, but no compression has occurred. In fact, with the Fourier transform there is an expansion in the amount of data. This occurs because the transform generates coefficients that are complex-valued. To achieve compression, the transform coefficients must be quantized and then coded to remove any remaining redundancy.

Two important issues in transform coding are the choice of transformation and the allocation of bits in the quantizer. The most commonly used transform is the discrete cosine transform (DCT). In fact, many of the proposed image and video standards utilize the DCT. The reasons for choosing a DCT include: its performance is superior to the other fast transforms and is very close to the optimal Karhunen–Loeve transform, it produces real-valued transform coefficients, and it has good symmetry properties, thus reducing the blocking artifacts inherent in block-based algorithms. One way to reduce these artifacts is by using a transform whose basis functions are even, i.e., the DCT, and another is to use overlapping blocks. For bit allocation, one can determine the variance of the transform coefficients and then assign the bits so the distortion is minimized. An example of a typical bit allocation map is shown in Figure 18.14.

Motion Estimation Techniques

Frame-to-frame changes in luminance are generated when objects move in video sequences. The luminance changes can be used to estimate the displacement of the moving objects if an appropriate model of the motion is specified. A variety of motion models have been developed for dynamic scene analysis in machine vision and for video communications applications. In fact, motion estimates were first

```
8 7 6 5 3 3 4 4 4 1 1 1 1 1 0 0
7 6 5 4 3 3 2 2 1 1 1 1 1 0 0 0
6 5 4 3 3 2 2 2 1 1 1 1 1 0 0 0
5 4 3 3 3 2 2 2 1 1 1 1 1 0 0 0
3 3 3 3 2 2 2 1 1 1 1 1 0 0 0 0
3 3 2 2 2 2 2 1 1 1 1 1 0 0 0 0
2 2 2 2 2 2 1 1 1 1 1 0 0 0 0 0
2 2 2 2 1 1 1 1 1 1 1 0 0 0 0 0
2 1 1 1 1 1 1 1 1 1 0 0 0 0 0 0
1 1 1 1 1 1 1 1 1 0 0 0 0 0 0 0
1 1 1 1 1 1 1 1 0 0 0 0 0 0 0 0
1 1 1 1 1 1 0 0 0 0 0 0 0 0 0 0
1 1 1 1 0 0 0 0 0 0 0 0 0 0 0 0
1 0 0 0 0 0 0 0 0 0 0 0 0 0 0 0
0 0 0 0 0 0 0 0 0 0 0 0 0 0 0 0
0 0 0 0 0 0 0 0 0 0 0 0 0 0 0 0
```

FIGURE 18.14 A typical bit allocation for 16×16 block coding of an image using the DCT. (*Source:* A.K. Jain, *Fundamentals of Digital Image Processing*, Englewood Cliffs, NJ: Prentice-Hall, 1989, p. 506. With permission.)

used as a control mechanism for the efficient coding of a sequence of images in an effort to reduce the temporal redundancy. Motion estimation algorithms can be classified in two broad categories: gradient or differential-based methods and token matching or correspondence methods. The gradient methods can be further divided into pel recursive, block matching, and optical flow methods.

Pel Recursive Methods

Netravali and Robbins (1979) developed the first pel recursive method for television signal compression. The algorithm begins with an initial estimate of the displacement, then iterates recursively to update the estimate. The iterations can be performed at a single pixel or at successive pixels along a scan line. The true displacement **D** at each pixel is estimated by

$$\hat{\mathbf{D}}^i = \hat{\mathbf{D}}^{i-1} + \mathbf{U}^i$$

where $\hat{\mathbf{D}}^i$ is the displacement estimate at the ith iteration and $\mathbf{U}^i$ is the update term. $\mathbf{U}^i$ is an estimate of $\hat{\mathbf{D}} - \hat{\mathbf{D}}^{i-1}$. The displaced frame difference (DFD):

$$\text{DFD}(x, y, \hat{\mathbf{D}}^{i-1}) = I(x, y, t) - I(x - \hat{\mathbf{D}}^{i-1}, t - T_S)$$

was then used to obtain a relationship for the update term $\mathbf{U}^i$. In the previous equation, T_S is the temporal sample spacing. If the displacement estimate is updated from sample to sample using a steepest-descent algorithm to minimize the weighted sum of the squared displaced frame differences over a neighborhood, then $\hat{\mathbf{D}}^i$ becomes

$$\hat{\mathbf{D}}^i = \hat{\mathbf{D}}^{i-1} - \frac{\varepsilon}{2} \nabla \hat{\mathbf{D}}^i \left[\sum_j W_j [\text{DFD}(x_{k-j}, \hat{\mathbf{D}}^{i-1})]^2 \right]$$

where $W_j \geq 0$ and

$$\sum_j W_j = 1$$

A graphical representation of pel recursive motion estimation is shown in Figure 18.15.

A variety of methods to calculate the update term have been reported. The advantage of one method over another is mainly in the improvement in compression. It should be noted that pel recursive algorithms assume that the displacement to be estimated is small. If the displacement is large, the estimates will be poor. Noise can also affect the accuracy of the estimate.

Block Matching

Block matching methods estimate the displacement within an $M \times N$ block in an image frame. The estimate is determined by finding the best match between the $M \times N$ block in a frame at time t and its best match from frame at $t - T_S$. An underlying

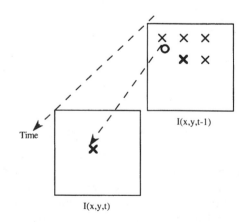

FIGURE 18.15 A graphical illustration of pel recursive motion estimation. The distance between the × and ○ pixels in the frame at $t - 1$ is $\hat{\mathbf{D}}^i$.

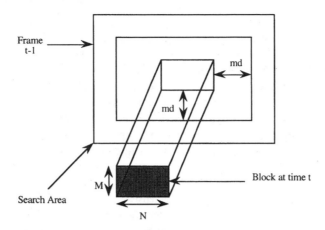

FIGURE 18.16 An illustration of block matching.

assumption in the block matching techniques is that each pixel within a block has the same displacement. A general block matching algorithm is given as follows:

1. Segment the image frame at time t into a fixed number of blocks of size $M \times N$.
2. Specify the size of the search area in the frame at time $t - 1$. This depends on the maximum expected displacement. If D_{max} is the maximum displacement in either the horizontal or vertical direction, then the size of the search area, SA, is

$$SA = (M + 2D_{max}) \times (N + 2D_{max})$$

Figure 18.16 illustrates the search area in the frame at time $t - 1$ for an $M \times N$ block at time t.
3. Using an appropriately defined matching criterion, e.g., mean-squared error or sum of absolute difference, find the best match for the $M \times N$ block.
4. Proceed to the next block in frame t and repeat Step 3 until displacement estimates have been determined for all blocks in the image.

Optical Flow Methods

The optical flow is defined as the apparent motion of the brightness patterns from one frame to the next. The optical flow is an estimate of the velocity field and hence requires two equations to solve for it. Typically a constraint is imposed on the motion model to provide the necessary equations. Optical flow can give useful information about the spatial arrangement of the objects in a scene, as well as the rate of change of those objects. Horn (1986) also defines a motion field, which is a two-dimensional velocity field resulting from the projection of the three-dimensional velocity field of an object in the scene onto the image plane. The motion field and the optical flow are not the same.

In general, the optical flow has been found difficult to compute because of the algorithm sensitivity to noise. Also, the estimates may not be accurate at scene discontinuities. However, because of its importance in assigning a velocity vector at each pixel, there continues to be research in the field.

The optical flow equation is based on the assumption that the brightness of a pixel at location (x, y) is constant over time; thus,

$$I_x = \frac{dx}{dt} + I_y \frac{dy}{dt} + I_t = 0$$

where dx/dt and dy/dt are the components of the optical flow. Several different constraints have been used with the optical flow equation to solve for dx/dt and dy/dt. A common constraint to impose is that the velocity field is smooth.

More details can be found in Tekalp (1995) or Wang et al. (2002).

Token Matching Methods

Token matching methods are often referred to as discrete methods since the goal is to estimate the motion only at distinct image features (tokens). The result is a sparse velocity field. The algorithms attempt to match the set of discrete features in the frame at time $t - 1$ with a set that best resembles them in the frame at time t. Most of the algorithms in this group assume that the estimation will be achieved in a two-step process. In the first step, the features are identified. The features could be points, corners, centers of mass, lines, or edges. This step typically requires segmentation and/or feature extraction. The second step determines the various velocity parameters. The velocity parameters include a translation component, a rotation component, and the rotation axis. The token matching algorithms fail if there are no distinct features to use.

All of the methods described in this subsection assume that the intensity at a given pixel location is reasonably constant over time. In addition, the gradient methods assume that the size of the displacements is small. Block matching algorithms have been used extensively in real systems because the computational complexity is not too great. The one disadvantage is that there is only one displacement estimate per block. To date, optical flow algorithms have found limited use because of their sensitivity to noise. Token matching methods work well for applications in which the features are well defined and easily extracted. They are probably not suitable for most video communications applications. See Tekalp (1995) for more details.

Hybrid Video Coding

The core of all the international video coding standards today makes use of a combination of block-based temporal prediction and transforms coding. Each video frame is divided into fixed-size blocks allowing for relatively independent processing of each block. Each block is then coded using a combination of transform coding and motion-compensated temporal prediction. Block-based motion estimation is used to predict each block from a previously encoded reference frame. A prediction error block, the difference between the actual block data and a motion-compensated prediction of the block, is generated and coded. The coding process includes transforming by a DCT, quantizing the DCT coefficients, and converting the coefficients into binary codewords using variable length coding. In practice, different block sizes may be needed for the motion estimation and the transform coding. Generally, motion estimation will be performed on a larger block, which is then subdivided into small blocks which are transformed using the DCT. Furthermore, the motion estimation may need to include multiple frames for increased accuracy. More details can be found in Wang et al. (2002).

Image Quality and Visual Perception

An important factor in designing video signal processing algorithms is that the final receiver of the video information is typically a human observer. This has an impact on how the quality of the final signal is assessed and how the processing should be performed. If our objective is video transmission over a limited bandwidth channel, we do not want to waste unnecessary bits on information that cannot be seen by the human observer. In addition, it is undesirable to introduce artifacts that are particularly annoying to the human viewer. Unfortunately, there are no perfect quantitative measures of visual perception. The human visual system is quite complicated. Despite the advances that have been made, no complete model of human perception exists. Therefore, we often have to rely on subjective testing to evaluate picture quality. Although no comprehensive model of human vision exists, certain functions can be characterized and then used in designing improved solutions. For more information, see Netravali and Haskell (1988).

Subjective Quality Ratings

There are two primary categories of subjective testing: category-judgment (rating-scale) methods and comparison methods. Category-judgment methods ask the subjects to view a sequence of pictures and assign each picture (video sequence) to one of several categories. Categories may be based on overall quality or on visibility of impairment (see Table 18.1).

TABLE 18.1   Quality and Impairment Ratings

5 Excellent	5 Imperceptible	3 Much better
4 Good	4 Perceptible but not annoying	2 Better
3 Fair	3 Slightly annoying	1 Slightly better
2 Poor	2 Annoying	0 Same
1 Bad	1 Very annoying	−1 Slightly worse
		−2 Worse
		−3 Much worse

Source: A.N. Netravali and B.G. Haskell, *Digital Pictures: Representation and Compression*, New York: Plenum Press, 1988, p. 247. With permission.

Comparison methods require the subjects to compare a distorted test picture with a reference picture. Distortion is added to the test picture until both pictures appear of the same quality to the subject. Viewing conditions can have a great impact on the results of such tests. Care must be taken in the experimental design to avoid biases in the results.

Visual Perception

In this subsection, a review of the major aspects of human psychophysics that have an impact in video signal processing is given. The phenomena of interest include light adaptation, visual thresholding and contrast sensitivity, masking, and temporal phenomena.

Light Adaptation

The human visual system (HVS) has two major classes of photoreceptors: the rods and the cones. Because these two types of receptors adapt to light differently, two different adaptation time constants exist. Furthermore, these receptors respond at different rates going from dark to light than from light to dark. It should also be noted that although the HVS has an ability to adapt to an enormous range of light intensity levels, on the order of 10^{10} in millilamberts, it does so adaptively. The simultaneous range is on the order of 10^3.

Visual Thresholding and Contrast Sensitivity

Determining how sensitive an observer is to small changes in luminance is important in the design of video systems. One's sensitivity will determine how visible noise will be and how accurately the luminance must be represented. The contrast sensitivity is determined by measuring the just-noticeable difference (JND) as a function of the brightness. The JND is the amount of additional brightness needed to distinguish a patch from the background. It is a visibility threshold. What is significant is that the JND is dependent on the background and surrounding luminances, the size of the background and surrounding areas, and the size of the patch, with the primary dependence being on the luminance of the background.

Masking

The response to visual stimuli is greatly affected by what other visual stimuli are in the immediate neighborhood (spatially and temporally). An example is the reduced sensitivity of the HVS to noise in areas of high spatial activity. Another example is the masking of details in a new scene by what was present in the previous scene. In both cases, the masking phenomenon can be used to improve the quality of image compression systems.

Temporal Effects

One relevant temporal phenomenon is the flicker fusion frequency. This is a temporal threshold that determines the point at which the HVS fuses the motion in a sequence of frames. Unfortunately, this frequency varies as a function of the average luminance. The HVS is more sensitive to flicker at high luminances than at low luminances. The spatial–temporal frequency response of the HVS is important in determining the

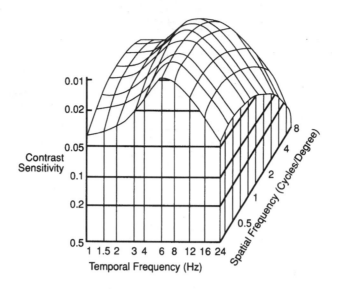

FIGURE 18.17   A perspective view of the spatio-temporal threshold surface. (*Source:* A.N. Netravali and B.G. Haskell, *Digital Pictures: Representation and Compression*, New York: Plenum Press, 1988, p. 273. With permission.)

sensitivity to small-amplitude stimuli. In both the temporal and spatial directions, the HVS responds as a bandpass filter (see Figure 18.17). Also significant is the fact that the spatial and temporal properties are not independent of one another, especially at low frequencies.

For more details on image quality and visual perception, see Netravali and Haskell (1988), Gonzalez and Woods (2002), and Wang et al. (2002).

Video Compression Standards

With potential applications of digital video including digital movies, digital television, videophones, multimedia, video mail, personal computers, video conferencing, and video games, it is important to have digital video compression standards. Use of standardized formats for digital video compresses facilitates the exchange or reception of information among vendors and across applications. The two international standards organizations that have the most significant impact on digital video compression are the International Telecommunications Union (ITU) (www.itu.int) headquartered in Geneva, Switzerland and the International Organization for Standards (ISO) (www.iso.ch).

Today video coding standards exist for a range of video applications. H.261 was adopted by ITU as a standard for ISDN video teleconferencing in 1990. About five years later ITU adopted H.263, a standard that enables video communications over analog telephone lines and in more recent phases video communication among desktop and mobile terminals connected to the Internet. ISO developed the MPEG standards. MPEG-1 was developed for progressively scanned video used in multimedia applications and includes standardization for both audio and video. MPEG-2 was developed as a standard for TV and HDTV. The goal was to have MPEG-2 systems somewhat compatible with MPEG-1, have error resilience, allow transmission over ATM networks, and transport more than one TV program in a stream. With the evolution of highly interactive multimedia applications, the need arose for a new standard. These applications not only require efficient compression, but also scalability of contents, interactivity with individual objects, and a high degree of error resilience. MPEG-4 was developed to allow for object-based coding of natural and synthetic audio and video, along with graphics. Recent efforts are focused on the development of MPEG-7, a set of tools to facilitate finding the content you need in multimedia data.

For more details on the development and content of the standards, see Tekalp (1995) and Wang et al. (2002).

Defining Terms

Aliasing: Distortion introduced in a digital signal when it is undersampled.

Compression: Process of compactly representing the information contained in a signal.

Motion estimation: Process of estimating the displacement of moving objects in a scene.

Quantization: Process of converting a continuous-valued signal into a discrete-valued signal.

Sampling: Process of converting a continuous-time/space signal into a discrete-time/space signal.

Scanning system: System used to capture a new image at periodic intervals in time and to convert the image into a digital representation.

References

R.C. Gonzalez and R.E. Woods, *Digital Image Processing*, Upper Saddle River, NJ: Prentice-Hall, 2002.

B.P. Horn, *Robot Vision*, Cambridge, MA: MIT Press, 1986.

A.K. Jain, *Fundamentals of Digital Image Processing*, Englewood Cliffs, NJ: Prentice-Hall, 1989.

A.N. Netravali and B.G. Haskell, *Digital Pictures: Representation and Compression*, New York: Plenum Press, 1988.

A.V. Oppenheim, A.S. Willsky, and I.T. Young, *Signals and Systems*, Englewood Cliffs, NJ: Prentice-Hall, 1983.

M. Tekalp, *Digital Video Processing*, Upper Saddle River, NJ: Prentice-Hall, 1995.

Y. Wang, J. Ostermann, and Y.Q. Zhang, *Video Processing and Communications*, Upper Saddle River, NJ: Prentice Hall, 2002.

Further Information

Two recent textbooks by Tekalp (1995) and Wang et al. (2002) provide broad coverage of the field of video signal processing. Other recommended sources of information include *IEEE Transactions on Circuits and Systems for Video Technology*, *IEEE Transactions on Image Processing*, *IEEE Signal Processing Magazine*, *IEEE Proceedings*, *Journal of Electronic Imaging*, and *Signal Processing: Image Communication*.

18.3 Sensor Array Processing

N.K. Bose and L.H. Sibul

Multidimensional signal processing tools apply to aperture and sensor array processing. Planar sensor arrays can be considered to be sampled apertures. Three-dimensional or volumetric arrays can be viewed as multidimensional spatial filters. Therefore, the topics of sensor array processing, aperture processing, and multidimensional signal processing can be studied under a unified format. The basic function of the receiving array is transduction of propagating waves in the medium into electrical signals. Propagating waves are fundamental in radar, communication, optics, sonar, and geophysics. In electromagnetic applications, basic transducers are antennas and arrays of antennas. The large body of literature that exists on antennas and antenna arrays can be exploited in the areas of aperture and sensor array processing. Much of the antenna literature deals with transmitting antennas and their radiation patterns. Because of the reciprocity of transmitting and receiving transducers, key results that have been developed for transmitters can be used for analysis of receiver aperture and/or array processing. Transmitting transducers radiate energy in desired directions, whereas receiving apertures/arrays act as spatial filters that emphasize signals from a desired look direction while discriminating against interferences from other directions. The spatial filter **wavenumber** response is called the receiver beam pattern. Transmitting apertures are characterized by their radiation patterns.

Conventional beamforming deals with the design of fixed beam patterns for given specifications. Optimum beamforming is the design of beam patterns to meet a specified optimization criterion. It can be compared to optimum filtering, detection, and estimation. Adaptive **beamformers** sense their operating environment (for example, noise covariance matrix) and adjust beamformer parameters so that their performance is optimized (Monzingo and Miller, 1980). Adaptive beamformers can be compared with adaptive filters.

Multidimensional signal processing techniques have found wide application in seismology—where a group of identical seismometers, called seismic arrays, are used for event location, studies of the Earth's sedimentation structure, and separation of coherent signals from noise, which sometimes may also propagate coherently across the array but with different horizontal velocities—by employing **velocity filtering** (Claerbout, 1976). Velocity filtering is performed by multidimensional filters and allows also for the enhancement of signals that may occupy the same wavenumber range as noise or undesired signals do. In a broader context, beamforming can be used to separate signals received by sensor arrays based on frequency, wavenumber, and velocity (speed as well as direction) of propagation. Both the transfer and unit impulse–response functions of a velocity filter are two-dimensional functions in the case of one-dimensional arrays. The transfer function involves frequency and wavenumber (due to spatial sampling by equally spaced sensors) as independent variables, whereas the unit impulse response depends upon time and location within the array. Two-dimensional filtering is not limited to velocity filtering by means of seismic array. Two-dimensional spatial filters are frequently used, for example, in the interpretation of gravity and magnetic maps to differentiate between regional and local features. Input data for these filters may be observations in the survey of an area conducted over a planar grid over the Earth's surface. Two-dimensional wavenumber digital filtering principles are useful for this purpose. Velocity filtering by means of two-dimensional arrays may be accomplished by properly shaping a three-dimensional response function $H(k_1, k_2, \omega)$. Velocity filtering by three-dimensional arrays may be accomplished through a four-dimensional function $H(k_1, k_2, k_3, \omega)$ as explained in the following subsection.

Spatial Arrays, Beamformers, and FIR Filters

A propagating plane wave, $s(\mathbf{x}, t)$, is, in general, a function of the three-dimensional space variables $(x_1, x_2, x_3)\Delta = \mathbf{x}$ and the time variable t. The 4-D Fourier transform of the stationary signal $s(\mathbf{x}, t)$ is

$$S(\mathbf{k}, \omega) = \int_{-\infty}^{\infty} \int_{-\infty}^{\infty} \int_{-\infty}^{\infty} \int_{-\infty}^{\infty} s(\mathbf{x}, t) e^{-j\left(\omega t - \sum_{i=1}^{3} k_i x_i\right)} dx_1 dx_2 dx_3 dt \tag{18.3}$$

which is referred to as the wavenumber–frequency spectrum of $s(\mathbf{x}, t)$, and $(k_1, k_2, k_3) \triangleq \mathbf{k}$ denotes the wavenumber variables in radians per unit distance and ω is the frequency variable in radians per second. If c denotes the velocity of propagation of the plane wave, the following constraint must be satisfied:

$$k_1^2 + k_2^2 + k_3^2 = \frac{\omega^2}{c^2}$$

If the 4-D Fourier transform of the unit impulse response $h(\mathbf{x}, t)$ of a 4-D linear shift-invariant (LSI) filter is denoted by $H(k, \omega)$, then the response $y(\mathbf{x}, t)$ of the filter to $s(\mathbf{x}, t)$ is the 4-D linear convolution of $h(\mathbf{x}, t)$ and $s(\mathbf{x}, t)$, which is uniquely characterized by its 4-D Fourier transform:

$$Y(\mathbf{k}, \omega) = H(\mathbf{k}, \omega)S(\mathbf{k}, \omega) \tag{18.4}$$

The inverse 4-D Fourier transform, which forms a 4-D Fourier transform pair with Equation (18.3), is

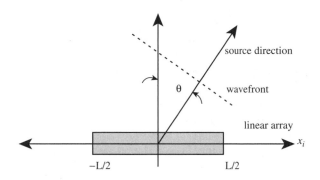

FIGURE 18.18 Uniformly weighted linear array.

$$s(\mathbf{x}, t) = \frac{1}{(2\pi)^4} \int_{-\infty}^{\infty} \int_{-\infty}^{\infty} \int_{-\infty}^{\infty} \int_{-\infty}^{\infty} S(\mathbf{k}, \omega) e^{j\left(\omega t - \sum_{i=1}^{3} k_i x_i\right)} dk_1 dk_2 dk_3 d\omega \qquad (18.5)$$

It is noted that $S(\mathbf{k}, \omega)$ in Equation (18.3) is product separable, i.e., expressible in the form:

$$S(\mathbf{k}, \omega) = S_1(k_1) S_2(k_2) S_3(k_3) S_4(\omega) \qquad (18.6)$$

where each function on the right-hand side is a univariate function of the respective independent variable, if and only if $s(\mathbf{x}, t)$ in Equation (18.3) is also product separable. In beamforming, $S_i(k_i)$ in Equation (18.6) would be the far-field beam pattern of a linear array along the x_i axis. For example, the normalized beam pattern of a uniformly weighted (shaded) linear array of length L is

$$S(k, \theta) = \frac{\sin\left(\dfrac{kL \sin \theta}{2}\right)}{\left(\dfrac{kL}{2} \sin \theta\right)}$$

where $\lambda = (2\pi/k)$ is the wavelength of the propagating plane wave and θ is the angle of arrival at the array site as shown in Figure 18.18. Note that θ is explicitly admitted as a variable in $S(k, \theta)$ to allow for the possibility that for a fixed wavenumber, the beam pattern could be plotted as a function of the angle of arrival. In that case, when θ is zero, the wave impinges the array broadside and the normalized beam pattern evaluates to unity.

The counterpart in aperture and sensor array processing of the use of window functions in spectral analysis for reduction of sidelobes is the use of aperture shading. In aperture shading, one simply multiplies a uniformly weighted aperture by the shading function. The resulting beam pattern is, then, simply the convolution of the beam pattern of the uniformly shaded volumetric array and the beam pattern of the shading function. The Fourier transform relationship between the stationary signal $s(\mathbf{x}, t)$ and the wavenumber frequency spectrum $S(\mathbf{k}, \omega)$ allows one to exploit high-resolution spectral analysis techniques for the high-resolution estimation of the direction of arrival (Pillai, 1989). The superscript *, t, and H denote, respectively, complex conjugate, transpose, and conjugate transpose.

Discrete Arrays for Beamforming

An array of sensors could be distributed at distinct points in space in various ways. Line arrays, planar arrays, and volumetric arrays could be either uniformly spaced or nonuniformly spaced, including the possibility of placing sensors randomly according to some probability distribution function. Uniform spacing along each coordinate axis permits one to exploit the well-developed multidimensional signal processing techniques concerned with filter design, DFT computation via FFT, and high-resolution spectral analysis of sampled signals (Dudgeon, 1977). Nonuniform spacing sometimes might be useful for reducing the number of sensors, which otherwise might be constrained to satisfy a maximum spacing between uniformly placed sensors to avoid **grating lobes** due to aliasing, as explained later. A discrete array, uniformly spaced, is convenient for the synthesis of a digital filter or beamformer by the performing of digital signal processing operations (namely delay, sum, and multiplication or weighting) on the signal received by a collection of sensors distributed in space. The sequence of the nature of operations dictates the types of beamformer. Common beamforming systems are of the straight summation, delay-and-sum, and weighted delay-and-sum types. The geometrical distribution of sensors and the weights w_i associated with each sensor are crucial factors in the shaping of the filter characteristics. In the case of a linear array of N equispaced sensors, which are spaced D units apart, starting at the origin $x_1 = 0$, the function:

$$W(k_1) = \frac{1}{N} \sum_{n=0}^{N-1} w_n e^{-jk_1 nD} \tag{18.8}$$

becomes the **array pattern**, which may be viewed as the frequency response function for a finite impulse response (FIR) filter, characterized by the unit impulse response sequence $\{w_n\}$. In the case when $w_n = 1$, Equation (18.8) simplifies to

$$W(k_1) = \frac{1}{N} \frac{\sin\left(\frac{k_1 ND}{2}\right)}{\sin\left(\frac{k_1 D}{2}\right)} \exp\left\{-j\frac{(N-1)k_1 D}{2}\right\}$$

If the N sensors are symmetrically placed on both sides of the origin, including one at the origin, and the sensor weights are $w_n = 1$, then the linear array pattern becomes

$$W(k_1) = \frac{1}{N} \frac{\sin\left(\frac{k_1 ND}{2}\right)}{\sin\left(\frac{k_1 D}{2}\right)}$$

For planar arrays, direct generalizations of the preceding linear array results can be obtained. To wit, if the sensors with unity weights are located at coordinates (kD, lD), where $k = 0, \pm1, \pm2, \ldots, \pm[(N-1)/2]$, and $l = 0, \pm1, \pm2, \ldots, \pm[(M-1)/2]$, for odd integer values of N and M, then the array pattern function becomes

$$W(k_1, k_2) = \frac{1}{NM} \sum_{k=-\left(\frac{N-1}{2}\right)}^{\left(\frac{N-1}{2}\right)} \sum_{l=-\left(\frac{M-1}{2}\right)}^{\left(\frac{M-1}{2}\right)} \exp\{-j(k_1 kD + k_2 lD\}$$

$$= \frac{1}{NM} \frac{\sin\left(\frac{k_1 ND}{2}\right) \sin\left(\frac{k_2 MD}{2}\right)}{\sin\left(\frac{k_1 D}{2}\right) \quad \sin\left(\frac{K_2 D}{2}\right)} \tag{18.10}$$

Routine generalizations to 3D spatial arrays are also possible. The array pattern functions for other geometrical distributions may also be routinely generated. For example, if unit weight sensors are located at the six vertices and the center of a regular hexagon, each of whose sides is D units long, then the array pattern function can be shown to be

$$W(k_1, k_2) = \frac{1}{7}\left[1 + 2\cos k_1 D + 4\cos\frac{k_1 D}{2}\cos\frac{\sqrt{3}k_2 D}{2}\right] \tag{18.11}$$

The array pattern function reveals how selective a particular beamforming system is. In the case of a typical array function shown in Equation (18.9), the beamwidth, which is the width of the main lobe of the array pattern, is inversely proportional to the array aperture. Because of the periodicity of the array pattern function, the main lobe is repeated at intervals of $2\pi/D$. These repetitive lobes are called grating lobes, whose existence may be interpreted in terms of spatial frequency aliasing resulting from a sampling interval D due to the N receiving sensors located at discrete points in space. If the spacing D between sensors satisfies

$$D \leq \frac{\lambda}{2} \tag{18.12}$$

where λ is the smallest wavelength component in the signal received by the array of sensors, then the grating lobes have no effect on the received signal. A plane wave of unit amplitude, which is incident upon the array at bearing angle θ degrees, as shown in Figure 18.18, produces outputs at the sensors given by the vector:

$$\mathbf{s}(\theta) \overset{\Delta}{=} \mathbf{s}_\theta = [\exp(j0) \exp(jk_1 D\sin\theta)\ldots\exp(jk_1(N-1)D\sin\theta)]^t \tag{18.13}$$

where $k_1 = 2\pi/\lambda$ is the wavenumber. In array processing, the array output y_θ may be viewed as the inner product of an array weight vector $\mathbf{w}$ and the steering vector $\mathbf{s}_\theta$. Thus, the beamformer response along a direction characterized by the angle θ is, treating $\mathbf{w}$ as complex:

$$y_\theta = \langle \mathrm{w}(\theta), \mathbf{s}_\theta \rangle = \sum_{k=0}^{N-1} w_k^* \exp(jk_1 kD\sin\theta) \tag{18.14}$$

The beamforming system is said to be robust if it performs satisfactorily despite certain perturbations (Ahmed and Evans, 1982). It is possible for each component $s_{k\theta}$ of $\mathbf{s}_\theta$ to belong to an interval $[s_{k\theta} - \phi_{k\theta}, s_{k\theta} + \phi_{k\theta}]$, and a robust beamformer will require the existence of at least one weight vector $\mathbf{w}$ which will guarantee the output y_θ to belong to an output envelope for each $\mathbf{s}_\theta$ in the input envelope. The robust beamforming problem can be translated into an optimization problem, which may be tackled by minimizing the value of the array output power:

$$P(\theta) = \mathbf{w}^{\mathrm{H}}(\theta)R\mathbf{w}(\theta) \tag{18.15}$$

when the response to a unit amplitude plane wave incident at the steering direction θ is constrained to be unity, i.e., $\mathbf{w}^{\mathrm{H}}(\theta)\mathbf{s}(\theta) = 1$, and R is the additive noise-corrupted signal autocorrelation matrix. The solution is called the minimum variance beamformer and is given by

$$\mathbf{w}_{\mathrm{MV}}(\theta) = \frac{R^{-1}\mathbf{s}(\theta)}{\mathbf{s}^{\mathrm{H}}(\theta)R^{-1}\mathbf{s}(\theta)} \tag{18.16}$$

and the corresponding power output is

$$P_{\mathrm{MV}}(\theta) = \frac{1}{s^{\mathrm{H}}(\theta)R^{-1}\mathbf{s}(\theta)} \tag{18.17}$$

The minimum variance power as a function of θ can be used as a form of the data-adaptive estimate of the directional power spectrum. However, in this mode of solution, the coefficient vector is unconstrained except at the steering direction. Consequently, a signal tends to be regarded as an unwanted interference and is, therefore, suppressed in the beamformed output unless it is almost exactly aligned with the steering direction. Therefore, it is desirable to broaden the signal acceptance angle while at the same time preserving the optimum beamformer's ability to reject noise and interference outside this region of angles. One way of achieving this is by the application of the principle of superdirectivity.

Discrete Arrays and Polynomials

It is common practice to relate discrete arrays to polynomials for array synthesis purposes (Steinberg, 1976). For volumetric equispaced arrays (it is only necessary that the spacing be uniform along each coordinate axis so that the spatial sampling periods D_i and D_j along, respectively, the ith and jth coordinate axes could be different for $i \neq j$), the weight associated with sensors located at coordinate $(i_1 D_1, i_2 D_2, i_3 D_3)$ is denoted by $w(i_1, i_2, i_3)$. The function in the complex variables (z_1, z_2, and z_3) that is associated with the sequence $\{w(i_1, i_2, i_3)\}$ is the generating function for the sequence and is denoted by

$$W(z_1, z_2, z_3) = \sum_{i_1} \sum_{i_2} \sum_{i_3} w(i_1, i_2, i_3) z_1^{i_1} z_2^{i_2} z_3^{i_3} \tag{18.18}$$

In the electrical engineering and geophysics literature, the generating function $W(z_1, z_2, z_3)$ is sometimes called the z-transform of the sequence $\{w(i_1, i_2, i_3)\}$. When there are a finite number of sensors, a realistic assumption for any physical discrete array, $W(z_1, z_2, z_3)$ becomes a trivariate polynomial. In the special case when $w(i_1, i_2, i_3)$ is product separable, the polynomial $W(z_1, z_2, z_3)$ is also product separable. Particularly, this separability property holds when the shading is uniform, i.e., $w(i_1, i_2, i_3) = 1$. When the support of the uniform shading function is defined by $i_1 = 0, 1, \ldots, N_1 - 1$, $i_2 = 0, 1, \ldots, N_2 - 1$, and $i_3 = 0, 1, \ldots, N_3 - 1$, the associated polynomial becomes

$$W(z_1, z_2, z_3) = \sum_{i_1=0}^{N_1-1} \sum_{i_2=0}^{N_2-1} \sum_{i_3=0}^{N_3-1} z_1^{i_1} z_2^{i_2} z_3^{i_3} = \prod_{i=1}^{3} \frac{z_i^{N_i} - 1}{z_i - 1} \tag{18.19}$$

In this case, all results developed for the synthesis of linear arrays become directly applicable to the synthesis of volumetric arrays. For a linear uniform discrete array composed of N sensors with intersensor spacing D_1 starting at the origin and receiving a signal at a known fixed wavenumber k_1 at a receiving angle θ, the far-field beam pattern:

$$S(k_1, \theta) \triangleq S(\theta) = \sum_{r=0}^{N-1} e^{jk_1 r D_1 \sin\theta}$$

may be associated with a polynomial $\sum_{r=0}^{N-1} z_1^r$, by setting $z_1 = e^{jk1D1\sin\theta}$. This polynomial has all its zeros on the unit circle in the z_1-plane. If the array just considered is not uniform but has a weighting factor w_r, for $r = 0, 1, \ldots, N_1 - 1$, the space factor:

$$Q(\theta) \triangleq \sum_{r=0}^{N_1-1} w_r e^{jk_1 D_1 r \sin\theta}$$

may again be associated with a polynomial $\sum_{r=0}^{N_1-1} w_r z_1^r$. By the pattern multiplication theorem, it is possible to get the polynomial associated with the total beam pattern of an array with weighted sensors by multiplying the polynomials associated with the array element pattern and the polynomial associated with the space factor $Q(\theta)$. The array factor $(\theta)|^2$ may also be associated with the polynomial spectral factor:

$$|Q(\theta)|^2 \leftrightarrow \sum_{r=0}^{N_1-1} w_r z_1^r \sum_{r=0}^{N_1-1} w_r^*(z_1^*)^r \tag{18.20}$$

where the weighting (shading) factor is allowed to be complex. Uniformly distributed apertures and uniformly spaced volumetric arrays which admit product separable sensor weightings can be treated by using the well-developed theory of linear discrete arrays and their associated polynomial. When the product separability property does not hold, scopes exist for applying results from multidimensional systems theory (Bose, 1982) concerning multivariate polynomials to the synthesis problem of volumetric arrays.

Velocity Filtering

The combination of individual sensor outputs in a more sophisticated way than the delay-and-sum technique leads to the design of multichannel velocity filters for linear and planar as well as spatial arrays. Consider first a linear (1D) array of sensors, which will be used to implement velocity discrimination. The pass and rejection zones are defined by straight lines in the (k_1, ω)-plane, where

$$k_1 = \frac{\omega}{V} = \frac{\omega}{(v/\sin\theta)}$$

is the wavenumber, ω the angular frequency in rad/sec, V the apparent velocity on the Earth's surface along the array line, v the velocity of wave propagation, and θ the horizontal arrival direction. The transfer function:

$$H(\omega, k_1) = \begin{cases} 1, & -\dfrac{|\omega|}{V} \leq k_1 \leq \dfrac{|\omega|}{V} \\ 0, & \text{otherwise} \end{cases}$$

of a "pie-slice" or "fan" velocity filter (Bose, 1985) rejects totally wavenumbers outside the range $-|\omega|/V \leq k_1 \leq |\omega|/V$ and passes completely wavenumbers defined within that range. Thus, the transfer function defines a high-pass filter which passes signals with apparent velocities of magnitude greater than V at a fixed frequency ω. If the equispaced sensors are D units apart, the spatial sampling results in a periodic wavenumber response with period $k_1 = 1/(2D)$. Therefore, for a specified apparent velocity V, the resolvable wavenumber and frequency bands are, respectively, $-1/(2D) \leq k_1 \leq 1/(2D)$ and $-V/(2D) \leq \omega \leq V/(2D)$ where $\omega/(2D)$ represents the folding frequency in rad/sec.

Linear arrays are subject to the limitation that the source is required to be located on the extended line of sensors so that plane wavefronts approaching the array site at a particular velocity excite the individual sensors, assumed equispaced, at arrival times which are also equispaced. In seismology, the equispaced interval between successive sensor arrival times is called a move-out or step-out and equals $(D \sin \theta)/v = D/V$. However, when the sensor-to-source azimuth varies, two or more independent signal move-outs may be present. Planar (2D) arrays are then required to discriminate between velocities as well as azimuth. Spatial (3D) arrays provide additional scope to the enhancement of discriminating capabilities when sensor/source locations are arbitrary. In such cases, an array origin is chosen and the mth sensor location is denoted by a vector $(x_{1m} x_{2m} x_{3m})^t$ and the frequency wavenumber response of an array of sensors is given by

$$H(\omega, k_1, k_2, k_3) = \frac{1}{N} \sum_{m=1}^{N} H_m(\omega) \exp\left[\sum_{i=1}^{3} -j2\pi k_i x_{im} \right]$$

where $H_m(\omega)$ denotes the frequency response of a filter associated with the mth recording device (sensor). The sum of all N filters provides flat frequency response so that waveforms arriving from the estimated directions of arrival at estimated velocities are passed undistorted and other waveforms are suppressed. In the planar specialization, the 2D array of sensors leads to the theory of 3D filtering involving a transfer function in the frequency wavenumber variables f, k_1, and k_2. The basic design equations for the optimum, in the least-mean-square error sense, frequency wavenumber filters have been developed (Burg, 1964). This procedure of Burg can be routinely generalized to the 4D filtering problem mentioned above.

Acknowledgment

N.K. Bose and L.H. Sibul acknowledge the support provided by the Office of Naval Research under, respectively, Contract N00014–92-J-1755 and the Fundamental Research Initiatives Program.

Defining Terms

Array pattern: Fourier transform of the receiver weighting function taking into account the positions of the receivers.

Beamformers: Systems commonly used for detecting and isolating signals that are propagating in a particular direction.

Grating lobes: Repeated main lobes in the array pattern interpretable in terms of spatial frequency aliasing.

Velocity filtering: Means for discriminating signals from noise or other undesired signals because of their different apparent velocities.

Wavenumber: 2π (spatial frequency in cycles per unit distance).

References

K.M. Ahmed and R.J. Evans, "Robust signal and array processing," *IEE Proceedings, F: Communications, Radar, and Signal Processing,* vol. 129, no. 4, pp. 297–302, 1982.

N.K. Bose, *Applied Multidimensional Systems Theory,* New York: Van Nostrand Reinhold, 1982.

N.K. Bose, *Digital Filters,* New York: Elsevier Science North-Holland, 1985. Reprint ed., Malabar, FL: Krieger Publishing, 1993.

J.P. Burg, "Three-dimensional filtering with an array of seismometers," *Geophysics,* vol. 23, no. 5, pp. 693–713, 1964.

J.F. Claerbout, *Fundamentals of Geophysical Data Processing,* New York: McGraw-Hill, 1976.

D.E. Dudgeon, "Fundamentals of digital array processing," *Proc. IEEE,* vol. 65, pp. 898–904, 1977.

R.A. Monzingo and T.W. Miller, *Introduction to Adaptive Arrays,* New York: Wiley, 1980.

S.M. Pillai, *Array Signal Processing*, New York: Springer-Verlag, 1989.
B.D. Steinberg, *Principles of Aperture and Array System Design*, New York: Wiley, 1976.

Further Information

Adaptive Signal Processing, edited by Leon H. Sibul, includes papers on adaptive arrays, adaptive algorithms and their properties, as well as other applications of adaptive signal processing techniques (IEEE Press, New York, 1987).

Adaptive Antennas:Concepts and Applications, by R. T. Compton, Jr., emphasizes adaptive antennas for electromagnetic wave propagation applications (Prentice-Hall, Englewood-Cliffs, NJ, 1988).

Array Signal Processing: Concepts and Techniques, by D. H. Johnson and D. E. Dudgeon, incorporates results from discrete-time signal processing into array processing applications such as signal detection, estimation of direction of propagation, and frequency content of signals (Prentice-Hall, Englewood Cliffs, NJ, 1993).

Neural Network Fundamentals with Graphs, Algorithms, and Applications, by N. K. Bose and P. Liang, contains the latest information on adaptive-structure networks, growth algorithms, and adaptive techniques for learning and capability for generalization (McGraw-Hill, New York, 1996).

19

Real-Time Digital Signal Processing

Cameron H.G. Wright
University of Wyoming

Thad B. Welch
United States Naval Academy

Michael G. Morrow
University of Wisconsin-Madison

An underlying assumption of most digital signal processing operations is that we have a sampled signal, our *digital signal*, that we wish to process. When time is not critical, these signals are often stored for subsequent retrieval or synthesized when needed. While this storage or synthesis method is very convenient, it does not allow for real-time processing of a signal.

19.1 Real-Time Processing

We use the term *real-time processing* to mean that the processing of a given sample must occur within a given time or the system will not operate properly. In a *hard real-time* system, the system will fail if the processing is not done in a timely manner. For example, in a gasoline engine control system, the calculations of fuel injection and spark timing must be completed in time for the next cycle or the engine will not operate. In a *soft real-time* system, the system will tolerate some failures to meet real-time targets and still continue to operate, but with some degradation in performance. For example, in a hand-held audio player, if the decoding for the next output sample is not completed in time, the system could simply repeat the previous sample instead. As long as this happened infrequently, it would be imperceptible to the user. Although general purpose microprocessors can be employed in many situations, the performance demands and power constraints of real-time systems often mandate specialized hardware. This may include specialized microprocessors optimized for signal processing (digital signal processors or DSPs), programmable logic devices, application-specific integrated circuits (ASICs), or a combination of any or all of them as required to meet system constraints. Another chapter of this book discussed specific examples of specialized DSP hardware devices.

19.2 Real-Time Hardware

This chapter is written to allow someone with a basic understanding of DSP theory to rapidly transition from the familiar MATLAB environment to performing real-time DSP operations on an industry-standard hardware target. Typically, real-time DSP hardware has to communicate with the "outside" world. This is usually accomplished on the input side with an analog-to-digital converter (ADC) and on the output side with a digital-to-analog converter (DAC). Integrated circuit (IC) chips which combine the ADC and DAC functions in one device are called *codec* chips, which is an acronym for "*coder* and *decoder*." While the C-code examples later in this chapter were written for a Texas Instruments TMS320C6713 processor and a specific codec, the same techniques apply to any manufacturer's hardware.

19.3   Real-Time DSP Programming

In this section, we will assume an example signal processing system that has been designed based on a programmable DSP, and that it has a single analog input and a single analog output. There are two common ways to program for real-time DSP: on a sample-by-sample basis or on a frame basis. While it is easier to understand most DSP theory and operations on a sample-by-sample basis, in reality this is often an inefficient way to configure the actual input and output of data. Just as a computer hard disk transfers data in blocks of many bytes rather than one byte or word at a time, many DSP systems process data in "blocks" known as *frames*. The discussion in this chapter is applicable to either form of processing. See Figure 19.1 for a pictorial comparison of sample-based vs. frame-based processing.

In general, there are three ways to structure the hardware and software that determines the system's operation:

- The DSP continually *polls* the input ADC to see if it has data available, and then processes the new data sample to determine the next output. The DSP then returns to polling the input until the next sample is available. Polling is an inherently inefficient use of DSP resources, so it is almost never used in practice.
- The system is designed to be *interrupt driven*. That is, when the next incoming sample is available (or the next output sample is required), a processor interrupt is triggered that causes the DSP to automatically switch to a short, fast subprogram called an *interrupt service routine* (ISR) which takes care of getting the sample into or out of the processor. When the ISR finishes execution, control is returned to the main program. This leads directly to the idea of the *real-time schedule*. For a given

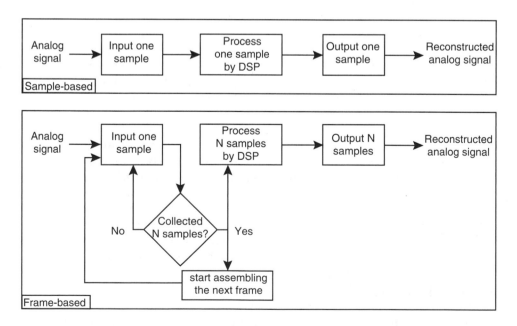

FIGURE 19.1   A comparison of a generic sample-based (top) vs. frame-based (bottom) processing system. Not all systems will require the input or output conversions from/to analog as shown here.

sample frequency F_s, the time between samples is $T_s = 1/F_s$. If the program is running on a sample-by-sample basis, the ISR has roughly T_s to do its processing before the next sample arrives. If the program is operating frame-based, the ISR will simply place the data in a buffer until a full frame is acquired, then signal the main program that a frame is available for processing. While the interrupt driven system eliminates the need for polling, the processor is still interrupted for every sample. In the interrupt service process, the processor context must be saved, control transferred to the ISR, and the processor context restored; the net result is *interrupt overhead* that represents lost processing time which can severely degrade the overall DSP performance.

- Use frame-based processing, but instead of the DSP responding to the sample interrupts, a peripheral device (typically integrated into the DSP) called a *direct memory access* (DMA) controller is used to automatically transfer the samples between the input/output devices and the DSP's memory. When a full frame is available, the DMA controller signals the DSP (usually through an interrupt). This eliminates using the high-performance DSP core to do the mundane sample data transfers, permits the DSP to be put into a low-power sleep state when it is not actively processing frames, and reduces the DSP interrupt overhead to a single interrupt per frame. For n-sample frames, a time of nearly nT_s is thus available for processing the samples. This DMA frame-based method is used in most commercial systems, and although simple in concept, is fraught with programming details that can be quite daunting to an engineer new to DSP systems.

Properly configuring the peripheral devices and interrupts, external hardware, and other configuration steps for DSP processors is different for each device and manufacturer. This is often the hardest part of getting started with real-time DSP, as often small configuration errors result in complete system failure. For simplicity of the discussion, we will use an FIR filter to demonstrate the process of transforming a nonreal-time DSP program in MATLAB to a real-time DSP program running on dedicated hardware. Previous chapters in this book explained FIR filters and explained their use in the MATLAB environment. Note there are many options for filter design, depending upon what optional Toolboxes you own. For example, the MATLAB Signal Processing Toolbox comes with versatile graphical filter design programs such as SPTool and FDATool. There is also a separate MATLAB Filter Design Toolbox.

19.4 MATLAB Implementation

MATLAB has a number of ways of performing the filtering operation; for simplicity we will only discuss two. The first is the built-in filter function and the second is to build your own routine to perform the FIR filtering operation. The built-in function allows us to filter signals almost immediately, but does very little to prepare us for the realities of real-time filtering using DSP hardware.

Built-in Approach

The MATLAB built-in function filter.m can be used to implement both an FIR filter (using only the numerator $[B]$ coefficients) and an IIR filter (using both the denominator $[A]$ and the numerator $[B]$ coefficients). The first few lines of the on-line help associated with the filter command are provided below:

```
>>help filter
```

```
FILTER One-dimensional digital filter.
   Y = FILTER (B, A, X) filters the data in vector X with the
   filter described by vectors A and B to create the filtered
   data Y. The filter is a ''Direct Form II Transposed''
   implementation of the standard difference equation:

   a(1)*y(n) = b(1)*x(n) + b(2)*x(n-1) + ... + b(nb+1)*x(n-nb)
            -a(2)*y(n-1) - ... - a(na+1)*y(n-na)
```

Notice that in the difference equation discussion of the MATLAB filter command, the A and B coefficient vector indices start at 1 instead of at 0. MATLAB does not allow for an index equal to zero. While this may only seem like a minor inconvenience, improper vector indices account for a significant number of the errors that occur during MATLAB algorithm development. To remain consistent with most DSP literature, we create another vector, say n, which is composed of integers with the first element equal to zero (i.e., $n = 0:15$ creates $n = \{0, 1, 2, 3, \ldots, 15\}$), and use this n vector to "fool" MATLAB into counting from zero. See the code given below for an example of this technique.

The MATLAB code shown below will filter the input vector x using the FIR filter coefficients in vector B. Notice that the input vector x is zero padded to flush the filter. This technique differs slightly from the direct implementation of the MATLAB filter command in which for M input values there will be M output values. Our technique assumes that the input vector is both preceded and followed by a large number of zeros. This implies that the filter is initially at rest (no initial conditions) and will relax or flush any remaining values at the end of the filtering operation.

Listing 19.1 Simple MATLAB FIR filter example.

```
%  This m-file is used to convolve x[n] and B[n]
%
%  Assumes that both x[n] and B[n] start at n = 0
%
%  written by Dr. Thad B. Welch, PE {t.b.welch@ieee.org}
%  copyright 2001
%  completed on 13 December 2001 revision 1.0


% Simulation inputs
x = [1 2 3 0 1 -3 4 1];                 % input vector x
B = [0.25 0.25 0.25 0.25];              % FIR filter coefficients B


% Calculated terms
PaddedX = [x zeros(1,length(B)-1)];  % zeros pad x to flush the filter
n = 0:(length(x) + length(B) - 2);   % plotting index for the output
y = filter(B, 1, PaddedX);           % performs the convolution


% Simulation outputs
stem(n, y)                              % output plot generation
ylabel('output values')
xlabel('sample number')
```

The output for this example follows:

```
Y =
Columns 1 through 8
  0.2500    0.7500    1.5000    1.5000    1.5000    0.2500    0.5000    0.7500
Column 9 through 11
  0.5000    1.2500    0.2500
```

The stem plot from the example is shown in Figure 19.2.

In this example, eight input samples were filtered and the results were returned all at once. Notice that when an eight-element vector x is filtered by a four-element vector B that 11 elements were returned $(8 + 4 - 1 = 11)$. This is an example of the general result that states that the length L of the sequence resulting from the convolution (filtering) of x and B is $L = \text{length}(x) + \text{length}(B) - 1$.

The FIR filter coefficients associated with this filter were $B = [0.25\,0.25\,0.25\,0.25]$. Since there are four coefficients for the filter, this is a third-order FIR filter (i.e., $N = 3$). The filtering effect produced is that it

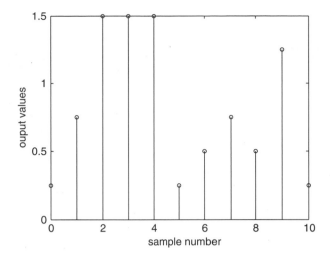

FIGURE 19.2　Stem plot of the filtering of *x* with *B*.

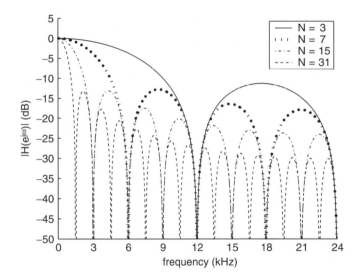

FIGURE 19.3　Magnitude of the frequency response for MA filters of order 3, 7, 15, and 31.

averages the most recent four input samples (i.e., the current sample and the previous three samples). This type of filter is called a moving average (MA) filter and is a type of lowpass FIR filter. Figure 19.3 shows the frequency response associated with MA filters of order $N = 3$, 7, 15, and 31, for a sample frequency of 48 kHz.

All of the MA filters shown in Figure 19.3 have a 0 Hz (dc) gain of 1 (which equals 0 dB). To insure that any filter has a dc gain of 0 dB, the impulse response $h[n]$ must sum to 1.

Creating Your Own Filter Algorithm

The last example helped us with MATLAB-based filtering, but the built-in function `filter.m` is of little use to us in performing real-time FIR filtering with DSP hardware. The next MATLAB example more closely implements the algorithm needed for a real-time process. This code will calculate a single output value based upon the current input value and the three previous input values.

Listing 19.2   FIR filter adjusted for real-time processing.

```
   %  This m-file is used to convolve x[n] and B[n] without
 2 %  using the MATLAB filter command.  This is one of the
   %  first steps toward being able to implement a real-time
 4 %  FIR filter in DSP hardware.
   %
 6 %  In sample-by-sample filtering, you are only trying to
   %  accomplish 2 things,
 8 %
   %  1.  Calculate the current output value, y(0), based on
10 %      just having received a new input sample, x(0).
   %  2.  Setup for the arrival of the next input sample.
12 %
   %  This is a BRUTE FORCE approach!
14 %
   %  written by Dr. Thad B. Welch, PE {t.b.welch@ieee.org}
16 %  copyright 2001
   %  completed on 13 December 2001 revision 1.0
18
   % Simulation inputs
20 x = [1 2 3 0];               % input vector x = x(0) x(-1) x(-2) x(-3)
   N = 3;                       % order of the filter = length(B) - 1
22 B = [0.25 0.25 0.25 0.25];   % FIR filter coefficients B

24 % Calculated terms
   y = 0;                       % initializes the output value y(0)
26 for i = 1:N+1                 % performs the dot product of B and x
       y = y + B(i)*x(i);
28 end

30 for i = N:-1:1               % shift stored x samples to the right so
       x(i+1) = x(i);           % the next x value, x(0), can be placed
32 end                          % in x(1)

34 % Simulation outputs
   x                            % notice that x(1) = x(2)
36 y                            % average of the last 4 input values
```

The input and output vectors from this FIR moving average filter program are shown below. Note that the four input samples result in a single output sample, as expected of a third-order filter:

```
x =
    1    1    2    3

y =
    1.5000
```

A few items need to be discussed concerning this example.

1. The filter order N was declared (line 21) despite the fact that MATLAB can determine the filter order based solely on the length of the vector B (that is, $N = \text{length}(B) - 1$). Declaring both the filter order N

and the FIR filter coefficients B (line 22), increases the portability of the code when you convert it to C/C++. Increased code portability may also be thought of as decreased machine dependance, which is generally a sought-after code attribute.

2. Only four values of x were stored (line 20). FIR filtering involves the dot product of only $N + 1$ terms. Since in this example $N = 3$, only four x terms are required.

3. The example is called a "brute force" approach based largely on the shifting of the stored x values within the x vector (lines 30 to 32). This unnecessary operation wastes resources that may be needed for other operations. Since our ultimate goal is efficient real-time implementation in DSP hardware, the more elegant and efficient solutions to this problem will be discussed in a later section.

19.5   Implementation in C

Several modifications to the MATLAB thought process are needed as we transition towards efficient real-time programming.

1. A semantic change is required since in MATLAB B is a vector, but in the C/C++ programming language, B is called an array.

2. The zero memory index, which does not exist in a MATLAB vector, does exist in the C/C++ programming language and it is routinely used in array notation.

3. The DSP hardware must process the data from the ADC in real time, therefore, we cannot wait for all message samples to be received prior to beginning the algorithmic process.

4. Real-time DSP is inherently an interrupt driven process and the input samples should only be processed using interrupt service routines (ISRs). Given this observation, it is incumbent upon the DSP programmer to ensure that the time requirements associated with periodic sampling are met. More bluntly, if you do not complete the algorithm's calculation before another input sample arrives, you have not met your real-time schedule and your system will fail. This leads to the observation that "the correct answer, if it arrives late, is wrong!"

5. Even though the DSP hardware has a phenomenal amount of processing power, this power should not be indiscriminately wasted.

6. The input and output ISRs are not magically linked! Nothing will come out of your DSP hardware unless you program the device to do so.

7. The digital portion of both an ADC and a DAC are inherently integer in nature. No matter what the ADC's input range is, the analog input voltage is mapped to an integer value. For a 16-bit converter, the possible values range from $+32,767$ to $-32,768$.

8. For clarity and understandability, declarations and assignments of variables, e.g., FIR filter coefficients, can be moved into .c and .h files.

Brute Force FIR Filtering – Part 1

The first version of the FIR implementation code, similar to the last MATLAB example, takes a brute force approach. The intention of this first approach is understandability, which comes at the expense of efficiency.

For simple programs, the actual FIR filtering operation code can be placed in the ISR itself. Many codecs are stereo devices, and your program should actually implement independent left and right channel filters. For clarity, only the left channel will be discussed here. In the code shown below, N is the filter order, the B array holds the FIR filter coefficients, the *xLeft* array holds the current input value $x[0]$, and past values of x, namely, $x[-1]$, $x[-2]$ and $x[-3]$, and *yLeft* is the current output value of the filter, $y[0]$. The integer i is used as an index counter in the loops.

Listing 19.3   Brute force FIR filter declarations.

```
#define N 3

float B[N+1] = {0.25, 0.25, 0.25, 0.25};

float xLeft[N+1];

float yLeft;

int i;
```

The code shown below performs the actual filtering operation. The five main steps involved in this operation are discussed below the code listing.

Listing 19.4   Brute force FIR filtering for real-time.

```
/* I added my routine here */
xLeft[0] = CodecDataIn.Channel[LEFT];    // current input value
yLeft = 0;                               // initialize the output value

for (i = 0; i <= N; i++) {
    yLeft += xLeft[i]*B[i]; // perform the dot-product
    }

for (i = N; i > 0; i--) {
    xLeft[i] = xLeft[i-1]; // setup for the next input
    }

CodecDataOut.Channel[LEFT] = yLeft; // output the value
/* end of my routine */
```

The Five Real-Time Steps involved in Brute Force FIR Filtering

An explanation of Listing 19.4 follows:

1. (Line 2): This code receives the next sample from the receive ISR and assigns it to the current input array element, xLeft[0].
2. (Line 3): The current output of this filter is given the name yLeft. Since this same variable will be used in the calculation of each output value of the filter, it must be reinitialized before each dot product is performed.
3. (Lines 5–7): These three lines of code perform the dot product of x and B. The equivalent operation is:

$$yLeft = xLeft[0]B[0] + xLeft[-1]B[1] + xLeft[-2]B[2] + xLeft[-3]B[3].$$

4. (Lines 9–11): These three lines of code shift all of the values in the *x* array one element to the right. The equivalent operation is

$$xLeft[2] \rightarrow xLeft[3]$$
$$xLeft[1] \rightarrow xLeft[2]$$
$$xLeft[0] \rightarrow xLeft[1].$$

After the shift to the right is complete, the next incoming sample, x[0], can be written into the xLeft[0] memory location without a loss of information. Also notice that xLeft[3] was overwritten by xLeft[2].

The expected operation of xLeft[3] → xLeft[4] is not needed since there is no xLeft[4] (xLeft only contains four elements). In summary, xLeft[3] is no longer needed and is, therefore, overwritten.

5. (Line 13): This line of code completes the filtering operation by transferring the result of the dot product, yLeft, to the DAC via the transmit ISR.

Brute Force FIR Filtering – Part 2

The previous section introduced a *brute force* approach to FIR filtering. While this implementation is straightforward and relatively easy to understand, it suffers from two major problems.

1. Most FIR filters use a considerably higher order than the filters discussed in the previous examples. Most filters also require a great deal of numerical precision to accurately specify the B coefficients. These facts make manual entry of the B coefficients very inconvenient.
2. Step 4 in the five real-time steps involved in the brute force FIR filtering section discussed above shifted all of the values in the *x* array one element to the right after each dot product operation. This shifting is a very inefficient use of the DSK's computational resources.

Entering a large number of filter coefficients by hand would be tedious and error prone. What many people do is use MATLAB to generate the filter coefficients, then copy and paste the values to a separate .c file where they are defined for the C/C++ DSP program. Using an .h header file in which the coefficients are declared would also be used, assuming separate compilation (the filter program in one .c file, the coefficients defined in another .c file, and the coefficient array variable declared in a .h file).

Listing 19.5 An example `coeff.h` file for a 30th-order FIR filter.

```
   /* coeff.h                          */
2  /* FIR filter coefficients          */

4  #define N 30

6  extern float B[];
```

Within the `coeff.h` file, line 4 is used to define the filter order and line 6 allows the B coefficients to be defined in another file. In this case, the coefficients are defined in the file `coeff.c`.

Listing 19.6 An example `coeff.c` file showing only the first 11 coefficients defined for the 30th-order FIR filter. Don't forget the closing brace } after the last coefficient!

```
   float B[N+1] = {
2  {-0.031913481327},   /* h[0] */
   {0.000000000000},    /* h[1] */
4  {-0.026040505746},   /* h[2] */
   {-0.000000000000},   /* h[3] */
6  {-0.037325855883},   /* h[4] */
   {0.000000000000},    /* h[5] */
8  {-0.053114839831},   /* h[6] */
   {-0.000000000000},   /* h[7] */
10 {-0.076709627018},   /* h[8] */
   {0.000000000000},    /* h[9] */
12 {-0.116853446730},   /* h[10]*/
```

Once you become familiar with these procedures and some of the MATLAB filter design techniques, FIR filters can be designed and implemented extremely quickly. You can easily modify the code shown in Listing 19.4 to use the more flexible separate compilation technique.

Circular Buffered FIR Filtering

As previously stated, shifting all of the values in the x array one element to the right after each dot product operation is a very inefficient use of the DSK's computational resources. The need to perform this shift is based on the assumption that the physical memory is linear. Given linear memory, with the inherent static labeling of each memory location, this shifting of values would seem to be an absolute requirement.

Figure 19.4 shows a linear memory model for the input to the filter x. As expected, to buffer the $N + 1$ elements in the x array, there are memory locations labeled $x[0], x[-1], \ldots, x[-N]$, but there is also an $x[-(N + 1)]$. While this location was not declared, it does physically exist, and any attempt to access the x array beyond its declared bounds will result in *something* being retrieved and used in any subsequent calculations. The results of this indexing error may be catastrophic, e.g., a run-time error, or more subtle, e.g., the program runs, but gives inaccurate results. Either way, this type of indexing error must be avoided.

An alternative to linear memory is circular memory. As shown in Figure 19.5, the circular memory concept wraps the memory location labeled $x[-N]$ back to the memory location labeled $x[0]$. Since the purpose of this circular memory is to store or buffer x, this concept is routinely referred to as circular buffering.

If instead of using static memory location labels, a pointer is used to point to and insert the most recent sample, $x[0]$, into the memory location containing the oldest sample, $x[N]$; thus, a circular buffer has been created. No shifting of the x values is required since the pointer will always point to the most recent sample value. As the pointer advances, the oldest sample in the buffer is replaced by the most recent sample. This process can be continued indefinitely. The result of inserting the next sample into the buffer is shown in Figure 19.6.

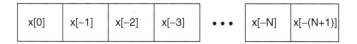

FIGURE 19.4   The linear memory concept with static memory location labeling.

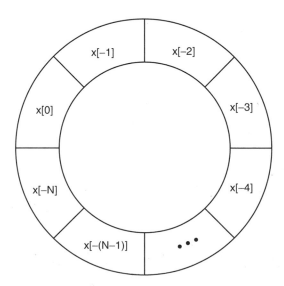

FIGURE 19.5   The circular buffer concept with static memory location labeling.

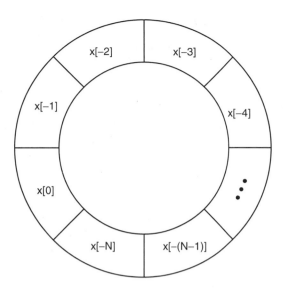

FIGURE 19.6 The circular buffer concept with dynamic memory location labeling.

Notice how the values shifted their effective position by just changing the pointer. To implement the circular buffer, a pointer must be established to the array xLeft. The required code to create this pointer is shown below.

```
float xLeft[N+1], *pLeft = xLeft;
```

The remainder of the circular buffered FIR filter code is shown below. The code comments explain the algorithm.

Listing 19.7 FIR filter using a circular buffer.

```
   *pLeft = CodecDataIn.Channel[LEFT];     // store LEFT input value
2
   output = 0;                             // set up for LEFT channel
4  p = pLeft;                              // save current sample pointer
   if(++pLeft > &xLeft[N])                 // update pointer, wrap if necessary
6     pLeft = xLeft;                       // and store
   for (i = 0; i <= N; i++) {              // do LEFT channel FIR
8     output += *p-- * B[i];               // do MAC (multiply and accumulate)
      if(p < &xLeft[N])                    // check for pointer wrap around
10       p = &xLeft[N];
      }
12 CodecDataOut.Channel[LEFT] = output; // store filtered LEFT value
```

The circular buffering technique is commonly used in many real-time applications. Some DSP processors include special commands to establish and manipulate circular buffers in the hardware.

19.6 Conclusion

Real-time DSP can be one of the "trickiest" topics to master. Even if your algorithm is perfectly valid, the actual implementation in real-time may suffer from problems that have more to do with computer engineering and software engineering principles than anything related to signal processing. Despite this, getting started in real-time DSP for many is the next logical step beyond MATLAB-based signal processing. While becoming an *expert* in real-time DSP typically requires many years of experience and learning, such skills are in very high demand.

References

D.A. Patterson and J.L. Hennessy, *Computer Organization and Design: The Hardware/Software Interface*, 2nd ed. San Francisco, CA: Morgan Kaufmann Publishers, 1997.

C.H.G. Wright, T.B. Welch, D.M. Etter, and M.G. Morrow, Teaching hardware-based DSP: Theory to practice. In *Proceedings of the IEEE International Conference on Acoustics, Speech, and Signal Processing* (May 2002), vol. IV, pp. 4148–4151. Paper 4024 (invited).

C.H.G. Wright, T.B. Welch, D.M. Etter, and M.G. Morrow, Teaching DSP: Bridging the gap from theory to real-time hardware. *ASEE Comput. Educ. J. XIII*, 3 (July 2003), 14–26.

Further Information

- Texas Instuments DSP resources:
  See `http://dspvillage.ti.com`
- Analog Devices DSP resources:
  See `http://www.analog.com/processors/index.html`
- DSP daughterboards and other resources for TI DSKs: Educational DSP (eDSP), L.L.C:
  See `http://www.educationaldsp.com/`
- For free DSP software such as winDSK6:
  See `http://eceserv0.ece.wisc.edu/ morrow/software/`

20

VLSI for Signal Processing

Keshab K. Parhi
University of Minnesota

Rulph Chassaing
Roger Williams University

Bill Bitler
InfiMed

20.1 Special Architectures

Keshab K. Parhi

Digital signal processing (DSP) is used in numerous applications. These applications include telephony, mobile radio, satellite communications, speech processing, video and image processing, biomedical applications, radar, and sonar. Real-time implementations of DSP systems require design of hardware that can match the application sample rate to the hardware processing rate (which is related to the clock rate and the implementation style). Thus, real-time does not always mean high speed. Real-time architectures are capable of processing samples as they are received from the signal source, as opposed to storing them in buffers for later processing as done in batch processing. Furthermore, real-time architectures operate on an infinite time series (since the number of the samples of the signal source is so large that it can be considered infinite). While speech and sonar applications require lower sample rates, radar and video image processing applications require much higher sample rates. The sample rate information alone cannot be used to choose the architecture. The algorithm complexity is also an important consideration. For example, a very complex and computationally intensive algorithm for a low-sample-rate application and a computationally simple algorithm for a high-sample-rate application may require similar hardware speed and complexity. These ranges of algorithms and applications motivate us to study a wide variety of architecture styles.

Using very large scale integration (VLSI) technology, DSP algorithms can be prototyped in many ways. These options include (1) single or multiprocessor programmable digital signal processors, (2) the use of core programmable digital signal processor with customized interface logic, (3) semicustom gate-array implementations, and (4) full-custom dedicated hardware implementations. The DSP algorithms are implemented in the programmable processors by translating the algorithm to the processor assembly code. This can require an extensive amount of time. On the other hand, high-level compilers for DSP can be used to generate the assembly code. Although this is currently feasible, the code generated by the compiler is not as efficient as hand-optimized code. Design of DSP compilers for generation of efficient code is still an active research topic. In the case of dedicated designs, the challenge lies in a thorough understanding of the DSP algorithms and theory of architectures. For example, just minimizing the number of multipliers in an algorithm may not lead to a better dedicated design. The area saved by the number of multipliers may be offset by the increase in control, routing, and placement costs.

Off-the-shelf programmable digital signal processors can lead to faster prototyping. These prototyped systems can prove very effective in fast simulation of computation-intensive algorithms (such as those encountered in speech recognition, video compression, and seismic signal processing) or in benchmarking and standardization. After standards are determined, it is more useful to implement the algorithms using dedicated circuits.

Design of dedicated circuits is not a simple task. Dedicated circuits provide limited or no programming flexibility. They require less silicon area and consume less power. However, the low production volume, high design cost, and long turnaround time are some of the difficulties associated with the design of dedicated systems. Another difficulty is the availability of appropriate computer-aided design (CAD) tools for DSP systems. As time progresses, however, the architectural design techniques will be better understood and can be incorporated into CAD tools, thus making the design of dedicated circuits easier. Hierarchical CAD tools can integrate the design at various levels in an automatic and efficient manner. Implementation of standards for signal and image processing using dedicated circuits will lead to higher volume production. As time progresses, dedicated designs will be more acceptable to customers of DSP.

Successful design of dedicated circuits requires careful algorithm and architecture considerations. For example, for a filtering application, different equivalent realizations may possess different levels of concurrency. Thus, some of these realizations may be suitable for a particular application while other realizations may not be able to meet the sample rate requirements of the application. The lower-level architecture may be implemented in a word-serial or word-parallel manner. The arithmetic functional units may be implemented in bit-serial or digit-serial or bit-parallel manner. The synthesized architecture may be implemented with a dedicated data path or shared data path. The architecture may be systolic or nonsystolic.

Algorithm transformations play an important role in the design of dedicated architectures (Parhi, 1989). This is because the transformed algorithms can be made to operate with better performance (where the performance may be measured in terms of speed, area, or power). Examples of these transformations include pipelining, parallel processing, retiming, unfolding, folding, look-ahead, associativity, and distributivity. These transformations and other architectural concepts are described in detail in subsequent sections.

Pipelining

Pipelining can increase the amount of concurrency (or the number of activities performed simultaneously) in an algorithm. Pipelining is accomplished by placing latches at appropriate intermediate points in a data flow graph that describes the algorithm. Each latch also refers to a storage unit or buffer or register. The latches can be placed at *feed-forward cutsets* of the data flow graph. In synchronous hardware implementations, pipelining can increase the clock rate of the system (and therefore the sample rate). The drawbacks associated with pipelining are the increase in system latency and the increase in the number of registers. To illustrate the speed increase using pipelining, consider the second-order three-tap finite impulse response (FIR) filter shown in Figure 20.1(a). The signal $x(n)$ in this system can be sampled at a rate limited by the throughput of one multiplication and two additions. For simplicity, if we assume the multiplication time to be two times the addition time (T_{add}), the effective sample or clock rate of this system is $1/4T_{add}$. By placing latches as shown in Figure 20.1(b) at the cutset shown by the dashed line, the sample rate can be improved to the rate of one multiplication or two additions. While pipelining can be easily applied to all algorithms with no feedback loops by the appropriate placement of latches, it cannot easily be applied to algorithms with feedback loops. This is because the cutsets in feedback algorithms contain feed-forward and feedback data flow and cannot be considered as feed-forward cutsets.

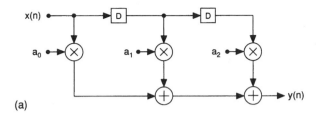

(a)

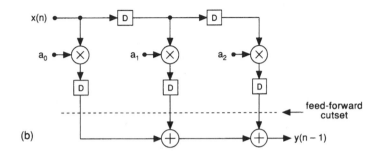

(b)

FIGURE 20.1   (a) A three-tap, second-order nonrecursive digital filter; (b) the equivalent pipelined digital filter obtained by placing storage units at the intersection of the signal wires and the feed-forward cutset. If the multiplication and addition operations require 2 and 1 units of time, respectively, then the maximum achievable sampling rates for the original and the pipelined architectures are 1/4 and 1/2 units, respectively.

Pipelining can also be used to improve the performance in software programmable multiprocessor systems. Most software programmable DSP processors are programmed using assembly code. The assembly code is generated by high-level compilers that perform scheduling. Schedulers typically use the acyclic precedence graph to construct schedules. The removal of all edges in the signal (or data) flow graph containing delay elements converts the signal flow graph to an acyclic precedence graph. By placing latches to pipeline a data flow graph, we can alter the acyclic precedence graph. In particular, the critical path of the acyclic precedence graph can be reduced. The new precedence graph can be used to construct schedules with lower iteration periods (although this may often require an increase in the number of processors).

Pipelining of algorithms can increase the sample rate of the system. Sometimes, for a constant sample rate, pipelining can also reduce the power consumed by the system. This is because the data paths in the pipelined system can be charged or discharged with lower supply voltage. Since the capacitance remains almost constant, the power can be reduced. Achieving low power can be important in many battery-powered applications (Chandrakasan et al., 1992).

Parallel Processing

Parallel processing is related to pipelining but requires replication of hardware units. Pipelining exploits concurrency by breaking a large task into multiple smaller tasks and by separating these smaller tasks by storage units. However, parallelism exploits concurrency by performing multiple larger tasks simul-taneously in separate hardware units.

To illustrate the speed increase due to parallelism, consider the parallel implementation of the second-order, three-tap FIR filter of Figure 20.1(a) shown in Figure 20.2. In the architecture of Figure 20.2, two input samples are processed and two output samples are generated in each clock cycle period of four addition times. Because each clock cycle processes two samples, however, the effective sample rate is $1/2T_{add}$, which is the same as that of Figure 20.1(b). The parallel architecture leads to the speed increase with significant hardware overhead. The entire data flow graph needs to be replicated with an increase in the amount of parallelism. Thus, it is more desirable to use pipelining as opposed to parallelism. However, parallelism may be useful if

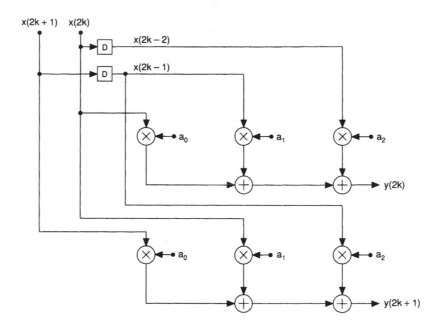

FIGURE 20.2   Twofold parallel realization of the three-tap filter of Figure 20.1(a).

pipelining alone cannot meet the speed demand of the application or if the technology constraints (such as limitations on the clock rate by the I/O technology) limit the use of pipelining. In obvious ways, pipelining and parallelism can be combined also. Parallelism, like pipelining, can also lead to power reduction but with significant overhead in hardware requirements. Achieving pipelining and parallelism can be difficult for systems with feedback loops. Concurrency may be created in these systems by using the look-ahead transformation.

Retiming

Retiming is similar to pipelining but yet different in some ways (Leiserson et al., 1983). Retiming is the process of moving the delays around in the data flow graph. Removal of one delay from all input edges of a node and insertion of one delay to each outgoing edge of the same node is the simplest example of retiming. Unlike pipelining, retiming does not increase the latency of the system. However, retiming alters the number of delay elements in the system. Retiming can reduce the critical path of the data flow graph. As a result, it can lead to clock period reduction in hardware implementations or critical path of the acyclic precedence graph or the iteration period in programmable software system implementations.

The single host formulation of the retiming transformation preserves the latency of the algorithm. The retiming formulation with no constraints on latency (i.e., with separate input and output hosts) can also achieve *pipelining with no retiming* or *pipelining with retiming*. Pipelining with retiming is the most desirable transformation in DSP architecture design. Pipelining with retiming can be interpreted to be identical to retiming of the original algorithm with a large number of delays at the input edges. Thus, we can increase the system latency arbitrarily and remove the appropriate number of delays from the inputs after the transformation.

The retiming formulation assigns retiming variables $r(.)$ to each node in the data flow graph. If $i(U \rightarrow V)$ is the number of delays associated with the edge $U \rightarrow V$ in the original data flow graph and $r(V)$ and $r(U)$, respectively, represent the retiming variable value of the nodes V and U, then the number of delays associated with the edge $U \rightarrow V$ in the retimed data flow graph is given by

$$i_r(U \rightarrow V) = i(U \rightarrow V) + r(V) - r(U)$$

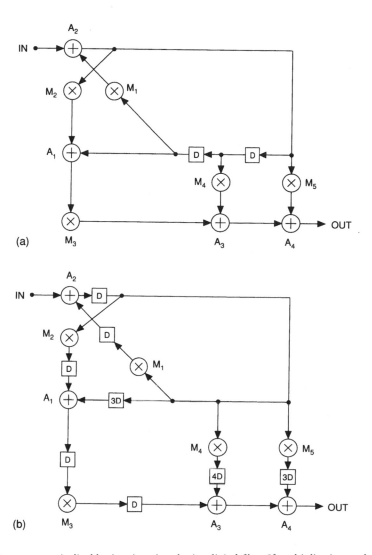

FIGURE 20.3   (a) A two-stage pipelinable time-invariant lattice digital filter. If multiplication and addition operations require 2 and 1 time units, respectively, then this data flow graph can achieve a sampling period of 10 time units (which corresponds to the critical path $M_1 \rightarrow A_2 \rightarrow M_2 \rightarrow A_1 \rightarrow M_3 \rightarrow A_3 \rightarrow A_4$). (b) The pipelined/retimed lattice digital filter can achieve a sampling period of 2 time units.

For the data flow graph to be realizable, $i_r(U \rightarrow V) \geq 0$ must be satisfied. The retiming transformation formulates the problem by calculating path lengths and by imposing constraints on certain path lengths. These constraints are solved as a shortest-path problem.

   To illustrate the usefulness of retiming, consider the data flow graph of a two-stage pipelined lattice digital filter graph shown in Figure 20.3(a) and its equivalent pipelined-retimed data flow graph shown in Figure 20.3(b). If the multiply time is 2 units and the add time is 1 unit, the architecture in Figure 20.3(a) can be clocked with period 10 units, whereas the architecture in Figure 20.3(b) can be clocked with period 2 units.

Unfolding

The **unfolding** transformation is similar to loop unrolling. In *J*-unfolding, each node is replaced by *J* nodes and each edge is replaced by *J* edges. The *J*-unfolded data flow graph executes *J* iterations of the original algorithm (Parhi, 1991).

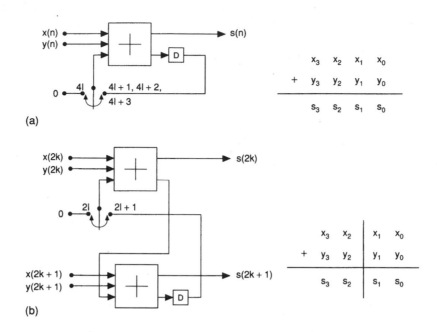

FIGURE 20.4   (a) A least-significant-bit first bit-serial adder for word length of 4; (b) a digit-serial adder with digit size 2 obtained by two-unfolding of the bit-serial adder. The bit position 0 stands for least significant bit.

The unfolding transformation can unravel the hidden concurrency in a data flow program. The achievable iteration period for a J-unfolded data flow graph is $1/J$ times the critical path length of the unfolded data flow graph. By exploiting interiteration concurrency, unfolding can lead to a lower iteration period in the context of a software programmable multiprocessor implementation.

The unfolding transformation can also be applied in the context of hardware design. If we apply an unfolding transformation on a (word-serial) nonrecursive algorithm, the resulting data flow graph represents a **word-parallel** (or simply parallel) algorithm that processes multiple samples or words in parallel every clock cycle. If we apply two-unfolding to the three-tap FIR filter in Figure 20.1(a), we can obtain the data flow graph of Figure 20.2.

Because the unfolding algorithm is based on graph theoretic approach, it can also be applied at the bit level. Thus, unfolding of a **bit-serial** data flow program by a factor of J leads to a **digit-serial** program with digit size J. The *digit size* represents the number of bits processed per clock cycle. The digit-serial architecture is clocked at the same rate as the bit-serial (assuming that the clock rate is limited by the communication I/O bound much before reaching the computation bound of the bit-serial program). Because the digit-serial program processes J bits per clock cycle the effective bit rate of the digit-serial architecture is J times higher. A simple example of this unfolding is illustrated in Figure 20.4, where the bit-serial adder in Figure 20.4(a) is unfolded by a factor of 2 to obtain the digit-serial adder in Figure 20.4(b) for digit size 2 and word length 4. In obvious ways, the unfolding transformation can be applied to both word level and bit level simultaneously to generate word-parallel, digit-serial architectures. Such architectures process multiple words per clock cycle and process a digit of each word (not the entire word).

Folding Transformation

The **folding** transformation is the reverse of the unfolding transformation. While the unfolding transformation is simpler, the folding transformation is more difficult (Parhi et al., 1992).

The folding transformation can be applied to fold a bit-parallel architecture to a digit-serial or bit-serial one or to fold a digit-serial architecture to a bit-serial one. It can also be applied to fold an algorithm data flow graph to a hardware data flow for a specified folding set. The folding set indicates the processor in which and the time partition at which a task is executed. A specified folding set may be infeasible, and this needs to be detected first. The folding transformation performs a preprocessing step to detect feasibility and in the feasible case transforms the algorithm data flow graph to an equivalent pipelined/retimed data flow graph that can be folded. For the special case of regular data flow graphs and for linear space–time mappings, the folding tranformation reduces to **systolic** array design.

In the folded architecture, each edge in the algorithm data flow graph is mapped to a communicating edge in the hardware architecture data flow graph. Consider an edge $U \rightarrow V$ in the algorithm data flow graph with associated number of delays $i(U \rightarrow V)$. Let the tasks U and V be mapped to the hardware units H_U and H_V, respectively. Assume that N time partitions are available, i.e., the iteration period is N. A modulo operation determines the time partition. For example, the time unit 20 for $N = 4$ corresponds to time partition 20 modulo 4 or 2. Let the tasks U and V be executed in time partitions u and v, i.e., the lth iterations of tasks U and V are executed in time units $Nl + u$ and $Nl + v$, respectively. The $i(U \rightarrow V)$ delays in the edge $U \rightarrow V$ implies that the result of the lth iteration of U is used for the $(l + i)$th iteration of V. The $(l + i)$th iteration of V is executed in time unit $N(l + i) + v$. Thus the number of storage units needed in the folded edge corresponding to the edge $U \rightarrow V$ is

$$D_F(U \rightarrow V) = N(l + i) + v - Nl - u - P_u = Ni + v - u - p_u$$

where P_u is the level of pipelining of the hardware operator H_U. The $D_F(U \rightarrow V)$ delays should be connected to the edge between H_U and H_V, and this signal should be switched to the input of H_V at time partition v. If the $D_F(U \rightarrow V)$'s as calculated here were always nonnegative for all edges $(U \rightarrow V)$, then the problem would be solved. However, some $D_F()$'s would be negative. The algorithm data flow graph needs to be pipelined and retimed such that all the $D_F()$'s are nonnegative. This can be formulated by simple inequalities using the retiming variables. The retiming formulation can be solved as a path problem, and the retiming variables can be determined if a solution exists. The algorithm data flow graph can be retimed for folding and the calculation of the $D_F()$'s can be repeated. The folded hardware architecture data flow graph can now be completed. The folding technique is illustrated in Figure 20.5. The algorithm data flow graph of a two-stage pipelined lattice recursive digital filter of Figure 20.3(a) is folded for the folding set shown in Figure 20.5. Figure 20.5(a) shows the pipelined/retimed data flow graph (preprocessed for folding) and Figure 20.5(b) shows the hardware architecture data flow graph obtained after folding.

As indicated before, a special case of folding can address systolic array design for regular data flow graphs and for linear mappings. The systolic architectures make use of extensive pipelining and local communication and operate in a synchronous manner (Kung, 1988). The systolic processors can also be made to operate in an asynchronous manner, and such systems are often referred to as wavefront processors. Systolic architectures have been designed for a variety of applications, including convolution, matrix solvers, matrix decomposition, and filtering.

Look-Ahead Technique

The **look-ahead** technique is a very powerful technique for pipelining of recursive signal processing algorithms (Parhi and Messerschmitt, 1989). This technique can transform a sequential recursive algorithm into an equivalent concurrent one, which can then be realized using pipelining or parallel processing or both. This technique has been successfully applied to pipeline many signal processing algorithms, including recursive digital filters (in direct form and lattice form), adaptive lattice digital filters, two-dimensional recursive digital filters, Viterbi decoders, Huffman decoders, and finite-state machines. This research demonstrated that the recursive signal processing algorithms can be operated at high speed. This is an important result since modern signal processing applications in radar and image processing and particularly in high-definition and super-high-definition television video signal processing require very high throughput. Traditional algorithms and topologies cannot be used for such high-speed applications because of the inherent speed bound of the algorithm created by the feedback loops. The look-ahead transformation creates

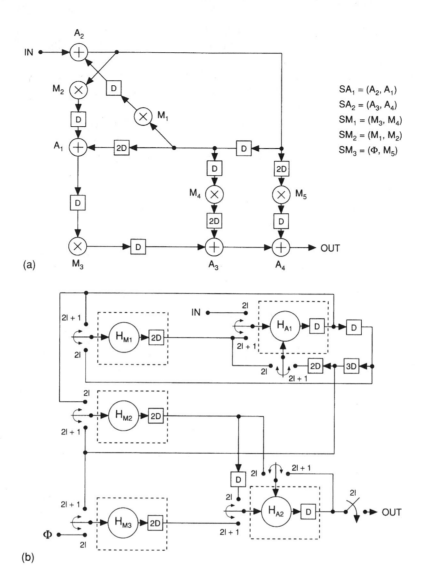

FIGURE 20.5 (a) A pipelined/retimed data flow graph obtained from Figure 20.3(a) by preprocessing for folding; (b) the folded hardware architecture data flow graph. In our folding notation, the tasks are ordered within a set and the ordering represents the time partition in which the task is executed. For example, $SA_1 = (A_2, A_1)$ implies that A_2 and A_1 are, respectively, executed in even and odd time partitions in the same processor. The notation Φ represents a null operation.

additional concurrency in the signal processing algorithms and the speed bound of the transformed algorithms is increased substantially. The look-ahead transformation is not free from drawbacks. It is accompanied by an increase in the hardware overhead. This difficulty has encouraged us to develop inherently pipelinable topologies for recursive signal processing algorithms. Fortunately, this is possible to achieve in adaptive digital filters using relaxations on the look-ahead or by the use of relaxed look-ahead (Shanbhag and Parhi, 1992).

To begin, consider a time-invariant, one-pole recursive digital filter transfer function:

$$H(z) = \frac{X(z)}{U(z)} = \frac{1}{1 - az^{-1}}$$

described by the difference equation:

$$x(n) = ax(n-1) + u(n)$$

and shown in Figure 20.6(a). The maximum achievable speed in this system is limited by the operating speed of one multiply–add operation. To increase the speed of this system by a factor of 2, we can express $x(n)$ in terms of $x(n-2)$ by substitution of one recursion within the other:

$$x(n) = a[ax(n-2) + u(n-1)] + u(n) = a^2 x(n-2) + au(n-1) + u(n)$$

The transfer function of the emulated second-order system is given by

$$H(z) = \frac{1 + az^{-1}}{1 - a^2 z^{-2}}$$

and is obtained by using a pole-zero cancellation at $-a$. In the modified system, $x(n)$ is computed using $x(n-2)$ as opposed to $x(n-1)$; thus we *look ahead*. The modified system has two delays in the multiply–add feedback loop, and these two delays can be distributed to pipeline the multiply–add operation by two stages. Of course, the additional multiply–add operation that represents one zero would also need to be pipelined by two stages to keep up with the sample rate of the system. To increase the speed by 4 times, we can rewrite the transfer function as

$$H(z) = \frac{(1 + az^{-1})(1 + a^2 z^{-2})}{(1 - a^4 z^{-4})}$$

This system is shown in Figure 20.6(b). Arbitrary speed increase is possible. However, for power-of-two speed increase the hardware complexity grows logarithmically with speed-up factor. The same technique can be applied to any higher-order system. For example, a second-order recursive filter with transfer function:

$$H(z) = \frac{1}{1 - 2r\cos\theta z^{-1} + r^2 z^{-2}}$$

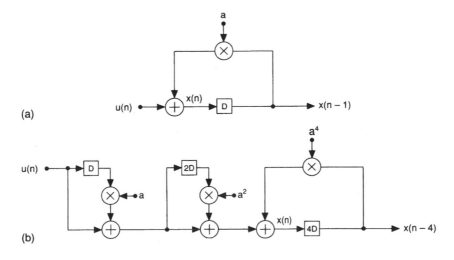

FIGURE 20.6   (a) A first-order recursive digital filter; (b) a four-stage pipelinable equivalent filter obtained by look-ahead computation.

can be modified to

$$H(z) = \frac{1 + 2r\cos\theta z^{-1} + r^2 z^{-2}}{1 - 2r\cos 2\theta z^{-1} + r^4 z^{-4}}$$

for a twofold increase in speed. In this example, the output $y(n)$ is computed using $y(n-2)$ and $y(n-4)$; thus, it is referred to as *scattered look-ahead*.

While look-ahead can transform any recursive digital filter transfer function to pipelined form, it leads to a hardware overhead proportional to $N \log_2 M$, where N is the filter order and M is the speed-up factor. Instead of starting with a sequential digital filter transfer function obtained by traditional design approaches and transforming it for pipelining, it is more desirable to use a constrained filter design program that can satisfy the filter spectrum and the pipelining constraint. The pipelining constraint is satisfied by expressing the denominator of the transfer function in scattered look-ahead form. Such filter design programs have now been developed in both time domain and frequency domain. The advantage of the constrained filter design approach is that we can obtain pipelined digital filters with marginal or zero hardware overhead compared with sequential digital filters. The pipelined transfer functions can also be mapped to pipelined lattice digital filters. The reader might note that the data flow graph of Figure 20.3(a) was obtained by this approach.

The look-ahead pipelining can also be applied to the design of transversal and adaptive lattice digital filters. Although look-ahead transformation can be used to modify the adaptive filter recursions to create concurrency, this requires a large hardware overhead. The adaptive filters are based on weight update operations, and the weights are adapted based on the current error. Finally, the error becomes close to zero and the filter coefficients have been adapted. Thus, making relaxations on the error can reduce the hardware overhead substantially without degradation of the convergence behavior of the adaptive filter. Three types of relaxations of look-ahead are possible. These are referred to as *sum relaxation*, *product relaxation*, and *delay relaxation*. To illustrate these three relaxations, consider the weight update recursion:

$$w(n+1) = a(n)w(n) + f(n)$$

where the term $a(n)$ is typically 1 for transversal least mean square (LMS) adaptive filters and of the form $(1 - \varepsilon(n))$ for lattice LMS adaptive digital filters, and $f(n) = \mu e(n)u(n)$ where μ is a constant, $e(n)$ is the error, and $u(n)$ is the input. The use of look-ahead transforms the above recursion to

$$w(n+M) = \prod_{i=0}^{M-1} a(n+M-i-1)w(n)$$

$$+ \left[1a(n+M-1)\prod_{i=0}^{1} a(n+M-i-1) \ldots \prod_{i=0}^{M-2} a(n+M-i-1) \right] \begin{bmatrix} f(n+M-1) \\ f(n+M-2) \\ \cdot \\ \cdot \\ \cdot \\ f(n) \end{bmatrix}$$

In sum relaxation, we only retain the single term dependent on the current input for the last term of the look-ahead recursion. The relaxed recursion after sum relaxation is given by

$$w(n+M) = \prod_{i=0}^{M-1} a(n+M-i-1)w(n) + f(n+M-1)$$

In lattice digital filters, the coefficient $a(n)$ is close to 1 for all n, since it can be expressed as $(1 - \varepsilon(n))$ and $\varepsilon(n)$ is close to zero for all n and is positive. The product relaxation on the above equation leads to

$$w(n + M) = (1 - M\varepsilon(n + M - 1))w(n) + f(n + M - 1)$$

The delay relaxation assumes the signal to be slowly varying or to be constant over D samples and replaces the look-ahead by

$$w(n + M) = (1 - M\varepsilon(n + M - 1))w(n) + f(n + M - D - 1)$$

These three types of relaxations make it possible to implement pipelined transversal and lattice adaptive digital filters with a marginal increase in hardware overhead. Relaxations on the weight update operations change the convergence behavior of the adaptive filter, and we are forced to examine carefully the convergence behavior of the relaxed look-ahead adaptive digital filters. It has been shown that the relaxed look-ahead adaptive digital filters do not suffer from degradation in adaptation behavior. Futhermore, when coding, the use of pipelined adaptive filters could lead to a dramatic increase in pixel rate with no degradation in signal-to-noise ratio of the coded image and no increase in hardware overhead (Shanbhag and Parhi, 1992).

The concurrency created by look-ahead and relaxed look-ahead transformations can also be exploited in the form of parallel processing. Furthermore, for a constant speed, concurrent architectures (especially the pipelined architectures) can also lead to low power consumption.

Associativity Transformation

The addition operations in many signal processing algorithms can be interchanged since the add operations satisfy associativity. Thus, it is possible to move the add operations outside the critical loops to increase the maximum achievable speed of the system. As an example of the associative transformation, consider the realization of a second-order recursion $x(n) = 5/8x(n-1) - 3/4x(n-2) + u(n)$. Two possible realizations are shown in Figure 20.7(a). The realization on the left contains one multiplication and two add operations in the critical inner loop, whereas the realization on the right contains one multiplication and one add operation in the critical inner loop. The realization on the left can be transformed to the realization on the right using the associativity transformation. Figure 20.7(b) shows a bit-serial implementation of this second-order recursion for the realization on the right for a word length of 8. This bit-serial system can be operated in a functionally correct manner for any word length greater than or equal to 5 since the inner loop computation latency is Five cycles. However, if associativity were not exploited, then the minimum realizable word length would be 6. Thus, associativity can improve the achievable speed of the system.

Distributivity

Another local transformation that is often useful is distributivity. In this transformation, a computation $(A \times B) + (A \times C)$ may be reorganized as $A \times (B + C)$. Thus, the number of hardware units can be reduced from two multipliers and one adder to one multiplier and one adder.

Arithmetic Processor Architectures

In addition to algorithms and architecture designs, it is also important to address implementation styles and arithmetic processor architectures.

Most DSP systems use fixed-point hardware arithmetic operators. While many number system representations are possible, the two's complement number system is the most popular number system. The other number systems include the residue number system, the redundant or signed-digit number system, and the logarithmic number system. The residue and logarithmic number systems are rarely used or are used in very special cases such as nonrecursive digital filters. Shifting or scaling and division are

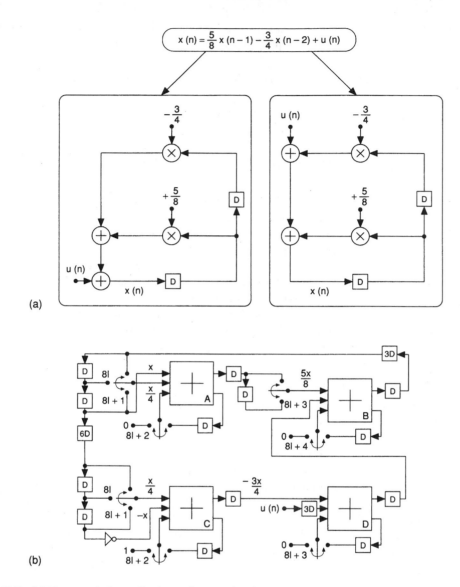

FIGURE 20.7   (a) Two associative realizations of a second-order recursion; (b) an efficient bit-serial realization of the recursion for a word length of 8.

difficult in the residue number system. Difficulty with addition and the overhead associated with logarithm and antilogarithm converters reduce the attractiveness of the logarithm number system. The use of the redundant number system leads to carry-free operation but is accompanied by the overhead associated with redundant-to-two's complement conversion. Another approach often used is distributed arithmetic. This approach has recently been used in a few video transformation chips.

The simplest arithmetic operation is addition. Multiplication can be realized as a series of add–shift operations, and division and square-root can be realized as a series of controlled add–subtract operations. The conventional two's complement adder involves carry ripple operation. This limits the throughput of the adder operation. In DSP, however, the combined multiply–add operation is most common. Carry–save operations have been used to realize pipelined multiply–adders using fewer pipelining latches. In conventional pipelined two's complement multiplier, the multiplication time is approximately 2 times the bit-level addition time. Recently, a technique has been proposed to reduce the multiplication time from $2W$ bit-level binary adder times to $1.25W$ bit-level binary adder times, where W is the word length. This technique is based on the use of hybrid number system representation, where one input operand is in two's complement number representation and the other in redundant number representation (Srinivas and Parhi, 1992). Using an efficient sign-select redundant-to-two's complement conversion technique, this multiplier can be made to operate faster and, in the pipelined mode, would require fewer pipelining latches and less silicon area.

Computer-Aided Design

With progress in the theory of architectures, the computer-aided design (CAD) systems for DSP application also become more powerful. In early 1980, the first silicon compiler system for signal processing was developed at the University of Edinburgh and was referred to as the FIRST design system. This system only addressed the CAD of bit-serial signal processing systems. Since then more powerful systems have been developed. The Cathedral I system from Katholieke Universiteit Leuven and the bit-serial silicon compiler (BSSC) from GE Research Center in Schenectady, New York, also addressed synthesis of bit-serial circuits. The Cathedral system has now gone through many revisions, and the new versions can systhesize parallel multiprocessor data paths and can perform more powerful scheduling and allocation. The Lager design tool at the University of California at Berkeley was developed to synthesize the DSP algorithms using parametrizable macro building blocks (such as ALU, RAM, ROM). This system has also gone through many revisions. The Hyper system also developed at the University of California at Berkeley and the MARS design system developed at the University of Minnesota at Minneapolis perform higher-level transformations and perform scheduling and allocation. These CAD tools are crucial to rapid prototyping of high-performance DSP integrated circuits.

Future VLSI DSP Systems

Future VLSI systems will make use of a combination of many types of architectures such as dedicated and programmable. These systems can be designed successfully with proper understanding of the algorithms, applications, theory of architectures, and with the use of advanced CAD systems.

Defining Terms

Bit serial: Processing of one bit per clock cycle. If word length is W, then one sample or word is processed in W clock cycles. In contrast, all W bits of a word are processed in the same clock cycle in a bit-parallel system.

Digit serial: Processing of more than one but not all bits in one clock cycle. If the digit size is W_1 and the word length is W, then the word is processed in W/W_1 clock cycles. If $W_1 = 1$, then the system is referred to as a bit-serial system and if $W_1 = W$, then the system is referred to as a bit-parallel system. In general, the digit size W_1 need not be a divisor of the word length W since the least and most significant bits of consecutive words can be overlapped and processed in the same clock cycle.

Folding: The technique of mapping many tasks to a single processor.

Look-ahead: The technique of computing a state $x(n)$ usng previous state $x(n - M)$ without requiring the intermediate states $x(n - 1)$ through $x(n - M + 1)$. This is referred to as an M-step look-ahead. In the case of higher-order computations, there are two forms of look-ahead: clustered look-ahead and scattered look-ahead. In clustered look-ahead, $x(n)$ is computed using the clustered states $x(n - M - N + 1)$ through $x(n - M)$ for an Nth-order computation. In scattered look-ahead, $x(n)$ is computed using the scattered states $x(n - iM)$ where i varies from 1 to N.

Parallel processing: Processing of multiple tasks independently by different processors. This also increases the throughput.

Pipelining: A technique to increase throughput. A long task is divided into components, and each component is distributed to one processor. A new task can begin even though the former tasks have not been completed. In the pipelined operation, different components of different tasks are executed at the same time by different processors. Pipelining leads to an increase in the system latency, i.e., the time elapsed between the starting of a task and the completion of the task.

Retiming: The technique of moving the delays around the system. Retiming does not alter the latency of the system.

Systolic: Flow of data in a rhythmic fashion from a memory through many processors, returning to the memory just as blood flows.

Unfolding: The technique of transforming a program that describes one iteration of an algorithm to another equivalent program that describes multiple iterations of the same algorithm.

Word parallel: Processing of multiple words in the same clock cycle.

References

A.P. Chandrakasan, S. Sheng, and R.W. Brodersen, "Low-power CMOS digital design," *IEEE J. Solid State Circuits*, vol. 27(4), pp. 473–484, April 1992.

S.Y. Kung, *VLSI Array Processors*, Englewood Cliffs, NJ: Prentice-Hall, 1988.

E.A. Lee and D.G. Messerschmitt, "Pipeline interleaved programmable DSP's," *IEEE Trans. Acoustics, Speech, Signal Processing*, vol. 35(9), pp. 1320–1345, September 1987.

C.E. Leiserson, F. Rose, and J. Saxe, "Optimizing synchronous circuitry by retiming," *Proc. 3rd Caltech Conf. VLSI*, Pasadena, CA, pp. 87–116, March 1983.

K.K. Parhi, "Algorithm transformation techniques for concurrent processors," *Proc. IEEE*, vol. 77(12), pp. 2079–2095, December 1989.

K.K. Parhi, "Systematic approach for design of digit-serial processing architectures," *IEEE Trans. Circuits Systems*, vol. 38(4), pp. 358–375, April 1991.

K.K. Parhi and D.G. Messerschmitt, "Pipeline interleaving and parallelism in recursive digital filters," *IEEE Trans. Acoustics, Speech, Signal Processing*, vol. 37(7), pp. 1099–1135, July 1989.

K.K. Parhi, C.Y. Wang, and A.P. Brown, "Synthesis of control circuits in folded pipelined DSP architectures," *IEEE J. Solid State Circuits*, vol. 27(1), pp. 29–43, January 1992.

N.R. Shanbhag, and K.K. Parhi, "A pipelined adaptive lattice filter architecture," *Proc. 1992 IEEE Int. Symp. Circuits and Systems*, San Diego, CA, May 1992.

H.R. Srinivas and K.K. Parhi, "High-speed VLSI arithmetic processor architectures using hybrid number representation," *J. VLSI Signal Processing*, vol. 4(2/3), pp. 177–198, 1992.

Further Information

A detailed video tutorial on "Implementation and Synthesis of VLSI Signal Processing Systems" presented by K.K. Parhi and J.M. Rabaey in March 1992 can be purchased from the customer service department of IEEE, 445 Hoes Lane, P.O. Box 1331, Piscataway, NJ 08855-1331.

Special architectures for video communications can be found in the book *VLSI Implementations for Image Communications*, published as the fourth volume of the series *Advances in Image Communications* (edited by Peter Pirsch) by Elsevier Science Publishing Co. in 1993. The informative article "Research on VLSI for Digital Video Systems in Japan," published by K.K. Parhi in the fourth volume of the *1991 Office of Naval Research Asian Office Scientific Information Bulletin* (pp. 93–98), provides examples of video codec designs using special architectures. For video programmable digital signal processor approaches, see I. Tamitani, H. Harasaki, and T. Nishitani, "A Real-Time HDTV Signal Processor: HD-VSP," published in *IEEE Transactions on Circuits and Systems for Video Technology*, March 1991, and T. Fujii, T. Sawabe, N. Ohta, and S. Ono, "Implementation of Super High-Definition Image Processing on HiPIPE," published in *1991 IEEE International Symposium on Circuits and Systems*, held in June 1991 in Singapore (pp. 348–351).

The *IEEE Design and Test of Computers* published three special issues related to computer-aided design of special architectures; these issues were published in October 1990 (addressing high-level synthesis), December 1990 (addressing silicon compilations), and June 1991 (addressing rapid prototyping).

Descriptions of various CAD systems can be found in the following references. The description of the FIRST system can be found in the article "A Silicon Compiler for VLSI Signal Processing," by P. Denyer et al. in the

Proceedings of the ESSCIRC conference held in Brussels in September 1982 (pp. 215–220). The Cathedral system has been described in R. Jain et al., "Custom Design of a VLSI PCM-FDM Transmultiplexor from System Specifications to Circuit Layout Using a Computer Aided Design System," published in *IEEE Journal of Solid State Circuits* in February 1986 (pp. 73–85). The Lager system has been described in "An Integrated Automatic Layout Generation System for DSP Circuits," by J. Rabaey, S. Pope, and R. Brodersen, published in the July 1985 issue of the *IEEE Transactions on Computer Aided Design* (pp. 285–296). A description of the MARS Design System can be found in C.-Y. Wang and K.K. Parhi, "High-Level DSP Synthesis Using MARS System," published in *Proceedings of the 1992 IEEE International Symposium on Circuits and Systems* in San Diego, May 1992. A tutorial article on high-level synthesis can be found in "The High-Level Synthesis of Digital Systems," by M.C. McFarland, A. Parker, and R. Composano, published in the February 1990 issue of the *Proceedings of the IEEE* (pp. 310–320).

Articles on pipelined multipliers can be found in T.G. Noll et al., "A Pipelined 330 MHZ Multiplier," *IEEE Journal of Solid State Circuits*, June 1986 (pp. 411–416) and in M. Hatamian and G. Cash, "A 70-MHz 8-Bit × 8-Bit-Parallel Pipelined Multiplier in 2.5 μm CMOS," *IEEE Journal of Solid State Circuits*, 1986.

Technical articles on special architectures and chips for signal and image processing appear in different places, including proceedings of conferences such as IEEE Workshop on VLSI Signal Processing; IEEE International Conference on Acoustics, Speech, and Signal Processing; IEEE International Symposium on Circuits and Systems; IEEE International Solid State Circuits Conference; IEEE Customs Integrated Circuits Conference; IEEE International Conference on Computer Design; ACM/IEEE Design Automation Conference; ACM/IEEE International Conference on Computer Aided Design; International Conference on Application Specific Array Processors; and journals such as *IEEE Transactions on Signal Processing, IEEE Transactions on Image Processing, IEEE Transactions on Circuits and Systems: Part II: Analog and Digital Signal Processing, IEEE Transactions on Computers, IEEE Journal of Solid State Circuits, IEEE Signal Processing Magazine, IEEE Design and Test Magazine,* and *Journal of VLSI Signal Processing.*

20.2   Signal Processing Chips and Applications

Rulph Chassaing and Bill Bitler

Recent advances in very large scale integration (VLSI) have contributed to the current digital signal processors. These processors are just special-purpose fast microprocessors characterized by architectures and instructions suitable for real-time digital signal processing (DSP) applications. The commercial DSP processor, a little more than a decade old, has emerged because of the ever-increasing number of signal processing applications. DSP processors are now being utilized in a number of applications from communications and controls to speech and image processing. They have found their way into talking toys and music synthesizers. A number of texts (such as Chassaing, 1992) and articles (such as Ahmed and Kline, 1991) have been written, discussing the applications that use DSP processors and the recent advances in DSP systems.

DSP Processors

Digital signal processors are currently available from a number of companies, including Texas Instruments, Inc. (Texas), Motorola, Inc. (Arizona), Analog Devices, Inc. (Massachusetts), AT&T (New Jersey), and NEC (California). These processors are categorized as either **fixed-point** or **floating-point processors.** Several companies are now supporting both types of processors. **Special-purpose digital signal processors,** designed for a specific signal processing application such as for fast Fourier transform (FFT), have also emerged. Currently available digital signal processors range from simple, low-cost processing units through high-performance units such as Texas Instruments' (TI) TMS320C40 (Chassaing and Martin, 1995) and TMS320C80, and Analog Devices[1] ADSP-21060 SHARC (Chassaing and Ayers, 1996).

One of the first-generation digital signal processors is the (N-MOS technology) TMS32010, introduced by Texas Instruments in 1982. This first-generation, fixed-point processor is based on the Harvard architecture, with a fast on-chip hardware multiplier/accumulator, and with data and instructions in separate memory spaces, allowing for concurrent accesses. This type of **pipelining feature** enables the processor to execute one instruction while fetching at the same time the next instruction. Other features include 144 (16-bit) words of on-chip data RAM and a 16-bit by 16-bit multiply operation in one instruction cycle time of 200 nsec. Since many instructions can be executed in one single cycle, the TMS32010 is capable of executing 5 million instructions per second (MIPS). Major drawbacks of this first-generation processor are its limited **on-chip memory** size and much slower execution time for accessing external memory. Improved versions of this first-generation processor are now available in C-MOS technology, with a faster instruction cycle time of 160 nsec.

The second-generation fixed-point processor TMS32020, introduced in 1985 by TI, was quickly followed by an improved C-MOS version TMS320C25 (Chassaing and Horning, 1990) in 1986. Features of the TMS320C25 include 544 (16-bit) words of on-chip data RAM, separate program and data memory spaces (each 64 K words), and an instruction cycle time of 100 nsec, thus enabling the TMS320C25 to execute 10 MIPS. A faster version, TI's fixed-point TMS320C50 processor, is available with an instruction cycle time of 35 ns.

The third-generation TMS320C30 (by TI) supports fixed- as well as floating-point operations (Chassaing, 1992). Features of this processor include 32-bit by 32-bit floating-point multiply operations in one instruction cycle time of 60 nsec. Since a number of instructions, such as load and store, multiply and add, can be performed in parallel (in one cycle time), the TMS320C30 can execute a pair of parallel instructions in 30 nsec, allowing for 33.3 MIPS. The Harvard-based architecture of the fixed-point processors was abandoned for one allowing four levels of pipelining with three subsequent instructions being consequently fetched, decoded, and read while the current instruction is being executed. The TMS320C30 has 2 K words of on-chip memory and a total of 16 million words of addressable memory spaces for program, data, and input/output. Specialized instructions are available to make common DSP algorithms such as filtering and spectral analysis execute fast and efficiently. The architecture of the TMS320C30 was designed to take advantage of higher-level languages such as C and ADA. The TMS320C31 and TMS320C32, recent members of the third-generation floating-point processors, are available with a 40-nsec instruction cycle time.

DSP starter kits (DSK) are inexpensive development systems available from TI and based on both the fixed-point TMS320C50 and the floating-point TMS320C31 processors. We will discuss both the fixed-point TMS320C25 and the floating-point TMS320C30 digital signal processors, including the development tools available for each of these processors and DSP applications.

Fixed-Point TMS320C25-Based Development System

TMS320C25-based development systems are now available from a number of companies such as Hyperception Inc., Texas, and Atlanta Signal Processors, Inc., Georgia. The Software Development System (SWDS), available from TI, includes a board containing the TMS320C25, which plugs into a slot on an IBM compatible PC. Within the SWDS environment, a program can be developed, assembled, and run. Debugging aids supported by the SWDS include single-stepping, setting of breakpoints, and display/modification of registers.

A typical workstation consists of:

1. An IBM compatible PC. Commercially available DSP packages (such as from Hyperception or Atlanta Signal Processors) include a number of utilities and filter design techniques.
2. The SWDS package, which includes an assembler, a linker, a debug monitor, and a **C compiler.**
3. Input/output alternatives such as TI's analog interface board (AIB) or analog interface chip (AIC).

The AIB includes a 12-bit analog-to-digital converter (ADC) and a 12-bit digital-to-analog converter (DAC). A maximum sampling rate of 40 kHz can be obtained. With (input) antialiasing and (output) reconstruction filters mounted on a header on the AIB, different input/output (I/O) filter bandwidths can be achieved. Instructions such as **IN** and **OUT** can be used for input/output accesses. The AIC, which provides an inexpensive I/O alternative, includes 14-bit ADC and DAC, antialiasing/reconstruction filters, all on a single C-MOS chip. Two inputs and one output are available on the AIC. (A TMS320C25/AIC interface diagram and communication routines can be found in Chassaing and Horning, 1990.)

The TLC32047 AIC is a recent member of the TLC32040 family of voiceband analog interface circuits, with a maximum sampling rate of 25 kHz.

Implementation of a Finite Impulse Response Filter with the TMS320C25

The convolution equation

$$
\begin{aligned}
y(n) = \sum_{k=0}^{N-1} h(k)x(n-k) \\
= h(0)x(n) + h(1)x(n-1) + \cdots + h(N-2)x(n-(N-2)) \\
+ h(N-1)x(n-(N-1))
\end{aligned}
\tag{20.1}
$$

represents a finite impulse response (FIR) filter with length N. The memory organization for the coefficients $h(k)$ and the input samples $x(n-k)$ is shown in Table 20.1. The coefficients are placed within a specified internal program memory space and the input samples within a specified data memory space. The program counter (PC) initially points at the memory location that contains the last coefficient $h(N-1)$, for example, at memory address FF00 h (in hex). One of the (eight) auxiliary registers points at the memory address of the last or least recent input sample. The most recent sample is represented by $x(n)$. The following program segment implements Equation (20.1):

> LARP   AR1
> RPTK   N − 1
> MACD   FF00h, *-
> APAC

The first instruction selects auxiliary register AR1, which will be used for indirect addressing. The second instruction RPTK causes the subsequent MACD instruction to execute N times (repeated $N-1$ times). The MACD instruction has the following functions:

1. Multiplies the coefficient value $h(N-1)$ by the input sample value $x(n-(N-1))$
2. Accumulates any previous product stored in a special register (TR)
3. Copies the data memory sample value into the location of the next-higher memory; this "data move" is to model the input sample delays associated with the next unit of time $n+1$

The last instruction APAC accumulates the last multiply operation $h(0)x(n)$.

TABLE 20.1 TMS320C25 Memory Organization for Convolution

Coefficients	Input Samples		
	Time n	Time $n+1$	Time $n+2$
PC → $h(N-1)$	$x(n)$	$x(n+1)$	$x(n+2)$
$h(N-2)$	$x(n-1)$	$x(n)$	$x(n+1)$
.	.	.	.
.	.	.	.
$h(2)$	$x(n-(N-3))$	$x(n-(N-4))$	$x(n-(N-5))$
$h(1)$	$x(n-(N-2))$	$x(n-(N-3))$	$x(n-(N-4))$
$h(0)$	AR1 → $x(n-(N-1))$	$x(n-(N-2))$	$x(n-(N-3))$

At time $n + 1$, the convolution Equation (20.1) becomes

$$y(n + 1) = h(0)x(n + 1) + h(1)x(n) + \ldots$$
$$+ h(N - 2)x(n - (N - 3)) + h(N - 1)x(n - (N - 2)) \qquad (20.2)$$

The previous program segment can be placed within a loop, with the PC and the auxiliary register AR1 reinitialized (see the memory organization of the samples $x(k)$ associated with time $n + 1$ in Table 20.1). Note that the last multiply operation is $h(0)x(\cdot)$, where $x(\cdot)$ represents the newest sample. This process can be continuously repeated for time $n + 2$, $n + 3$, and so on.

The characteristics of a frequency-selective FIR filter are specified by a set of coefficients that can be readily obtained using commercially available filter design packages. These coefficients can be placed within a generic FIR program. Within 5 to 10 minutes, an FIR filter can be implemented in real-time. This includes finding the coefficients; assembling, linking, and downloading the FIR program into the SWDS; and observing the desired frequency response displayed on a spectrum analyzer. A different FIR filter can be quickly obtained since the only necessary change in the generic program is to substitute a new set of coefficients.

The approach for modeling the sample delays involves moving the data. A different scheme is used with the floating-point TMS320C30 processor with a circular mode of addressing.

Floating-Point TMS320C30-Based Development System

TMS320C30-based DSP development systems are also currently available from a number of companies. The following are available from Texas Instruments:

1. *An evaluation module (EVM).* The EVM is a powerful, yet relatively inexpensive 8-bit card that plugs into a slot on an IBM AT compatible. It includes the third-generation TMS320C30, 16 K of user RAM, and an AIC for I/O. A serial port connector available on the EVM can be used to interface the TMS320C30 to other input/output devices (the TMS320C30 has two serial ports). An additional AIC can be interfaced to the TMS320C30 through this serial port connector. A very powerful, yet inexpensive analog evaluation fixture, available from Burr-Brown (Arizona), can also be readily interfaced to the serial port on the EVM. This complete two-channel analog evaluation fixture includes an 18-bit DSP102 ADC, an 18-bit DSP202 DAC, antialiasing, and reconstruction filters. The ADC has a maximum sampling rate of 200 kHz.
2. *An XDS1000 emulator*—powerful but quite expensive. A module can be readily built as a target system to interface to the XDS1000 (Chassaing, 1992). This module contains the TMS320C30 and 16 K of static RAM. Two connectors are included on this module, for interfacing to either an AIC module or to a second-generation analog interface board (AIB). The AIC was discussed in conjunction with the TMS320C25. The AIB includes Burr-Brown's 16-bit ADC and DAC with a maximum sampling rate of 58 kHz. An AIC is also included on this newer AIB version.

EVM Tools

The EVM package includes an assembler, a linker, a simulator, a C compiler, and a C source debugger. The second-generation TMS320C25 fixed-point processor is supported by C with some degree of success. The architecture and instruction set of the third-generation TMS320C30 processor facilitate the development of high-level language compilers. An optimizer option is available with the C compiler for the TMS320C30. A C-code program can be readily compiled, assembled, linked, and downloaded into either a simulator or the EVM for real-time processing. A runtime support library of C functions, included with the EVM package, can be used during linking. During simulation, the input data can be retrieved from a file and the output data written into a file. Input and output port addresses can be appropriately specified. Within a real-time processing environment with the EVM, the C source debugger can be used. One can single-step through a C-code program while observing the equivalent step(s) through the assembly code. Both the C code and the corresponding assembly code can be viewed through the EVM windows. One can also monitor at the same time the contents of registers, memory locations, and so on.

Implementation of a Finite Impulse Response Filter with the TMS320C30

Consider again the convolution equation, Equation (20.1), which represents an FIR filter. Table 20.2 shows the TMS320C30 memory organization used for the coefficients and the input samples. Initially, all the input samples can be set to zero. The newest sample $x(n)$, at time n, can be retrieved from an ADC using the following instructions:

$$\begin{array}{ll} \text{FLOAT} & *AR3, R3 \\ \text{STF} & R3, *AR1{+}{+}\% \end{array}$$

These two instructions cause an input value $x(n)$, retrieved from an input port address specified by auxiliary register AR3, to be loaded into a register R3 (one of eight 40-bit-wide extended precision registers), then stored in a memory location pointed by AR1 (AR1 would be first initialized to point at the "bottom" or higher-memory address of the table for the input samples). AR1 is then postincremented in a circular fashion, designated with the modulo operator %, to point at the oldest sample $x(n-(N-1))$, as shown in Table 20.2. The size of the circular buffer must first be specified. The following program segment implements Equation (20.1):

$$\begin{array}{lll} & \text{RPTS} & \text{LENGTH} - 1 \\ & \text{MPYF} & *AR0{+}{+}\%, *AR1{+}{+}\%, R0 \\ \| & \text{ADDF} & R0, R2, R2 \\ & \text{ADDF} & R0, R2 \end{array}$$

The repeat "single" instruction RPTS causes the next (multiply) floating-point instruction MPYF to be executed LENGTH times (repeated LENGTH − 1), where LENGTH is the length of the FIR filter. Furthermore, since the first ADDF addition instruction is in parallel (designated by ‖) with the MPYF instruction, it is also executed LENGTH times. From Table 20.2, AR0, one of the eight available auxiliary registers, initially points at the memory address (a table address) which contains the coefficient $h(N-1)$, and a second auxiliary register AR1 now points to the address of the oldest input sample $x(n-(N-1))$. The second indirect addressing mode instruction multiplies the content in memory (address pointed by AR0) $h(N-1)$ by the content in memory (address pointed by AR1) $x(n-N-1)$), with the result stored in R0. Concurrently (in parallel), the content of R0 is added to the content of R2, with the result stored in R2. Initially, R0 and R2 are set to zero; hence, the resulting value in R2 is *not* the product of the first multiply operation. After the first multiply operation, both AR0 and AR1 are incremented, and $h(N-2)$ is multiplied by $x(n-(N-2))$. Concurrently, the result of the first multiply operation (stored in R0) is accumulated into R2. The second addition instruction, executed only once, accumulates the last product $h(0)x(n)$ (similar to the APAC

TABLE 20.2   TMS320C30 Memory Organization for Convolution

Coefficients	Time n	Time $n + 1$	Time $n + 2$
AR0 → $h(N-1)$	AR1 → $x(n-(N-1))$	$x(n+1)$	$x(n+1)$
$h(N-2)$	$x(n-(N-2))$	AR1 → $x(n-(N-2))$	$x(n+2)$
$h(N-3)$	$x(n-(N-3))$	$x(n-(N-3))$	AR1 → $x(n-(N-3))$
.	.	.	.
.	.	.	.
.	.	.	.
$h(1)$	$x(n-1)$	$x(n-1)$	$x(n-1)$
$h(0)$	$x(n)$	$x(n)$	$x(n)$

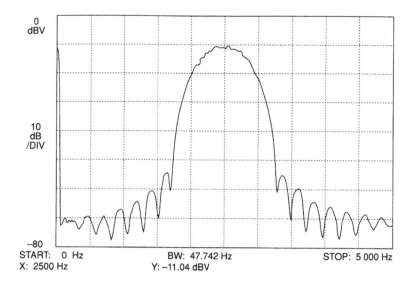

FIGURE 20.8   Frequency response of 41-coefficient FIR filter.

instruction associated with the fixed-point TMS320C25). The overall result yields an output value $y(n)$ at time n. After the last multiply operation, both AR0 and AR1 are post-incremented to point at the "top" or lower-memory address of each circular buffer. The process can then be repeated for time $n + 1$ in order to obtain a second output value $y(n + 1)$. Note that the newest sample $x(n + 1)$ would be retrieved from an ADC using the FLOAT and STF instructions, then placed at the top memory location of the buffer (table) containing the samples, overwriting the initial value $x(n - (N - 1))$. AR1 is then incremented to point at the address containing $x(n - (N - 2))$, and the previous four instructions can be repeated. The last multiply operation involves $h(0)$ and $x(\cdot)$, where $x(\cdot)$ is the newest sample $x(n + 1)$, at time $n + 1$. The foregoing procedure would be repeated to produce an output $y(n + 2)$, $y(n + 3)$, and so on. Each output value would be converted to a fixed-point equivalent value before being sent to a DAC. The frequency response of an FIR filter with 41 coefficients and a center frequency of 2.5 kHz, obtained from a signal analyzer, is displayed in Figure 20.8.

FIR and IIR Implementation Using C and Assembly Code

A real-time implementation of a 45-coefficient bandpass FIR filter and a sixth-order IIR filter with 345 samples, using C code and TMS320C30 code, is discussed in Chassaing and Bitler (1991). Table 20.3 and Table 20.4 show a comparison of execution times of these two filters. The C language FIR filter, implemented without the modulo operator % and compiled with a C compiler V4.1, executed two times slower[1] than an equivalent assembly language filter (which has a similar execution time as one implemented with a filter routine in assembly, called by a C program). The C language IIR filter ran 1.3 times slower than the corresponding assembly language IIR filter. These slower execution times may be acceptable for many applications. Where execution speed is crucial, a time-critical function may be written in assembly and called from a C program. In applications where speed is not absolutely crucial, C provides a better environment because of its portability and maintainability.

Real-Time Applications

A number of applications are discussed in Chassaing and Horning (1990) using TMS320C25 code and in Chassaing (1992) using TMS320C30 and C code. These applications include multi-rate and adaptive filtering,

---

[1] 1.5 times slower using a newer C compiler V4.4.

TABLE 20.3 Execution Time and Program Size of
FIR Filter

FIR (45 samples)	Execution Time (msec)	Size (words)
C with modulo	4.16	122
C without modulo	0.338	116
C-called assembly	0.1666	74
Assembly	0.1652	27

TABLE 20.4 Execution Time and Program Size
of Sixth Order IIR Filter

IIR (345 samples)	Execution Time (msec)	Size (words)
C	1.575	109
Assembly	1.18	29

modulation techniques, and graphic and parametric equalizers. Two applications are briefly discussed here: a ten-band multi-rate filter and a video line rate analysis.

1. The functional block diagram of the multi-rate filter is shown in Figure 20.9. The multi-rate design provides a significant reduction in processing time and data storage, compared to an equivalent single-rate design. With multi-rate filtering, we can use a decimation operation to obtain a sample rate reduction or an interpolation operation (as shown in Figure 20.9) in order to obtain a sample rate increase (Crochiere and Rabiner, 1983). A pseudorandom noise generator implemented in software provides the input noise to the ten octave band filters. Each octave band filter consists of three 1/3-octave filters (each with 41 coefficients), which can be individually controlled. A controlled noise source can be obtained with this design. Since each 1/3-octave band filter can be turned *on* or *off*, the noise spectrum can be shaped accordingly. The interpolation filter is a low-pass FIR filter with a 2:1 data-rate increase, yielding two sample outputs for each input sample. The sample rate of the highest octave-band filter is set at 32,768 samples per second, with each successively lower band processing at half the rate of the next-higher band. The multi-rate filter (a nine-band version) was implemented with the TMS320C25 (Chassaing et al., 1990). Figure 20.10 shows the three 1/3-octave band filters of band 10 implemented with the EVM in conjunction with the two-channel analog fixture (made by Burr-Brown). The center frequency of the middle 1/3-octave band 10 filter is at approximately 8 kHz since the coefficients were designed for a center frequency of 1/4 the sampling rate (the middle 1/3-octave band 9 filter would be centered at 4 kHz, the middle 1/3-octave band 8 filter at 2 kHz, and so on). Note that the center frequency of the middle 1/3-octave band 1 filter would be at 2 Hz if the highest sampling rate is set at 4 kHz. Observe from Figure 20.10 that the crossover frequencies occur at the 3-dB points. Since the main processing time of the multi-rate filter

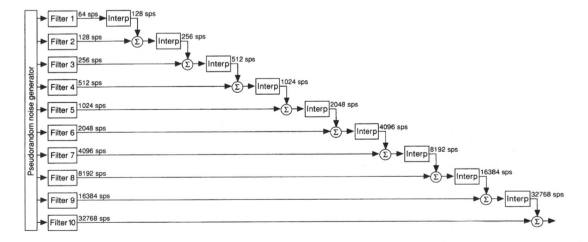

FIGURE 20.9 Multi-rate filter functional block diagram.

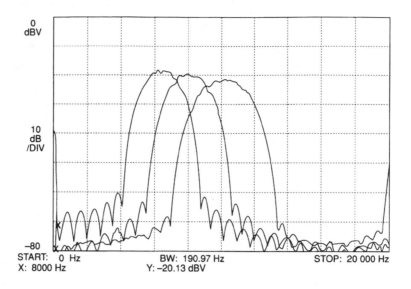

FIGURE 20.10   Frequency responses of the 1/3-octave band 10 filters.

(implemented in assembly code) was measured to be 8.8 msec, the maximum sampling rate was limited to 58 kspsec.

2. A video line rate analysis implemented entirely in C code is discussed in Chassaing and Bitler (1992). A module was built to sample a video line of information. This module included a 9.8-MHz clock, a high sampling rate 8-bit ADC and appropriate support circuitry (comparator, FIFO buffer, etc.). Interactive features allowed for the selection of one (out of 256) horizontal lines of information and the execution of algorithms for digital filtering, averaging, and edge enhancement, with the resulting effects displayed on the PC screen. Figure 20.11 shows the display of a horizontal line (line #125) of information obtained from a test chart with a charge-coupled device (CCD) camera. The function key **F3** selects the 1-MHz low-pass filter resulting in the display shown in Figure 20.12. The 3-MHz filter (with **F4**) would pass more of the higher-frequency components of the signal but with less noise reduction. **F5** implements the noise averaging algorithm. The effect of the edge enhancement algorithm (with **F7**) is displayed in Figure 20.13.

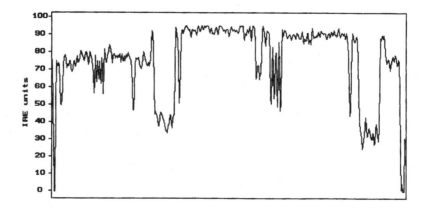

FIGURE 20.11   Display of a horizontal line of video signal.

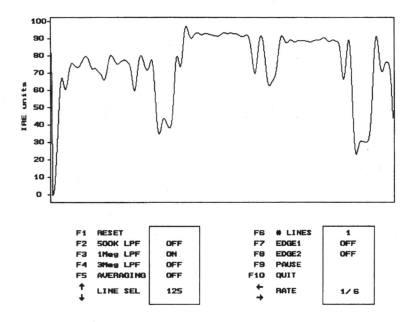

FIGURE 20.12 Video line signal with 1-MHz filtering.

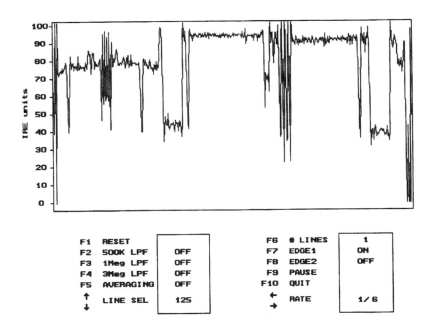

FIGURE 20.13 Video line signal with edge enhancement.

Conclusions and Future Directions

Digital signal processors have been used extensively in a number of applications, even in non-DSP applications such as graphics. The fourth-generation floating-point TMS320C40, code compatible with the TMS320C30, features an instruction cycle time of 40 nsec and six serial ports. The fifth-generation fixed-point TMS320C50, code compatible with the first two generations of fixed-point processors, features an instruction cycle time of 35 nsec and 10 K words (16-bit) of on-chip data and program memory. Currently, both the fixed-point and floating-point processors are being supported by TI.

Defining Terms

C compiler: Program that translates C code into assembly code.

Digital signal processor: Special-purpose microprocessor with an architecture suitable for fast execution of signal processing algorithms.

Fixed-point processor: A processor capable of operating on scaled integer and fractional data values.

Floating-point processor: Processor capable of operating on integers as well as on fractional data values without scaling.

On-chip memory: Internal memory available on the digital signal processor.

Pipelining feature: Feature that permits parallel operations of fetching, decoding, reading, and executing.

Special-purpose digital signal processor: Digital signal processor with a special feature for handling a specific signal processing application, such as FFT.

References

H.M. Ahmed and R.B. Kline, "Recent advances in DSP systems," *IEEE Communications Magazine,* 1991.

R. Chassaing, *Digital Signal Processing with C and the TMS320C30,* New York: Wiley, 1992.

R. Chassaing and R. Ayers, "Digital signal processing with the SHARC," in *Proceedings of the 1996 ASEE Annual Conference,* 1996.

R. Chassaing and B. Bitler, "Real-time digital filters in C," in *Proceedings of the 1991 ASEE Annual Conference,* 1991.

R. Chassaing and B. Bitler, "A video line rate analysis using the TMS320C30 floating-point digital signal processor," in *Proceedings of the 1992 ASEE Annual Conference,* 1992.

R. Chassaing and D.W. Horning, *Digital Signal Processing with the TMS320C25,* New York: Wiley, 1990.

R. Chassaing and P. Martin, "Parallel processing with the TMS320C40," in *Proceedings of the 1995 ASEE Annual Conference,* 1995.

R. Chassaing, W.A. Peterson, and D.W. Horning, "A TMS320C25-based multirate filter," *IEEE Micro,* 1990.

R.E. Crochiere and L.R. Rabiner, *Multirate Digital Signal Processing,* Englewood Cliffs, NJ: Prentice-Hall, 1983.

K.S. Lin (Ed.), *Digital Signal Processing Applications with the TMS320 Family. Theory, Algorithms, and Implementations,* vol. 1, Texas Instruments Inc., TX, 1989.

A.V. Oppenheim and R. Schafer, *Discrete-Time Signal Processing,* Englewood Cliffs, NJ: Prentice-Hall, 1989.

P. Papamichalis (Ed.), *Digital Signal Processing Applications with the TMS320 Family. Theory, Algorithms, and Implementations,* vol. 3, Texas Instruments, Inc., TX, 1990.

21

Acoustic Signal Processing

Juergen Schroeter
AT&T Labs

Gary W. Elko
Agere Systems

M. Mohan Sondhi
AT&T Labs

Vyacheslav Tuzlukov
Ajou University

Won-Sik Yoon
Ajou University

Yong Deak Kim
Ajou University

21.1 Digital Signal Processing in Audio and Electroacoustics

Juergen Schroeter, Gary W. Elko, and M. Mohan Sondhi

Introduction

In this section we will focus on some of the algorithms and technologies in digital signal processing (DSP) that are used in audio and electroacoustics (A&E). Because A&E embraces a wide range of topics, it is impossible for us to go here into any depth in any one of them. Instead, this section will try to give a compressed overview of the topics the authors judge to be most important.

In the following, we will look into steerable microphone arrays, digital hearing aids, spatial processing, audio coding, echo cancelation, and active noise and sound control. We will not cover basic techniques in digital recording [1] and computer music [2].

Steerable Microphone Arrays

Steerable microphone arrays have controllable directional characteristics. One important application is in teleconferencing. Here, sound pickup can be highly degraded by reverberation and room noise. One solution to this problem is to utilize highly directional microphones. Instead of pointing such a microphone manually to a desired talker, steerable microphone arrays can be used for reliable automatic tracking of speakers as they move around in a noisy room or auditorium, if combined with a suitable speech detection algorithm.

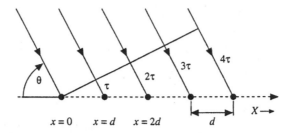

FIGURE 21.1   A linear array of N microphones (here, $N = 5$; $\tau = d/c \cos \theta$).

Figure 21.1 depicts the simplest kind of steerable array using N microphones that are uniformly spaced with distance d along the linear x-axis. It can be shown that the response of this system to a plane wave impinging at an angle θ is

$$H(j\omega) = \sum_{n-0}^{N-1} a_n e^{-j(\omega/c)nd \cos \theta} \tag{21.1}$$

Here, $j = \sqrt{-1}$, ω is the radian frequency, and c is the speed of sound. Equation (21.1) is a spatial filter with coefficients a_n and the delay operator $z^{-1} = \exp(-jd\omega/c \cos \theta)$. Therefore, we can apply finite impulse response (FIR) filter theory. For example, we could taper the weights a_n to suppress sidelobes of the array. We also have to guard against spatial aliasing, that is, grating lobes that make the directional characteristic of the array ambiguous. The array is steered to an angle θ_0 by introducing appropriate delays into the N microphone lines. In Equation (21.1), we can incorporate these delays by letting

$$a_n = e^{-j\omega\tau_0} e^{+j(\omega/c)nd \cos \theta_0} \tag{21.2}$$

Here, τ_0 is an overall delay equal to or larger than $Nd/c \cos \theta_0$ that ensures causality, while the second term in Equation (21.2) cancels the corresponding term in Equation (21.1) at $\theta = \theta_0$. Because of the axial symmetry of the one-dimensional (linear, 1D) array, the directivity of the array is a figure of revolution around the x-axis. Therefore, in case we want the array to point to a single direction in space, we need a 2D array.

Since most of the energy of typical room noise and the highest level of reverberation in a room is at low frequencies, one would like to use arrays that have their highest directivity (i.e., narrowest beamwidth) at low frequencies. Unfortunately, this need collides with the physics of arrays: the smaller the array relative to the wavelength, the wider the beam. (Again, the corresponding notion in filter theory is that systems with shorter impulse responses have wider bandwidth.) One solution to this problem is to superimpose different-size arrays and filter each output by the appropriate bandpass filter, similar to a crossover network used in two- or three-way loudspeaker designs. Such a superposition of three five-element arrays is shown in Figure 21.2. Note that we only need nine microphones in this example, instead of $5 \times 3 = 15$.

Another interesting application is the use of an array to mitigate discrete noise sources in a room. For this, we need to attach an FIR filter to each of the microphone signal outputs. For any given frequency, one can show that N microphones can produce $N-1$ nulls in the directional characteristic of the array. Similarly,

FIGURE 21.2   Three superimposed linear arrays depicted by large, midsize, and small circles. The largest array covers the low frequencies, the midsize array covers the midrange frequencies, and the smallest covers the high frequencies.

attaching an M-point FIR filter to each of the microphones, we can get these zeros at $M-1$ frequencies. The weights for these filters have to be adapted, usually under the constraint that the transfer function (frequency characteristic) of the array for the desired source is optimally flat. In practical tests, systems of this kind work nicely in (almost) anechoic environments. Their performance degrades, however, with increasing reverberation.

More information on microphone arrays can be found in Ref. [3]; in particular, Flanagan et al. describe there how to make arrays adapt to changing talker positions in a room by constantly scanning the room with a moving search beam and by switching the main beam accordingly. Interesting research issues are, among others, 3D arrays and how to take advantage of low-order wall reflections.

Digital Hearing Aids

Commonly used hearing aids attempt to compensate for both conductive (mechanical response) and sensorineural (cochlear) hearing loss by delivering an amplified acoustic signal to the external ear canal. As will be pointed out below, the most important problem is how to find the best aid for a given patient.

Historically, technology has been the limiting factor in hearing aids. Early on, carbon microphone-based hearing aids provided limited gain and a narrow, peaky frequency response. Nowadays, hearing aid transducers have broader bandwidth, a flatter frequency response, and higher sound output power. Consequently, more people can benefit from the improved technology. With the recent wide adoption of digital technology, there is renewed promise that even more people will be able to benefit from using hearing aids. Current digital designs are just now starting to add features that are offering the hearing impaired significant advantages over earlier digitally controlled analog and analog hearing aids. Analog hearing aids contain only (low-power) pre-amp, filter(s), (optional) automatic gain control (AGC) or compressor, power amplifier, and output limiter. Digitally controlled aids have certain additional components: one kind adds a digital controller to monitor and adjust the analog components of the aid. Another kind contains switched-capacitor circuits that represent sampled signals in analog form, in effect allowing simple discrete-time processing (e.g., filtering). Aids with switched-capacitor circuits have a lower power consumption compared to digital aids. Digital aid products are growing rapidly and now account for more that one-half of the products produced by major hearing aid vendors. Digital aids contain A/D and D/A converters and at least one programmable digital signal processing (DSP) chip, allowing for the use of sophisticated DSP algorithms, (small) two- or three-element differential microphone arrays for beamforming (see above), speech enhancement and noise suppression, feedback suppression, etc. Clinical results have shown that the addition of directional pickup in hearing aids has been a significant improvement in hearing aid performance as rated by users. For noise suppression and speech enhancement, however, experts disagree as to the usefulness of these techniques. To date, the most successful approach for signal enhancement after initial linear beamforming seems to be to ensure that all parts of the signal get amplified so that they are clearly audible but not too loud and to "let the brain sort out signal and noise."

Hearing aids pose a tremendous challenge for the DSP engineer, as well as for the audiologist and acoustician. Because of the continuing progress in chip technology, the physical size of a digital aid is no longer a serious problem; however, power consumption will still be a design parameter for quite some time. Besides the obvious necessity of avoiding howling (acoustic feedback), for example, by employing sophisticated models of the electroacoustic transducers, acoustic leaks, and ear canal to control the aid accordingly, there is a much more fundamental problem: since DSP allows complex schemes of splitting, filtering, compressing, and (re)combining the signal, hearing aid performance is no longer limited by bottlenecks in technology. It is still limited, however, by the lack of basic knowledge about how to map an arbitrary input signal (i.e., speech from a desired speaker) onto the reduced capabilities of the auditory system of the targeted wearer of the aid. Hence, the *selection and fitting of an appropriate aid* becomes the most important issue. This serious problem is illustrated in Figure 21.3.

It is important to note that for speech presented at a constant level, a linear (no compression) hearing aid can be tuned to do as well as a hearing aid with compression. However, if parameters like signal and

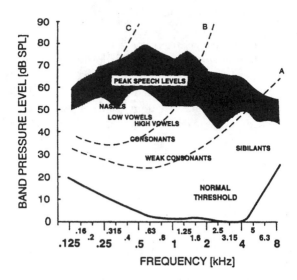

FIGURE 21.3 Peak third-octave band levels of normal to loud speech (hatched) and typical levels/dominant frequencies of speech sound (identifiers). Both can be compared to the third-octave threshold of normal-hearing people (solid line), thresholds for a mildly hearing-impaired person A, for a severely hearing-impaired person B, and for a profoundly hearing-impaired person C. For example, for person A, sibilants and some weak consonants in a normal conversation cannot be perceived. (*Source:* H. Levitt, "Speech discrimination ability in the hearing impaired: spectrum considerations," in *The Vanderbilt Hearing-Aid Report: State of the Art-Research Needs*, G.A. Studebaker and F.H. Bess (Eds.), Monographs in Contemporary Audiology, Upper Darby, PA, 1982, p. 34. With permission.)

background noise levels change dynamically, compression aids, in particular those with two bands or more, should have an advantage.

While a patient usually has no problem telling whether setting A or B is "clearer," adjusting more than just two to three (usually interdependent) parameters is very time consuming. For a multiparameter aid, an efficient fitting procedure that maximizes a certain objective is needed. Possible objectives are, for example, *intelligibility maximization* or *loudness restoration*. The latter objective is assumed in the following.

It is known that an impaired ear has a reduced dynamic range. Therefore, the procedure for fitting a patient with a hearing aid could estimate the so-called loudness-growth function (LGF) that relates the sound pressure level of a specific (band-limited) sound to its loudness. An efficient way of measuring the LGF is described by Allen et al. [4]. Once the LGF of an impaired ear is known, a multiband hearing aid can implement the necessary compression for each band [5]. Note, however, that this assumes that interactions between the bands can be neglected (problem of summation of partial loudnesses). This might not be valid for aids with a large number of bands. Other open questions include the choice of widths and filter shape of the bands, and optimization of dynamic aspects of the compression (e.g., time constants). For aids with just two bands, the crossover frequency is a crucial parameter that is difficult to optimize.

Spatial Processing

In spatial processing, audio signals are modified to give them new spatial attributes such as for example, the perception of having been recorded in a specific concert hall. The auditory system—using only the two ears as inputs—is capable of perceiving the direction and distance of a sound source with a high degree of accuracy by exploiting *binaural* and *monaural* spectral cues. Wave propagation in the ear canal is essentially one-dimensional. Hence, the 3D spatial information is coded by sound diffraction into spectral information before the sound enters the ear canal. The sound diffraction is caused by the head/torso (in the order of 20-dB and 600-μsec *interaural* level difference and delay, respectively) and at the two pinnae (auriculae) [6]. Binaural

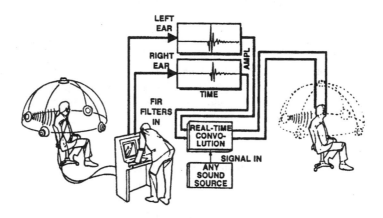

FIGURE 21.4 Measuring and using transfer functions of the external ear for binaural mixing (FIR = finite impulse response). (*Source:* E.M. Wenzel, Localization in virtual acoustic displays, *Presence*, vol. 1, p. 91, 1992. With permission.)

techniques like the one discussed below can be used for evaluating room and concert-hall acoustics (optionally in reduced-scale model rooms using a miniature dummy head), for noise assessment (e.g., in cars), and for "Kunstkopfstereophonie" (dummy-head stereophony). In addition, there are techniques for *loudspeaker reproduction* (like "Q-Sound") that try to extend the range in horizontal angle of traditional stereo speakers by using interaural cross cancelation. Largely an open question is how to reproduce spatial information for large audiences, for example, in movie theaters.

Figure 21.4 illustrates the technique for filtering a single-channel source using measured head-related transfer functions, in effect creating a virtual sound source in a given direction of the listener's auditory space (assuming plane waves, i.e., infinite source distance). On the left in this figure, the measurement of head-related transfer functions is shown. Focusing on the *left* ear for a moment (subscript l), we need to estimate the so-called free-field transfer function (subscript ff) for given angles of incidence in the horizontal plane (azimuth ϕ) and vertical plane (elevation δ):

$$H_{\mathrm{ff},1}(j\omega, \varphi, \delta) = P_{\mathrm{probe},1}(j\omega, \varphi, \delta)/P_{\mathrm{ref}}(j\omega) \qquad (21.3)$$

where $P_{\mathrm{probe},1}$ is the Fourier transform of the sound pressure measured in the subject's left ear, and P_{ref} is the Fourier transform of the pressure measured at a suitable reference point in the free field without the subject being present (e.g., at the midpoint between the two ears). (Note that this is independent of the direction of sound incidence since we assume an anechoic environment.) The middle of Figure 21.4 depicts the convolution of any "dry" (e.g., mono, low reverberation) source with the stored $H_{\mathrm{ff},1}(j\omega, \varphi, \delta)$'s and corresponding $H_{\mathrm{ff},r}(j\omega, \varphi, \delta)$'s. On the right side in the figure, the resulting binaural signals are reproduced via equalized headphones. The equalization ensures that a sound source with a flat spectrum (e.g., white noise) does not suffer any perceivable coloration for any direction (ϕ, δ).

Implemented in a real-time "binaural mixing console," the above scheme can be used to create "virtual" sound sources. When combined with an appropriate scheme for interpolating head-related transfer functions, moving sound sources can be mimicked realistically. Furthermore, it is possible to superimpose early reflections of a hypothetical recording room, each filtered by the appropriate head-related transfer function. Such inclusion of a room in the simulation makes the spatial reproduction more robust against individual differences between "recording" and "listening" ears, in particular if the listener's head movements are fed back to the binaural mixing console. (Head movements are useful for disambiguating spatial cues.) Finally, such a system can be used to create "virtual acoustic displays," for example, for pilots and astronauts [7]. Other research issues are, for example, the required accuracy of the head-related transfer functions, intersubject variability, and psychoacoustic aspects of room simulations.

Audio Coding

Audio coding is concerned with compressing (reducing the bit rate) of audio signals. The uncompressed digital audio of compact disks (CDs) is recorded at a rate of 705.6 kbit/sec for each of the two channels of a stereo signal (i.e., 16 bit/sample, 44.1-kHz sampling rate; 1411.2 kbit/sec total). This is too high a bit rate for digital audio broadcasting (DAB) or for transmission via end-to-end digital telephone connections (integrated services digital network, ISDN). Current audio coding algorithms provide at least "better than FM" quality at a combined rate of 128 kbit/sec for the two stereo channels (two ISDN B channels!), "transparent coding" at rates of 96 to 128 kbit/sec per mono channel, and "studio quality" at rates between 128 and 196 kbit/sec per mono channel. (While a large number of people will be able to detect distortions in the first class of coders, even so-called "golden ears" should not be able to detect any differences between original and coded versions of known "critical" test signals; the highest quality category adds a safety margin for editing, filtering, and/or recoding.)

To compress audio signals by a factor as large as 11 while maintaining a high quality requires sophisticated algorithms for reducing the *irrelevance and redundancy* in a given signal. A large portion (but usually less than 50%) of the bit-rate reduction in an audio coder is due to the first of the two mechanisms. Eliminating irrelevant portions of an input signal is done with the help of psychoacoustic models. It is obvious that a coder can eliminate portions of the input signal that, when played back, will be below the threshold of hearing. More complicated is the case when we have multiple signal components that tend to cover each other, that is, when weaker components cannot be heard due to the presence of stronger components. This effect is called *masking*. To let a coder take advantage of masking effects, we need to use good masking models. Masking can be modeled in the time domain where we distinguish so-called simultaneous masking (masker and maskee occur at the same time), forward masking (masker occurs *before* maskee), and backward masking (masker occurs *after* maskee). Simultaneous masking usually is modeled in the frequency domain. This latter case is illustrated in Figure 21.5.

Audio coders that employ common frequency-domain models of masking start out by splitting and subsampling the input signal into different frequency bands (using filterbanks such as subband filterbanks or time-frequency transforms). Then, the masking threshold (i.e., predicted masked threshold) is determined, followed by quantization of the spectral information and (optional) noiseless compression using variable-length coding. The encoding process is completed by multiplexing the spectral information with side information, adding error protection, etc.

The first stage, the filter bank, has the following requirements. First, decomposing and then simply reconstructing the signal should not lead to distortions ("perfect reconstruction filterbank"). This results in the advantage that all distortions are due to the quantization of the spectral data. Since each quantizer works on band-limited data, the distortion (also band-limited due to refiltering) is controllable by using the masking

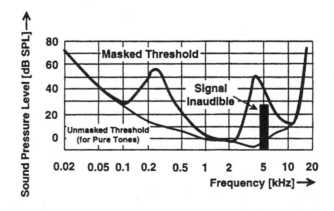

FIGURE 21.5 Masked threshold in the frequency domain for a hypothetical input signal. In the vicinity of high-level spectral components, signal components below the current masked threshold cannot be heard.

models described above. Second, the bandwidths of the filters should be narrow to provide sufficient coding gain. However, the length of the impulse responses of the filters should be short enough (time resolution of the coder!) to avoid so-called pre-echoes, that is, backward spreading of distortion components that result from sudden onsets (e.g., castanets). These two contradictory requirements, obviously, have to be worked out by a compromise. *Critical band* filters have the shortest impulse responses needed for coding of transient signals. On the other hand, the optimum frequency resolution (i.e., the one resulting in the highest coding gain) for a typical signal can be achieved by using, for example, a 2048-point modified discrete cosine transform (MDCT).

In the second stage, the (time-varying) masking threshold as determined by the psychoacoustic model usually controls an iterative analysis-by-synthesis quantization and coding loop. It can incorporate rules for masking of tones by noise and of noise by tones, though little is known in the psychoacoustic literature for more general signals. Quantizer step sizes can be set and bits can be allocated according to the known spectral estimate, by block companding with transmission of the scale factors as side information or iteratively in a variable-length coding loop (Huffman coding). In the latter case, one can low-pass filter the signal if the total required bit rate is too high.

The decoder has to invert the processing steps of the encoder, that is, do the error correction, perform Huffman decoding, and reconstruct the filter signals or the inverse-transformed time-domain signal. More information can be found, for example, in [8].

Echo Cancelation

Echo cancelers were first deployed in the U.S. telephone network in 1979. Today, they are virtually ubiquitous in long-distance telephone circuits where they cancel so-called line echoes (i.e., *electrical* echoes) resulting from nonperfect hybrids (the devices that couple local two-wire to long-distance four-wire circuits). In satellite circuits, echoes bouncing back from the far end of a telephone connection with a round-trip delay of about 600 msec are very annoying and disruptive. Acoustic echo cancelation—where the echo path is characterized by the transfer function $H(z)$ between a loudspeaker and a microphone in a room (e.g., in a speakerphone)—is crucial for teleconferencing where two or more parties are connected via full-duplex links. Here, echo cancelation can also alleviate acoustic feedback ("howling").

The principle of acoustic echo cancelation is depicted in Figure 21.6(a). The echo path $H(z)$ is canceled by modeling $H(z)$ by an adaptive filter and subtracting the filter's output $\hat{y}(t)$ from the microphone signal $y(t)$. The adaptability of the filter is necessary since $H(z)$ changes appreciably with movement of people or objects in the room and because periodic measurements of the room would be impractical. *Acoustic* echo cancelation is more challenging than canceling line echoes for several reasons. First, room impulse responses $h(t)$ are longer than 200 msec compared to less than 20 msec for line echo cancelers. Second, the echo path of a room $h(t)$ is likely to change constantly (note that even small changes in temperature can cause significant changes in h). Third, teleconferencing eventually will demand larger audio bandwidths (e.g., 7 kHz) compared to standard telephone connections (about 3.2 kHz) and multi-channel transmission (e.g., stereo; extension to more than two channels is fairly straightforward). We will address the stereo echo cancelation problem later in this section.

It is obvious that the initially unknown echo path $H(z)$ has to be "learned" by the canceller. It is also clear that for adaptation to work, there needs to be a nonzero input signal $x(t)$ that excites all the eigenmodes of the system (resonances, or "peaks" of the system magnitude response $|H(j\omega)|$). Another important problem is how to handle double-talk (speakers at both ends are talking simultaneously). In such a case, the canceller could easily get confused by the speech from the near end that acts as an uncorrelated noise in the adaptation. Finally, the convergence rate, that is, how fast the canceller adapts to a change in the echo path, is an important measure to compare different algorithms.

Adaptive filter theory suggests several algorithms for use in echo cancelation. The most popular one is the so-called *least-mean square* (LMS) algorithm that models the echo path by an FIR filter with an impulse response $\hat{h}(t)$. Using vector notation $\mathbf{h}$ for the true echo path impulse response, $\hat{\mathbf{h}}$ for its estimate, and $\mathbf{x}$ for the excitation time signal, an estimate of the echo is obtained by $\hat{y}(t) = \hat{\mathbf{h}}'\mathbf{x}$, where the prime denotes vector

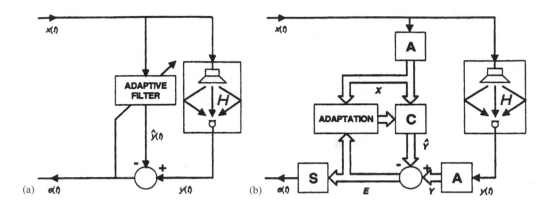

FIGURE 21.6 (a) Principle of using an echo canceler in teleconferencing. (b) Realization of the echo canceler in subbands. (After M.M. Sondhi and W. Kellermann, "Adaptive echo cancelation for speech signals," in *Advances in Speech Signal Processing*, S. Furui and M.M. Sondhi, Eds., New York: Marcel Dekker, 1991. Courtesy of Marcel Dekker, Inc.)

transpose. A reasonable objective for a canceler is to minimize the instantaneous squared error $e^2(t)$, where $e(t) = y(t) - \hat{y}(t)$. The time derivative of $\hat{\mathbf{h}}$ can be set to

$$\frac{d\hat{h}}{dt} = -\mu \nabla_{\hat{h}} e^2(t) = -2\mu e(t) \nabla_{\hat{h}} e(t) = 2\mu e(t) x \tag{21.4}$$

resulting in the simple update equation $\hat{h}_{k+1} = \hat{h}_k + \alpha e_k x_k$, where α (or μ) control the rate of change. In practice, whenever the far-end signal $x(t)$ is low in power, it is a good idea to freeze the canceler by setting $\alpha = 0$. Sophisticated logic is needed to detect double talk. When it occurs, then also set $\alpha = 0$. It can be shown that the spread of the eigenvalues of the autocorrelation matrix of $x(t)$ determines the convergence rate, where the slowest-converging eigenmode corresponds to the smallest eigenvalue. Since the eigenvalues themselves scale with the power of the predominant spectral components in $x(t)$, setting $\alpha = 2\mu/(\mathbf{x}'\mathbf{x})$ will make the convergence rate independent of the far-end power. This is the *normalized* LMS method. Even then, however, all eigenmodes will converge at the same rate only if $x(t)$ is white noise. Therefore, pre-whitening the far-end signal will help in speeding up convergence.

The LMS method is an iterative approach to echo cancelation. An example of a noniterative, block-oriented approach is the *least-squares* (LS) algorithm. Solving a system of equations to get $\hat{\mathbf{h}}$, however, is computationally more costly. This cost can be reduced considerably by running the LS method on a sample-by-sample basis and by taking advantage of the fact that the new signal vectors are the old vectors with the oldest sample dropped and one new sample added. This is the *recursive least-squares* (RLS) algorithm. It also has the advantage of normalizing $\mathbf{x}$ by multiplying it with the inverse of its autocorrelation matrix. This, in effect, equalizes the adaptation rate of all eigenmodes.

Another interesting approach is outlined in Figure 21.6(b). As in subband coding (discussed earlier), splitting the signals $\mathbf{x}$ and $\mathbf{y}$ into subbands with analysis filterbanks $\mathbf{A}$, doing the cancelation in bands, and resynthesizing the outgoing ("error") signal e through a synthesis filterbank $\mathbf{S}$ also reduces the eigenvalue spread of each bandpass signal compared to the eigenvalue spread of the fullband signal. This is true for the eigenvalues that correspond to the "center" (i.e., unattenuated) portions of each band. It turns out, however, that the slowly converging "transition-band" eigenmodes get attenuated significantly by the synthesis filter $\mathbf{S}$. The main advantage of the subband approach is the reduction in computational complexity due to the down-sampling of the filterbank signals. The drawback of the subband approach, however, is the introduction of the combined delay of $\mathbf{A}$ and $\mathbf{S}$. Eliminating the analysis filterbank on $y(t)$ and moving the synthesis filterbank into the adaptation branch $\hat{Y}$ will remove this delay, with the result that the canceler will not be able to model the earliest portions of the echo-path impulse response $h(t)$. To alleviate this problem, we could add in parallel a fullband echo canceler with a short filter. Further information and an extensive bibliography can be found in [9].

Finally, let us return to the stereo echo cancelation problem. In addition to teleconferencing, there are several other potential applications for stereophonic echo cancelers. One example is a so-called desktop teleconference in which the conferees are all at different locations. Each conferee has a desktop with pictures or videos of all the other conferees displayed on the screen. Here, we would like to associate stereo sound images with each conferee's image on the screen (left to right). Other potential applications are in on-line gaming, where friends and foes need to communicate and should speak from different apparent directions.

In order to understand the fundamental problem of stereo acoustic echo cancelation [10], consider the schematic of a stereophonic conference shown in Figure 21.7. We will consider echo cancelation for one end of the circuit (room B, the receiving room), it being understood that analogous configurations are used for echo cancelation in room A (in our discussion, the transmitting room). Each room has two loudspeakers and two microphones. The microphones in the transmitting room A pick up signals $x_1(n)$ and $x_2(n)$. These two signals are transmitted to loudspeakers in the receiving room B. Each microphone in room B picks up an echo of *both* signals, linearly combined via the acoustic echo paths to the microphone from the two loudspeakers. For simplicity we will consider the cancelation of the echo picked up by only one of the microphones, as shown in Figure 21.7, and denote the impulse responses from the loudspeakers to that microphone by $h_1(n)$ and $h_2(n)$. A similar discussion applies to cancelation of the echo at the other microphone, with $h_1(n)$ and $h_2(n)$ replaced by the responses appropriate to that microphone. (The impulse response of an acoustic path is assumed to include the responses of the loudspeaker and microphone in that path.)

Neglecting ambient noise and signals generated in the receiving room, the signal picked up by the microphone is

$$y(n) = \hat{\mathbf{h}}_1^T(n)\mathbf{x}_1(n) + \hat{\mathbf{h}}_2^T(n) \tag{21.5}$$

and the error signal is

$$e(n) = y(n) - \hat{y}(n) = \mathbf{r}_1^T(n)\mathbf{x}_1(n) + \mathbf{r}_2^T(n)\mathbf{x}_2(n) \tag{21.6}$$

where $\mathbf{r}_{1,2}^T(n) = \mathbf{h}_{1,2}^T(n) - \hat{\mathbf{h}}_{1,2}^T(n)$ are the misalignments in impulse responses. Let us define the "stacked" vector $\mathbf{x}(n)$ as the concatenation of $\mathbf{x}_1(n)$ and $\mathbf{x}_2(n)$ (i.e., $\mathbf{x}(n) = [\mathbf{x}_1^T(n)\mathbf{x}_2^T(n)]^T$). Similarly define the stacked vectors $\mathbf{h}$, $\hat{\mathbf{h}}(n)$, and $\mathbf{r}(n)$. In terms of these vectors, the expressions for $y(n)$, $\hat{y}(n)$, and $e(n)$ are exactly the same as the ones for the single channel canceler discussed above. Therefore, one would expect to be able to use the algorithm of Equation (21.4) in the ideal case to drive the misalignment vector $\mathbf{r}(n) \rightarrow 0$, i.e., $\hat{\mathbf{h}}_1(n) \rightarrow \mathbf{h}_1(n)$ and $\hat{\mathbf{h}}_2(n) \rightarrow \mathbf{h}_2(n)$. This turns out not to be true. In the ideal case, the algorithm does drive the error to zero.

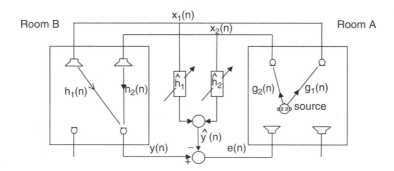

FIGURE 21.7 Schematic diagram of stereophonic echo cancelation. (*Source*: J. Benesty, D.R. Morgan, and M.M. Sondhi, "A better understanding and an improved solution to the specific problems of stereophonic acoustic echo cancellation," *IEEE Trans Speech and Audio*, vol. 6, no. 2, pp. 156–165, March 1998. With permission.)

However, the misalignment does not necessarily converge to zero. This is because the condition of zero error can be achieved by infinitely many nonzero vectors $\mathbf{r}(n)$. This nonuniqueness becomes apparent when Equation (21.6) is written in terms of the signals and impulse responses as time series. The error $e(n)$ can be written as

$$e(n) = r_1(n) * x_1(n) + r_2(n) * x_2(n) \tag{21.7}$$

where $*$ denotes convolution, and $r_1(n)$ and $r_2(n)$ are the misalignments in the two channels. If a single source is active in the transmitting room, the source signal $s(n)$ is convolved with the two transmitting room impulse responses $g_1(n)$ and $g_2(n)$ to generate the two microphone signals $x_1(n)$ and $x_2(n)$. That is, $x_1(n) = s(n) * g_1(n)$ and $x_2(n) = s(n) * g_2(n)$. Note that, since convolution $*$ is commutable,

$$g_2(n) * x_1(n) = g_2(n) * g_1(n) * s(n) = g_1(n) * x_2(n) \tag{21.8}$$

In view of Equation (21.8), the expression for the error in Equation (21.7) can be rewritten as

$$e(n) = [r_1(n) + \alpha g_2(n)] * x_1(n) + [r_2(n) + \alpha g_1(n)] * x_2(n) \tag{21.9}$$

where α is an arbitrary constant. Clearly the settings of the adaptive filters $\hat{h}_1(n)$ and $\hat{h}_2(n)$ are not unique.

Thus, although $\mathbf{r}(n) = 0$ implies $e(n) = 0$, the converse is not true. The condition $e(n) = 0$ only implies a relationship between the impulse responses of the receiving room and the transmission room. This nonuniqueness is the basic difference between single- and two-channel cancelers. It arises when only one source is active in the transmitting room, and is due to the fact that in such a case the signals $x_1(n)$ and $x_2(n)$ are highly correlated.

Examination of Equation (21.7) suggests a possible solution, first proposed in [10]. If the signals $x_1(n)$ and $x_2(n)$ were uncorrelated they would not be able to compensate for each other. The minimum of the error (zero in the ideal case) would then be obtained only when the two misalignments $r_1(n)$ and $r_2(n)$ were both zero.

Thus, one could try decorrelating the two channel signals. Of course, $x_1(n)$ and $x_2(n)$ cannot be completely decorrelated because then the stereo effect itself would be lost. The challenge, therefore, is to decorrelate them by an amount adequate to make the adaptive algorithm converge, yet small enough to be perceptually negligible.

Several attempts have been made to exploit this idea. However, the simplest and most effective proposal is to *distort* the signals $x_1(n)$ and $x_2(n)$ by passing each through a zero memory nonlinearity [11], thereby reducing their coherence. The simple nonlinear transformation:

$$\tilde{x}_{1,2} = x_{1,2}(n) + \beta |x_{1,2}(n)| \tag{21.10}$$

allows convergence of the misalignment. Yet, somewhat surprisingly, for speech signals it is hardly perceptible for β even as large as 0.2. Although this nonlinear transformation provides a good solution to the nonuniqueness problem for teleconferencing, it does not provide a completely satisfactory solution for music signals. For complex musical signals consisting of several instruments playing together, the distortion is almost imperceptible, as in the case of speech signals. However, for certain music signals, the nonlinearity introduces an unacceptable degradation. The signal from a flute, for instance, is almost a pure sinusoid. The distortion products for such a signal are quite perceptible.

Active Noise and Sound Control

Active noise control (ANC) is a way to reduce the sound pressure level of a given noise source through electroacoustic means. ANC and echo cancelation are somewhat related. While even *acoustic* echo cancelation is actually done on electrical signals, ANC could be labeled "wave cancelation," since it involves using one or

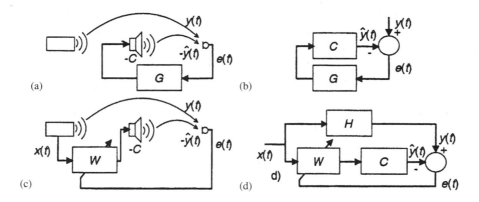

FIGURE 21.8 Two principles of active noise control. Feedback control system (a) and (b); feedforward control system (c) and (d). Physical block diagrams (a) and (c), and equivalent electrical forms (b) and (d). (After P.A. Nelson and S.J. Elliott, *Active Control of Sound*, London: Academic Press, 1992. With permission.)

more secondary acoustic or vibrational sources. Another important difference is the fact that in ANC one usually would like to cancel a given noise in a whole *region* in space, while echo cancelation commonly involves only one microphone picking up the echo signal at a single *point* in space. Finally, the transfer function of the transducer used to generate a cancelation ("secondary source") signal needs to be considered in ANC.

Active sound control (ASC) can be viewed as an offspring of ANC. In ASC, instead of trying to cancel a given sound field, one tries to control specific spatial and temporal characteristics of the sound field. One application is in adaptive sound reproduction systems. Here, ASC aims at solving the large-audience spatial reproduction problem mentioned in the section "Spatial Processing" in this chapter.

Two important principles of ANC are depicted in Figure 21.8. In the upper half (Figure 21.8(a) and (b)), a feedback loop is formed between the controller $G(s)$ and the transfer function $C(s)$ of the secondary source, and the acoustic path to the error microphone. Control theory suggests that $E/Y = 1/[1 + C(s)G(s)]$, where $E(s)$ and $Y(s)$ are Laplace transforms of $e(t)$ and $y(t)$, respectively. Obviously, if we could make C a real constant and $G \rightarrow \infty$, we would get a "zone of quiet" around the error microphone. Unfortunately, in practice $C(s)$ will introduce at least a delay, thus causing stability problems for too large a magnitude $|G|$ at high enough frequencies. The system can be kept stable, for example, by including a low-pass filter in G and by positioning the secondary source in close vicinity to the error microphone. A highly successful application of the feedback control in ANC is in active hearing protective devices (HPDs) and high-quality headsets and "motional-feedback" loudspeakers. *Passive* HPDs offer little or no noise attenuation at low frequencies due to inherent physical limitations. Since the volume enclosed by earmuffs is rather small, HPDs can benefit from the increase in low-frequency attenuation brought about by feedback-control ANC. Finally, note that the same circuit can be used for high-quality reproduction of a communications signal $s(t)$ fed into a headset by subtracting $s(t)$ electrically from $e(t)$. The resulting transfer function is $E/S = C(s)G(s)/[1 + C(s)G(s)]$ assuming $Y(s) = 0$. Thus, a high loop gain $|G(s)|$ will ensure both a high noise attenuation at low frequencies and a faithful bass reproduction of the communications signal.

The principle of the feedforward control method in ANC is outlined in Figure 21.8(c) and (d). The obvious difference from the feedback control method is that the separate reference signal $x(t)$ is used. Here, cancelation is achieved for the filter transfer function $W = H(s)/C(s)$, which is most often implemented by an adaptive filter. The fact that $x(t)$ reaches the ANC system earlier than $e(t)$ allows for a causal filter, needed in broadband systems. However, a potential problem with this method is the possibility of feedback of the secondary source signal $\hat{y}(t)$ into the path of the reference signal $x(t)$. This is obviously the case when $x(t)$ is picked up by a microphone in a duct just upstream of the secondary source C. An elegant solution for ANC in a duct without explicit feedback cancelation is to use a recursive filter W.

Single error signal/single secondary source systems cannot achieve *global* cancelation or sound control in a room. An intuitive argument for this fact is that one needs at least as many secondary sources and error microphones as there are orthogonal wave modes in the room. Since the number of wave modes in a room below a given frequency is approximately proportional to the third power of this frequency, it is clear that ANC (and ASC) is practical only at low frequencies. In practice, using small (point-source) transducers, it turns out that one should use more error microphones than secondary sources. Examples of such multidimensional ANC systems are employed for canceling the lowest few harmonics of the engine noise in an airplane cabin and in a passenger car. In both of these cases, the adaptive filter matrix is controlled by a multiple-error version of the LMS algorithm. Further information can be found in [12].

Summary and Acknowledgment

In this section, we have touched upon several topics in audio and electroacoustics. The reader may be reminded that the authors' choice of these topics was biased by their background in communication acoustics. Furthermore, ongoing efforts in integrating different communication modalities into systems for teleconferencing [13] had a profound effect in focusing this contribution. Experts in topics covered in this contribution, like Jont Allen, David Berkley, Joe Hall, Jim Johnston, Mead Killion, Harry Levitt, and Dennis Morgan, are gratefully acknowledged for their patience and help.

Defining Terms

Audio: Science of processing signals that are within the frequency range of hearing, that is, roughly between 20 Hz and 20 kHz. Also the name for this kind of signal.

Critical bands: Broadly used to refer to psychoacoustic phenomena of limited frequency resolution in the cochlea. More specifically, the concept of critical bands evolved in experiments on the audibility of a tone in noise of varying bandwidth, centered around the frequency of the tone. Increasing the noise bandwidth beyond a certain critical value has little effect on the audibility of the tone.

Electroacoustics: Science of interfacing between acoustical waves and corresponding electrical signals. This includes the engineering of transducers (e.g., loudspeakers and microphones), but also parts of the psychology of hearing, following the notion that it is not necessary to present to the ear signal components that cannot be perceived.

Intelligibility maximization and loudness restoration: Two different objectives in fitting hearing aids. Maximizing intelligibility involves conducting laborious intelligibility tests. Loudness restoration involves measuring the mapping between a given sound level and its perceived loudness. Here, we assume that recreating the loudness a normal hearing person would perceive is close to maximizing the intelligibility of speech.

Irrelevance and redundancy: In audio coding, irrelevant portions of an audio signal can be removed without perceptual effect. Once removed, however, they cannot be regenerated in the decoder. Contrary to this, redundant portions of a signal that have been removed in the encoder can be regenerated in the decoder. The "lacking" irrelevant parts of an original signal constitute the major cause for a (misleadingly) low signal-to-noise ratio (SNR) of the decoded signal while its subjective quality can still be high.

Monaural/interaural/binaural: *Monaural* attributes of ear input signals (e.g., timbre, loudness) require, in principle, only one ear to be detected. *Interaural* attributes of ear input signals (e.g., localization in the horizontal plane) depend on differences between, or ratios of measures of, the two ear input signals (e.g., delay and level differences). Psychoacoustic effects (e.g., cocktail-party effect) that depend on the fact that we have two ears are termed *binaural*.

References

1. K.C. Pohlmann, *Principles of Digital Audio*, 2nd ed., Carmel, IN: SAMS/Macmillan Computer Publishing, 1989.
2. F.R. Moore, *Elements of Computer Music*, Englewood Cliffs, NJ: Prentice-Hall, 1990.
3. J.L. Flanagan, D.A. Berkley, G.W. Elko, J.E. West, and M.M. Sondhi, "Autodirective microphone systems," *Acustica*, 73, 58–71, 1991.
4. J.B. Allen, J.L. Hall, and P.S. Jeng, "Loudness growth in 1/2-octave bands (LGOB) — a procedure for the assessment of loudness," *J. Acoust. Soc. Am.*, vol. 88, no. 2, pp. 745–753, 1990.
5. E. Villchur, "Signal processing to improve speech intelligibility in perceptive deafness," *J. Acoust. Soc. Am.*, vol. 53, no. 6, pp. 1646–1657, 1973.
6. E.A.G. Shaw, "The acoustics of the external ear," in *Acoustical Factors Affecting Hearing Aid Performance*, G.A. Studebaker and I. Hochberg, Eds., Baltimore, MD: University Park Press, 1980.
7. E.M. Wenzel, "Localization in virtual acoustic displays," *Presence*, 1, 80–107, 1992.
8. J.D. Johnston and K. Brandenburg, "Wideband coding — perceptual considerations for speech and music," in *Advances in Speech Signal Processing*, S. Furui and M.M. Sondhi, Eds., New York: Marcel Dekker, 1991.
9. E. Haensler, "The hands-free telephone problem — an annotated bibliography," *Signal Process.*, 27, 259–271, 1992.
10. M.M. Sondhi, D.R. Morgan, and J.L. Hall, "Stereophonic acoustic echo cancelation — an overview of the fundamental problem," *IEEE Sig. Proc. Lett.*, vol. 2, no. 8, pp. 148 –151, 1995.
11. J. Benesty, D.R. Morgan, and M.M. Sondhi, "A better understanding and an improved solution to the specific problems of stereophonic acoustic echo cancelation," *IEEE Trans. Speech Audio*, vol. 6, no. 2, pp. 156–165, 1998.
12. P.A. Nelson and S.J. Elliott, *Active Control of Sound*, London: Academic Press, 1992.
13. J.L. Flanagan, D.A. Berkley, and K.L. Shipley, "Integrated information modalities for human/machine communication: HuMaNet, an experimental system for conferencing," *J. Vis. Commun. Image Represent.*, vol. 1, no. 2, pp. 113–126, 1990.

Further Information

A highly informative article that is complementary to this contribution is the one by P.J. Bloom, "High-quality digital audio in the entertainment industry: An overview of achievements and challenges," *IEEE-ASSP Magazine*, October 1985. An excellent introduction to the fundamentals of audio, including music synthesis and digital recording, is contained in the 1992 book *Music Speech Audio*, by W.J. Strong and G.R. Plitnik, available from Soundprint, 2250 North 800 East, Provo, UT 84604 (ISBN 0-9611938-2-4). *Oversampling Delta-Sigma Data Converters* is a 1992 collection of papers edited by J.C. Candy and G.C. Temes. It is available from IEEE Press (IEEE order number PC0274-1). Specific issues of the *Journal of Rehabilitation Research and Development* (ISSN 007-506X), published by the Veterans Administration, are a good source of information on hearing aids, in particular the Fall 1987 issue. *Spatial Hearing* is the title of a 1982 book by J. Blauert, available from MIT Press (ISBN 0-262-02190-0). Anyone interested in *Psychoacoustics* should look into the 1990 book of this title by E. Zwicker and H. Fastl, available from Springer-Verlag (ISBN 0-387-52600-5).

The Institute of Electrical and Electronics Engineers (IEEE) *Transactions on Speech and Audio Processing* is keeping up-to-date on algorithms in audio. Every two to three years, a workshop on applications of signal processing to audio and electroacoustics covers the latest advances in areas introduced in this section. IEEE can be found on the web at http://www.ieee.org. *The Journal of the Audio Engineering Society* (AES) is another useful source of information on audio (http://www.aes.org). *The Journal of the Acoustical Society of America* (ASA) contains information on physical, psychological, and physiological acoustics, as well as on acoustic signal processing, among other things. ASA's "Auditory Demonstrations" CD contains examples of signals demonstrating hearing-related phenomena ranging from "critical bands" over "pitch" to "binaural beats." ASA can be found on the web at http://asa.aip.org.

21.2   Underwater Acoustical Signal Processing

Vyacheslav Tuzlukov, Won-Sik Yoon, and Yong Deak Kim

Background of Underwater Acoustics

Underwater acoustics entails the development and employment of acoustical methods to image underwater features, to communicate information via the oceanic wave-guide, or to measure oceanic properties. Underwater acoustics is the active or passive use of sound to study physical parameters and processes, as well as biological species and targets (for example, ships, submarines, mines, fishes, phyto- and zooplankton, etc.) at sea. In some cases, a specifically designed sound source is used to learn about the ocean and its boundaries or targets (active underwater acoustics). In other research, a natural sound or a sound generated by targets in the sea is analyzed to reveal the physical or biological characteristics of the sound source (passive underwater acoustics). Light, radar, microwaves, and other electromagnetic waves attenuate very rapidly and do not propagate any significant distance through salt water. Because sound suffers very much less attenuation than electromagnetics, it has become the preeminent tool for sensing, detection, identifying, and communicating under the ocean surface. And yet, for decades, inadequate oceanographic information about the extraordinary spatial and temporal variability of this medium has hindered underwater acousticians in their desire to predict sound propagation. It was necessary to learn more about those ocean characteristics that the traditional oceanographic instruments measure rather crudely, with great difficulty, and at great expense.

Acoustical researchers invert the problem; they use the complex nature of sound propagation to learn about the ocean. The many successes of this young science range from the identification and counting of physical and biological inhomogeneties — such as microbubbles, fragile zooplankton, fish, and mammals — to the remote measurement of distant rainfall and sea surface roughness, deep sea mountains, rocks, consolidated and suspensed sediments, as well as the shape and strength of internal waves, ocean frontal systems, and immense churning ocean eddies, hundreds of kilometers in extent. All of these unknowns can be measured by underwater acoustical techniques.

In retrospect, underwater acoustics started in 1912, when the steamship *Titanic* struck an iceberg. The subsequent loss of hundreds of lives triggered man's use of sound to sense scatterers in the sea. Within a month, two patent applications were filed by L.R. Richardson in the United Kingdom for "detecting the presence of large objects under water by means of the echo of compressional waves…directed in a beam…by a projector." The basic idea was that a precise knowledge of the speed of sound in water, and the time of travel of the sound, permits the calculation of the distance to the scatterer [1]. By 1935, acoustical sounding was used to determine the ocean depth as well as to hunt for fish schools. Much more recently it has been realized that the physical and spatial character of the scatterers can be inferred from the statistical characteristics of the scattered sound and that high-resolution images can be obtained at long range in optically opaque, turbid water.

Knowing the sound speed in water is critical to many of the applications of underwater acoustics. A value of 1435 m/sec was found, but it was soon realized that the speed in saline water is somewhat greater than this, and that in general the temperature of the water is an even more important parameter. Numerous laboratory and field measurements have now shown that the sound speed increases in a complicated way with increasing temperature, hydrostatic pressure, and the amount of dissolved salts in the water. A simplified formula for the speed in m/sec, accurate to 0.1 m/sec but good only to 1 km depth, was given by Medwin [2]:

$$c = 1449.2 + 4.6T - 0.055T^2 + 0.00029T^3 + (1.34 - 0.01T){\cdot}(S - 35) + 0.016z \qquad (21.11)$$

In the above, temperature T is in degrees centigrade, salinity S is parts per thousand of dissolved weight of salts, and the depth z is in meters. The effect of salinity is quite small except near estuaries or in polar regions, where fresh water enters the sea, but microbubbles have a very large effect on the speed of propagation near the ocean surface; frequency-dependent sound speed deviations of tens of meters per second are common in the upper ocean.

The propagation of sound in the sea has been studied intensely since the beginning of World War II when it was recognized that an understanding of this phenomenon was essential to the successful conduct of anti-submarine warfare (ASW) operations. These early measurements were quickly transformed into effective, albeit primitive, prediction tools. The study of sound propagation in the sea is fundamental to the understanding and prediction of all other underwater acoustic phenomena. Both marine seismologists and underwater acousticians have achieved advances in sound propagation, although the motivating factors have been quite different. Marine seismologists have traditionally used Earth-borne propagation or elastic waves to study the solid Earth beneath the oceans. Underwater acousticians have concentrated on the study of water-borne, compressional-wave propagation phenomena in the ocean as well as in the shallow sub-bottom layers. As research in underwater acoustics has extended to frequencies below several hundred Hertz, it has overlapped with the spectral domain of marine seismologists. Moreover, marine seismologists have become more interested in exploring the velocity–depth structure of the uppermost layers of the sea floor using higher frequencies. This area of overlapping interests has been recognized as a sub-discipline of both communities and is referred to as "ocean seismo-acoustics."

What Is Underwater Acoustical Signal Processing?

The use of acoustical signals that have propagated through water to detect, classify, and localize underwater objects is referred to as underwater acoustical signal processing or sonar. Sonar stands for "Sound Navigation Ranging" and is a method or device for detecting and locating objects, especially underwater, by means of sound waves transmitted or reflected by an object. Specific sonar application problems are submarine detection, mine hunting, torpedo homing, and bathymetric sounding. There are active and passive sonars. Active sonars transmit acoustic energy and detect targets by echolocation. Passive sonars operate by listening for acoustic emissions and can function in a covert manner.

Sonar Systems

Consider briefly the types of sonar systems. Sonar systems are used to remotely sense the interior of the ocean, what is in it, its surface, its bottom, and the structure beneath the bottom. Many specialized systems have been developed to do this. Data interpretation methods range from a simple display of "what is there and what is it doing" to statistical analysis of the pressure signals received by a system. In display and analysis there are two signal-processing classes: resolved and unresolved scatterers or reverberation. With *resolved signals*, the sources, scatterers, and so forth are separately displayed or imaged in time and space. Decades of development in sonar systems have improved the time and spatial resolution of the systems. With *unresolved signals*, the pressure signals from the sources, scatterers, and so forth are not separated in time or space and are called "reverberation." Statistical analysis, spectral analysis, and directional scattering are used in the study of unresolved scatterers in the reverberation.

Many sonar systems are almost one-task devices. The introduction of digital recording and data analysis has broadened the range of usefulness of an instrument so that a single instrument may be able to do several related tasks. Digital software has replaced many of analog operations in sonar systems, and digital signal processing has improved the adaptability of a system to new tasks. Sonar hardware and transducer configurations tend to be specialized to measurement tasks. Starting with simple sonar to echo sounder, we describe sonar configurations and their relation to remote sensing tasks. Generally, acoustic pings are used.

Echo Sounder. The most common sonar system is the echo sounder (see Figure 21.9). It employs an electrical signal generator and amplifier, called a *transmitter*, a transducer to convert an electrical signal to sound; a transducer to convert sound to an electrical signal; an electrical receiving circuit; and a display. Separate transmitting and receiving transducers are shown. A trigger from the display or transmitter starts

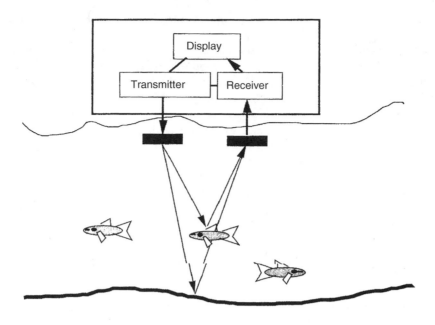

FIGURE 21.9   Echo sounder.

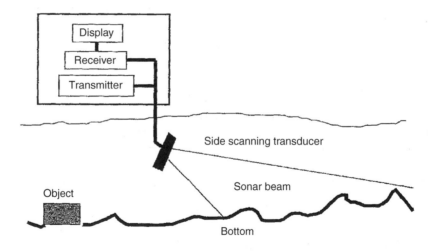

FIGURE 21.10   Side-scanning sonar.

the cycle. Many sonar systems use the same transducer for transmission and reception. Systems range in complexity from the *fish finders* that are sold in sporting goods departments to multibeam systems that are used by commercial fishermen and navies. The multibeam systems are basically combinations of many single-beam systems.

   Side-Scanning Sonar. The side-scanning sonar is an echo sounder that is pointed sideways (see Figure 21.10). The sonar looks to the side of the ship and makes an echo sounding record as the ship moves. The time of return of a pulse is interpreted as the range to the bottom feature that caused the scatter. Display software converts the *raw* image to a map of features on the bottom. However, although the design concepts are the same as the simple echo sounder, the sending transducer produces a fan-shaped beam, and the receiver has a time-variable gain to compensate for range. Side-scanning sonars are used in geological studies to give images of rough features on the sea floor. The instruments are also used to locate objects such as sunken ships on the sea floor.

Multibeam Sonar. Comparisons of mapping and object location operations that use radar in air and sonar in water demonstrate the large differences between the use of electromagnetic waves in air and sound waves in water. Radar (electromagnetic wave velocity $= 3 \times 10^8$ m/sec): the radar pulse travel time for a range of 30 km is 2×10^{-4} sec, and a simple radar systems can send, receive, and display in a very short time. The time required to make 360° image at 1° increments can be less than 0.1 sec. Thus, radar systems can use a single rotating dish to give good images. Consider the airborne radar. In 0.1 sec, an aircraft moving at a little less than the speed of sound in air (about 600 mph or 1000 kph) moves only about 30 m. The attenuation of electromagnetic waves in seawater is very large, and radar does not have a useful working range in the ocean. However, the attenuation of electromagnetic waves in glacial ice is small enough that radar soundings are used. Sonar (sound speed $= 1500$ m/sec): the time required for sonar to range to 30 km is 40 sec. In a sequential data acquisition system that takes one echo measurement at a time, several hours at one location would be needed to make one 360° image. A ship moving at 9 kph (2.5 m/sec) moves 100 m during the time for a single echo ranging measurement. A technological solution is to acquire sonar data in parallel by transmitting and receiving in many directions at the same time. Figure 21.11 shows an example of a multibeam sonar for sea-floor mapping. A cross-section of the ship is shown. The transmission is a broad beam. By adjusting time delays of the receiving elements, the multi-element receiving array is preformed to a set of narrow beams that look from port to starboard and measure the depths to various positions such as those shown at 1 to 7 in the figure. As the ship moves, the computer makes a contour plot of the depths. Using color coding, one gets a highly revealing picture. This system is intended to map a swath of depths along the ship track. Since these systems are usually mounted on the hull of the ship, the receiving array points in different directions as the ship

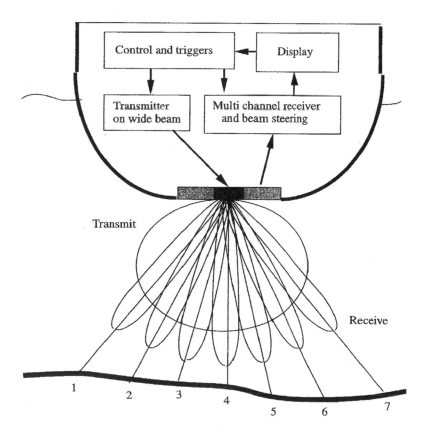

FIGURE 21.11 Multichannel sonar system using preformed beams.

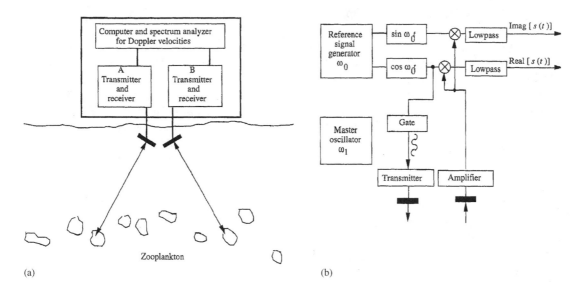

FIGURE 21.12 Doppler sonar system: (a) pings from the transmitter are backscattered from zooplankton; (b) block diagram of one channel.

pitches and rolls. The data-reduction system must compensate for the ship motions and the direction in which the receiving array is pointing when the echoes arrive.

Doppler Sonars. Doppler sonars are used to measure the velocities of ships relative to the water or the sea floor (see Figure 21.12). Commonly, Doppler sonars have four channels — two look fore and aft, and two look to starboard and port. They may also be used to measure the motion of the ocean surface or swimming objects, or internal waves, within the volume. Another application of the Doppler phenomenon is the ocean-going *portable* Doppler velocimeter [1].

Passive Sonar. Passive sonars listen to sounds in the ocean. A system may range in complexity from a single hydrophone to an elaborate, steered array of hydrophones similar to the multibeam system in Figure 21.11. There are the following noises at sea:

- Natural physical sounds: wave–turbulence interactions and oscillating bubble clouds (20 to 500 Hz); near shipping lanes the noise in the 10 to 150 Hz band is due largely to the machinery of distant ships; ocean sound on the band 500 to 20,000 Hz has been called *wind sea*, *sea state noise*, or *Knudsen noise*, because, during World War II, Vern O. Knudsen discovered that it correlated very well with wind speed [3]; the depth dependence owing to the attenuation of sea surface sound by near-surface bubble layers and bubble plumes cannot be ignored; rainfall sound.
- Natural biological sounds: noise generated by marine animals.
- Shipping noise: this noise can exhibit both spatial and temporal variabilities; the spatial variability is largely governed by the distribution of shipping routes in the oceans; the temporal variability can be introduced, for example, by the seasonal activities of fishing fleets.
- Seismoacoustic noise: microseismic band (80 mHz to 3 Hz) contains high-level microseismic noise resulting from nonlinear wave–wave interactions; noise-notch band (20 to 80 mHz) contains noise controlled by currents and turbulence in the boundary layer near the sea floor; ultra-low-frequency band (>20 mHz) contains noise resulting from surface gravity waves.

Steered Array Sonars. Transmitting or receiving arrays of transducers are steered by adding the signals from each transducer with proper time delays. The same analysis applies to send or to receive; we give the analysis for a receiving array. Consider the array of transducers in a line perpendicular to the direction $\psi = 0$ (see Figure 21.13). To steer the array, the elements at positions y_0, y_1, and so forth are given time delays τ_0, τ_1, and so forth that depend on the angle. The combination of the transducer sensitivity, analog-to-digital conversion,

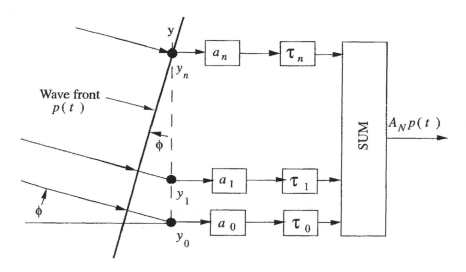

FIGURE 21.13 Electronically/digitally steered array for a plane wave entering the line of transducer array elements at angle ψ.

and amplifiers for the individual elements has the value a_n. The concept works for sources as well as receivers. To electronically/digitally steer the array, we insert appropriate time delays in each y_n-channel. Let the signal at the 0th hydrophone be $p(t)$ and the channel amplification factor be a_0. From the geometry in Figure 21.13, the plane wave front arrives at the nth hydrophone at advance Δt_n before reaching the 0th hydrophone, where

$$\Delta t_n = \frac{y_n \sin \psi}{c} \tag{21.12}$$

The signal at hydrophone n is $p(t - \Delta t_n)$. The analog/digital conversion and amplification is in the a_n. The time delay τ_n is inserted to give the signal $p(t - \Delta t_n + \tau_n)$. The sum signal for N channels is

$$A_N p_N(t) = \sum_{n=0}^{N-1} A_n p(t - \Delta t_n + \tau_n) \tag{21.13}$$

where A_n is an amplitude factor. Now, if τ_n is chosen to equal Δt_n, then the signals add in phase for that direction ψ, and we have

$$A_N p_N(t) = p(t) \sum_{n=0}^{N-1} A_n \quad \text{for } \Delta t_n = \tau_n \tag{21.14}$$

This method of array steering is called *delay and sum*. The only assumption is that the signals in each channel are the sum except for their time delays. Delay and sum processing works for any $p(t)$. The directional response of a steered array in other directions can be computed by choosing an incoming angle ψ' and letting

$$\tau_n = \frac{y_n \sin \psi'}{c} \tag{21.15}$$

Then

$$\Delta t_n - \tau_n = \frac{y_n(\sin\psi - \sin\psi')}{c} \tag{21.16}$$

The directional response of the array as a function of ψ depends on the value of ψ'. We have assumed that the incident sounds are plane waves. This is equivalent to assuming that the curvature of the wave front is small over the dimensions of the array, i.e., less than 0.125λ, where λ is the acoustic wavelength. The plane-wave assumption is effective for small arrays and distant sources. Arrays are built in many configurations: cylinders, spheres, and so on. The multibeam sonar described before is one example. By using time delays, almost any shape can be steered to receive signals of any curvature from any direction. However, when the arrays are built around a structure, diffraction effects can cause the performance to deteriorate.

Quantitative measurements of sound scattered from an object require that we understand the specifications and use of the sonar system as well as the physics of the scattering process. Sonars are often designed and adapted to the physics of a particular type of measurements. However, the operating characteristics of the sonar and the physics of the reflection and scattering processes are really independent. Now consider a generic sonar system.

Generic Sonar. The generic sonar, shown in Figure 21.14, is a combination of analog and digital components. The generic sonar system has an automatic transmit/receive switch. Some systems have separate transducers for sending and receiving. The trigger, from a clock that is internal and/or external, initiates the transmission and reception cycle. The receiver includes the electronic and digital signal processing. The transducers may be mounted on the ship or in a tow body, the *fish*. The signals may be recorded on an analog tape, a digital tape, or a compact disc. Typical displays are paper chart recorders and video display terminals. Sonars that are used for surveys and research usually include control of ping duration, choice of the time-varying-gain (TVG) function, calibration signals, displays, and analog signal outputs. TVG as a function of range for two settings for a generic sonar is shown in Figure 21.15. The operation is used to compensate for range dependence. Unresolved overlapping echoes (volume scatter) have a pressure amplitude proportional to t^{-2}, and the amplitude compensation is proportional to t and $20\log R$, where R is a range. Isolated echoes from individuals have echo amplitudes proportional to t^{-2} and compensation is t^2 or $40\log R$. As shown in Figure 21.15, for each transmission and reception cycle, the TVG starts at low gain and increases as a function of time. All sonar receivers have a minimum output that is related to the noise and a maximum output/limit when the amplifier overloads. Digital systems have equivalent minimum and maximum limits. The TVG is chosen to keep the output electrical signal amplitudes approximately the same for near and distant scatterers and to keep the signals above the minimum and less than the maximum limits.

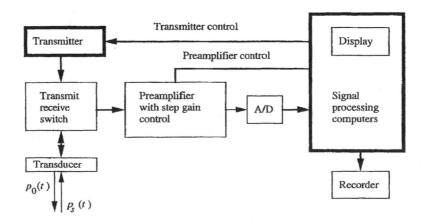

FIGURE 21.14   Block diagram of a generic sonar.

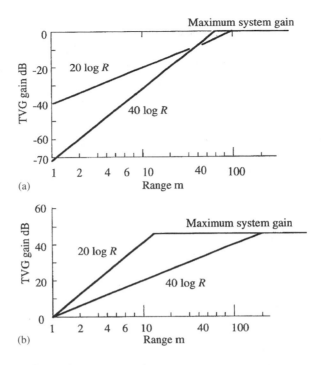

FIGURE 21.15 TVG as a function of range for two settings for a generic sonar: (a) the final TVG reference gain is unity (0 dB); (b) the initial reference gain is unity (0 dB). The maximum TVG action is limited by the maximum gain of the amplifier.

It is preferable to choose a TVG function that keeps the electrical signals in a good recording range rather than trying to match some preconceived notion of what the TVG ought to be. Standard preprogrammed choices are gain proportional to t^2 or $40 \log R$ and gain proportional to t or $20 \log R$. How the TVG operates on a voltage signal in an instrument depends on the engineering design. Some designs start with a very low gain and then increase to the amplifiers limit (see Figure 21.15a). Others start with an initial gain of unity (0 dB) and increase to the amplifier's limit (see Figure 21.15b). If one has individual echoes from many small, isolated targets, then the echo pressure amplitudes (the echoes) decrease as $\approx t^{-2}$, and a TVG of $40 \log R$ compensates for the spherical divergence.

When the objects are close together, as in a cloud of scatterers, and their echoes overlap, the sum of all of the unresolved echo pressures tends to decrease as t^{-1}, and a TVG of $20 \log R$ gives compensation. This is a characteristic of volume scatter. In acoustical surveys, both clouds of fish and individuals can be present, and neither choice of TVG fits all. It is better to use one TVG choice that keeps the voltage levels in a good range for recording. Whatever the TVG choice, an appropriate range dependence can be included in digital signal processing. Digital signal processing technology has enabled the mass manufacture of small sports-fisherman's sonars. These inexpensive instruments contain preprogrammed computers and are actually very sophisticated. The small, portable sonars can identify echoes from individual fish, display relative fish sizes, look sideways and separate echoes of large fish from reverberation, and show water depth.

Sonar with Band-Shifting or Heterodyning Operations. The ping from sonar may have a carrier frequency of 100 kHz and duration of 1 msec. Examples of pings having the same envelope and different carrier frequencies are sketched in Figure 21.16 and Figure 21.17. Since the carrier frequency f_c is known, sampling the envelope and measuring the relative phase of the carrier frequency can describe the ping. This can simplify the signal processing operations and greatly reduce memory requirements in sonar systems.

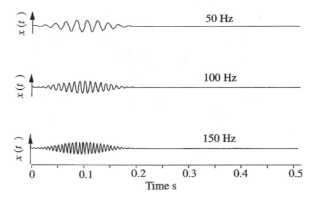

FIGURE 21.16 Signals and their spectra. All of the pings have the same duration. Time-domain presentation of pings $x(t)$ with carrier frequencies of 50, 100, and 150 Hz.

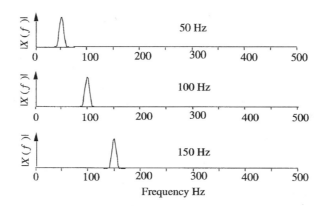

FIGURE 21.17 Signals and their spectra. All of the pings have the same duration. Modulus of the spectral amplitudes $|X(f)|$.

The ping is determined by:

$$x(t) = \begin{cases} e(t)\sin 2\pi f_c t & \text{for } 0 < t < t_p \\ 0 & \text{otherwise} \end{cases} \tag{21.17}$$

where the envelope of the ping is

$$e(t) = 0.5 \left| 1 - \cos \frac{2\pi t}{t_p} \right| \tag{21.18}$$

The spectrum of $x(t)$ is $X(f)$, which is sketched in Figure 21.18. The two multiplication operations, shown in Figure 21.19, are the first step:

$$x_H(t) = e(t) \sin 2\pi f_c t \cos 2\pi f_H t \tag{21.19}$$

and

$$x_Q(t) = e(t) \sin 2\pi f_c t \sin 2\pi f_H t \tag{21.20}$$

We make the approximation that the time dependence of the envelope $1 - \cos \dfrac{2\pi}{t_p}$ can be ignored because

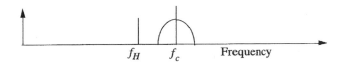

FIGURE 21.18   Band-shift or heterodyne operations: spectrum of the input signal.

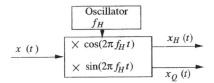

FIGURE 21.19   Band-shift or heterodyne operations: diagram of the band-shift operation for both cosine (*H*) and the sine quadrature (*Q*) multiplications.

the frequencies are very small compared with f_c. Expansion of the products of the sine and cosine terms gives

$$2 \sin 2\pi f_c t \cos 2\pi f_H t = \sin[2\pi(f_c - f_H)t] - \sin[2\pi(f_c + f_H)t] \tag{21.21}$$

and

$$2 \sin 2\pi f_c t \sin 2\pi f_H t = \cos[2\pi(f_c - f_H)t] - \cos[2\pi(f_c + f_H)t] \tag{21.22}$$

and the *H* and *Q* components of the band-shifted pings are

$$x_H(t) \approx 0.5 e(t)\{\sin[2\pi(f_c - f_H)t] + \sin[2\pi(f_c + f_H)t]\} \tag{21.23}$$

and

$$x_Q(t) \approx 0.5 e(t)\{\cos[2\pi(f_c - f_H)t] - \cos[2\pi(f_c + f_H)t]\} \tag{21.24}$$

The spectra of the band-shifted pings are sketched in Figure 21.20. Band-shifting operations are also known as *heterodyning*. The final steps are to low-pass-filter the results to select the $f_c - f_H$ bands. It is sufficient to sample the heterodyned signal at more than twice the frequency bandwidth of the envelope. For example, let a ping have the duration of 1 msec and a carrier frequency of 100 kHz. The bandwidth of the 1 msec ping is approximately 1 kHz, and thus the minimum sampling frequency is 2 kHz. At 4 kHz, the envelope would be sampled four times.

   The relative phases of heterodyned signals are preserved in the heterodyning operation. Let the ping given by Equation (21.17) have a relative phase η:

$$x(t, \eta) = e(t) \sin(2\pi f_c t + \eta) \tag{21.25}$$

The shift of the envelope is very small. Repeating the steps that gave Equation (21.21) to Equation (21.24), the

FIGURE 21.20   Band-shift or heterodyne operations: output spectra.

low-pass heterodyned signals with a phase shift are

$$x_H(t, \eta) \approx 0.5e(t)\sin(2\pi f_d t + \eta) \tag{21.26}$$

where $f_d = f_c - f_H$, and

$$x_Q(t, \eta) \approx 0.5e(t)\cos(2\pi f_d t + \eta) \tag{21.27}$$

The reference signal is $x(t)$, the phase-shifted signal is $x(t, \eta)$, and we want to measure the relative phase. The cross-correlations of the signals are less sensitive to noise than direct comparisons of the phases. Consider the cross-correlation of $x_H(t)$ and $x_Q(t, \eta)$:

$$<x_H(t)x_Q(t, \eta)> \equiv \frac{1}{\text{Norm}} \int_0^{t_p} e^2(t) \sin 2\pi f_d t \cos(2\pi f_d t + \eta)dt \tag{21.28}$$

where

$$\text{Norm} \equiv \int_0^{t_p} e^2(t) \sin^2 2\pi f_d t dt \tag{21.29}$$

The expansions of the product of the sine and cosine are

$$\sin 2\pi f_d t \cos(2\pi f_d t + \eta) = -0.5 \sin \eta + 0.5 \sin(4\pi f_d t + \eta) \tag{21.30}$$

and

$$\sin^2 2\pi f_d t = 0.5 - 0.5 \cos 4\pi f_d t \tag{21.31}$$

The substitution of Equation (21.30) and Equation (21.31) into Equation (21.28) gives the sum of two integrals. The integral that includes the $\sin(4\pi f_d t + \eta)$ term tends to 0. The remaining integral is

$$<x_H(t)x_Q(t, \eta)> \approx -\frac{1}{2\text{Norm}} \int_0^{t_p} e^2(t) \sin \eta dt \tag{21.32}$$

The evaluation of *Norm*, using Equation (21.30) and Equation (21.31), reduces Equation (21.32) to

$$<x_H(t)x_Q(t, \eta)> \approx -\sin \eta \tag{21.33}$$

The other cross-correlations are determined as follows:

$$<x_Q(t)x_H(t,\eta)> \approx \sin\eta \qquad (21.34)$$

$$<x_H(t)x_H(t,\eta)> \approx \cos\eta \qquad (21.35)$$

$$<x_Q(t)x_Q(t,\eta)> \approx \cos\eta \qquad (21.36)$$

The following expression gives the tg η:

$$\text{tg}\,\eta = -\frac{<x_H(t)x_Q(t,\eta)> - <x_Q(t)x_H(t,\eta)>}{<x_H(t)x_H(t,\eta)> + <x_Q(t)x_Q(t,\eta)>} \qquad (21.37)$$

The cross-correlation method of measuring the phase difference is useful when two pressure signals (pings) are received on a pair of transducers and their phase difference is used to calculate the direction of the incident pressure signal [4].

Echo Identification Rules. The echo amplitude display from echo sounders shows an interesting phenomenon as the instrument moves slowly over isolated fish. Normally, TVG compensates for range so that the echo amplitudes are the same for the same sizes of fish at different ranges. Figure 21.21 shows the details of echoes from one fish in a sonar beam. First for simplicity of drawing, the sonar is fixed and the fish swims through the center of the sonar beam. Pings measure the sequence of ranges to the fish at positions "a" to "i." The graphic recorder plots the echoes beneath the time of the ping and the sequence of echoes form a crescent. The echo amplitudes are given by the position of the fish in the transducer-response pattern. The graphic record of echoes would be the same if the fish was fixed and the sonar transducer was moved from left to right ("a" to "i") over the fish. An echo crescent is formed as the scattering object (fish) moves through the echo sounder's beam pattern. Starting with the object at the left edge of the beam, the echo is weak, and, from the geometry, the range is a little larger than at position "e." As the object moves into the center of the beam ("e"), the echo is larger, and the range is smaller. The width of the crescent and its amplitude depend on depth and whether the object goes through the center of the beam or is off to one side. The procedures to use these effects in the analysis of echoes from fish are suggested in [5]. The computer identification of an echo requires acceptance rules. The envelopes of a single echo, reverberations, and several echoes are shown in Figure 21.22. The sonar return has been compensated for spherical spreading. The echo envelope is $e(t)$. A single echo has shape parameters. The ping has a duration of t_p, and ideally the duration of the echo is t_p. A specific example

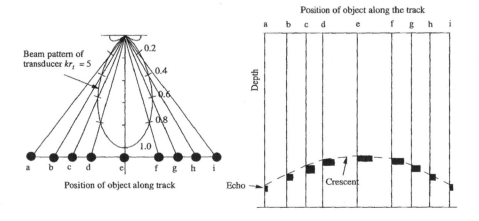

FIGURE 21.21 Simulation of a graphic display of echoes from a single fish.

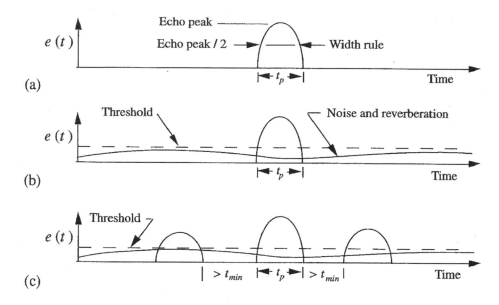

FIGURE 21.22 Identification of echoes: (a) at half-echo amplitude, the width must be less than 1.5 t_p and greater than 0.5 t_p; (b) the echo amplitude must be greater than a threshold and the reverberation noise; (c) the echo is less than the threshold before and after the echo and the minimum time t_{min} is greater than the reciprocal frequency bandwidth of the receiver.

of acceptance rules follows. Single echo identification and acceptance rules are important parts of the signal processing codes that are used in data analysis. Acceptance rules are used in the fish-finding sonars for sportsmen.

Signal Processing

Signals are the messages that we want to receive at our hydrophone. Noises are everything that we do not want to receive. The types of signal messages include impulses and continuous wave (CW) tones of short or long duration and constant or varying frequency; they also include complicated coded messages and random sequences. The form of noise can run the gamut. We all know the popular saying "Beauty (or ugliness) is in the eye of the beholder." One can propose a comparable acoustical maxim: "Signals (or noise) are in the ear of the listener." There are many examples: the sonar operator searching for the *signal* of a submarine will call the sounds of whales and dolphins *noise*. Needless to say, the marine mammals seeking to communicate or locate food would characterize man-made sounds as *noise*. Some sounds called *noise* for many years are now recognized as bearing information that qualifies them as *signals*; for example, the sound of rainfall at sea is now used to measure the size and number of raindrops per square meter per second. Flow noise at a transducer, electrical circuit noises, and the 60- or 50-Hz electrical interferences from power lines are generally regarded as noise by everybody.

Traditionally, underwater ambient noise has been specified in terms of the sound measured at a convenient hydrophone, some distance from the sources. The origins of the sound are often a mystery. We initiate a different approach. We will survey the acoustic power, source pressure, directionality, and intermittency of physical, biological, and man-made ocean sounds at their source. When this information is known, one's knowledge of propagation in the ocean allows us to calculate the ambient sound at any location. In addition to our survey of many of the more common sound sources at sea, we need operations that allow the listener to sort out the signal from the noise

Sampling Rules

Practically all underwater sound signals and noise are recorded digitally, and the results of analysis are displayed on computer terminals. The acoustician uses signal acquisition and digitizing equipment, signal processing algorithms, and graphic display software to make this happen. The noisy signal that comes from our hydrophone is an electrical voltage that is a continuous function of time. It is called an analog signal. Hydrophone signals must be sampled to convert them from analog to digital format in order to enter a digital computer. We must sample signals properly or we get garbage. The sampling rules are general. They apply to either temporal or spatial sampling of the oceanic environment. When the rules are obeyed, the original signal can be recovered from the sampled signal with the aid of an interpolation procedure. If the rules are not obeyed, and sampling is too sparse, the original signal cannot be recovered. The Nyquist sampling rules are [6]:

- *Space-Domain Rule.* The spatial sampling interval must be less than half of the shortest wavelength of the spatial variation. Spatial sampling is sometimes described in terms of the spatial wave number, $k_s = \dfrac{2\pi}{\lambda_s}$, where λ_s is the distance between samples.
- *Time-Domain Rule.* The time interval between samples must be less than half of the shortest period in the signal. Sampling is defined in terms of the sampling frequency $f_s = t_0^{-1}$, where t_0 is the time between samples. Otherwise stated, the sampling frequency must be greater than twice the highest frequency component in the signal.

Spatial sampling. In traditional marine biology, samples are taken by towing a net through the water. A net is lowered to the depth, opened, and towed for a specified distance. Net sampling takes a lot of work.

Temporal sampling. The electrical signal $x(t)$ is sampled by an analog-to-digital converter to create a sequence of numbers (see Figure 21.23). The clock (Figure 21.23(a)) gives the sampling instruction. The sample is the

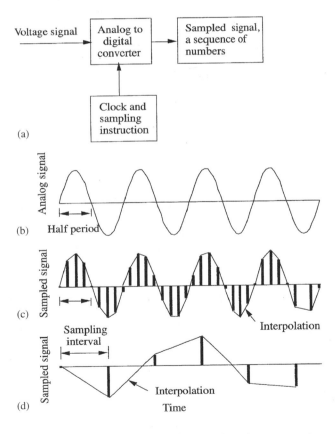

FIGURE 21.23 Temporal sampling of a simple signal: (a) sampling system to change an analog voltage into a sequence of numbers; (b) input analog sinusoidal voltage; (c) the result of sampling four times during each half-period (the vertical lines represent the magnitudes of the sampled voltages and the straight lines between ends of the vertical lines are interpolations); (d) the result of sampling at times greater than the half-period.

instantaneous value of the signal voltage at the clock time. No information is recordered about the signal voltage between samples where straight lines are drawn. Examples of data taken at two different sampling intervals are shown. In Figure 21.23(c) there are four samples in a half-period. In Figure 21.23(d) the sampling interval is larger than the half-period. Reconstruction of the inadequately sampled signal in Figure 21. 23(d) does not resemble the original signal, whereas Figure 21.23(c) does. A practical rule of thumb is to sample at intervals less than the period/3 for approximate reconstruction of the original signal.

Filter Operations

Electrical filters were originally introduced into electronic systems by radio and telephone engineers to separate the signals they wanted from those that they did not. Instrumentation manufacturers built analog filters or *black boxes* for research laboratories. These filters had switches on the front to make frequency bandpass choices. We use the filters that are built into our radio or television set when we select a channel that tunes in our desired station and rejects others. Audio amplifiers have filters (equalizers, bass, and treble controls) to modify the amplitudes of the input frequencies and to enhance the quality of the sound coming from the speakers. Digital communication engineers have developed the digital equivalent of the black box analog filter. An analog-to-digital converter digitizes the incoming analog electrical signal, and a computer does the filter operations. In many systems, the filtered digital sequence of numbers is converted back to an analog signal for listening and display. For simplicity in our discussion of operations on signals and noise, we use the neutral symbols x, h, and y to represent source, filters, and filter outputs, respectively.

Finite Fourier Transformations. Digital computers, digitized data, and efficient algorithms have made the numerical computations of Fourier transformations practical. The Fourier transformation of a finite number of data points is called the finite Fourier transformation (FFT) or discrete Fourier transformation (DFT). The time intervals between all digitally sampled data points are t_0. For N data points, the Fourier transformation pairs are

$$X_{\text{fft}}(m) = \sum_{n=0}^{N-1} x(n) \cdot e^{-2j\frac{\pi mn}{N}} \tag{21.38}$$

$$x(n) = \frac{1}{N} \sum_{m=0}^{N-1} X_{\text{fft}}(m) \cdot e^{2j\frac{\pi mn}{N}} \tag{21.39}$$

where $x(n)$ is the nth digital input signal amplitudes, and $X_{\text{fft}}(m)$ is the mth spectral component.

Equation (21.38) changes a time-dependent series of terms into a frequency spectrum. The companion Equation (21.39) or inverse finite Fourier transformation or (IFFT), changes a spectrum into a time-dependent expression. Real $x(n)$ transforms to complex $X_{\text{fft}}(m)$ and vice versa. A simple example demonstrates the periodic properties. Let $N = 64$ and suppose that the original digitized signal is real and exists between $n = 0$ and $n = 63$ as shown in Figure 21.24(a). The digitized signal is assumed to be an isolated event (see Figure 21.24(a)). The FFT method assumes that the signal has the period $N = 64$, as shown in Figure 21.24(b). Evaluations of the FFT give real and imaginary components of $X_{\text{fft}}(m)$. Inspection of Figure 21.24(c) shows that the real components are symmetric about 0, 0.5N, and N. In Figure 21.24(d) the imaginary components are antisymmetric about 0, 0.5N, and N. The modulus $|X_{\text{fft}}(m)|$, Figure 21.24(e), is

$$|X_{\text{fft}}(m)| = \sqrt{\{\text{Re}X_{\text{fft}}(m)\}^2 + \{\text{Im}X_{\text{fft}}(m)\}^2} \tag{21.40}$$

which is symmetric about 0, 0.5N, and N. If we had started with $\text{Re}X_{\text{fft}}(m)$ and $\text{Im}X_{\text{fft}}(m)$, we would have the periodic $x(n)$. Figure 21.24 shows most of the important properties of the FFT.

Filter-Response Measurements. As shown in Figure 21.25, the frequency response of the filter is the ratio of the output/input voltages for a long-duration sinusoidal input signal (the oscillator). Two measurements are sketched in Figure 21.25. To record signals digitally, we need a low-pass filter to prepare the signal for the digitization operation. The filter shown in Figure 21.25(b) is a *low-pass, antialiasing filter*. It is adjusted to pass

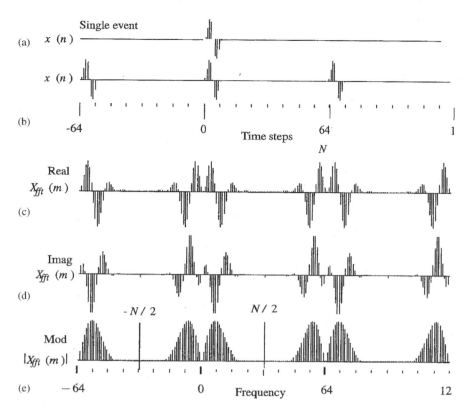

FIGURE 21.24 Signal, periodic Fourier series and its spectrum. Even if the orthogonal signal is not periodic, expansion in the Fourier series creates a new signal that is periodic. In applications, many zeros are appended to the signal to move the next cycle out of the way: (a) original signal; (b) periodic signal; (c) real component of $X_{fft}(m)$; (d) imaginary component of $X_{fft}(m)$; and (e) modulus of $X_{fft}(m)$, $|X_{fft}(m)|$.

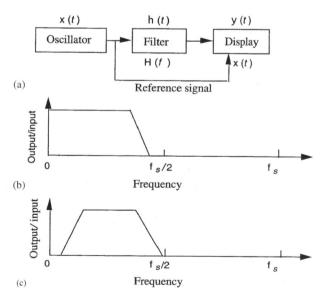

FIGURE 21.25 Filters and their responses: (a) block diagram for a typical filter response measurement; (b) response of an antialiasing, low-pass filter that is used ahead of analog-to-digital conversion at sampling frequency f_s; (c) response of a bandpass filter.

frequencies that are less than half the sampling frequency of the analog-to-digital converter and to reject higher frequencies. It thereby can prevent higher-frequency components from appearing as *alias* signals. The action of a bandpass filter is sketched in Figure 21.25(c). It is used to pass a signal and to reject unwanted signals and noise within the designed frequency range.

Time-Domain View of Bandpass Filtering. Figure 21.26(a) shows a short-duration signal, $x_1(t)$, which then passes through an appropriate bandpass filter to give the output $y_1(t)$ in Figure 21.26(b). The high- and low-pass settings of the filter were chosen to pass the signal with the signal with an acceptable amount of distortion of waveform. A longer-duration, low-frequency whale song, $x_2(t)$, is emitted during this same time so that the sum of the two signals at the input (see Figure 21.26(c)) is

$$x(t) = x_1(t) + x_2(t) \tag{21.41}$$

The output of the bandpass filter $y(t)$ is shown in Figure 21.26(d). The bandpass filter effectively removes the interfering whale song and reveals the short-duration 150-Hz ping.

Filter Operations in the Frequency Domain. Operations in the frequency domain are intuitively simple. The frequency-dependent functions, $X(f)$ and $Y(f)$, are the amplitude spectral densities of the input and output signals, and $H(f)$ is the filter-frequency response. The input–output expressions for analog signals are

$$Y(f) = H(f) \cdot X(f) \tag{21.42}$$

and

$$Y_{\text{fft}}(f) = H_{\text{fft}}(f) \cdot X_{\text{fft}}(f) \tag{21.43}$$

The output signals $Y(f)$ or $Y_{\text{fft}}(f)$ have the frequency components that are passed by the filters. Our discussion uses the amplitude spectral densities for brevity.

Figure 21.27 represents a spectrum analyzer that was constructed of many bandpass filters. The complex signal is an input to each of the filters. The spectral output of the jth filter is

$$Y_j(f) = H_j(f) \cdot X_j(f) \tag{21.44}$$

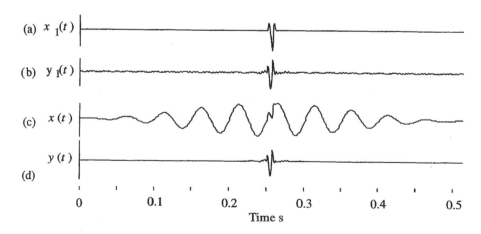

FIGURE 21.26   Filter operation shown in the time-domain: (a) signal input is a 150-Hz ping having a duration of 0.01sec; (b) signal out of a 50 to 150 Hz bandpass filter; (c) input 150-Hz ping and a 20-Hz whale song; (d) filtered signal output using the 50 to 150 Hz bandpass filter.

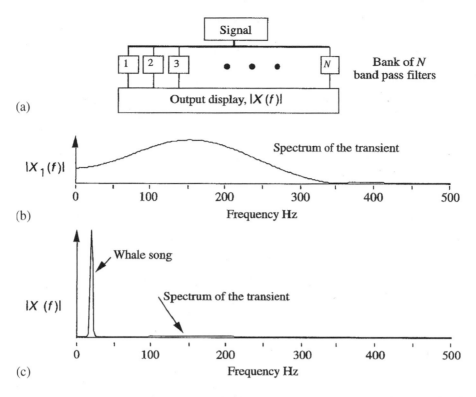

FIGURE 21.27 Spectral analysis of the signals of Figure 21.26 using a digital spectral analysis. The bandwidths of the equivalent bandpass filters are 2 Hz: (a) an analog spectrum analyzer that uses a bank of bandpass filters; (b) the digitally calculated spectrum of the 150-Hz ping in Figure 21.26(a); (c) the digitally calculated spectrum of the ping and whale song in Figure 21.26(c). The spectral amplitude factor of the ping is 0.025 that of the whale song. The ping does not show the detail of Figure 21.26(c) because of the change of scale.

Parseval's theorem gives the equivalence between the integral squares of signals in the time- and frequency-domains:

$$\int_{-\infty}^{\infty} y^2(t)\mathrm{d}t = \int_{-\infty}^{\infty} |Y(f)|^2\mathrm{d}f = 2\int_{0}^{\infty} |Y(f)|^2\mathrm{d}f \tag{21.45}$$

where the time integral of $y^2(t)$ is finite, and the absolute squares of $|Y(f)|$ and $|Y(-f)|$ are equal. For signals that start at 0 time, the limits of the doubly infinite integral become 0 to ∞. Using the filter input and output, Equation (21.44), and Parseval's theorem [7], the integral square output of the jth filter is

$$\int_{0}^{\infty} y_j^2(t)\mathrm{d}t = 2\int_{0}^{\infty} |Y_j(f)|^2\mathrm{d}f \tag{21.46}$$

The substitution of Equation (21.44) into Equation (21.46) gives the filter output:

$$\int_{0}^{\infty} y_j^2(t)\mathrm{d}t = 2\int_{0}^{\infty} |X_j(f)\cdot H_j(f)|^2\mathrm{d}f \tag{21.47}$$

For an approximation, let $H_j(f)$ be a *boxcar* filter defined by

$$H_j(f) = \begin{cases} 1 & \text{for} \quad f_j - 0.5\Delta f \leqslant f \leqslant f_j + 0.5\Delta f \\ 0 & \text{otherwise} \end{cases} \tag{21.48}$$

If $X(f)$ is approximately constant in the pass band (Equation (21.48)), then the time integral square (Equation (21.47)) is approximately:

$$\int_0^\infty y_j^2(t)dt \approx 2|X(f_j)|^2 \Delta f \tag{21.49}$$

This is the raw output of the boxcar spectrum analyzer. The spectral density in a 1-Hz bandwidth is obtained by dividing both sides of Equation (21.49) by Δf:

$$\frac{1}{\Delta f}\int_0^\infty y_j^2(t)dt \approx 2|X(f_j)|^2 \tag{21.50}$$

The quantity in Equation (21.50) is sometimes called an *energy spectral density*, $E_{xx}(f_j)$, when the signal amplitude is a voltage, because the *integral square* over time is proportional to the energy of the electrical signal during that time:

$$E_{xx}(f_j) = \frac{1}{\Delta f}\int_0^\infty y_j^2(t)dt \approx 2|X(f_j)|^2 \tag{21.51}$$

The subscript $_{xx}$ indicates that $x(t)$ is the function being analyzed.

The $x(t)$ and $y(t)$ have units of Pa, the units of $E_{xx}(f_j)$ are Pa^2 sec/Hz. Although the quantity $E_{xx}(f)$ is often called an *energy spectral density*, it is actually proportional to the energy spectral density of the wave. As shown in the discussion of intensity [1], true expressions for energy spectral density (joules/m^2 Hz) require that Pa^2 sec/Hz be divided by $\rho_A c$, where ρ_A is the ambient density of the medium and c is the sound speed or velocity. Most spectrum analyzers are digital and use computers to do the spectral analysis. The digital spectrum analyzers often can have the equivalent of more than 1000 very narrow band-pass filters. Examples of spectrum analysis are shown in Figure 21.27(b) and (c). The spectrum of the 150-Hz ping (see Figure 21.27(b)) and the spectrum of the short-duration ping and longer-duration whale song are shown in Figure 21.27(c).

Gated Signals

The spectrum of a signal depends on its time-domain waveform. Consider some pings and their spectra. These comparisons display the relation of periodicity and duration, in the time-domain, to the peak frequency and bandwidth of the spectrum. For these examples, the ping has a slow turn-on and turn-off. The signal $x(t)$ is

$$x(t) = \begin{cases} 0.5\left[1 - \cos\dfrac{2\pi t}{t_p}\right]\sin 2\pi f_c t & \text{for} \quad 0 < t < t_p \\ 0 & \text{otherwise} \end{cases} \tag{21.52}$$

where t_p is the total nonzero ping duration and f_c is the carrier frequency. The amplitude factor in the brackets gives a spectrum with very small side lobes. This signal is similar to the sound-pressure signal radiated by many sonar transducers and some marine animals. The envelope of the sine wave is tapered from zero to a maximum and then back to zero.

Dependence of Spectrum on Ping Carrier Periodicity. Figure 21.16 and Figure 21.17 show the dependence of the peak of the spectrum on the carrier frequency of the pings. The durations of the pings were chosen to be long so that the widths of spectral peaks are narrow. The 50-Hz signal has a spectral peak at approximately 50 Hz. The other signals have their spectral peaks at 100 and 150 Hz. The period of the signal can be measured to estimate the peak or central frequency of the spectrum. This is the first rule of thumb.

Dependence of Spectrum on Ping Duration. Figure 21.28 shows the dependence of the widths of the spectral peaks on the durations of the pings. The effective durations of the signals t_d are a little less than the t_p in Equation (21.52) because the turn-ons and turn-offs are very gradual. The same frequency, 100 Hz, was used for all examples. To define the bandwidth Δf, we use the half-power width given by the two frequencies where the amplitude is 0.707 of the peak amplitude. The spectra shown in Figure 21.28(b) are the module of the absolute amplitudes. The widths of the spectra decrease as the signal duration increases and are approximately the reciprocals of the durations of the signals. The comparisons are in Table 21.1. These comparisons give a second rule of thumb:

$$\Delta f \cdot t_d \geqslant 1 \tag{21.53}$$

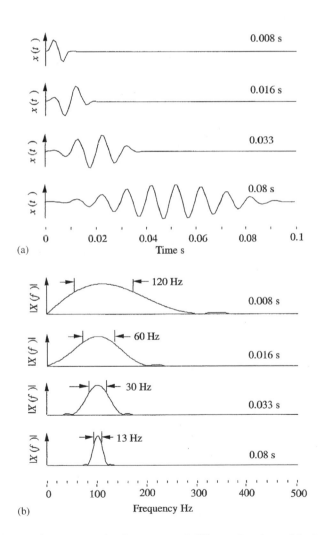

(a)

(b)

FIGURE 21.28 Signals having the same carrier frequency and different durations: (a) signals in the time-domain; (b) module of spectral amplitudes in the frequency-domain. The bandwidths were measured at the half-power points (i.e., at 0.707 of the peak amplitudes).

TABLE 21.1 Comparison between the Signal Duration
and Spectrum Width

t_p, sec	t_d, sec	$\frac{1}{t_d}$, Hz	Δf, Hz
0.01	0.0085	118.0	120
0.02	0.017	58.8	60
0.04	0.034	29.4	30
0.10	0.08	12.5	13

We use the $\geqslant$ sign for $\Delta f \cdot t_d$ because many signals have durations greater than Δf^{-1}. The time t_d gives the minimum duration of a signal for a sonar system to have a bandwidth Δf.

Power Spectra of Random Signals

Sound pressures that have random characteristics are often called noise, whether they are cleverly created as such or are the result of random and uncontrolled processes in the ocean. The spectral analysis of both is the same. As inputs to a spectrum analyzer, they are signals, and their spectral descriptions are to be determined. For a short name, these are called *random signals*.

Signals Having Random Characteristics. In their simplest form, signals that have random characteristics are the result of some process that is not predictable. In honest games, the toss of a coin and the roll of a die give sequences of random events. In the Earth, processes that range from the occurrence and location of earthquakes to rainfall at sea are generators of random signals. For simulations and laboratory tests, we use computers and function generators to make sequences or sets of random numbers. Many of the algorithms generate sequences that repeat, and these algorithms are known as *pseudorandom number generators*. The numerical recipe books give random number-generating algorithms. Programming languages usually include a function call such as *rng* (.) in the library of functions.

Spectral Density and Correlation Methods. The correlation or covariance method of analyzing random signals is discussed in detail by Blackman and Tukey in their monograph *The Measurement of Power Spectra* [1958]. The random signal is the sequence of numbers $x(n)$, and the sequence has $N + k_{max}$ numbers. The covariance of the random signal is the summation:

$$c_{xx}(k) = \begin{cases} \dfrac{1}{N} \displaystyle\sum_{n=0}^{N-1} x(n) \cdot x(n+k); \\ 0 \qquad \text{otherwise} \end{cases} \tag{21.54}$$

The covariance $c_{xx}(k)$ is symmetric, and $c_{xx}(k) = c_{xx}(-k)$. The Fourier transformation of $c_{xx}(k)$ is, using Equation (21.38):

$$C_{xx,\text{fft}}(m) = \sum_{k=0}^{N-1} c_{xx}(k) \cdot e^{-2j\frac{\pi km}{N}} \tag{21.55}$$

The substitution of Equation (21.54) in Equation (21.55) gives:

$$C_{\text{fft}}(m) = \frac{1}{N} \sum_{k=0}^{N-1} \sum_{n=0}^{N-1} x(n) \cdot x(n+k) \cdot e^{-2j\frac{\pi mk}{N}} \tag{21.56}$$

Change variables by letting $j = n + k$, and Equation (21.56) becomes:

$$C_{\text{fft}}(m) = \frac{1}{N} \sum_{n=0}^{N-1} x(n) \cdot e^{2j\frac{\pi mn}{N}} \sum_{k=0}^{N-1} x(k) \cdot e^{-2j\frac{\pi mk}{N}} \tag{21.57}$$

The first summation is the complex conjugate $X^*_{\text{fft}}(m)$ and the second summation is $X_{\text{fft}}(m)$. The spectrum is

$$C_{xx,\text{fft}}(m) = \frac{X^*_{\text{fft}}(m) \cdot X_{\text{fft}}(m)}{N} \tag{21.58}$$

and, using Equation 6.2.27 in [1], the spectral density is

$$C_{xx}(m) = C_{xx,\text{fft}}(m) \cdot t_0 \tag{21.59}$$

where t_0 is the time between samples. Frequency domain expressions for the autocovariance are

$$c_{xx}(\tau) = \int\limits_{-\infty}^{\infty} C_{xx}(f) \cdot e^{2j\pi f \tau} df \tag{21.60}$$

and

$$C_{xx}(f) = \int\limits_{-\infty}^{\infty} c_{xx}(\tau) \cdot e^{-2j\pi f \tau} d\tau \tag{21.61}$$

where $C_{xx}(f)$ has positive and negative frequencies, and $C_{xx}(f) = C_{xx}(-f)$, or

$$c_{xx}(\tau) = 2 \int\limits_{0}^{\infty} C_{xx}(f) \cos(2\pi f \tau) df \tag{21.62}$$

This pair of transformations, Equation (21.61) and Equation (21.62), is known as the Wiener–Khinchine theorem. The *power spectral density* of $x(t)$ is the sum of the positive and negative frequency components:

$$\Pi_{xx}(f) = 2C_{xx}(f) \tag{21.63}$$

Random Signal Simulations: Intensity Spectral Density

In the simulation of a random signal, the random function generator gives a sequence of random numbers: $x(0), x(1)$, and so on. Figure 21.29(a) shows a sequence where the numbers have been connected by interpolation lines. The result of bandpass filtering the input random signal gives a new random signal, the output $y(n)$ in Figure 21.29(b). The operations of bandpass filtering, squaring the signal, and summing or integrating the squared signal are indicated in Figure 21.30. The effective number of independent trials is

$$N_{\text{it}} = t_d \cdot \Delta f_m \tag{21.64}$$

where t_d is the duration of the signal and Δf_m is the filter bandwidth:

$$\Pi_{xx}(f_m) = \frac{1}{N \Delta f_m} \sum_{n=0}^{N-1} y_m^2(n) \tag{21.65}$$

where N is the number of samples.

The filtered signal is $y_m(n)$, where the subscript is added to indicate the filtering by the mth filter. The signal is squared, summed, and averaged over t_d to give the power. Since the mean square output (e.g., volt^2) is

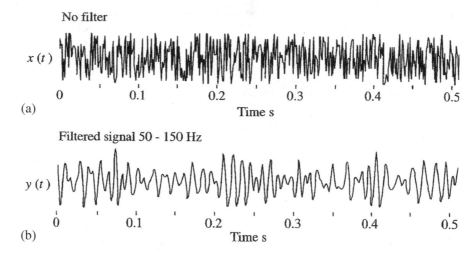

FIGURE 21.29 (a) Random signal created by using a random number generator (the sampling interval was 1 msec). (b) Result of bandpass filtering the signal in (a) through a 50- to 150-Hz bandpass filter and the duration of the signal is t_d.

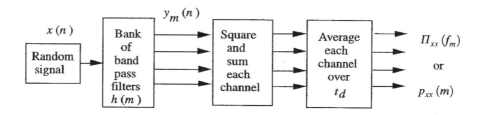

FIGURE 21.30 Block diagram of spectral analysis of a random signal in a computer or a single dedicated instrument; the set of filters $h(m)$ have center frequencies f_m and filter widths Δf_m; the duration of the random signal is t_d.

proportional to the filter bandwidth and the duration of the signal, it is customary to define the power spectral density by Equation (21.65). The first step in deriving an equivalent integral expression for continuous functions of time uses the multiplication and division by t_0:

$$\Pi_{xx}(f_m) = \frac{1}{Nt_0\Delta f_m} \sum_{n=0}^{N-1} y_m^2(n)t_0 \tag{21.66}$$

Let Nt_0 become t_d, the duration of the signal, and t_0 become dt. The summation becomes the integral:

$$\Pi_{xx}(f_m) = \frac{1}{t_d\Delta f_m} \int_0^{t_d} y_m^2(n)\mathrm{d}t \tag{21.67}$$

If $x(n)$ has the units of volts, the so-called power spectral density has units of (volts)^2/Hz. True power spectral density would require division by a load resistance in an electrical circuit to give watts/Hz. Since a hydrophone output in volts is proportional to the acoustic pressure, $x(n)$ has the units of Pa, the spectral density has units of (Pa)^2/Hz, and the true *intensity spectral density* requires division by $\rho_A c$, to give

$(\text{Pa})^2/\rho_A c\text{Hz} = \text{watts/m}^2\text{Hz}$. Acoustic spectra are often reported in dB relative to one $(\mu\text{Pa})^2/\text{Hz}$, so that the intensity spectrum level (ISL):

$$\text{ISL} = 10\,\log_{10}\left\{\frac{\Pi_{xx}(f_m)}{\dfrac{(\mu\text{Pa})^2}{\text{Hz}}}\right\} \tag{21.68}$$

The spectrum levels depend on the reference sound pressure, which is sometimes unclear. It is better to use *Pascal* units such as $(\text{Pa})^2/\text{Hz}$ or $\text{watts/m}^2\text{Hz}$.

Spectral Smoothing. Consider the following example of spectrum analysis. A random signal is constructed of 512 magnitudes at separation $t_0 = 0.001$ sec and duration 0.512 sec. Figure 21.31 shows the results of processing the signal by the equivalents of very narrow, wide, and very wide bandpass filters. The output of the narrow 2-Hz filter (Figure 21.31(a)) is extremely rough. The number of independent samples given by Equation (21.54) in the 2-Hz bandpass filter is 1. Figure 21.31(b) shows the result of using a wider filter, $\Delta f = 64$ Hz. Here the number of independent samples is 32. The spectrum is much smoother and has less detail. An increase in the filter width to $\Delta f = 128$ Hz and the number of independent samples to 64 is shown in Figure 21.31(c). Another random signal would have a different spectrum. These examples show the basic trade-off between resolution and reduction of roughness or variance of the estimate of the spectral density. The importance of smoothing power spectra and the trade-off between the reduction of frequency resolution and the reduction of fluctuations is given in detail by Blackman and Tukey in their monograph *The Measurement of Power Spectra* [1958].

Traditional Measures of Sound Spectra. The measurement of underwater sounds has inherited the instrumentation and the vocabulary that were developed for measurements of sounds heard by humans in air.

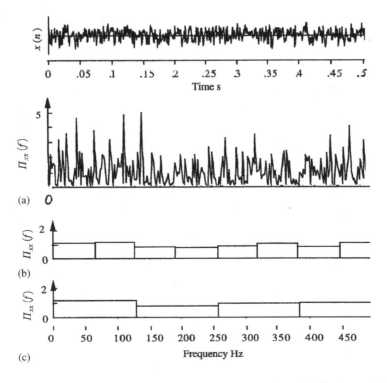

FIGURE 21.31 Smoothing of power spectra by filtering. The top trace is a random signal $x(n)$ or $x(t)$. Filter bandwidths are (a) $\Delta f = 2$ Hz, (b) $\Delta f = 64$ Hz, and (c) $\Delta f = 128$ Hz.

The principal areas of interest to humans have been acoustic pressure threshold for hearing; acoustic threshold of damage to hearing; threshold for speech communication in the presence of noise; and community response to annoying sounds. The vast amount of data required to evaluate human responses, and then to communicate the recommendations to laymen, forced psychoacousticians and noise-control engineers to adopt simple instrumentation and a simple vocabulary that would provide simple numbers for complex problems. Originally this was appropriate to the analog instrumentation. But even now digital measurements are reported according to former constraints. For example, the *octave band*, which is named for the eight notes of musical notation that corresponds to the 2:1 ratio of the top of the frequency band to the bottom, remains common in noise-control work. For finer analysis, one-third octave band instruments are used; they have an upper-to-lower-band frequency ratio of $2^{0.33}$, so that three bands span one octave.

The use in water of instruments and references that were designed for air has caused great confusion. The air reference for acoustic pressure level in dB was logically set at the threshold of hearing (approximately $20\,\mu$Pa at 1000 Hz) for the average adult human. This is certainly not appropriate for underwater measurements, where the chosen reference is $1\,\mu$Pa or 1 Pa. Furthermore, plane-wave intensity of CW is calculated from Equation 2.5.16 in Reference 1, where Intensity $= P_{rms}^2/\rho_A c$ (where P_{rms}^2 is the mean squared pressure; ρ_A is the water density; and c is the speed of sound in water). Therefore, the dB reference for sound intensity in water is clearly different from that in air because the specific acoustic impedance $\rho_A c$ is about 420 kg/m^{2}sec for air compared with 1.5×10^6 for water. This ratio corresponds to about 36 dB, if one insists on using the decibel as a reference.

The potential for confusion in describing the effects of sound on marine animals is aggravated when physical scientists use the decibel notation in talking to biological scientists. Confusion will be minimized if psychoacoustical characteristics of marine mammals—such as thresholds of pain, hearing communication perception, and so forth—are described by the use of SI units (i.e., pascals; acoustic pressure at a receiver), watts/m^2 (acoustic intensity for CW at a receiver) and joules/m^2 (impulse energy/area at a receiver). Likewise, only SI units should be used for sources—that is, watts (power output of a continuous source) and joules (energy output of a transient impulse source). The directivity of the source should always be part of its specification. All of these quantities are functions of sound frequency and can be expressed as *spectral* densities (i.e., per 1-Hz frequency band).

Matched Filters and Autocorrelation

The coded signal and its matched filter and associated concepts have become very important in the applications to sound transmission in the ocean. The simple elegance of the original paper [8] is well worth a trip to the library. The generality of their concepts was far ahead of the then-existing signal processing methods. Digital signal processing facilitates the design of many types of filters for processing sonar signals. Each definition of an optimum condition also defines a class of optimum filter. We limit our discussion to the simplest of the optimum filters, the *matched filter* [8]. An example is shown in Figure 21.32.

An example of a coded signal $x(n)$ is shown in Figure 21.32(b). Recalling the convolution summation, Equation 6.2.29 in Reference 1, the convolution of $h_M(n)$ and $x(n)$ is

$$y_M(j) = \sum_{m=0}^{m_1} x(m) \cdot h_M(j-m) \qquad (21.69)$$

where the subscript $_M$ means the matched filter. The matched filter uses the criterion that the square of the peak output value $y_M(0)$ is a maximum. To maximize the square of $y_M(0)$, we use Cauchy's inequality [9]:

$$y_M^2(0) \leqslant \sum_{m=0}^{m_1} h_M^2(m) \times \sum_{m=0}^{m_1} x^2(m) \qquad (21.70)$$

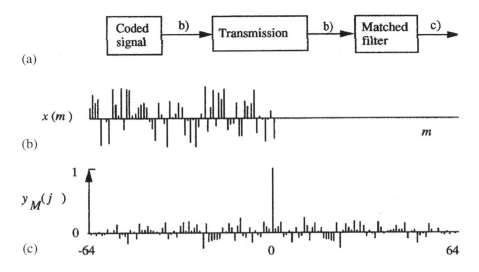

FIGURE 21.32 Coded signal, matched filter, and output: (a) simple matched filter system; (b) coded signal (here, the coded signal is a short sequence from a random-number generator); (c) output of the matched filter.

The filter produces a maximum output (the equal sign in Equation (21.70)) when $x(m)$ is time-reversed to $x(m) = x(-m)$:

$$h_{\mathrm{M}}(m) = Ax(-m) \quad \text{or} \quad h_{\mathrm{M}}(-m) = Ax(m) \tag{21.71}$$

where A is a constant of proportionality. In our examples we use $A = \dfrac{1}{m_1 + 1}$. The filter defined by Equation (21.71) is called the *matched filter*. In computations, it is convenient to shift the indices by m_1 and to shift the time by $m_1 t_0$:

$$h_{\mathrm{M}}(m_1 - m) = h_{\mathrm{M}}(-m) \tag{21.72}$$

to give a causal filter. Except for constants and normalization, $y_{\mathrm{M}}(j)$ has the same form as the autocorrelation (see Equation (21.54)):

$$y_{\mathrm{M}}(j) = A \sum_{m=0}^{m_1} x(m) \cdot x(m + j) \tag{21.73}$$

The output of the matched filter is shown in Figure 21.32(c). Here, the output of the matched filter is the autocorrelation or covariance of $x(n)$. The signal-to-noise amplitude ratio gain is proportional to the square root of the number of independent samples of the coded signal. While we do not prove it, ignoring noise in our derivation is equivalent to assuming that the noise is an uncorrelated sequence of random numbers having a mean value of zero. This kind of noise is often called *white noise*.

In 1962, Parvulescu obtained a classified patent for the use of the *matched equivalent signal*, for measuring the reproducibility of signal transmissions over large ranges in the ocean. In this first use of the matched filter technique in the ocean, the multipath received signals were regarded as a coded signal. An analog tape recorder was employed, with the tape direction reversed to convert the multipath arrivals into a matched filter [10,11].

Summary

Development and employment of acoustical techniques allow us to image underwater features, communicate information via the oceanic wave-guide or measure oceanic properties. Representative applications of these techniques can be summarized in the following form:

- *Image underwater features*: detection, classification and localization of objects in the water column and in the sediments using monostatic or bistatic sonars; obstacle avoidance using forward-looking sonars; navigation using echo sounders or sidescan sonars to recognize sea-floor topographic reference features.
- *Communicate information via the oceanic wave-guide*: acoustic transmission and reception of voice or data signals in the oceanic wave-guide; navigation and docking guided by acoustic transponders; release of moored instrumentation packages using acoustically activated mechanisms.
- *Measure oceanic properties*: measurement of ocean volume and boundaries using either direct or indirect acoustical methods; acoustical monitoring of the marine environment for regulatory compliance; acoustical surveying of organic and inorganic marine resources.

Nomenclature

Symbol	Quantity	Symbol	Quantity
ASE	Amplified spontaneous emission	IFFT	Inverse finite Fourier transformation
ASW	Anti submarine warfare		
CW	Continuous waves	ISL	Intensity spectrum level
DFT	Discrete Fourier transformation	SI	Sound intensity
FFT	Finite Fourier transformation	TVG	Time-varying gain

References

1. H. Medwin and C. Clay, *Fundamentals of Acoustical Oceanography*, London: Academic Press, 1998.
2. H. Medwin, "Speed of sound in water: a simple equation for realistic parameters," *J. Acoust. Soc. Am.*, 58, 1318–1319, 1975.
3. V. Knudsen, R. Alford, and J. Emliing, "Underwater ambient noise," *J. Marine Res.*, 7, 410–429, 1948.
4. J. Traynor and J. Ehrenberg, "Fish and standard-sphere, target-strength measurements obtained with a dual-beam and split-beam echo-sounding system," *Rapp. P.-V. Reun. Cons. Int. Explor. Mer.*, 189, 325–335, 1990.
5. R. Craig and S. Forbes, "Design of a sonar for fish counting," *Fisk. Dir. Ser. Havunders.*, 15, 210–219, 1969.
6. P. Etter, *Underwater Acoustic Modeling and Simulation*, 3rd ed., London: Spon Press, 2003.
7. L. Ziomek, *Acoustic Field Theory and Space–Time Signal Processing*, Boca Raton, FL: CRC Press, 1995.
8. J. Van Vleck and D. Middleton, "A theoretical comparison of the visual, aural, and meter reception of pulsed signals in the presence of noise," *J. Appl. Phys.*, 17, 940–971, 1946.
9. M. Abramowitz and I. Stegun, *Handbook of Mathematical Functions*, National Bureau of Standard Applied Mathematical Series 55, Washington, D.C.: Government Printing Office, 1964.
10. A. Parvulescu, "Matched-signal (MESS) processing by the ocean," *J. Acoust. Soc. Am.*, 98, 943–960, 1995.
11. I. Tolstoy and C. Clay, *Ocean Acoustics*, New York: McGraw-Hill, 1966. Reprinted by the Acoustical Society of America, Woodbury, NY, 1987.

22

Neural Networks and Adaptive Signal Processing

Jose C. Principe
University of Florida

Mohamed Ibnkahla
Queen's University

Ahmad Iyanda Sulyman
Queen's University

Yu Cao
Queen's University

22.1   Artificial Neural Networks

Jose C. Principe

Definitions and Scope

Introduction

Artificial neural networks are one of the newest signal processing technologies in the engineer's toolbox. The field is highly interdisciplinary, but our approach will restrict the view to the engineering perspective. In engineering, neural networks serve two important functions: pattern classifiers and nonlinear adaptive filters. We will provide a brief overview of the theory, learning rules, and applications of the most important neural network models.

Definitions and Style of Computation

An artificial neural network (ANN) is an adaptive, most often nonlinear system that learns to perform a function (an input–output map) from data. The input output map is also called the *neural network topology*, and it includes the perceptron, the multilayer perceptron, the radial basis function network, or the recurrent topologies, to name just a few. Adaptive means that the system parameters are changed during operation, normally called the *training phase*. After the training phase, the ANN parameters are fixed and the system is deployed to solve the problem at hand (the *testing phase*). The ANN includes a systematic step-by-step procedure to minimize a performance criterion or to follow some implicit internal constraint, which is commonly referred to as the *learning rule*. The input–output training data are fundamental in neural network technology because they convey the necessary information to "discover" the optimal operating point. The nonlinear nature of the neural network processing elements provides the system with lots of flexibility to achieve practically any desired input–output map, i.e., some ANNs are *universal mappers*.

There is a style in neural computation that is worth describing (Figure 22.1). An input is presented to the network and a corresponding desired or target response set at the output (when this is the case, the training is

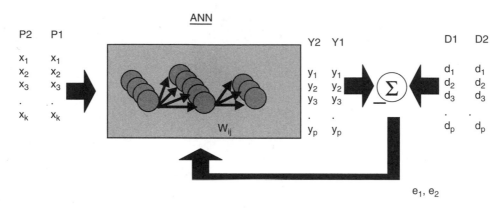

FIGURE 22.1 The style of neural computation. (*Source:* Principe, J. et al., *Neural and Adaptive Systems: Fundamentals through Simulation*, New York: John Wiley & Sons, 2000.)

called *supervised*). An error is composed from the difference between the desired response and the system output. This error information is fed back to the system and adjusts the system parameters in a systematic fashion (the learning rule). The process is repeated until the performance is acceptable. It is clear from this description that the performance hinges heavily on the data. If one does not have data that covers a significant portion of the operating conditions, or if they are noisy, then neural network technology is probably not the right solution. However, if there are plenty of data and the problem is poorly understood to derive an approximate model, then neural network technology is a good choice.

This operating procedure should be contrasted with the traditional engineering design, made of exhaustive subsystem specifications, and inter-communication protocols. In ANNs, the designer chooses the network topology, the performance function, the learning rule, and the criterion to stop the training phase, but the system automatically adjusts the parameters. So it is difficult to bring in *a priori* information into the design, and when the system does not work properly, it is also hard to incrementally refine the solution. However, ANN-based solutions are extremely efficient in terms of development time and resources, and in many difficult problems ANNs provide performance that are difficult to match with other technologies. Denker, 15 years ago, said that "ANNs are the second best way to implement a solution" exactly due to the simplicity of design and their universality, only shadowed by the traditional design obtained by studying the physics of the problem, but which becomes too difficult for complex problems. Presently, ANNs are emerging as the technology of choice for many applications as pattern recognition, data mining, time-series analysis, system identification, and control.

ANNs Types and Applications

It is always risky to establish a taxonomy of a technology, but our motivation is one of providing a quick overview of the application areas and the most popular topologies and learning paradigms.

Application	Topology	Supervised Learning	Unsupervised Learning
Association	Hopfield [1,2] Multilayer perceptron [1,2,3] Linear Associative. Mem. [1,2]	Backpropagation [1,2,3]	Hebbian [1,2,4] Hebbian
Pattern recognition	Multilayer Perceptron [1,2,3] Radial Basis Functions [1,3]	Backpropagation LMS [1]	k-Means [3]
Feature extraction	Competitive [1,2] Kohonen [1,2] Multilayer Perceptron [4] Principal Comp. Anal. [1,4]	Backpropagation	Competitive Kohonen Oja's [1,4]
Prediction system ID	Time Lagged Networks [1,4,5] Fully Recurrent Networks [1]	Backpropagation through time [1]	

It is clear that *multilayer perceptrons* (*MLPs*) and the *backpropagation algorithm* and its extensions (time lagged networks and backpropagation through time [BPTT], respectively) hold a prominent position in the ANN technology. It is therefore only natural to spend most of our overview presenting the theory and tools of backpropagation learning. It is also important to notice that *Hebbian learning* (and its extension, the Oja rule) is also a very useful (and biologically plausible) learning mechanism. It is an *unsupervised learning* method since there is no need to specify the desired or target response to the ANN.

Multilayer Perceptrons

Multilayer perceptrons are a layered arrangement of nonlinear processing elements (PEs), as shown in Figure 22.2. The layer that receives the input is called the *input layer*, and the layer that produces the output is the *output layer*. The layers that do not have direct access to the external world are called *hidden layers*. A layered network with just the input and output layers is called the *perceptron*. Each connection between PEs is weighted by a scalar w_i, called a *weight*, which is adapted during learning.

The PEs in the MLP are composed of an adder followed by a smooth saturating nonlinearity of the sigmoid type (Figure 22.3). The most common saturating nonlinearities are the logistic function and the hyperbolic tangent. The threshold is used in other nets.

The importance of the MLP is that it has been shown to be a universal mapper (implements arbitrary input–output maps) when the topology has at least two hidden layers (Haykin, 1994). Even MLPs with a single layer are able to approximate continuous input–output maps. This means that rarely we will need to choose topologies with more than two hidden layers. However, these are existence proofs, so the issue that we must solve as engineers is to choose how many layers and how many PEs in each layer are required to produce good results.

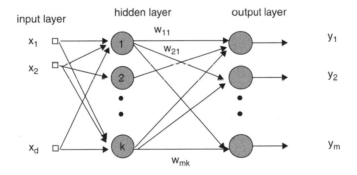

FIGURE 22.2 Multilayer perceptron (MLP) with one hidden layer (d–k–m). (*Source:* Principe, J. et al., *Neural and Adaptive Systems: Fundamentals through Simulation*, New York: John Wiley & Sons, 2000.)

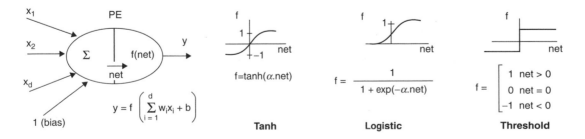

FIGURE 22.3 A processing element (PE) and most common nonlinearities. (*Source:* Principe, J. et al., *Neural and Adaptive Systems: Fundamentals through Simulation*, New York: John Wiley & Sons, 2000.)

Many problems in engineering can be thought of as a transformation of an input space, containing the input to an output space where the desired response exists. For instance, dividing data into classes can be thought of as transforming the input into 0 and 1 responses that will code the classes (Bishop, 1995). Likewise, identification of an unknown system can be also be framed as a mapping (function approximation) from the input to the system output (Kung, 1993). The MLP is highly recommended for these applications.

The Function of Each PE

Let us study briefly the function of a single PE with two inputs (Zurada, 1992). If the nonlinearity is the threshold nonlinearity, we can immediately see that the output is simply 1 and −1. The surface that divides these sub-spaces is called a *separation surface* and in this case it is a line of equation:

$$y(w_1, w_2) = w_1 x_1 + W_2 x_2 + b = 0 \tag{22.1}$$

i.e., the PE weights and the bias control the orientation and position of the separation line, respectively (Figure 22.4). In many dimensions the separation surface becomes a hyperplane of dimension one less than the dimensionality of the input space. So, each PE creates a dichotomy in input space. For smooth nonlinearities, the separation surface is not crisp; it becomes fuzzy but the same principles apply. In this case, the size of the weights control the with of the fuzzy boundary (larger weights shrink the fuzzy boundary).

The perceptron input–output map is built from a justaposition of linear separation surfaces so, as a classifier, the perceptron gives zero classification error only for *linearly separable classes* (i.e., classes that can be exactly classified by hyperplanes).

When one adds one layer to the perceptron creating a one-hidden-layer MLP, the type of separation surfaces changes drastically. It can be shown that this learning machine is able to create "bumps" in the input space, i.e., an area of high response surrounded by low responses (Principe et al., 2000). The function of each PE is always the same, no matter if the PE is part of a perceptron or MLP. However, notice that the output layer in the MLP works with the result of the hidden layer activations, creating an embedding of functions and producing more complex separation surfaces. The one-hidden-layer MLP is able to produce *nonconvex separation surfaces*, which can be interpreted as an *universal mapper*. If one adds an extra layer (i.e., two hidden layers), the learning machine can now create and combine at will "bumps," i.e., areas that correspond to one class surrounded by areas that belong to the other class. One important aspect to remember is that changing a single weight in the MLP can drastically change the location of the separation surfaces, i.e., the MLP achieves the input–output map through the interplay of all its weights.

How to Train MLPs

One fundamental issue is how to adapt the weights w_i of the MLP to achieve a given input–output map. The core ideas have been around for many years in optimization, and they are extensions of well-known

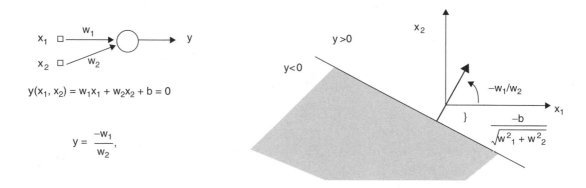

FIGURE 22.4 A two-input PE and its separation surface.

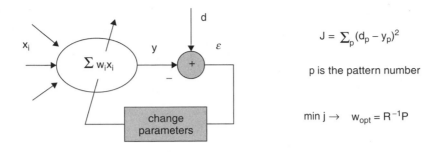

FIGURE 22.5 Computing analytically optimal weights for the linear PE.

engineering principles such as the *least mean square (LMS) algorithm* of adaptive filtering (Principe et al., 2000). Let us review the theory here. Assume that we have a linear PE (f(net)=net), and that one wants to adapt the weights as to minimize the square difference between the desired signal and the PE response (Figure 22.5).

This problem has an analytic solution known as *least squares* (Haykin, 1994). The optimal weights are obtained as the product of the inverse of the input autocorrelation function (R^{-1}) and the cross-correlation vector (P) between the input and the desired response. The analytical solution is equivalent to a search for the minimum of the quadratic performance surface $J(w_i)$ using gradient descent, i.e., where the weights at each iteration k are adjusted by

$$w_i(k+1) = w_i(k) - \eta \Delta J_i(k) \quad \Delta J_i = \frac{\partial J}{\partial w_i} \quad (22.2)$$

where η is a small constant called the *step size*, and $\nabla J(k)$ is the gradient of the performance surface at iteration k. Bernard Widrow in the late 1960s proposed a very efficient estimate to compute the gradient at each iteration:

$$\nabla J_i(k) = \frac{\partial}{\partial w_i} J(k) \sim \frac{1}{2} \frac{\partial}{\partial w_i} (\varepsilon^2(k)) = -\varepsilon(k) x_i(k) \quad (22.3)$$

which when substituted into Equation (22.2) produces the so-called LMS algorithm. He showed that the LMS converged to the analytic solution provided the step size η is small enough. Since it is a steepest descent procedure, the largest step size is limited by the inverse of the largest eigenvalue of the input autocorrelation matrix. The larger the step size (below this limit), the faster is the convergence, but the final values will "rattle" around the optimal value in a basin that has a radius proportional to the step size. Hence, there is a fundamental trade-off between speed of convergence and accuracy in the final weight values. One great appeal of the LMS algorithm is that it is very efficient (just one multiplication per weight), and that it requires only local quantities to be computed.

The LMS algorithm can be framed as a computation of partial derivatives of the cost with respect to the unknowns, i.e., the weight values. In fact, if one writes

$$\frac{\partial J}{\partial w_i} = \frac{\partial J}{\partial y} \frac{\partial y}{\partial w_i} = \frac{\partial}{\partial y} \left(\sum (d-y)^2 \right) \frac{\partial}{\partial w_i} \left(\sum w_i x_i \right) = -\varepsilon x_i \quad (22.4)$$

we obtain the LMS algorithm for the linear PE. What happens if the PE is nonlinear? If the nonlinearity is differentiable (smooth), we still can apply the same method, because of the chain rule, which prescribes

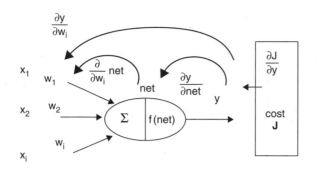

FIGURE 22.6 How to extend LMS to nonlinear PEs with the chain rule. (*Source:* Principe, J. et al., *Neural and Adaptive Systems: Fundamentals through Simulation*, New York: John Wiley & Sons, 2000.)

that (Figure 22.6):

$$\frac{\partial J}{\partial w_i} = \frac{\partial J}{\partial y} \frac{\partial y}{\partial \text{net}} \frac{\partial}{\partial w_i} \text{net} = -(d-y)\dot{f}(\text{net})x_i = -\varepsilon \dot{f}(\text{net})x_i \tag{22.5}$$

where $f'(\text{net})$ is the derivative of the nonlinearity computed at the operating point. Equation (22.5) is known as the *delta rule*, and it will train the perceptron (Principe et al., 2000). Note that throughout the derivation we skipped the pattern index p for simplicity, but this rule is applied for each input pattern. However, the delta rule cannot train MLPs since it requires the knowledge of the error signal at each PE.

The principle of the ordered derivatives can be extended to multilayer networks provided we organize the computations in flows of activation and error propagation. The principle is very easy to understand, but a little complex to formulate in equation form (Principe et al., 2000).

Suppose that we want to adapt the weights connected to a hidden layer PE, the ith PE (Figure 22.7).

One can decompose the computation of the partial derivative of the cost with respect to the weight w_{ij} as:

$$\frac{\partial J}{\partial w_{ij}} = \underbrace{\frac{\partial J}{\partial y_i}}_{1} \underbrace{\frac{\partial y_i}{\partial \text{net}_i} \frac{\partial}{\partial w_{ij}} \text{net}_i}_{2} \tag{22.6}$$

i.e., the partial with respect to the weight is the product of the partial with respect to the PE state (part 1 in Equation (22.6)) times the partial of the local activation to the weights (part 2 in Equation (22.6)). This last quantity is exactly the same as for the nonlinear PE ($f'(\text{net}_i)x_j$), so the big issue is the computation of $\partial J/\partial y$. For an output PE, $\partial J/\partial y$ becomes the injected error ε (Equation (22.4)). For the hidden ith PE, $\partial J/\partial y$ is evaluated by summing all the errors that reach the PE from the top layer through the topology when the

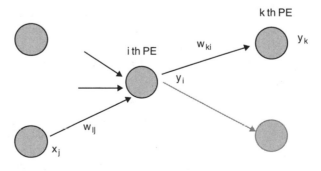

FIGURE 22.7 How to adapt the weights connected to ith PE.

injected errors ε_k are clamped at the top layer, or in an equation:

$$\frac{\partial J}{\partial y_i} = \left(\sum_k \frac{\partial J \partial y_k}{\partial y_k \partial \text{net}_k} \frac{\partial}{\partial y_i} \text{net}_k \right) = \sum_k \varepsilon_k f'(\text{net}_k) w_{ki} \qquad (22.7)$$

Substituting back in Equation (22.6), we finally get

$$\frac{\partial J}{\partial w_{ij}} = - \underbrace{x_j f'(\text{net}_i)}_{1} \underbrace{\sum_k \varepsilon_k f'(\text{net}_k) w_{ki}}_{2} \qquad (22.8)$$

This equation embodies the *backpropagation training algorithm* (Haykin, 1994; Bishop, 1995). It can be rewritten as the product of a local activation (part 1) and a local error (part 2), exactly as the LMS and the delta rules. However, now the local error is a composition of errors that flow through the topology, which becomes equivalent to the existence of a desired response at the PE.

There is an intrinsic flow in the implementation of the backpropagation algorithm: first, inputs are applied to the net, and activations computed everywhere to yield the output activation. Second, the external errors are computed by subtracting the net output from the desired response. Third, these external errors are utilized in Equation (22.8) to compute the local errors for the layer immediately preceding the output layer, and the computations chained up to the input layer. Once all the local errors are available, Equation (22.2) can be used to update every weight. These three steps are then repeated for other training patterns until the error is acceptable.

Step three is equivalent to injecting the external errors in the *dual or adjoint topology* and backpropagate them up to the input layer (Principe et al., 2000). The dual topology is obtained from the original one by reversing data flow, and substituting summing junctions by splitting nodes and vice versa. The error at each PE of the dual topology is then multiplied by the activation of the original network to compute the weight updates. So, effectively, the dual topology is being used to compute the local errors, which makes the procedure highly efficient. This is the reason backpropagation trains a network of N PEs with a number of multiplications proportional to N, (O(N)), instead of (O(N²)) for previous methods of computing partial derivatives known in control theory. Using the dual topology to implement backpropagation is the best and most general method to program the algorithm in a digital computer.

Applying Backpropagation in Practice

Now that we know how to train MLPs, let us see what are the practical issues to apply it. We will address the following aspects: size of training set versus weights, search procedures, how to stop training, how to set the topology for maximum generalization.

Size of Training Set. The size of the training set is very important for good performance. Remember that the ANN gets its information from the training set. If the training data does not cover the full range of operating conditions, the system may perform badly when deployed. Under no circumstances should the training set be less than the number of weights in the ANN. A good size for the training data is ten times the number of weights in the network, with the lower limit being set around three times the number of weights (these values should be taken as an indication, subject to experimentation for each case) (Haykin, 1994).

Search Procedures. Going in the direction of the gradient is fine if the performance surface is quadratic. However, in ANNs rarely is this the case due to the use of nonlinear PEs and topologies with several layers. So, gradient descent can be caught in *local minima* and make the search very slow in regions of small curvature. One efficient way to speed up the search in regions of small curvature and at the same time stabilize it in narrow valleys is to include a momentum term in the weight adaptation:

$$w_{ij}(n+1) = w_{ij}(n) + \eta\delta(n)x_j(n) + \alpha(w_{ij}(n) - w_{ij}(n-1)) \tag{22.9}$$

The value of momentum α should be set experimentally between 0.5 and 0.9. There are many more modifications to the conventional gradient search, such as adaptive step sizes, annealed noise, conjugate gradients, and second-order methods (using information contained in the Hessian matrix), but the simplicity of momentum learning is hard to beat (Principe et al., 2000).

How to Stop Training. The stop criterion is a fundamental aspect of training. The simple ideas of capping the number of iterations or letting the system train until the error reaches a preset value are not recommended. The reason is that we want the ANN to perform well in the test set data, i.e., we would like the system to perform well in data it never saw before (good *generalization*) (Bishop, 1995). The error in the training set tends to decrease with iteration when the ANN has enough degrees of freedom to represent the input–output map. However, the system may be remembering the training patterns (*overfitting*) instead of finding the underlying mapping rule. This is called *overtraining*. To avoid overtraining, the performance in a *validation set*, i.e., a set of input data that the system never saw before, must be checked regularly during training (i.e., once every 50 passes over the training set). The training should be stopped when the performance in the validation set starts to increase, albeit that the performance in the training set continues to decrease. This method is called *early stopping* based on cross-validation. The validation set should be 10% of the training set and distinct from it.

The Size of the Topology. The size of the topology should also be carefully selected. If the number of layers or the size of each layer is too small, the network does not have enough degrees of freedom to classify the data or to approximate the function, and the performance suffers.

However, if the size of the network is too large, performance may also suffer. This is the phenomenon of *overfitting* that we mentioned above, but one alternate way to control it is to reduce the size of the network. There are basically two procedures to set the size of the network: either one starts small and adds new PEs, or one starts with a large network and prunes PEs (Haykin, 1994). One quick way to prune the network is to impose a penalty term in the performance function on a *regularizing term* — such as limiting the slope of the input–output map (Bishop, 1995). A regularization term that can be implemented locally is

$$w_{ij}(n+1) = w_{ij}(n)\left(1 - \frac{\lambda}{(1 + w_{ij}(n))^2}\right) + \eta\delta_i(n)x_j(n) \tag{22.10}$$

where λ is the *weight decay* parameter and δ the local error. Weight decay tends to drive unimportant weights to zero.

A *Posteriori* Probabilities

We will finish the discussion of the MLP by noting that this topology is able to estimate directly at its outputs *a posteriori probabilities*, i.e., the probability that a given input pattern belongs to a given class (Bishop, 1995). This property is very useful because the MLP outputs can be interpreted as probabilities. In order to guarantee this property, one has to make sure that each class is attributed to one output PE, that the topology is sufficiently large to represent the mapping, that the training has converged to the absolute minimum, and that the outputs are normalized between 0 and 1. The first requirements are met by good design, while the last can be easily enforced if the *softmax activation* is used as the output PE (Bishop, 1995):

$$y = \frac{\exp(\text{net})}{\sum_j \exp(\text{net}_j)} \tag{22.11}$$

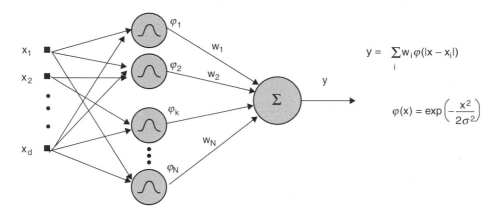

FIGURE 22.8   Radial basis function (RBF) network. (*Source:* Principe, J. et al., *Neural and Adaptive Systems: Fundamentals through Simulation*, New York: John Wiley & Sons, 2000.)

Radial Basis Function Networks

The radial basis function (RBF) network constitutes another way of implementing arbitrary input–output mappings. The most significant difference between the MLP and RBF is in the PE nonlinearity. While the PE in the MLP responds to the full input space, the PE in the radial basis function is local, normally a Gaussian kernel in the input space. Hence, it only responds to inputs that are close to its center, i.e., it has basically a *local response.*

The radial basis function network is also a layered net with the hidden layer built from Gaussian kernels and a linear (or nonlinear) output layer (Figure 22.8). Training of the RBF network done normally in two stages (Principe et al., 2000): first, the centers x_i are adaptively placed in the input space using competitive learning or k means clustering (Bishop, 1995), which are unsupervised procedures. Competitive learning is explained later in the chapter. The variances of each Gaussian is chosen as a percentage (30 to 50%) to the distance to the nearest center. The goal is to cover adequately the input data distribution. Once the RBF is located, the second layer weights w_i are trained using the LMS procedure.

RBF networks are easy to work with, they train very fast, and have shown good properties both for function approximation and classification. The problem is that they require lots of Gaussian kernels in high-dimensional spaces.

Time Lagged Networks (TLNs)

The MLP is the most common neural network topology, but it can only handle instantaneous information, since the system has no memory and it is feedforward. In engineering, the processing of signals that exist in time require systems with memory, i.e., linear filters. Another alternative to implement memory is to use feedback, which gives rise to *recurrent networks*. Fully recurrent networks are difficult to train and to stabilize, so it is preferable to develop topologies based on MLPs but where explicit subsystems to store the past information are included. These subsystems are called *short-term memory structures* (Principe et al., 2000). The combination of an MLP with short-term memory structures is called a *time lagged network* (*TLN*). The memory structures can be eventually recurrent, but the feedback is local so stability is still easy to guarantee. Here, we will cover just one TLN topology called *focused* where the memory is at the input layer. The most general TLN has memory added anywhere in the network, but requires other more involved training strategies (back propagation through time (Haykin, 1994)). The interested reader is referred to Principe et al. (2000) for further details, or to NeuroSolutions (1993) for simulation software.

The function of a short-term memory in the focused TLN is to represent the past of the input signal, while the nonlinear PEs provide the mapping as in the MLP (Figure 22.9).

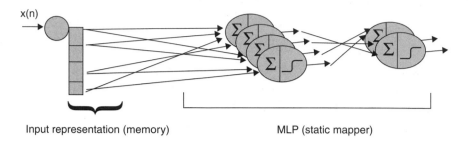

FIGURE 22.9 A focused time lagged network.

Memory Structures

The simplest memory structure is built from a tap delay line (Figure 22.10). The memory by delays is a single-input, multiple-output system that has no free parameters except its size K. The tap delay memory is the memory utilized in the time delay neural network (TDNN) and has been utilized successfully in speech recognition and system identification (Principe et al., 2000).

A different mechanism for linear memory is feedback (Figure 22.11). Feedback allows the system to remember past events because of the exponential decay of the response.

This memory has limited resolution due to the lowpass required for long memories, but notice that unlike memory by delay, memory by feedback provides the learning system with a free parameter μ that controls the length of the memory. Memory by feedback has been used in Elman and Jordan networks (Haykin, 1994).

It is possible to combine the advantages of memory by feedback with the ones of the memory by delays in linear systems called *dispersive delay lines*. The most studied of these memories is a cascade of lowpass functions called the *gamma memory* (deVries and Principe, 1992).

The gamma memory has a free parameter μ that controls and decouples memory depth from resolution of the memory. *Memory depth D* is defined as the first moment of the impulse response from the input to the last tap K, while *memory resolution R* is the number of taps per unit time. For the gamma memory $D = K/\mu$, and $R = \mu$, i.e., changing μ modifies the memory depth and resolution inversely. This recursive parameter μ can be adapted with the output MSE as the other network parameters, i.e., the ANN is able to choose the best memory depth to minimize the output error, which is unlike tap delay memory.

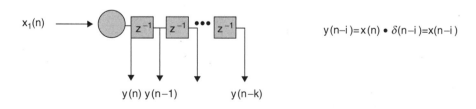

FIGURE 22.10 Tap delay line memory.

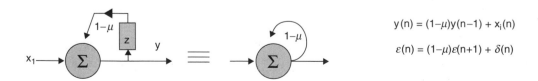

FIGURE 22.11 Memory by feedback (context PE). (*Source:* Principe, J. et al., *Neural and Adaptive Systems: Fundamentals through Simulation*, New York: John Wiley & Sons, 2000.)

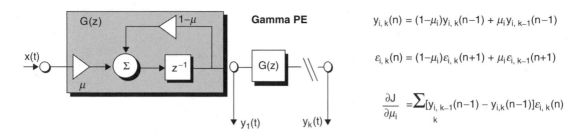

$$y_{i,k}(n) = (1-\mu_i)y_{i,k}(n-1) + \mu_i y_{i,k-1}(n-1)$$

$$\varepsilon_{i,k}(n) = (1-\mu_i)\varepsilon_{i,k}(n+1) + \mu_i \varepsilon_{i,k-1}(n+1)$$

$$\frac{\partial J}{\partial \mu_i} = \sum_k [y_{i,k-1}(n-1) - y_{i,k}(n-1)]\varepsilon_{i,k}(n)$$

FIGURE 22.12   Gamma memory (dispersive delay line).

Training Focused TLN Architectures.   The appeal of the focused architecture is that the MLP weights can be still adapted with backpropagation. However, the input–output mapping produced by these networks is static. The input memory layer is bringing in past input information to establish the value of the mapping. As we know in engineering, the size of the memory is fundamental to identify, for instance, an unknown plant or to perform prediction with a small error. However, note now that with the focused TLN, the models for system identification become nonlinear (i.e., nonlinear moving average [NMA]).

When the tap delay implements the short-term memory, straight backpropagation can be utilized since the only adaptive parameters are the MLP weights. When the gamma memory is utilized (or the context PE), the recursive parameter is adapted in a total adaptive framework (or preset the parameter by some external consideration). The equations to adapt the context PE and the gamma memory are shown in Figure 22.11 and Figure 22.12, respectively. For the context PE $\delta(n)$ refers to the total error that is backpropagated from the MLP and that reaches the dual-context PE.

Hebbian Learning and Principal Component Analysis (PCA) Networks

Hebbian learning is an unsupervised learning rule that *captures similarity* between an input and an output through *correlation*. To adapt a weight w_i using Hebbian learning, we adjust the weights according to $\Delta w_i = \eta x_i y$ or in an equation (Haykin, 1994):

$$w_i(n+1) = w_i(n) + \eta x_i(n)y(n) \tag{22.12}$$

where η is the step size, x_i is the ith input and y is the PE output.

The output of the single PE is an inner product between the input and the weight vector (formula in Figure 22.13). It measures the similarity between the two vectors; i.e., if the input is close to the weight vector, the output y is large, otherwise it is small. The weights are computed by an outer product of the input X and output Y, i.e., $W = XY^T$ where T means transpose. The problem of Hebbian learning is that it is unstable, i.e., the weights will keep on growing with the number of iterations (Principe et al., 2000).

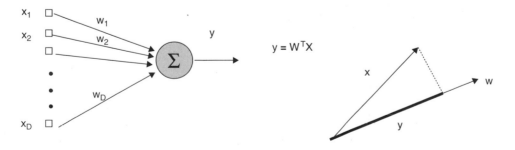

FIGURE 22.13   Hebbian PE.

Oja proposed to stabilize the Hebbian rule by normalizing the new weight by its size, which gives the rule (Principe et al., 2000):

$$w_i(n+1) = w_i(n) + \eta y(n)[x_i(n) - y(n)w_i(n)] \tag{22.13}$$

The weights now converge to finite values. They still define in the input space the direction where the data cluster has its largest projection, which corresponds to the largest eigenvector of the input correlation matrix (Kung, 1993). The output of the PE provides the largest eigenvalue of the input correlation matrix.

Principal Component Analysis

Principal component analysis is a well-known technique in signal processing that is used to project a signal into a signal-specific basis. The importance of PCA is that it provides *the best linear projection* to a subspace in terms of preserving the signal energy (Haykin, 1994). Normally, PCA is computed analytically through a singular value decomposition. PCA networks offer an alternative to this computation by providing an iterative implementation that may be preferred for real-time operation in embedded systems.

The PCA network is a one-layer network with linear processing elements (Figure 22.14). One can extend Oja's rule for many output PEs (less or equal to the number of input PEs), according to the formula shown in Figure 22.14, which is called Sanger's rule (Haykin, 1994). The weight matrix rows (that contain the weights connected to the output PEs in descending order) are the eigenvectors of the input correlation matrix. If we set the number of output PEs equal to $M < D$, we will be projecting the input data onto the M largest principal components. Their outputs will be proportional to the M largest eigenvalues. Note that we are performing an eigendecomposition through an iterative procedure.

Associative Memories

Hebbian learning is also the rule to create *associative memories* (Zurada, 1992). The most utilized associative memory implements *hetero-association*, where the system is able to associate an input X to a designated output Y, which can be of a different dimension (Figure 22.15). So, in hetero-association, the signal Y works as the desired response.

We can train such a memory using Hebbian learning or LMS, but LMS provides a more efficient encoding of information. Associative memories differ from conventional computer memories in several respects. First, they are content addressable and the information is distributed throughout the network, so they are robust to noise in the input. With nonlinear PEs or recurrent connections (as in the famous Hopfield network (Haykin, 1994)), they display the important property of *pattern completion*; i.e., when the input is distorted or only partially available, the recall can still be perfect.

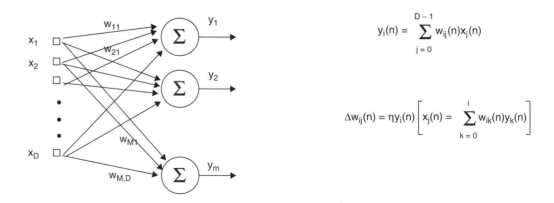

$$y_i(n) = \sum_{j=0}^{D-1} w_{ij}(n)x_j(n)$$

$$\Delta w_{ij}(n) = \eta y_i(n)\left[x_j(n) = \sum_{k=0}^{i} w_{ik}(n)y_k(n)\right]$$

FIGURE 22.14 Principal component analysis network. (*Source:* Principe, J. et al., *Neural and Adaptive Systems: Fundamentals through Simulation*, New York: John Wiley & Sons, 2000.)

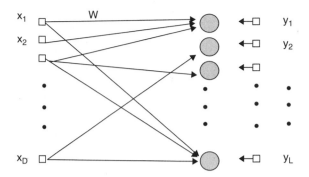

FIGURE 22.15 Associative memory (hetero-association).

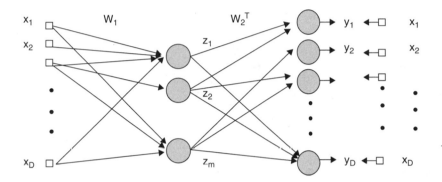

FIGURE 22.16 Auto-associator. (*Source:* Principe, J. et al., *Neural and Adaptive Systems: Fundamentals through Simulation*, New York: John Wiley & Sons, 2000.)

A special case of associative memories is called the *auto-associator* (Figure 22.16), where the training output of size D is equal to the input signal (also a size D) (Kung, 1993). Note that the hidden layer has less PEs ($M \ll D$) than the input (bottleneck layer). $W_1 = W_2^T$ is enforced.

The function of this network is one of *encoding or data reduction*. The training of this network (W_2 matrix) is done with LMS. It can be shown that this network also implements PCA with M components, even when the hidden layer is built from nonlinear PEs.

Competitive Learning and Kohonen Networks

Competition is a very efficient way to divide the computing resources of a network. Instead of having each output PE more or less sensitive to the full input space as in the associative memories, in a competitive network each PE specializes into a piece of the input space, and represents it (Principe et al., 2000). Competitive networks are linear, single-layer nets (Figure 22.17).

Their functionality is directly related to the competitive learning rule, which belongs to the unsupervised category. First, only the PE that has the largest output gets its weights updated. The weights of the winning PE are updated according to the formula in Figure 22.17 in such a way that they approach the present input. The step size exactly controls how much is this adjustment (see Figure 22.17).

Notice that there is an intrinsic nonlinearity in the learning rule: only the PE that has the largest output (the winner) has its weights updated. All the other weights remain unchanged. This is the mechanism that allows the competitive net PEs to specialize.

Competitive networks are used for clustering, i.e., an M-output PE net will seek M clusters in the input space. The weights of each PE will correspond to the centers of mass of one of the M clusters of input samples.

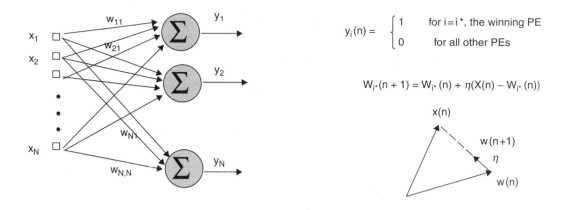

FIGURE 22.17   Competitive neural network.

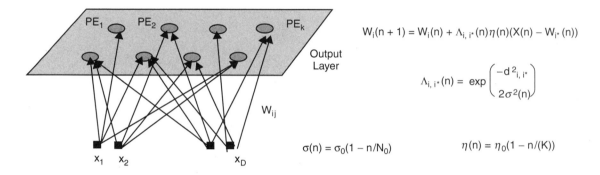

FIGURE 22.18   Kohonen self-organizing feature map (SOFM).

When a given pattern is shown to the trained net, only one of the outputs will be active and can be used to *label* the sample as belonging to one of the clusters. No more information about the input data is preserved.

Competitive learning is one of the fundamental components of the Kohonen self-organizing map (SOM) network, also a single layer network with linear PEs (Haykin, 1994). Kohonen learning creates annealed competition in the output space by adapting not only the winner PE weights but also their spatial neighbors using a Gaussian neighborhood function Λ. The output PEs are arranged in linear or 2D neighborhoods (Figure 22.18).

Kohonen SOM networks produce a mapping between the continuous input space to the discrete output space preserving topological properties of the input space (i.e., local neighbors in the input space are mapped to neighbors in the output space). During training, both the spatial neighborhoods and the learning constant are decreased slowly by starting with a large neighborhood σ_0, and decreasing it (N_0 controls the scheduling). The initial step size η_0 also needs to be scheduled (by K).

The Kohonen SOM network is useful to project the input to a subspace in alternative PCA networks. The topological properties of the output space provide more information about the input than straight clustering.

References

Bishop C. M., *Neural Networks for Pattern Recognition*, Oxford: Oxford University Press, 1995.

deVries and Principe J.,"The gamma model – A new neural model for temporal processing", *Neural Networks*, 5, 565–576, 1992.

Haykin S., *Neural Networks: A Comprehensive Foundation*, New York: MacMillan, 1994.

Kung S.Y., *Digital Neural Networks*, Englewood Cliffs, NJ: Prentice-Hall, 1993.

Principe J., Euliano N., Lefebvre C., *Neural and Adaptive Systems: Fundamentals through Simulation*, New York: John Wiley & Sons, 2000.

Zurada J. M., *Artificial Neural Systems*, St Paul, MN: West Publishing, 1992.

Further Information

The literature in this field is voluminous. We decided to limit the references to textbooks for an engineering audience, with different levels of sophistication. Zurada and Principe are the most accessible texts, Haykin the most comprehensive. Kung provides interesting applications of both PCA networks and nonlinear signal processing and system identification. Bishop stresses the links to statistical pattern recognition. Principe has the appeal that it is an electronic book that comes with a simulator and over 200 computer examples to illustrate the important concepts in this exciting field.

Interested readers are directed to the following journals for more information: *IEEE Trans. on Signal Processing, IEEE Trans. on Neural Networks, Neural Networks, Neural Computation*, and *Proceedings of the Neural Information Processing System Conference (NIPS)*.

22.2   Adaptive Signal Processing for Wireless Communications

Mohamed Ibnkahla, Ahmad Iyanda Sulyman, and Yu Cao

Adaptive signal processing plays an important role in wireless communications because it allows the wireless communication system to adapt itself to the channel changes and to the wireless network conditions. Adaptation can be applied to the receiver (e.g., channel equalization), to the transmitter side (e.g., adaptive modulation and coding), or, more generally, to the different levels of the wireless layers (e.g., adaptive call admission control). Furthermore, for new wireless communication technologies (such as multiple-input multiple-output [MIMO] systems) the use of efficient adaptive signal processing techniques is essential in order to achieve high data rates and enable quality of service (QoS) applications. This section reviews several adaptive techniques and their impact on current and future wireless communications.

Introduction

Extensive research efforts are being made worldwide in order to enable high data rate transmissions over wireless communications channels while keeping the required QoS. This is expected to allow users, regardless of their geographic location, to have equal access to leading-edge healthcare, education, government services, interactive multimedia, Internet banking, e-commerce, online entertainment, etc. However, wireless channels suffer from performance degradation due to time-varying multipath propagation and fading. Adaptive signal processing plays a central role in current and future wireless communication systems as it tries to adaptively overcome these problems and optimize the system parameters according to the available resources, as well as the network and channel conditions.

Adaptive signal processing which started in the mid-twentieth century from very basic adaptive filtering algorithms, today covers almost all aspects of wireless communications, such as equalization, channel identification and estimation, modulation, coding, multiple access, networking, cross-layer design, etc. In each of these areas, researchers are paying more and more attention to adaptive techniques as alternatives to classical nonadaptive approaches.

This chapter reviews the basic principles of adaptive signal processing. The section "Linear Adaptive Filtering and the LMS Algorithm" presents the basics of linear adaptive filtering and the least mean square (LMS) algorithm. The section "Channel Identification, Modeling and Tracking" reviews adaptive filtering and adaptive channel identification, including single-input single-output (SISO) systems, satellite mobile channels, and MIMO systems. The section "Channel Equalization" covers adaptive equalization in both SISO and MIMO systems. The section "Other Aspects of Adaptive Processing in Wireless Communications" presents adaptive modulation and coding (AMC) and cross-layer design approaches in wireless communications.

Linear Adaptive Filtering and the LMS Algorithm

Figure 22.19 represents the block diagram of the adaptive filtering problem, in which $x(n)$ is the input signal to the adaptive filter at time n, $y(n)$ is the output, and $d(n)$ (desired signal) is a reference signal. The error between the desired signal and the filter output is computed as

$$e(n) = d(n) - y(n) \tag{22.14}$$

The adaptive filter has a finite set of parameters, called the parameter vector. The parameter vector is denoted at time n, as $\underline{w}(n) = [w_0(n)\, w_1(n) \ldots w_{L-1}(n)]^{\mathrm{T}}$ (where ^{T} denotes the transpose). This has to be updated so that the output of the adaptive filter becomes a better match to the desired output. In the case of finite impulse response (FIR) filtering, the output is expressed as

$$y(n) = \sum_{k=0}^{L-1} w_k(n)x(n-k) = \underline{w}^{\mathrm{T}}(n)\underline{x}(n) \tag{22.15}$$

where $\underline{x}(n) = [x(n)\, x(n-1)\ldots x(n-L+1)]^{\mathrm{T}}$ denotes the input signal vector.

When $d(n)$ is available, the adaptation is called supervised learning. In wireless communication applications, $d(n)$ is not available all the time. In such situations, adaptation occurs typically when $d(n)$ is available. When $d(n)$ is not available, an estimated value is used. In some applications, $d(n)$ is never available. In such cases, some additional information can be found (or assumptions can be made) with regard to $d(n)$ (such as its predicted statistical behavior) to form suitable estimates of $d(n)$ from the signals available to the adaptive filter. These methods are called blind adaptive algorithms.

The mean squared error (MSE) is usually taken as the cost function to be minimized:

$$J(n) = \frac{1}{2} E(e^2(n)) \tag{22.16}$$

where $E(.)$ is the expectation operator.

The Wiener solution is the value of the weight vector that minimizes the cost function. In our case, it is expressed as

$$\underline{w}_{\mathrm{opt}} = R_x^{-1}(n)\underline{p}_{dx}(n) \tag{22.17}$$

where $R_x(n)$ is the correlation matrix of the input signal vector (i.e., $R_x(n) = E(\underline{x}(n)\underline{x}^{\mathrm{T}}(n))$), and $\underline{p}_{dx}(n) = E(d(n)\underline{x}(n))$. The method of steepest descent adjusts the weights according to

$$\underline{w}(n+1) = \underline{w}(n) - \mu(n)\,\mathrm{grad}(J(n)) \tag{22.18}$$

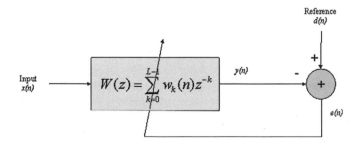

FIGURE 22.19   Principle of adaptive filtering.

where $\mu(n)$ is a positive step-size parameter (called learning rate), and grad($J(n)$) denotes the ordinary gradient with respect to the weight vector $\underline{w}(n)$, which can be expressed here as

$$\text{grad}(J(n)) = \left[\frac{\partial J(n)}{\partial w_k(n)} \quad \frac{\partial J(n)}{\partial w_2(n)} \cdots \cdots \frac{\partial J(n)}{\partial w_L(n)} \right]^t = -\left(\underline{p}_{dx}(n) - R_x^{-1}(n)\underline{w}(n) \right) \qquad (22.19)$$

Therefore, the weights are updated as

$$\underline{w}(n+1) = \underline{w}(n) + \mu(n)\left(\underline{p}_{dx}(n) - R_x^{-1}(n)\underline{w}(n) \right) \qquad (22.20)$$

Thus, the steepest descent approach depends on the statistical quantities $R_x^{-1}(n)$ and $\underline{p}_{dx}(n)$. In practice, these quantities are unavailable and need to be estimated from the measurements of $x(n)$ and $d(n)$. The LMS procedure uses an approximate version of the method of steepest descent and minimizes the instantaneous squared errors (instead of the MSE):

$$J(n) = \frac{1}{2}e^2(n) \qquad (22.21)$$

and the weights are updated as

$$\underline{w}(n+1) = \underline{w}(n) - \mu(n)\,\text{grad}(J(n)) = \underline{w}(n) + \mu(n)e(n)\underline{x}(n) \qquad (22.22)$$

Under some statistical assumptions, it can be shown that the weight vector converges to the Wiener solution.

There are several other approaches to update the filter weights, such as the sliding window LMS algorithm, normalized LMS algorithm, transform domain LMS algorithm, recursive least squares (RLS) algorithm, affine projection algorithm, etc. [13].

Channel Identification, Modeling, and Tracking

Overview

Channel identification, modeling, and tracking are very important research areas in wireless communications. For example, for the purpose of receiver design, the effect of the channel on the transmitted symbols has to be estimated in order to successfully detect the transmitted information. Similarly, for transmitter design, a channel model is needed, e.g., for the choice of the modulation and coding schemes to be employed, or to choose the transmission rate and power, etc. In cross-layer design approaches, a channel model is needed in order to optimize the networking and medium access control (MAC) protocols.

However, in wireless communications, the channel is generally time-varying. Therefore, adaptive approaches are needed for channel modeling and identification. An example of adaptive identification is illustrated in Figure 22.20 where both the unknown channel and the adaptive filter are driven by the same input signal. The channel identification and tracking scheme may include a parameter measurement block that estimates some useful parameters about the channel which can be used by the adaptive algorithms to improve the learning and tracking processes.

In Figure 22.20, the adaptive channel model (e.g., adaptive FIR filter) adjusts its parameters (e.g., filter weights) such that its output is a best least squares fit to that of the unknown channel. For successful identification or tracking, the structure and parameter values of the adaptive system may or may not be similar to those of the unknown channel. However, the input–output relationships should be close. The ability to track the channel variations depends on the algorithm structure and design and on how fast the channel changes in time. The capabilities of the LMS and RLS algorithms for time-varying system tracking have been largely studied in the literature [13,14]. The general conclusions drawn from these studies are that the LMS algorithm exhibits better tracking behavior than the RLS algorithm. In [14], an extended version of the

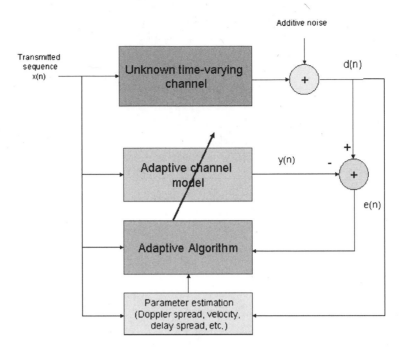

FIGURE 22.20   Block diagram of adaptive channel modeling and identification.

RLS algorithm has been proposed exploiting a Kalman-based formulation of the RLS algorithm. This extended version of the RLS algorithm has shown good tracking capabilities compared to the regular RLS and LMS algorithms.

Since the transmitted data sequence is unknown to the receiver, the wireless system designers include in the transmitted sequence, in addition to the data sequence, a set of training and/or pilot symbols which are used by the adaptive algorithms. The learning algorithm has, in general, two modes: The *training sequence (TS) mode* and the *decision-directed (DD)* mode. In the TS mode, a training sequence (which is a known transmitted sequence $x(n)$) is used at the adaptive system input. In the DD mode, a training sequence is unavailable, and the adaptive algorithm uses an estimation of the transmitted sequence, $\hat{x}(n)$, as input, instead of the exact transmitted sequence.

Figure 22.21 shows an example of a time division multiple access (TDMA) time slot structure used in global system for mobile communications (GSM). There is a training sequence of 26 known symbols in the middle of the time slot. For relatively high transmission rates and low mobile speeds, the channel can be considered as constant during the time slot and as slowly varying when moving from one time slot to another. The training sequence can be used, for instance, by the tracking algorithm and the channel parameter estimator. Figure 22.22 shows an example of the time slot structure used in the IS-136 system. The slot duration is relatively long and the channel can significantly change during the time slot. Therefore, in addition

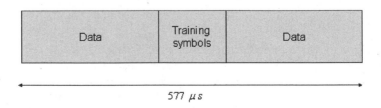

FIGURE 22.21   GSM time slot structure.

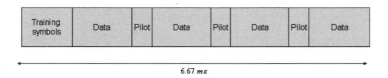

FIGURE 22.22 IS-136 time slot structure.

to the training sequence at the beginning of the slot, clusters of known pilot symbols are placed through the time slot. These clusters can be used by the adaptive algorithm for training (this is also called pilot-aided mode [PA]). They may also be used to estimate the channel parameters (such as fading gain or Doppler spread). An interpolation technique can then be used to predict the channel state at the data locations (i.e., when pilot or training sequences are unavailable). In this case, the channel gain carried by the data symbols can be interpolated as a weighted sum of the channel gain on the nearest K pilot symbols. Obviously, using a larger interpolation order K or frequently inserting pilot symbols, will yield higher estimation accuracy. However, a larger interpolation order K will increase the computational complexity and cause longer delay for symbol detection, while frequently inserting pilot symbols will decrease the system bandwidth efficiency. Therefore, in practice, choosing these interpolation parameters is a trade-off between the bandwidth redundancy, system complexity, and estimation accuracy. The interpolation weights, corresponding to different interpolation methods, also need to be carefully designed to minimize the estimation error. Several interpolation methods have been suggested in the literature. Cavers [6] derived the optimum Wiener interpolation that yields MMSE estimation. This Wiener filtering interpolator, however, requires some prior information of the statistical property of the channel and is very computationally complex. Another interpolator worth to mention is a *Sinc* interpolator [21], which gives near-optimum performance and is easy to implement. Therefore, it is considered to be suitable for practical use. Other interpolation techniques include Gaussian [29], linear interpolation, and Nyquist filtering [4]. The above conventional fading estimation techniques use only pilot symbols for channel estimation. A novel idea was proposed [19] that the channel information can be estimated by utilizing both the pilot and decided data symbols (this is also called DD mode). In this case, decision feedback and noise smoothing filters are used for fading estimation. The interpolation scheme was later improved [24]. The challenge here is to carefully design the interpolation algorithm to avoid the large error caused by wrongly detected symbols.

It is very important that the adaptive algorithm uses the available information about the channel in order to use it during the learning/tracking process. Doppler spread, delay spread, and angle spread are important channel parameters that can be used and included in the adaptation process. For example, the step size, $\mu(n)$, of an LMS channel tracker can be changed according to the online estimation of the Doppler spread, i.e., increase $\mu(n)$ when the Doppler spread is high and decrease it when the Doppler spread is low. In a sliding-window-based channel tracking algorithm, the window size can be adjusted according to the Doppler spread and SNR information [3].

Example of Satellite Mobile Channel Identification Using Adaptive Neural Networks

Modeling and identification of satellite mobile channels represents a very good illustration of the capabilities of adaptive algorithms. A typical satellite mobile channel model is represented in Figure 22.23.

In this model, the transmitted signal is modulated and amplified by a nonlinear high-power amplifier (HPA), it is then affected by multi-path fading caused by the downlink satellite mobile channel [17].

We assume here that the transmitted signal is multilevel quadrature amplitude modulation (M-QAM), modulated and expressed in the complex form as:

$$x(n) = r(n)e^{j\phi_0(n)} \tag{22.23}$$

where $r(n)$ and $\phi_0(n)$ are the amplitude and phase, respectively.

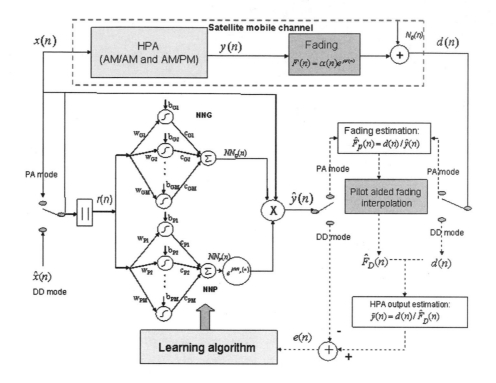

FIGURE 22.23 Neural network channel estimator structure.

The equivalent baseband HPA output, $y(n)$, can be expressed as:

$$y(n) = A(r(n)) \exp\{j(\phi(r(n)) + \phi_0(n))\} \tag{22.24}$$

where $A(r)$ and $\phi(r)$ are, respectively, the amplitude to amplitude (AM/AM) and amplitude to phase (AM/PM) conversions of the on-board HPA, which are taken in this section as:

$$A(r) = \frac{2r}{1 + r^2} \quad \text{and} \quad \phi(r) = \frac{\pi}{3} \times \frac{r^2}{1 + r^2} \tag{22.25}$$

The channel output is given by:

$$d(n) = F(n) \times y(n) + N_0(n) \tag{22.26}$$

where $F(n) = \alpha(n)e^{j\psi(n)}$ is the complex fading gain caused by the downlink propagation channel, and N_0 is an additive white Gaussian noise.

The system identification scheme is composed of a neural network (NN) [15] and a linear fading estimator (Figure 22.23). The NN serves aims at identifying the HPA nonlinearity while the fading estimator aims at tracking the time-varying fading gain. (See also [16] and [18] for the case where the propagation channel is fixed but has memory).

The NN is composed of two subnetworks: the gain network (NNG) and the phase network (NNP). The gain and phase networks aim at identifying and tracking the AM/AM and AM/PM conversions, respectively. When the transmitted sequence, $x(n)$, is available at the receiver (i.e., in the TS mode), the NN output is expressed as

$$\hat{y}(n) = x(n)NN_{\mathrm{G}}(r(n))e^{jNN_{\mathrm{P}}(r(n))} = r(n)NN_{\mathrm{G}}(r(n))e^{jNN_{\mathrm{P}}(r(n))}e^{j\Phi_0(n)} \tag{22.27}$$

$$NN_{\mathrm{G}}(r(n)) = \sum_{i=1}^{M} c_{\mathrm{G}i}f(w_{\mathrm{G}i}r(n) + b_{\mathrm{G}i}) \tag{22.28}$$

$$NN_{\mathrm{P}}(r(n)) = \sum_{i=1}^{M} c_{\mathrm{P}i}f(w_{\mathrm{P}i}r(n) + b_{\mathrm{P}i}) \tag{22.29}$$

where $x(n)$ is the input to the NN, M is the number of neurons in the hidden layer, $c_{\mathrm{G}i}, b_{\mathrm{G}i}, w_{\mathrm{G}i}$ (respectively, $c_{\mathrm{P}i}, b_{\mathrm{P}i}, w_{\mathrm{P}i}$) are the neural weights in the gain network (respectively, phase network), and $f(.)$ is the tanh(.) function. Subscripts G and P are referred to as the gain part and phase part, respectively.

Note that in the DD mode, the detected symbol, $\hat{x}(n)$, is taken as input instead of $x(n)$.

Learning Process. The goal of this adaptive system is to estimate the fading gain and to update the NN weights in order to identify the static nonlinearity of the HPA. To achieve this goal, two modes are employed:

Pilot aided mode: During the PA mode, a known pilot, $x(n)$, is transmitted and the fading gain, $\hat{F}_{\mathrm{p}}(n)$, is estimated according to

$$\tilde{F}_{\mathrm{p}}(n) = d(n)/\hat{y}(n) \tag{22.30}$$

Decision directed mode: In this mode, the pilot symbol is unavailable, therefore we use the decided symbol $\hat{x}(n)$ (that can be obtained for instance by a maximum likelihood (ML) detector [18]) as input to the NN. In this case, the fading gain, $\hat{F}_{\mathrm{D}}(n)$, is estimated using an interpolation method.

During the DD mode, the NN weights are updated in order to identify the unknown nonlinearity of the HPA. In this case, an estimated HPA output is calculated using the true channel output and the fading gain that was already given by the interpolation method:

$$\tilde{y}(n) = \frac{d(n)}{\hat{F}_{\mathrm{D}}(n)} \tag{22.31}$$

The NN weights are updated in order to minimize the loss function $J(n)$ between the estimated HPA output and the NN output:

$$J(n) = \frac{1}{2}\|e(n)\|^2 = \frac{1}{2}\|\tilde{y}(n) - \hat{y}(n)\|^2 \tag{22.32}$$

where $\tilde{y}(n)$ is the estimated HPA output, $\hat{y}(n)$ is the NN output, and $e(n)$ is the error between the NN output and the estimated HPA output.

Let θ represent the set of the adaptive weights:

$$\underline{\theta} = [w_{\mathrm{G1}} \ldots w_{\mathrm{GM}}, b_{\mathrm{G1}} \ldots b_{\mathrm{GM}}, c_{\mathrm{G1}} \ldots c_{\mathrm{G1}}, w_{\mathrm{P1}} \ldots w_{\mathrm{PM}}, b_{\mathrm{P1}} \ldots b_{\mathrm{PM}}, c_{\mathrm{P1}} \ldots c_{\mathrm{PM}}]^{\mathrm{T}}$$

A gradient descent-based algorithm updates the NN weights according to

$$\underline{\theta}(n+1) = \underline{\theta}(n) - \mu\nabla_{\underline{\theta}(n)}J(n) \tag{22.33}$$

where ∇ represents a gradient operator. It can be the ordinary gradient, in which case we deal with the classical backpropagation (BP) algorithm; or the natural gradient (NG) [2,18], in which case the algorithm follows the steepest descent. The NG algorithm shows, in general, a faster convergence speed than the BP algorithm, but it is less stable [16].

Simulation Examples.   A satellite mobile channel has been simulated using the HPA characteristics given in Equation (22.25). The transmitted signal is a 16-QAM modulated signal. The fading was assumed to be Rayleigh distributed with a normalized Doppler frequency of 0.01. For the pilot aided fading estimation approach, one pilot symbol has been sent after every four consecutive data symbols. Figure 22.24 and Figure 22.25 show the learning curves (i.e., MSE error vs. iteration number) of the NG and BP algorithms, respectively. It can be seen that as the learning rate μ increases, the algorithms are faster. Figure 22.26 shows the MSE (obtained after 50,000 iterations) vs. the learning rate. For each algorithm, there is an optimal value of the learning rate that gives the lowest MSE. The lowest MSE yielded by the NG algorithm is smaller than that of the BP algorithm. Figure 22.27 and Figure 22.28 show that the amplifier AM/AM and AM/PM curves have been successfully identified by both algorithms. Finally, Figure 22.29 shows that the estimated fading gain fits very well with the true fading gain.

   This tracking and identification scheme has been used for ML detection. The symbol error rate (SER) performance of the ML detector is presented in Figure 22.30 for the NG and BP algorithms (Rayleigh fading). In addition, two other cases have been used for comparison: the linear fading channel case, and the case of nonlinear fading assuming perfect knowledge of the nonlinearity. It can be seen that the NG algorithm performs better than the BP algorithm, and is very close to the case where a perfect knowledge of the nonlinearity is assumed. This result is expected since the NG approach allows a very good approximation of the nonlinearity.

   Figure 22.31 displays the results for the Ricean fading case (Ricean factor $K = 6$ dB). The SER performances of the NG and BP algorithms are close to each other and are better than those obtained in the Rayleigh fading case. This is expected since Ricean fading is less severe than Rayleigh fading. This allows the NN to better approximate the nonlinearity.

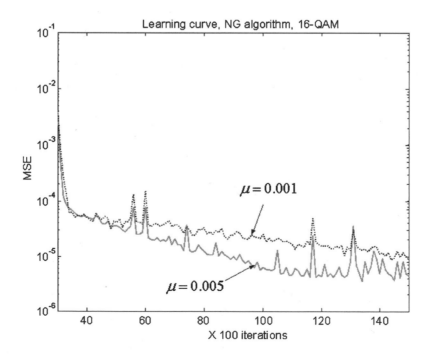

FIGURE 22.24   Learning curve for the NG algorithm.

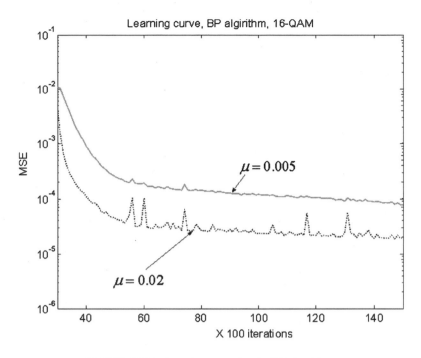

FIGURE 22.25 Learning curve for the BP algorithm.

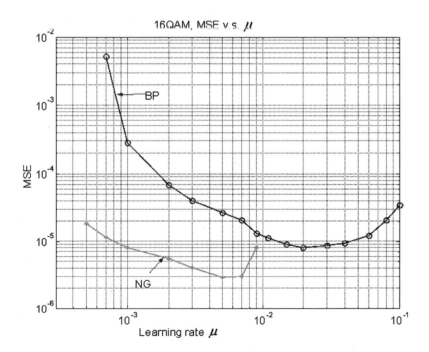

FIGURE 22.26 MSE vs. learning rate.

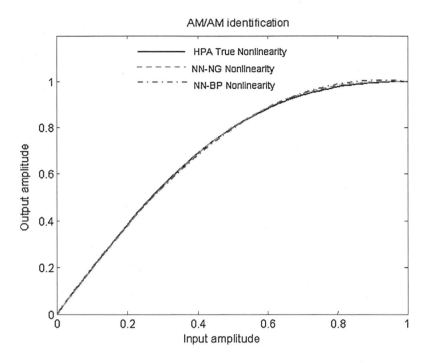

FIGURE 22.27   Nonlinearity AM/AM identification (NG vs. BP algorithm).

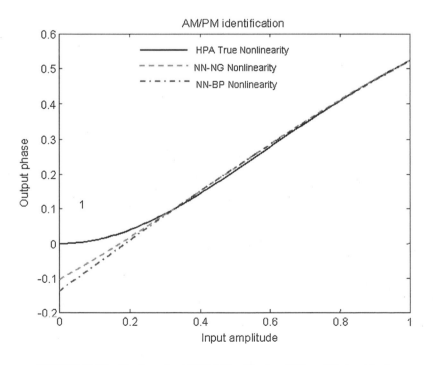

FIGURE 22.28   Nonlinearity AM/PM identification (NG vs. BP algorithm).

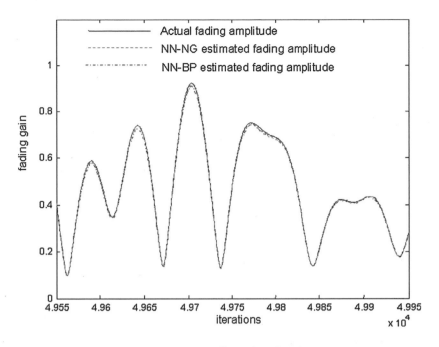

FIGURE 22.29   Fading gain estimation.

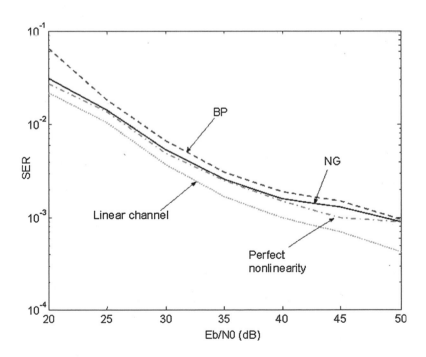

FIGURE 22.30   SER performance of the ML detector, Rayleigh fading.

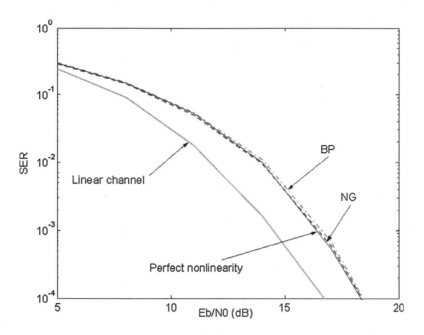

FIGURE 22.31   SER performance of the ML detector, Ricean fading.

MIMO Channel Tracking and Identification

Most MIMO [27] signal processing algorithms, as well as theoretical analyses on performance of MIMO systems, have traditionally relied on the assumption of the availability of the MIMO channel coefficients (or MIMO channel state information [CSI]) at the receiver. In this section, we illustrate this presumption by discussing the space–time encoding and decoding operations. The various problems that arise in the tracking of MIMO channel coefficients are then explained.

Space–Time Coding.   In this section we discuss the idea behind the space–time codes presented in [33], and illustrate the tracking problems in MIMO channels. Consider a MIMO transmission, with N transmitting and L receiving antennas, over a wireless communication channel illustrated in Figure 22.32. At any time instant n, let the information signals be encoded by the space-time encoder into $N \times 1$ code vector $\underline{c}(n) = [c^1(n)c^2(n) \cdots c^N(n)]^t$, and let each code symbol be transmitted simultaneously from a different antenna. Assume that l consecutive code vectors, $\{\underline{c}(n)\}_{n=1}^l$, have been transmitted.

At the receiver side, signals arriving at the different receiving antennas undergo independent fading. The received signal is a linear combination of the transmitted signal and the MIMO channel coefficients h_{ij}, $(i = 1, \ldots, N)$, $(j = 1, \ldots, L)$, corrupted by additive noise. At the nth transmission period, the $L \times 1$ received signal vector is therefore given by

$$\underline{y}(n) = H(n)\underline{c}(n) + \underline{N}_0(n) \quad n = 1, \ldots, l \tag{22.34}$$

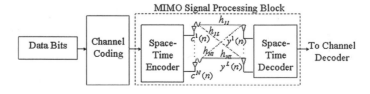

FIGURE 22.32   MIMO communications systems.

where $H(n)$ is the $L \times N$ matrix whose elements, h_{ij}, are the channel fading coefficients from the jth transmit antenna to the ith receive antenna. $\underline{N}_0(n)$ is an $L \times 1$ additive white Gaussian noise (AWGN) vector, and is assumed to be spatially and temporally white (i.e., $\underline{N}_0(n)$ is a zero mean complex Gaussian vector with covariance matrix $\sigma^2 I$).

Maximum likelihood decoding (MLD) of the transmitted data $\underline{c}(n)$ from the received signal sequences is then carried out. The maximum likelihood (ML) decoder can be realized using Viterbi algorithm with the ML metric given, in form of minimum Euclidean distance, as

$$
\begin{aligned}
\underline{c} &= \underset{[\underline{\tilde{c}}(1),\underline{\tilde{c}}(2),\cdots,\underline{\tilde{c}}(l)]}{\operatorname{argmin}} \; \|(\underline{y}(1),\underline{y}(2),\cdots\underline{y}(l)) - (\tilde{H}(1)\underline{c}(1),\tilde{H}(2)\underline{c}(2),\cdots,\tilde{H}(l)\underline{c}(l))\|^2 \\
&= \underset{[\underline{\tilde{c}}(1),\underline{\tilde{c}}(2),\cdots,\underline{\tilde{c}}(l)]}{\operatorname{argmin}} \; \sum_{n=1}^{l} \|\underline{y}(n) - \tilde{H}(n)\underline{c}(n)\|^2
\end{aligned}
\tag{22.35}
$$

where $\tilde{H}(n)$ is the MIMO channel estimate at any time instant n. Therefore, it is clear from Equation (22.35) that ML detection of space–time codes require the provision of the MIMO CSI at the receiver. In practice, some form of channel tracking or estimation will have to be employed in order to obtain the MIMO channel estimate prior to the detection process.

Tracking and Identification of Frequency-Selective MIMO Channels.

When MIMO technology is deployed over mobile radio channels, the memory and time-varying nature of the mobile radio channels have to be taken into account. In the encoding and decoding operation of space-time codes in the section "Space–Time Coding" the channel coefficient has been assumed time-varying. In the sequel, however, we will need to modify the system equations in Equation (22.34) and Equation (22.35) and account for the memory nature of the mobile radio channels.

Let $H_k(n)$ ($k = 0, 1, 2 \ldots, m-1$) be the $L \times N$ complex channel matrix representing the kth tap of the channel matrix response with $\underline{c}(n)$ as the input and $\underline{y}_n = [y^1(n)y^2(n) \cdots y^L(n)]^t$ as the output at time instant n. The received signal is given by

$$
\underline{y}(n) = \sum_{k=0}^{m-1} H_k(n)\underline{c}(n-k) + \underline{N}_0(n)
\tag{22.36}
$$

Finally, modifying this equation to accommodate a block or frame of l consecutive symbols, we obtain:

$$
Y(n) = \hat{H}(n)C + N(n)
\tag{22.37}
$$

where $\hat{H}(n) = [H_1(n)H_2(n) \cdots H_m(n)]$ is the $L \times Nm$ MIMO-FIR channel matrix, C is the $Nm \times l$ convolution matrix obtained from the l input sequences so that the ith column of C is $[\underline{c}_{i-1}(n)^t, \underline{c}_{i-2}(n)^t, \cdots, \underline{c}_{i-m}(n)^t]^t$. $Y(n) = [\underline{y}_1(n), \underline{y}_2(n) \cdots \underline{c}_l(n),]$ is the stacked received l samples of the input sequences. $N(n)$ has the same structure as $Y(n)$. Next we stack the Nm columns of the matrix $H(n)$ into the vector $\underline{h}(n) = \text{vec}\{H(n)\}$ and the l columns of the matrix $Y(n)$ into the vector $\hat{y}(n) = \text{vec}\{Y(n)\}$ to obtain the final system equation as

$$
\hat{\underline{y}}(n) = (C^t \otimes I_L)\underline{h}(n) + \underline{N}_0(n) = X\underline{h}(n) + \underline{N}_0(n)
\tag{22.38}
$$

where $\underline{N}_0(n) = \text{vec}\{N(n)\}$, X is the result of a Kronecker product operation, $\otimes$, between the transposed matrix of the input sequences C, and the identity matrix I_L. The channel coefficient vector $\underline{h}(n)$ in Equation (22.38) can then be tracked using the conventional adaptive system identification techniques [13,32].

Adaptive Algorithms for MIMO Channel Tracking. As evident from Equation (22.34) to Equation (22.38) MIMO receivers require the knowledge of CSI for the detection process. The CSI can first be estimated and then tracked at each transmission block by using training sequences/pilot symbols inserted in each block [1]. In the following, we describe a simple channel estimation and tracking technique [1] that can be employed to track the MIMO channel coefficients in the system Equation (22.38). At initialization, an initial channel estimate $\underline{\tilde{h}}(0)$ can be obtained by transmitting a full training block. Subsequent transmission blocks will contain pilot tones that are then used to acquire new estimates $\underline{h}'(n)$. The MIMO channel estimate for the nth transmission is obtained using a forgetting factor α as follows:

$$\underline{\tilde{h}}(n) = \alpha\underline{\tilde{h}}(n-1) + (1-\alpha)\underline{h}'(n) \tag{22.39}$$

Obviously, the performance quality of a MIMO receiver employing this estimation and tracking method will be at its best when frequent retraining (re-initialization) processes are made. This, however, is at the expense of training symbols overhead. In practice therefore, a compromise will have to be made between the quality of channel estimates obtained in Equation (22.39) and the affordable training overhead.

An alternative to this two-step approach is to employ an adaptive algorithm for the MIMO channel tracking. Adaptive receivers do not explicitly estimate the CSI at the receiver but rather employ adaptive filtering techniques, using any of the classical algorithms developed for adaptive system identifications, to adaptively identify the MIMO channel coefficients. However, this method still requires training overhead for the coefficients of the adaptive filter to converge to their optimum weights. Among the conventional adaptive algorithms, the LMS algorithm is widely used in the SISO system today due to its low implementation complexity. Its main drawback, however, is its slow convergence and its performance degradation (relative to the performance achieved with the optimum weights) when used in channels with large eigenvalue spread. MIMO communications channels are known to have large eigenvalue spreads due to the temporal correlation introduced in the transmitted signals via the space-time encoding, and therefore LMS adaptations are unsuitable for tracking MIMO channels [1]. Faster convergence in MIMO channel tracking can be achieved using the family of RLS algorithms (including the Kalman algorithm). However, the RLS algorithms suffer from two major drawbacks: high computational complexity (compared to the LMS algorithm) and instability problems. For these reasons, the RLS algorithms have traditionally not been as much embraced in real-time applications compared to the LMS algorithm.

Recently however, the use of the Kalman algorithm for tracking MIMO channel coefficients was given some attention [22]. This is largely due to the fact that a low-order autoregressive model approximates the MIMO channel variation and this facilitates tracking via a Kalman filter. Also, it is known that Kalman algorithms exhibit reasonable measure of tracking robustness. This is, however, without regard to their relatively heavy implementation complexities. The authors in [22] also show that the Kalman algorithm offers good tracking behavior for multi-user fading ISI channels at the expense of higher complexity than conventional adaptive algorithms.

In [8], the authors study the effect of ambiguities in correct determination of the phase, ϕ, of the ith entry, $h_i e^{j\phi}$, of the complex MIMO channel coefficients. Phase ambiguities arise from a random rotation of one constellation point to another. These ambiguities cause error propagation when data detection and channel estimation are jointly done at the MIMO receiver. The authors in [8] discussed an enhanced channel tracking with speed estimation method for MIMO applications. They have shown that their improved tracking algorithm estimates not only the CSI (amplitude and phase) accurately, but also the speed of variations of the CSI. This additional information on speed of CSI variations enhances further the identification of the trajectory of the CSI. Using such speed information does not only improve the performance of the channel tracking, but also reduces the probability of random rotation of the estimated parameters — an event that leads to the phase ambiguity problem. Consequently, error propagation becomes less probable with such an improved tracking and speed estimation algorithm.

Channel Equalization

Adaptive channel equalization [28] is a very important signal processing technique for transmission over frequency-selective channels that cause intersymbol interference (ISI). A multi-path propagation channel is considered frequency-selective when the coherence bandwidth of the channel is smaller than the signal bandwidth. The purpose of the equalizer is to remove or considerably reduce the ISI, so that the overall propagation channel, plus the receiver, behaves like a flat (or frequency-nonselective) channel (e.g., a pure delay). In the case where the channel is fixed, analytical approaches give the appropriate linear equalizer structures to overcome the ISI. However, when the channel is time-varying, an adaptive equalization method is required. This is so that, for each iteration, the equalizer will be able to track the channel changes and adapt itself to the new frequency response of the channel.

The Linear FIR Equalizer

The basic scheme of an adaptive linear equalizer is shown in Figure 22.33. At each iteration, the equalizer parameters are updated in order to minimize the error between a known sequence (training symbols) and the equalizer output:

$$J(n) = \frac{1}{2} \|e(n)\|^2 = \frac{1}{2} \|x(n - \Delta) - y(n)\|^2 \tag{22.40}$$

where Δ is a delay and $y(n)$ is the equalizer output. In the case of real-valued signals, the LMS algorithm updates the FIR filter weights as

$$\underline{w}(n+1) = \underline{w}(n) - \mu(n)\,\mathrm{grad}(J(n)) = \underline{w}(n) + \mu(n)e(n)\underline{d}(n) \tag{22.41}$$

where $d(n) = [d(n)\,d(n-1)\ldots d(n-L+1)]^{\mathrm{T}}$, and $d(n)$ is the channel output and equalizer input at time n. In the case of complex-valued signals, this equation becomes

$$\underline{w}(n+1) = \underline{w}(n) + \mu(n)e(n)\underline{d}^{*}(n) \tag{22.42}$$

where * denotes the complex conjugate.

Note that, with the help of an accurate estimation of the channel parameters, we can improve the equalizer convergence speed and performance. For example, the learning rate can be chosen as a function of the Doppler spread. Similarly, an estimation of the RMS delay spread or the maximum delay spread can be used to optimize the equalizer length L, etc.

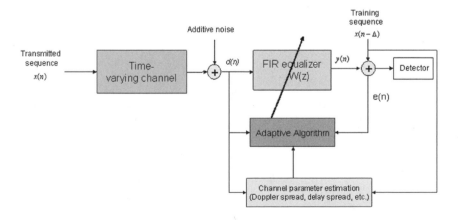

FIGURE 22.33 Basic structure of an adaptive FIR equalizer.

When no training sequence is available, the DD mode can be activated, and the estimated value after detection, $\hat{x}(n)$, is taken instead of $x(n)$.

The Decision Feedback Equalizer (DFE)

The basic scheme of the DFE is shown in Figure 22.34. This equalizer uses previously detected symbols to eliminate the ISI that affects the current symbol to be detected. The feedforward filter acts as a linear equalizer. The feedback filter has its input from the previously detected symbols. The input to the detector is the difference between the outputs of the feedforward and feedback filters. Hence, the feedback section removes the lagging ISI that is beyond the reach of the forward section.

The tap weights of the feedforward and feedback filters are updated so that the MSE error between the input to the detector and the training sequence is minimized. However, the DFE suffers from error propagation, since if a wrong decision is made, it is fed back to the equalizer input.

The DFE is a nonlinear equalizer because it uses the hard decisions made by the detector. More generally, the feedback filter can be replaced by a nonlinear process (e.g., a neural network) to perform the feedback task [35].

Other Equalization Techniques for Fading Channels

In [23], the authors present a survey of non adaptive equalization techniques for fading channels. Two major structures are presented: block equalizers and serial equalizers, both including linear and DFE equalizers. The authors present closed-form expressions for the minimum MSE (MMSE) block and serial equalizers based on the channel knowledge. In practical cases, however, the channel is unknown and has to be determined. This can be done, for example, through adaptive estimation approaches. See [34] for some examples of other joint channel estimation-equalization approaches.

MIMO Channel Equalization

For MIMO transmission over frequency-selective channels, the channel output is given by the expression in Equation (22.23) above and has the Z-transform given by

$$\underline{y}(z) = \check{H}(z)\underline{c}(z) + \underline{N}_0(z) \tag{22.43}$$

where $\check{H}(z) = \sum\limits_{k=0}^{m-1} H_k(n)z^{-k}$.

An adaptive equalizer employed at the MIMO receiver has the functionality of carrying out a reverse operation of the frequency-selective MIMO channel actions in Equation (22.43) in order to recover the information bits from the noisy observation $y(z)$. For the equalizer to function effectively, however, it has to know the nature of the underlying MIMO channel. Channel tracking (see the section "MIMO Channel Tracking and Identification") can be employed in order to "learn" the MIMO channel state information (MIMO CSI) adaptively, and make this information available to the equalizer for effective equalization of the channel. In the rest of this section, we will assume knowledge of the MIMO CSI at the receiver.

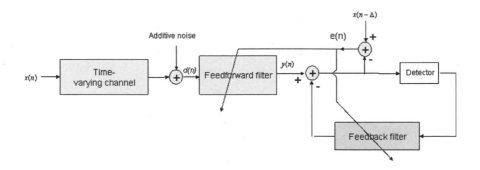

FIGURE 22.34 Block diagram of a DFE equalizer.

Assuming perfect knowledge of the channel coefficients, it is well known that the optimum receiver is a maximum likelihood sequence estimator (MLSE). For transmission over frequency-selective MIMO channels therefore, the best performance in terms of error rate can be achieved through trellis equalization of the space-time codes based on MLSE or symbol-by-symbol maximum *a posteriori* probability (MAP) estimation [30]. However, the complexity of these methods is proportional to the number of states of the trellis which grows exponentially with the product of the channel memory and the number of transmit antennas. When the channel memory becomes large and high-order constellations are used, the algorithm becomes impractical. On the other hand, increasing the number of states of the trellis, the constellation size, and the number of antennas are effective means to achieve high bit rates. It then becomes imperative to develop reduced-complexity equalization methods without significant performance degradation (i.e., without the need to significantly reduce these three parameters). In the next section, we review two families of such suboptimum equalizers achieving a good performance–complexity tradeoff, which have been employed in MIMO channels. The first of these is the family of block linear and decision-feedback equalizers, and the second is the family of list-type equalizers.

Block Linear and Decision-Feedback Equalizers. Because block linear and decision-feedback equalizers are optimized for block transmission systems [20], they are easily adapted for MIMO systems. In the following, we discuss the performances of four structures. These include:

- Two types of block linear equalizers: zero-forcing block linear equalizer (ZF-BLE) and MMSE block linear equalizer (MMSE-BLE).
- Two types of block decision-feedback equalizers: ZF block decision-feedback equalizer (ZF-BDFE) and MMSE block decision-feedback equalizer (MMSE-BDFE).

Block Linear Equalizers. The expression for the signal estimate at the output of ZF-BLE can be written in the form:

$$\hat{\underline{d}}_{\text{ZF-BLE}} = \underline{d} + \kappa \underline{N_0} \tag{22.44}$$

where $\underline{d}$ is the $Nl \times 1$ vector that stacks the transmitted symbols (from the N transmit antennas) during the transmission of a block of length l. The matrix κ is an amplification factor that represents noise enhancements due to the zero-forcing operation, and $\underline{N_0}$ is the noise vector.

A similar expression for the MMSE-BLE can be written as

$$\hat{\underline{d}}_{\text{MMSE-BLE}} = W \hat{\underline{d}}_{\text{ZF-BLE}} \tag{22.45}$$

where the elements of W can be seen as coefficients of a Weiner filter. The estimate from an MMSE-BLE can then be interpreted as the output of the ZF-BLE followed by a Weiner filter. The Weiner filter reduces the performance degradation caused by noise enhancement in ZF-BLE. Therefore, the SNR at the output of the MMSE-BLE per symbol is, in general, larger than that of the ZF-BLE.

Block Decision-Feedback Equalizers. Figure 22.35 shows the block diagram of a block decision-feedback equalizer employed in a MIMO setup. At any time instant, n, the received signal vector $\underline{y}(n)$ is filtered by the equalizer's feedforward filter (FFF), with coefficients $W(n)$, to obtain the filtered signal vector $\underline{y}'(n)$. Previous data estimates are processed through a feedback filter (FBF), with coefficients $B(n)$, and subtracted from $\underline{y}'(n)$. The resultant signals are then fed into threshold detectors from where estimates of the transmitted data $\hat{\underline{c}}(n - \Delta)$, are obtained, where Δ is the delay in the equalizer and $\underline{c}(n - \Delta)$ corresponds to the input signals at time $n - \Delta$. Similar analysis, analogous to Equation 22.44 and Equation 22.45 for the MMSE-BDFE and ZF-BDFE in [30], shows that the SNR at the output of MMSE-BDFE is in general larger than the SNR at the output of the counterpart ZF-BDFE. Therefore, both block linear and block decision-feedback equalization of a MIMO channel based on MMSE criteria will yield better performance than their counterpart zero-forcing schemes. This conclusion is quite consistent with what is known for the SISO channel case.

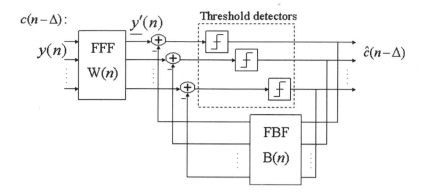

FIGURE 22.35   MIMO DFE block diagram.

List-Type Equalizers.   These equalizers employ a state reduction (complexity reduction) algorithm in the Viterbi or MAP equalizer, using the concept of per-survivor processing (PSP) [12]. The equalizers consider a reduced number of taps of the channel to construct the trellis, leading to a reduced number of states, and an adaptive equalization of the channel is carried out based on the reduced states. To ensure that the best suboptimum performance is achieved, it is desirable to use a receiver filter that concentrates the channel energy on the first taps in order to ensure that the first few taps chosen for the trellis construction have the strongest energy. In the MIMO channel case, this is achieved by using a multidimensional whitened matched filter (WMF) as a prefilter for the equalizer.

Comparison between the performance of the block equalizers and the prefiltered list-type MAP equalizers in MIMO channel [30] shows that the prefiltered list-type MAP equalizer achieves better performance than the block equalizer. However, the list-type MAP equalizer is much more complex to implement. Hence, the regular trade-off between performance and complexity has to be part of the criteria for selecting any of these structures for MIMO applications.

Other Aspects of Adaptive Processing in Wireless Communications

New generations of mobile communication systems must achieve the goal of high data rate applications under spectrum and power constraints, while maintaining the required QoS. The system must provide higher capacity and performance through better use of the available resources. Therefore, adaptation techniques have become popular for optimizing mobile radio transmission and reception, not only at the physical layer but also at the higher layers of the network stack. Here, we briefly introduce adaptive modulation and coding, time-frequency-space link adaptation, and cross-layer design.

Adaptive Modulation and Coding

Classical wireless designs are based, in general, on the worst-case scenario. This leads to nonefficient use of resources. Wireless link adaptation can be defined as any alteration of the transmitter parameters based on information about the channel and network conditions. Methods of adaptation can be classified by the type of adaptation performed (e.g., power, modulation, code rate, etc.). Similar to what has been discussed in previous sections, adaptation can be exploited because of the time-varying nature of the channel, traffic changes, user needs, and QoS requirements. However, this time-varying nature makes the efficiency of the adaptation process dependent on the quality of the measurements and the delays in transmitting the necessary information to the transmitter. Figure 22.36 displays the principle of adaptive modulation and coding schemes.

In uncoded systems, a set of candidate constellations is used. For example, for M-QAM transmissions, the following schemes can be used: 0-QAM (no transmission), 2-QAM, 4-QAM, 16-QAM, and 64-QAM. Depending on the application, a target bit error rate is chosen for a given average SNR. The adaptive algorithm

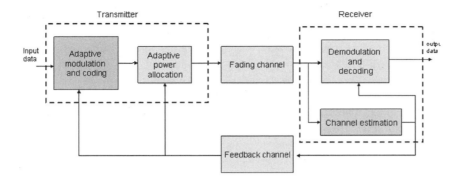

FIGURE 22.36 Adaptive modulation and coding scheme.

calculates the predicted BER based on the fading conditions and chooses the maximum M (i.e., the highest rate) that fulfills the BER requirements. Once a constellation has been chosen, the system is essentially a fixed-rate system. In this case, the goal of the system becomes that of minimizing transmitted power while maintaining the target BER.

Note that in coded schemes, the memory complicates the design of the adaptive system [10]. If the current channel fading is accurately known at the transmitter, the effective channel given the outdated estimate is AWGN. Therefore, coding structure designed for AWGN should be employed [11], where a base trellis coded modulation tuned to the average received SNR can be selected and uncoded bits can be added or deleted based on the channel estimate. With a small amount of prediction error, the conditional channel becomes more Rician (if the channel gain is high) and the same scheme can be used. When the path gain is small, the channel is rather Rayleigh distributed, and a code appropriate to this fading channel should be employed.

When there is a large amount of channel prediction error, the use of uncoded bits is no longer possible. In this case, the instantaneous rate is adapted and all bits are coded [9,26].

Space–Time–Frequency Adaptation

In multicarrier systems (such as orthogonal frequency division multiplexing [OFDM]), where multipath channels introduce frequency selectivity, link adaptation can be made over the frequency domain. Transmission over multicarrier systems maps the information bits over the various carriers. The idea is to choose the modulation scheme based on the channel condition on each subband. For example, avoid transmission over deeply faded subcarriers, while using high-level modulation over carriers that offer good channel conditions. This technique requires, however, high computational load, since it requires information about the channel in each subcarrier. Alternate solutions based on adapting the modes on a per-subband (as opposed to subcarrier) basis offer less overhead [5].

In MIMO transmission, link adaptation can be made over space. Space selectivity occurs when the fading gain depends on the spatial location of the antennas as well as on the spread of angles at the transmitter and receiver sides. Therefore, the performance depends on a number of parameters. This includes the mapping scheme used to map the signals into the transmitting antennas, antenna polarization and location, processing scheme at the receiver, etc. Space–time adaptation selects the best way of combining antennas (e.g., choosing the number of antennas, space–time coding scheme, etc.).

In a multi-carrier MIMO system, link adaptation can be exploited in all domains: space, time, and frequency. This leads to an optimal use of the available resources.

We should mention, however, that there are practical limitations and implementation issues in link adaptation. These include the additional overhead, the feedback channel required to send back the channel state information to the transmitter, the different delays, the determination of the best adaptation threshold and rate, etc. [5].

Cross-Layer Design

The adaptive techniques discussed thus far take place in the physical layer of the wireless communications network. Future generations of wireless systems will have facilities to allow inter-layer adaptive strategies among the various network layers: application, transport, network (IP), medium access control (MAC), and physical layers. This strategy is known as cross-layer design. In cross-layer design, the physical and MAC layers' knowledge of the wireless medium is shared with higher layers, in order to provide efficient methods of allocating network resources. Thus, optimization of the network layer functionality is performed by means of incorporating unconventional lower-layer parameters into the network layer's traditional function. Since the parameters of each layer change from time to time in any real-time system, this kind of joint-layer optimization will therefore have to be done adaptively.

Joint optimization of data transmissions across various layers right from the application layer down to the physical layer has been considered recently [7]. An observation harnessed for such cross-layer design is the fact that, while the IP or MAC layer does not know the context of a data packet, the application layer does have such information. The application layer knows whether a given packet is the beginning, the middle, or end of a long data stream, or whether it is all alone. However, the physical layer has the knowledge about the amount of link capacity currently available, the BER performance of the wireless link, etc. If joint coordination functions or management interfaces (management I/F) are employed, for example at the application and physical layers, such that cross-layer exchange of such vital information is achieved, the overall performance of the network will be improved. The idea is as illustrated in Figure 22.37. It is important to mention that the management I/F does not only span the application and physical layers. This function can be present in all other OSI protocol entities existing between the application and the physical layers. The main challenge in cross-layer design, however, is how to communicate the auxiliary information through the management I/Fs across the layers, with minimum impact on the standard network protocol stack, and using as far as possible already existing route reservation protocols. Another challenge for cross-layer design consideration is that of interoperability of separate networks implementing various architectures of the design. In addressing this challenge, standardization efforts are already in progress to ensure smooth interoperations of the various architectures [31]. In the following, some examples of cross-layer design schemes proposed in the literature are reviewed.

Cross-layer design approaches involving joint network and physical layer optimizations [7], or joint MAC and physical layer optimizations [25] have been proposed. QoS guarantees for CDMA networks are provided by means of cross-layer optimization across the physical and network layers [7]. At the physical layer, the QoS requirements are specified in terms of a target signal-to-interference ratio (SIR) requirement, and optimal target powers are dynamically adjusted according to the current number of users in the system. At the network layer, both the blocking probabilities as well as call connection delay constraints are considered. A reservation-based MAC scheme where users reserve data channels through a slotted-ALOHA procedure has been

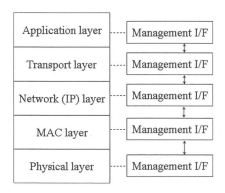

FIGURE 22.37 Principle of cross-layer design.

introduced [25]. Users request transmissions by sending a signature randomly chosen from a pool of orthogonal codes representing the set of available channels. The base station grants or denies access to users based on (signal strength) measurements at the physical layer and the system information at the MAC layer. Such random access schemes have been proposed for the UMTS-WCDMA.

One of the difficulties of joint-layer designs in general is the lack of analytical expressions that relate performance parameters across the various layers. In [25], a novel analytical framework was presented for joint physical and MAC layer designs.

Due to the heterogeneity and, thus, complexity of 4G/4G+ wireless networks, it is expected that application-layer adaptation to network and access conditions is crucial. QoS support becomes an increasingly challenging issue in the presence of external networks. A time-dependent QoS support is an example of the dependency of QoS on variable lower-layer performance.

Lower-layer adaptation based on higher-layer information is also an important research direction to be explored. For example, physical channel sensing mechanisms need to be combined with efficient random access strategies to ensure that channel utilization is maximized, or at least enhanced. Here, a measure of the intra-layer network design can be defined as the ability of two protocol entities belonging to different wireless stations to exchange information of relevance with their internal operation. This would be significantly useful for medium access control and regulation in which such information may affect the queuing behavior of the MAC entity.

Conclusion

The chapter has given a survey of the most popular adaptive signal processing techniques used in wireless communications. We have discussed, in particular, channel identification and equalization, including satellite communication channels and MIMO channels. Other applications of adaptive techniques such as adaptive modulation and coding, space–time–frequency adaptation, as well as cross layer design, have been also covered.

Acknowledgments

This work has been supported in part by the Natural Sciences and Engineering Research Council of Canada (NSERC), Communications and Information Technology Ontario (CITO), and the Ontario Premier's Research Excellence Award (PREA). The authors would like to thank Dr. Quazi M. Rahman and Ms. Suzan Eren for their help.

References

1. N. Al-Dhahir, "Space–time coding and signal processing for broadband wireless communications," in *Signal Processing for Mobile Communications Handbook*, M. Ibnkahla, Ed., Boca Raton, FL: CRC Press, 2004 (Chapter 13).
2. S.I. Amari, H. Park, and K. Fukumizu, "Adaptive method for realizing natural gradient learning for multi-layer perceptrons," *Neural Comput.*, 12, 1399–1409, 2000.
3. H. Arslan, "Adaptation techniques and enabling parameter estimation algorithms for wireless communications systems," in *Signal Processing for Mobile Communications Handbook*, M. Ibnkahla Ed., Boca Raton, FL: CRC Press, 2004 (Chapter 28).
4. H. Arslan and G. Bottomley, "Channel estimation in narrowband wireless communication systems," *Wireless Commun. Mobile Comput. J.*, 1, 201–219, 2001.
5. S. Catreux, V. Erceg, D. Gesbert, and R.W. Heath, "Adaptive modulation and MIMO coding for broadband wireless data networks," *IEEE Commun. Mag.*, vol. 40, no. 6, pp. 108–115, 2002.
6. J.K. Cavers, "An analysis of pilot symbol assisted modulation for Rayleigh fading channels," *IEEE Trans. Veh. Technol.*, vol. 40, no. 4, pp. 686–693, 1991.
7. C. Comaniciu and H.V. Poor, "Jointly optimal power and admission control for delay sensitive traffic in CDMA networks with LMMSE receivers," *IEEE Trans. Signal Proc.*, 51, 2031–2042, 2003.

8. S. Gazor and H.S. Rad, "Space-time coding ambiguities in joint adaptive channel estimation and detection," *IEEE Trans. Signal Process.*, vol. 52, no. 2, pp. 372–384, 2004.

9. D. Goeckel, "Adaptive coding for time-varying channels using outdated fading estimates," *IEEE Trans. Commun.*, 47, 844–855, 1999.

10. D. Goeckel, "Adaptive coded modulation for transmission over fading channels," in M. Ibnkahla Ed., *Signal Processing for Mobile Communications Handbook*, Boca Raton, FL: CRC Press, 2004 (Chapter 6).

11. A. Golsmith and S. Chua, "Adaptive coded modulation for fading channels," *IEEE Trans. Commun.*, 46, 595–602, 1998.

12. T. Hashimoto, "A list-type reduced-constraint generalization of the Viterbi algorithm," *IEEE Trans. Inform. Theory*, IT-33, 866–876, 1987.

13. S. Haykin, *Adaptive Filter Theory*, Englewood Cliffs, NJ: Prentice-Hall, 1996.

14. S. Haykin, "Adaptive tracking of linear time-variant systems by extended RLS algorithm," *IEEE Trans. Signal Process.*, vol. 45, no. 5, pp. 1118–1128, 1997.

15. M. Ibnkahla, "Applications of neural networks to digital communications — A survey," *EURASIP Signal Processing*, vol. 80, Amsterdam: Elsevier, 2000, pp. 1185–11215.

16. M. Ibnkahla, "Nonlinear system identification using neural networks trained with natural gradient descent," *EURASIP J. Appl. Signal Process.*, 1229–1237, 2003 (December).

17. M. Ibnkahla, Q. Rahman, A. Sulyman, H. Al-Asady, Y. Jun, and A. Safwat, "High speed satellite mobile communications: technologies and challenges," *Proc. IEEE, Spec. Issue Gigabit Wireless Commun.: Technol. Challenges*, 312–339, 2004 (February).

18. M. Ibnkahla and J. Yuan, "A neural network MLSE receiver based on natural gradient descent: application to satellite communications," *EURASIP J. Appl. Signal Process.*, pp. 2580–2591, 2004.

19. G.T. Irvine and P.J. McLane, "Symbol-aided plus decision-directed reception for PSK/TCM modulation on shadowed mobile satellite fading channels," *IEEE J. Sel. Areas Commun.*, vol. 10, no. 8, pp. 1289–1299, 1992 (October).

20. G. Kaleh, "Channel equalization for block transmission systems," *IEEE J. Selected Areas Commun.*, vol. 13, no. 1, pp. 110–121, 1995.

21. Y.-S. Kim, C.-J. Kim, G.-Y. Jeong, Y.-J. Bang, H.-K. Park, and S.S. Choi, "New Rayleigh fading channel estimator based on PSAM channel sounding technique," in *Proc. IEEE (ICC'97)*, 1518–1520, 1997.

22. C. Komninakis, C. Fragouli, A.H. Sayed, and R.D. Wesel, "Multiple input-multiple output fading channel tracking and equalization using Kalman estimation," *IEEE Trans. Signal Process.*, vol. 50, no. 5, pp. 1065–1076, 2002.

23. G. Leus and M. Moonen, "Equalization techniques for fading channels," in M. Ibnkahla Ed., *Signal Processing for Mobile Communications Handbook*, Boca Raton, FL: CRC Press, 2004 (Chapter 16).

24. Y. Liu and S.D. Blostein, "Identification of frequency non-selective fading channels using decision feedback and adaptive linear prediction," *IEEE Trans. Commun.*, vol. 43, no. 234, pp. 1484–1492, 1995.

25. A. Maharshi, L. Tong, and A. Swami, "Cross-layer designs of multichannel reservation MAC under Rayleigh fading," *IEEE Trans. Signal Process.*, vol. 51, no. 8, pp. 2054–2067, 2003.

26. P. Ormeci, X. Liu D. Goeckel, and R.D. Wesel, "Adaptive bit-interleaved coded modulation," *IEEE Trans. Commun.*, 49, 1572–1581, 2001.

27. A.J. Paulraj, D.A. Gore, R.U. Nabar, and H. Bolcskei, "An overview of MIMO systems — a key to gigabit wireless," *Proc. IEEE, Spec. Issue Gigabit Wireless Commun.: Technol. Challenges*, 198–218, 2004 (February).

28. S. Qureshi, "Adaptive equalization," *IEEE Proc.*, vol. 73, no. 9, pp. 1349–1387, 1985.

29. S. Sampei and T. Sunaga, "Rayleigh fading compensation for QAM in land mobile radio communications," *IEEE Trans. Veh. Technol.*, vol. 42, no. 2, pp. 137–147, 1993.

30. N. Sellami, I. Fijalkow, and M. Siala, "Overview of equalization techniques for MIMO fading channels," in M. Ibnkahla Ed., *Signal Processing for Mobile Communications Handbook*, Boca Raton, FL: CRC Press, 2004 (Chapter 18).

31. S. Shakkottai, T.S. Rappaport, and P.C. Karlsson, "Cross-layer design for wireless networks," *IEEE Commun. Mag.*, 74 – 80, 2003 (October).

32. D.T.M. Slock, "On the convergence behavior of the LMS and the normalized LMS algorithms," *IEEE Trans. Signal Process.*, vol. 41, no. 9, pp. 2811–2825, 1993.

33. V. Tarokh, N. Seshadri, and A.R. Calderbank, "Space-time codes for high data rate wireless communication: performance criterion and code construction," *IEEE Trans. Inform. Theory*, 744–765, 1998 (March).

34. J. Tugnait, "Modeling and estimation of mobile channels," in M. Ibnkahla Ed., *Signal Processing for Mobile Communications Handbook*, Boca Raton, FL: CRC Press, 2004 (Chapter 3).

35. A. Zerguine and A. Shafi, "Performance of the multilayer perceptron-based decision feedback equalizer with lattice structure in nonlinear channels," in L. Wang, Ed., *Soft Computing in Communications*, Berlin: Springer, pp. 31–53, 2003.

23

Computing Environments for Digital Signal Processing

Robert W. Ives
*United States Naval
Academy*

Delores M. Etter
*United States Naval
Academy*

Computing environments provided by software tools allow users to design, simulate, and implement digital signal processing (DSP) techniques with speed, accuracy, and confidence. With access to libraries of high-performance algorithms and to advanced visualization capabilities, we can design and analyze systems using the equations and notations that we use to think about signal processing problems; we do not have to translate the equations and techniques into a different notation and syntax. The graphics interface provides an integral part of this design environment, and is accessible from any point within our algorithms. Within this type of computing environment, we are more productive. However, even more importantly, we develop better solutions because we have so many more tools for analyzing solutions, for experimenting with "what if" questions, and for developing extensive simulations to test our solutions. To illustrate the power of these environments, we present a brief description of MATLAB, one of the most popular technical computing environments in both industry and academia, and then present five examples that use MATLAB.

23.1 MATLAB Environment

MATLAB is an integrated technical environment designed to provide accelerated DSP design capabilities. In addition to the basic software package that contains powerful functions for numeric computations, advanced graphics and visualization capabilities, a high-level programming language, and tools for designing a graphical user interface (GUI), MATLAB also provides a number of application-specific toolboxes that contain specialized libraries of functions. The discussion and examples contained in this chapter use capabilities from the Signal Processing and Image Processing toolboxes. Other toolboxes that are applicable to solving signal processing problems include the following: Communications, Control Systems, Data Acquisition, Fuzzy Logic, Higher Order Spectral Analysis, Image Acquisition, Neural Networks, Nonlinear Control, Optimization,

Partial Differential Equations, Quantization Feedback Control, μ-Analysis and Synthesis, Statistics, Symbolic Math, System Identification, and Wavelets.

An interactive environment for modeling, analyzing, and simulating a wide variety of dynamic systems is also provided by MATLAB through SIMULINK—a graphical user interface designed to construct block diagram models using "drag-and-drop" operations. Simulations of the block diagrams can be used to test a number of "what if" questions. Special purpose block libraries are available for DSP algorithm development, and include a Communications Blockset, a DSP Blockset, a Fixed-Point Blockset, and a Nonlinear Control Design Blockset.

In order to bridge the gap between interactive prototyping and embedded systems, MATLAB includes a compiler to generate optimized C code from MATLAB code. Automatic C code generation eliminates manual coding and algorithm recoding, thus providing a hierarchical framework for designing, simulating, and prototyping DSP solutions.

23.2   Example 1: DTMF Signal Analysis (Stationary Signal)

In this first example, we generate a signal that represents a dial tone from a telephone network that uses dual-tone multi-frequency (DTMF) signaling. In this system, pairs of tones (or sinusoids) are used to signal each character on the telephone keypad as shown in Figure 23.1. For example, the digit 1 is represented by tones at 697 Hz and 1209 Hz. Figure 23.2 is a plot of the time domain representation of the signal that

Frequencies	1209 Hz	1336 Hz	1477 Hz
697 Hz	1	2	3
770 Hz	4	5	6
852 Hz	7	8	9
941 Hz	*	0	#

FIGURE 23.1   Dual-tone multi-frequency (DTMF) signaling.

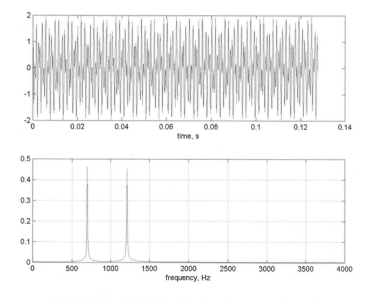

FIGURE 23.2   DTMF signal (top) and spectrum (bottom).

```
%  Example 1:  These statements create and analyze a DTMF tone.
%
clear, clf                          % clear memory and figure window
N = 1024;                           % set number of samples of DTMF signal
Fs = 8000;                          % set sample frequency
T = 1/Fs;                           % set sample interval
t = (0:N-1)*T;                      % specify the time signal in seconds
%
x1 = cos(2*pi*697*t)+cos(2*pi*1209*t);  % create the DTMF signal for digit "1"
y1 = abs(fft(x1))*1/N;              % compute and scale the FFT magnitude
y1 = y1(1:N/2);                     % select the first half of y1
f = (0:length(y1)-1)*Fs/N;          % compute the FFT axis in Hz
%
figure(1)                           % plot signal in time and frequency
subplot(2,1,1),plot(t,x1),xlabel('time, s'),
   title('DTMF Signal and Spectrum')
subplot(2,1,2),plot(f,y1),xlabel('frequency, Hz'),grid
```

FIGURE 23.3 MATLAB code for Example 1.

represents the digit 1, as well as this signal's frequency content computed using the MATLAB fast Fourier transform function. The MATLAB code that generated these plots is shown in Figure 23.3. This code illustrates some of the important characteristics of high-level computational tools. The fundamental data structure is a matrix, and all operations and functions are designed to work with matrices. Hence, loops are rarely necessary, and thus the code is generally much shorter, more readable, and more self-documenting.

23.3   Example 2: Speech Signal Analysis (Nonstationary Signal)

A common DSP application is the analysis of signals that have been collected from experiments or from a physical environment. These signals are typically stored in data files, and often need preprocessing steps applied to them before we are able to extract the desired information. Preprocessing can include removing means or linear trends, filtering noise, removing anomalies, and interpolating for missing data. Once the data are ready to analyze, we are usually interested in statistical information (mean, median, variance, autocorrelation, etc.) along with an estimate of the distribution of the values (uniform, Gaussian, etc.). The frequency content of a signal is also important to determine; if the signal is nonstationary, the frequency content needs to be determined using relatively short time windows.

To illustrate the use of MATLAB in computing some of the steps mentioned above, we use a speech signal collected at 8 kHz. After loading the signal from a data file, we will remove any linear trend that might have been introduced in the collection process (this also removes any constant term). Figure 23.4 contains a plot of the signal, which clearly shows the time-varying nature of the signal. Figure 23.5 contains a histogram of the distribution of the values, showing that the values are closer to a Laplacian or Gamma distribution than to a uniform or Gaussian distribution. Figure 23.6 contains a spectrogram which displays the frequency content of the signal computed using short overlapping time windows. The MATLAB code that generated these plots is shown in Figure 23.7.

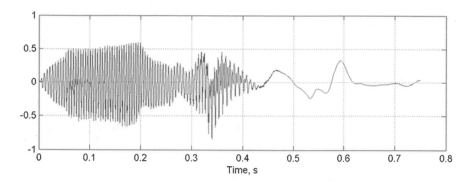

FIGURE 23.4   Speech signal.

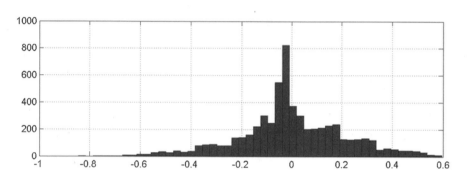

FIGURE 23.5   Histogram of speech values.

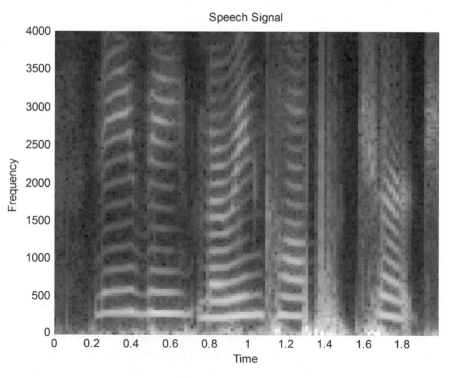

FIGURE 23.6   Spectrogram of speech signal.

```
% Example 2:  These statements read and process a speech file.
%
clear, clf                  % clear memory and figure window
load speech.dat;            % load speech data file
T = 1/8000;                 % set sampling interval
s = detrend(speech);        % remove mean and linear trend from speech
N = length(s);              % determine number of speech samples
t = (0:N-1)*T;              % specify time signal in seconds
%
figure(1)                   % plot the speech signal
subplot(2,1,1),plot(t,s),title('Speech Signal'),
    xlabel('Time, s'),grid,
%
figure(2)                   % plot the speech histogram using 50 bins
subplot(2,1,1),hist(s,50),title('Distribution of Speech Values'),grid
%
figure(3)                   % plot a spectrogram using windows of 256 points
specgram(s,256,8000),title('Speech Signal'),pause
```

FIGURE 23.7   MATLAB code for Example 2.

23.4   Example 3: Filter Design and Analysis

MATLAB gives us a number of different options for designing both IIR and FIR digital filters. We can design classical IIR filters (Butterworth, Chebyshev type I, Chebyshev type II, and elliptic) that are low-pass, high-pass, bandpass, or bandstop filters. We can also use other techniques, such as the Yule–Walker technique, to design IIR filters with arbitrary passbands. Several techniques allow us to design FIR filters using windowed least squares techniques. The Parks–McClellan algorithm uses the Remez exchange algorithm to design filters with an optimal fit to an arbitrary desired response. Once a filter is designed, it can be easily translated to other forms, including transfer functions, impulse responses, and poles/zeros.

Assume that we are going to analyze the dial tones from a telephone network that uses dual-tone multi-frequency (DTMF) signaling as discussed in Example 1. All of the tones are between 697 Hz and 1477 Hz. Thus, before analyzing the signal to determine the two tones that it contains, we might want to remove signals outside of the band that contains all possible tones in order to increase the signal-to-noise ratio. In this example, we design a bandpass filter with a passband between 500 Hz and 1800 Hz. Designs are compared using an elliptic IIR filter of order 8 and a causal FIR filter of order 70. Figure 23.8 contains magnitude plots

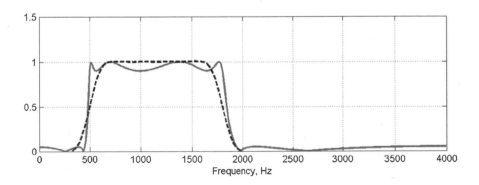

FIGURE 23.8   Comparison of IIR (dashed line) and FIR (solid line) filters.

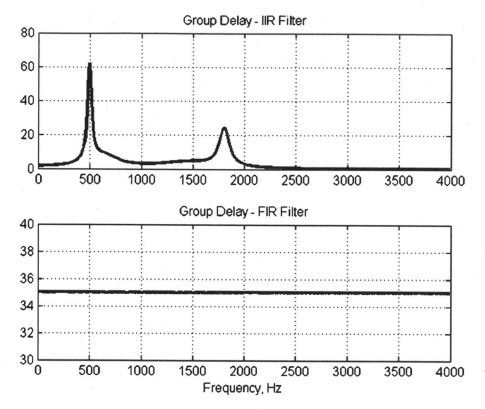

FIGURE 23.9   Group delays of IIR (top) and FIR (bottom) filters.

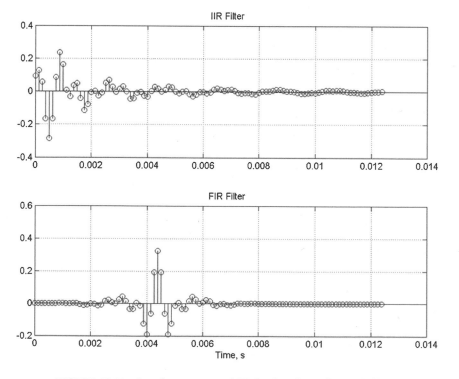

FIGURE 23.10   Impulse responses of IIR (top) and FIR (bottom) filters.

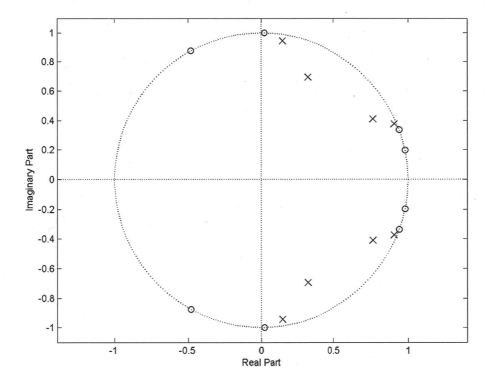

FIGURE 23.11   Poles/zeros for IIR filter.

of the two filters, and clearly shows the characteristics of the filters. The elliptic filter has sharp transitions with ripple in the passband and in the stopband, while the FIR filter (which also uses a Hamming window) is flat in the passband and the stopband, but has wider transition bands. Figure 23.9 contains the group delays for the two filters. The FIR filter has a linear phase response, and thus the group delay is a fixed value of 35 samples; the IIR filter has a nonlinear phase, but has a relatively constant delay in the passband. Figure 23.10 contains the corresponding impulse responses, illustrating the finite impulse response of the FIR filter and the infinite impulse response of the IIR filter. Figure 23.11 contains the pole/zero plots for the IIR solution. The code for performing the designs and generating all of the plots is shown in Figure 23.12.

23.5   Example 4: Multi-Rate Signal Processing

Given a signal that has been collected or computed using a process that eliminates or minimizes aliasing from components above the Nyquist frequency (half the sampling frequency), we have a great deal of flexibility in modifying the sampling rate. For example, if the frequency content of the signal is much lower than the Nyquist frequency, then the sampling rate can be reduced without losing any of the signal content. This "decimation" process allows us to compress the signal into a form that requires less memory. An "interpolation" process can be used to interpolate new data points between points of the decimated signal in such a way that the frequency content of the new signal is essentially the same as the original signal. The decimation process requires a reduction of data points by an integer factor, M, such as a factor of 3. The interpolation process requires that an integral number of points, $L-1$, be interpolated between existing points, such as interpolation of five new points between existing pairs of points. The decimation process increases a sampling interval by M, and the interpolation process

```
%  Example 3:  These statements design and analyze IIR and FIR filters.
%
clear, clf                        % clear memory and figure window
Fs = 8000;                        % specify sampling frequency
band = [500/4000 1800/4000];      % specify passband in normalized freq.
Rp = -20*log10(0.9);              % specify passband ripple
Rs = -20*log10(0.05);             % compute stopband ripple
[B1,A1] = ellip(4,Rp,Rs,band);    % design elliptic passband filter
B2 = fir1(70,band);               % design causal FIR filter
[H1,f] = freqz(B1,A1,512,Fs);     % compute frequency content of filters
[H2,f] = freqz(B2,1,512,Fs);
mag_H1 = abs(H1);                 % compute magnitude of filters
mag_H2 = abs(H2);
[gd1,f] = grpdelay(B1,A1,512,Fs);   % compute group delay of filters
[gd2,f] = grpdelay(B2,1,512,Fs);
%
figure(1)                         % plot filter magnitudes
subplot(2,1,1),plot(f,mag_H1,f,mag_H2,'--'),xlabel('Frequency, Hz'),
   title('Comparision of IIR and FIR Filters'),grid
%
figure(2)                         % plot filter group delays
subplot(2,1,1),plot(f,gd1),title('Group Delay - IIR Filter'),grid
subplot(2,1,2),plot(f,gd2),title('Group Delay - FIR Filter'),grid,
   xlabel('Frequency, Hz')
%
figure(3)                         % plot impulse responses
[h1,t] = impz(B1,A1,100,Fs);
[h2,t] = impz(B2,1,100,Fs);
subplot(2,1,1),stem(t,h1),title('Impulse Response - IIR Filter'),grid
subplot(2,1,2),stem(t,h2),title('Impulse Response - FIR Filter'),grid,
   xlabel('Time, s')
%
figure(4)                         % plot poles and zeros of the IIR filter
zplane(B1,A1),title('Poles/Zeros for IIR FIlter')
```

FIGURE 23.12   MATLAB code for Example 3.

decreases a sampling interval by a factor of *L*. MATLAB contains functions for decimation and interpolation, as well as a function for a resampling of a signal using a noninteger factor of *P/Q* where *P* and *Q* are integers.

Consider a signal that is one sinusoid modulated by another sinusoid. The signal has been sampled at a frequency chosen to provide efficient storage of the data. However, when plotting the data for further analysis, we want to interpolate by a factor of 8 so that the signal looks smoother. Therefore, we use the MATLAB interpolation function. Figure 23.13 contains plots of the original and interpolated time signals. Figure 23.14 contains frequency plots to confirm that the interpolation did not significantly affect the frequency content. Figure 23.15 contains the MATLAB code for this process.

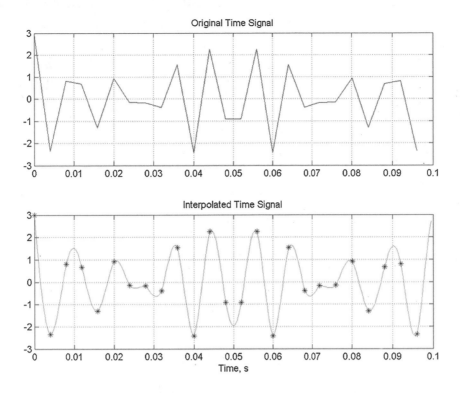

FIGURE 23.13 Original (top) and interpolated (bottom) time signals.

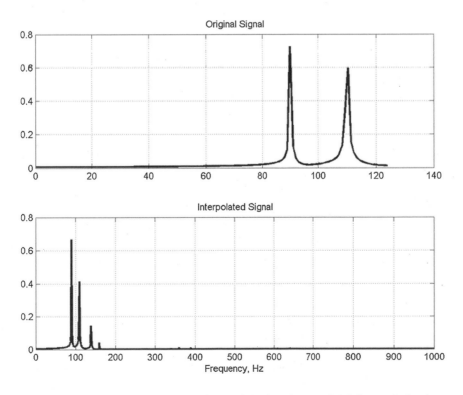

FIGURE 23.14 Frequency content of original (top) and interpolated (bottom) signals.

```
%  Example 4:  These statements interpolate a signal by a factor of 8.
%
clear, clf                           % clear memory and figure window
N1 = 256; N2 = 2048;                 % specify numbers of points
T1 = 0.004; T2 = 0.0005;             % specify time intervals
k1 = 0:255; k2 = 0:2047;             % specify time index
t1 = k1*T1; t2 = k2*T2;              % specify time signals
x1 = 3*cos(20*pi*t1).*cos(200*pi*t1);   % generate original signal
x2 = interp(x1,8);                   % interpolate by a factor of 8
%
figure(1)                            % plot original & interpolated signals
subplot(2,1,1),plot(t1(1:25),x1(1:25)),grid,
    title('Original Time Signal'),
subplot(2,1,2),plot(t2(1:200),x2(1:200),t1(1:25),x1(1:25),'*'),grid,
    title('Interpolated Time Signal'),xlabel('Time, s')
%
X1 = fft(x1);                        % compute frequency content of signals
X2 = fft(x2);
f1 = k1/(N1*T1);                     % determine axis in Hz
f2 = k2/(N2*T2);
figure(2)                            % plot frequency content
subplot(2,1,1),plot(f1(1:128),abs(X1(1:128)/N1)),grid,
    title('Frequency Content of Original Signal'),
subplot(2,1,2),plot(f2(1:1024),abs(X2(1:1024)/N2)),grid
    title('Frequency Content of Interpolated Signal'),
    xlabel('Frequency, Hz')
```

FIGURE 23.15   MATLAB code for Example 4.

23.6   Example 5: Image Analysis and Processing

MATLAB is a matrix-oriented programming language, so it is designed to handle multi-dimensional signals with the same ease with which one-dimensional signals are handled. In fact, many of the same commands can be used with 2D signals, or images, that could be used with one-dimensional signals. Two-dimensional signals are typically stored in formatted image files. In many cases, it is necessary to preprocess the images to reduce the effects of noise or illumination conditions, or simply to enhance the image for viewing. Similar to one-dimensional signals, we are usually interested in statistical information along with an estimate of the statistical distribution of the pixel values. The frequency content of an image may also be important to determine.

Consider a dark photographic image that is contaminated by "impulse" or "salt and pepper" noise, which appears as black and white dots superimposed on a grayscale image, as shown in Figure 23.16(a). Suppose it is desired to use this image on a website, and thus its visual clarity must be improved. A median filter is an order-statistics filter that is particularly effective in the presence of impulse noise. Once the noise is reduced, the darkness can be addressed by a number of methods, one being contrast-limited adaptive histogram equalization (CLAHE). Figure 23.16 contains the original image, the image after denoising, and the image after denoising/equalization. These operations were all performed using MATLAB functions. The effects of this processing on the histogram of the original image are shown in Figure 23.17. The histogram of the original image indicates a large number of low values (darker regions). After denoising, the number of the very brightest and the very darkest values is somewhat reduced, but the shape of the histogram has not changed much. However, after histogram equalization, the range of pixel values is more uniformly distributed, which is reflected in more apparent detail and better contrast in the end result in Figure 23.16(c). Figure 23.18 contains

(a) (b) (c)

FIGURE 23.16 Image processing example: (a) original; (b) denoised; (c) histogram equalized.

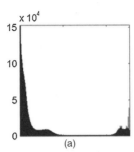
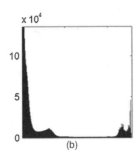
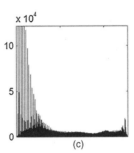

(a) (b) (c)

FIGURE 23.17 Histograms of the corresponding images in Figure 23.16. (a) Original; (b) denoised; (c) histogram equalized.

```
%  Example 5:  These statements process an image.
%
clear, clf                     % clear memory and figure window
a = imread('Chapel-old.jpg');  % read in the input image
b = medfilt2(a);               % apply median filter
c = adapthisteq(b);            % Contrast-Limited Adaptive equalization
%
figure(1)                      % plot original, denoised, denoised/equalized images
subplot(1,3,1),image(a),axis off,colormap(gray(256))
   title('Original'),axis image
subplot(1,3,2),image(b),axis off,colormap(gray(256))
   title('Denoised'),axis image
subplot(1,3,3),image(c),axis off,colormap(gray(256))
   title('Histogram Equalized'),axis image
%
figure(2)                      % plot histogram-equalized image and its histogram
subplot(1,3,1),imhist(a),
   title('Original'),axis square
subplot(1,3,2),imhist(b),
   title('Denoised'),axis square
subplot(1,3,3),imhist(c),axis square
   title('Hist. Equalized'),
%
imwrite(c,'Chapel-new.jpg','jpg');   % write new image as a jpeg file
```

FIGURE 23.18 MATLAB code for Example 5.

the MATLAB code for performing this processing and creating the images displayed. Note that the denoising and equalization required only two function references. Most of the code is generating and plotting the results.

23.7   Conclusions

These examples have demonstrated the ease with which MATLAB can be used to perform signal processing techniques. These techniques included the analysis of stationary and nonstationary signals in both the time domain and frequency domain; the sampling time conversion of a signal; the design and implementation of both FIR and IIR filters; and the analysis and modification of a 2D signal. The MATLAB code for each of the examples was presented to illustrate the ease with which digital signal processing techniques can be implemented with MATLAB functions.

Defining Terms

Drag and drop operation: Graphical operation for building diagrams by selecting, copying, and moving icons using a mouse or track ball.

Graphical user interface (GUI):   Interface using pull-down menus, push buttons, sliders, and other point-and-click icons.

Toolbox: Library of specialized functions.

"What if" question: Question that allows a user to determine the effect of parameter changes in a problem solution.

References

Buck, J., Daniel, M., and Singer, A., *Computer Explorations in Signals and Systems Using MATLAB*, Englewood Cliffs, NJ: Prentice-Hall, 1997.

Burris, C., McClellan, J., and Oppenheim, A., *Computer-Based Exercises for Signal Processing*, Englewood Cliffs, NJ: Prentice-Hall, 1994.

Etter, D., *Engineering Problem Solving with MATLAB*, 2nd ed., Englewood Cliffs, NJ: Prentice-Hall, 1997.

Etter, D., *Introduction to MATLAB for Engineers and Scientists*, Englewood Cliffs, NJ: Prentice-Hall, 1996.

Garcia, A., Numerical Methods for Physics, Englewood Cliffs, NJ: Prentice-Hall, 1994.

Gonzales, R., Woods, R., and Eddins, S., *Digital Image Processing Using MATLAB*, Englewood Cliffs, NJ: Prentice-Hall, 2004.

Hanselman, D., and Kuo, B., *MATLAB Tools for Control System Analysis and Design*, 2nd ed., Englewood Cliffs, NJ: Prentice-Hall, 1995.

Jang, J., Sun, C., and Mizutani, E., *Neuro-Fuzzy and Soft Computing: A Computational Approach to Learning and Machine Intelligence*, Englewood Cliffs, NJ: Prentice-Hall, 1997.

Kamen, E., and Heck, B., *Fundamentals of Signals and Systems Using MATLAB*, 2nd ed., Englewood Cliffs, NJ: Prentice-Hall, 1997.

Marcus, M., *Matrices and MATLAB: A Tutorial*, Englewood Cliffs, NJ: Prentice-Hall, 1993.

Polking, J., *Ordinary Differential Equations Using MATLAB*, Englewood Cliffs, NJ: Prentice-Hall, 1995.

Roberts, M., *Signals and Systems: Analysis Using Transform Methods and MATLAB*, Boston, MA: McGraw-Hill Higher Education, 2004.

Stearns, S., *Digital Signal Processing with Examples in MATLAB*, Boca Raton, FL: CRC Press, 2002.

Van Loan, C., *Introduction to Scientific Computing: A Matrix Vector Approach Using MATLAB*, Englewood Cliffs, NJ: Prentice-Hall, 1997.

Further Information

For further information on MATLAB, here are some e-mail addresses and Internet sites:

E-mail addresses:

news-notes@mathworks.com (MATLAB News and Notes editor)
support@mathworks.com (technical support for all products)
info@mathworks.com (general information)

Web sites:

http://www.mathworks.com (the MathWorks home page)
http://www.mathworks.com/academia (educational products and services)
http://www.mathworks.com/matlabcentral (MATLAB and Simulink user community)

24

An Introduction to Biometrics

Robert W. Ives
United States Naval Academy

Delores M. Etter
United States Naval Academy

24.1 Introduction

Biometrics is a relatively new area of technology that uses unique and measurable physical, biological, or behavioral traits of people to establish or to verify their identification. Identification and verification are two separate operations. In identification, the biometric system asks and attempts to answer the question "Who is this person?" by collecting and comparing biometric samples from an individual and comparing it to the information contained in its database. This is a *one-to-many* search. With verification, the biometric system asks and attempts to answer the question "Is this Joe?". Here, Joe claims he is Joe and submits biometric samples that are compared to the information contained in the database for Joe. This is a *one-to-one* search. Biometrics is most often used to perform identity verification for authorized access to computer networks or secure facilities. The physical attributes typically used include face, iris, fingerprints, hand geometry, handwriting, and voice. Compared to common identification methods, such as identification (ID) cards, personal identification numbers (PINs), or passwords, biometrics is more convenient for users, has lower costs for businesses, reduces fraud, and is more secure.

There is a need for biometrics in federal, state, and local governments, in the military, and in commercial applications. Biometrics has been used in the criminal justice system, in U.S. immigration and naturalization services, and in place of passwords or keys for e-commerce. In 2001, the *MIT Technology Review* named biometrics "one of the top ten emerging technologies that will change the world." Since September 11, 2001, a heightened awareness of security issues is driving the adoption of biometrics within numerous application environments. Some considerations for choosing a biometric system to use, as well as a description of some of the more widely used biometrics, are presented in the following sections.

24.2 Biometric Systems

One way to differentiate types of biometric technologies is based on the level of the users' involvement and cooperation in providing biometric samples. In this case, they can be termed either *active* or *passive*. The active biometric requires users to submit to some form of measurement such that they are aware a sample is being taken. This form of biometric would be used more in a verification role, where individuals wish to prove their identity, such as for log-on to a computer network or in making a banking transaction. Systems that use active biometrics are able to control the environment in which the samples are taken, which facilitates processing. Active biometrics includes iris scanning, fingerprint recognition, and hand geometry recognition.

In contrast, passive biometrics are those that can be sampled without user cooperation or knowledge. These technologies are very dependent on the environment in which the samples are taken. A noncooperative subject, with reliance on the environment, complicates the processing. For example, if the subject is not facing the camera, or is wearing a beard or floppy hat, or if the lighting is dim, facial recognition can fail. Since they do not require cooperation, passive biometrics are more commonly used in identification (as opposed to verification) applications. As a case in point, facial recognition was used to scan the crowds entering turnstiles at the Superbowl in Tampa Bay, Florida, in 2001, searching for known criminals. Common passive biometrics includes face and voice. Note that there are some applications that result in some overlap in the categories of active and passive biometrics; for example, voice can be used as an active biometric for entry to a lab, or as a passive biometric when collected without the speaker's knowledge.

Like many automatic target recognition systems, the performance of biometric systems is based on the accuracy with which the system can correctly match or reject previously unseen samples to the templates contained in its database. Some of the measures of performance can be expressed in terms of the false acceptance rate (FAR) and the false rejection rate (FRR). The FAR measures the percentage of individuals who are incorrectly identified, and is defined as

$$\text{FAR}(\%) = \frac{\text{number of incidents of false acceptance}}{\text{total number of samples presented}} \times 100\%. \tag{24.1}$$

The FRR measures the percentage of individuals who should be identified, but are not, and is defined as

$$\text{FRR}(\%) = \frac{\text{number of incidents of false rejections}}{\text{total number of samples presented}} \times 100\%. \tag{24.2}$$

In addition, a biometric system can be judged by its failure to enroll rate (FTER). The FTER measures the percentage of persons who attempt to enroll, but for various reasons, the system was not able to acquire samples of sufficient quality to create a user template in its database. For example, people who engage in sailing as a sport handle ropes frequently and tend to have their fingerprints worn down to an extent that a fingerprint device may not be able to capture a valid print. The FTER is defined as

$$\text{FTER}(\%) = \frac{\text{number of incidents of unsuccessful enrollment}}{\text{total number of enrollment attempts}} \times 100\%. \tag{24.3}$$

Using system performance in these terms can determine the suitability of a particular system for a desired application.

In addition, there are several other factors that should be considered in implementing a biometric system. These include (in no particular order):

- *Cost.* This includes the cost of the system itself, the cost of deploying the system, and the cost to support and maintain it. Many commercial biometric systems are now affordable to smaller organizations.
- *Ease of use.* Some systems are more user-friendly than others.
- *User acceptance/nonacceptance.* This relates to whether those who would use a particular system feel that it might be of benefit to them, or at least is better than other alternatives.
- *Individual privacy protection.* With identity theft increasing, people are concerned with providing a biometric sample and its susceptibility to theft for criminal use.
- *Invasive measurements.* This relates to the extent to which an individual must cooperate in the collection of a biometric sample, or the perceived intrusiveness on their privacy. A picture of a person

taken in public may not be invasive, but collecting an electronic fingerprint can require physical contact with a collection system.

- *Stable technology.* How long a technology has been on the market is indicative of its reliability, and how fast it is evolving relates to its maturity. A stable technology tends to be well supported.
- *Spoofing the system.* There are those who would actively try to "beat" a biometrics identification system; an example would be playing a recorded voice with a tape recorder for a voice recognition system. One means to reduce the chances of spoofing incorporates some type of "liveness" testing to ensure the submitted sample is from a live human. Some systems are harder to beat than others.

All of these points typify biometric systems in general. Some of the more popular technologies are introduced in the following section.

24.3 Common Biometric Signals

There are a number of physical, biological, or behavioral traits that can be used for verification or identification, with varying degrees of success depending on the application. Some of the common biometrics in reasonably widespread use include: fingerprint, iris, face, and voice. Other biometrics used for identification or verification include gait, handwriting, hand geometry, retinal scan, and ear, among others. The following paragraphs provide a brief description of the more common biometrics.

Fingerprints

The practice of using fingerprints as a means of identification is an indispensable aid to modern law enforcement, and has been used for over a century. Every person has raised ridges of skin on the inside surfaces of their fingers that form interesting patterns of loops, arches, ridge endings, and ridge bifurcations. The local ridge characteristics that occur where a ridge ends or bifurcates (splits into two ridges) are known as minutiae (Figure 24.1), and do not change naturally during a person's life. Identification algorithms that use fingerprints typically extract information about the location, type, and direction of significant minutiae that appear in a fingerprint image and place that information into a template where it can be compared to other templates for matching. Fingerprint recognition begins with preprocessing in order to segment the ridges from the background, and to thin the ridges to a one-pixel width. There are two major categories of

FIGURE 24.1 A fingerprint image.

fingerprint recognition algorithms: minutiae-based and correlation-based. Minutiae-based algorithms seek to determine the presence and relative locations of the minutiae on the finger, including flow direction. The disadvantages of this method are that it relies on a fairly good quality print in order to locate the minutiae and does not take into account a more global picture of the ridges and furrows. Correlation-based methods are sensitive to translation and rotation of the finger, and require both fingerprints to be compared to be precisely registered. Fingerprint recognition systems usually include a sensor to electronically acquire fingerprints (instead of an ink pad) and software for fingerprint analysis and recognition.

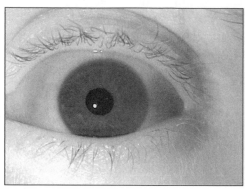 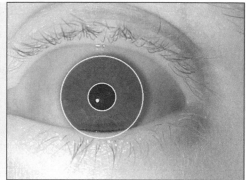

FIGURE 24.2   An iris image. (a) Original image. (b) With iris boundaries delineated by concentric circles.

Iris Scans

The iris is the colored portion of the eye that surrounds the pupil. It is an internal organ whose texture is random, stable, and very unique to an individual throughout their lifetime. The randomness of the pits, striations, filaments, rings, dark spots, and freckles within the colored membrane allows for high confidence recognition in very large databases. Typical iris recognition systems use a near infrared (NIR) camera (to reduce reflections and penetrate glasses and contact lenses) to capture an iris image, then preprocessing extracts the iris portion of the image from the pupil, eyelid, and eyelash pixels, which can then be further analyzed. Figure 24.2 (a) is a NIR image of an eye that could be used for iris recognition and Figure 24.2 (b) shows the iris pattern (within the concentric circles) that might be processed in a recognition algorithm. Several approaches to using the iris pattern in a matching algorithm have been documented. A two-dimensional (2D) Gabor wavelet approach for iris pattern analysis and recognition is the basis for commercially available iris systems today.

Face Images

Face recognition is the primary means by which people recognize each other. It has also become one of the major areas of biometric research because of its noninvasive nature. Images of the face are captured by photographs or video. The fundamental principle of face recognition uses a special mathematical model to measure the dissimilarity of features in the face. Currently, both 2D and three-dimensional (3D) facial recognition systems are available, and some research has gone into a new area called "5D," which is a combination of 2D and 3D technologies.

Two-dimensional facial recognition is based on comparing two digital face images in such a manner as to be invariant to the subject's facial hair or expression, pose, and whether or not they are wearing glasses when the images were captured. The comparison involves the inherent features of the face, such as nose, eyes, lips, chin, and ears, and their relative position to each other in the two-dimensional plane of the captured image. The technologies work best under controlled conditions; as the camera angle or lighting conditions vary, the performance can diminish. Examples of two different facial images are shown in Figure 24.3. The image on the right would be much more difficult to match to an "enrollment" image.

Three-dimensional facial recognition takes the 2D technology a step further and introduces depth by generating a 3D face model. In particular, it addresses the two most critical and complicating factors that affect 2D performance: illumination conditions and pose variation. An example of a 3D system is the A4 Vision system, where structured light is used to acquire the 3D geometry of the face. Three-dimensional reconstruction algorithms are used to formulate a 3D mesh of the face, which is then used for identification.

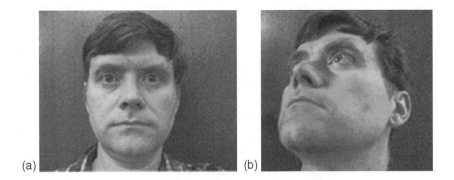

FIGURE 24.3 A 2D facial image. (a) Nearly ideal conditions for recognition. (b) A much harder problem.

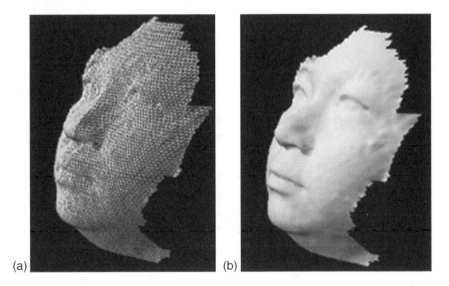

FIGURE 24.4 A 3D facial image. (a) 3D mesh template of a face. (b) Reconstructed 3D mesh w/surface added.

Figure 24.4 (left) shows an example of a 3D mesh of a face used in recognition. Further processing can create a 3D surface of the face (right).

Voice Recognition

Speech is produced via the vocal tract. It is the shape of the vocal tract that makes the voice unique and suitable for use in speaker identification. The vibration of the vocal cords, as well as the positions, shapes, and sizes of the various articulators (lips, tongue, etc.) change over time to produce the sound. The characteristics of the sound vary from person to person, and can be used to identify an individual. For example, Figure 24.5 shows the speech waveform from three different people speaking the word "Honolulu." A person's voice is not necessarily stable over a lifespan, varying with age and in the presence of disease. It can also vary over the short term, in the presence of stress, colds, and allergies. Voice recognition is occasionally confused with the technology of speech recognition. In the latter, an algorithm translates what a user is saying, whereas voice recognition technology verifies the identity of the individual who is speaking.

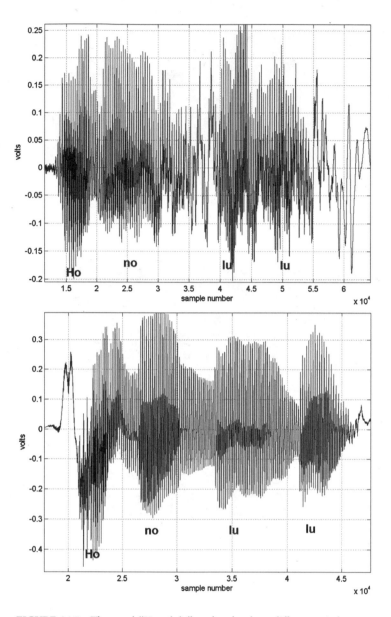

FIGURE 24.5   The word "Honolulu" spoken by three different people.

24.4   Conclusions

The use of biometrics for security applications is becoming more common. Any particular biometric may prove most suitable for a given application, but the performance of any can be improved using *multimodal* biometrics. Multimodal biometrics is the use of a combination of different biometrics in the identification/ verification process. For example, by requiring both a fingerprint and hand geometry samples, the possibility of an incorrect match is more unlikely than when using only a fingerprint.

Improvements in the technology have made biometric systems more accurate, more convenient, and more secure than the more widely accepted means of identification such as ID cards. In an age where identity theft is becoming more prevalent and terrorist threats substantiate the need for automatic identification of people from a distance to identify potential terrorists, biometrics appears to be an important part of solutions to these problems.

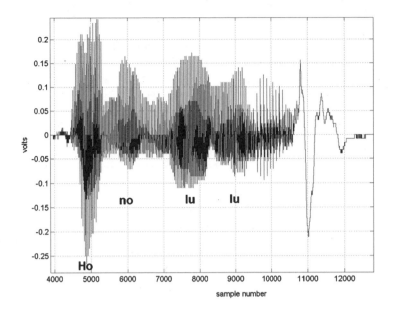

FIGURE 24.5 (Continued).

Defining Terms

Biometric: A unique and measurable physical, biological, or behavioral characteristic of a person that can be used to establish or to verify their identification.

Identification: The process of comparing a person's biometric sample to a database to establish their identity; a one-to-many identity check.

Verification: The process of certifying that an individual is who they say by using biometric samples; a one-to-one identity check.

Iris recognition: Automatically establishing identity using an infrared image of the iris.

Facial recognition: Automatically establishing identity using a photo or video of a person.

Fingerprint recognition: Automatically establishing identity using a fingerprint image.

Voice recognition: Automatically establishing identity using a speech sample.

Multimodal biometrics: Automatically establishing identity using samples from multiple biometrics.

References

Reid, *Biometrics for Network Security*, Upper Saddle River, NJ: Prentice-Hall, 2004.

Woodward, Orlans, and Higgins, *Identity Assurance in the Information Age*, Berkeley, CA: McGraw-Hill, 2003.

Biometrics for Identification and Authentification: Advice on Product Selection, *BiometricsAdvice.pdf*, http://www.cesg.gov.uk/site/ast/biometrics/media/BiometricsAdvice.pdf, 2002.

Prabhakar and Jain, *Fingerprint Identification*, http://biometrics.cse.msu.edu/fingerprint.html, 2004.

Further Information

For further information on biometrics, here are some Internet sites and other resources.

Web sites

http://www.biometrics.org: The Biometrics Consortium
http://www.biometricgroup.com: International Biometrics Group
http://www.eubiometricforum.com: European Biometrics Forum
http://www.biometricdomains.com: Biometrics Domain (a link to other biometric resources)
http://www.iapr.org/: The International Association for Pattern Recognition

Other resources

Interested readers are directed to the following journals for more information:

IEEE Transactions on Pattern Analysis and Machine Intelligence

IEEE Transactions on Signal Processing

IEEE Transactions on Circuits and Systems for Video Technology

IEEE Transactions on Image Processing

International Journal of Pattern Recognition and Artificial Intelligence (IJPRAI)

Pattern Recognition

25

Iris Recognition

Yingzi Du
Indiana University/
Purdue University

Robert W. Ives
United States Naval Academy

Delores M. Etter
United States Naval Academy

The iris provides one of the most stable biometric signals for use in identification, with a distinctive texture that is formed before birth and remains constant throughout life unless there is an injury to the eye. The striations, filaments, and rings that make up the iris pattern are unique to each person, and the left eye differs from the right eye. Because of its uniqueness to an individual, an iris can provide identification with very high confidence, even with large databases. Compared with other biometric features such as face and fingerprint, iris patterns are more stable and reliable. Iris recognition systems are noninvasive, but require a cooperative subject.

25.1 The Iris

The iris (Figure 25.1) is the round, pigmented tissue that lies behind the cornea. It gives color to the eye and controls the amount of light entering the eye by varying the size of the black pupillary opening. The iris contracts and expands its dilator muscles depending on the surrounding light conditions. By regulating the size of the pupil, the iris directs light on to the retina.

There are four layers in the iris: the anterior border layer, the stroma, the dilator pupillae muscle, and the posterior pigment epithelium. These layers together determine eye color and produce the pits, striations, filaments, rings, dark spots, and freckles (which make up the iris patterns through a combination of scattering effects and pigmentation). The natural substance that gives color (pigment) to the human hair, skin, and iris is melanin. Brown eyes are due to heavy pigmentation of the anterior border layer with eumelanin. Blue/green eyes are due to pigmentation with pheomelanin. The dimensions of the iris vary slightly between individuals, with an average size of 12 mm in diameter.

25.2 Iris Recognition Technology

Ophthalmologists first noted from clinical experience that every iris had a highly detailed and unique texture, which remained unchanged after the first year of human life. Ophthalmologists Flom and Safir (1987) hold

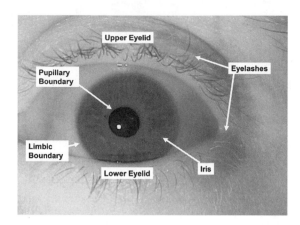

FIGURE 25.1   An iris image.

a patent for describing methods and apparatus for iris recognition based on visible iris features. They were the first to propose the use of the iris as a biometric trait for individual identification. In their 1987 patent, their technique required adjustment of the intensity of a light source until the pupil reached a predetermined size and then compared it with stored image information from an eye with the same pupil size. The requirement of the same pupil size made this iris recognition algorithm somewhat cumbersome.

Daugman (1994) was awarded a patent in 1994 for a biometric personal identification system based on iris analysis. It was the first automatic iris recognition system. He used the two-dimensional (2D) Gabor wavelet to transform iris images into iris codes. Daugman defines iris recognition as the "real-time, high confidence recognition of a person's identity by mathematical analysis of the random patterns that are visible within the iris of an eye from some distance." The iris has proven to be the most stable and reliable means of biometric identification. The block diagram in Figure 25.2 depicts the principal steps of the process of iris recognition, and is described in the following sections. The iris recognition system has two subsystems: the iris enrollment system and the iris identification system. The iris enrollment system is used to enroll the iris patterns in the database for further identification. The iris identification system compares a newly input iris pattern with the known iris patterns in the database and decides if it is in the database.

Image Acquisition

One of the major challenges of an iris recognition system is to capture a high-quality iris image while remaining noninvasive to the user. In the image acquisition step, near-infrared (NIR) cameras are used to acquire iris images and save them in digital format (see Figure 25.1b). A NIR light source may be required to provide ample illumination. In the visible light wavelengths, the amount of detail visible in the iris varies from person to person and is correlated to eye color. Lighter color eyes (such as light blue eyes) tend to show more detailed patterns than darker color eyes (such as dark brown eyes). However, at NIR wavelengths (700 to 900 nm), even dark irises reveal rich and complex features. The actual focal length to acquire an iris image depends on the iris camera, typically 5 to 20 in. resolving a minimum of 70 pixels in iris radius. There is research in progress to enable longer-distance iris scanning.

Preprocessing

The preprocessing step locates the various components of the iris boundary. In particular, it defines the limbic (outer) boundary of the iris, the pupillary (inner) boundary of the iris, and the eyelids. Edge detection can be used in this step to find the iris boundaries. The pupillary and limbic boundaries are approximated by circles,

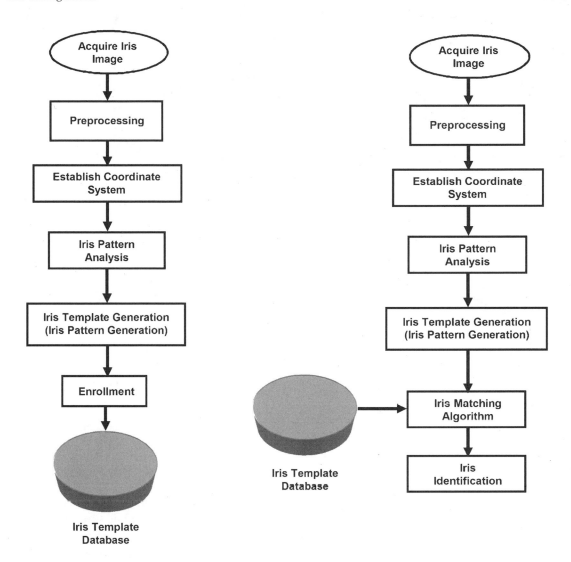

FIGURE 25.2   Iris recognition system: (a) iris enrollment system; (b) iris identification system.

and after defining the pupillary boundary, the center of the pupil can be estimated using this circular model. Then, with this as the center location of the circles, the iris image can be transformed into polar coordinates.

Establish Coordinate System

The size of the same iris captured at different times may be variable as a result of changes in the distance from the camera to the face. Also, due to stimulation by light or for other reasons, the pupil may be constricted or dilated. These factors will change the iris resolution and the actual distance between the pupillary and limbic boundaries. Instead of fixing the pupil size as ophthalmologists Flom and Safir (1987) proposed, in an automatic iris recognition system, the image is processed to ensure the accurate location of the boundary and to fix the resolution.

Figure 25.3 illustrates the rectangular to polar transformation. Using (x_0, y_0) as the location of the center of the pupil, for each pixel in the original iris image located at rectangular coordinates (x_i, y_i), we compute its

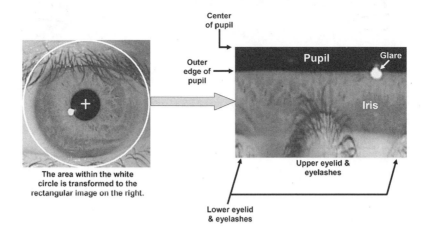

FIGURE 25.3 Rectangular to polar transformation.

polar coordinates (r_i, θ_i) as

$$r_i = \sqrt{(x_i - x_0)^2 + (y_i - y_0)^2},$$

$$\theta_i = \begin{cases} \arcsin\left(\frac{y_i - y_0}{x_i - x_0}\right) & y_i \geq y_0 \\ \pi + \arcsin\left(\frac{y_i - y_0}{x_i - x_0}\right) & y_i < y_0 \end{cases} \tag{25.1}$$

This transformation takes the circular iris and reforms it into a rectangular shape that can be further analyzed using image processing.

Iris Pattern Analysis

Iris pattern analysis is a key step in the process, and there are different approaches. Daugman (1994, 2004) used a quadrature 2D Gabor wavelet method to analyze both coherent and incoherent detailed texture of the iris. Only the phase information is used for iris recognition, as the phase can provide optimal information about the orientation, spatial frequency content, and 2D position of the local image structure while the magnitude provides little discrimination. Du et al. (2004) designed a local texture analysis algorithm to calculate the local variances of iris images. Some other approaches include the Laplacian parameter approach by Wildes et al. (1996), zero-crossings of the one-dimensional (1D) wavelet transform at various resolution levels by Boles and Boashash (1998), the independent component analysis (ICA) approach by Huang et al. (2002), the texture analysis using multichannel Gabor filtering and wavelet transform by Zhu et al. (2000), the circular symmetric filter approach by Ma et al. (2002) and the self-organizing neural network approach by Liam et al. (2002). Among them, Daugman's 2D Gabor wavelet approach has been successfully tested using a large scale iris database and has been commercialized by Iridian.

Iris Template Generation

After the iris pattern has been analyzed, iris template generation extracts the unique iris patterns and stores them as a template that will be used for identification. Again, there are various approaches to template generation. As an example, Daugman's (1994, 2004) method transforms the 2D Gabor phase information into binary 0s and 1s using a zero-crossing method. The iris template is encoded in 2048 bits, while at the same time an equal number of masking bits are also used to denote non-iris artifacts to prevent their interference in the comparison. Many other iris recognition methods transfer the iris patterns into 2D codes. However, Du et al. (2004) generate 1D iris signatures (templates) from the 2D local texture patterns (LTP) of each

(a)

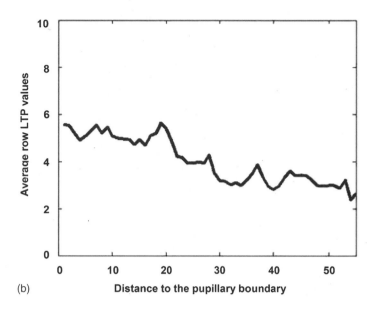

(b)

FIGURE 25.4 An example of different iris templates: (a) iris phase code; (b) 1D iris signature.

row of the rectangular images, after removing the non-iris pixels. Figure 25.4 shows a sample of a 2D iris code using Daugman's method, and a 1D signature generated via the method of Du et al.

Enrollment

For an iris pattern to be recognizable in the system, we first need to enroll the iris pattern in the database. The enrollment process usually takes multiple iris images of the same iris to register and generate the enrollment iris patterns. Iris recognition systems usually require a higher quality of iris images for enrollment purposes. For example, to enroll, the users usually cannot wear eyeglasses or contact lenses. However, these restrictions could be relaxed for identification purposes.

Iris Template Database

The enrolled iris templates are stored in the iris template database for further identification. The security of this iris template database must be protected, and the templates are usually encrypted.

Iris Matching Algorithm

When an iris image is presented for identification, an iris template is generated by the template generation algorithm. The iris recognition system then attempts to match it with the enrolled iris templates stored in the database. Here again, there are various matching algorithms in use. Since the eyes may tilt during the acquisition step, the iris matching algorithm should take into consideration any rotation of the iris patterns. Instead of rotating the iris image, Daugman and many other researchers compare templates in the database to the newly created template by cyclic scrolling of its angular variable (for Daugman's algorithm, this is the 2D iris phase code). The match score is the best match after numerous relative rotations of the two iris templates. To compute the match score, Daugman used the Hamming distance (HD) as the measure of the dissimilarity between two irises. For 1D iris signatures by Du et al. (2004), this rotation is not necessary, since the 1D iris templates are orientation invariant. Here, the *Du measure* was used to generate the 1D iris template matching scores.

Iris Identification

Based on the matching score generated by the matching algorithm, this step will decide if the input iris is identified as the one in the database. The identification step involves statistical analysis of a threshold applied to the match scores. According to Daugman, if using a Hamming distance threshold of 0.35 (that is, 65% of the bits match), the false match rate could be as high as 1 in 133,000. However, using 0.26 as the threshold, the false identification rate could be as high as 1 in 10^{13} while the false rejection rate will be increased dramatically.

Speed Performance of Iris Recognition

Different iris recognition methods vary in the speed with which they are able to carry out the recognition process. Daugman showed the execution time for critical steps on a 300 MHz Sun workstation using optimized integer code. The demodulation and IrisCode creation step (i.e., the iris pattern analysis and iris template generation steps in Figure 25.2) takes only 102 msec, while the preprocessing step (assess image focus, scrub specula reflections, localize eye and iris, detect pupillary boundary, detect and fit both eyelids, and remove eyelashes and contact lens edges) takes a much longer time, 344 msec. For commercialized iris recognition systems, iris recognition just takes approximately 2 sec.

Iris recognition can be used to search large databases in real time. A 2001 study conducted by the U.K.'s National Physical Laboratory found that iris technology was capable of nearly 20 times more matches per minute than fingerprints (Forrester et al., 2001).

25.3   Applications of Iris Recognition

Iris recognition is noninvasive to the user, avoiding any physical contact with the device. The iris is an extremely difficult characteristic to spoof. In an iris recognition system, a "liveness" test is usually deployed to test if the iris is from a living person rather than an artificial or lifeless eye. The natural pupillary response (changing pupil size in response to changes in illumination) can be used to confirm the liveness of an iris. Some iris recognition systems also use the "hippus" (the constant movement of the pupil) and "saccade" (movement of the eye) as criteria for the liveness test.

Compared to other biometrics, such as face, fingerprint, and voice, the iris has a number of advantages. The iris is a highly protected, internal organ of the eye, and yet it is externally visible so patterns can be imaged from a distance. The iris has six times as many distinct, identifiable features as a fingerprint and a high information density (approximately 3.2 bits per square millimeter). Like fingerprints, iris patterns develop randomly. No two iris patterns are alike, even those of identical twins, and even between the right and left eye of the same person. The iris remains unchanged throughout a person's life, although damage to the cornea, disease, or other ailments might hinder or prevent its use for identification. Iris recognition has the highest accuracy level of all biometrics, with a near 0% false accept rate (FAR) and very low false reject rate (FRR). It can provide fast, scalable authentication in large database environments.

Iris technologies have provided important assets for security in terms of positive human identification. Currently, there are three general categories of commercialized iris recognition systems. These are described in the following sections.

PC Iris Recognition System

This kind of iris recognition system uses a low-end iris camera, such as the Panasonic Authenticam. The iris camera is linked directly with a PC (Figure 25.5), commonly with a USB connection. This system typically can hold very small databases (fewer than 200 users). The enrollment and identification are performed with the

FIGURE 25.5 PC iris recognition system (the iris camera is a Panasonic Authenticam).

same iris camera. The advantage is the ease of installation and relatively low cost, but this kind of system is more vulnerable to spoofing compared to the more sophisticated iris systems. The applications are usually seen in computer access control, Internet authentication, and e-commerce applications.

Walkup Iris Recognition System

The walkup system (Figure 25.6) usually uses a server–client distribution system and comprises multiple iris cameras, such as the LG IrisAcess System. The enrollment is done within a secure environment with a separate enrollment iris camera. The identification is performed in a less secured environment, such as outside a door or gate that leads to a secure area.

The walkup iris recognition system usually includes an administrative server, an enrollment server, a database server, and the remote access units. The administrative server manages the user accounts and the access schedule (the schedule each user is allowed to access the door/gate) for each user, and monitors system events. The enrollment server connects to an iris enrollment camera. The database server contains the entire

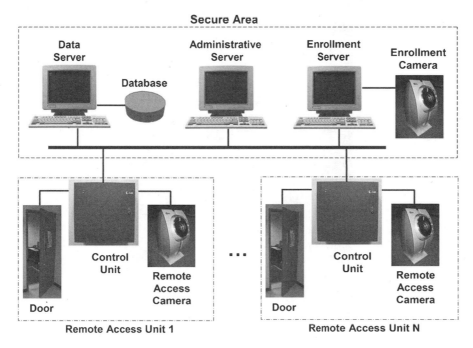

FIGURE 25.6 Walkup iris recognition system architecture.

enrolled iris database. These three servers should be located in a secured area. In a smaller user group, these three server functions could be performed on a single computer. For large user groups, there could be multiple database and enrollment servers.

A remote access unit has an access control iris camera, a control unit, and a door control unit. The access control iris camera acquires iris images, and provides audio and visual cues that indicate whether or not the user is identified. No biometric templates are stored on the optical unit. The control unit is a customized computer that processes the iris image acquired from the iris camera into iris templates and compares these with stored templates. If a match is found, the unit generates a signal to the door control unit to open or unlock the door. The control unit should be installed in a protected area. The iris recognition system can have multiple remote access units. The remote access unit is typically connected to the servers via a computer network. Many walkup iris recognition systems are using the Internet (TCP/IP) for connection purposes.

Walkup iris recognition systems are found in government, financial institutions, key research areas, and other corporations for door/gate access control and positive identification.

Stand-Alone/Portable Iris Recognition System

A portable iris recognition system is a fully self-contained iris enrollment and recognition system. It is a stand-alone handheld device. SecuriMetrics Inc. developed the first stand-alone iris recognition system named PIER (portable iris enrollment and recognition) as shown in Figure 25.7. The portable/stand-alone iris recognition

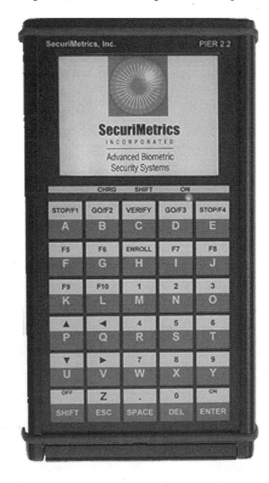

FIGURE 25.7 An example of a portable/stand-alone iris recognition system (from PIER Version 2.2 at http://www.securimetrics.com/).

system integrates an iris camera, a computer processor, a database, and a keyboard in a single device. On top of the PIER is a small monitor that shows the status of iris recognition. On the top back portion is an iris camera. The bottom part of the PIER is the keyboard that enables users to input user names and allows interaction with the iris recognition process.

The portable and standalone features of this kind of system enable the application of iris recognition in the field. According to the *Homeland Defense Journal* in March 2004, U.S. Special Operations Forces have deployed PIER Version 2.2 in Iraq in 2003. It has been deployed worldwide in support of Operation Enduring Freedom for numerous missions, including port security.

25.4 Conclusion

The iris provides a unique stable biometric signal that is the basis of iris recognition algorithms. Most of the algorithms employ 2D signal processing techniques, but some recent algorithms use 1D techniques. Commercial systems provide reliable accurate identification, and range from inexpensive PC devices to more expensive multiple access-point systems. Iris recognition systems are noninvasive, but require a cooperative subject.

Defining Terms

Iris: Round pigmented tissue that lies behind the cornea and gives color to the eye (e.g., blue eyes). It controls the amount of light entering the eye by varying the size of the black pupillary opening.

Iris recognition: Real-time, high confidence recognition of a person's identity by mathematical analysis of the random patterns that are visible within the iris of an eye from some distance.

Iris liveness test: Checks if the iris image is taken from a living person.

References

W.W. Boles and B. Boashash, "A human identification technique using images of the iris and wavelet transform," *IEEE Trans. Signal Process.*, vol. 46, no. 4, 1998.

J. Daugman, United States Patent No. 5,291,560, *Biometric Personal Identification System Based on Iris Analysis*, Washington D.C.: U.S. Government Printing Office, 1994 (issued March 1).

J. Daugman, "How iris recognition works," *IEEE Trans. Circ. Syst. Video Technol.*, vol. 14, no. 1, pp. 21–30, 2004.

Y. Du, C.-I. Chang, H. Ren, F.M. D'Amico, and J. Jensen, "A new hyperspectral discrimination measure for spectral characterization," *Optical Engineering*, vol. 43, no. 8, pp. 1777–1786.

Y. Du, R.W. Ives, D.M. Etter, T.B. Welch, and C.-I Chang, "One dimensional approach to iris recognition", *Proc. SPIE's Def. Secur. Symp.*, 2004.

L. Flom and A. Safir, United States Patent No. 4,641,349, *Iris Recognition System*, Washington D.C.: U.S. Government Printing Office, 1987 (issued February 3).

J. Forrester, A. Dick, P. McMenamin, and W. Lee, *The Eye: Basic Sciences in Practice*, London: W. B. Saunders, 2001.

Y.-P. Huang, S.-W. Luo, and E.-Y. Chen, "An efficient iris recognition system", *Proc. First Int. Conf. Mach. Learn. Cybern.*, 450–454, 2002.

Iris (anatomy), http://www.worldhistory.com/wiki/I/Iris-(anatomy).htm, 2004.

L.W. Liam, A. Chekima, L.C. Fan, and J.A. Dargham, "Iris recognition using self-organizing neural network," *Student Conf. Res. Dev. Proc.*, 169–172, 2002.

L. Ma, Y. Wang, and T. Tan, "Iris recognition using circular symmetric filters," *16th Int. Conf. Pattern Recognit.*, 2, 414–417, 2002.

R.P. Wildes, J.C. Asmuth, G.L. Green, S.C. Hsu, R.J. Kolczynski, J.R. Matey, and S.E. McBride, "A machine vision system for Iris recognition," *Mach. Vis. Appl.*, 9, 1–8, 1996.

J.D. Woodward, N.M. Orlans, and P.T. Higgins, *Biometrics*, California: McGraw-Hill Company, 2002.

Y. Zhu, T. Tan, and Y. Wnag, "Biometric personal identification based on iris patterns," *15th Int. Conf. Pattern Recognit.*, 2, 801–804, 2000.

Further Information

For further information on iris recognition technology and systems, some Internet sites and other resources locations are listed below.

Web Sites

http://www.cl.cam.ac.uk/users/jgd1000 (Dr. John Daugman's homepage)

http://www.iridiantech.com (Iridian Technologies)

http://www.sinobiometrics.com (Center for Biometric Authentication and Testing, Chinese Academy of Sciences. You can download very good iris images for your research)

http://www.biometrics.org (Biometrics Organization Webpage)

Other Resources

Currently, many researchers are working on iris recognition technologies. Interested readers are directed to the following journals for more information:

IEEE Transactions on Pattern Analysis and Machine Intelligence

IEEE Transactions on Signal Processing

IEEE Transactions on Circuits and Systems for Video Technology

IEEE Transactions on Image Processing

Optical Engineering

Pattern Recognition

Additionally, readers are encouraged to attend the following conferences to update information about the current state of iris recognition:

Biometric Consortium (held annually in Arlington, VA)

SPIE Defense and Security Symposium (held annually in Orlando, FL)

IEEE International Conference on Pattern Recognition (held annually)

26

Liveness Detection in Biometric Devices

Stephanie A.C. Schuckers
Clarkson University

Reza Derakhshani
University of Missouri

Sujan T.V. Parthasaradhi
Bioscrypt, Inc.

Lawrence Hornak
West Virginia University

26.1 Introduction

Biometric devices have been suggested for use in applications from access to personal computers, automated teller machines, credit card transactions, and electronic transactions to access control for airports, nuclear facilities, and border control. Given this diverse array of potential applications, biometric devices have the potential to provide additional security over traditional security means such as passwords, keys, signatures, picture identification, etc. Examples of biometrics include recognition of a fingerprint, face, voice, iris, and handprint [1,2]. The following are the basic components and function of a system, as shown in Figure 26.1.

First, a sensor measures the biometric sample and converts it into a digital signal or image. An algorithm then extracts features from the signal or image, that are used for the recognition algorithm. For example, with fingerprint recognition, features commonly include ridge endings and bifurcations, called minutiae. The algorithm then compares these features to a template, which are the features stored previously at an enrollment visit. In the case of verification, the sample is compared to the enrolled template of the person that they claim to be. In the case of identification, the sample is compared to all templates in the database. A match score quantifies the degree of match between the sample and the template, and a threshold is set to decide if it is a match or nonmatch. Typical statistics that define biometric system performance include false accept/ false reject error rates and receiver operating characteristic curves, which balance the tradeoff between false accept and false reject error rates for a variety of thresholds [3].

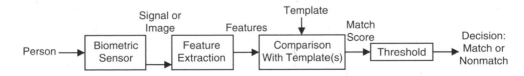

FIGURE 26.1 Basic structure of a biometric system.

While biometrics may improve security, biometric systems also have vulnerabilities. System vulnerabilities include attacks at the biometric sensor level, replay attacks on the data communication stream, and attacks on the database, among others [4]. One vulnerability includes attempts to gain unauthorized access at the biometric sensor with artificial fingers created from fingerprints of authorized users, called *spoofs*, or, in the worst-case scenario, dismembered fingers. Previous work has shown that it is possible to spoof a variety of fingerprint technologies through relatively simple techniques using casts of a finger with molds made of materials including silicon, Play-Doh, clay, and gelatin (gummy finger). It has been demonstrated that spoof molds can be scanned and verified when compared to a live enrolled finger [5–10]. In addition, our research has shown that cadaver fingers can be scanned and verified against enrolled cadaver fingers [10].

Liveness, i.e., determination of whether the introduced biometric is coming from a live source, has been suggested as a means to circumvent attacks that use spoof fingers. The goal of liveness testing is to determine if the biometric being captured is an actual measurement from the authorized, live person who is present at the time of capture. Typically, liveness is a secondary measure after biometric authentication, which must be met in order to achieve a positive response. Liveness and antispoofing methods are covered in detail in the following summaries [10–12]. Other techniques that can make spoofing more difficult include the use of biometrics in combination with a challenge response, passwords, tokens, smart cards, and multiple biometrics.

Liveness falls into three main approaches. The first approach is the utilization of additional hardware in conjunction with the biometric sensor. Examples of this approach include thermal sensing of finger temperature [13], ECG [13], impedance and electrical conductivity of the skin (dielectric response) [14], and pulse oximetry [13,15]. Pulse oximetry is a standard medical monitoring tool, which utilizes the variation over time of the absorption of light due to oxygenated and deoxygenated blood (typically, 660 and 940 nm) [16]. While effective in determining liveness, these methods require additional hardware, which is costly and, unless integrated properly, may be spoofed with an unauthorized live person.

The second approach is to further process the biometric sample to gather liveness information. Examples include quantifying saccade movements in the eye for iris recognition, lip-reading [17], or perspiration in the fingerprint, illustrated in an example of liveness detection discussed in the next section [9]. The advantage of this approach is that the biometric and liveness information are linked, such that it would be more difficult to spoof with an unauthorized live person.

Finally, for some biometrics, liveness is an inherent aspect of the biometric, i.e., "liveness" is a prerequisite even to capture the biometric. Examples include ECG [18], the electrical measurement of the heart, spectroscopy [19], and the reflection of different wavelengths of light measuring characteristics of underlying skin and tissue structure. While the most difficult to circumvent, these methods as biometrics are relatively new to the field and will need to be evaluated both as a biometric and a liveness measure.

Although many liveness detection methods have been suggested and some have been implemented, no independent testing has been performed to report on the effectiveness of these methods for detecting liveness. Liveness detection is a stage in the authentication process. Therefore, it must be treated as part of the biometric system, in that it has an impact on the false reject ratio, false accept ratio, failure to acquire, and other statistics. In addition, other characteristics for evaluating biometrics systems such as ease of use, universality, and user acceptance [1] need to be considered before implementing a liveness algorithm. Liveness algorithms are not spoof-proof and therefore will also have varying degrees of spoofing vulnerability.

26.2 Use of Signal Processing for Liveness Detection in Biometric Devices

In this section, an example of signal processing is given where liveness is determined in fingerprint devices based on a time series of fingerprint images [9,20]. The method uses the physiological process of perspiration to determine the liveness of a fingerprint. In brief, when in contact with the fingerprint sensor surface, live fingers, as opposed to cadaver or spoof, demonstrate a distinctive spatial moisture pattern that evolves in time due to the physiological perspiration process. Optical, electro-optical, and solid-state fingerprint sensors are sensitive to the skin's moisture changes on the contacting ridges of the fingertip skin and can capture the time-dependent, spatial pattern (Figure 26.2).

To quantify the perspiration phenomenon, our algorithm maps a two-dimensional fingerprint image to a "signal" that represents the gray-level values along the ridges (Figure 26.3). Variations in gray levels in the signal correspond to variations in moisture, both statically (on one image) and dynamically (difference between consecutive images). The static feature measures periodic variability in gray level along the ridges

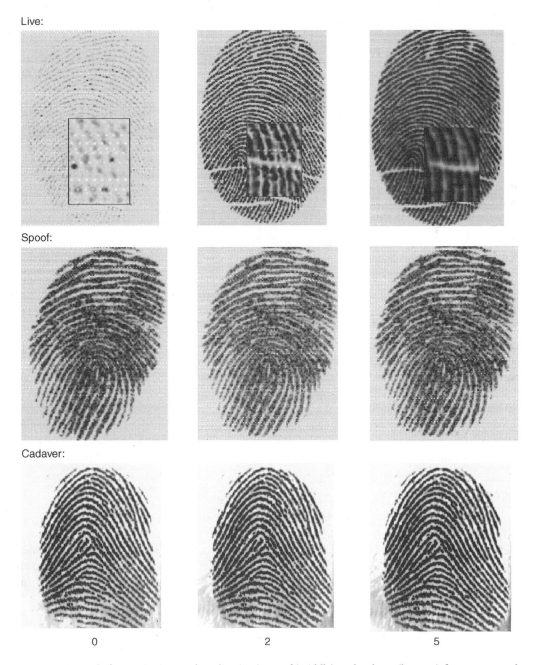

FIGURE 26.2 Example fingerprint images from live (top), spoof (middle), and cadaver (bottom) fingers captured at 0, 2, and 5 sec (left to right) after placement on scanner. Live fingerprint images demonstrate changing moisture over time due to perspiration. (*Source:* Parthasaradhi, S.T.V. et al., *IEEE Trans. Syst., Man and Cybernetics*, Part C, Vol. 35, no. 3, pp. 335–343, 2005.)

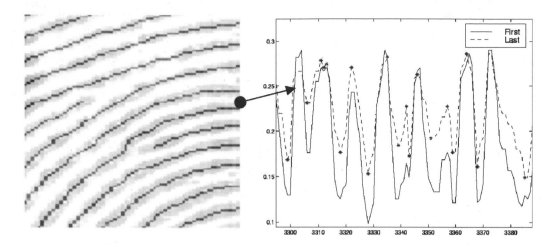

FIGURE 26.3   Ridge mask (dark lines) superimposed over the original grayscale fingerprint image (left) and an example of the resulting ridge signal for the first (solid) and second (dashed) image captures (right). The ridge signal is the gray levels of the image under the mask, where a larger amplitude indicates a higher gray level. (*Source:* Parthasaradhi, S.T.V. et al., *IEEE Trans. Syst., Man and Cybernetics*, Part C, Vol. 35, no. 3, pp. 335–343, 2005.)

due to the presence of perspiration around the pores. In this measure, the small, highly moisturized areas around the perspiring pores (and not the pores themselves) are detected as a sign of active perspiration and thus liveness. The dynamic features quantify the temporal change of the ridge signal due to propagation of this moisture between pores in the initial image relative to image captures 2 (or 5) sec later, while the pore areas remain saturated and almost unchanged. The cadaver and spoof fingers fail to provide the mentioned static and dynamic patterns due to the lack of active pore-emanated perspiration.

26.3   Algorithm Details

Two fingerprint images captured within a 2- (or 5-) sec interval (referred to as first and second capture) are the inputs to the algorithm. The algorithm can be divided into five main components, outlined in Figure 26.4: (1) preprocessing to remove noise, (2) conversion of images to "ridge signal," (3) extraction of static-based feature on first image, (4) extraction of dynamic features based on first and second image, and (5) classification (neural network) [9,20].

Preprocessing

A program developed to clean the image subtracts the permanent irregularities in the scanner by comparing it to a "blank" capture taken for each individual case. It also removes the background static by discarding those pixels that change only within 2% of the blank scan. Next, a median filter is applied here to "cover" the white pixels in the middle of the pores (3×3 for 300 dpi device and scaled accordingly for other devices). This also smoothes the image further and eliminates "salt-and-pepper" noise, if any. Next, a software module transforms the image to binary.

Conversion to Ridge Signal

A copy of the last captured image is thinned to locate the ridges. Y-junctions are removed using a simple nonoverlapping neighbor operation (3×3 for 300 dpi and scaled accordingly for other devices). Ridges that are not long enough to cover at least 12 mm (approximately two pores) are discarded. Using the thinned ridge locations as a mask, the gray levels of the original image underneath these ridge paths are recorded.

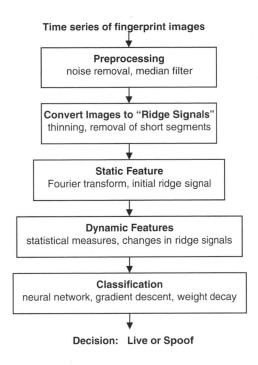

Time series of fingerprint images

Preprocessing
noise removal, median filter

Convert Images to "Ridge Signals"
thinning, removal of short segments

Static Feature
Fourier transform, initial ridge signal

Dynamic Features
statistical measures, changes in ridge signals

Classification
neural network, gradient descent, weight decay

Decision: Live or Spoof

FIGURE 26.4 General flowchart of algorithms for detection of liveness in biometric fingerprint devices.

The resulting signals for the first and the last capture are representative of the moisture level along the ridges for a given image in the time series. Figure 26.5 illustrates these steps by showing a portion of the ridge signals derived from the first and last captures from a live, spoof, and cadaver source along the mentioned mask.

Extraction of Static Feature (Fourier Transform)

Static feature (*SM*) quantifies variations in the gray levels of the first ridge signal, where the peaks represent the moist pores and the valleys the dryer regions between each of two pores. The average Fourier transform of the signal segments is calculated from the first capture using a 256-point FFT for each ridge signal segment. The total energy is summed across the spatial frequency range between the typical pore spacing of 0.4 to 1.2 mm (8 to 24 pixel distance for a 300 dpi device or 11 and 33 number of FFT points/spatial period). Before taking the FFT, in order to eliminate the spike around zero frequency, the dc of the signal is removed by subtracting the mean. The procedure can be mathematically expressed as

$$\mathrm{SM} = \sum_{k=11}^{33} f(k)^2$$

where

$$f(k) = \frac{\sum_{j=1}^{n} \left| \sum_{i=1}^{256} C1_{j,i}^{a} \, e^{-2\pi(k-1)(i-1)/256} \right|}{n} \qquad C1_{j}^{a} = C1_{j} - \mathrm{mean}\,(C1_{j})$$

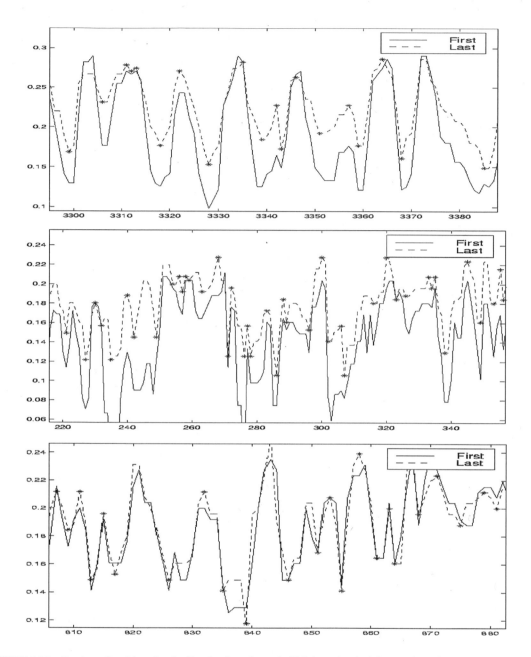

FIGURE 26.5 Portion of a ridge signals: live (top), cadaver (middle), and spoof (bottom). ✳ denotes minimums and maximums. The ridge signal for the live finger has peaks indicating moist regions around the pores and valleys indicating the dry region between pores. Over time, the valleys increase in live images as the perspiration diffuses between the ridges. The peaks and valleys of the spoof and cadaver images are random without a specific time-domain change. (*Source:* R. Derakhshani et al., *Pattern Recognition*, vol. 36, pp. 386–396, 2003.)

where $C1_{j,i}$ is the gray level for the jth ridge signal of the first capture and the ith point, j is equal to 1 to the number of ridge signals (n), i equal to 1 to the length of each ridge signal (m), and $f(k)$ is the average FFT of the strings. Figure 26.6 plots the average FFT for the first image for live, cadaver, and spoof fingerprints. The total energy (SM) for cadaver and spoof is very low compared to live.

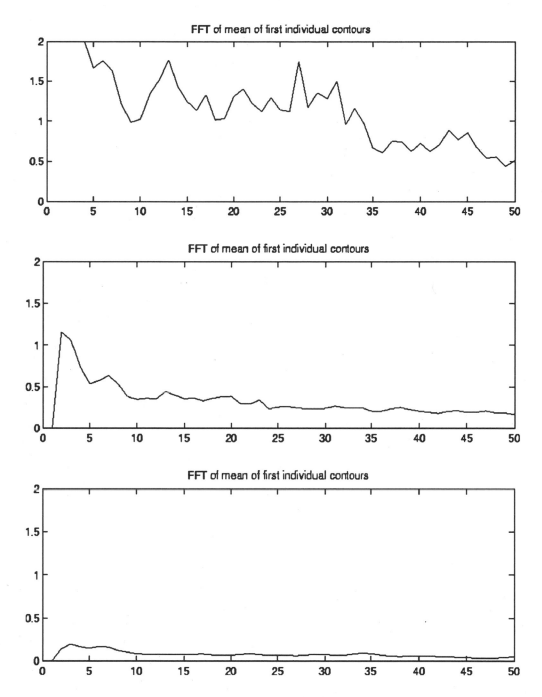

FIGURE 26.6 Average FFTs calculated from signal segments from the live (top), cadaver (middle), and spoof (bottom) fingerprint images. *X*-axis is the FFT index where 11 to 33 indicates 8 to 24 pixel spacing. *Y*-axis is magnitude. Live images have more total energy than spoof/cadaver for the first capture for frequencies related to the moist pore spacing. (*Source:* R. Derakhshani et al., *Pattern Recognition*, vol. 36, pp. 386–396, 2003.)

Extraction of Dynamic Features

Each of the dynamic measures uses the ridge signals from both the first and last capture. Each dynamic measure is an average across the n ridge signals:

- *Total swing ratio of first to last fingerprint signal (DM1)*: The fluctuation of the live fingerprint signal is more in the first capture, where there are moist pores and drier regions in between pores, compared to last fingerprint signal, where sweat has diffused into drier regions. In mathematical terms,

$$DM1 = \frac{1}{n}\sum_{j=1}^{n} \frac{\sum_{i=1}^{m} |C1_{j,i} - C1_{j,i-1}|}{\sum_{i=1}^{m} |C2_{j,i} - C2_{j,i-1}|}$$

where $C2_{j,i}$ is second capture. The number of points for each ridge signal (m) is the same for $C1$ and $C2$ since the same mask was used.

- *Min/max growth ratio of first to last fingerprint signal (DM2)*: For the live fingerprint signal, the heights of the maximums do not increase as fast as the minimums as the perspiring pores are already saturated. In mathematical terms,

$$DM2 = \frac{1}{n}\sum_{j=1}^{n} \frac{\sum_{i} (C2_{j,i}^{\min} - C1_{j,i}^{\min})}{\sum_{k} (C2_{j,k}^{\max} - C1_{j,k}^{\max})}$$

where $C1_{j,i}^{\min}$ and $C2_{j,i}^{\min}$ are the local minimums for the first and last scan, respectively, and $C1_{j,k}^{\max}$ and $C2_{j,k}^{\max}$ are the local maximums. Location of minimums and maximums were determined from the second capture and applied to both.

- *Last-first fingerprint signal difference mean (DM3)*: The first ridge signal is subtracted from the second. In mathematical terms,

$$DM3 = \frac{1}{n}\sum_{j=1}^{n} \frac{\sum_{i=1}^{m} (C2_{j,i} - C1_{j,i})}{m}$$

- *Percentage change of standard deviations of first and last fingerprint signals (DM4)*: This measure is the percentage change in standard deviation of last and first fingerprint signals. In mathematical terms,

$$DM4 = \frac{1}{n}\sum_{j=1}^{n} \frac{SD(C1_j) - SD(C2_j)}{SD(C1_j)}$$

where SD is the standard deviation operator

$$SD(C_j) = \sqrt{\frac{\sum_{i=1}^{m} (C_{j,i} - \text{mean}(C_j))^2}{m - 1}}$$

- *Dry saturation percentage change (DM5)*: This fifth dynamic measure indicates how fast the low cut-off region of the ridge signal is disappearing where higher DM5 corresponds to faster disappearance of dry saturation. In mathematical terms,

$$DM5 = \frac{1}{n}\sum_{j=1}^{n} \frac{\sum_{i=1}^{m} \delta(C1_{j,i} - LT) - \delta(C2_{j,i} - LT)}{0.1 + \sum_{i=1}^{m} \delta(C2_{j,i} - LT)}$$

LT is the low-cutoff threshold of the ridge signal [$\min(C_i)$], δ is the discrete delta function, and 0.1 is added to the denominator to avoid division by zero.

- *Wet saturation percentage change (DM6)*: The sixth dynamic measure indicates how fast the high cut-off region of the ridge signal is appearing. In mathematical terms,

$$DM6 = \frac{1}{n}\sum_{j=1}^{n} \frac{\sum_{i=1}^{m} \delta(C2_{j,i} - HT) - \delta(C1_{j,i} - HT)}{0.1 + \sum_{i=1}^{m} \delta(C1_{j,i} - HT)}$$

HT is the high-cutoff (saturation) threshold of the ridge signal [$\max(C_i)$].

Classification (Neural Network)

One static and six dynamic measures are used as input features to the neural network. Classification of images is divided into live and spoof, where spoof, includes images from fingerprint molds and cadavers. A single hidden layer neural network with gradient descent learning in conjunction with weight decay and momentum learning was used [21,22]. The hidden layer utilizes five nodes with unipolar logsig activation functions. The output layer is composed of one node with bipolar tansig activation function. For training, desired output values were set to +1 for live and −1 for spoof. In the weight-decay method, a constant value (0.0001) is subtracted from all the weights during each iteration of the training phase, such that insignificant weights are driven towards zero and eventually eliminated. This reduces the classifier's variance by reducing the number of its free parameters, resulting in robustness and better generalization.

26.4 Example Results

This section details some example results [20]. A dataset of approximately 30 live, 30 spoof (Play-Doh finger molds created from dental material casts of 30 subjects), and 15 cadaver sets of fingerprint images were used to test the algorithms from three fingerprint scanner technologies: optical (Secugen, EyeD hamster #HFDUO1A), capacitive DC (Precise Biometrics, 100 sc), and electro-optical (Ethentica, Ethenticator USB 2500). Of the data, 75% was used for training and 25% for testing. Classification was performed separately for two different time windows (2 and 5 sec). Various thresholds (with 0.05 steps) were tested on the output of the neural network to achieve 100% live classification rate along with the maximum possible spoof classification rate. Results are shown in Figure 26.7, where 100% detection of live fingerprints is possible, with 80 to 100% spoof classification depending on device technology.

26.5 Conclusion

An example of signal processing applied to liveness detection in biometric devices has been described. Given biometric recognition is an emerging technology, the vulnerability of these devices to simple spoofing techniques will need to be addressed using liveness detection or other antispoofing methods depending on the risk assessment for the application.

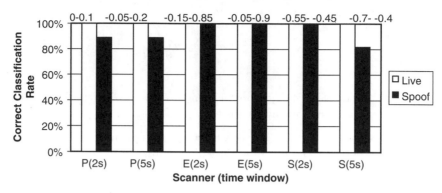

FIGURE 26.7 Live and spoof classification rates using weight decay neural network for a set of thresholds (top) for two time windows (2 and 5 sec) over a variety of biometric devices (P: Precise Biometrics, E: Ethentica, S: Secugen). (*Source:* S.A.C. Schuckers et al., *Proc. SPIE Def. Secur. Symp., Biometric Technology for Human Identification*, Orlando, FL, April 2004.)

Acknowledgments

Part of this work was funded by the Center for Identification Technology Research (CITeR), an NSF Industry/ University Cooperative Research Center. Special thanks to my graduate research assistants, Reza Derakhshani, Sujan Parthnasardi, and Aditya Abhyankar.

References

1. A. K. Jain, R. Bolle, and S. Pankanti, Eds., *Biometrics: Personal Identification in Networked Society*, Dordrecht: Kluwer Academic Publishers, 1999.
2. J. D. Woodward, Jr., N. M. Orlans, and R. T. Higgins, Eds., *Biometrics*, New York: McGraw-Hill Osborne Media, 2002.
3. A.J. Mansfield and J.L. Wayman, "Best practices in testing and reporting performance of biometric devices: version 2.01," *National Physical Laboratory Report, CMSC 14/02*, United Kingdom, August 2002.
4. N.K. Ratha, "Enhancing security and privacy in biometrics-based authentication systems," *IBM Syst. J.*, vol. 40, no. 3, pp. 614–634, 2001.
5. T. Matsumoto, H. Matsumoto, K. Yamada, and S. Hoshino, "Impact of artificial 'gummy' fingers on fingerprint systems," *Proc. SPIE*, 4677, 2002 (January).
6. L. Thalheim and J. Krissler, "Body check: biometric access protection devices and their programs put to the test", *c't magazine*, November 2002.
7. D. Willis and M. Lee, "Biometrics under our thumb," *Network Computing*, June 1, 1998.
8. T. van der Putte and J. Keuning, "Biometrical fingerprint recognition: don't get your fingers burned," *Proceedings of the Fourth Working Conference on Smart Card Research and Advanced Applications*, Dordrecht: Kluwer Academic Publishers, 2000, pp. 289–303.
9. R. Derakhshani, S.A.C. Schuckers, L.A. Hornak, and L. O'Gorman, "Determination of vitality from a non-invasive biomedical measurement for use in fingerprint scanners," *Pattern Recognition*, vol. 36, pp. 386–396, 2003.
10. S.A.C. Schuckers, "Spoofing and anti-spoofing measures," *Inf. Secur. Tech. Rep.*, vol. 7, no. 4, pp. 56–62, 2002.
11. V. Valencia and C. Horn, "Biometric liveness testing," in *Biometrics*, J. D. Woodward, Jr., N. M. Orlans, and R. T. Higgins, Eds., New York: McGraw-Hill Osborne Media, 2002.

12. *Liveness Detection in Biometric Systems*, International Biometric Group white paper, available at http://www.ibgweb.com/reports/public/reports/liveness.html.

13. D. Osten, H.M. Carim, M.R. Arneson, and B.L. Blan, "Biometric, personal authentication system," Minnesota Mining and Manufacturing Company, U.S. Patent #5,719,950, February 17, 1998.

14. P. Kallo, I. Kiss, A. Podmaniczky, and J. Talosi, "Detector for recognizing the living character of a finger in a fingerprint recognizing apparatus," Dermo Corporation, Ltd. U.S. Patent #6,175,64, January 16, 2001.

15. P.D. Lapsley, J.A. Less, D.F. Pare, Jr., and N. Hoffman, "Anti-fraud biometric sensor that accurately detects blood flow," SmartTouch, LLC, U.S. Patent #5,737,439, April 7, 1998.

16. J.G. Webster, Ed., *Design of Pulse Oximeters*, Bristol and Philadelphia: Institute of Physics Publishing, 1997.

17. C.C. Broun, X. Zhang, R.M. Mersereau, and M.A. Clements, "Automatic speechreading with application to speaker verification," *Proc. ICASSP*, Orlando, FL, p I/685-I/688, 2002 (May 13–17).

18. L. Biel, O. Pettersson, L. Philipson, and P. Wide, "ECG analysis: a new approach in human identification," *IEEE Trans. Instrum. Meas.*, vol. 50, no. 3, pp. 808–812, 2001.

19. K.A. Nixon, R.K. Rowe, J. Allen, S. Corcoran, L. Fang, D. Gabel, D. Gonzales, R. Harbour, S. Love, R. McCaskill, B. Ostrom, D. Sidlaukus, and K. Unruh, "Novel spectroscopy-based technology for biometric and liveness verification," *Proc. SPIE Def. Secur. Symp., Biometric Technology for Human Identification*, Orlando, FL, April, 2004.

20. S.A.C. Schuckers, R. Derakhshani, S. Parthasaradhi, and L. Hornak, "Improvement of an algorithm for recognition of liveness using perspiration in fingerprint devices," *Proc. SPIE Def. Secur. Symp., Biometric Technology for Human Identification*, Orlando, FL, April, 2004.

21. J.C. Principe, N.R. Euliano, and W.C. Lefebvre, *Neural and Adaptive Systems: Fundamentals through Simulations*, London: Wiley, 2000.

22. S. Haykin, *Neural Networks, A Comprehensive Foundation*, 2nd ed., Englewood Cliffs, NJ: Prentice-Hall, 1999.

27

Human Identification Using Gait and Face

Rama Chellappa
University of Maryland

Amit K. Roy-Chowdhury
University of California

Shaohua Kevin Zhou
University of Maryland

27.1 Introduction

One of the main goals of computer vision research is to develop methods to recognize objects and events. A subclass of these topics is the recognition of humans and their activities. In this chapter, we summarize some recent methods for human recognition using face and gait.

Gait recognition is related to the broader problem of human motion modeling and has important implications in such areas as surveillance, medical diagnoses, the entertainment industry, and video communications. Traditionally, there has been a keen interest in studying human motion in various disciplines. In psychology, Johansson conducted classic experiments by attaching light displays to body parts and showing that humans can identify motion when presented with only a small set of these moving dots [32]. Muybridge captured the first photographic recordings of humans and animals in motion in his famous publication on animal locomotion published at the end of the 19th century [52]. In kinesiology, the goal has been to understand human motion and apply that knowledge to applications in sports, medicine, elderly care, and the early detecting of movement disorders [28]. Gait recognition is a relatively new area for computer vision researchers. However, significant progress has been made and reasonably good performance on large datasets under controlled circumstances has been achieved. Problems in this area include poor performance in uncontrolled outdoor situations and the effects of time. Some progress has been made in recognizing people walking in arbitrary directions to the camera.

The problem in face recognition can be defined as follows. A database of a large number of faces is available as the gallery. The faces can be represented as a single image or by a set of images, either as a video sequence or as a collection of discrete poses. These images are usually referred to as training images since they train the parameters of a recognition algorithm. Given an image or a set of images of an individual (known as test images), the issue is to identify the individual from the gallery or decide that the person is not part of the gallery. The main challenges in face recognition are working with:

- varying conditions of illumination between training and test images
- changes in appearance, makeup, and clothing between training and test images
- changes due to differences in time between the recording of training and test images
- different face poses in various instances of recording

All these issues make face recognition a challenging problem. However, considerable progress has been made in the past decade, and face recognition technologies are under consideration for deployment at public facilities.

27.2   Review of Existing Work

Human Recognition Using Gait

Human gait is a spatio-temporal phenomenon concerning the motion of an individual. When identification is attempted in natural settings for surveillance applications, biometrics such as fingerprint or iris processes are not applicable. Furthermore, night vision capability (an important surveillance component) is usually not possible with these biometrics. Even though an infrared (IR) camera would reveal the presence of people, the facial features are far from discernible at long distances. The attractiveness of gait as a biometric comes from the fact that it is nonintrusive and can be detected and measured even in low-resolution video. Furthermore, it is harder to disguise than static appearance features such as face and it does not require a cooperating subject.

Although the study of human gait is a relatively new area for computer vision researchers, extensive work has been performed in the psychophysics community on the ability of humans to recognize others by their walking style. In the computer vision community, gait research has concentrated mainly on recognition algorithms. These methods can be divided into two groups—appearance-based and model-based. Appearance-based methods can be further divided into deterministic or stochastic methods. Deterministic, appearance-based methods are discussed in References 10, 35 and 55, while well-known stochastic methods in the same category use hidden Markov models (HMM) [41,70]. Model-based methods are less popular largely because of the difficulty of obtaining accurate three-dimensional (3D) models of the human body.

The belief that humans can distinguish between gait patterns of different individuals is widely held. These gait-related qualities include stride length, bounce, rhythm, and speed, and are even found in swagger and body swing. The suggestion that humans can identify people by their gait was investigated in a series of early studies by Johansson [33]. He presented participants with images reduced to point-light displays. His experiments suggested that we have some implicit notion of human movement and can recognize temporal data within this context. Later work using point-light displays went further, demonstrating that not only could a walking figure rapidly be extracted from moving lights but also that a perceiver could distinguish between different sorts of biological motions. These included walking, climbing stairs, and jumping [15]. Attempts to address the question of identification from gait have proceeded in small steps. Kozlowski and Cutting [38] first investigated whether perceivers could identify the gender of a point-light walker. Their results indicated an accuracy rate of 65 to 70% when the walker was viewed from the side. In [14], it was suggested that gender could be identified indirectly through a determination of the walker's "center of moment." The demonstration that gender could be extracted from gait provided insight into how perceivers might discriminate between the gait patterns of different individuals. Cutting and Kozlowski [13] demonstrated that perceivers could reliably recognize themselves and their friends from dynamic point-light displays. Barclay et al. [2] suggested that individual walking styles might be captured by differences in a basic series of pendular limb motions. Interestingly, Beardsworth and Buckner [4] have shown that the ability to recognize oneself from a point-light display is greater than the ability to recognize friends, despite the fact that we rarely see our own gait from a third-person perspective. Stevenage et al. [69] also explored the ability of people to identify others using gait information alone. He found that even with a brief exposure time and unfamiliarity with the walking subjects, perceivers could identify the target correctly at a rate that was greater than chance.

A recent study by Schollhorn et al. [62] studied the gait of 15 subjects to ascertain the presence of gait identity information. This study found that kinetic variables (captured using a force platform), as well as kinematic variables (captured by reflective markers on the thigh, shank, and hip), were necessary for gait identification. Simply using the leg portion of the body was adequate for good identification performance.

Approaches to gait recognition problems can be broadly classified as being either model-based or model-free. Both methodologies follow the general framework of feature extraction, feature correspondence, and high-level processing. The main difference is in feature correspondence between two consecutive frames. Methods assuming *a priori* models match the 2D image sequences to the model data. Feature correspondence is automatically achieved once a match between the images and model data is established. Examples of this approach include the work of Lee et al. [42], where several ellipses are fitted to different parts of the person's binarized silhouette. The parameters of these ellipses, including the location of its centroid and eccentricity, are used to represent the gait of a person. Recognition is achieved by template matching. Cunado et al. [12] extracted a gait signature by fitting the movement of the thighs to an articulated, pendulum-like motion model. The idea is somewhat similar to an early work by Murray [51], who modeled the hip-rotation angle as a simple pendulum, the motion of which was described by simple harmonic motion. Model-free methods establish correspondence between successive frames based upon the prediction or estimation of features related to position, velocity, shape, texture, and color. Alternatively, they assume some implicit notion of what is being observed. Examples of this approach include the work of Huang et al. [31], where optical flow derives a motion image sequence for a walk cycle. Principal components analysis is then applied to the binarized silhouette to derive what are called eigen gaits. Benabdelkader et al. [8] use image self-similarity plots as a gait feature. Little and Boyd [45] extract frequency and phase features from moments of the motion image derived from optical flow and use template matching to recognize different people by their gait. A dynamic time-warping (DTW) [22] algorithm for gait recognition was proposed in [35]. This algorithm matches two gait sequences (probe and gallery) by computing the distance as a function of time between two feature sets representing the data. This approach can be used even when substantial training data are not available. It can also account for modest variation in walking speed. Two of the most successful approaches to date in gait recognition are shown in [70,72]. In Sundaresan, the authors used a HMM [58] to represent an individual's gait. This algorithm will be described in detail in the next section. A method for identifying individuals by shape, which is automatically extracted from a cluster of similar poses obtained from a spectral partitioning framework, was proposed in [72]. Most of the above methods rely on the availability of a side view to extract the gait parameters. Two approaches that address the problem of view-invariant recognition are Shakhnarovich et al. [27] and Kale et al. [34]. A study of the role of kinematics and shape in computer vision-based gait recognition problems was presented in Veeraraghavan et al. [75].

Similar to FERET evaluations for face recognition, a HumanID GaitChallenge Problem was introduced to measure the progress of different recognition algorithms (http://www.gaitchallenge.org) [56]. The challenge problem consists of a baseline algorithm, a set of 12 experiments, and a dataset of 122 people. The baseline algorithm estimates silhouettes by background subtraction and performs recognition by the temporal correlation of silhouettes. Twelve experiments examine the effects of five covariates: change of viewing angle, change in shoe type, change in walking surface, whether a briefcase is carried, and temporal differences. A description of the different experiments (probe sets) is shown in Table 27.1. Identification and verification scores for all the experiments are reported using the baseline algorithm. The relative performance of the HMM-based method with the baseline is presented in the next section.

Human Recognition Using Face

Chronologically speaking, face recognition started with still images. Popular methods proposed for study are principal components analysis or eigenfaces [50,73], linear discriminant analysis or Fisherfaces [5,80], elastic graph matching [77], local feature analysis [53], morphable models [6], and many others. The reader is referred to a recent survey paper [81] for additional details on recognition.

TABLE 27.1   Probe Sets for the GaitChallenge Data

| | Description of GaitChallenge Data | |
Experiment	Probe Description (Surface C/G, Shoe A/B, Camera L/R, Carry NB/BF, Time)	Number of Subjects
A	(G, A, L, NB, T1+T2)	122
B	(G, B, R, NB, T1+T2)	54
C	(G, B, L, NB, T1+T2)	54
D	(C, A, R, NB, T1+T2)	121
E	(C, B, R, NB, T1+T2)	60
F	(C, A, L, NB, T1+T2)	121
G	(C, B, L, NB, T1+T2)	60
H	(G, A, R, BF, T1+T2)	120
I	(G, B, R, BF, T1+T2)	60
J	(G, A, L, BF, T1+T2)	120
K	(G, A/B, R, NB, T2)	33
L	(G, A/B, R, NB, T2)	33

The gallery is (G,A,R,NB,T1+T2). C/G represents concrete/grass surface, L/R represents left or right camera, NB/BF represents carrying a briefcase or not, T1 and T2 represent the data collected at two different time instants.

Statistical approaches to face modeling have been popular since Turk and Pentland's work on eigenface in 1991 [73]. In this statistical approach, the 2D appearance of face image is treated as a vector by scanning the image in lexicographical order, with the vector dimension referring to the number of pixels in the image. In the eigenface approach [73], all face images consist of a distinctive face subspace. This subspace is linear and spanned by the eigenvectors of the covariance matrix found using PCA. Typically, the number of eigenvectors is kept at less than the true dimension of the vector space. The task of face recognition is to find the closest match in this face subspace. However, PCA might not be efficient for recognition accuracy, since the construction of the face subspace does not capture class separability between humans. This motivated the use of LDA [5] and its variants. In LDA, the linear subspace is constructed so that the within-class scatter is minimized and the between-class scatter is maximized. This idea is further generalized in Bayesian face recognition [49], where intra-personal space (IPS) and extra-personal space (EPS) are used instead of within-class scatter and between-class scatter measures. The IPS models variations in the appearance of the same individual and EPS models variations in the appearance due to a difference in identity. Probabilistic subspace density is then fitted on each space. A Bayesian decision is taken using a *maximum a posteriori* (MAP) rule to determine identity. In the famous EGM [77] algorithm, the face is represented as a labeled graph. The graph nodes are located at facial landmarks, such as the pupils and the tip of the nose. Each node is labeled with jets derived from responses obtained by convolving the image with a family of Gabor functions. Edges in the graph represent the geometric distance between two nodes. Face recognition is then formalized as a graph matching problem. The above approaches are based on 2D appearances and perform poorly when significant pose and illumination variations are present [81].

While recognition rates under controlled indoor situations are reasonably good, work must continue before these technologies can be deployed in outdoor situations. Many researchers believe that the use of video sequences, as opposed to single images, will lead to better recognition rates. This is based on the idea that integrating the recognition performance over a sequence would give a better result than considering a single image from that sequence. Most of the current research in this area is focused on exploiting video sequence.

However, nearly all video-based recognition systems apply still-image-based recognition to selected good frames. In Kozlowski and Cutting [38], McKenna and Gong [48], and Wechsler et al. [76], radial basis function (RBF) networks are used for track and recognition purposes. In Howell and Buxton [30], the system uses an RBF network for recognition. Because no warping is done, the RBF network must learn individual variations and possible transformations. Performance appears to vary widely, depending on the size of the training data. Wechsler et al. [76] present a fully automatic person-authentication system. The system uses video break, face detection, and authentication modules and cycles over successive video images until high-recognition confidence is reached. This system was tested on three image sequences: the first was taken indoors with one subject present, the second was taken outdoors with two subjects, and the third was taken outdoors with one subject in stormy conditions. Perfect results were reported on all three sequences when verified against a database of 20 still-face images. A multimodal-based person recognition system is described in Choudhury et al. [9]. This system consists of a face recognition module, a speaker identification module, and a classifier fusion module. The most reliable video frames and audio clips are selected for recognition. Three-dimensional information from the head is used to detect the presence of an actual person as opposed to an image of that person. Recognition and verification rates of 100% were achieved for 26 registered clients.

27.3 Gait Recognition Using Hidden Markov Models

The HMM approach is suitable because the gait of an individual can be visualized as the person adopting postures from a set in a sequence that has an underlying structured-probabilistic nature. The postures that the individual adopts can be regarded as the states of the HMM and are typical of that individual. They also provide a means of discrimination. This approach assumes that, during a walk cycle, the individual transitions among N discrete postures or states. An adaptive filter automatically detects the cycle boundaries. The method is not dependent on the particular feature vector to represent the gait information contained in the postures. The statistical nature of the HMM provides robustness to the model. In the method described below, the binarized background-subtracted image is used as the feature vector. Different distance metrics, such as those based on the L_1 and L_2 norms of the vector difference and the normalized inner product of the vectors, are used to measure the similarity between feature vectors.

Overview of the HMM Method

Let the database consists of video sequences of P persons. The model for the pth person is given by $\lambda_p = (A_p, B_p, \pi_p)$, with N number of states. The model, λ_p, is built from the observation sequence for the pth person, using the sequence of feature vectors given by $\mathcal{O}_p = \{\mathbf{O}_1^p, \mathbf{O}_2^p, \ldots, \mathbf{O}_{T_p}^p\}$, where T_p is the number of frames in the sequence of the pth person. A_p is the transition matrix, and π_p is the initial distribution. The B_p parameter consists of the probability distributions for a feature vector conditioned on the state index, or, the set $\{P_1^p(.), P_2^p(.), \ldots, P_N^p(.)\}$. The probability distributions are defined in terms of *exemplars*, where the jth exemplar is a typical realization of the jth state. The exemplars for the pth person are given by $\mathcal{E}_p = \{\mathbf{E}_1^p, \mathbf{E}_2^p, \ldots, \mathbf{E}_N^p\}$. The superscript denoting the index of the person will be dropped for simplicity. The reason behind using an exemplar-based model is that recognition can be based on the distance measure between the observed feature vector and the exemplars. The distance metric is evidently a key factor in the performance of the algorithm. $P_j(\mathbf{O}_t)$ is defined as a function of $D(\mathbf{O}_t, \mathbf{E}_j)$, the distance of the feature vector $\mathbf{O}_t$ from the jth exemplar:

$$P_j(\mathbf{O}_t) = \alpha e^{-\alpha D(\mathbf{O}_t, \mathbf{E}_j)} \tag{27.1}$$

During the **training** phase, a model is built for all the subjects, indexed by $p = 1, 2, \ldots, P$, in the gallery. An initial estimate of $\mathcal{E}_p$ and λ_p is formed from $\mathcal{O}_p$, and these estimates are refined iteratively. Note that B is completely defined by $\mathcal{E}$ if α is fixed previously. We can iteratively estimate A and π by using the Baum–Welch algorithm, keeping $\mathcal{E}$ fixed. The algorithm to reestimate ε is determined by the choice of the distance metric. During **testing**, given a gallery $\mathcal{L} = \{\lambda_1, \lambda_2, \ldots, \lambda_P\}$ and the probe sequence of length T, $\mathcal{X} = \{X_1, X_2, \ldots, X_T\}$ traversing the path $\mathcal{Q} = \{q_1, q_2, \ldots, q_T\}$, q_t being the state index at time t, we obtain the ID of the probe

FIGURE 27.1   Part of an observation sequence.

sequence as

$$ID = \arg_p \max_{\mathcal{Q},p} \Pr[\mathcal{Q}|\mathcal{X}, \lambda_p] \tag{27.2}$$

The feature vector used is the binarized version of the background subtracted images. The images are scaled and aligned to the center of the frame as in Figure 27.1. This figure features part of a sequence of feature vectors. We now describe the methods to obtain initial estimates of the HMM parameters, the training algorithm, and identification results using USF data described in [56].

Initial Estimate of HMM Parameters

To obtain a good estimate of the exemplars and the transition matrix, we first obtain an initial estimate of an ordered set of exemplars from the sequence and the transition matrix, and successively refine the estimate. The initial estimate for the exemplars, $\mathcal{E}^0 = \{\mathbf{E}_1^0, \mathbf{E}_2^0, \ldots, \mathbf{E}_N^0\}$ allows only for transitions from the jth state to either the jth or the state. A corresponding initial estimate of the transition matrix, A^0 (with $A_{j,j}^0 = A_{j,j\,\mathrm{mod}\,N+1}^0 = 0.5$, and all other $A_{j,k}^0 = 0$) is also obtained. The initial probabilities π_j are set to be equal to $1/N$.

We observe that the gait sequence is quasi-periodic, and we use this fact to obtain the initial estimate ε^0. We can divide the sequence into "cycles," where a cycle is defined as that segment of the sequence bounded by those silhouettes where the subject has arms on each side and legs approximately aligned with each other. We can further divide each cycle into N temporally adjacent clusters of approximately equal size. We visualize the frames of the pth cluster of all cycles to be generated from the jth state. Thus we can get a good initial estimate of E_j from the feature vectors belonging to the jth cluster. For example, assume that the training sequence is given by $\mathcal{Y} = \{\mathbf{Y}_1, \mathbf{Y}_2, \ldots, \mathbf{Y}_T\}$. We can partition the sequence into K cycles, with the kth cycle given by frames in the set $\mathcal{Y}_k = \{\mathbf{Y}_{S_k}, \mathbf{Y}_{S_k+1}, \ldots, \mathbf{Y}_{S_k+L_k-1}\}$, where S_k and L_k are the index of the first frame of the kth cycle, and the length of the kth cycle, respectively. We define the first cluster as comprising of frames with indices $S_k, S_k + 1, \ldots, S_k + \frac{1}{2}L_k/N, S_k + L_k - \frac{1}{2}L_k/N, S_k + L_k - \frac{1}{2}L_k/N + 1, \ldots, S_k + L_k - 1$. The jth cluster $(j = 2, 3, \ldots, N)$ consists of frames with indices $S_k + (j - \frac{3}{2}) L_k/N, S_k + (j - \frac{3}{2})L_k/N + 1, \ldots, S_k + (j - \frac{1}{2})L_k/N$. We must robustly estimate the cycle boundaries so that we can partition the sequence into N clusters and obtain the initial estimates of the exemplars. If the sums of the foreground pixels of each image are plotted with time then, following our definition of a cycle, the minima should correspond to the cycle boundaries. We denote the sum of the foreground silhouette pixels in the nth frame as $s[n]$. This signal is noisy and may contain several spurious minima. However we can exploit the signal's quasi-periodicity and filter the signal to remove the noise before identifying the minima. Median filtering and differential smoothing of $s[n]$ are not robust because they do not take into account the gait frequency.

The specifications of the bandpass filter allow frequencies that are typical for a fast walk. The video is captured at 30 frames per second and the sampling frequency, $f_s = 1/30$ and $T_s = 30a$. The maximum gait frequency is assumed to be $f_m = 0.1$, corresponding to a cycle period of $T_m = 10$. A Hamming window of length L is used. The extended sequence $x[n]$ is obtained by symmetrically extending $s[n]a$ in both directions by $L/2a$. Therefore, the sequence $x[n]$ has length $M = N + L$. The resultant sequence is filtered using a bandpass filter (with upper cutoff frequency $f_{uc} = f_m$) in both directions to remove phase delay. The distances between the minima of the filtered sequence provide an estimate of the cycle period. The cycle frequency is estimated as the inverse of the median of cycle periods. Using this revised estimate of the gait frequency, $\hat{f}$, a new filter is constructed with upper cutoff frequency $f_{uc} = \hat{f} + 0.02$. A manual examination of all the sequences in the gallery in the GaitChallenge database revealed a 100% detection rate with hardly any false detection of cycle boundaries.

Training the HMM Parameters

The iterative refining of the estimates is performed in two steps. In the first step, a Viterbi evaluation [58] of the sequence is performed using the current values for the exemplars and the transition matrix. Feature vectors are clustered according to the most likely state where they originated. The exemplars for the states are newly estimated from these clusters. Using the current exemplar values, $\mathcal{E}^{(i)}$, and the transition matrix, $A^{(i)}$, Viterbi decoding is performed on the sequence $\mathcal{Y}$ to obtain the most probable path $\mathcal{Q} = \{q_1^{(i)}, q_2^{(i)}, \ldots, q_T^{(i)}\}$, where $q_t^{(i)}$ is the state at time t. The set of observation indices, whose corresponding observation is estimated to have been generated from state j, is given by $\mathcal{T}_j^{(i)} = \{t : q_t^{(i)} = j\}$. We now have a set of frames for each state and we would like to select the exemplars to maximize the probability in Equation (27.3). If we use the definition in Equation (27.1), Equation (27.4) follows:

$$\mathbf{E}_j^{(i+1)} = \arg_{\mathbf{E}} \max \prod_{t \in \mathcal{T}_j^{(i)}} P(\mathbf{Y}_t | \mathbf{E}) \tag{27.3}$$

$$\mathbf{E}_j^{(i+1)} = \arg_E \min \sum_{t \in \mathcal{T}_j^{(i)}} D(\mathbf{Y}_t, \mathbf{E}) \tag{27.4}$$

The actual method for minimizing the distance in Equation (27.4) depends on the distance metric used. We have experimented with three distance measures, namely the Euclidean (EUCLID) distance, the inner product (IP) distance, and the sum of absolute difference (SAD) distance. These are given by Equation (27.5), Equation (27.6), and Equation (27.7), respectively. Note that though $\mathbf{Y}_t$ and $\mathbf{E}$ are two-dimensional images, they are represented as vectors of dimension $D \times 1$ for ease of notation. $1_{D \times 1}$ is a vector of D ones:

$$D_{\text{EUCLID}}(\mathbf{Y}, \mathbf{E}) = (\mathbf{Y} - \mathbf{E})^T (\mathbf{Y} - \mathbf{E}) \tag{27.5}$$

$$D_{\text{IP}}(\mathbf{Y}, \mathbf{E}) = 1 - \frac{\mathbf{Y}^T \mathbf{E}}{\sqrt{\mathbf{Y}^T \mathbf{Y} \mathbf{E}^T \mathbf{E}}} \tag{27.6}$$

$$D_{\text{SAD}}(\mathbf{Y}, \mathbf{E}) = |\mathbf{Y} - \mathbf{E}|^T 1_{D \times 1} \tag{27.7}$$

The equations for updating the jth element of the exemplars in the EUCLID distance, IP distance, and the SAD distance cases are presented in Equation (27.8), Equation (27.9), and Equation (27.10), respectively. $\tilde{\mathbf{Y}}$ denotes the normalized vector $\mathbf{Y}$ and $|\mathcal{T}_j^{(i)}|$ denotes the cardinality of the set $\mathcal{T}_j^{(i)}$:

$$\mathbf{E}_j^{(i+1)}(j) = \frac{1}{|\mathcal{T}_j^{(i)}|} \sum_{t \in \mathcal{T}_j^{(i)}} \mathbf{Y}_t(j) \tag{27.8}$$

$$\mathbf{E}_j^{(i+1)}(j) = \sum_{t \in \mathcal{T}_j^{(i)}} \tilde{\mathbf{Y}}_t(j) \tag{27.9}$$

$$\mathbf{E}_j^{(i+1)}(j) = \text{median}_{t \in \mathcal{T}_j^{(i)}} \{\mathbf{Y}_t(j)\} \tag{27.10}$$

Given $\mathcal{E}^{(i+1)}$ and $A^{(i)}$, we can calculate $A^{(i+1)}$ using the Baum–Welch algorithm [58]. Then we can successively refine estimates of the HMM parameters. This usually only takes a few iterations to obtain an acceptable estimate.

Identifying from a Test Sequence

Identifying a sequence involves deciding which model parameters to use for discrimination parameters. Given gallery models $\mathcal{L} = \{\lambda_1, \lambda_2, \ldots, \lambda_P\}$ and the probe sequence $\mathcal{X} = \{\mathbf{X}_1, \mathbf{X}_2, \ldots, \mathbf{X}_T\}$, we find the model and the path that maximizes the probability of the path, given the probe sequence. The ID is obtained as in Equation (27.2).

We do not need to use the trained parameter set, λ, as a whole. For example, if we believe that the transition matrix is predominantly indicative of the speed that the subject walks and is therefore not suitable as a discriminant of the ID of the subject, then we have the option of using only part of the parameter set given by $\gamma_p = (B_p, \pi_p)$ instead of the entire HMM parameter set. In this case, the conditional probability of the sequence, given the ID, is given as follows. The Baum–Welch algorithm could be used to obtain $A_p^{\mathcal{X}}$ recursively in Equation (27.12):

$$\Pr[\mathcal{Q}|\mathcal{X}, \gamma_p] = \Pr[\mathcal{Q}|\mathcal{X}, A_p^{\mathcal{X}}, \gamma_p] \tag{27.11}$$

$$A_p^{\mathcal{X}} = \arg_A \max \Pr[\mathcal{X}|A, \gamma_p] \tag{27.12}$$

Experimental Results

The objective of our experiments was to evaluate the algorithm's performance and compare the efficacy of the different distance measures to gauge the similarity between two images for posture. As described before, the GaitChallenge or USF database contains video sequences of 122 individuals, a subset featuring sequences collected under each of 12 conditions. The sequences are labeled gallery and probe A–L. We trained our parameters on the sequences from the gallery set. In each experiment, we attempted to identify the sequences in each of the seven probe sets from the parameters obtained from the gallery set, using the inner product-distance measure. The ID was calculated using Equation (27.2). The experiments were repeated with different distance measures. The results of the experiment, using the IP distance measure between feature vectors in the form of cumulative match scores (CMS) plots [55], are in Figure 27.2(a). We observe that the distance

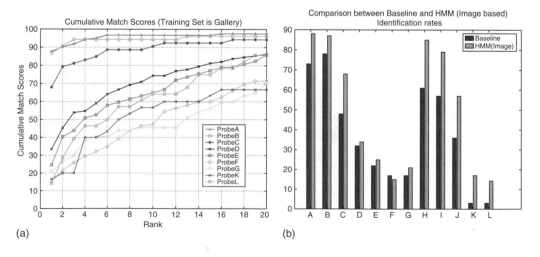

(a)   (b)

FIGURE 27.2   (a) CMS plots of probes A–L tested against gallery. (b) Comparison of identification rates of HMM and baseline algorithm.

measure that works best and is the simplest to implement is the inner product distance. The performance comparison with the baseline [55] is illustrated in Figure 27.2(b).

From the experiments, we note that the biggest performance drop occurs due to changes in surface types and when there is a time difference between the gallery and the probe. Reasons for the sudden drop are not yet fully understood. Probable causes may be changes in the silhouette (especially the lower part for surface change) and a change in clothing due to time differences. Note that the performance does not change greatly with small changes in viewing direction. In summary, gait recognition under arbitrary conditions is still an open research problem.

27.4 View Invariant Gait Recognition

A person's gait is best reflected from a side view (referred to in this chapter as a canonical view) to the camera. Most of the above gait recognition algorithms rely on the availability of the subject's side view. The situation is analogous to face recognition, where it is desirable to have frontal views of the person's face. In realistic scenarios, gait recognition algorithms must work in a situation where the person walks at an arbitrary angle to the camera. The most general solution to this problem is to estimate the three-dimensional (3D) model for the person. Features extracted from the 3D model can provide the gait model for a person. This problem requires the structure's solution from motion (SfM) or stereo reconstruction problems [18,29], which are known to be difficult for articulating objects. In the absence of methods for the robust recovery of accurate 3D models, a simple way to exploit existing appearance-based methods is to synthesize the canonical views of a walking person. Shakhnarovich et al. [27] compute an image-based visual hull from a set of monocular views, used to render virtual canonical views for tracking and recognition. Gait recognition is achieved by matching a set of image features, based on moments extracted from the silhouettes of the synthesized probe video to the gallery. An alternative to synthesizing canonical views is the work of Bobick and Johnson [7]. In this work, two sets of activity-specific static and stride parameters are extracted for different individuals. The expected confusion for each set is computed to guide the choice of parameters under different imaging conditions (indoor vs. outdoor, side view vs. angular view). A cross-view mapping function accounts for changes in viewing direction. The set of stride parameters (which is smaller than the set of static parameters) exhibits greater resilience to viewing direction. A method for recognizing an individual's gait using joint angle trajectories was presented in Tanswonsuwan and Bobick [71]. Representation using such a small set of parameters may not give good recognition rates on large databases.

We have developed a view-invariant gait recognition algorithm for the single camera case by synthesizing a canonical view from an arbitrary one without explicitly computing the 3D depth. Consider a person walking along a straight line subtending an angle θ with the image plane (AC in Figure 27.4). If the distance, Z_0, of the person from the camera is much larger than the width, ΔZ, of the person, then it is reasonable to replace the scaling factor $\dfrac{f}{Z_0 + \Delta Z}$ for perspective projection by an average scaling factor $\dfrac{f}{Z_0}$. In other words, for human identification at a distance, we can approximate the actual 3D human as a planar object. Assume that we are given a video of a person walking at a fixed angle θ (Figure 27.4). By tracking the direction of motion, α, in the video sequence, one can estimate the 3D angle θ. This can be done by using the optical flow-based SfM equations. Under the assumption of planarity, knowing angle θ and the calibration parameters, we can synthesize side views of the sequence of images of an unknown walking person without explicitly computing the 3D model of the person. We refer to this as the "implicit SfM" approach. Where there is no real translation of the person, such as a person walking on a treadmill, an alternative approach obtains the synthesized views of the person. Given a set of point correspondences for a planar surface between the canonical and noncanonical views in a set of training images, we compute a homography. This homography is applied to the person's binary silhouette to obtain the synthesized views. We refer to this as the "homography approach."

An overview of our gait recognition framework is given in Figure 27.3. We have reported recognition experiments [34] using two publicly available gait databases (NIST and CMU). The implicit SfM approach is used for the NIST databases, while the homography approach is used for the CMU database. Keeping in mind

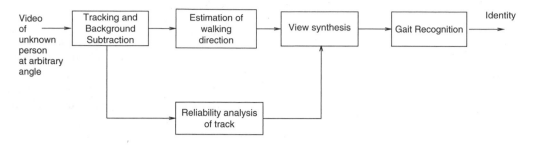

FIGURE 27.3 Framework for view invariant gait recognition.

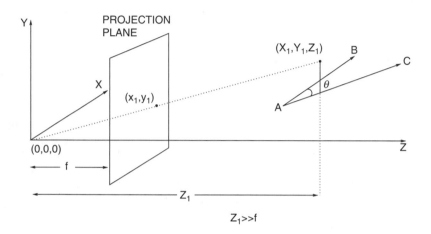

FIGURE 27.4 Imaging geometry.

the limited quantity of training data, the DTW algorithm [35] is used for gait recognition. Acceptable recognition results are obtained for θ less than 30°. A byproduct of the above method is a simple algorithm to synthesize novel views of a planar scene.

27.5 Human Recognition Using Face

Though face recognition has been intensively investigated for more than 10 years, state-of-the-art face recognition systems yield satisfactory performance only when confronted with controlled conditions. Unconstrained conditions, such as illumination/pose variations and surveillance video scenarios, impose significant challenges to existing recognition systems. Below we summarize emerging techniques dealing with face recognition under illumination/pose variations and from videos. Because different approaches experimented on different datasets, comparison of recognition performance is not appropriate and no performance is actually reported below.

Face Recognition under Illumination Variation

Recent works on face recognition under illumination variations [3,23,66,84] employ a Lambertian reflectance model with a varying albedo field. A pixel h^s under a distant illuminant s is formulated as

$$h^s = p\mathbf{n}^T\mathbf{s} = \mathbf{t}^T\mathbf{s};\ \mathbf{n}_{3\times1} \doteq [n_x, n_y, n_z]^T;\ \mathbf{t}_{3\times1} \doteq p\mathbf{n} \qquad (27.13)$$

where p is the albedo at the pixel, $\mathbf{n}$ is the unit surface normal vector at the pixe, and $\mathbf{s}$ (a 3×1 unit vector multiplied by its intensity) specifies the distant illuminant s. For an image $\mathbf{h}^s$, a collection of d pixels $\{h_i^s, i = 1, ..., d\}$, by stacking all the pixels into a column vector, we have

$$\mathbf{h}^s \doteq [h_1^s, h_2^s, ..., h_d^s]^T = [\mathbf{t}_1, \mathbf{t}_2, ..., \mathbf{t}_d]^T \mathbf{s} = \mathbf{T} \, \mathbf{s}, \qquad (27.14)$$

where $\mathbf{T} \doteq [\mathbf{t}_1, \mathbf{t}_2, ..., \mathbf{t}_d]^T$ contains complete albedo and shape information for the object and is called the *object-specific albedo-shape* matrix [84]. The Lambertian reflectance model, when the attached and cast shadows are ignored, implies a rank-3 subspace [65] where the appearances are located under different illuminations.

If attached shadows are considered [3], the rank grows to infinity but the energy is largely packed in a few harmonics components. This enables a low-dimensional subspace approximation. However, in Refs. [3,23], for one object to be recognized, multiple (≥ 3) images must be stored in a gallery set. This is inconvenient in practice. Generalization across illumination variation is offered by the illumination model, but no generalization between identities is available.

The requirement of storing multiple images is relaxed in Refs. [66,84]: Only the training set stores multiple observations for multiple objects, and the gallery set stores only one image per object. Here, a continuous-valued identity signature is used, and a linear generalization from the training set to the gallery and probe set is assumed. The difference between Shashua and Raviv [66] and Zhou et al. [84] lies in how the linear blending coefficients are learned. Once learned, the blending coefficients offer an illumination-invariant signature of the identity. Both approaches recognize probe images under illumination different from gallery images. The quotient image approach in Shashua and Jacobs [66] assumes that shapes of all objects are the same and the albedo field of an unknown object lies in the rational span of the training set. The approach in Zhou et al. [84] poses a rank constraint on the product of the albedo and surface normal. Below, we briefly review the approach in Zhou et al. [84].

It states that any $\mathbf{T}$ matrix can be represented as a linear combination of some basis matrices $\{\mathbf{T}_1, \mathbf{T}_2, \ldots, \mathbf{T}_m\}$ coming from some m hypothetical base objects. Mathematically, coefficients f_i are present, such that

$$\mathbf{T} = \sum_{i=1}^{m} f_i \mathbf{T}_i = [\mathbf{T}_1, \mathbf{T}_2, ..., \mathbf{T}_m](\mathbf{f} \otimes \mathrm{I}_3) = \mathbf{W}(\mathbf{f} \otimes \mathrm{I}_3), \qquad (27.15)$$

where $\mathbf{f}_{m \times 1} \doteq [f_1, f_2, ..., f_m]^T$, $\mathbf{W}_{d \times 3m} \doteq [\mathbf{T}_1, \mathbf{T}_2, ..., \mathbf{T}_m]$, and $\otimes$ denotes the matrix Kronecker (tensor) product. So, an image $\mathbf{h}^s$ can be re-expressed as

$$\mathbf{h}^s = \mathbf{T}\mathbf{s} = \mathbf{W}(\mathbf{f} \otimes \mathrm{I}_3)\mathbf{s} = \mathbf{W}(\mathbf{f} \otimes \mathbf{s}). \qquad (27.16)$$

Since the coefficient vector $\mathbf{f}$ only relates the albedos and surface normals of the basis matrices, it has no relationship with the illumination $\mathbf{s}$. Thus, $\mathbf{f}$ is an illumination-invariant description of the identity and is an appropriate quantity for face recognition under the illumination variation. Equation (27.16) presents a bilinear relationship between $\mathbf{f}$ and $\mathbf{s}$. Once the $\mathbf{W}$ matrix is given, the $\mathbf{f}$ vector can be easily recovered using a bilinear algorithm.

However, learning the $\mathbf{W}$ matrix from a set of training images is not trivial. In [20], the recovered $\mathbf{W}$ minimizes the approximation error in the mean square sense and need not satisfy the integrability constraint. In other words, the hypothetical base objects in $\mathbf{W}$ are not integrable. In Zhou et al. [84], the recovered $\mathbf{W}$ minimizes the above approximation error, as well as a cost function that enforces the integrability constraint. As a consequence, [20] can only process the image ensemble, consisting of different objects under the same set of lighting sources (the case considered here) while Zhou et al. [84] can process the image ensemble consisting of different objects under completely different lighting condition. Figure 27.5 shows the recovered $\mathbf{W}$ using the algorithm developed in Zhou et al. [84].

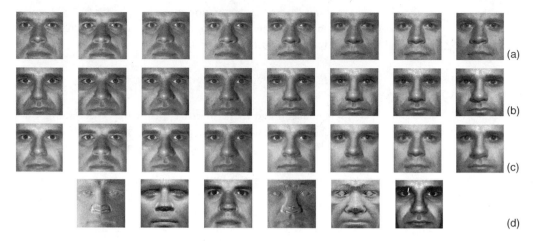

FIGURE 27.5   (a) The first basis object under eight different illuminations. (b) The second basis object under the same set of eight different illuminations. (c) Eight images (constructed by random linear combinations of two basis objects) illuminated by eight different lighting sources. (d) Recovered class-specific albedo-shape matrix **W** showing the product of varying albedos and surface normals of two basis objects (i.e., the three columns of T_1 and T_2) using the algorithm in [84].

Face Recognition under Pose Variation

The issue of pose essentially amounts to a correspondence problem. If dense correspondences across poses are available and if a Lambertian reflectance model is further assumed, a rank-1 constraint is implied because, theoretically, a 3D model can be recovered and used to render novel poses. However, recovering a 3D model from 2D images is a difficult task. There are two types of approaches: model-based and image-based. Model-based approaches [6,21,60,64] require explicit knowledge of prior 3D models, while image-based approaches [39,43,47,57] do not use prior 3D models. In general, model-based approaches [6,21,60,64] register the 2D face image to 3D models that are given beforehand. In Fox [21] and Shan et al. [64], a generative face model is deformed through bundle adjustment to fit 2D images. In Chowdhury and Chellappa [60], a generative face model is used to regularize the 3D model recovered using the SfM algorithm. In Blanz and Vetter [6], 3D morphable models are constructed based on many prior 3D models. There are mainly three types of image-based approaches: Structure from motion (SfM) [57], visual hull [39,47], and light field rendering [24,43] methods. The SfM approach [57], using sparse correspondence, does not reliably recover the 3D model amenable for practical use. Recently developed methods [61] using optical flow and FFT computation show promise. The visual hull methods [39,47] assume that the shape of the object is convex, a pattern that is not always satisfied by the human face, and also require accurate calibration information. The light field rendering methods [24,43] relax the calibration requirement of calibration by a fine quantization of the pose space and recover a novel view by sampling the captured data that forms the so-called light field. Figure 27.6 illustrates the concept using a simple example of the 2D light field of a 2D object.

As mentioned earlier, pose variation essentially amounts to a correspondence problem. If dense correspondences across poses are available and a Lambertian reflectance is assumed, then a rank-1 constraint is implied. Unfortunately, finding correspondences is a difficult task. There are no subspaces based on an appearance representation when confronted with pose variation. Approaches to face recognition under pose variation [23,26,54] avoid the correspondence problem by sampling the continuous pose space into a set of poses. This process stores multiple images at different poses for each person, at least in the training set. In Pentland et al. [54], view-based "Eigenfaces" are learned from the training set and used for recognition. In Georghiades et al. [23], a denser sampling covers the pose space. However, Georghiades et al. [23] use object-specific images, and appearances belonging to a novel object (or not in the training set) cannot be handled. In Gross et al. [26], the concept of light field [43] characterizes the continuous pose space. Eigen light fields are learned from the training set. However, the implementation of Gross [26] makes the pose space

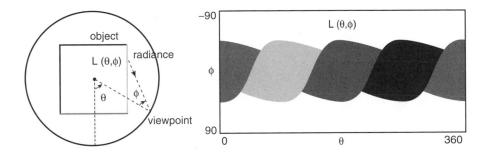

FIGURE 27.6 This figure illustrates the 2D light field of a 2D object (a square with four differently colored sides) placed within a circle. The angles θ and ϕ relate the viewpoint with the radiance from the object. The right image shows the actual light field for the square object. See another illustration in [26].

discrete. Recognition can be based on probe images at poses in the discrete set. The light field is not related to variations in illumination.

Face Recognition under Illumination and Pose Variations

Approaches to handling illumination and pose variations include Blanz and Vetter [6], Gross [25], Vasilescu and Terzopoulos [74], Zhou and Chellappa [83], and Zhou and Chellappa [82]. The approach in Ref. [6] uses morphable 3D models to characterize human faces. Both geometry and texture are linearly spanned by the training ensemble, consisting of 3D prior models. It is able to handle both illumination and pose variations. Its only weakness is a complicated fitting algorithm. Recently, a fitting algorithm more efficient than Blanz and Vetter [6] was proposed in Romdhani and Vetter [59]. In Gross et al. [25], the Fisher light field is proposed to handle illumination and pose variations, where the light field covers the pose variation and Fisher discriminant analysis covers the illumination variation. Discriminant analysis is a statistical analysis tool that minimizes the within-class scatter while maximizing between-class scatter and has no relationship with any physical illumination model. It is questionable that discriminant analysis can generalize to new lighting conditions. Instead, this generalization may be inferior because discriminant analysis tends to overly tune to the lighting conditions in the training set. The "Tensorface" approach [74] uses multilinear analysis to handle factors such as identity, illumination, pose, and expression. The factors of identity and illumination are suitable for linear analysis, as evidenced by the "Eigenface" approach (assuming a fixed illumination and a fixed pose), and the subspace induced by the Lambertian model, respectively. However, the factor of expression is arguably amenable for linear analysis and the factor of pose is not amenable for linear analysis. In Zhou and Chellappa [83], preliminary results are reported by first warping the albedo and surface normal fields at the desired pose and then carrying on recognition as usual. The approach in Zhou and Chellappa [82] extends the algorithm in [84] that is for illumination variation to deal with pose variation. In Zhou et al. [84], face images are in frontal view and no pose variances are present. In Zhou and Chellappa [82], we consider a finite set of views that uniformly cover the left profile to the right profile. By treating the images consisting of all views of the same individual, illuminated by the same lighting source as an "augmented" image, we can apply the algorithm developed in Zhou et al. [84] to the augmented image, since it is illuminated by one source. Figure 27.7 shows the part of the **W** matrix learned using images in the PIE (pose, illumination, and expression) database [67].

Face Recognition from Videos

Various approaches for performing face recognition based on video sequences have been proposed. However, most approaches [9] are still-image-based and treat each video frame separately. Typically, they first perform

FIGURE 27.7 The first nine columns of the learned **W** matrix.

face tracking and then perform recognition based on one or several tracked face regions satisfying certain criteria. In the above strategy, two important characteristics of a video sequence are disregarded:

1. *Multiple looks.* A video sequence provides a large amount of observations. This is attractive considering the projective nature of the imaging geometry and the uncontrolled lighting distribution. These factors, coupled with personal variations such as facial expression, make the 2D face appearances of one individual possess infinitely many possibilities. However, the above approach makes recognition decisions based only on a sparse set of observations.

2. *Temporal continuity.* Frames of a video sequence come in a sequential fashion and possess certain smoothness between successive frames that is referred as temporal continuity. Such continuity is often exploited in developing tracking algorithms. However, there is psychophysical evidence [37] suggesting that temporal continuity also is important for recognition.

We now highlight some recent approaches to face recognition from videos, utilizing video characteristics.

In Yamaguchi et al. [79], two mutual view subspaces are prelearned. In testing, for each frame of a video sequence, the algorithm computes a similarity score with both subspaces and takes the higher ones. A time-evolving curve of similarity score is plotted. If the temporal information is stripped, this method is similar to view-based and modular Eigenspaces proposed by Pentland et al. [54]. Evidence integration is also performed in Edwards et al. [17] and Li et al. [44]. Both Edwards and Li based their concepts on the framework of active appearance models [17,44]. In Li et al. [44], the kernel discriminant analysis features are extracted from the image, warped to a frontal view for recognition.

Another line of research effort summarizes the appearances presented in the video sequence. Examples of such a method are Shaknarovich et al. [63], Fitzgibbon and Zisserman [19], and Wolf and Shashua [78]. In Shakhnarovich et al. [63], a multivariate Gaussian density is fitted for a video sequence, where all facial images cropped out by a separate tracker are assumed to be samples from that distribution. Recognition is performed by comparing the Kullback–Leibler divergence distance [11] between the gallery and probe videos. However, a multivariate Gaussian density has difficulty in modeling significant variations caused by pose and illumination. In Fitzgibbon and Zisserman [19], manifolds are formed for multiple images. Recognition is performed by computing the shortest distance between two manifolds. The manifold takes a certain parameterized form and the parameters are directly learned from the visual appearances. In Wolf and Shashua [78], principal subspaces are learned for multiple images, and the principal angle between two subspaces is used for recognition. The computation of principal angle is carried out on the feature space embedded by kernel functions. One common disadvantage of the two approaches is that they assume that the face regions have been cropped previously, either from a detector or a tracker.

When the above approaches utilize the multiple appearances provided by the video sequence, they do not take into account temporal continuity between successive video frames. We now show approaches that utilize the temporal coherence embedded in the video sequences.

In Zhou et al. [85], still-to-video recognition is solved, where the gallery consists of still images and the probes are video sequences. Since the detected face might be moving in the video sequence, we have to deal with uncertainty in tracking and in recognition. Rather than resolving these two uncertainties separately, our strategy is to perform simultaneous tracking and recognition of human faces from a video sequence. A time-series–state-space model is proposed to fuse temporal information in a probe video, which simultaneously characterizes kinematics and identity using a motion vector and an identity variable, respectively. The joint posterior distribution of the motion vector and the identity variable is estimated at each time instant and propagated to the next time instant. Marginalization over the motion vector yields a robust estimate of the posterior distribution of the identity variable. A computationally efficient sequential importance sampling (SIS) algorithm [16] estimates the posterior distribution. Empirical results demonstrate that, due to the propagation of the identity variable over time, a degeneracy in posterior probability of the identity variable is achieved.

The gallery can be extended to have videos as inputs. A learning algorithm to automatically extract exemplars from the gallery video sequences has been described in Zhou et al. [85]. To represent each gallery object, multiple exemplars are extracted. During testing, the SIS is adopted to use temporal coherence to boost recognition performance.

In Liu and Chen [46], hidden Markov models are used to learn the dynamics before successive appearances. Matching video sequences is equivalent to comparing two Markov models. In Lee et al. [40], pose variations are learned through view-discretized appearance manifolds from the training ensemble. Transition probabilities from one view to another regularize the search space. However, in Liu and Chen [46] and Lee et al. [40], the cropped images are used for testing. Recently, a linear dynamical system model [68] has been used to model the video sequence, and system model coefficients that are used in face recognition [1].

27.6 Conclusions

In this chapter, we have described some of the methods developed toward the goal of human recognition using biometrics such as face and gait. While face recognition has been a subject for research in computer vision for many years, the use of gait for recognition is a more recent phenomenon. There are a number of limitations in these techniques for uncontrolled environments, such as outdoors under various lighting conditions. Progress in computer vision research and allied fields like image/video processing, pattern recognition, and machine learning will allow us to develop more realistic algorithms in the future. Most of the present face recognition algorithms work with a frontal view of the face, while most gait recognition methods assume a side view of the person. The full potential of face recognition using video sequences still must be explored and its advantages compared to still images need to be studied. The problems that affect the performance of these techniques include the effects of variable illumination, elapsed time, and pose invariance. Today, multimodal biometrics are becoming more popular in achieving high recognition rates. An example of fusing face and gait signatures can be found in Kale et al. [36]. The concept of multimodal biometrics is to combine different cues like face, gait, fingerprint, iris, and ear to develop an individual's identifying signature. Finding efficient means of automatically combining some or all of these different biometrics is an open question.

References

1. G. Aggarwal, A. Roy-Chowdhury, and R. Chellappa, "A system identification approach for video-based face recognition," *International Conference on Pattern Recognition*, Cambridge, UK, August 2004.

2. C.D. Barclay, J.E. Cutting, and L.T. Kozlowski, "Temporal and spatial factors in gait perception that influence gender recognition," *Percept. Psychophys.*, 23, 145–152, 1978.

3. R. Basri and D.W. Jacobs, "Lambertian reflectance and linear subspaces," *IEEE Trans. Pattern Anal. Mach. Int.*, vol. 25, no. 2, pp. 218–233, 2003.

4. T. Beardsworth and T. Buckner, "The ability to recognize oneself from a video recording of ones movements without seeing ones body," *Bull. Psychon. Soc.*, vol. 18, no. 1, pp. 19–22, 1981.

5. P.N. Belhumeur, J.P. Hespanha, and D.J. Kriegman, "Eigenfaces vs. fisherfaces: recognition using class specific linear projection," *IEEE Trans. Pattern Anal. Mach. Int.*, 19, 711–720, 1997.

6. V. Blanz and T. Vetter, "Face recognition based on fitting a 3D morphable model," *IEEE Trans. Pattern Anal. Mach. Int.*, vol. 25, no. 9, pp. 1063–1074, 2003.

7. A.F. Bobick and A. Johnson, "Gait recognition using static activity-specific parameters," in *Proc. IEEE Conf. Comput. Vision Pattern Recognit.*, Hawaii, December 2001.

8. R. Cutler C. Benabdelkader and L.S. Davis, "Motion based recognition of people in eigengait space," in *Proc. IEEE Conf. Face Gesture Recognit.*, Washington, DC, May 2002, pp. 267–272.

9. T. Choudhury, B. Clarkson, T. Jebara, and A. Pentland, "Multimodal person recognition using unconstrained audio and video," in *Proc. Int. Conf. Audio-Video-Based Person Authentication*, Washington, 1999, pp. 176–181.

10. R. Collins, R. Gross, and J. Shi, "Silhouette-based human identification from body shape and gait," in *Proc. IEEE Conf. Face Gesture Recognit.*, May 2002.

11. T.M. Cover and J.A. Thomas, *Elements of Information Theory*, New York: Wiley, 1991.

12. D. Cunado, J.M. Nash, M.S. Nixon, and J.N. Carter, "Gait extraction and description by evidence-gathering," in *Proc. Int. Conf. Audio Video Based Biometric Person Authentication*, 1999, pp. 43–48.

13. J. Cutting and L. Kozlowski, "Recognizing friends by their walk: gait perception without familiarity cues," *Bull. Psychon. Soc.*, 9, 353–356, 1977.

14. J.E. Cutting and D.R. Proffitt, *Gait Perception as an Example of How We Perceive Events*, London: Plenum Press, 1981.

15. W.H. Dittrich, "Action categories and the perception of biological motion," *Perception*, 22, 15–22, 1993.

16. A. Doucet, N. de Freitas, and N. Gordon, Eds., *Sequential Monte Carlo Methods in Practice*, Berlin: Springer, 2001.

17. G. Edwards, C. Taylor, and T. Cootes, "Improving identification performance by integrating evidence from sequences," in *Proc. IEEE Comp. Soc. Conf. Comp. Vision Pattern Recognit.*, 1999.

18. O.D. Faugeras, *Three-Dimensional Computer Vision: A Geometric Viewpoint*, Cambridge, MA: MIT Press, 1993.

19. A. Fitzgibbon and A. Zisserman, "Joint manifold distance: a new approach to appearance based clustering," in *Proc. IEEE Comput. Soc. Conf. Comput. Vision Pattern Recognit.*, 2003.

20. W.T. Freeman and J.B. Tenenbaum, "Learning bilinear models for two-factor problems in vision," in *Proc. IEEE Comput. Soc. Conf. Comput. Vision Pattern Recognit.*, 1997.

21. P. Fua, "Regularized bundle adjustment to model heads from image sequences without calibrated data," *Int. J. Comput. Vision*, 38, 153–157, 2000.

22. S. Furui, "Cepstral analysis technique for automatic speaker verification," *IEEE Trans. Acoust. Speech Signal Process.*, vol. 29, no. 2, pp. 254–272, 1981.

23. A. Georghiades, P. Belhumeur, and D. Kriegman, "From few to many: illumination cone models for face recognition under variable lighting and pose," *IEEE Trans. Pattern Anal. Mach. Int.*, vol. 23, no. 6, pp. 643 –660, 2001.

24. S.J. Gortler, R. Grzeszczuk, R. Szeliski, and M. Cohen, "The lumigraph," in *Proc. SIGGRAPH*, New Orleans, LA, 1996, pp. 43–54.

25. R. Gross, I. Matthews, and S. Baker, "Fisher light-fields for face recognition across pose and illumination," in *Proc. 24th Symp. German Assoc. Pattern Recognit. (DAGM)*, 2002.

26. R. Gross, I. Matthews, and S. Baker, "Eigen light-fields and face recognition across pose," in *Proc. Face Gesture Recognit.*, Washington, DC, May 2002.

27. G. Shakhnarovich, L. Lee, and T. Darrell, "Integrated face and gait recognition from multiple views," in *Proc. IEEE Conf. Comput. Vision Pattern Recognit.*, December 2001.

28. G.F. Harris and P.A. Smith, Eds., *Human Motion Analysis: Current Applications and Future Directions*, New York: IEEE Press, 1996.

29. R.I. Hartley and A. Zisserman, *Multiple View Geometry in Computer Vision*, Cambridge, MA: Cambridge University Press, 2000.

30. A. Howell and H. Buxton, "Face recognition using radial basis function neural networks," in *Proc. British Mach. Vision Conf.*, 1996, pp. 455–464.

31. P.S. Huang, C.J. Harris, and M.S. Nixon, "Recognizing humans by gait via parametric canonical space," *Artif. Int. Eng.*, vol. 13, no. 4, pp. 359–366, 1999.

32. G. Johansson, "Visual perception of biological motion and a model for its analysis," *PandP*, vol. 14, no. 2, pp. 201–211, 1973.

33. G. Johansson, "Visual motion perception," *Sci. Am.*, 232, 76–88, 1975.

34. A. Kale, A.K. Roy-Chowdhury, and R. Chellappa, "Towards a view invariant gait recognition algorithm," in *Proc. IEEE Conf. Adv. Video Signal Based Surveillance*, 2003, pp. 143–150.

35. A. Kale, N. Cuntoor, B. Yegnanarayana, A.N. Rajagopalan, and R. Chellappa, "Gait analysis for human identification," in *Proc. Audio Video Biometric Person Authentication*, 2003.

36. A. Kale, A. Roy-Chowdhury, and R. Chellappa, "Fusion of gait and face for human identification," *International Conference on Acoustics, Speech and Signal Processing*, Montreal, Canada, May 2004.

37. B. Knight and P. Johnston, "The role of movement in face recognition," *Visual Cognition*, 4, 265–274, 1997.

38. L. Kozlowski and J. Cutting, "Recognizing the sex of a walker from a dynamic point display," *Percept. Psychophys.*, 21, 575–580, 1977.

39. A. Laurentini, "The visual hull concept for silhouette-based image understanding," *IEEE Trans. Pattern Anal. Mach. Int.*, vol. 16, no. 2, pp. 150–162, 1994.

40. K. Lee, J. Ho, M. Yang, and D. Kriegman, "Video-based face recognition using probabilistic appearance manifolds," in *Proc. IEEE Comput. Soc. Conf. Comput. Vision Pattern Recognit.*, 2003.

41. L. Lee and G. Dalley, "Learning pedestrian models for silhouette refinement," *International Conference on Computer Vision*, Nice, France, 2003.

42. L. Lee and W.E.L. Grimson, "Gait analysis for recognition and classification," in *Proc. IEEE Conf. Face Gesture Recognit.*, 2002, pp. 155–161.

43. M. Levoy and P. Hanrahan, "Light field rendering," in *Proc. SIG-GRAPH*, 1996.

44. Y. Li, S. Gong, and H. Liddell, "Constructing facial identity surfaces for recognition," *Int. J. Comput. Vision*, vol. 53, no. 1, pp. 71–92, 2003.

45. J. Little and J. Boyd, "Recognizing people by their gait: the shape of motion," *Videre*, vol. 1, no. 2, pp. 1–32, 1998.

46. X. Liu and T. Chen, "Video-based face recognition using adaptive hidden Markov models," in *Proc. IEEE Comput. Soc. Conf. Comput. Vision Pattern Recognit.*, 2003.

47. W. Matusik, C. Buehler, R. Raskar, L. Gortler, and S. McMillan, "Image-based visual hulls," in *Proc. SIGGRAPH*, 2000, pp. 369–374.

48. S. McKenna and S. Gong, "Non-intrusive person authentication for access control by visual tracking and face recognition," in *Proc. Int. Conf. Audio-Video-Based Biometric Person Authentication*, 1997, pp. 177–183.

49. B. Moghaddam, "Principal manifolds and probabilistic subspaces for visual recognition," *IEEE Trans. PAMI*, vol. 24, no. 6, pp. 780–788, 2002.

50. B. Moghaddam and A.P. Pentland, "Probabilistic visual learning for object representation," *IEEE Trans. Pattern Anal. Mach. Int.*, 19, 696–710, 1997.

51. M.P. Murray, A.B. Drought, and R.C. Kory, "Walking patterns of normal men," *J. Bone Joint Surg.*, vol. 46-A, no. 2, pp. 335–360, 1964.

52. E. Muybridge, *The Human Figure in Motion*, New York: Dover Publications, 1901.

53. P. Penev and J. Atick, "Local feature analysis: a general statistical theory for objecct representation," *Network: Comput. Neural Sys.*, 7, 477–500, 1996.

54. A.P. Pentland, B. Moghaddam, and T. Starner, "View-based and modular eigenspaces for face recognition," *IEEE Conference on Computer Vision and Pattern Recognition*, 1994.

55. P.J. Phillips, S. Sarkar, I. Robledo, P. Grother, and K.W. Bowyer, "Baseline results for the challenge problem of human ID using gait analysis," in *Proc. fifth IEEE Int. Conf. Autom. Face Gesture Recognit.*, 2002.

56. P.J. Phillips, S. Sarkar, I. Robledo, P. Grother, and K.W. Bowyer, "The gait identification challenge problem: data sets and baseline algorithm," in *Proc. Int. Conf. Pattern Recognit.*, 2002.

57. G. Qian and R. Chellappa, "Structure from motion using sequential Monte Carlo methods," in *Proc. IEEE Int. Conf. Comput. Vision*, 2, 614–621, Vancouver, Canada, 2001.

58. L.R. Rabiner, "A tutorial on Hidden Markov Models and selected applications in speech recognition," in *Proc. IEEE*, vol. 77, no. 2, pp. 257–285, 1989.

59. S. Romdhani and T. Vetter, "Efficient, robust and accurate fitting of a 3D morphable model," in *Proc. IEEE Int. Conf. Comput. Vision*, Nice, France, 2003, pp. 59–66.

60. A.R. Chowdhury and R. Chellappa, "Face reconstruction from video using uncertainty analysis and a generic model," *Comput. Vision Image Understanding*, 91, 188–213, 2003.

61. A.R. Chowdhury and R. Chellappa, "Stochastic approximation and rate distortion analysis for robust structure and motion estimation," *Int. J. Comput. Vision*, vol. 55, no. 1, pp. 27–53, 2003.

62. W.I Scholhorn, B.M. Nigg , D.J. Stephanshyn, and W. Liu, "Identification of individual walking patterns using time discrete and time continuous data sets," *Gait Posture*, 15, 180–186, 2002.

63. G. Shakhnarovich, J. Fisher, and T. Darrell, "Face recognition from long-term observations," *European Conference on Computer Vision*, Copenhagen, Denmark, May 2002.

64. Y. Shan, Z. Liu, and Z. Zhang, "Model-based bundle adjustment with application to face modeling," in *Proc. Int. Conf. Comput. Vision*, Vancouver, Canada, 2001, pp. 645–651.

65. A. Shashua, "On photometric issues in 3D visual recognition from a single 2D image," *Int. J. Comput. Vision*, 21, 99–122, 1997.

66. A. Shashua and T.R. Raviv, "The quotient image: class based re-rendering and recognition with varying illuminations," *IEEE Trans. Pattern Anal. Mach. Int.*, 23, 129–139, 2001.

67. T. Sim, S. Baker, and M. Bsat, "The CMU pose, illumination, and expression (PIE) database," in *Proc. Face Gesture Recognit.*, 2002.

68. S. Soatto, G. Doretto, and Y.N. Wu, "Dynamic textures," *International Conference on Computer Vision*, 2001.

69. S.V. Stevenage, M.S. Nixon, and K. Vince, "Visual analysis of gait as a cue to identity," *Appl. Cogn. Psychol.*, 13, 513–526, 1999.

70. A. Sundaresan, A.R. Chodhury, and R. Chellappa, "A Hidden Markov Model based framework for recognition of humans from gait sequences," in *Proc. Int. Conf. Image Process.*, September 2003.

71. R. Tanawongsuwan and A.F. Bobick, "Gait recognition from time-normalized joint-angle trajectories in the walking plane," in *IEEE Comput. Soci. Conf. Vision Pattern Recognit.*, 2001, pp. II:726–731.

72. D. Tolliver and R. Collins, "Gait shape estimation for identification," in *Proc. AVBPA*, 2003, pp. 734–742.

73. M. Turk and A. Pentland, "Eigenfaces for recognition," *J. Cogn. Neurosci.*, 3, 72–86, 1991.

74. M.A.O. Vasilescu and D. Terzopoulos, "Multilinear analysis of image ensembles: tensorfaces," *European Conference on Computer Vision*, Copenhagen, Denmark, vol. 2350, May 2002, pp. 447–460.

75. A. Veeraraghavan, A.R. Chowdhury, and R. Chellappa, "Role of shape and kinematics in human movement analysis," *IEEE Comput. Soci. Conf. Comp. Vision Pattern Recognit.*, Washington, DC, 2004.

76. H. Wechsler, V. Kakkad, J. Huang, S. Gutta, and V. Chen, "Automatic video-based person authentication using the RBF network," in *Proc. Int. Conf. Audio- Video-Based Biometric Person Authentication*, 1997, pp. 85–92.

77. L. Wiskott, J.M Fellous, and C. von der Malsburg, "Face recognition by elastic bunch graph matching," *IEEE Trans. Pattern Anal. Mach. Int.*, 19, 775–779, 1997.

78. L. Wolf and A. Shashua, "Kernel principal angles for classification machines with applications to image sequence interpretation," in *Proc. IEEE Comput. Soc. Conf. Comput. Vision Pattern Recognit.*, Madison, WI, 2003.

79. O. Yamaguchi, K. Fukui, and K. Maeda, "Face recognition using temporal image sequence," in *Proc. Int. Conf. Autom. Face Gesture Recognit.*, Nara, Japan, October, 1998.

80. W. Zhao, R. Chellappa, and A. Krishnaswamy, "Discriminant analysis of principal components for face recognition," in *Proc. Int. Conf. Autom. Face Gesture Recognit.*, 1998, pp. 336–341.

81. W. Zhao, R. Chellappa, A. Rosenfeld, and J. Phillips, "Face recognition: a literature survey," *ACM Comput. Surv.*, 12, 399–458, 2003.

82. S. Zhou and R. Chellappa, "Illuminating light field: image-based face recognition across illuminations and poses," *IEEE Int. Conf. Autom. Face Gesture Recognit.*, Korea, May 2004.

83. S. Zhou and R. Chellappa, "Rank constrained recognition under unknown illumination," *IEEE Int. Workshop Anal. Modeli. Faces Gestures*, Nice, France, 2003.

84. S. Zhou, R. Chellappa, and D. Jacobs, "Characterization of human faces under illumination variations using rank, integrability, and symmetry constraints," *European Conference on Computer Vision*, Prague, The Czech Republic, May 2004.

85. S. Zhou, V. Krueger, and R. Chellappa, "Probabilistic recognition of human faces from video," *Comput. Vision Image Understand.*, 91, 214–245, 2003.

Mathematics, Symbols, and Physical Constants

Ronald J. Tallarida
Temple University

THE GREAT ACHIEVEMENTS in engineering deeply affect the lives of all of us and also serve to remind us of the importance of mathematics. Interest in mathematics has grown steadily with these engineering achievements and with concomitant advances in pure physical science. Whereas scholars in nonscientific fields, and even in such fields as botany, medicine, geology, etc., can communicate most of the problems and results in nonmathematical language, this is virtually impossible in present-day engineering and physics. Yet it is interesting to note that until the beginning of the twentieth century, engineers regarded calculus as something of a mystery. Modern students of engineering now study calculus, as well as differential equations, complex variables, vector analysis, orthogonal functions, and a variety of other topics in applied analysis. The study of systems has ushered in matrix algebra and, indeed, most engineering students now take linear algebra as a core topic early in their mathematical education.

This section contains concise summaries of relevant topics in applied engineering mathematics and certain key formulas, that is, those formulas that are most often needed in the formulation and solution of engineering problems. Whereas even inexpensive electronic calculators contain tabular material (e.g., tables of trigonometric and logarithmic functions) that used to be needed in this kind of handbook, most calculators do not give symbolic results. Hence, we have included formulas along with brief summaries that guide their use. In many cases we have added numerical examples, as in the discussions of matrices, their inverses, and their use in the solutions of linear systems. A table of derivatives is included, as well as key applications of the derivative in the solution of problems in maxima and minima, related rates, analysis of curvature, and finding approximate roots by numerical methods. A list of infinite series, along with the interval of convergence of each, is also included.

Of the two branches of calculus, integral calculus is richer in its applications, as well as in its theoretical content. Though the theory is not emphasized here, important applications such as finding areas, lengths, volumes, centroids, and the work done by a nonconstant force are included. Both cylindrical and spherical polar coordinates are discussed, and a table of integrals is included. Vector analysis is summarized in a separate section and includes a summary of the algebraic formulas involving dot and cross multiplication, frequently needed in the study of fields, as well as the important theorems of Stokes and Gauss. The part on special functions includes the gamma function, hyperbolic functions, Fourier series, orthogonal functions, and both Laplace and z-transforms. The Laplace transform provides a basis for the solution of differential equations and is fundamental to all concepts and definitions underlying analytical tools for describing feedback control systems. The z-transform, not discussed in most applied mathematics books, is most useful in the analysis of discrete signals as, for example, when a computer receives data sampled at some prespecified time interval. The Bessel functions, also called cylindrical functions, arise in many physical applications, such as the heat transfer in a "long" cylinder, whereas the other orthogonal functions discussed—Legendre, Hermite, and Laguerre polynomials—are needed in quantum mechanics and many other subjects (e.g., solid-state electronics) that use concepts of modern physics.

The world of mathematics, even applied mathematics, is vast. Even the best mathematicians cannot keep up with more than a small piece of this world. The topics included in this section, however, have withstood the test of time and, thus, are truly *core* for the modern engineer.

This section also incorporates tables of physical constants and symbols widely used by engineers. While not exhaustive, the constants, conversion factors, and symbols provided will enable the reader to accommodate a majority of the needs that arise in design, test, and manufacturing functions.

Mathematics, Symbols, and Physical Constants

Greek Alphabet

Greek Letter		Greek Name	English Equivalent	Greek Letter		Greek Name	English Equivalent
A	α	Alpha	a	N	ν	Nu	n
B	β	Beta	b	Ξ	ξ	Xi	x
Γ	γ	Gamma	g	O	o	Omicron	$\breve{o}$
Δ	δ	Delta	d	Π	π	Pi	P
E	ε	Epsilon	$\breve{e}$	P	ρ	Rho	r
Z	ζ	Zeta	z	Σ	σ	Sigma	s
H	η	Eta	$\bar{e}$	T	τ	Tau	t
Θ	$\theta\ \vartheta$	Theta	th	Y	υ	Upsilon	u
I	ι	Iota	i	Φ	$\phi\ \varphi$	Phi	ph
K	κ	Kappa	k	X	χ	Chi	ch
Λ	λ	Lambda	l	Ψ	ψ	Psi	ps
M	μ	Mu	m	Ω	ω	Omega	$\bar{o}$

International System of Units (SI)

The International System of units (SI) was adopted by the 11th General Conference on Weights and Measures (CGPM) in 1960. It is a coherent system of units built form seven *SI base units,* one for each of the seven dimensionally independent base quantities: they are the meter, kilogram, second, ampere, kelvin, mole, and candela, for the dimensions length, mass, time, electric current, thermodynamic temperature, amount of substance, and luminous intensity, respectively. The definitions of the SI base units are given below. The *SI derived units* are expressed as products of powers of the base units, analogous to the corresponding relations between physical quantities but with numerical factors equal to unity.

In the International System there is only one SI unit for each physical quantity. This is either the appropriate SI base unit itself or the appropriate SI derived unit. However, any of the approved decimal prefixes, called *SI prefixes,* may be used to construct decimal multiples or submultiples of SI units.

It is recommended that only SI units be used in science and technology (with SI prefixes where appropriate). Where there are special reasons for making an exception to this rule, it is recommended always to define the units used in terms of SI units. This section is based on information supplied by IUPAC.

Definitions of SI Base Units

Meter: The meter is the length of path traveled by light in vacuum during a time interval of 1/299,792,458 of a second (17th CGPM, 1983).

Kilogram: The kilogram is the unit of mass; it is equal to the mass of the international prototype of the kilogram (3rd CGPM, 1901).

Second: The second is the duration of 9,192,631,770 periods of the radiation corresponding to the transition between the two hyperfine levels of the ground state of the cesium-133 atom (13th CGPM, 1967).

Ampere: The ampere is that constant current which, if maintained in two straight parallel conductors of infinite length, of negligible circular cross-section, and placed 1 m apart in vacuum, would produce between these conductors a force equal to 2×10^{-7} newton per meter of length (9th CGPM, 1948).

Kelvin: The kelvin, unit of thermodynamic temperature, is the fraction 1/273.16 of the thermodynamic temperature of the triple point of water (13th CGPM, 1967).

Mole: The mole is the amount of substance of a system which contains as many elementary entities as there are atoms in 0.012 kg of carbon-12. When the mole is used, the elementary entities must be specified and may be atoms, molecules, ions, electrons, or other particles or specified groups of such particles (14th CGPM, 1971).

Examples of the use of the mole:

1 mol of H_2 contains about 6.022×10^{23} H_2 molecules, or 12.044×10^{23} H atoms.

1 mol of HgCl has a mass of 236.04 g.

1 mol of Hg_2Cl_2 has a mass of 472.08 g.

1 mol of Hg_2^{2+} has a mass of 401.18 g and a charge of 192.97 kC.

1 mol of $Fe_{0.91}S$ has a mass of 82.88 g.

1 mol of e^- has a mass of 548.60 μg and a charge of -96.49 kC.

1 mol of photons whose frequency is 10^{14} Hz has energy of about 39.90 kJ.

Candela: The candela is the luminous intensity in a given direction of a source that emits monochromatic radiation of frequency 540×10^{12} hertz and that has a radiant intensity in that direction of (1/683) watt per steradian (16th CGPM, 1979).

Names and Symbols for the SI Base Units

Physical Quantity	Name of SI Unit	Symbol for SI Unit
Length	meter	m
Mass	kilogram	kg
Time	second	s
Electric current	ampere	A
Thermodynamic temperature	kelvin	K
Amount of substance	mole	mol
Luminous intensity	candela	cd

SI Derived Units with Special Names and Symbols

Physical Quantity	Name of SI Unit	Symbol for SI Unit	Expression in Terms of SI Base Units	
Frequency[1]	hertz	Hz	s^{-1}	
Force	newton	N	$m\ kg\ s^{-2}$	
Pressure, stress	pascal	Pa	$N\ m^{-2}$	$= m^{-1}\ kg\ s^{-2}$
Energy, work, heat	joule	J	$N\ m$	$= m^2\ kg\ s^{-2}$
Power, radiant flux	watt	W	$J\ s^{-1}$	$= m^2\ kg\ s^{-3}$
Electric charge	coulomb	C	$A\ s$	
Electric potential, electromotive force	volt	V	$J\ C^{-1}$	$= m^2\ kg\ s^{-3}\ A^{-1}$
Electric resistance	ohm	Ω	$V\ A^{-1}$	$= m^2\ kg\ s^{-3}\ A^{-2}$
Electric conductance	siemens	S	Ω^{-1}	$= m^{-2}\ kg^{-1}\ s^3\ A^2$
Electric capacitance	farad	F	$C\ V^{-1}$	$= m^{-2}\ kg^{-1}\ s^4\ A^2$
Magnetic flux density	tesla	T	$V\ s\ m^{-2}$	$= kg\ s^{-2}\ A^{-1}$
Magnetic flux	weber	Wb	$V\ s$	$= m^2\ kg\ s^{-2}\ A^{-1}$
Inductance	henry	H	$V\ A^{-1}\ s$	$= m^2\ kg\ s^{-2}\ A^{-2}$
Celsius temperature[2]	degree Celsius	°C	K	

(continued)

SI Derived Units with Special Names and Symbols (continued)

Physical Quantity	Name of SI Unit	Symbol for SI Unit	Expression in Terms of SI Base Units	
Luminous flux	lumen	lm	cd sr	
Illuminance	lux	lx	cd sr m^{-2}	
Activity (radioactive)	becquerel	Bq	s^{-1}	
Absorbed dose (of radiation)	gray	Gy	J kg^{-1}	$= $ m^2 s^{-2}
Dose equivalent (dose equivalent index)	sievert	Sv	J kg^{-1}	$= $ m^2 s^{-2}
Plane angle	radian	rad	1	$= $ m m^{-1}
Solid angle	steradian	sr	1	$= $ m^2 m^{-2}

[1]For radial (circular) frequency and for angular velocity the unit rad s^{-1}, or simply s^{-1}, should be used, and this may not be simplified to Hz. The unit Hz should be used only for frequency in the sense of cycles per second.

[2]The Celsius temperature θ is defined by the equation:

$$\theta/°C = T/K - 273.15$$

The SI unit of Celsius temperature interval is the degree Celsius, °C, which is equal to the kelvin, K. °C should be treated as a single symbol, with no space between the ° sign and the letter C. (The symbol °K and the symbol ° should no longer be used.)

Units in Use Together with the SI

These units are not part of the SI, but it is recognized that they will continue to be used in appropriate contexts. SI prefixes may be attached to some of these units, such as milliliter, ml; millibar, mbar; megaelectronvolt, MeV; kilotonne, ktonne.

Physical Quantity	Name of Unit	Symbol for Unit	Value in SI Units
Time	minute	min	60 s
Time	hour	h	3600 s
Time	day	d	86,400 s
Plane angle	degree	°	$(\pi/180)$ rad
Plane angle	minute	′	$(\pi/10,800)$ rad
Plane angle	second	″	$(\pi/648,000)$ rad
Length	ångstrom[1]	Å	10^{-10} m
Area	barn	b	10^{-28} m^2
Volume	liter	l, L	dm^3 $=$ 10^{-3} m^3
Mass	tonne	t	Mg $=$ 10^3 kg
Pressure	bar[1]	bar	10^5 Pa $=$ 10^5 N m^{-2}
Energy	electronvolt[2]	eV ($= e \times$ V)	$\approx 1.60218 \times 10^{-19}$ J
Mass	unified atomic mass unit[2,3]	u ($= m_a(^{12}$C$)/12$)	$\approx 1.66054 \times 10^{-27}$ kg

[1]The ångstrom and the bar are approved by CIPM for "temporary use with SI units," until CIPM makes a further recommendation. However, they should not be introduced where they are not used at present.

[2]The values of these units in terms of the corresponding SI units are not exact, since they depend on the values of the physical constants e (for the electronvolt) and N_a (for the unified atomic mass unit), which are determined by experiment.

[3]The unified atomic mass unit is also sometimes called the dalton, with symbol Da, although the name and symbol have not been approved by CGPM.

Conversion Constants and Multipliers

Recommended Decimal Multiples and Submultiples

Multiples and Submultiples	Prefixes	Symbols	Multiples and Submultiples	Prefixes	Symbols
10^{18}	exa	E	10^{-1}	deci	d
10^{15}	peta	P	10^{-2}	centi	c
10^{12}	tera	T	10^{-3}	milli	m
10^{9}	giga	G	10^{-6}	micro	μ (Greek mu)
10^{6}	mega	M	10^{-9}	nano	n
10^{3}	kilo	k	10^{-12}	pico	p
10^{2}	hecto	h	10^{-15}	femto	f
10	deca	da	10^{-18}	atto	a

Conversion Factors—Metric to English

To Obtain	Multiply	By
Inches	centimeters	0.3937007874
Feet	meters	3.280839895
Yards	meters	1.093613298
Miles	kilometers	0.6213711922
Ounces	grams	$3.527396195 \times 10^{-2}$
Pounds	kilogram	2.204622622
Gallons (U.S. liquid)	liters	0.2641720524
Fluid ounces	milliliters (cc)	$3.381402270 \times 10^{-2}$
Square inches	square centimeters	0.155003100
Square feet	square meters	10.76391042
Square yards	square meters	1.195990046
Cubic inches	milliliters (cc)	$6.102374409 \times 10^{-2}$
Cubic feet	cubic meters	35.31466672
Cubic yards	cubic meters	1.307950619

Conversion Factors—English to Metric*

To Obtain	Multiply	By
Microns	mils	**25.4**
Centimeters	inches	**2.54**
Meters	feet	**0.3048**
Meters	yards	**0.9144**
Kilometers	miles	**1.609344**
Grams	ounces	28.34952313
Kilograms	pounds	**0.45359237**
Liters	gallons (U.S. liquid)	**3.785411784**
Millimeters (cc)	fluid ounces	29.57352956
Square centimeters	square inches	**6.4516**
Square meters	square feet	**0.09290304**
Square meters	square yards	**0.83612736**
Milliliters (cc)	cubic inches	**16.387064**
Cubic meters	cubic feet	$2.831684659 \times 10^{-2}$
Cubic meters	cubic yards	0.764554858

*Boldface numbers are exact; others are given to ten significant figures where so indicated by the multiplier factor.

Conversion Factors—General*

To Obtain	Multiply	By
Atmospheres	feet of water @ 4°C	2.950×10^{-2}
Atmospheres	inches of mercury @ 0°C	3.342×10^{-2}
Atmospheres	pounds per square inch	6.804×10^{-2}
BTU	foot-pounds	1.285×10^{-3}
BTU	joules	9.480×10^{-4}
Cubic feet	cords	**128**
Degree (angle)	radians	57.2958
Ergs	foot-pounds	1.356×10^{7}
Feet	miles	**5280**
Feet of water @ 4°C	atmospheres	33.90
Foot-pounds	horsepower-hours	1.98×10^{6}
Foot-pounds	kilowatt-hours	2.655×10^{6}
Foot-pounds per min	horsepower	3.3×10^{4}
Horsepower	foot-pounds per sec	1.818×10^{-3}
Inches of mercury @ 0°C	pounds per square inch	2.036
Joules	BTU	1054.8
Joules	foot-pounds	1.35582
Kilowatts	BTU per min	1.758×10^{-2}
Kilowatts	foot-pounds per min	2.26×10^{-5}
Kilowatts	horsepower	0.745712
Knots	miles per hour	0.86897624
Miles	feet	1.894×10^{-4}
Nautical miles	miles	0.86897624
Radians	degrees	1.745×10^{-2}
Square feet	acres	**43,560**
Watts	BTU per min	17.5796

*Boldface numbers are exact; others are given to ten significant figures where so indicated by the multiplier factor.

Temperature Factors

$$°F = 9/5 \ (°C) + 32$$
$$\text{Fahrenheit temperature} = 1.8 \ (\text{temperature in kelvins}) - 459.67$$
$$°C = 5/9 \ [(°F) - 32)]$$
$$\text{Celsius temperature} = \text{temperature in kelvins} - 273.15$$
$$\text{Fahrenheit temperature} = 1.8 \ (\text{Celsius temperature}) + 32$$

Conversion of Temperatures

From	To	
°Celsius	°Fahrenheit	$t_F = (t_C \times 1.8) + 32$
	Kelvin	$T_K = t_C + 273.15$
	°Rankine	$T_R = (t_C + 273.15) \times 18$
°Fahrenheit	°Celsius	$t_C = \dfrac{t_F - 32}{1.8}$
	Kelvin	$T_k = \dfrac{t_F - 32}{1.8} + 273.15$
	°Rankine	$T_R = t_F + 459.67$
Kelvin	°Celsius	$t_C = T_K - 273.15$
	°Rankine	$T_R = T_K \times 1.8$
°Rankine	Kelvin	$T_K = \dfrac{T_R}{1.8}$
	°Fahrenheit	$t_F - T_R - 459.67$

Physical Constants

General

Equatorial radius of the Earth $=$ 6378.388 km $=$ 3963.34 miles (statute)

Polar radius of the Earth, 6356.912 km $=$ 3949.99 miles (statute)

1 degree of latitude at $40°$ $=$ 69 miles

1 international nautical mile $=$ 1.15078 miles (statute) $=$ 1852 m $=$ 6076.115 ft

Mean density of the earth $=$ 5.522 g/cm^3 $=$ 344.7 lb/ft^3

Constant of gravitation $(6.673 \pm 0.003) \times 10^{-8}$ cm^3 gm^{-1} s^{-2}

Acceleration due to gravity at sea level, latitude $45°$ $=$ 980.6194 cm/s^2 $=$ 32.1726 ft/s^2

Length of seconds pendulum at sea level, latitude $45°$ $=$ 99.3575 cm $=$ 39.1171 in.

1 knot (international) $=$ 101.269 ft/min $=$ 1.6878 ft/s $=$ 1.1508 miles (statute)/h

1 micron $=$ 10^{-4} cm

1 ångstrom $=$ 10^{-8} cm

Mass of hydrogen atom $=$ $(1.67339 \pm 0.0031) \times 10^{-24}$ g

Density of mercury at $0°C$ $=$ 13.5955 g/ml

Density of water at $3.98°C$ $=$ 1.000000 g/ml

Density, maximum, of water, at $3.98°C$ $=$ 0.999973 g/cm^3

Density of dry air at $0°C$, 760 mm $=$ 1.2929 g/l

Velocity of sound in dry air at $0°C$ $=$ 331.36 m/s $-$ 1087.1 ft/s

Velocity of light in vacuum $=$ $(2.997925 \pm 0.000002) \times 10^{10}$ cm/s

Heat of fusion of water $0°C$ $=$ 79.71 cal/g

Heat of vaporization of water $100°C$ $=$ 539.55 cal/g

Electrochemical equivalent of silver 0.001118 g/s international amp

Absolute wavelength of red cadmium light in air at $15°C$, 760 mm pressure $=$ 6438.4696 Å

Wavelength of orange-red line of krypton 86 $=$ 6057.802 Å

π Constants

$$\pi = 3.14159\ 26535\ 89793\ 23846\ 26433\ 83279\ 50288\ 41971\ 69399\ 37511$$
$$1/\pi = 0.31830\ 98861\ 83790\ 67153\ 77675\ 26745\ 02872\ 40689\ 19291\ 48091$$
$$\pi^2 = 9.8690\ \ 44010\ 89358\ 61883\ 44909\ 99876\ 15113\ 53136\ 99407\ 24079$$
$$\log_e\pi = 1.14472\ 98858\ 49400\ 17414\ 34273\ 51353\ 05871\ 16472\ 94812\ 91531$$
$$\log_{10}\pi = 0.49714\ 98726\ 94133\ 85435\ 12682\ 88290\ 89887\ 36516\ 78324\ 38044$$
$$\log_{10}\sqrt{2\pi} = 0.39908\ 99341\ 79057\ 52478\ 25035\ 91507\ 69595\ 02099\ 34102\ 92128$$

Constants Involving e

$$e = 2.71828\ 18284\ 59045\ 23536\ 02874\ 71352\ 66249\ 77572\ 47093\ 69996$$
$$1/e = 0.36787\ 94411\ 71442\ 32159\ 55237\ 70161\ 46086\ 74458\ 11131\ 03177$$
$$e^2 = 7.38905\ 60989\ 30650\ 22723\ 04274\ 60575\ 00781\ 31803\ 15570\ 55185$$
$$M = \log_{10}e = 0.43429\ 44819\ 03251\ 82765\ 11289\ 18916\ 60508\ 22943\ 97005\ 80367$$
$$1/M\cdot = \log_e10 = 2.30258\ 50929\ 94045\ 68401\ 79914\ 54684\ 36420\ 67011\ 01488\ 62877$$
$$\log_{10}M = 9.63778\ 43113\ 00536\ 78912\ 29674\ 98645\ -10$$

Numerical Constants

$$\sqrt{2} = 1.41421\ 35623\ 73095\ 04880\ 16887\ 24209\ 69807\ 85696\ 71875\ 37695$$
$$3\sqrt{2} = 1.25992\ 10498\ 94873\ 16476\ 72106\ 07278\ 22835\ 05702\ 51464\ 70151$$
$$\log_e2 = 0.69314\ 71805\ 59945\ 30941\ 72321\ 21458\ 17656\ 80755\ 00134\ 36026$$
$$\log_{10}2 = 0.30102\ 99956\ 63981\ 19521\ 37388\ 94724\ 49302\ 67881\ 89881\ 46211$$

$$\sqrt{3} = 1.73205\ 08075\ 68877\ 29352\ 74463\ 41505\ 87236\ 69428\ 05253\ 81039$$
$$\sqrt[3]{3} = 1.44224\ 95703\ 07408\ 38232\ 16383\ 10780\ 10958\ 83918\ 69253\ 49935$$
$$\log_e 3 = 1.09861\ 22886\ 68109\ 69139\ 52452\ 36922\ 52570\ 46474\ 90557\ 82275$$
$$\log_{10} 3 = 0.47712\ 12547\ 19662\ 43729\ 50279\ 03255\ 11530\ 92001\ 28864\ 19070$$

Symbols and Terminology for Physical and Chemical Quantities

Name	Symbol	Definition	SI Unit
Classical Mechanics			
Mass	m		kg
Reduced mass	μ	$\mu = m_1 m_2/(m_1 + m_2)$	kg
Density, mass density	ρ	$\rho = M/V$	kg m^{-3}
Relative density	d	$d = \rho/\rho^{\theta}$	1
Surface density	ρ_A, ρ_S	$\rho_A = m/A$	kg m^{-2}
Momentum	p	$p = mv$	kg m s^{-1}
Angular momentum, action	L	$l = r ¥ p$	J s
Moment of inertia	I, J	$I = \Sigma m_i r_i^2$	kg m^2
Force	F	$F = d\mathbf{p}/dt = ma$	N
Torque, moment of a force	$T, (M)$	$T = r \times \mathbf{F}$	N m
Energy	E		J
Potential energy	E_p, V, Φ	$E_p = Fds$	J
Kinetic energy	E_k, T, K	$e_k = (1/2)mv^2$	J
Work	W, w	$w = Fds$	J
Hamilton function	H	$H(q, p) = T(q, p) + V(q)$	J
Lagrange function	L	$L(q, \dot{q})T(q, \dot{q}) - V(q)$	J
Pressure	p, P	$p = F/A$	Pa, N m^{-2}
Surface tension	γ, σ	$\gamma = dW/dA$	N m^{-1}, J m^{-2}
Weight	$G, (W, P)$	$G = mg$	N
Gravitational constant	G	$\Gamma = Gm_1 m_2/r^2$	N m^2 kg^{-2}
Normal stress	υ	$\sigma = F/A$	Pa
Shear stress	τ	$\tau = F/A$	Pa
Linear strain, relative elongation	ε, e	$\varepsilon = \Delta l/l$	1
Modulus of elasticity, Young's modulus	E	$E = \sigma/\varepsilon$	Pa
Shear strain	γ	$\gamma = \Delta x/d$	1
Shear modulus	G	$G = \tau/\gamma$	Pa
Volume strain, bulk strain	θ	$\theta = \Delta V/V_0$	1
Bulk modulus, compression modulus	K	$K = -V_0(dp/dV)$	Pa
Viscosity, dynamic viscosity	η, μ	$\tau_{x,z} = \eta(dv_x/dz)$	Pa s
Fluidity	ϕ	$\phi = 1/\eta$	m kg^{-1} s
Kinematic viscosity	v	$v = \eta/\rho$	m^2 s^{-1}
Friction coefficient	$\mu, (f)$	$F_{frict} = \mu F_{norm}$	1
Power	P	$P = dW/dt$	W
Sound energy flux	P, P_a	$P = dE/dt$	W
Acoustic factors			
Reflection factor	ρ	$\rho = P_t/P_0$	1
Acoustic absorption factor	$\alpha_a, (\alpha)$	$\alpha_a = 1 - \rho$	1
Transmission factor	τ	$\tau = P_{tr}/P_0$	1
Dissipation factor	δ	$\delta = \alpha_a - \tau$	1

(continued)

Symbols and Terminology for Physical and Chemical Quantities (continued)

Name	Symbol	Definition	SI Unit
Electricity and Magnetism			
Quantity of electricity, electric charge	Q		C
Charge density	ρ	$\rho = Q/V$	C m^{-3}
Surface charge density	σ	$\sigma = Q/A$	C m^{-2}
Electric potential	V, ϕ	$V = dW/dQ$	V, J C^{-1}
Electric potential difference	$U, \Delta V, \Delta\phi$	$U = V_2 - V_1$	V
Electromotive force	E	$E = (F/Q)ds$	V
Electric field strength	$\mathbf{E}$	$\mathbf{E} = \mathbf{F}/Q = -\text{grad } V$	V m^{-1}
Electric flux	Ψ	$\Psi = \mathbf{D}d\mathbf{A}$	C
Electric displacement	$\mathbf{D}$	$\mathbf{D} = \varepsilon\mathbf{E}$	C m^{-2}
Capacitance	C	$C = Q/U$	F, C V^{-1}
Permittivity	ε	$D = \varepsilon E$	F m^{-1}
Permittivity of vacuum	ε_0	$\varepsilon_0 = \mu_0^{-1} c_0^{-2}$	F m^{-1}
Relative permittivity	ε_r	$\varepsilon_\text{r} = \varepsilon/\varepsilon_0$	1
Dielectric polarization (dipole moment per volume)	$\mathbf{P}$	$\mathbf{P} = \mathbf{D} - \varepsilon_0\mathbf{E}$	C m^{-2}
Electric susceptibility	χ_e	$\chi_\text{e} = \varepsilon_\text{r} - 1$	1
Electric dipole moment	$\mathbf{p}, \mu$	$\mathbf{p} = Q\mathbf{r}$	C m
Electric current	I	$I = dQ/dt$	A
Electric current density	$\mathbf{j}, \mathbf{J}$	$I = \mathbf{j}dx\mathbf{A}$	A m^{-2}
Magnetic flux density, magnetic induction	$\mathbf{B}$	$\mathbf{F} = Qv \times \mathbf{B}$	T
Magnetic flux	Φ	$\Phi = \mathbf{B}dA$	Wb
Magnetic field strength	$\mathbf{H}$	$\mathbf{B} = \mu\mathbf{H}$	A M^{-1}
Permeability	μ	$\mathbf{B} = \mu\mathbf{H}$	$\text{N A}^{-2}, \text{H m}^{-1}$
Permeability of vacuum	μ_0		H m^{-1}
Relative permeability	μ_r	$\mu_\text{r} = \mu/\mu_0$	1
Magnetization (magnetic dipole moment per volume)	$\mathbf{M}$	$\mathbf{M} = \mathbf{B}/\mu_0 - \mathbf{H}$	A m^{-1}
Magnetic susceptibility	$\chi, \kappa, (\chi_\text{m})$	$\chi = \mu_\text{r} - 1$	1
Molar magnetic susceptibility	χ_m	$\chi_\text{m} = V_\text{m}\chi$	$\text{m}^3 \text{ mol}^{-1}$
Magnetic dipole moment	$\mathbf{m}, \mu$	$E_\text{p} = -\mathbf{m} \cdot \mathbf{B}$	$\text{A m}^2, \text{J T}^{-1}$
Electrical resistance	R	$\mathbf{P} = \mathbf{Y}/\mathbf{I}$	Ω
Conductance	G	$G = 1/R$	S
Loss angle	δ	$\delta = (\pi/2) + \phi_I - \phi_U$	1, rad
Reactance	X	$X = (U/I)\sin \delta$	Ω
Impedance (complex impedance)	Z	$Z = R + iX$	Ω
Admittance (complex admittance)	Y	$Y = 1/Z$	S
Susceptance	B	$Y = G + iB$	S
Resistivity	ρ	$\rho = E/j$	$\Omega \text{ m}$
Conductivity	κ, γ, σ	$\kappa = 1/\rho$	S m^{-1}
Self-inductance	L	$E = -L(dI/dt)$	H
Mutual inductance	M, L_{12}	$E_1 = L_{12}(Di_2/dt)$	H
Magnetic vector potential	$\mathbf{A}$	$\mathbf{B} = \mathbf{V} \times \mathbf{A}$	Wb m^{-1}
Poynting vector	$\mathbf{S}$	$\mathbf{S} = \mathbf{E} \times \mathbf{H}$	W m^{-2}
Electromagnetic Radiation			
Wavelength	λ		m
Speed of light			m s^{-1}
in vacuum	c_0		
in a medium	c	$c = c_0/n$	

(continued)

Symbols and Terminology for Physical and Chemical Quantities (continued)

Name	Symbol	Definition	SI Unit
Electromagnetic Radiation			
Wavenumber in vacuum	V	$V = V/c_0 = 1/n\lambda$	m^{-1}
Wavenumber (in a medium)	σ	$\sigma = 1/\lambda$	m^{-1}
Frequency	v	$v = c/\lambda$	Hz
Circular frequency, pulsatance	ω	$\omega = 2\pi v$	s^{-1}, rad s^{-1}
Refractive index	n	$n = c_0/c$	1
Planck constant	h		J s
Planck constant/2π	$\hbar$	$\hbar = h/2\pi$	J s
Radiant energy	Q, W		J
Radiant energy density	ρ, w	$\rho = Q/V$	$J\ m^{-3}$
Spectral radiant energy density			
in terms of frequency	ρ_v, w_v	$\rho_v = \delta\rho/dv$	$J\ m^{-3}\ Hz^{-1}$
in terms of wavenumber	$\rho_{\bar{v}}, w_{\bar{v}}$	$\rho_{\bar{v}} = d\rho/d\bar{v}$	$J\ m^{-2}$
in terms of wavelength	ρ_λ, w_λ	$\rho_\lambda = \delta\rho/d\lambda$	$J\ m^{-4}$
Einstein transition probabilities			
Spontaneous emission	A_{nm}	$dN_n/dt = -A_{nm}N_n$	s^{-1}
Stimulated emission	B_{nm}	$dn_n/dt = -\rho\bar{v}(\bar{V}_{nm}) \times B_{nm}N_n$	$s\ kg^{-1}$
Radiant power, radiant energy per time	Φ, P	$\Phi = dQ/dt$	W
Radiant intensity	I	$I = d\Phi/d\Omega$	$W\ sr^{-1}$
Radiant exitance (emitted radiant flux)	M	$M = d\Phi/dA_{source}$	$W\ m^{-2}$
Irradiance (radiant flux received)	$E, (I)$	$E = d\Phi/\delta A$	$W\ m^{-2}$
Emittance	ε	$\varepsilon = M/M_{bb}$	1
Stefan–Boltzmann constant	σ	$M_{bb} = \sigma T^4$	$W\ m^{-2}\ K^{-4}$
First radiation constant	c_1	$c_1 = 2\pi h c_0^2$	$W\ m^2$
Second radiation constant	c_2	$c_2 = hc_0/k$	K m
Transmittance, transmission factor	τ, T	$\tau = \Phi_{tr}/\Phi_0$	1
Absorptance, absorption factor	α	$\alpha = \phi_{abs}/\phi_0$	1
Reflectance, reflection factor	ρ	$\rho = \phi_{refl}/\Phi_0$	1
(Decadic) absorbance	A	$A = \lg(1 - \alpha_i)$	1
Napierian absorbance	B	$B = \ln(1 - \alpha_i)$	1
Absorption coefficient			
(Linear) decadic	a, K	$a = A/l$	m^{-1}
(Linear) napierian	α	$\alpha = B/l$	m^{-1}
Molar (decadic)	ε	$\varepsilon = a/c = A/cl$	$m^2\ mol^{-1}$
Molar napierian	κ	$\kappa = \alpha/c = B/cl$	$m^2\ mol^{-1}$
Absorption index	k	$k = \alpha/4\pi\bar{v}$	1
Complex refractive index	$\hat{n}$	$\hat{n} = n + ik$	1
Molar refraction	R, R_m	$R = \frac{(n^2-1)}{(n^2+2)}V_m$	$m^3\ mol^{-1}$
Angle of optical rotation	α		1, rad
Solid State			
Lattice vector	$\mathbf{R}, \mathbf{R}_0$		m
Fundamental translation vectors for the crystal lattice	$\mathbf{a}_1; \mathbf{a}_2; \mathbf{a}_3, \mathbf{a}; \mathbf{b}; \mathbf{c}$	$R = n_1\mathbf{a}_1 + n_2\mathbf{a}_2 + n_3\mathbf{a}_3$	m
(Circular) reciprocal lattice vector	$\mathbf{G}$	$\mathbf{G} \cdot \mathbf{R} = 2\pi m$	m^{-1}

(continued)

Symbols and Terminology for Physical and Chemical Quantities (continued)

Name	Symbol	Definition	SI Unit
Solid State			
(Circular) fundamental translation vectors for the reciprocal lattice	$\mathbf{b}_1; \mathbf{b}_2; \mathbf{b}_3, \mathbf{a}^\star; \mathbf{b}^\star; \mathbf{c}^\star$	$\mathbf{a}_i \cdot \mathbf{b}_k = 2\pi\delta_{ik}$	m^{-1}
Lattice plane spacing	d		m
Bragg angle	θ	$n\lambda = 2d \sin \theta$	l, rad
Order of reflection	n		l
Order parameters			
Short range	σ		l
Long range	s		l
Burgers vector	b		m
Particle position vector	r, R_j		m
Equilibrium position vector of an ion	R_o		m
Displacement vector of an ion	$\mathbf{u}$	$\mathbf{u} = \mathbf{R} - \mathbf{R}_0$	m
Debye–Waller factor	B, D		l
Debye circular wavenumber	q_D		m^{-1}
Debye circular frequency	ω_D		s^{-1}
Grüneisen parameter	γ, Γ	$\gamma = \alpha V/\kappa C_V$	l
Madelung constant	$\alpha, \mathscr{M}$	$E_{coul} = \frac{\alpha N_A z_+ z_- e^2}{4\pi\varepsilon_0 R_0}$	l
Density of states	N_E	$N_E = dN(E)/dE$	$J^{-1}\,m^{-3}$
(Spectral) density of vibrational modes	N_ω, g	$N_\omega = dN(\omega)/d\omega$	$s\,m^{-3}$
Resistivity tensor	ρ_{ik}	$E = \rho \cdot j$	$\Omega\,m$
Conductivity tensor	σ_{ik}	$\sigma = \rho^{-1}$	$S\,m^{-1}$
Thermal conductivity tensor	λ_{ik}	$J_q = -\lambda \cdot \text{grad } T$	$W\,m^{-1}\,K^{-1}$
Residual resistivity	ρ_R		$\Omega\,m$
Relaxation time	τ	$\tau = l/v_F$	s
Lorenz coefficient	L	$L = \lambda/\sigma T$	$V^2\,K^{-2}$
Hall coefficient	A_H, R_H	$\mathbf{E} = \rho \cdot \mathbf{j} + R_H(\mathbf{B} \times \mathbf{j})$	$m^3\,C^{-1}$
Thermoelectric force	E		V
Peltier coefficient	Π		V
Thomson coefficient	$\mu,(\tau)$		$V\,K^{-1}$
Work function	Φ	$\Phi = E_\infty - E_F$	J
Number density, number concentration	$n, (p)$		m^{-3}
Gap energy	E_γ		J
Donor ionization energy	E_δ		J
Acceptor ionization energy	E_α		J
Fermi energy	E_Φ, ε_F		J
Circular wave vector, propagation vector	$\mathbf{k, q}$	$k = 2\pi/\lambda$	m^{-1}
Bloch function	$u_k(\mathbf{r})$	$\psi(\mathbf{r}) = u_k(\mathbf{r}) \exp(i\mathbf{k} \cdot \mathbf{r})$	$m^{-3/2}$
Charge density of electrons	ρ	$\rho(\mathbf{r}) = -e\psi^\star(\mathbf{r})\psi(\mathbf{r})$	$C\,m^{-3}$
Effective mass	$m^\star$		kg
Mobility	μ	$\mu = v_{drift}/E$	$m^2\,V^{-1}\,s^{-1}$
Mobility ratio	b	$b = \mu_n/\mu_p$	l
Diffusion coefficient	D	$dN/dt = -DA(dn/dx)$	$m^2\,s^{-1}$
Diffusion length	L	$L = \sqrt{D\tau}$	m
Characteristic (Weiss) temperature	ϕ, ϕ_W		K
Curie temperature	T_C		K
Néel temperature	T_N		K

Credits

Material in Section III was reprinted from the following sources:

D. R. Lide, Ed., *CRC Handbook of Chemistry and Physics*, 76th ed., Boca Raton, FL: CRC Press, 1992: International System of Units (SI), conversion constants and multipliers (conversion of temperatures), symbols and terminology for physical and chemical quantities, fundamental physical constants, classification of electromagnetic radiation.

D. Zwillinger, Ed., *CRC Standard Mathematical Tables and Formulae*, 30th ed., Boca Raton, FL: CRC Press, 1996: Greek alphabet, conversion constants and multipliers (recommended decimal multiples and submultiples, metric to English, English to metric, general, temperature factors), physical constants, series expansion.

Probability for Electrical and Computer Engineers

Charles W. Therrien

The Algebra of Events

The study of probability is based upon experiments that have uncertain outcomes. Collections of these outcomes comprise *events* and the collection of all possible outcomes of the experiment comprise what is called the *sample space*, denoted by S. Outcomes are members of the sample space and events of interest are represented as *sets* of outcomes (see Figure III.1).

The algebra $\mathcal{A}$ that deals with representing events is the usual set algebra. If A is an event, then A^c (the *complement* of A) represents the event that "A did not occur." The complement of the sample space is the *null event*, $\emptyset = S^c$. The event that *both* event A_1 and event A_2 have occurred is the intersection, written as "$A_1 \cdot A_2$" or "$A_1 A_2$" while the event that *either* A_1 or A_2 *or both* have occurred is the union, written as "$A_1 + A_2$."[1]

Table III.1 lists the two postulates that define the algebra $\mathcal{A}$, while Table III.2 lists seven axioms that define properties of its operations. Together these tables can be used to show all of the properties of the algebra of events. Table III.3 lists some additional useful relations that can be derived from the axioms and the postulates.

Since the events "$A_1 + A_2$" and "$A_1 A_2$" are included in the algebra, it follows by induction that for any finite number of events $A_1 + A_2 + \cdots + A_N$ and $A_1 \cdot A_2 \cdots A_N$ are also included in the algebra. Since problems often involve the union or intersection of an *infinite* number of events, however, the algebra of events must be defined to include these infinite intersections and unions. This extension to infinite unions and intersections is known as a sigma algebra.

A set of events that satisfies the two conditions:

1. $A_i A_j = \emptyset \neq$ for $\neq i \neq j$
2. $A_1 + A_2 + A_3 + \cdots = S$

is known as a *partition* and is important for the solution of problems in probability. The events of a partition are said to be *mutually exclusive* and *collectively exhaustive*. The most fundamental partition is the set outcomes defining the random experiment, which comprise the sample space by definition.

Probability

Probability measures the likelihood of occurrence of events represented on a scale of 0 to 1. We often estimate probability by measuring the *relative frequency* of an event, which is defined as

$$\text{relative frequency} = \frac{\text{number of occurrences of the event}}{\text{number of repetitions of the experiment}}$$

(for a large number of repetitions). Probability can be defined formally by the following axioms:

   (I) The probability of any event is nonnegative:

$$\Pr[A] \geq 0 \tag{III.1}$$

   (II) The probability of the universal event (i.e., the entire sample space) is 1:

$$\Pr[S] = 1 \tag{III.2}$$

---

[1]Some authors use $\cap$ and $\cup$ rather than $\cdot$ and $+$, respectively.

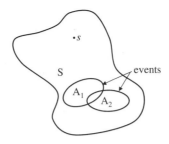

FIGURE III.1 Abstract representation of the sample space S with outcome s and sets A_1 and A_2 representing events.

(III) If A_1 and A_2 are mutually exclusive, i.e., $A_1 A_2 = \varnothing$, then

$$\Pr[A_1 + A_2] = \Pr[A_1] + \Pr[A_2] \tag{III.3}$$

(IV) If $\{A_i\}$ represent a countably infinite set of mutually exclusive events, then

$$\Pr[A_1 + A_2 + A_3 + \cdots] = \sum_{i=1}^{\infty} \Pr[A_i] \quad (\text{ if } A_i A_j = \varnothing \quad i \neq j) \tag{III.4}$$

Note that although the additivity of probability for any finite set of disjoint events follows from (III), the property has to be stated explicitly for an infinite set in (IV). These axioms and the algebra of events can be used to show a number of other important properties which are summarized in Table III.4. The last item in the table is an especially important formula since it uses probabilistic information about

TABLE III.1 Postulates for an Algebra of Events

1.	If $A \in \mathcal{A}$ then $A^c \in \mathcal{A}$
2.	If $A_1 \in \mathcal{A}$ and $A_2 \in \mathcal{A}$ then $A_1 + A_2 \in \mathcal{A}$

TABLE III.2 Axioms of Operations on Events

$A_1 A_1^c = \varnothing$	Mutual exclusion
$A_1 S = A_1$	Inclusion
$(A_1^c)^c = A_1$	Double complement
$A_1 + A_2 = A_2 + A_1$	Commutative law
$A_1 + (A_2 + A_3) = (A_1 + A_2) + A_3$	Associative law
$A_1(A_2 + A_3) = A_1 A_2 + A_1 A_3$	Distributive law
$(A_1 A_2)^c = A_1^c + A_2^c$	DeMorgan's law

TABLE III.3 Additional Identities in the Algebra of Events

$S^c = \varnothing$	
$A_1 + \varnothing = A_1$	Inclusion
$A_1 A_2 = A_2 A_1$	Commutative law
$A_1(A_2 A_3) = (A_1 A_2) A_3$	Associative law
$A_1 + (A_2 A_3) = (A_1 + A_2)(A_1 + A_3)$	Distributive law
$(A_1 + A_2)^c = A_1{}^c A_2{}^c$	DeMorgan's law

TABLE III.4   Some Corollaries Derived from the Axioms
of Probability

$\Pr[A^c] = 1 - \Pr[A]$
$0 \leqslant \Pr[A] \leqslant 1$
If $A_1 \subseteq A_2$ then $\Pr[A_1] \leqslant \Pr[A_2]$
$\Pr[\varnothing] = 0$
If $A_1 A_2 = \varnothing -$ then $= \Pr[A_1 A_2] = 0$
$\Pr[A_1 + A_2] = \Pr[A_1] + \Pr[A_2] - \Pr[A_1 A_2]$

individual events to compute the probability of the union of two events. The term $\Pr[A_1 A_2]$ is referred to as the *joint probability* of the two events. This last equation shows that the probabilities of two events add as in Equation (III.3) only if their joint probability is 0. The joint probability is 0 when the two events have no intersection ($A_1 A_2 = \varnothing$).

Two events are said to be statistically *independent* if and only if

$$\Pr[A_1 A_2] = \Pr[A_1] \cdot \Pr[A_2] \quad \text{(independent events)} \tag{III.5}$$

This definition is not derived from the earlier properties of probability. An argument to give this definition intuitive meaning can be found in Ref. [1]. Independence occurs in problems where two events are not influenced by one another and Equation (III.5) simplifies such problems considerably.

A final important result deals with partitions. *A partition* is a finite or countably infinite set of events $A_1, A_2, A_3, \ldots$ that satisfy the two conditions:

$$A_i A_j = \varnothing \text{ for } i \neq j$$

$$A_1 + A_2 + A_3 + \cdots = S$$

The events in a partition satisfy the relation:

$$\sum_i \Pr[A_i] = 1 \tag{III.6}$$

Further, if B is *any* other event, then

$$\Pr[B] = \sum_i \Pr[A_i B] \tag{III.7}$$

The latter result is referred to as the *principle of total probability* and is frequently used in solving problems. The principle is illustrated by a Venn diagram in Figure III.2. The rectangle represents the sample space and other events are defined therein. The event B is seen to be comprised of all of the pieces

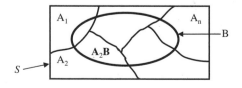

FIGURE III.2   Venn diagram illustrating the principle of total probability.

that represent intersections or overlap of event B with the events A_i. This is the graphical interpretation of Equation (III.7).

An Example

Simon's Surplus Warehouse has large barrels of mixed electronic components (parts) that you can buy by the handful or by the pound. You are not allowed to select parts individually. Based on your previous experience, you have determined that in one barrel, 29% of the parts are bad (faulted), 3% are bad resistors, 12% are good resistors, 5% are bad capacitors, and 32% are diodes. You decide to assign probabilities based on these percentages. Let us define the following events:

Event	Symbol
Bad (faulted) component	B
Good component	G
Resistor	R
Capacitor	C
Diode	D

A Venn diagram representing this situation is shown below along with probabilities of various events as given:

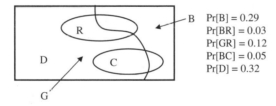

$$Pr[B] = 0.29$$
$$Pr[BR] = 0.03$$
$$Pr[GR] = 0.12$$
$$Pr[BC] = 0.05$$
$$Pr[D] = 0.32$$

Note that since any component must be a resistor, capacitor, or diode, the region labeled D in the diagram represents everything in the sample space which is not included in R or C.

We can answer a number of questions.

1. What is the probability that a component is a resistor (either good *or* bad)?

   Since the events B and G form a partition of the sample space, we can use the principle of total probability Equation (III.7) to write:

$$Pr[R] = Pr[GR] + Pr[BR] = 0.12 + 0.03 = 0.15$$

2. Are bad parts and resistors independent?

   We know that $Pr[BR] = 0.03$ and we can compute:

$$Pr[B] \cdot Pr[R] = (0.29)(0.15) = 0.0435$$

   Since $Pr[BR] \neq Pr[B] \cdot Pr[R]$, the events are *not* independent.

3. You have no use for either bad parts or resistors. What is the probability that a part is either bad and/or a resistor?

Using the formula from Table III.4 and the previous result we can write:

$$Pr[B + R] = Pr[B] + Pr[R] - Pr[BR] = 0.29 + 0.15 - 0.03 = 0.41$$

4. What is the probability that a part is useful to you?
   Let U represent the event that the part is useful. Then (see Table III.4):

$$Pr[U] = 1 - Pr[U^c] = 1 - 0.41 = 0.59$$

5. What is the probability of a bad diode?
   Observe that the events R, C, and D form a partition, since a component has to be one and only one type of part. Then using Equation (III.7) we write:

$$Pr[B] = Pr[BR] + Pr[BC] + Pr[BD]$$

Substituting the known numerical values and solving yields

$$0.29 = 0.03 + 0.05 + Pr[BD] \text{ or } Pr[BD] = 0.21$$

Conditional Probability and Bayes' Rule

The *conditional* probability of an event A_1 given that an event A_2 has occurred is defined by

$$Pr[A_1|A_2] = \frac{Pr[A_1A_2]}{Pr[A_2]} \tag{III.8}$$

($Pr[A_1|A_2]$ is read "probability of A_1 *given* A_2.") As an illustration, let us compute the probability that a component in the previous example is bad given that it is a resistor:

$$Pr[B|R] = \frac{Pr[BR]}{Pr[R]} = \frac{0.03}{0.15} = 0.2$$

(The value for $Pr[R]$ was computed in question 1 of the example.) Frequently the statement of a problem is in terms of conditional probability rather than joint probability, so Equation (III.8) is used in the form:

$$Pr[A_1A_2] = Pr[A_1|A_2] \cdot Pr[A_2] = Pr[A_2|A_1] \cdot Pr[A_1] \tag{III.9}$$

(The last expression follows because $Pr[A_1A_2]$ and $Pr[A_2A_1]$ are the same thing.) Using this result, the principle of total probability Equation (III.7) can be rewritten as

$$Pr[B] = \sum_j Pr[B|A_j] Pr[A_j] \tag{III.10}$$

where B is any event and $\{A_j\}$ is a set of events that forms a partition.

Now, consider any one of the events A_i in the partition. It follows from Equation (III.9) that

$$Pr[A_i|B] = \frac{Pr[B|A_i] \cdot Pr[A_i]}{Pr[B]}$$

Then substituting in Equation (III.10) yields:

$$Pr[A_i|B] = \frac{Pr[B|A_i] \cdot Pr[A_i]}{\sum_j Pr[B|A_j]\,Pr[A_j]} \tag{III.11}$$

This result is known as *Bayes' theorem* or *Bayes' rule*. It is used in a number of problems that commonly arise in electrical engineering. We illustrate and end this section with an example from the field of communications.

Communication Example

The transmission of bits over a binary communication channel is represented in the drawing below:

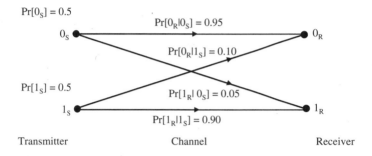

where we use notation like 0_S, 0_R ... to denote events "0 sent," "0 received," etc. When a 0 is transmitted, it is correctly received with probability 0.95 or incorrectly received with probability 0.05. That is, $Pr[0_R|0_S] = 0.95$ and $Pr[1_R|0_S] = 0.05$. When a 1 is transmitted, it is correctly received with probability 0.90 and incorrectly received with probability 0.10. The probabilities of sending a 0 or a 1 are denoted by $Pr[0_S]$ and $Pr[1_S]$. It is desired to compute the *probability of error* for the system.

This is an application of the principle of total probability. The two events 0_S and 1_S are mutually exclusive and collectively exhaustive and thus form a partition. Take the event B to be the event that an error occurs. It follows from Equation (III.10) that

$$Pr[error] = Pr[error|0_S]\,Pr[0_S] + Pr[error|1_S]\,Pr[1_S]$$

$$= Pr[1_R|0_S]Pr[0_S] + Pr[0_R|1_S]\,Pr[1_S]$$

$$= (0.05)(0.5) + (0.10)(0.5) = 0.075$$

Next, given that an error has occurred, let us compute the probability that a 1 was sent or a 0 was sent. This is an application of Bayes' rule. For a 1, Equation (III.11) becomes

$$Pr[1_S|error] = \frac{Pr[error|1_S]\,Pr[1_S]}{Pr[error|1_S]\,Pr[1_S] + Pr[error|0_S]\,Pr[0_S]}$$

Substituting the numerical values then yields:

$$Pr[1_S|error] = \frac{(0.10)(0.5)}{(0.10)(0.5) + (0.05)(0.5)} \approx 0.667$$

For a 0, a similar analysis applies:

$$\Pr[0_S|\text{error}] = \frac{\Pr[\text{error}|0_S]\,\Pr[0_S]}{\Pr[\text{error}|1_S]\,\Pr[1_S] + \Pr[\text{error}|0_S]\,\Pr[0_S]}$$

$$= \frac{(0.05)(0.5)}{(0.10)(0.5) + (0.05)(0.5)} \approx 0.333$$

The two resulting probabilities sum to 1 because 0_S and 1_S form a partition for the experiment.

Reference

1.   C. W. Therrien and M. Tummala, *Probability for Electrical and Computer Engineers*. Boca Raton, FL: CRC Press, 2004.

Indexes

Author Index

Subject Index

Page on which a term is defined is indicated in bold.